HANDBOOK OF HEAVY TAILED DISTRIBUTIONS IN FINANCE

HANDBOOKS IN FINANCE

BOOK 1

Series Editor

WILLIAM ZIEMBA

Advisory Editors

KENNETH J. ARROW
GEORGE CONSTANTINIDES
HARRY M. MARKOWITZ
ROBERT C. MERTON
STEWART MYERS
PAUL A. SAMUELSON
WILLIAM F. SHARPE

N·H

ELSEVIER

AMSTERDAM • BOSTON • LONDON • NEW YORK • OXFORD • PARIS
SAN DIEGO • SAN FRANCISCO • SINGAPORE • SYDNEY • TOKYO

HANDBOOK OF HEAVY TAILED DISTRIBUTIONS IN FINANCE

Edited by

SVETLOZAR T. RACHEV
University of California

2003

ELSEVIER

AMSTERDAM • BOSTON • LONDON • NEW YORK • OXFORD • PARIS
SAN DIEGO • SAN FRANCISCO • SINGAPORE • SYDNEY • TOKYO

ELSEVIER SCIENCE B.V.
Sara Burgerhartstraat 25
P.O. Box 211, 1000 AE Amsterdam, The Netherlands

First edition 2003

Library of Congress Cataloging in Publication Data
A catalog record from the Library of Congress has been applied for.

British Library Cataloguing in Publication Data
Handbook of heavy tailed distributions in finance. –
(Handbooks in finance; v. 1)
1. Finance – Statistical methods
I. Rachev, Svetlozar T. (Svetlozar Todorov) II. Heavy tailed distributions in finance
332'.015195

ISBN: 0 444 50896 1
ISSN (Series): 1568-4997

Transferred to Digital Printing 2009

INTRODUCTION TO THE SERIES

The *Handbooks in Finance* are intended to be a definitive source for comprehensive and accessible information in the field of finance. Each individual volume in the series should present an accurate self-contained survey of a sub-field of finance, suitable for use by finance and economics professors and lecturers, professional researchers, graduate students and as a teaching supplement. The goal is to have a broad group of outstanding volumes in various areas of finance.

CONTENTS OF THE HANDBOOK

PREFACE

"... some scholars now believe that stock prices have been more volatile (with Black Monday and Terrible Tuesday, October 19–20, 1987, being prime examples) that any existing theory would suggest."
William F. Sharpe and Cordon J. Alexander, Investment, Prentice-Hall, 1992, Preface.

"Most of the models assume that stock options follow a log-normal distribution, in fact, I found out that the actual price distributions of virtually all financial markets tend to have fatter tails than suggested by the log-normal distribution."
Blair Hull, Jack D. Schwager, The New Market Wizards, Harper Business, 1992, p. 337.

"... However, it has often been argued that the true distributions returns (even after standardizing by the volatility) implies a larger probability of extreme returns than that implied from the normal distribution. Although we could try to specify a distribution that fits returns better, it would be a daunting task, especially if we consider that the new distribution would have to provide a good fit across all asset classes."
Risk Metrics Group, Return to RiskMetrics: The Evolution of a Standard, Technical Document, April 2001, p. 25.

Modern finance relies heavily on the assumption that the random variables under investigation follow a normal distribution. Distributional assumptions for financial processes have important theoretical implications, given that financial decisions are commonly based on expected returns and risk of alternative investment opportunities. Hence, solutions to such problems like portfolio selection, option pricing, and risk management depend critically on distributional specifications.

However, time series observed in finance often deviate from the Gaussian model in that their marginal distributions are heavy-tailed and, possibly, asymmetric. In such situations, the appropriateness of the commonly adopted normal assumption is highly questionable. This is especially true in financial modeling.

Benoit Mandelbrot's fundamental work in the 1960s strongly rejected normality as a distributional model for asset returns. Examining various time series on commodity returns and interest rates, he conjectured that financial return processes behave like non-Gaussian stable processes.

Mandelbrot's contributions give rise to a new probabilistic foundation for financial theory and empirics, and are of similar importance as the fundamental contributions of Louis Bachelier (1900) and Paul Samuelson (1955). His early investigations on asset returns were carried further by Eugene Fama and Richard Roll, amongst others, and led to a consolidation of the Stable Paretian Hypothesis.

The Mandelbrot Stable Paretian Hypothesis involves more than simply fitting marginal asset-return distributions. Stable Paretian laws and other heavy-tailed infinitely divisable laws describe the fundamental "building blocks" (e.g., innovations) that drive asset return processes. In addition to describing these "building blocks", a complete financial model should be rich enough to encompass relevant stylized facts such as:

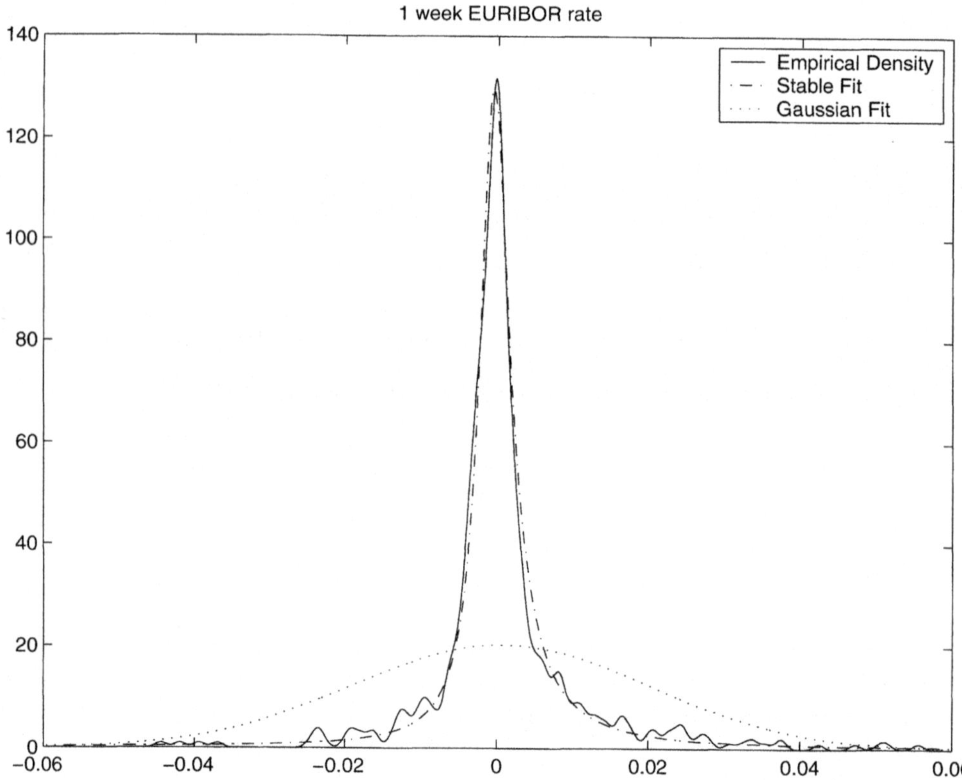

Fig. 1. The probability density fit of a Gaussian, and stable (non-Gaussian) distribution to the empirical (sample) distribution 1 week EURIBOR log returns. One-week EURIBOR (EURo InterBank Offered Rate) spot rate. Daily log-returns (sample size 567, from January 1, 1999 to March 5, 2001). (European Central Bank and Datastream.)

- Non-Gaussian, heavy-tailed and skewed distributions;
- Volatility clustering (ARCH-effects);
- Temporal dependence of the tail behavior;
- Short and long-range dependence.

An attractive feature of stable models – not shared by other distributional models – is that they allow us to generalize Gaussian-based financial theories and thus, to build a coherent and more general framework for financial modeling. The generalizations are only possible because of specific probabilistic properties that are unique to (Gaussian and non-Gaussian) stable laws, namely, the stability property, the Central Limit Theorem, and the Invariance Principle for Lévy-stable processes.

Although stable and other heavy-tailed distributions and processes have been studied intensively in recent years, many theoretical and practical financial aspects have yet to be explored. This volume presents current research focusing on heavy-tailed distributions in finance. The contributions cover methodological issues, i.e., probabilistic, statistical and econometric modeling under non-Gaussian assumptions, as well as the application of stable and other non-Gaussian models in finance and risk management.

As editor, I would like to thank the contributors for their fruitful collaboration. I am also grateful for the editorial assistance I received. In particular, I would like to thank Carlo Marinelli (Columbia University) for his skillful organizational help and Thomas Plum (University of Karslruhe) for his excellent technical word-processing support.

SVETLOZAR T. RACHEV
Karlsruhe and Santa Barbara, 2002

CONTENTS

Chapter 1

HEAVY TAILS IN FINANCE FOR INDEPENDENT OR MULTIFRACTAL PRICE INCREMENTS

BENOIT B. MANDELBROT

Sterling Professor of Mathematical Sciences, Yale University, New Haven, CT 065020-8283, USA

Contents

Handbook of Heavy Tailed Distributions in Finance, Edited by S.T. Rachev

Abstract

This chapter has two goals. Section 1 sketches the history of heavy tails in finance through the author's three successive models of the variation of a financial price: mesofractal, unifractal and multifractal. The heavy tails occur, respectively, in the marginal distribution only (Mandelbrot, 1963), in the dependence only (Mandelbrot, 1965), or in both (Mandelbrot, 1997). These models increase in the scope of the "principle of scaling invariance", which the author has used since 1957.

The mesofractal model is founded on the stable processes that date to Cauchy and Lévy. The unifractal model uses the fractional Brownian motions introduced by the author. By now, both are well-understood.

To the contrary, one of the key features of the multifractals (Mandelbrot, 1974a, b) remains little known. Using the author's recent work, introduced for the first time in this chapter, the exposition can be unusually brief and mathematically elementary, yet covering all the key features of multifractality. It is restricted to very special but powerful cases: (a) the Bernoulli binomial measure, which is classical but presented in a little-known fashion, and (b) a new two-valued "canonical" measure. The latter generalizes Bernoulli and provides an especially short path to negative dimensions, divergent moments, and divergent (i.e., long range) dependence. All those features are now obtained as separately tunable aspects of the same set of simple construction rules.

My work in finance is well-documented in easily accessible sources, many of them reproduced in Mandelbrot (1997 and also in 2001a, b, c, d). That work having expanded and been commented upon by many authors, a survey of the literature is desirable, but this is a task I cannot undertake now. However, it was a pleasure to yield to the entreaties of this Handbook's editors by a text in which a new technical contribution is preceded by an introductory sketch followed by a simple new presentation of an old feature that used to be dismissed as "technical", but now moves to center stage.

The history of heavy tails in finance began in 1963. While acknowledging that the successive increments of a financial price are interdependent, I assumed independence as a first approximation and combined it with the principle of scaling invariance. This led to (Lévy) stable distributions for the price changes. The tails are very heavy, in fact, power-law distributed with an exponent $\alpha < 2$.

The multifractal model advanced in Mandelbrot (1997) extends scale invariance to allow for dependence. Readily controllable parameters generate tails that are as heavy as desired and can be made to follow a power-law with an exponent in the range $1 < \alpha < \infty$. This last result, an essential one, involves a property of multifractals that was described in Mandelbrot (1974a, b) but remains little known among users. The goal of the example described after the introduction is to illustrate this property in a very simple form.

1. Introduction: A path that led to model price by Brownian motion (Wiener or fractional) of a multifractal trading time

Given a financial price record $P(t)$ and a time lag $\mathrm{d}t$, define $L(t, \mathrm{d}t) = \log P(t + \mathrm{d}t) - \log P(t)$. The 1900 dissertation of Louis Bachelier introduced Brownian motion as a model of $P(t)$. In later publications, however, Bachelier acknowledged that this is a very rough first approximation: he recognized the presence of heavy tails and did not rule out dependence. But until 1963, no one had proposed a model of the heavy tails' distribution.

1.1. From the law of Pareto to infinite moment "anomalies" that contradict the Gaussian "norm"

All along, search for a model was inspired by a finding rooted in economics outside of finance. Indeed, the distribution of personal incomes proposed in 1896 by Pareto involved tails that are heavy in the sense of following a power-law distribution $Pr\{U > u\} = u^{-\alpha}$.

However, almost nobody took this income distribution seriously. The strongest "conventional wisdom" argument against Pareto was that the value $\alpha = 1.7$ that he claimed leads to the variance of U being infinite.

Infinite moments have been a perennial issue both before my work and (unfortunately) ever since. Partly to avoid them, Pareto volunteered an exponential multiplier, resulting in

$$Pr\{U > u\} = u^{-\alpha} \exp(-\beta u).$$

Also, Herbert A. Simon expressed a universally held view when he asserted in 1953 that infinite moments are (somehow) "improper". But in fact, the exponential multipliers are not needed and infinite moments are perfectly proper and have important consequences. In multifractal models, depending on specific features, variance can be either finite or infinite. In fact, all moments can be finite, or they can be finite only up to a critical power q_{crit} that may be 3, 4, or any other value needed to represent the data.

Beginning in the late 1950s, a general theme of my work has been that the uses of statistics must be recognized as falling into at least two broad categories. In the "normal" category, one can use the Gaussian distribution as a good approximation, so that the common replacement of the term, "Gaussian", by "normal" is fully justified. To the contrary, in the category one can call "abnormal" or "anomalous", the Gaussian is very misleading, even as an approximation.

To underline this distinction, I have long suggested – to little effect up to now – that the substance of the so-called ordinary *central limit* theorem would be better understood if it is relabeled as the *center* limit theorem. Indeed, that theorem concerns the *center* of the distribution, while the anomalies concern the *tails*. Following up on this vocabulary, the *generalized central* limit theorem that yields Lévy stable limits would be better understood if called a *tail* limit theorem. This distinction becomes essential in Section 8.5.

Be that as it may, I came to believe in the 1950s that the power-law distribution and the associated infinite moments are key elements that distinguish economics from classical physics. This distinction grew by being extended from independent to highly dependent random variables. In 1997, it became ready to be phrased in terms of randomness and variability falling in one of several distinct "states". The "mild" state prevails for classical errors of observation and for sequences of near-Gaussian and near-independent quantities. To the contrary, phenomena that present deep inequality necessarily belong to the "wild" state of randomness.

1.2. A scientific principle: scaling invariance in finance

A second general theme of my work is the "principle" that financial records are invariant by dilating or reducing the scales of time and price in ways suitably related to each other. There is no need to believe that this principle is exactly valid, nor that its exact validity could ever be tested empirically. However, a proper application of this principle has provided the basis of models or scenarios that can be called good because they satisfy all the following properties:

(a) they closely model reality,

(b) they are exceptionally *parsimonious*, being based on very few very general a priori assumptions, and

(c) they are *creative* in the following sense: extensive and correct predictions arise as consequences of a few assumptions; when those assumptions are changed the consequences also change. By contrast, all too many financial models start with Brownian motion, then build upon it by including in the input every one of the properties that one wishes to see present in the output.

1.3. Analysis alone versus statistical analysis followed by synthesis and graphic output

The topic of multifractal functions has grown into a well-developed analytic theory, making it easy to apply the multifractal formalism blindly. But it is far harder to understand it and draw consequences from its output. In particular, statistical techniques for handling multifractals are conspicuous by their near-total absence. After they become actually available, their applicability will have to be investigated carefully.

A chastening example is provided by the much simpler question of whether or not financial series exhibit global (long range) dependence. My claim that they do was largely based on *R/S* analysis which at this point relies heavily on graphical evidence. Lo (1991) criticized this conclusion very severely as being subjective. Also, a certain alternative test Lo described as "objective" led to a mixed pattern of "they do" and "they do not". This pattern being practically impossible to interpret, Lo took the position that the simpler outcome has not been shown wrong, hence one can assume that long range dependence is absent.

Unfortunately, the "objective test" in question assumed the margins to be Gaussian. Hence, Lo's experiment did not invalidate my conclusion, only showed that the test is not robust and had repeatedly failed to recognize long range dependence.

The proper conclusion is that careful graphic evidence has not yet been superseded. The first step is to attach special importance to models for which sample functions can be generated.

1.4. Actual implementation of scaling invariance by multifractal functions: it requires additional assumptions that are convenient but not a matter of principle, for example, separability and compounding

By and large, an increase in the number and specificity in the assumptions leads to an increase in the specificity of the results. It follows that generality may be an ideal unto itself in mathematics, but in the sciences it competes with specificity, hence typically with simplicity, familiarity, and intuition.

In the case of multifractal functions, two additional considerations should be heeded. The so-called multifractal formalism (to be described below) is extremely important. But it does not by itself specify a random function closely enough to allow analysis to be followed by synthesis. Furthermore, multifractal functions are so new that it is best, in a first stage, to be able to rely on existing knowledge while pursuing a concrete application. For these and related reasons, my study of multifractals in finance has relied heavily on two special cases.

One is implemented by the recursive "cartoons" investigated in Mandelbrot (1997) and in much greater detail in Mandelbrot (2001c).

The other uses compounding. This process begins with a random function $F(\theta)$ in which the variable θ is called an "intrinsic time". In the key context of financial prices, θ is called "trading time". The possible functions $F(\theta)$ include all the functions that have been previously used to model price variation. Foremost is the Wiener Brownian motion $B(t)$

postulated by Bachelier. The next simplest are the fractional Brownian motion $B_H(t)$ and the Lévy stable "flight" $L(t)$.

A separate step selects for the intrinsic trading time a scale invariant random functions of the physical "clock time" t. Mandelbrot (1972) recommended for the function $\theta(t)$ the integral of a multifractal measure. This choice was developed in Mandelbrot (1997) and Mandelbrot, Calvet and Fisher (1997).

In summary, one begins with two statistically independent random functions $F(\theta)$ and $\theta(t)$, where $\theta(t)$ is non-decreasing. Then one creates the "compound" function $F[\theta(t)] = \varphi(t)$. Choosing $F(\theta)$ and $\theta(t)$ to be scale-invariant insures that $\varphi(t)$ will be scale-invariant as well. A limitation of compounding as defined thus far is that it demands independence of F and θ, therefore restricts the scope of the compound function.

In a well-known special case called Bochner subordination, the increments of $\theta(t)$ are independent. As shown in Mandelbrot and Taylor (1967), it follows that $B[\theta(t)]$ is a Lévy stable process, i.e., the mesofractal model. This approach has become well-known. The tails it creates are heavy and do follow a power law distribution but there are at least two drawbacks. The exponent α is at most 2, a clearly unacceptable restriction in many cases, and the increments are independent.

Compounding beyond subordination was introduced because it allows α to take any value > 1 and the increments to exhibit long term dependence. All this is discussed elsewhere (Mandelbrot, 1997 and more recent papers).

The goal of the remainder of this chapter is to use a specially designed simple case to explain how multifractal measure suffices to create a power-law distribution. The idea is that $L(t, \mathrm{d}t) = \mathrm{d}\varphi(t)$ where $\varphi = B_H[\theta(t)]$. Roughly, $\mathrm{d}\mu(t)$ is $|\mathrm{d}B_H|^{1/H}$. In the Wiener Brownian case, $H = 1/2$ and $\mathrm{d}\mu$ is the "local variance". This is how a price that fluctuates up and down is reduced to a positive measure.

2. Background: the Bernoulli binomial measure and two random variants: shuffled and canonical

The prototype of all multifractals is nonrandom: it is a Bernoulli binomial measure. Its well-known properties are recalled in this section, then Section 3 introduces a random "canonical" version. Also, all Bernoulli binomial measures being powers of one another, a broader viewpoint considers them as forming a single "class of equivalence".

2.1. Definition and construction of the Bernoulli binomial measure

A multiplicative nonrandom cascade. A recursive construction of the Bernoulli binomial measures involves an "initiator" and a "generator". The initiator is the interval [0, 1] on which a unit of mass is uniformly spread. This interval will recursively split into halves, yielding dyadic intervals of length 2^{-k}. The generator consists in a single parameter u, variously called *multiplier* or *mass*. The first stage spreads mass over the halves of every dyadic interval, with unequal proportions. Applied to [0, 1], it leaves the mass u in [0, 1/2] and the mass v in $[1/2, 1]$. The $(k+1)$-th stage begins with dyadic intervals of length 2^{-k}, each split in two subintervals of length 2^{-k-1}. A proportion equal to u goes to the left subinterval and the proportion v, to the right.

After k stages, let φ_0 and $\varphi_1 = 1 - \varphi_0$ denote the relative frequencies of 0's and 1's in the finite binary development $t = 0.\beta_1\beta_2 \ldots \beta_k$. The "pre-binomial" measures in the dyadic interval $[\mathrm{d}t] = [t, t + 2^{-k}]$ takes the value

$$\mu_k(\mathrm{d}t) = u^{k\varphi_0} v^{k\varphi_1},$$

which will be called "pre-multifractal". This measure is distributed uniformly over the interval. For $k \to \infty$, this sequence of measures $\mu_k(\mathrm{d}t)$ has a limit $\mu(\mathrm{d}t)$, which is the Bernoulli binomial multifractal.

Shuffled binomial measure. The proportion equal to u now goes to either the left or the right subinterval, with equal probabilities, and the remaining proportion v goes to the remaining subinterval. This variant must be mentioned but is not interesting.

2.2. *The concept of canonical random cascade and the definition of the canonical binomial measure*

Mandelbrot (1974a, b) took a major step beyond the preceding constructions.

The random multiplier M. In this generalization every recursive construction can be described as follows. Given the mass m in a dyadic interval of length 2^{-k}, the two subintervals of length 2^{-k-1} are assigned the masses $M_1 m$ and $M_2 m$, where M_1 and M_2 are *independent* realizations of a random variable M called multiplier. This M is equal to u or v with probabilities $p = 1/2$ and $1 - p = 1/2$.

The Bernoulli and shuffled binomials both impose the constraint that $M_1 + M_2 = 1$. The canonical binomial does not. It follows that the canonical mass in each interval of duration 2^{-k} is multiplied in the next stage by the sum $M_1 + M_2$ of two independent realizations of M. That sum is either $2u$ (with probability p^2), or 1 (with probability $2(1-p)p$), or $2v$ (with probability $1 - p^2$).

Writing p instead of 1/2 in the Bernoulli case and its variants complicates the notation now, but will soon prove advantageous: the step to the TVCM will simply consist in allowing $0 < p < 1$.

2.3. *Two forms of conservation: strict and on the average*

Both the Bernoulli and shuffled binomials repeatedly redistribute mass, but within a dyadic interval of duration 2^{-k}, the mass remains exactly conserved in all stages beyond the k-th. That is, the limit mass $\mu(t)$ in a dyadic interval satisfies $\mu_k(\mathrm{d}t) = \mu(\mathrm{d}t)$.

In a canonical binomial, to the contrary, the sum $M_1 + M_2$ is not identically 1, only its expectation is 1. Therefore, canonical binomial construction preserve mass on the average, but not exactly.

The random variable Ω. In particular, the mass $\mu([0,1])$ is no longer equal to 1. It is a basic random variable denoted by Ω and discussed in Section 4.

Within a dyadic interval dt of length 2^{-k}, the cascade is simply a reduced-scale version of the overall cascade. It transforms the mass $\mu_k(dt)$ into a product of the form $\mu(dt) = \mu_k(dt)\Omega(dt)$ where all the $\Omega(dt)$ are independent realizations of the same variable Ω.

2.4. *The term "canonical" is motivated by statistical thermodynamics*

As is well known, statistical thermodynamics finds it valuable to approximate large systems as juxtapositions of parts, the "canonical ensembles", whose energy only depends on a common temperature and not on the energies of the other parts. Microcanonical ensembles' energies are constrained to add to a prescribed total energy. In the study of multifractals, the use of this metaphor should not obscure the fact that the multiplication of canonical factors introduces strong dependence among $\mu(dt)$ for different intervals dt.

2.5. *In every variant of the binomial measure one can view all finite (positive or negative) powers together, as forming a single "class of equivalence"*

To any given real exponent $g \neq 1$ and multipliers u and v corresponds a multiplier M_g that can take either of two values $u_g = \psi u^g$ with probability p, and $v_g = \psi v^g$ with probability $1-p$. The factor ψ is meant to insure $pu_g + (1-p)v_g = 1/2$. Therefore, $\psi[pu^g + (1-p)v^g] = 1/2$, that is, $\psi = 1/[2EM^g]$. The expression $2EM^g$ will be generalized and encountered repeatedly especially through the expression

$$\tau(q) = -\log_2\left[pu^q + (1-p)v^q\right] - 1 = -\log_2\left(2EM^q\right).$$

This is simply a notation at this point but will be justified in Section 5. It follows that $\psi = 2^{-\tau(g)}$, hence

$$u_g = u^g 2^{\tau(g)} \quad \text{and} \quad v_g = v^g 2^{\tau(g)}.$$

Assume $u > v$. As g ranges from 0 to ∞, u_g ranges from 1/2 to 1 and v_g ranges from 1/2 to 0; the inequality $u_g > v_g$ is preserved. To the contrary, as g ranges from 0 to ∞, $v_g < u_g$. For example, $g = -1$ yields

$$u_g = \frac{1/u}{1/u + 1/v} = v \quad \text{and} \quad v_g = \frac{1/v}{1/v + 1/v} = u.$$

Thus, inversion leaves both the shuffled and the canonical binomial measures unchanged. For the Bernoulli binomial, it only changes the direction of the time axis.

Altogether, every Bernoulli binomial measure can be obtained from any other as a reduced positive or negative power. If one agrees to consider a measure and its reduced powers as equivalent, *there is only one Bernoulli binomial measure.*

In concrete terms relative to non-infinitesimal dyadic intervals, the sequences representing $\log \mu$ for different values of g are mutually affine. Each is obtained from the special case $g = 1$ by a multiplication by g followed by a vertical translation.

2.6. *The full and folded forms of the address plane*

In anticipation of TVCM, the point of coordinates u and v will be called the *address* of a binomial measure in a *full address space*. In that plane, the locus of the Bernoulli measures is the interval defined by $0 < v$, $0 < u$, and $u + v = 1$.

The *folded address space* will be obtained by identifying the measures (u, v) and (v, u), and representing both by one point. The locus of the Bernoulli measures becomes the interval defined by the inequalities $0 < v < u$ and $u + v = 1$.

2.7. *Alternative parameters*

In its role as parameter added to $p = 1/2$, one can replace u by the ("information-theoretical") fractal dimension $D = -u \log_2 u - v \log_2 v$ which can be chosen at will in this open interval $]0, 1[$. The value of D characterizes the "set that supports" the measure. It received a new application in the new notion of multifractal concentration described in Mandelbrot (2001c). More generally, the study of all multifractals, including the Bernoulli binomial, is filled with fractal dimensions of many other sets. All are unquestionably positive. One of the newest features of the TVCM will prove to be that they also allow negative dimensions.

3. Definition of the two-valued canonical multifractals

3.1. *Construction of the two-valued canonical multifractal in the interval* $[0, 1]$

The TVCM are called two-valued because, as with the Bernoulli binomial, the multiplier M can only take 2 possible values u and v. The novelties are that p need not be $1/2$, the multipliers u and v are not bounded by 1, and the inequality $u + v \neq 1$ is acceptable.

For $u + v \neq 1$, the total mass cannot be preserved exactly. Preservation on the average requires

$$EM = pu + (1 - p)v = \frac{1}{2},$$

hence $0 < p = (1/2 - v)/(u - v) < 1$.

The construction of TVCM is based upon a recursive subdivision of the interval $[0, 1]$ into equal intervals. The point of departure is, once again, a uniformly spread unit mass. The first stage splits $[0, 1]$ into two parts of equal lengths. On each, mass is poured uniformly, with the respective densities M_1 and M_2 that are independent copies of M. The second stage continues similarly with the interval $[0, 1/2]$ and $[1/2, 1]$.

3.2. *A second special two-valued canonical multifractal: the unifractal measure on the canonical Cantor dust*

The identity $EM = 1/2$ is also satisfied by $u = 1/2p$ and $v = 0$. In this case, let the lengths and number of non-empty dyadic cells after k stages be denoted by $\Delta t = 2^{-k}$ and N_k. The random variable N_k follows a simple birth and death process leading to the following alternative.

When $p > 1/2$, $EN_k = (EN_1)^k = (2p)^k = (\mathrm{d}t)^{\log(2p)}$. To be able to write $EN_k = (\mathrm{d}t)^{-D}$, it suffices to introduce the exponent $D = -\log(2p)$. It satisfies $D > 0$ and defines a fractal dimension.

When $p < 1/2$, to the contrary, the number of non-empty cells almost surely vanishes asymptotically. At the same time, the formal fractal dimension $D = -\log(2p)$ satisfies $D < 0$.

3.3. *Generalization of a useful new viewpoint: when considered together with their powers from $-\infty$ to ∞, all the TVCM parametrized by either p or $1 - p$ form a single class of equivalence*

To take the key case, the multiplier M^{-1} takes the values

$$u_{-1} = \frac{1/u}{2(p/u + (1-p)/v)} = \frac{v}{2(v+u) - 1} \quad \text{and} \quad v_{-1} = \frac{u}{2(v+u) - 1}.$$

It follows that $pu_{-1} + (1-p)v_{-1} = 1/2$ and $u_{-1}/v_{-1} = v/u$. In the full address plane, the relations imply the following: (a) the point (u_{-1}, v_{-1}) lies on the extension beyond $(1/2, 1/2)$ of the interval from (u, v) to $(1/2, 1/2)$ and (b) the slopes of the intervals from 0 to (u, v) and from 0 to (u_{-1}, v_{-1}) are inverse of one another. It suffices to fold the full phase diagram along the diagonal to achieve $v > u$. The point (u_{-1}, v_{-1}) will be the intersection of the interval corresponding to the probability $1 - p$ and of the interval joining 0 to (u, v).

3.4. *The full and folded address planes*

In the full address plane, the locus of all the points (u, v) with fixed p has the equation $pu + (1-p)v = 1/2$. This is the negatively sloped interval joining the points $(0, 1/2p)$ and $([1/2(1-p)], 0)$. When (u, v) and (v, u) are identified, the locus becomes the same interval plus the negatively sloped interval from $[0, 1/2(1-p)]$ to $(1/2p, 0)$.

In the folded address plane, the locus is made of two shorter intervals from $(1, 1)$ to both $(1/2p, 0)$ and $([1/2(1-p)], 0)$. In the special case $u + v = 1$ corresponding to $p = 1/2$, the two shorter intervals coincide.

Those two intervals correspond to TVCM in the same class of equivalence. Starting from an arbitrary point on either interval, positive moments correspond to points to the same interval and negative moments, to points of the other. Moments for $g > 1$ correspond to points to the left on the same interval; moments for $0 < g < 1$, to points to the right on the same interval; negative moments to points on the other interval.

For $p \neq 1/2$, the class of equivalence of p includes a measure that corresponds to $u = 1$ and $v = [1/2 - \min(p, 1-p)]/[\max(p, 1-p)]$. This novel and convenient universal point of reference requires $p \neq 1/2$. In terms to be explained below, it corresponds to $\alpha_{\min} = -\log u = 0$.

3.5. Background of the two-valued canonical measures in the historical development of multifractals

The construction of TVCM is new but takes a well-defined place among the three main approaches to the development of a theory of multifractals.

General mathematical theories came late and have the drawback that they are accessible to few non-mathematicians and many are less general than they seem.

The heuristic presentation in Frisch and Parisi (1985) and Halsey et al. (1986) came after Mandelbrot (1974a, b) but before most of the mathematics. Most importantly for this paper's purpose, those presentations fail to include significantly random constructions, hence cannot yield measures following the power law distribution.

Both the mathematical and the heuristic approaches seek generality and only later consider the special cases. To the contrary, a third approach, the first historically, began in Mandelbrot (1974a, b) with the careful investigation of a variety of special random multiplicative measures. I believe that each feature of the general theory continues to be best understood when introduced through a special case that is as general as needed, but no more. The general theory is understood very easily when it comes last.

In pedagogical terms, the "third way" associates with each distinct feature of multifractals a special construction, often one that consists of generalizing the binomial multifractal in a new direction. TVCM is part of a continuation of that effective approach; it could have been investigated much earlier if a clear need had been perceived.

4. The limit random variable $\Omega = \mu([0, 1])$, its distribution and the star functional equation

4.1. The identity $EM = 1$ implies that the limit measure has the "martingale" property, hence the cascade defines a limit random variable $\Omega = \mu([0, 1])$

We cannot deal with martingales here, but positive martingales are mathematically attractive because they converge (almost surely) to a limit. But the situation is complicated because the limit depends on the sign of $D = 2[-pu \log_2 u - (1-p)v \log_2 v]$.

Under the condition $D > 0$, which is discussed in Section 9, what seemed obvious is confirmed: $Pr\{\Omega > 0\} > 0$, conservation on the average continues to hold as $k \to \infty$, and Ω is either non-random, or is random and satisfies the identity $E\Omega = 1$.

But if $D < 0$, one finds that $\Omega = 0$ almost surely and conservation on the average holds for finite k but fails as $k \to \infty$. The possibility that $\Omega = 0$ arose in mathematical esoterica and seemed bizarre, but is unavoidably introduced into concrete science.

4.2. *Questions*

(A) Which feature of the generating process dominates the tail distribution of Ω? It is shown in Section 6 to be the sign of $\max(u, v) - 1$.
(B) Which feature of the generating process allows Ω to have a high probability of being either very large or very small? Section 6 will show that the criterion is that the function $\tau(q)$ becomes negative for large enough q.
(C) Divide [0, 1] into 2^k intervals of length 2^{-k}. Which feature of the generating process determines the relative distribution of the overall Ω among those small intervals? This relative distribution motivated the introduction of the functions $f(\alpha)$ and $\rho(\alpha)$, and is discussed in Section 8.
(D) Are the features discussed under (B) and (C) interdependent? Section 10 will address this issue and show that, even when Ω has a high probability of being large, its value does not affect the distribution under (C).

4.3. *Exact stochastic renormalizability and the "star functional equation" for Ω*

Once again, the masses in [0, 1/2] and [1/2, 1] take, respectively, the forms $M_1\Omega_1$ and $M_2\Omega_2$, where M_1 and M_2 are two independent realizations of the random variable M and Ω_1, and Ω_2 are two independent realizations of the random variable Ω. Adding the two parts yields

$$\Omega \equiv \Omega_1 M_1 + \Omega_2 M_2.$$

This identity in distribution, now called the "star equation", combines with $E\Omega = 1$ to determine Ω. It was introduced in Mandelbrot (1974a, b) and has since then been investigated by several authors, for example by Durrett and Liggett (1983). A large bibliography is found in Liu (2002).

In the special case where M is non-random, the star equation reduces to the equation due to Cauchy whose solutions have become well-known: they are the Cauchy–Lévy stable distributions.

4.4. *Metaphor for the probability of large values of Ω, arising in the theory of discrete time branching processes*

A growth process begins at $t = 0$ with a single cell. Then, at every integer instant of time, every cell splits into a random non-negative number of N_1 cells. At time k, one deals with a clone of N_k cells. All those random splittings are statistically independent and identically distributed. The normalized clone size, defined as N_k/EN_1^k has an expectation equal to 1. The sequence of normalized sizes is a positive martingale, hence (as already mentioned) converges to a limit random variable.

When $EN > 1$, that limit does not reduce to 0 and is random for a very intuitive reason. As long as clone size is small, its growth very much depends on chance, therefore the normalized clone size is very variable. However, after a small number of splittings, a law of large numbers comes into force, the effects of chances become negligible, and the clone grows near-exponentially. That is, the randomness in the relative number of family members can be very large but acts very early.

4.5. *To a large extent, the asymptotic measure Ω of a TVCM is large if, and only if, the pre-fractal measure $\mu_k([0, 1])$ has become large during the very first few stages of the generating cascade*

Such behavior is suggested by the analogy to a branching process, and analysis shows that such is indeed the case. After the first stage, the measures $\mu_1([0, 1/2])$ and $\mu_1([1/2, 1])$ are both equal to u^2 with probability p^2, uv with probability $2p(1-p)$, and v^2 with probability $(1-p)^2$. Extensive simulations were carried out for large k in "batches", and the largest, medium, and smallest measure was recorded for each batch. Invariably, the largest (resp., smallest) Ω started from a high (resp., low) overall level.

5. The function $\tau(q)$: motivation and form of the graph

So far $\tau(q)$ was nothing but a notation. It is important as it is the special form taken for TVCM by a function that was first defined for an arbitrary multiplier in Mandelbrot (1974a, b). (Actually, the little appreciated Figure 1 of that original paper did not include $q < 0$ and worked with $-\tau(q)$, but the opposite sign came to be generally adopted.)

5.1. *Motivation of $\tau(q)$*

After k cascade stages, consider an arbitrary dyadic interval of duration $\mathrm{d}t = 2^{-k}$. For the k-approximant TVCM measure $\mu_k(\mathrm{d}t)$ the q-th power has an expected value equal to $[pu^q + (1-p)v^q]^k = \{EM^q\}^k$. Its logarithm of base 2 is

$$\log_2\{[pu^q + (1-p)v^q]^k\} = k \log_2\{pu^q + (1-p)v^q\}$$
$$= \log_2(\mathrm{d}t)[\tau(q) + 1].$$

Hence

$$E\mu_k^q(\mathrm{d}t) = (\mathrm{d}t)^{\tau(q)+1}.$$

5.2. *A generalization of the role of Ω: middle- and high-frequency contributions to microrandomness*

Exactly the same cascade transforms the measure in $\mathrm{d}t$ from $\mu_k(\mathrm{d}t)$ to $\mu(\mathrm{d}t)$ and the measure in $[0, 1]$ from 1 to Ω. Hence, one can write

$$\mu(\mathrm{d}t) = \mu_k(\mathrm{d}t)\Omega(\mathrm{d}t).$$

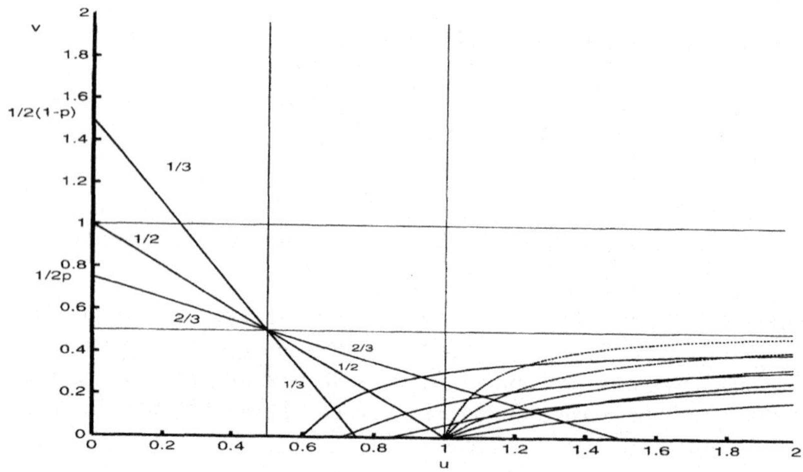

Fig. 1. The full phase diagram of TVCM with coordinates u and v. The isolines of the quantity p are straight intervals from $(1/\{2(1-p)\}, 0)$ to $(0, 1/\{2p\})$. The values p and $1-p$ are equivalent and the corresponding isolines are symmetric with respect to the main bisector $u=v$. The acceptable part of the plane excludes the points (u, v) such that either $\max(u, v) < 1/2$ or $\min(u, v) > 1/2$. Hence, the relevant part of this diagram is made of two infinite halfstrips reducible to one another by folding along the bisector. The folded phase diagram of TVCM corresponds to $v < 0.5 < u$. It shows the following curves. The isolines of $1-p$ and p are straight intervals that start at the point $(1, 1)$ and end at the points $(1/\{2p\}, 0)$ and $(1/\{2(1-p)\}, 0)$. The isolines of D start on the interval $1/2 < u < 1$ of the u-axis and continue to the point $(\infty, 0)$. The isolines of q_{crit} start at the point $(1, 0)$ and continue to the point $(\infty, 0)$. The Bernoulli binomial measure corresponds to $p = 1/2$ and the canonical Cantor measure corresponds to the half line $v = 0$, $u > 1/2$.

In this product, frequencies of wavelength $> \mathrm{d}t$, to be described as "low", contribute $\mu_k([0, 1])$, and frequencies of wavelength $< \mathrm{d}t$, to be described as "high", contribute Ω.

5.3. *The expected "partition function"* $\sum E\mu^q(\mathrm{d}_i t)$

Section 6 will show that $E\Omega^q$ need not be finite. But if it is, the limit measure $\mu(\mathrm{d}t) = \mu_k(\mathrm{d}t)\Omega(\mathrm{d}t)$ satisfies

$$E\mu^q(\mathrm{d}t) = (\mathrm{d}t)^{\tau(q)+1} E\Omega^q.$$

The interval $[0, 1]$ subdivides into $1/\mathrm{d}t$ intervals $\mathrm{d}_i t$ of common length $\mathrm{d}t$. The sum of the q-th moments over those intervals takes the form

$$E\chi(\mathrm{d}t) = \sum E\mu^q(\mathrm{d}_i t) = (\mathrm{d}t)^{\tau(q)} E\Omega^q.$$

Estimation of $\tau(q)$ from a sample. It is affected by the prefactor Ω insofar as one must estimate both $\tau(q)$ and $\log E\Omega^q$.

5.4. Form of the $\tau(q)$ graph

Due to conservation on the average, $EM = pu + (1 - p)v = 1/2$, hence $\tau(1) = -\log_2[1/2] - 1 = 0$. An additional universal value is $\tau(0) = -\log_2(1) - 1 = -1$. For other values of q, $\tau(q)$ is a cap-convex continuous function satisfying $\tau(q) < -1$ for $q < 0$.

For TVCM, a more special property is that $\tau(q)$ is asymptotically linear: assuming $u > v$, and letting $q \to \infty$:

$$\tau(q) \sim -\log_2 p - 1 - q \log u \quad \text{and} \quad \tau(-q) \sim -\log_2(1-p) - 1 + q \log v.$$

The sign of $u - 1$ affects the sign of $\log u$, a fact that will be very important in Section 6.

Moving as little as possible beyond these properties. The very special tau function of the TVCM is simple but Figure 2 suffices to bring out every one of the delicate possibilities first reported in Mandelbrot (1974a), where $-\tau(q)$ is plotted in that little appreciated Figure 1.

Other features of τ that deserve to be mentioned. Direct proofs are tedious and the short proofs require the multifractal formalism that will only be described in Section 11.

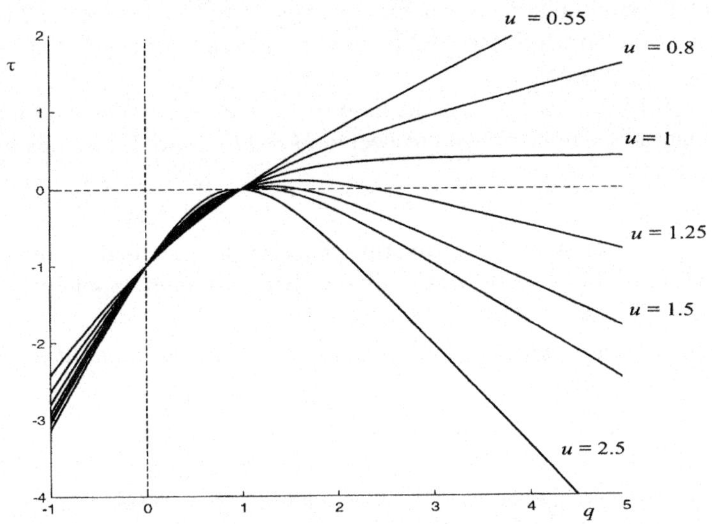

Fig. 2. The function $\tau(q)$ for $p = 3/4$ and varying g. By arbitrary choice, the value $g = 1$ is assigned $u = 1$, from which follows that $g = -1$ is assigned to the case $v = 1$. Behavior of $\tau(q)$ for the value $g > 0$: as $q \to -\infty$, the graph of $\tau(q)$ is asymptotically tangent to $\tau = -q \log_2 v$, as $q \to \infty$, the graph of $\tau(q)$ is asymptotically tangent to $\tau = -q \log_2 u$. Those properties are widely believed to describe the main facts about $\tau(q)$. But for TVCM they do not. Thus, $\tau(q)$ is also tangent to $\tau = q\alpha^*_{\max}$ and $\tau = q\alpha^*_{\min}$. Beyond those points of tangency, f becomes < 0. For $g > 1$, that is, for $u > 1$, $\tau(q)$ has a maximum. Values of q beyond this maximum correspond to $\alpha_{\min} < 0$. Because of the capconvexity of $\tau(q)$, the equation $\tau(q) = 0$ may, in addition to the "universal" value $q = 1$, have a root $q_{\text{crit}} > 1$. For $u > 2.5$, one deals with a very different phenomenon also first described in Mandelbrot (1974a, b). One finds that the construction of TVCM leads to a measure that degenerates to 0.

The quantity $D(q) = \tau(q)/(q-1)$. This popular expression is often called a "generalized dimension", a term too vague to mean anything. $D(q)$ is obtained by extending the line from (q, τ) to $(1, 0)$ to its intercept with the line $q = 0$. It plays the role of a critical embedding codimension for the existence of a finite q-th moment. This topic cannot be discussed here but is treated in Mandelbrot (2003).

The ratio $\tau(q)/q$ *and the "accessible" values of* q. Increase q from $-\infty$ to 0 then to $+\infty$. In the Bernoulli case, $\tau(q)/q$ increases from $\alpha_{\max}$ to ∞, jumps down to $-\infty$ for $q = 0$, then increases again from $-\infty$ to $\alpha_{\min}$. For TVCM with $p \neq 1/2$, the behavior is very different. For example, let $p < 1/2$. As q increases from 1 to ∞, $\tau(q)$ increases from 0 to a maximum $\alpha^*_{\max}$, then decreases. In a way explored in Section 10, the values of $\alpha > \alpha^*_{\max}$ are not "accessible".

5.5. *Reducible and irreducible canonical multifractals*

Once again, being "canonical" implies conservation on the average. When there exists a microcanonical (conservative) variant having the same function $f(\alpha)$, a canonical measure can be called "reducible". The canonical binomial is reducible because its $f(\alpha)$ is shared by the Bernoulli binomial. Another example introduced in Mandelbrot (1989b) is the "Erice" measure, in which the multiplier M is uniformly distributed on [0, 1]. But the TVCM with $p \neq 1/2$ is not reducible.

In the interval [0, 1] subdivided in the base $b = 2$, reducibility demands a multiplier M whose distribution is symmetric with respect to $M = 1/2$. Since $u > 0$, this implies $u < 1$.

6. When $u > 1$, the moment $E\Omega^q$ diverges if q exceeds a critical exponent q_{crit} satisfying $\tau(q) = 0$; Ω follows a power-law distribution of exponent q_{crit}

6.1. *Divergent moments, power-law distributions and limits to the ability of moments to determine a distribution*

This section injects a concern that might have been voiced in Sections 4 and 5. The canonical binomial and many other examples satisfy the following properties, which everyone takes for granted and no one seems to think about: (a) $\Omega = 1$, $E\Omega^q < \infty$, (b) $\tau(q) > 0$ for all $q > 0$, and (c) $\tau(q)/q$ increases monotonically as $q \to \pm\infty$.

Many presentations of fractals take those properties for granted in all cases. In fact, as this section will show, the TVCM with $u > 1$ lead to the "anomalous" divergence $E\Omega^q = \infty$ and the "inconceivable" inequality $\tau(q) < 0$ for $q_{\text{crit}} < q < \infty$. Also, the monotonicity of $\tau(q)/q$ fails for all TVCM with $p \neq 1/2$.

Since Pareto in 1897, infinite moments have been known to characterize the power-law distributions of the form $Pr\{X > x\} = x^{-q_{\text{crit}}}$. But in the case of TVCM and other canonical multifractals, the complicating factor $L(x)$ is absent. One finds that when $u > 1$, the overall measure Ω follows a power law of exponent q_{crit} determined by $\tau(q)$.

6.2. Discussion

The power-law "anomalies" have very concrete consequences deduced in Mandelbrot (1997) and discussed, for example, in Mandelbrot (2001c).

But does all this make sense? After all, $\tau(q)$ and $E\Omega^q$ are given by simple formulas and are finite for all parameters. The fact that those values cannot actually be observed raises a question. Are high moments lost by being unobservable? In fact, they are "latent" but can be made "actual" by a process is indeed provided by the process of "embedding" studied elsewhere.

An additional comment is useful. The fact that high moments are non-observable does not express a deficiency of TVCM but a limitation of the notion of moment. Features ordinarily expressed by moments must be expressed by other means.

6.3. An important apparent "anomaly": in a TVCM, the q-th moment of Ω may diverge

Let us elaborate. From long past experience, physicists' and statisticians' natural impulse is to define and manipulate moments without envisioning or voicing the possibility of their being infinite. This lack of concern cannot extend to multifractals. The distribution of the TVCM within a dyadic interval introduces an additional critical exponent q_{crit} that satisfies $q_{\text{crit}} > 1$. When $1 < q_{\text{crit}} < \infty$, which is a stronger requirement that $D > 0$, the q-th moment of $\mu(\mathrm{d}t)$ diverges for $q > q_{\text{crit}}$.

A stronger result holds: the TVCM cascade generates a measure whose distribution follows the power law of exponent q_{crit}.

Comment. The heuristic approach to non-random multifractals fails to extend to random ones, in particular, it fails to allow $q_{\text{crit}} < \infty$. This makes it incomplete from the viewpoint of finance and several other important applications.

The finite q_{crit} has been around since Mandelbrot (1974a, b) (where it is denoted by α) and triggered a substantial literature in mathematics. But it is linked with events so extraordinarily unlikely as to appear incapable of having any perceptible effect on the generated measure. The applications continue to neglect it, perhaps because it is ill-understood. A central goal of TVCM is to make this concept well-understood and widely adopted.

6.4. An important role of $\tau(q)$: if $q > 1$ the q-th moment of Ω is finite if, and only if, $\tau(q) > 0$; the same holds for $\mu(\mathrm{d}t)$ whenever $\mathrm{d}t$ is a dyadic interval

By definition, after k levels of iteration, the following symbolic equality relates independent realizations of M and μ. That is, it does not link random variables but distributions

$$\mu_k([0,1]) = M\mu_{k-1}([0,1]) + M\mu_{k-1}([0,1]).$$

Conservation on the average is expressed by the identity $E\mu_{k-1}([0,1]) = 1$. In addition, we have the following recursion relative to the second moment.

$$E\mu^2([0,1]) = 2EM^2\left[E\mu_{k-1}^2([0,1])\right] + 2EM^2\left[E\mu_{k-1}([0,1])\right]^2.$$

The second term to the right reduces to 1/2. Now let $k \to \infty$. The necessary and sufficient condition for the variance of $\mu_k([0,1])$ to converge to a finite limit is

$$2(EM^2) < 1 \quad \text{in other words} \quad \tau(2) = -\log_2(EM^2) - 1 > 0.$$

When such is the case, Kahane and Peyrière (1976) gave a mathematically rigorous proof that there exists a limit measure $\mu([0,1])$ satisfying the formal expression

$$E\mu^2([0,1]) = \frac{1}{2(1-2^{\tau(2)})}.$$

Higher integer moments satisfy analogous recursion relations. That is, knowing that all moments of order up to $q-1$ are finite, the moment of order q is finite if and only if $\tau(q) > 0$.

The moments of non-integer order q are more delicate to handle, but they too are finite if, and only if, $\tau(q) > 0$.

6.5. *Definition of q_{crit}; proof that in the case of TVCM q_{crit} is finite if, and only if, $u > 1$*

Section 5.4 noted that the graph of $\tau(q)$ is always cap-convex and for large $q > 0$,

$$\tau(q) \sim -\log_2(pu^q) + -1 \sim -\log_2 p - 1 - q\log_2 u.$$

The dependence of $\tau(q)$ on q is ruled by the sign of $u - 1$, as follows.

- *The case when $u < 1$, hence $\alpha_{\min} > 0$.* In this case, $\tau(q)$ is monotone increasing and $\tau(q) > 0$ for $q > 1$. This behavior is exemplified by the Bernoulli binomial.
- *The case when $u > 1$, hence $\alpha_{\min} < 0$.* In this case, one has $\tau(q) < 0$ for large q. In addition to the root $q = 1$, the equation $\tau(q) = 1$ has a second root that is denoted by q_{crit}.

Comment. In terms of the function $f(\alpha)$ graphed on Figure 3, the values 1 and q_{crit} are the slopes of the two tangents drawn to $f(\alpha)$ from the origin (0, 0).

Within the class of equivalence of any p and $1-p$; the parameter g can be "tuned" so that q_{crit} begins by being > 1 then converges to 1; if so, it is seen that D converges to 0.

- Therefore, the conditions $q_{\text{crit}} = 1$ and $D = 0$ describe the same "anomaly".

In Figure 1, isolines of q_{crit} are drawn for $q_{\text{crit}} = 1, 2, 3$, and 4. When $q = 1$ is the only root, it is convenient to say that $q_{\text{crit}} = \infty$. This isoset $q_{\text{crit}} = \infty$ is made of the half-line $\{v = 1/2 \text{ and } u > 1/2\}$ and of the square $\{0 < v < 1/2, 1/2 < u < 1\}$.

6.6. *The exponent q_{crit} can be considered as a macroscopic variable of the generating process*

Any set of two parameters that fully describes a TVCM can be called "microscopic". All the quantities that are directly observable and can be called macroscopic are functions of those two parameters.

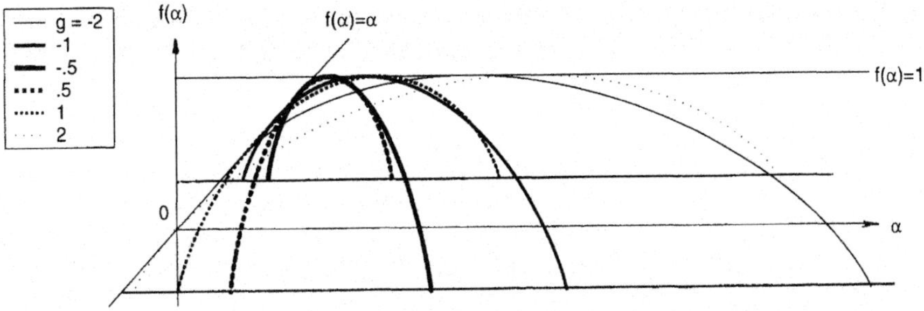

Fig. 3. The functions $f(\alpha)$ for $p = 3/4$ and varying g. All those graphs are linked by horizontal reductions or dilations followed by translation and further self-affinity. It is widely anticipated that $f(\alpha) > 0$ holds in all cases, but for the TVCM this anticipation fails, as shown in this figure. For $g > 0$ (resp., $g < 0$) the left endpoint of $f(\alpha)$ (resp., the right endpoint) satisfies $f(\alpha) < 0$ and the other endpoint, $f(\alpha) > 0$.

For the general canonical multifractal, a full specification requires a far larger number of microscopic quantities but the same number of macroscopic ones. Some of the latter characterize each sample, but others, for example q_{crit}, characterize the population.

7. The quantity α: the original Hölder exponent and beyond

The multiplicative cascades – common to the Bernoulli and canonical binomials and TVCM – involve successive multiplications. An immediate consequence is that both the basic $\mu(\mathrm{d}t)$ and its probability are most intrinsically viewed through their logarithms. A less obvious fact is that a normalizing factor $1/\log(\mathrm{d}t)$ is appropriate in each case. An even less obvious fact is that the normalizations $\log\mu/\log \mathrm{d}t$ and $\log P/\log \mathrm{d}t$ are of far broader usefulness in the study of multifractals. The exact extend of their domain of usefulness is beyond the goal of this chapter, but we keep some special cases that can be treated fully by elementary arguments.

7.1. The Bernoulli binomial case and two forms of the Hölder exponent: coarse-grained (or coarse) and fine-grained

Recall that due to conservation, the measure in an interval of length $\mathrm{d}t = 2^{-k}$ is the same after k stages and in the limit, namely, $\mu(\mathrm{d}t) = \mu_k(\mathrm{d}t)$. As a result, the coarse-grained Hölder exponent can be defined in either of two ways,

$$\alpha(\mathrm{d}t) = \frac{\log\mu(\mathrm{d}t)}{\log(\mathrm{d}t)} \quad \text{and}$$

$$\tilde{\alpha}(\mathrm{d}t) = \frac{\log\mu_k(\mathrm{d}t)}{\log(\mathrm{d}t)}.$$

The distinction is empty in the Bernoulli case but prove prove essential for the TVCM. In terms of the relative frequencies φ_0 and φ_1 defined in Section 2.1,

$$\alpha(\mathrm{d}t) = \tilde{\alpha}(\mathrm{d}t) = \alpha(\varphi_0, \varphi_1) = -\varphi_0 \log_2 u - \varphi_1 \log_2 v$$
$$= -\varphi_0(\log_2 u - \log_2 v) - \log v.$$

Since $u > v$, one has $0 < \alpha_{\min} = -\log_2 u \leqslant \alpha = \tilde{\alpha} \leqslant \alpha_{\max} = -\log_2 v < \infty$. In particular, $\alpha > 0$, hence $\tilde{\alpha} > 0$. As $\mathrm{d}t \to 0$, so does $\mu(\mathrm{d}t)$, and a formal inversion of the definition of α yields

$$\mu(\mathrm{d}t) = (\mathrm{d}t)^{\alpha}.$$

This inversion reveals an old mathematical pedigree. Redefine φ_0 and φ_1 from denoting the finite frequencies of 0 and 1 in an interval, into denoting the limit frequencies at an instant t. The instant t is the limit of an infinite sequence of approximating intervals of duration 2^{-k}. The function $\mu([0, t])$ is non-differentiable because $\lim_{\mathrm{d}t \to 0} \mu(\mathrm{d}t)/\mathrm{d}t$ is not defined and cannot serve to define the local density of μ at the instant $\mathrm{d}t$.

The need for alternative measures of roughness of a singularity expression first arose around 1870 in mathematical esoterica due to L. Hölder. In fractal/multifractal geometry this expression merged with a very concrete exponent due to H.E. Hurst and is continually being generalized. It follows that for the Bernoulli binomial measure, it is legitimate to interpret the coarse αs as finite-difference surrogates of the local (infinitesimal) Hölder exponents.

7.2. *In the general TVCM measure, $\alpha \neq \tilde{\alpha}$, and the link between "$\alpha$" and the Hölder exponent breaks down; one consequence is that the "doubly anomalous" inequalities $\alpha_{\min} < 0$, hence $\tilde{\alpha} < 0$, are not excluded*

A Hölder (Hurst) exponent is necessarily positive. Hence negative $\tilde{\alpha}$s cannot be interpreted as Hölder exponents. Let us describe the heuristic argument that leads to this paradox and then show that $\tilde{\alpha} < 0$ is a serious "anomaly": it shows that the link between "some kind of α" and the Hölder exponent requires a searching look. The resolution of the paradox is very subtle and is associated with the finite q_{crit} introduced in Section 6.5.

Once again, except in the Bernoulli case, $\Omega \neq 1$ and $\mu(\mathrm{d}t) = \mu_k(\mathrm{d}t)\Omega(\mathrm{d}t)$, hence

$$\alpha(\mathrm{d}t) = \tilde{\alpha}(\mathrm{d}t) + \frac{\log \Omega(\mathrm{d}t)}{\log \mathrm{d}t}.$$

In the limit $\mathrm{d}t \to 0$ the factor $\log = \Omega/\log(\mathrm{d}t)$ tends to 0, hence it seems that $\alpha = \tilde{\alpha}$. Assume $u > 1$, hence $\alpha_{\min} < 0$ and consider an interval where $\tilde{\alpha}(\mathrm{d}t) < 0$. The formal equality

$$\text{“}\mu_k(\mathrm{d}t) = (\mathrm{d}t)^{\tilde{\alpha}}\text{”}$$

seems to hold and to imply that "the" mass in an interval increases as the interval length $\to 0$. On casual inspection, this is absurd. On careful inspection, it is not – simply because the variable $dt = 2^{-k}$ and the function $\mu_k(dt)$ both depend on k. For example, consider the point t for which $\varphi_0 = 1$. Around this point, one has $\mu_k = u\mu_{k-1} > \mu_{k-1}$. This inequality is not paradoxical.

Furthermore, Section 8 shows that the theory of the multiplicative measures introduces $\tilde{\alpha}$ intrinsically and inevitably and allows $\tilde{\alpha} < 0$.

Those seemingly contradictory properties will be reexamined in Section 9. Values of $\mu(dt)$ will be seen to have a positive probability but one so minute that they can never be observed in the way $\alpha > 0$ are observed. But they affect the distribution of the variable Ω examined in Section 4, therefore are observed indirectly.

8. The full function $f(\alpha)$ and the function $\rho(\alpha)$

8.1. *The Bernoulli binomial measure: definition and derivation of the box dimension function $f(\alpha)$*

The number of intervals of denumerator 2^{-k} leading to φ_0 and φ_1 is $N(k, \varphi_0, \varphi_1) = k!/(k\varphi_0)!(k\varphi_1)!$, and dt is the reduction ratio r from $[0, 1]$ to an interval of duration dt. Therefore, the expression

$$f(k, \varphi_0, \varphi_1) = -\frac{\log N(k, \varphi_0, \varphi_1)}{\log(dt)} = -\frac{\log[k!/(k\varphi_0)!(k\varphi_1)!]}{\log(dt)}$$

is of the form $f(k, \varphi_0, \varphi_1) = -\log N/\log r$. Fractal geometry calls this the "box similarity dimension" of a set. This is one of several forms taken by *fractal dimension*. More precisely, since the boxes belong to a grid, it is a *grid fractal dimension*.

The dimension function $f(\alpha)$. For large k, the leading term in the Stirling approximation of the factorial yields

$$\lim_{k\to\infty} f(k, \varphi_0, \varphi_1) = f(\varphi_0, \varphi_1) = -\varphi_0 \log_2 \varphi_0 - \varphi_1 \log_2 \varphi_1.$$

8.2. *The "entropy ogive" function $f(\alpha)$; the role of statistical thermodynamics in multifractals and the contrast between equipartition and concentration*

Eliminate φ_0 and φ_1 between the functions f and $\alpha = -\varphi_0 \log u - \varphi_1 \log v$. This yields in parametric form a function, $f(\alpha)$. Note that $0 \leqslant f(\alpha) \leqslant \min\{\alpha, 1\}$. Equality to the right is achieved when $\varphi_0 = u$. The value α where $f = \alpha$ is very important and will be discussed in Section 9. In terms of the reduced variable $\varphi_0 = (\alpha - \alpha_{\min})/(\alpha_{\max} - \alpha_{\min})$, the function $f(\alpha)$ becomes the "ogive"

$$\tilde{f}(\varphi_0) = -\varphi_0 \log_2 \varphi_0 - (1 - \varphi_0)\log_2(1 - \varphi_0).$$

This $\tilde{f}(\varphi_0)$ can be called a universal function. The $f(\alpha)$ corresponding to fixed p and varying g are affine transforms of $\tilde{f}(\varphi_0)$, therefore of one another. The ogive function $\tilde{f}$ first arose in thermodynamics as an entropy and in 1948 (with Shannon) entered communication theory as an information. Its occurrence here is the first of several roles the formalism of thermodynamics plays in the theory of multifractals.

An essential but paradoxical feature. Equilibrium thermodynamics is a study of various forms of *near-equality*, for example postulates the equipartition of states on a surface in phase space or of energy among modes. In sharp contrast, multifractals are characterized by extreme *inequality* between the measures in different intervals of common duration $\mathrm{d}t$. Upon more careful examination, the paradox dissolves by being turned around: the main tools of thermodynamics can handle phenomena well beyond their original scope.

8.3. *The Bernoulli binomial measure, continued: definition and derivation of a function $\rho(\alpha) = f(\alpha) - 1$ that originates as a rescaled logarithm of a probability*

The function $f(\alpha)$ never fully specifies the measure. For example, it does not distinguish between the Bernoulli, shuffled and canonical binomials. The function $f(\alpha)$ can be generalized by being deduced from a function $\rho(\alpha) = f(\alpha) - 1$ that will now be defined. Instead of dimensions, that deduction relies on probabilities. In the Bernoulli case, the derivation of ρ is a minute variant of the argument in Section 8.1, but, contrary to the definition of f, the definition of ρ easily extends to TVCM and other random multifractals.

In the Bernoulli binomial case, the probability of hitting an interval leading to φ_0 and φ_1 is simply $P(k, \varphi_0, \varphi) = N(k, \varphi_0, \varphi_1)2^{-k} = k!/(k\varphi_0)!(k\varphi_1)!2^{-k}$. Consider the expression

$$\rho(k, \varphi_0, \varphi_1) = -\frac{\log[P(k, \varphi_0, \varphi_1)]}{\log(\mathrm{d}t)},$$

which is a rescaled but not averaged form of entropy. For large k, Stirling yields

$$\begin{aligned}\lim_{k\to\infty} \rho(k, \varphi_0, \varphi_1) &= \rho(\varphi_0, \varphi_1) = -\varphi_0 \log_2 \varphi_0 - \varphi_1 \log_2 \varphi_1 - 1 \\ &= f(\alpha) - 1.\end{aligned}$$

8.4. *Generalization of $\rho(\alpha)$ to the case of TVCM; the definition of $f(\alpha)$ as $\rho(\alpha) + 1$ is indirect but significant because it allows the generalized f to be negative*

Comparing the arguments in Sections 8.1 and 8.2 link the concepts of fractal dimension and of minus log (probability). However, when $f(\alpha)$ is reported through $f(\alpha) = \rho(\alpha) + 1$, the latter is not a mysterious "spectrum of singularities". It is simply the peculiar but proper way a probability distribution must be handled in the case of multifractal measures. Moreover, there is a major a priori difference exploited in Section 10. Minus log (probability) is not subjected to any bound. To the contrary, every one of the traditional definitions of fractal dimension (including Hausdorff–Besicovitch or Minkowski–Bouligand) necessarily yields a positive value.

The point is that the dimension argument in Section 8.1 does not carry over to TVCM, but the probability argument does carry over as follows. The probability of hitting an interval leading to φ_0 and φ_1 now changes to $P(k, \varphi_0, \varphi_1) = p(\varphi_0 k)!/(k\varphi_0)!(k\varphi_1)!$ One can now form the expression

$$\rho(k, \varphi_0, \varphi_1) = -\frac{\log[P(k, \varphi_0, \varphi_1)]}{\log(\mathrm{d}t)}.$$

Stirling now yields

$$\begin{aligned}\rho(\varphi_0, \varphi_1) &= \lim_{k\to\infty} \rho(k, \varphi_0, \varphi_1) \\ &= \{-\varphi_0 \log_2 \varphi_0 - \varphi_1 \log_2 \varphi_1\} + \{\varphi_0 \log_2 p + \varphi_1 \log_2(1 - p)\}.\end{aligned}$$

In this sum of two terms marked by braces, we know that the first one transforms (by horizontal stretching and translation) into the entropy ogive. The second is a linear function of φ, namely $\varphi_0[\log_2 p - \log_2(1 - p)] + \log_2(1 - p)$. It transforms the entropy ogive by an affinity in which the line joining the two support endpoints changes from horizontal to inclined. The overall affinity solely depends on p, but φ_0 depends explicitly on u and v.

This affinity extends to all values of p. Another property familiar from the binomial extends to all values of p. For all u and v, the graphs of $\rho(\alpha)$, hence of $f(\alpha)$ have a vertical slope for $q = \pm\infty$.

Alternatively, $\rho(\varphi_0, \varphi_1) = -\varphi_0 \log_2[\varphi_0/p] - \varphi_1 \log_2[\varphi_1/(1 - p)]$.

8.5. *Comments in terms of probability theory*

Roughly speaking, the measure μ is a *product* of random variables, while the limit theorems of probability theory are concerned with *sums*. The definition of α as $\log \mu(\mathrm{d}t)/\log(\mathrm{d}t)$ replaces a product of random variables M by a weighted sum of random variables of the form $\log M$. Let us now go through this argument step by step in greater rigor and generality. One needs a cumbersome restatement of $\alpha_k(\mathrm{d}t)$.

The low frequency factor of $\mu_k(\mathrm{d}t)$ and the random variable H_{low}. Consider once again a dyadic cell of length 2^{-k} that starts at $t = 0.\beta_1\beta_2\ldots\beta_k$. The first k stages of the cascade can be called of *low frequency* because they involve multipliers that are constant over dyadic intervals of length $\mathrm{d}t = 2^{-}k$ or longer. These stages yield

$$\mu_k(\mathrm{d}t) = M(\beta_1)M(\beta_1, \beta_2) \cdots M(\beta_1, \ldots, \beta_k) = \prod M.$$

We transform $\mu_k(\mathrm{d}t)$ into the *low frequency* random variable

$$\begin{aligned}H_{\mathrm{low}} &= \frac{\log[\mu_k(\mathrm{d}t)]}{\log(\mathrm{d}t)} \\ &= \frac{1}{k}[-\log_2 M(\beta_1) - \log_2 M(\beta_1, \beta_2) - \cdots].\end{aligned}$$

We saw in Section 4.5 that the first few values of M largely determine the distribution of Ω. But the last expression involves an operation of averaging in which the first terms contributing to $\mu(\mathrm{d}t)$ are asymptotically washed out.

8.6. *Distinction between "center" and "tail" theorems in probability*

The quantity $\tilde{\alpha}_k(\mathrm{d}t) = \varphi_0 \log_2 u - \varphi_1 \log_2 v$ is the average of a sum of variables $-\log M$; but why is its distribution is not Gaussian and the graph of $\rho(\alpha)$ is an entropy ogive rather than a parabola? Why is this so? The law of large numbers tells us that $\tilde{\alpha}_k(\mathrm{d}t)$ almost surely converges to its expectation which tells us very little. A tempting heuristic argument continues as follows. The central limit theorem is believed to ensure that for small $\mathrm{d}t$, $H_{\mathrm{low}}(\mathrm{d}t)$ becomes Gaussian, therefore the graph of $\log p(\mathrm{d}t)$ should be expected to be a parabola. This being granted, why is it that the Stirling approximation yields an entropy ogive – not a parabola?

In fact, there is no paradox of any kind. While the central limit theorem is indeed central to probability theory, all it asserts in this context is that, asymptotically, the Gaussian rules the *center* of the distribution, its "bell". Renormalizations reduce this center to the immediate neighborhood of the top of the $\rho(\alpha)$ graph and the central limit theorem is correct in asserting that the top of the entropy ogive is locally parabolic. But in the present context this information is of little significance. We need instead an alternative that is only concerned with the tail behavior which it ought to blow up. For this and many other reasons, it would be an excellent idea to speak of *center,* not *central* limit theorem. The tail limit theorem is due to H. Cramer and asserts that the tail consisting in the bulk of the graph is not a parabola but an entropy ogive.

8.7. *The reason for the anomalous inequalities $f(\alpha) < 0$ and $\alpha < 0$ is that, by the definition of a random variable $\mu(\mathrm{d}t)$, the sample size is bounded and is prescribed intrinsically; the notion of supersampling*

The inequality $\rho(\alpha) < -1$ characterizes events whose probability is extraordinarily small. The finding that this inequality plays a significant role was not anticipated, remains difficult to understand and appreciate, and demands comment.

The common response is that even extremely low probability events are captured if one simply takes a sufficiently long sample of independent values. But this is impossible, even if one forgets that, in the present uncommon context, the values are extremely far from being statistically independent. Indeed, the choice the duration $\mathrm{d}t = 2^{-k}$ has two effects. Not only does it fix the distribution of $\mu(\mathrm{d}t)$, but it also sets the sample size at the value $N = 1/\mathrm{d}t = 2^k$. Roughly speaking, a sample of size N can only reveal values having a probability greater than $1/N$, which means $\rho(\alpha) > -1$.

In summary, it is true that decreasing $\mathrm{d}t$ to 2^{-k-1} increases the sample size. But it also changes the distribution and does so in such a way that the bound $\rho = -1$ remains untouched.

This bound excludes ∂u items of information that correspond to $f(\alpha) < 0$ (for example, the value of q_{crit} when finite). Those items remain hidden and latent in the sense that they cannot be inferred from one sample of values of $\mu(\mathrm{d}t)$. Ways of revealing those values, supersampling and embedding, are examined in Mandelbrot (1989b, 1995) and forthcoming Mandelbrot (2003).

Figure 3 shows, for $p = 3/4$, how the graph of $f(\alpha)$ depends on g.

8.8. *Excluding the Bernoulli case $p = 1/2$, TVCM faces either one of two major "anomalies": for $p > -1/2$, one has $f(\alpha_{\min}) = 1 + \log_2 p > 0$ and $f(\alpha_{\max}) = 1 + \log_2(1 - p) < 0$; for $p < 1/2$, the opposite signs hold*

The fact that the values of $\rho(\alpha_{\min}) = f(\alpha_{\min}) - 1$ and $\rho(\alpha_{\max}) = f(\alpha_{\max}) - 1$ are logarithms of probabilities confirms and extends the definition of $p(\alpha) = f(\alpha) - 1$ as a limit rescaled probability. Here, those endpoint values of $f(\alpha)$ are independent of g and the affinity that deduces them from the entropy ogive (with ends on the horizontal axis) characterizes the class of equivalence of p and $1-p$. If, and only if, $p = 1/2$ and $u+v = 1$, that is, in the familiar Bernoulli binomial case, one has $\rho(\alpha_{\min}) = \rho(\alpha_{\max}) = \log_2(1/2) = -1$ hence $f(\alpha_{\min}) = f(\alpha_{\max}) = 0$. When $u + v \neq 1$, one of the endpoints satisfies $f > 0$ and the other satisfies $f < 0$. Sections 8.9 and 10 shall examine the sharply differing consequences of those inequalities.

8.9. *The "minor anomalies" $f(\alpha_{\max}) > 0$ or $f(\alpha_{\min}) > 0$ lead to sample function with a clear "ceiling" or "floor"*

Suppose that $f(\alpha_{\min}) = 0$ and $f(\alpha_{\max}) = 0$, as is the case for $p = 1/2$. Then, using terms often applied to the printed page – but after it has been turned 90° to the side – the sample functions are "non-justified" or "ragged" for both high and low values. That is, the values tend to be unequal; one is clearly larger than all others, a second is clearly the second largest, etc.

To the contrary, TVCM with $p \neq 1/2$ yield either $f(\alpha_{\max}) > 0$ or $f(\alpha_{\min}) > 0$. Sample functions have a conspicuous "ceiling" (resp., a "floor"). That is, a largest (resp., smallest) value is attained repeatedly for values of t belonging to a set of positive dimension. To use the printers' vocabulary, when one side is "ragged" the other is "justified". On visual inspection of the data, the ceiling is always visible; the floor merges with the time axis, except when one plots $\log[\mu(\mathrm{d}t)]$.

9. The fractal dimension $D = \tau'(1) = 2[-pu \log_2 u - (1 - p)v \log_2 v]$ and multifractal concentration

The function $f(\alpha)$ satisfies $f(\alpha) \leqslant \alpha$, with equality $f(\alpha) = \alpha$ when $\alpha = D = \tau'(1)$. From the value of $\alpha = D$ follows one of the most important properties of multifractals. Mandelbrot (2001d) proposed to call it "multifractal concentration". This section will first examine its opposite, which is asymptotic negligibility.

9.1. In the Bernoulli binomial measures weak asymptotic negligibility holds but strong asymptotic negligibility fails

Recall that during construction, the total binomial measure of [0, 1] remains constant and equal to 1. But the first few stages of construction make its distribution become very unequal and a few values that stand out as sharp spikes. After k stages, the maximum measure is u^k, which is far larger than the minimum measure v^k. From the relations

$$2^{-k} = \mathrm{d}t, \quad 2^k = N, \quad -\log_2 u = \alpha_{\min} < 1, \quad \text{and} \quad -\log_2 v = \alpha_{\min} > 1,$$

it follows that

$$u^k = b^{(-\log_b u)(-k)} = (\mathrm{d}t)^{\alpha_{\min}} = N^{-\alpha_{\min}}.$$

In words: even the maximum u^k tends to 0. This is a *weak* form of asymptotic negligibility following a power-law.

The preceding result holds for every multifractal for which there is an $\alpha_{\min} > 0$ that plays the same role as in the binomial case. (In more general multifractals the same role is held by some $\alpha^*_{\min} > \max\{\alpha_{\min}, 0\}$.)

Similarly, the total contribution of any fixed number of largest spikes is asymptotically negligible.

9.2. For the Bernoulli or canonical binomials, the equation $f(\alpha) = \alpha$ has one and only one solution; that solution satisfies $D > 0$ and is the fractal dimension of the "carrier" of the measure

We now proceed to the total contribution of a number of spikes that is no longer fixed but increases with N. In the simplest of all possible worlds, many spikes would have been more or less equal to the largest, and the sum of all the other spikes would have been negligible. If so, the sum of $N^{\alpha_{\min}}$ spikes would have been of the order of $N^{\alpha_{\min}} N^{-\alpha_{\min}} = 1$.

While the world is actually more complicated there is an element of orderliness. The equality $\varphi_0 = u$ is achieved for $\alpha = f(\alpha) = -u \log u - v \log v = D$. For finite but large k, it follows that

$$\mu(k, \varphi_0, \varphi_1) \sim 2^{-k\alpha} = 2^{-kD} \quad \text{and} \quad N(k_1\varphi_0, \varphi_1) \sim 2^{kf(\alpha)} = 2^{kD}.$$

Hence,

$$\mu(k_1\varphi_0, \varphi_1)N(k_1\varphi_0\varphi_1) \text{ is approximately equal to } 1.$$

Actually, this product is necessarily $\leqslant 1$ but the difference tends to 0 as $k \to \infty$. That is, an increasingly overwhelming bulk of the measure tends to "concentrate" in the cells where $\alpha = D$. The remainder is small, but in the theory of multifractals even very small remainders are extremely significant for some purposes.

9.3. The notion of "multifractal concentration"

A key feature of multifractals is a subtle interaction between number and size that is elaborated upon in Mandelbrot (2001d). Section 9.2 showed that the contributions that are large are too few to matter. The small contributions are very numerous, but so extremely small that their total contribution is negligible as well. The bulk of the measure is found in a rather inconspicuous intermediate range one can call "mass carrying". Since $D > \alpha_{\min}$, the N^D spikes of size N^{-D} are far smaller than the largest one. Separately, each is asymptotically negligible. But their number N^D is exactly large enough to insure that their total contribution is nearly equal to the overall measure 1. When a sample is plotted, this range does not stand out but it makes a perfect match between size and frequency.

Practically, the number of visible peaks is so small compared to N^D that a combination of the peaks and the intermediate range is still of the order of N^D. The combined range has the advantage of simplicity, since it includes the N^D largest values. Note that the peaks tend to be located in the midst of stretches of values of intermediate size.

9.4. The case of TVCM with $p < 1/2$, allows D to be positive, negative, or zero

Using the alternative expression for $f(\alpha)$ given in Section 8.4, the identity $f(\alpha) = \alpha$ demands the equality of the two expressions

$$f(\alpha) = -\varphi_0 \log_2\left[\frac{\varphi_0}{p}\right] - \varphi_1 \log_2\left[\frac{\varphi_1}{1-p}\right] \quad \text{and} \quad \alpha = -\varphi_0 \log_2 u - \varphi_1 \log_2 v.$$

The solution is, obviously, $\varphi_0 = pu$ and $\varphi_1 = (1-p)v$. The sum $\varphi_1 + \varphi_1$ is 1, as it must. Hence, $D = -pu \log_2 u - (1-p)v \log_2 v$, as announced. The novelty is that TVCM allow $D > 0$, $D = 0$, and $D < 0$.

Familiar role of D under the inequality $D > 0$. Mandelbrot (1974a, b) obtained the following criterion, which has become widely known and includes the TVCM case. When positive, D is the fractal dimension of the "set that supports" the measure. Figure 1 shows isolines of D for $D = 0, 1/4, 1/2$, and $3/4$. The isoline for $D = 1$ is made of the interval $\{u = 1,\ 0 < v < 1\}$ and the half-line $\{v = 1,\ u \geqslant 1\}$. The key result is that, contrary to the Bernoulli binomial case, the half line $1 < q < \infty$ subdivides into up to three subranges of values.

Largely unfamiliar consequence of the inequality $D < 0$. For all non-random multifractals, $\tau\prime(1) > 0$. A casual acquaintance with multifractals takes for granted that this is not changed by randomness. But Mandelbrot (1974a, b) also allows for an alternative possibility, which has so far remained little known. The example of TVCM shows that, in a canonical case, the formally evaluated D can be negative. In the example of TVCM, D is negative when the point (u, v) falls in a domain to the bottom right of the folded phase diagram in Figure 1. The consequences of $D < 0$ are drastic: the multifractal reduces to 0 almost surely and is called degenerate.

A classical "pathological limit" as metaphor. This limit behavior of the distribution of μ seems incompatible with the fact that $E\mu = 1$ by definition. But in fact, no contradiction is observed. A convincing idea of the distribution is provided for each p, by the behavior of the $g \to \infty$ limit of the weights $u^g 2^{\tau(g)}$ and $v^g 2^{\tau(g)}$. This recalls a classical counterexample of analysis, namely, the behavior for $k \to \infty$ of the variable P_k defined as follows: $P_k = k$ with the probability $1/k$ and $P_k = 0$ with the probability $1 - 1/k$. For finite k, one has $EP_k = 1$. But in the limit $k \to \infty$, $P_\infty = 0$, hence $EP_\infty = 0$, so that in the limit the expectation drops discontinuously from 1 to 0. In practice, the preasymptotic measure is extremely small with a high probability and huge with a tiny probability.

The condition $D = 0$. It defines the threshold of degeneracy.

10. A noteworthy and unexpected separation of roles, between the "dimension spectrum" and the total mass Ω; the former is ruled by the accessible α for which $f(\alpha) > 0$, the latter, by the inaccessible α for which $f(\alpha) < 0$

Brought together, Sections 4, 7, 8, and 9 imply, in plain words, that what you do not necessarily see may affect you significantly. This section serves to underline that the notion of canonical multifractal is very subtle and deserves to be well-understood and further discussed.

10.1. *Definitions of the "accessible ranges" of the variables: qs from $q^*_{\min}$ to $q^*_{\max}$ and αs from $\alpha^*_{\min}$ to $\alpha^*_{\max}$; the accessible functions $\tau^*(q)$ and $f^*(\alpha)$*

Mandelbrot (1995) worked to introduce to the function $f^*(\alpha) = \max\{0, f(\alpha)\}$. That is,

- In the interval $[\alpha^*_{\min}, \alpha^*_{\max}]$ where $f(\alpha) > 0$, $f^*(\alpha) = f(\alpha)$;
- When $f(\alpha) \leqslant 0$, $f^*(\alpha) = 0$.

The graph of $f^*(\alpha)$ is identical to that of $f(\alpha)$ except that the "tails" with $f < 0$ are truncated so that $f^* > 0$. In terms of $\tau(q)$, the equality $f(\alpha) = 0$ corresponds to lines that are tangent to the graph of $\tau(q)$ and also go through $(0, 0)$. In the most general case, those lines' slopes are $\alpha^*_{\min}$ and $\alpha^*_{\max}$ and the points of contact are denoted by $q^*_{\max}$ (satisfying >0) and $q^*_{\min}$ (satisfying <0). Therefore, the function $f^*(\alpha)$ corresponds to the following truncated function $\tau^*(q)$.

- When $q < q^*_{\min}$, $\tau^*(q) = \alpha^*_{\max} q$;
- When $q^*_{\min} < q < q^*_{\max}$, $\tau^*(q) = \tau(q)$;
- When $q > q^*_{\max}$, $\tau^*(q) = \alpha^*_{\min} q$.

In other words, the graph of τ^* is identical to that of τ except that beyond $q^*_{\max}$ or $q^*_{\min}$ it follows the tangents that go through the origins. Therefore it is straight.

For the TVCM, one has either $\alpha^*_{\max} = \alpha_{\max}$ with $q^*_{\min} = -\infty$, or $\alpha^*_{\min} = \alpha_{\min}$ with $q^*_{\max} = \infty$.

10.2. *A confrontation*

Section 4 noted that the largest values of $\Omega([0, 1])$ are generated when a sample cascade begins with a few large values. Section 7 noted that the value of $\Omega([0, 1])$ – irrespective of size – ceases, for $k \to \infty$, to have any impact on α. Section 8 noted that, again for $k \to \infty$, values of α such that $f(\alpha) < 0$ have a vanishing probability of being observed. Section 9.1 followed up by defining the accessible function $f(\alpha)$. Section 9 returned to large values of $\Omega([0, 1])$ and noted their association with $q_{crit} < \infty$. The values of α they involve satisfy $\alpha < 0$, hence a fortiori $f(\alpha) < 0$. Those values do not occur in multifractal decomposition, yet they are extremely important.

10.3. *The simplest cases where $f(\alpha) > 0$ for all α, as exemplified by the canonical binomial*

Here, the large values of Ω are ruled by the left-most part of the graph of $f(\alpha)$. That is, the same graph controls those large values and the distribution of $\Omega([0, 1])$ among the $1/\mathrm{d}t$ intervals of length $\mathrm{d}t$.

10.4. *The extreme case where $f(\alpha) < 0$ and $\alpha < 0$ both occur, as exemplified by TVCM when $u > 1$*

Due to the inequality $f(\alpha) < \alpha$, the graph of $f(\alpha)$ never intersects the quadrant where $\alpha < 0$ and $f > 0$. The key unexpected fact is that the portions of $f(\alpha)$ within other quadrants play more or less separate roles. In the TVCM case, those quadrants are parts of one (analytically simple) function. But in general they are nearly independent of each other.

The function $f(\alpha)$ was defined as having a graph that lies in the non-anomalous quadrant $\alpha > 0$ and $f > 0$. This f determines completely the multifractal decomposition of our TVCM measure, in particular, the dimension D and the exponents q^*_{min}, q^*_{max}, α^*_{min} and α^*_{max}.

To the contrary, q_{crit} is entirely determined by the doubly anomalous left tail located in the quadrant characterized by $f(\alpha) < 0$ and $\alpha < 0$. A priori, it was quite unexpected that this quadrant should exist and play *any* role, least of all a central role, in the theory of multifractals. But in fact, q_{crit} has a major effect on the distribution, hence the value of the total measure in an interval.

10.5. *The intermediate case where $\alpha_{min} > 0$ but $f(\alpha) < 0$ for some values of α*

When $p < 1/2$, but $u < 1$ so that $q_{crit} = \infty$ and all moments are finite, large values of μ have a much lower probability than when $u > 1$. As always, however, their probability distribution continues to be determined by the left tail of the probability graph where $f < 0$.

11. A broad form of the multifractal formalism that allows $\alpha < 0$ and $f(\alpha) < 0$

The collection of rules that relate $\tau(q)$ to $f(\alpha)$ is called "multifractal formalism". TVCM was specifically designed to understand multifractals directly, thus avoiding all formalism.

However, general random multifractals more than TVCM demand their own broad multifractal formalism. Once again, the most widely known form of the multifractal formalism does not allow randomness and yields $f(\alpha) > 0$, but the broad formalism first introduced in Mandelbrot (1974a, b) concerns a generalized function for which $f(\alpha) < 0$ is allowed.

11.1. The broad "multifractal formalism" confirms the form of $f(\alpha)$ and allows $f(\alpha) < 0$ for some α

Through a point on the graph of coordinates q and $\tau(q)$, draw the tangent to that graph. Under wide conditions, the tangent's slope is $\alpha(q)$ and its intercept by the ordinate axis is $-f(q)$. Thus

$$\alpha(q) = \frac{d\tau(q)}{dq} \quad \text{and} \quad -f(q) = \tau(q) - q\frac{d\tau(q)}{dq}.$$

Through the quantities $\alpha(q)$ and $f(q)$, a function $f(\alpha)$ is defined by using q as parameter.

The slope $f'(\alpha)$ is the inverse of the function $\alpha(q)$. The tangent of slope $f'(\alpha)$ intersects the line $\alpha = 0$ at the point of ordinate $-\tau(q)$. The $D(q)$ tangent's equation being $-\tau(q) + q\alpha$, its intersection with the bisector satisfies the condition $-\tau + q = \alpha$, hence $D = \tau(q)/(q - 1)$. This is the critical embedding dimension discussed in Section 5.4.

11.2. The Legendre and inverse Legendre transforms and the thermodynamical analogy

The transforms that replace q and $\tau(q)$ by α and $f(\alpha)$, or conversely, are due to Legendre. They play a central role in thermodynamics, as does already the argument that yielded $f(\alpha)$ and $\rho(\alpha)$ in the original formalism introduced in Mandelbrot (1974a, b).

Acknowledgments

The figures in this chapter were prepared by Evgenyi Vilenkin, Yale class of 2003.

References

Chernoff, H., 1952. A measure of asymptotic efficiency for tests of a hypothesis based on the sum of observations. The Annals of Mathematical Statistics 23, 493–507.

Cootner, P.H. (Ed.), 1964. The Random Character of Stock Market Prices. MIT Press, Cambridge, MA.

Daniels, H.E., 1954. Saddlepoint approximations in statistics. The Annals of Mathematical Statistics 25, 631–649.

Durrett, R., Liggett, T.M., 1983. Fixed points of the smoothing transformation. Zeitschrift für Wahrscheinlichkeitstheorie 64, 275–301.

Frisch, U., Parisi, G., 1985. Fully developed turbulence and intermittency. In: Ghil, M. (Ed.), Turbulence and Predictability in Geophysical Fluid Dynamics and Climate Dynamics, North-Holland, pp. 84–86. Excepted in Mandelbrot (1999a).

Halsey, T.C., Jensen, M.H., Kadanoff, L.P., Procaccia, I., Shraiman, B.I., 1986. Fractal measures and their singularities: the characterization of strange sets. Physical Review A 33, 1141–1151.

Hentschel, H.G.E., Procaccia, I., 1983. The infinite number of generalized dimensions of fractals and strange attractors. Physica (Utrecht) 8D, 435–444.

Kahane, J. P., Peyrière, J., 1976. Sur certaines martingales de B. Mandelbrot. Advances in Mathematics 22, 131–145. Translated in Mandelbrot (1999a) as Chapter N17.

Kolmogorov, A.N., 1962. A refinement of previous hypotheses concerning the local structure of turbulence in a viscous incompressible fluid at high Reynolds number. Journal of Fluid Mechanics 13, 82–85.

Liu, Q.S., 2002. An extension of a fundamental equation of Poincaré and Mandelbrot. Asian Journal of Mathematics 6, 145–68.

Lo, A.W., 1991. Long-term memory in stock market prices. Econometrica 59, 1279–1313.

Mandelbrot, B.B., 1963. The variation of certain speculative prices. Journal of Business (Chicago) 36, 394–419. Reprinted in Cootner (1964), as Chapter E 14 of Mandelbrot (1997), in Telser (2000), and several other collections of papers on finance.

Mandelbrot, B.B., 1965. Une classe de processus stochastiques homothétiques à soi; application à la loi climatologique de H.E. Hurst. Comptes Rendus (Paris) 260, 3274–3277. Translated as Chapter H9 of Mandelbrot (2002).

Mandelbrot, B.B., 1967. The variation of some other speculative prices. Journal of Business (Chicago) 40, 393–413. Reprinted as Chapter E14 of Mandelbrot (1997), pp. 419–443, in Telser (2000), and several other collections of papers on finance.

Mandelbrot, B.B., 1972. Possible refinement of the lognormal hypothesis concerning the distribution of energy dissipation in intermittent turbulence. In: Rosenblatt, M., Van Atta, C. (Eds.), Statistical Models and Turbulence. Springer-Verlag, New York, pp. 333–351. Reprinted in Mandelbrot (1999a) as Chapter N14.

Mandelbrot, B.B., 1974a. Intermittent turbulence in self similar cascades; divergence of high moments and dimension of the carrier. Journal of Fluid Mechanics 62, 331–358. Reprinted in Mandelbrot (1999a) as Chapter N15.

Mandelbrot, B.B., 1974b. Multiplications aléa.atoires itérées et distributions invariantes par moyenne pondérée aléatoire, Comptes Rendus (Paris) A 278, 289–292 and 355–358. Reprinted in Mandelbrot (1999a) as Chapter N16.

Mandelbrot, B.B., 1982. The Fractal Geometry of Nature. Freeman, New York.

Mandelbrot, B.B., 1984. Fractals in physics: squid clusters, diffusions, fractal measures and the unicity of fractal dimension. Journal of Statistical Physics 34, 895–930.

Mandelbrot, B.B., 1989a. Multifractal measures, especially for the geophysicist. Pure and Applied Geophysics 131, 5–42.

Mandelbrot, B.B., 1989b. A class of multinomial multifractal measures with negative (latent) values for the "dimension" $f(\alpha)$. In: Pietronero, L. (Ed.), Fractals' Physical Origin and Properties. Plenum, New York, pp. 3–29.

Mandelbrot, B.B., 1990a. Negative fractal dimensions and multifractals. Physica A 163, 306–315.

Mandelbrot, B.B., 1990b. New "anomalous" multiplicative multifractals: left-sided $f(\alpha)$ and the modeling of DLA. Physica A 168, 95–111.

Mandelbrot, B.B., 1990c. Limit lognormal multifractal measures. In: Gotsman, E.A., Neéman, Y., Voronel, A. (Eds.), Frontiers of Physics: Landau Memorial Conference. Pergamon, New York, pp. 309–340.

Mandelbrot, B.B., 1995. Negative dimensions and Hölders, multifractals and their Hölder spectra, and the role of lateral preasymptotics in science. In: Bonami, A., Peyrière, J. (Eds.), J.P. Kahane's meeting, Paris, 1993. The Journal of Fourier Analysis and Applications, 409–432.

Mandelbrot, B.B., 1997. Fractals and Scaling in Finance: Discontinuity, Concentration, Risk (Selecta Volume E). Springer-Verlag.

Mandelbrot, B.B., 1999a. Multifractals and $1/f$ Noise: Wild Self-Affinity in Physics (Selecta Volume N). Springer-Verlag.

Mandelbrot, B.B., 1999b. A multifractal walk through Wall Street. Scientific American, February issue, 50–53.

Mandelbrot, B.B., 2001a. Scaling in financial prices, I: Tails and dependence. Quantitative Finance 1, 113–124. Reprint: Farmer, D., Geanakoplos, J. (Eds.), Beyond Efficiency and Equilibrium. Oxford University Press, UK, 2002.

Mandelbrot, B.B., 2001b. Scaling in financial prices, II: Multifractals and the star equation. Quantitative Finance 1, 124–130. Reprint: Farmer, D., Geanakoplos, J. (Eds.), Beyond Efficiency and Equilibrium. Oxford University Press, UK, 2002.

Mandelbrot, B.B., 2001c. Scaling in financial prices, III: Cartoon Brownian motions in multifractal time. Quantitative Finance 1, 427–440.

Mandelbrot, B.B., 2001d. Scaling in financial prices, IV: Multifractal concentration. Quantitative Finance 1, 558–559.

Mandelbrot, B.B., 2001e. Stochastic volatility, power-laws and long memory. Quantitative Finance 1, 427–440.

Mandelbrot, B.B., 2002. Gaussian Self-Affinity and Fractals (Selecta Volume H). Springer-Verlag.

Mandelbrot, B.B., 2003, forthcoming.

Mandelbrot, B.B., Calvet, L., Fisher, A., 1997. The multifractal model of asset returns. Large deviations and the distribution of price changes. The multifractality of the Deutschmark/US Dollar exchange rate. Discussion Papers numbers 1164, 1165, and 1166 of the Cowles Foundation for Economics at Yale University, New Haven, CT. Available on the web at the following addresses.
http://papers.ssrn.com/sol3/paper.taf? ABSTRACT ID=78588.
http://papers.ssrn.com/sol3/paper.taf? ABSTRACT ID=78606.
http://papers.ssrn.com/sol3/paper.taf? ABSTRACT ID=78628.

Mandelbrot, B.B., Taylor, H.M., 1967. On the distribution of stock price differences. Operations Research 15, 1057–1062.

Obukhov, A.M., 1962. Some specific features of atmospheric turbulence. Journal of Fluid Mechanics 13, 77–81.

Telser, L. (Ed.), 2000. Classic Futures: Lessons from the Past for the Electronic Age. Risk Books, London.

Chapter 2

FINANCIAL RISK AND HEAVY TAILS

BRENDAN O. BRADLEY and MURAD S. TAQQU

Department of Mathematics and Statistics, Boston University, 111 Cummington Street, Boston, MA 02215, USA
e-mail: bbradley@bu.edu, murad@bu.edu
web: http://math.bu.edu/people/murad

Contents

Handbook of Heavy Tailed Distributions in Finance, Edited by S.T. Rachev

Abstract

It is of great importance for those in charge of managing risk to understand how financial asset returns are distributed. Practitioners often assume for convenience that the distribution is normal. Since the 1960s, however, empirical evidence has led many to reject this assumption in favor of various heavy-tailed alternatives. In a heavy-tailed distribution the likelihood that one encounters significant deviations from the mean is much greater than in the case of the normal distribution. It is now commonly accepted that financial asset returns are, in fact, heavy-tailed. The goal of this survey is to examine how these heavy tails affect several aspects of financial portfolio theory and risk management. We describe some of the methods that one can use to deal with heavy tails and we illustrate them using the *NASDAQ* composite index.

1. Introduction

Financial theory has long recognized the interaction of risk and reward. The seminal work of Markowitz (1952) made explicit the trade-off of risk and reward in the context of a portfolio of financial assets. Others such as Sharpe (1964), Lintner (1965), and Ross (1976), have used equilibrium arguments to develop asset pricing models such as the capital asset pricing model (CAPM) and the arbitrage pricing theory (APT), relating the expected return of an asset to other risk factors. A common theme of these models is the assumption of normally distributed returns. Even the classic Black and Scholes option pricing theory (Black and Scholes, 1973) assumes that the return distribution of the underlying asset is normal. The problem with these models is that they do not always comport with the empirical evidence. Financial asset returns often possess distributions with tails heavier than those of the normal distribution. As early as 1963, Mandelbrot (1963) recognized the heavy-tailed, highly peaked nature of certain financial time series. Since that time many models have been proposed to model heavy-tailed returns of financial assets.

The implication that returns of financial assets have a heavy-tailed distribution may be profound to a risk manager in a financial institution. For example, 3σ *events* may occur with a much larger probability when the return distribution is heavy-tailed than when it is normal. Quantile based measures of risk, such as value at risk, may also be drastically different if calculated for a heavy-tailed distribution. This is especially true for the highest quantiles of the distribution associated with *very rare* but *very damaging* adverse market movements.

This chapter serves as a review of the literature. In Section 2, we examine financial risk from an historical perspective. We review risk in the context of the mean–variance portfolio theory, CAPM and the APT, and briefly discuss the validity of their assumption of normality. Section 3 introduces the popular risk measure called *value at risk* (*VaR*). The computation of *VaR* often involves estimating a scale parameter of a distribution. This scale parameter is usually the volatility of the underlying asset. It is sometimes regarded as constant, but it can also be made to depend on the previous observations as in the popular class of ARCH/GARCH models.

In Section 4, we discuss the validity of several risk measures by reviewing a proposed set of properties suggested by Artzner, Delbean, Eber and Heath (1999) that any sensible risk measure should satisfy. Measures satisfying these properties are said to be *coherent*. The popular measure *VaR* is, in general, not coherent, but the *expected shortfall* measure is. The expected shortfall, in addition to being coherent, gives information on the expected size of a large loss. Such information is of great interest to the risk manager.

In Section 5, we return to risk, portfolios and dependence. *Copulas* are introduced as a tool for specifying the dependence structure of a multivariate distribution separately from the univariate marginal distributions. Different measures of dependence are discussed including *rank correlations* and *tail dependence*. Since the use of linear correlation in finance is ubiquitous, we introduce the class of *elliptical distributions*. Linear correlation is shown to be the canonical measure of dependence for this class of multivariate distributions and the standard tools of risk management and portfolio theory apply.

Since the risk manager is concerned with extreme market movements we introduce *extreme value theory* (EVT) in Section 6. We review the fundamentals of EVT and argue that it shows great promise in quantifying risk associated with heavy-tailed distributions. Lastly, in Section 7, we examine the use of *stable distributions* in finance. We reformulate the mean–variance portfolio theory of Markowitz and the CAPM in the context of the multivariate stable distribution.

2. Historical perspective

2.1. Risk and utility

Perhaps the most cherished tenet of modern day financial theory is the trade-off between risk and return. This, however, was not always the case, as Bernstein's (1996) narrative on risk indicates. In fact, investment decisions used to be based primarily on expected return. The higher the expected return, the better the investment. Risk considerations were involved in the investment decision process, but only in a qualitative way, *stocks are more risky than bonds*, for example. Thus any investor considering only the expected payoff $\mathbb{E}X$ of a game (investment) would, in practice, be willing to pay a fee equal to $\mathbb{E}X$ for the right to play.

The practice of basing investment decisions solely on expected return is problematic, however. Consider the game known today as the *Saint Petersburg Paradox*, introduced in 1728 by Nicholas Bernoulli. The game involves flipping a fair coin and receiving a payoff of 2^{n-1} roubles[1] if the first head appears on the nth toss of the coin. The longer tails appears, the larger the payoff. While in this game the expected payoff is infinite, no one would be willing to wager an infinite sum to play, hence the paradox. Investment decisions cannot be made on the basis of expected return alone.

Daniel Bernoulli, Nicholas' cousin, proposed a solution to the paradox ten years later. He believed that, instead of trying to maximize their expected wealth, investors want to maximize their expected *utility* of wealth. The notion of utility is now widespread in economics.[2] A utility function $U : \mathbb{R} \to \mathbb{R}$ indicates how desirable is a quantity of wealth W. One generally agrees that the utility function U should have the following properties:

(1) U is continuous and differentiable over some domain D.
(2) $U'(W) > 0$ for all $W \in D$, meaning investors prefer more wealth to less.
(3) $U''(W) < 0$ for all $W \in D$, meaning investors are risk averse. Each additional dollar of wealth adds less to the investors utility when wealth is large than when wealth is small.

In other words, U is smooth and concave over D. An investor can use his utility function to express his level of risk aversion.

[1] In fact, it was *ducats* (Bernstein, 1996).

[2] For introductions to utility theory see for example Ingersoll (1987) or Huang and Litzenberger (1988).

2.2. Markowitz mean–variance portfolio theory

In 1952, while a graduate student at the University of Chicago, Harry Markowitz (1952) produced his seminal work on portfolio theory connecting risk and reward. He defined the reward of the portfolio as the expected return and the risk as its standard deviation or variance.[3] Since the expectation operator is linear, the portfolio's expected return is simply given by the weighted sum of the individual assets' expected returns. The variance operator, however, is not linear. This means that the risk of a portfolio, as measured by the variance, is not equal to the weighted sum of risks of the individual assets. This provides a way to quantify the benefits of diversification.

We briefly describe Markowitz' theory in its classical setting where we assume that the assets distribution is multivariate normal. We will relax this assumption in the sequel. For example, in Section 5.3, we will suppose that the distribution is elliptical and, in Section 7.1, that it is an infinite variance stable distribution.

Consider a universe with n risky assets with random rates of return $\mathbf{X} = (X_1, \ldots, X_n)$, with mean $\boldsymbol{\mu} = (\mu_1, \ldots, \mu_n)$, covariance matrix $\boldsymbol{\Sigma}$ and portfolio weights $\mathbf{w} = (w_1, \ldots, w_n)$. If $\mathbf{X}$ is assumed to have a multivariate normal distribution $\mathbf{X} \sim \mathcal{N}(\boldsymbol{\mu}, \boldsymbol{\Sigma})$, then the return distribution of the portfolio $X_p = \mathbf{w}^{\mathrm{T}}\mathbf{X}$ is also normally distributed, $X_p \sim \mathcal{N}(\mu_p, \sigma_p^2)$ where $\mu_p = \mathbf{w}^{\mathrm{T}}\boldsymbol{\mu}$ and $\sigma_p^2 = \mathbf{w}^{\mathrm{T}}\boldsymbol{\Sigma}\mathbf{w}$. The problem is to find the portfolio of minimum variance that achieves a minimum level a of expected return:

$$\begin{aligned} &\min_{\mathbf{w}} \mathbf{w}^{\mathrm{T}}\boldsymbol{\Sigma}\mathbf{w} \\ &\text{such that } \quad \mathbf{w}^{\mathrm{T}}\boldsymbol{\mu} \geqslant a, \\ &\qquad\qquad\quad \mathbf{e}^{\mathrm{T}}\mathbf{w} = 1. \end{aligned} \tag{1}$$

Here $\mathbf{e} = (1, \ldots, 1)$ and T denotes a transpose. The last condition in (1),

$$\mathbf{e}^{\mathrm{T}}\mathbf{w} = \sum_{i=1}^{n} w_i = 1,$$

indicates that the portfolio is fully invested. Additional restrictions are usually added on the weights[4] and the problem is generally solved through quadratic programming. By varying the minimum level a of expected return, a set of portfolios X_p is chosen, each of which is optimal in the sense that an investor cannot achieve a greater expected return, $\mu_p = \mathbb{E}X_p$, without increasing his risk, σ_p. The set of optimal portfolios corresponds to a convex curve $(\sigma_p, \mathbb{E}X_p)$ called the *efficient frontier*. Any rational investor making decisions based only on the mean and variance of the distribution of returns of a portfolio would only choose

[3] In practice, one minimizes the variance, but it is convenient to view risk as measured by the standard deviation.

[4] For example, $w_i \geqslant 0$, in other words no *short* selling. Without the additional constraints, the problem can be solved as a system of linear equations.

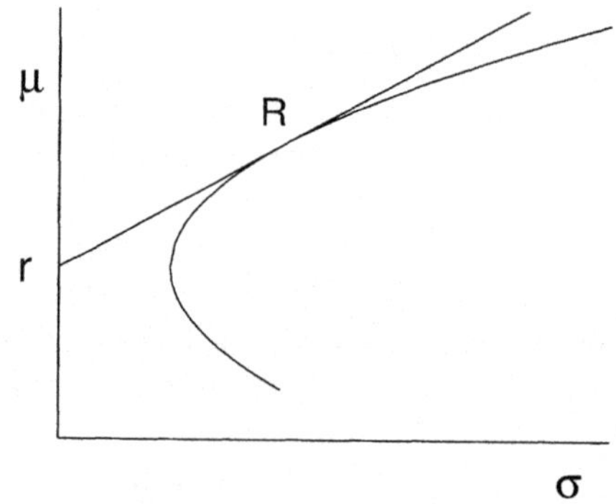

Fig. 1. The efficient frontier (σ_p, μ_p). In the case when only risky assets R are available, the frontier traces out a convex curve in risk-return space. The inclusion of a risk-free asset r, has a profound effect on the efficient set. In this case, all efficient portfolios will consist of linear combinations of r and some risky portfolio R, where (σ_R, μ_R) lies on the efficient frontier.

to own portfolios on this efficient frontier. The specific portfolio he chooses depends on his level of risk aversion.[5] If the universe of assets also includes a *risk-free* asset which the investor may borrow and lend without constraint, then the optimal portfolio is a linear combination of the risk-free asset r and a certain risky portfolio X_R on the efficient frontier. As shown in Figure 1, this line is tangent to the convex risky asset efficient frontier at the point $(\sigma_R, \mathbb{E}X_R)$. The risky portfolio therefore maximizes the slope of this linear combination,

$$\max_{\mathbf{w}} \frac{\mathbb{E}(X_R) - r}{\sigma_{X_R}}. \tag{2}$$

Again, the specific weights given to the risk-free and risky assets depend on the individual investors level of risk aversion.

2.3. CAPM and APT

The mean–variance portfolio theory of Markowitz describes the construction of an optimal portfolio, in the mean–variance sense, for an individual investor. It requires only estimates

[5] One can reconcile maximizing expected utility with the mean–variance portfolio theory of Markowitz, but one has to assume either a quadratic utility function or that returns are multivariate normal or, more generally, elliptical. (Elliptical distributions are introduced in Section 5.3.) For example, if returns are multivariate normal and if X_{p_1} and X_{p_2} are the returns of two linear portfolios with the same expected return, then for all utility functions U with properties listed in Section 2.1,

$$\mathbb{E}U(X_{p_1}) \geqslant \mathbb{E}U(X_{p_2}) \quad \text{if and only if} \quad \sigma^2_{p_1} \leqslant \sigma^2_{p_2}.$$

See for example Ingersoll (1987).

for each asset mean return, and the covariance between assets.[6] If all investors act in a way consistent with Markowitz' theory, then under additional assumptions, one will be able to learn something about the trade-off between risk and return in a market in equilibrium.[7] This is what the CAPM does.

The *capital asset pricing model* (CAPM) is an equilibrium pricing model [see Sharpe (1964) and Lintner (1965)] which relates the expected return of an asset to the risk-free return, to the *market's* expected return and to the covariance between the market and the asset. In addition to assuming that market participants use the mean–variance framework, the model makes two additional major assumptions. First, the market is assumed *frictionless*. This means that securities are infinitely divisible, there exist no transaction costs, no taxes, and there are no trading restrictions. Second, the investors beliefs are *homogeneous*. This means investors agree on mean returns and covariances for all assets in the market.

The efficient frontier in Figure 1 depended on the investors' belief. Under the CAPM assumptions, since all investors assume the same expected return and covariances for all assets in the market, they all have the same (risky) efficient frontier. However, the individual investors choice of the optimal risky portfolio still depends on the investors own level of risk aversion. Additionally, with the inclusion of a risk-free asset, we saw that the investors portfolios become dramatically more simple. Each investor can own only two assets: the risk-free asset and an optimal risky portfolio, with the relative weights depending on the investors appetite for risk. But since each investor holds the same optimal portfolio of risky assets, and since the market is assumed to be in equilibrium, this optimal risky portfolio must be the *market* portfolio. Thus Figure 1 applies with $R = M$, where M denotes the market portfolio. M consists of all risky assets held in proportion to their overall market capitalization. Letting X_M denote the return on the market portfolio, X_i denote the return of asset i, and r denote the risk-free return, the CAPM establishes the following relationship:

$$\mathbb{E}(X_i - r) = \beta_i \mathbb{E}(X_M - r), \tag{3}$$

where

$$\beta_i = \frac{\operatorname{Cov}(X_i, X_M)}{\operatorname{Var} X_M}. \tag{4}$$

The CAPM thus relates in a linear way the expected premium $\mathbb{E}X_i - r$ of holding the risky asset i over the risk-free asset to the expected premium $\mathbb{E}X_M - r$ of holding the market portfolio over the risk-free asset. The constant of proportionality is the asset's *beta*. The coefficient β_i is a measure of asset i's sensitivity to the market portfolio. The expected

6 For a universe of n assets it is necessary to compute $n(n-1)/2+n$ covariances. This means that if the universe under consideration consists of $n = 1000$ assets, it is necessary to estimate over 500000 covariances.

7 By market equilibrium, we mean a market place where security prices are set so that supply equals demand.

premium for asset i is greater than that of the market if $\beta_i > 1$ and less if $\beta_i < 1$. But if $\beta_i > 1$, then the risk will be greater. Indeed, if we assume that

$$X_i - r = \beta_i (X_M - r) + \varepsilon_i, \tag{5}$$

where ε_i is such that $\mathbb{E}\varepsilon_i = 0$ and $\mathrm{Cov}(\varepsilon_i, X_M) = 0$, then we have (3) and

$$\sigma_{X_i}^2 = \beta_i^2 \sigma_{X_M}^2 + \sigma_{\varepsilon_i}^2. \tag{6}$$

Equation (5) is often known as a *single factor model* for asset returns. Notice from (6) that the asset's risk is the sum of two terms, the *systematic* or *market* risk $\beta_i^2 \sigma_{X_M}^2$ and the *unsystematic* or *residual risk* $\sigma_{\varepsilon_i}^2$. For a portfolio X_p with weights $\mathbf{w} = (w_1, \ldots, w_n)$, one gets similarly $\sigma_{X_p}^2 = \beta_p^2 \sigma_{X_M}^2 + \sigma_{\varepsilon_p}^2$ where $\beta_p = \sum_{i=1}^n w_i \beta_i$. If one additionally assumes that $\mathrm{Cov}(\varepsilon_i, \varepsilon_j) = 0$ for all $i \neq j$ then the residual risk is

$$\sigma_{\varepsilon_p}^2 = \sum_{i=1}^n w_i^2 \sigma_{\varepsilon_i}^2. \tag{7}$$

It is bounded by c/n for some constant c, if for example, $w_i = 1/n$, and hence the portfolio's residual risk can be greatly reduced by diversification. The investor, for example, is only rewarded for bearing systematic or market risk, that is, he can expect a higher return than the market only by holding a portfolio which is riskier ($\beta_p > 1$) than the market.

In the CAPM, all assets are exposed to a single *common* source of randomness, namely the market. The *arbitrage pricing theory* (APT) model, due to Ross (1976), is a generalization of the CAPM in which assets are exposed to a larger number of common sources of randomness. The APT differs from the CAPM in that the mean–variance framework that led to (5) is now replaced by the assumption of a *multifactor* model

$$X_i = \alpha_i + \beta_{i1} f_1 + \cdots + \beta_{ik} f_k + \varepsilon_i \tag{8}$$

for generating security returns. All assets are exposed to the k sources of randomness f_j, $j = 1, \ldots, k$, called factors. Additionally, each asset i is exposed to its own specific source of randomness ε_i. The equilibrium argument used in the CAPM led to the central result (3). In the APT, the equilibrium assumption takes a slightly different form, namely, one assumes that the market is free of arbitrage. The major result of the APT then relates the expected premium of asset i to its exposure β_{ij} to factor j, and to each factor premium λ_j, $j = 1, \ldots, k$. Specifically

$$\mathbb{E}X_i = r + \beta_{i1}\lambda_1 + \cdots + \beta_{ik}\lambda_k, \tag{9}$$

where λ_j, $j = 1, \ldots, k$, is the expected premium investors demand for bearing the risk of factor j. Notice that the factor premiums λ_j are the same for each security, and it is the

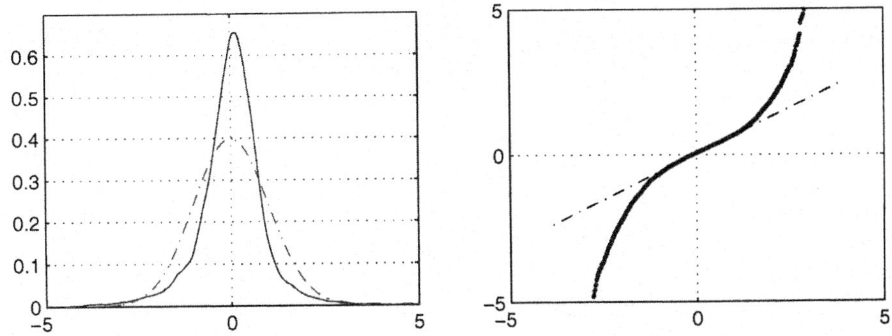

Fig. 2. Left: Empirical probability density function (pdf) for *NASDAQ* standardized returns (solid) versus the normal distribution (dot–dash) over the period February 1971 to February 2001. Right: Corresponding quantile–quantile (QQ) plot with quantiles of the normal distribution on the abscissa and empirical quantiles on the ordinate. Returns are expressed as a %.

exposure β_{ij} to each factor that depends on the security. Additionally if $k = 1$ in (8) and if we assume the existence of a risk-free asset r, $f_1 = X_M$ and that ε_i are uncorrelated with each other and the market, then $\lambda_1 = \mathbb{E}(X_M - r)$ and we get back the CAPM.

2.4. Empirical evidence

Markowitz's mean–variance portfolio theory, as well as the CAPM and APT models, rely either explicitly or implicitly on the assumption of normally distributed asset returns.[8] Today, with long histories of price/return data available for a great many financial assets, it is easy to see that this assumption is inadequate. Empirical evidence suggests that asset returns have distributions which are heavier-tailed than the normal distribution. Figure 2 illustrates this for the *NASDAQ*.[9] The quantile–quantile (QQ) plot[10] shows clearly that the distribution tails of the *NASDAQ* are heavier than the tails of the normal distribution. As early as 1963, Mandelbrot (1963) and Fama (1965) rejected the assumption of normality for other heavier-tailed distributions. In his 1963 paper, Mandelbrot not only confirmed the poor fit of the normal distribution, but proposed the model which is known today as the stable model for asset returns.

[8] As noted before, the multivariate normal assumption is consistent with maximizing expected utility.

[9] The daily *NASDAQ* time series, the corresponding returns and their maxima and minima are displayed in Figure 16. The time series starts in February 1971 and ends February 2001 (actually from February 08, 1971 to January 26, 2001). The corresponding empirical statistics can be found in Table 1.

[10] A quantile–quantile (QQ) plot is a graphical check to see if two distributions are of the same type. Two random variables X and Y are said to be of the same type if their distributions are the same up to a change in location and scale. That is $X \stackrel{d}{=} aY + b$ for some $a \in \mathbb{R}^+$, $b \in \mathbb{R}$. Since the QQ plot plots quantiles of two distributions, if they are of the same type, the plot should be linear. In this case we are checking whether the empirical distribution of *NASDAQ* standardized returns and the hypothesized normal distribution are of the same type.

Table 1
Empirical statistics for daily returns (as %) of several financial assets: the S&P 500 index, the USD/British pound exchange rate, the Thai Baht/USD exchange rate and the *NASDAQ* composite index

Asset	Period	Mean	Std. dev.	Skewness	Kurtosis[11]	Min	Max
S&P 500	01/51–03/2001	0.033	0.870	−1.61	43.9	−22.9	8.71
USD/GBP	02/1985–02/2001	0.006	0.677	0.043	3.40	−4.13	4.59
TB/USD	02/85–03/2001	0.011	0.663	4.22	158	−8.57	17.8
NASDAQ	02/1971–02/2001	0.044	1.08	−0.523	15.5	−12.0	13.3

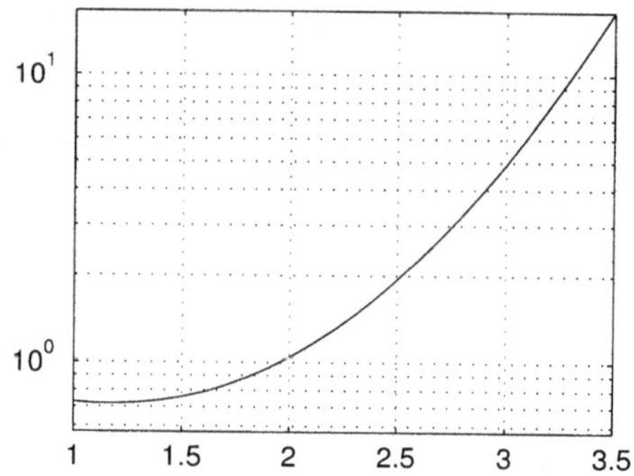

Fig. 3. Ratio of tail probabilities $\mathbb{P}(T > k\sigma)/\mathbb{P}(X > k\sigma)$ plotted in units of k. Here $T \sim t_4$ and X is normal, both with variance σ^2. T is more likely to take large values than X.

Recall that if the normal distribution is valid, then about 95% of the observations would lie within two standard deviations of the mean, and about 99% would lie within three standard deviations of the mean. In financial time series, large returns (both positive and negative) occur far too often to be compatible with the normal distribution assumption. The distribution of the financial return series are characterized not only by heavy tails, but also by a high peakedness at the center. In the Econometric terminology, they are said to be *leptokurtotic*.

To the risk manager trying to guard against large losses, the deviation from normality cannot be neglected. Suppose for example that daily returns are distributed as a stable distribution with 4 degrees of freedom (denoted t_4) and a variance given by σ^2. Since this distribution has a much heavier tail than a normal distribution with the same variance, as one moves farther out into the tail of the distribution, rare events occur much more frequently. Figure 3 shows how much more likely *rare* events occur under the t_4 assumption than under the normal, when rare is defined in terms of standard deviations.

[11] In this chapter we use as definition of kurtosis

$$K(X) = \frac{\mathbb{E}(X - \mu_X)^4}{(\operatorname{Var} X)^2} - 3,$$

so that the normal distribution has a kurtosis of zero. Heavy tails, therefore, will lead to positive kurtosis.

3. Value at risk

In the early 1990s, a number of financial institutions (J.P. Morgan, Bankers Trust, ...) proposed a new risk measure to quantify by a single number the firms aggregate exposure to market risk. This measure, commonly known today as *value at risk* (*VaR*), is now used to measure not only market risk but other forms of risk to which the firm is exposed, such as credit, operational, liquidity, and legal risk. *VaR* is defined as the loss of a financial position over a time horizon τ that would be exceeded with small probability $1-\alpha$, that is,

$$\mathbb{P}(\text{Loss} > \mathit{VaR}) \leqslant 1-\alpha. \tag{10}$$

The confidence level α is typically a large number[12] between 0.95 and 1.

To define *VaR* precisely, let X be the random variable whose cumulative distribution function F_X describes the negative profit and loss distribution (P&L) of the risky financial position at the specified horizon time τ. Negative values of X correspond now to profits and positive values of X correspond to losses. This is a useful convention in risk management since there is then no ambiguity when discussing large losses (*large* values of X correspond to *large* losses).

Formally, value at risk is a *quantile* of the probability distribution F_X, that is roughly, the x corresponding to a given value of $0 < \alpha = F_X(x) < 1$.

Definition 3.1. Let X be the random variable whose cumulative distribution function F_X describes the negative profit and loss distribution (P&L) of the risky financial position at the specified horizon time τ (so that losses are positive). Then, for a confidence level $0 < \alpha < 1$,

$$\mathit{VaR}_\alpha(X) = \inf\{x \mid F_X(x) \geqslant \alpha\}. \tag{11}$$

We set, avoiding technicalities

$$\mathit{VaR}_\alpha(X) = F_X^{-1}(\alpha),$$

where F_X^{-1} denotes the inverse function of F_X[13] (see Figure 4). Hence the value $\mathit{VaR}_\alpha(X)$ over the horizon time τ would be exceeded on the average $100(1-\alpha)$ times every 100τ time periods.

[12] In statistics, α and $1-\alpha$ are usually interchanged because α, in statistics, denotes typically the Type 1 hypothesis testing error and is chosen small. The corresponding confidence level is then $1-\alpha$.

[13] This is strictly correct when F_X is strictly increasing and continuous. Otherwise, one needs to use the *generalized inverse* of F_X, denoted $F_X^{\leftarrow}$, and defined as

$$F_X^{\leftarrow}(\alpha) = \inf\{x \mid F_X(x) \geqslant \alpha\}, \quad 0 < \alpha < 1.$$

The definition (11) of $\mathit{VaR}_\alpha(X)$ is then $\mathit{VaR}_\alpha(X) = F_X^{\leftarrow}(\alpha)$. Thus, if $F_X(x) = \alpha$ for $x_0 \leqslant x \leqslant x_1$, then $\mathit{VaR}_\alpha(X) = F_X^{\leftarrow}(\alpha) = x_0$.

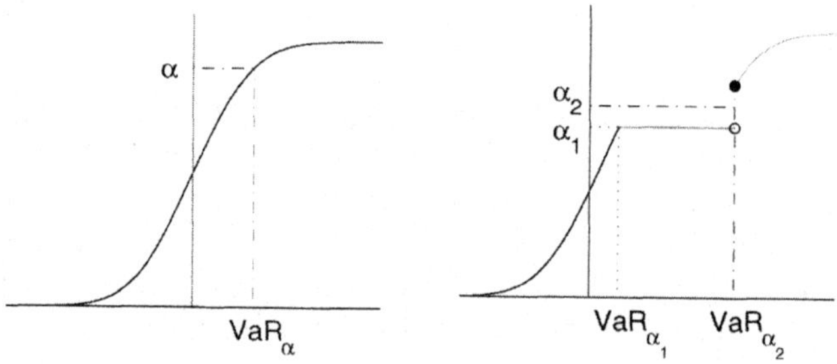

Fig. 4. $VaR_\alpha(X)$ for different cumulative distributions functions (cdfs) of the loss distribution X. The cdf on the right corresponds to an asset with discontinuous payoff, for example a binary option. See Definition 3.1.

Because of its intuitive appeal and simplicity, it is no surprise that *VaR* has become the *de facto* standard risk measure used around the world today. For example, today *VaR* is frequently used by regulators to determine minimum capital adequacy requirements. In 1995, the Basle Committee on Banking Supervision[14] suggested that banks be allowed to use their own internal *VaR* models for the purpose of determining minimum capital reserves. The internal models approach of the Basle Committee is a ten day *VaR* at the $\alpha = 99\%$ confidence level multiplied by a *safety factor* of at least 3. Thus if $VaR = 1M$, the institution is required to have at least $3M$ in reserve in a safe account.

The safety factor of three is an effort by regulators to ensure the solvency of their institutions. It has also been argued, see Stahl (1997) or Danielsson et al. (1998), that the safety factor of three comes from the heavy-tailed nature of the return distribution. Since most *VaR* calculations are based on the simplifying assumption that the distribution of returns are normal,[15] how bad does this assumption effect *VaR*? Assume that the Profit and Loss (P&L) distribution is symmetric and has finite variance σ^2. Then regardless of the actual distribution, if X represents the random loss over the specified horizon time with mean zero, Chebyshev's inequality gives

$$\mathbb{P}[X > c\sigma] \leqslant \frac{1}{2c^2}.$$

So if we are interested in *VaR* bounds for $\alpha = 0.99$, setting $1/2c^2 = 0.01$ gives $c = 7.071$, and this implies $VaR^{\max}_{\alpha=0.99}(X) = 7.071\sigma$. If the *VaR* calculation were done under the assumption of normality (Gaussian distribution) then $VaR^{\mathrm{Ga}}_{\alpha=0.99}(X) = 2.326\sigma$, and so if the true distribution is indeed heavy-tailed with finite variance then the correction for $VaR_{\alpha=0.99}$ of three is reasonable, since $3 \times 2.326\sigma = 6.978\sigma$.

[14] See Basle Committee on Banking Supervision (1995a, 1995b). Basle is a city in Switzerland. In French, Basle is Bâle, in German, it is Basel. Basle is the old name for the city. The accent in Bâle stands for the s that has been dropped from Basle.

[15] See for example the RiskMetrics manual (RiskMetrics, 1996).

3.1. Computation of VaR

Before we discuss how $VaR_\alpha(X)$ is computed, we need to say a few words about X. Typically X represents the risk of some aggregated position which is influenced by many underlying risk factors $Y_1, \ldots, Y_d$,

$$X = f(Y_1, \ldots, Y_d). \tag{12}$$

The functional form of the dependence of X on the factors $Y_1, \ldots, Y_d$ is usually never known exactly, but it may be approximated in several standard ways depending on the nature of the position. For example, f is linear in the case of a portfolio of straight equity positions. The function f is non-linear, for example, if the portfolio contains a call option on an equity since the value of the call changes non-linearly with respect to a change in the underlying asset. The usual procedure is to approximate the change in the calls value with respect to its underlying by the options delta. For small changes in the underlying such an approximation is reasonable. However for large changes in the underlying, the approximation can be quite bad. In an effort to improve the approximation, a second order term is sometimes added, the options gamma. This second order approximation is referred to as the delta–gamma approximation.

In practice, the *VaR* of a risky position X is calculated in one of three ways: through historical simulation, through a parametric model, or through some sort of Monte Carlo simulation. Each way involves assumptions and approximations and it is the responsibility of the user to be aware of them. The risk manager who blindly performs the model calculations does so at his or her peril. For a full treatment of the commonly used procedures for the calculation of *VaR*, see Jorion (2001), Dowd (1998) or Wilson (1998). See Duffie and Pan (1997) for a discussion of heavy tails and *VaR* calculations. We now describe the three ways of calculating *VaR*.

3.1.1. Historical simulation VaR

The historical simulation model uses the historical returns of assets currently held in the portfolio in order to calculate *VaR*.[16] First, returns over the horizon time τ are constructed for each asset in the portfolio using historical price information. Then portfolio returns are computed using the current weight distribution of assets as though the portfolio had been held during the whole historical period which is being sampled. The *VaR* is then read from the *historical sample* by using the order statistics. For example, if 1000 time periods are sampled, then 1000 portfolio returns are calculated, one for each time period. Let $X_p^{(1)} \geqslant X_p^{(2)} \geqslant \cdots \geqslant X_p^{(1000)}$ be the order statistics of these returns, where losses are positive. Then $VaR_{\alpha=0.95}(X_p) = X_p^{(50)}$. The size of the sample is chosen by the user, but may be constrained by the available data for some of the assets currently held.

[16] Over a fixed time horizon, *VaR* may be reported in units of rate of return (%) or of currency (profit and loss) since these are essentially the same, up to multiplication by the initial wealth/value.

The model is simple to implement and has several advantages. Since it is based on historical prices it allows for a non-linear dependence between assets in the portfolio and underlying risk factors. Also since it uses historical returns it allows for the presence of heavy tails without making assumptions on the probability distributions of returns of the assets in the portfolio. There is therefore no *model risk*. In addition, there is no need to worry about the dependence structure of assets within the portfolio since it is already reflected in the price and return data.

The drawbacks are typical of models involving historical data. There may not be enough data available and there may be no reason to believe that the future will look like the past. For example, if the user would like to compute *VaR* for regulatory requirements, then $\tau = 10$ days. With about 260 business days, there are only 26 such observations in each year, four years worth of data are required to get about 100 historical simulations. This is the absolute minimum necessary to calculate *VaR* with $\alpha = 0.99$, since with 100 data points, there is but a single observation in the tail. If one or several of the assets in the portfolio have insufficient histories then adjustments must be made. For example, some practitioners bootstrap from the shorter return histories in order to take advantage of the longer histories on other assets.

When working only with historical data it is important to realize that we are assuming that the future will look like the past. If this assumption is likely to be unrealistic, the *VaR* estimate may be dangerously off the mark. For instance, if the sample period or window is devoid of large price changes, then our historical *VaR* will be low. But it will be large if there were large price fluctuations during the sample period. As large price fluctuations leave the sample window, the *VaR* will change accordingly. This yields a highly variable estimate and one which does not take into account the current financial climate. The deficiencies of historical simulation notwithstanding, its ease of use makes it the most popular method for *VaR* calculations.

3.1.2. Parametric VaR

The parametric *VaR* model assumes that the returns possess a specific distribution, usually normal. The parameters of the distribution are estimated using either historical data or forward looking option data.

Example 3.1. Assume that over the desired time horizon τ the (negative) return distribution of a portfolio is given by $F_X \sim \mathcal{N}(\mu_\tau, \sigma_\tau^2)$. Then the value at risk of portfolio X for horizon τ and confidence level $\alpha > 0.5$ is given by

$$VaR_\alpha(X) = \inf\{x \mid F_X(x) \geqslant \alpha\} = F_X^{-1}(\alpha) = \mu_\tau + \sigma_\tau \Phi^{-1}(\alpha),$$

where $\Phi^{-1}(\alpha)$ is the α quantile of the standard normal distribution.

More generally, if the (negative) return distribution of X is any F_X with finite mean μ_τ and finite variance σ_τ^2, then

$$VaR_\alpha(X) = \mu_\tau + \sigma_\tau q_\alpha, \tag{13}$$

where q_α is the α quantile of the standardized version of X. In other words, $q_\alpha = F_{\widetilde{X}}^{-1}(\alpha)$ where $\widetilde{X} = (X - \mu_\tau)/\sigma_\tau$.

If the *VaR* is computed under the assumption that returns are light-tailed, say normal, when in fact they are heavy tailed, say t_ν (Student-t distribution with ν degrees of freedom), the risk may be seriously underestimated for high confidence levels. This is because for large α, $F_{\text{normal}}^{-1}(\alpha) \leqslant F_{t_\nu}^{-1}(\alpha)$, so that the value of x that achieves $F_{\text{normal}}(x) = \alpha$ is smaller than the value of x that achieves $F_{t_\nu}(x) = \alpha$. It is thus very important that the return distribution be modelled well. A wide variety of parametric distributions can be considered.

Within the portfolio context, the most easily implemented parametric model is the so called *delta-normal* method, where the joint distribution of the risk factor returns is multivariate normal and the returns of the portfolio are assumed to be a linear function of the returns of the underlying risk factors. In this case the portfolio returns are themselves normally distributed.

Example 3.2. Take a portfolio of equities whose (negative) returns are given by $X_p = w_1 X_1 + \cdots + w_n X_n$ where w_i is the weight given to asset i and X_i is the assets (negative) return over the horizon in question. Assume $(X_1, \ldots, X_n) \sim \mathcal{N}(\mathbf{0}, \boldsymbol{\Sigma})$. Then, for $\alpha \in (0.5, 1)$,

$$VaR_\alpha(X_p) = \Phi^{-1}(\alpha)\sqrt{\mathbf{w}^{\mathrm{T}}\Sigma\mathbf{w}} = \sqrt{\overrightarrow{VaR_\alpha}^{\mathrm{T}}\boldsymbol{\rho}\overrightarrow{VaR_\alpha}},$$

where $\overrightarrow{VaR_\alpha} = (VaR_\alpha(w_1X_1), \ldots, VaR_\alpha(w_nX_n))$ is the vector of the individual weighted asset *VaRs* and $\boldsymbol{\rho}$ is the asset return correlation matrix. See Dowd (1998) for details.

When the number of assets is large, the central limit theorem is often invoked in defense of the normal model. Even if the individual asset returns are non-normal, the central limit theorem tells us that the weighted sum of many assets should be approximately normal. This argument may be disposed of in various ways. Consider, for example, the empirical distribution of daily returns of a large diversified index such as the *NASDAQ*, which is clearly heavy-tailed (see Figure 2). From a probabilistic point of view it is not at all obvious that the assumptions of the central limit theorem are satisfied. For example, if the returns do not have finite variance, there may be convergence to the class of stable distributions.

The class of stable distributions (also known as α-stable or *stable Paretian*) may be defined in a variety of ways. More will be said about them in Section 7. We define, at this stage, a stable distribution as the only possible limiting distribution of appropriately normalized sums of independent random variables.

Definition 3.2. The random variable X has a stable distribution if there exists a sequences of i.i.d. random variables $\{Y_i\}$ and constants $\{a_n\} \in \mathbb{R}$ and $\{b_n\} \in \mathbb{R}^+$ such that

$$\frac{Y_1 + \cdots + Y_n}{b_n} - a_n \xrightarrow{d} X \quad \text{as } n \to \infty. \tag{14}$$

The stable distribution of X in (14) is characterized by four parameters $(\alpha, \sigma, \beta, \mu)$ and we write $X \sim S_\alpha(\beta, \sigma, \mu)$. The parameter $\alpha \in (0, 2]$ is called the *index of stability* or the *tail exponent* and controls the decay in the tails of the distribution. The remaining parameters σ, β, μ control scale, skewness, and location respectively. If the Y_i have finite variance (the case in the usual CLT) then $\alpha = 2$ and the distribution of X is Gaussian. For all $\alpha \in (0, 2)$ the distribution is non-Gaussian stable and possess heavy tails.

Example 3.3. Properties of weekly returns of the Nikkei 225 Index over a 12 year period are examined in Mittnik, Rachev and Paolella (1998). The authors fit the return distribution using a number of parametric distributions, including the normal, Student-t and stable. According to various measures of goodness of fit, the partially asymmetric Weibull, Student-t and the asymmetric stable provide the best fit. The fit by the normal is shown to be relatively poor. The stable distribution, in addition, fits best the tail quantiles of the empirical distribution, which is a result most relevant to the calculation of *VaR*.

The central limit theorem typically assumes independence. Although it has extensions to allow for mild dependence, this dependence must be sufficiently weak. In fact, for a given number of assets, the greater the dependence, the worse the normal approximation. This affects the speed of the convergence. Since a *VaR* calculation involves the tails of the distribution, it is most important that the approximation hold in the tails. However, even when the conditions for the central limit theorem hold, the convergence in the tail is known to be very slow. The normal approximation may then only be valid in the central part of the distribution. In this case, the return distribution may be better approximated by a heavier-tailed distribution such as the Student-t or *hyperbolic* whose use in finance is becoming more common.

The hyperbolic distribution is a subclass of the class of generalized hyperbolic distributions. The generalized hyperbolic distributions were introduced in 1977 by Barndorff-Nielsen (1977) in order to explain empirical findings in geology. Today these distributions are becoming popular in finance, and in particular in risk management. Two subclasses, the hyperbolic and the *inverse Gaussian*, are most commonly used. Both these subclasses may be shown to be mixtures of Gaussians. As such, they possess heavier tails than the normal distribution but not as heavy as the stable distribution. For an introduction to generalized hyperbolic distributions in finance, see for example Eberlein and Keller (1995), Eberlein and Prause (2002) or Shiryaev (1999).

3.1.3. Monte Carlo VaR

Monte Carlo procedures are perhaps the most flexible methods for computing *VaR*. The risk manager specifies a model for the underlying risk factors, which incorporates somehow their dependence. For example, the risk factors in (12) may be described by the *stochastic differential equation*

$$dY_t^{(i)} = Y_t^{(i)}\left(\mu_t^{(i)}\,dt + \sigma_t^{(i)}\,dW_t^{(i)}\right), \tag{15}$$

for $i = 1, \ldots, d$, where $\mathbf{W}_t = (W_t^{(1)}, \ldots, W_t^{(d)})$ is a multivariate Wiener process. Once parameters of the model are estimated, for example by using historical data, or option implied estimates, the risk factors paths are then computer generated, thousands of paths for each risk factor. Each set of simulated paths for the risk factors yields a portfolio path and the portfolio is priced accordingly. Each computed price of the portfolio represents a point on the portfolio's return distribution. After many such points are obtained the portfolio's *VaR* may then be read off the simulated distribution.

This method has the advantage of being extremely versatile. It allows for heavy tails, non-linear payoffs and a great many other user specifications. Within the Monte Carlo framework, risk managers may use their own pricing models to determine non-linear payoffs under many different scenarios for the underlying risk factors. The method has also the advantage of allowing for time varying parameters within the risk factor processes. See for example Broadie and Glasserman (1998).

There are two major drawbacks to Monte Carlo methods. First, they are computationally very expensive. Thousands of simulations of the risk factors may have to be carried out for results to be trusted. For a portfolio with a large number of assets this procedure may quickly become unmanageable, since each asset within the portfolio must be valued using these simulations. Second, the method is prone to model risk. The risk factors and the pricing models of assets with non-linear payoffs may both be mis-specified. And, as is the case of the parametric *VaR*, there is the risk of mis-specifying the model parameters.

3.2. *Parameter estimation*

The parametric and Monte Carlo *VaR* methods require parameters to be estimated. When one is interested in short time horizons, the primary goal is to estimate the volatility and covariance/correlation.[17] We outline some of the common estimation techniques here.

3.2.1. *Historical volatility*

There are two different approaches to modelling volatility and covariance using only historical data. The more common approach gives constant weights to each data point. It assumes that volatility and covariance are constant over time. The other approach attempts to address the fact that volatility and covariance are time dependent by giving more weight to the more recent data points in the sample window.

First assume that variances and covariances do not to change over time. Take a large window of length n in which historical data on the risk factors is available. Let Y_{i,t_k} be the return of factor i at time period t_k. The variance of factor i and covariance of factors i and j are then computed by giving equal weights to each data point in the past. The n-period estimates at time T for the variance and covariance

$$\hat{\sigma}_i^2 = \frac{1}{n-1} \sum_{t=T-n}^{T-1} (Y_{i,t} - \hat{\mu}_{Y_i})^2, \quad \text{where } \hat{\mu}_{Y_i} = \frac{1}{n} \sum_{t=T-n}^{T-1} Y_{i,t}, \tag{16}$$

[17] For example, over short time horizons, the mean return is usually assumed to be zero.

and

$$\hat{\sigma}_{i,j} = \frac{1}{n-1} \sum_{t=T-n}^{T-1} (Y_{i,t} - \hat{\mu}_{Y_i})(Y_{j,t} - \hat{\mu}_{Y_j}) \tag{17}$$

respectively.[18] Since equal weight is given to each data point in the sample, the estimated volatility and covariance change only slowly. If one keeps the window length fixed, the estimated values will rise or fall as new large returns enter the sample period and old large returns leave it. This means that even a single extreme return will affect the estimates in the same way, whether it occurred at time $T-1$ or time $T-n$. The estimated variance and covariance, therefore, are greatly influenced by the choice of the window size n.

Another *stylized fact* of financial time series, however, is that volatility itself is volatile. With this in mind, another historical estimate of variance and covariance uses a weighting scheme which gives more weight to more recent observations. The corresponding estimates of variance and covariance are

$$\hat{\sigma}_i^2(T) = \sum_{t=T-n}^{T-1} \alpha_t (Y_{i,t} - \hat{\mu}_{Y_i})^2,$$

$$\hat{\sigma}_{i,j}(T) = \sum_{t=T-n}^{T-1} \alpha_t (Y_{i,t} - \hat{\mu}_{Y_i})(Y_{j,t} - \hat{\mu}_{Y_j}),$$

where the weights α_t, $\sum_{t=T-n}^{T-1} \alpha_t = 1$, are chosen to reflect current volatility conditions. In particular, more weight is given to recent observations: $1 > \alpha_{T-1} > \alpha_{T-2} > \cdots > \alpha_{T-n} > 0$. The model using exponentially decreasing weights, such as that used by RiskMetrics, is probably the most popular. In RiskMetrics, the volatility estimator is given by

$$\hat{\sigma}_i(T) = \sqrt{(1-\lambda) \sum_{t=1}^{n} \lambda^{t-1} (Y_{i,T-t} - \hat{\mu}_{Y_i})^2}, \tag{18}$$

where the decay factor λ is chosen to best match a large group of assets.[19] The covariance estimate is similar. RiskMetrics chooses $\lambda = 0.94$ in the case of daily returns.

[18] The normalization constant $n-1$ gives an unbiased estimate. It is sometimes replaced by n in order to correspond to the maximum likelihood estimate.

[19] In this estimate it is assumed that the decay parameter λ and window length n are such that the approximation

$$\sum_{t=1}^{n} \lambda^{t-1} \cong \frac{1}{1-\lambda}$$

is valid.

The choice (18) allows the forecast of the next periods volatility given the current information, and hence to make parametric *VaR* calculations given the current information. To see this, assume that the time T (negative) return distribution X_T is being modelled by

$$X_T \stackrel{d}{=} \sigma_T Z_T, \tag{19}$$

where Z_t, $t \in \mathbb{Z}$, is an innovation process, that is a sequence of i.i.d. mean zero and unit variance random variables. Letting $\mathcal{F}_t$ denote the filtration[20] we have

$$\begin{aligned}\sigma^2_{T+1|\mathcal{F}_T} &= (1-\lambda)\sum_{t=0}^{\infty}\lambda^t X^2_{i,T-t}\\ &= (1-\lambda)X_T^2 + \lambda(1-\lambda)(X^2_{T-1} + \lambda X^2_{T-2} + \lambda^2 X^2_{T-3} + \cdots)\\ &= (1-\lambda)X_T^2 + \lambda\sigma^2_{T|\mathcal{F}_{T-1}}.\end{aligned}$$

This allows us to make our *VaR* calculation depend on the conditional return distribution $F_{X_{T+1}|\mathcal{F}_T}$. If $VaR_\alpha^{T+1}(X)$ denotes the estimated value at risk for X at confidence level α for the period $T+1$ at time T, then, by (19),

$$VaR_\alpha^{T+1}(X) = \sigma_{T+1|\mathcal{F}_T} q_\alpha,$$

where q_α is the α quantile of the innovation process Z_{t+1}. In RiskMetrics Z is $\mathcal{N}(0,1)$, in which case the return process X_t is *conditionally* normal.[21]

The modelling of the volatility using exponential weights and the assumption of conditional normality has two major effects. First, the volatility estimator, which is now truly time varying, attempts to account for the local volatility conditions by giving more weight to the most recent observations. It also has a second less obvious, but no less profound effect on the calculation of *VaR*. Even though the conditional return distribution may be assumed to be normal (thin-tailed) within the *VaR* calculation, the unconditional return distribution will typically have heavier tails than the normal. This result is not surprising since we may think of our time t return as being sampled from a normal distribution with changing variance. This means that our unconditional distribution is more likely to fit the empirical returns and thus to provide a better estimate of the true *VaR*.

3.2.2. ARCH/GARCH volatilities

The ARCH/GARCH class of conditional volatility models were first proposed by Engle (1982) and Bollerslev (1986) respectively. We will again assume that the (negative) return

[20] Conditioning over $\mathcal{F}_T$ means conditioning over all the observations $X_1, \ldots, X_T$.

[21] RiskMetrics allows the assumption of conditional normality to be relaxed in favor of heavier-tailed conditional distributions. For example the conditional distribution of returns may be mixture of normals or a generalized error distribution, that is, a double sided exponential.

process to be modelled is of the form (19) where Z_t are i.i.d. mean zero, unit variance random variables representing the innovations of the return process. In the GARCH(p, q) model,[22] the conditional variance is given by

$$\sigma_t^2 = \alpha_0 + \sum_{i=1}^{p} \alpha_i X_{t-i}^2 + \sum_{j=1}^{q} \beta_j \sigma_{t-j}^2.$$

In its most common form, $Z_t \sim \mathcal{N}(0, 1)$, so that the returns are conditionally normal. Just as in the exponentially weighted model for volatility (see Section 3.1.1), the GARCH model with a conditionally normal return distribution can lead to heavy tails in the unconditional return distribution. In the case of the GARCH$(1, 1)$ model

$$X_t = \sigma_t Z_t, \quad \text{where } Z_t \sim \mathcal{N}(0, 1) \text{ i.i.d.},$$
$$\sigma_t^2 = \alpha_0 + \alpha_1 X_{t-1}^2 + \beta_1 \sigma_{t-1}^2,$$

it is straightforward to show that under certain conditions[23] the unconditional centered kurtosis is given by

$$K = \frac{\mathbb{E}X_t^4}{(\mathbb{E}X_t^2)^2} - 3 = \frac{6\alpha_1^2}{1 - \beta_1^2 - 2\alpha_1\beta_1 - 3\alpha_1^2},$$

which for most financial return series will be greater than zero. For example, in the case of a stationary ARCH(1) model, $X_t = \sqrt{\alpha_0 + \alpha_1 X_{t-1}^2}\, Z_t$, with $\alpha_0 > 0$ and $\alpha_1 \in (0, 2\,\mathrm{e}^{\gamma})$, where γ is Euler's constant,[24] Embrechts, Klüppelberg and Mikosch (1997) show that the unconditional distribution is formally heavy-tailed, that is

$$\mathbb{P}(X > x) \sim cx^{-\alpha}, \quad x \to \infty, \tag{20}$$

where $\alpha/2 > 0$ is the unique solution to the equation $h(u) = \frac{(2\alpha_1)^u}{\sqrt{\pi}} \Gamma(u + \frac{1}{2}) = 1$.

The ARCH/GARCH models allow for both volatility clustering (periods of large volatility) and for heavy tails. The GARCH$(1, 1)$ estimated volatility process σ_t for the *NASDAQ* is displayed in Figure 5. The assumption of conditional normality can be checked, for ex-

[22] The ARCH(p) model first proposed by Engle is equivalent to the GARCH$(p, 0)$ model later proposed by Bollerslev. The advantage of the GARCH model over the ARCH model is that it requires fewer parameters to be estimated, because AR models (ARCH) of high order are often less parsimonious than ARMA models (GARCH) of lower order.

[23] These conditions are $\alpha_1 + \beta_1 < 1$ to guarantee stationarity, and $3\alpha_1^2 + 2\alpha_1\beta_1 + \beta_1^2 < 1$ for $K > 0$. Both are generally met in financial time series.

[24] Euler's constant γ is given by $\gamma = \lim_{n\to\infty}(\sum_{k=1}^{n} \frac{1}{k} - \ln n)$ and is approximately $\gamma \approx 0.577$.

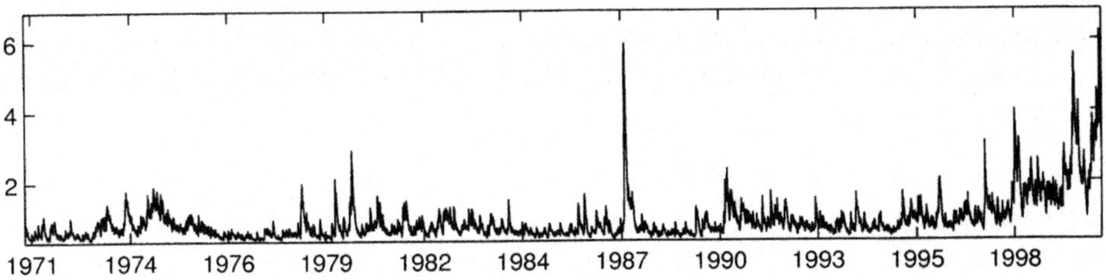

Fig. 5. GARCH(1, 1) volatilities σ_t for *NASDAQ*.

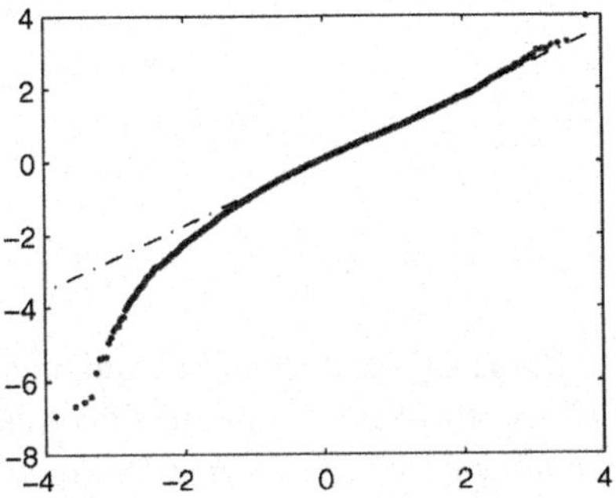

Fig. 6. Quantile–quantile (QQ) plot of the conditionally normal GARCH(1,1) standardized *ex post* innovations for *NASDAQ* with the $\mathcal{N}(0, 1)$ distribution.

ample, by examining a QQ plot of the *ex post* innovations, that is $\widehat{Z}_t = X_t/\hat{\sigma}_t$. Figure 6 displays the QQ plot of $\widehat{Z}_t$ in the traditional, conditionally normal GARCH(1,1) model for the *NASDAQ*. The fit of the GARCH(1,1) conditionally normal model in the lower tail is poor, showing the lower tail of $\widehat{Z}_t$ is heavier than the normal distribution.

If the distribution of the historical innovations $Z_{t-n}, \ldots, Z_t$ is heavier-tailed than the normal, one can modify the model to allow a heavy-tailed conditional distribution $F_{X_{t+1}|\mathcal{F}_t}$.[25] In Panorska, Mittnik and Rachev (1995) and Mittnik, Paolella and Rachev (1997), returns on the Nikkei index are modelled using an ARMA-GARCH model of the form

$$X_t = a_0 + \sum_{i=1}^{r} a_i X_{t-i} + \varepsilon_t + \sum_{j=1}^{s} b_j \varepsilon_{t-j} \tag{21}$$

(contrast with (19)), where $\varepsilon_t = \sigma_t Z_t$, with Z_t an i.i.d. location zero, unit scale heavy-tailed random variable. The conditional distribution of the return series $F_{X_t|\mathcal{F}_{t-1}}$ is given

25 For example the GARCH module in the statistical software package SPlus allows for three different non-Gaussian conditional distributions. As long as the user can estimate the GARCH parameters, usually through maximum likelihood, there are virtually no limits to the choice of the conditional distribution.

by the distribution type of Z_t. The ARMA structure in (21) is used to model the conditional mean $\mathbb{E}(X_t|\mathcal{F}_{t-1})$ of the return series X_t. The GARCH structure is imposed on the scale parameter[26] σ_t through

$$\sigma_t^2 = \alpha_0 + \sum_{i=1}^{p} \alpha_i \varepsilon_{t-i}^2 + \sum_{j=1}^{q} \beta_j \sigma_{t-j}^2.$$

Several choices for the distribution of Z_t are tested. In the case where Z_t are realizations from a stable distribution, the GARCH model used is

$$\sigma_t = \alpha_0 + \sum_{i=1}^{p} \alpha_i |\varepsilon_{t-i}| + \sum_{j=1}^{q} \beta_j \sigma_{t-j},$$

and the index of stability exponent α for the stable distribution is constrained to be greater than one.

Using several goodness of fit measures, the authors find that it is better to model the conditional distribution of returns for the Nikkei than the unconditional distribution, since the unconditional distribution cannot capture the observed temporal dependencies of the return series.[27] Within the tested models for Z_t, the partially asymmetric Weibull, the Student-t, and the asymmetric stable all outperform the normal. In order to perform reliable value at risk calculations one must model the tail of the distribution Z_t particularly well. The Anderson–Darling (AD) statistic can be used to measure goodness of fit in the tails. Letting $F_{\text{emp}}(x)$ and $F_{\text{hyp}}(x)$ denote the empirical and hypothesized parametric distributions respectively, the AD statistic

$$AD = \sup_{x \in \mathbb{R}} \frac{|F_{\text{emp}}(x) - F_{\text{hyp}}(x)|}{\sqrt{F_{\text{hyp}}(x)(1 - F_{\text{hyp}}(x))}}$$

gives more weight to the tails of the distribution. Using this statistic, as well as others, the authors propose the asymmetric stable distribution as the best of the tested models for performing *VaR* calculations at high quantiles.

The class of ARCH/GARCH models have become increasingly popular for computing *VaR*. The modelling of the conditional distribution has two immediate benefits. First, it allows for the predicted volatility (or scaling) to use local information, i.e., it allows for volatility clustering. Second, since volatility is allowed to be volatile, the unconditional distribution will typically not be thin-tailed. This is true, as we have seen, even when the conditional distribution is normal.

[26] In their model σ_t is to be interpreted as a scale parameter, not necessarily a volatility, since for some of the distributional choices for Z_t, the variance may not exist.

[27] The type of the conditional distribution is that of Z_t, the unconditional distribution is that of X_t.

There now exist many generalizations of the class of ARCH/GARCH models. Models such as EGARCH, HGARCH, AGARCH, and others, all attempt to use the local volatility structure to better predict future volatility while trying to account for other observed phenomenon. See Bollerslev, Chou and Kroner (1992) for a review. The time series of returns $\{X_t\}_{t\in\mathbb{Z}}$ in (19) is generally assumed to be stationary. In a recent paper, Mikosch and Stărică (2000) show that this assumption is not supported, at least globally, by the *S&P 500* from 1953 to 1990 and the DEM/USD foreign exchange rate from 1975 to 1982. The authors show that when using a GARCH model the parameters must be updated to account for changes of structure (changes in the unconditional variance) of the time series. A method for detecting these changes is also proposed. Additionally, they show that the long range dependence behavior associated with the absolute return series, another of the so called *stylized facts* of financial time series, may only be an artifact of structural changes in the series, that is, to non-stationarity.

Stochastic volatility models are not limited to the class of ARCH/GARCH models and their generalizations. Other models may involve additional sources of randomness. For example, the model of Hull and White (1987)

$$\mathrm{d}Y_t = \mu Y_t + \sigma_t Y_t \,\mathrm{d}W_t^{(1)},$$

$$\mathrm{d}V_t = \nu V_t + \xi V_t \,\mathrm{d}W_t^{(2)},$$

where $\sigma_t^2 = V_t$ and $(W_t^{(1)}, W_t^{(2)})$ is a bivariate Wiener process, introduces a second source of randomness through the volatility. The two sources of randomness $W_t^{(1)}$ and $W_t^{(2)}$ need not be uncorrelated. Again, the introduction of a stochastic scaling generally leads to an unconditional return distribution which is leptokurtotic. See Shiryaev (1999), for an introduction to stochastic volatility models in discrete and continuous time.

3.2.3. *Implied volatilities*

The parametric *VaR* calculation requires a forecast of the volatility. All of the models examined so far have used historical data. One may prefer to use a *forward* looking data set instead of historical data in the forecast of volatility, for example options data, which provide the market estimate of future volatility. To do so, one could use the *implied volatility* derived from the Black–Scholes model. In this model, European call options prices $C_t = C(S_t, K, r, \sigma, T - t)$ are an increasing function of the volatility σ. The stock price S_t at time t, the strike price K, the interest rate r and the time to expiration $T - t$ are known at time t. Since σ is the only unknown parameter/variable, we may then use the observed market price C_t to solve for σ. This estimate of σ is commonly called the (Black–Scholes) implied volatility. The Black–Scholes model, however is imperfect. While σ should be constant, one typically observes that σ depends on the time to expiration $T - t$ and on the strike price K. For fixed $T - t$, the implied volatility $\sigma = \sigma(T - t, K)$ as a function of the strike price K is often convex, a phenomenon known as the *volatility smile*. To obtain volatility estimates it is common to use at-the-money options, where $S_t = K$, since they are the most actively traded and hence are thought to provide the most accurate estimates.

3.2.4. *Extreme value theory*

Since *VaR* calculations are only concerned with the tails of a probability distribution, techniques from Extreme Value Theory (EVT) may be particularly effective. Proponents of EVT have made compelling arguments for its use in calculating *VaR* and for risk management in general. We will discuss EVT in Section 6.

4. Risk measures

We have considered two different measures of risk: standard deviation and value at risk. Standard deviation, used by Markowitz and others, is still commonly used in portfolio theory today. The second measure, *VaR*, is the standard measure used today by regulators and investment banks. We detailed some of the computational issues surrounding these measures but have not discussed their validity.

It is easy to criticize standard deviation and value at risk. Even in Markowitz's pioneering work on portfolio theory, the shortcomings of standard deviation as a risk measure were recognized. In Markowitz (1959), an entire chapter is devoted to *semi-variance*[28] as a potential alternative. In Artzner et al. (1997), for example, measures based on standard deviation are criticized based on their inability to describe rare events and *VaR* is criticized because of its inability to aggregate risks in a logical manner. In two now famous papers (Artzner et al., 1997, 1999) on financial risk, the authors propose a set of properties any reasonable risk measure should satisfy. Any risk measure which satisfies these properties is called *coherent*. We shall now introduce these properties and indicate why the risk measures described above are not coherent.

4.1. *Coherent risk measures*

Suppose that the financial position of an investor will lead at time T to a loss X,[29] which is a random variable. Let $\mathcal{G}$ be the set of all such X. A risk measure ρ is defined as a mapping from $\mathcal{G}$ to $\mathbb{R}$. Intuitively, for a given potential loss X in the future we may think of $\rho(X)$ as the minimum amount of cash that we need to invest prudently today (in a reference instrument) to be allowed to take the position X.[30] A risk measure ρ may be *coherent* or not.

Definition 4.1. Given a reference instrument with return r, possibly random, a risk measure ρ satisfying the following four axioms is said to be coherent:

[28] In order to put the accent on (negative) returns above the mean, semi-variance is defined as

$$\widetilde{\sigma}_X = \mathbb{E}[(X - \mathbb{E}X)1_{\{X > \mathbb{E}X\}}]^2.$$

[29] Losses are positive and profits negative. This is at odds with the authors' original notation.

[30] The authors refer to X as *risk* and axiomatically define *acceptance sets* which are sets of acceptable risks, and proceed to define measures of risk as describing the risks proximity to the acceptance set.

Translation invariance. For all $X \in \mathcal{G}$ and all $\alpha \in \mathbb{R}$, we have $\rho(X + \alpha r) = \rho(X) + \alpha$. This means that adding the amount α to the position, and investing it prudently, reduces the overall risk of the position by α.

Subadditivity. For all X_1 and $X_2 \in \mathcal{G}$, $\rho(X_1 + X_2) \leqslant \rho(X_1) + \rho(X_2)$. Hence *a merger does not create extra risk*. This is the basis for diversification.

Positive homogeneity. For all $\lambda \geqslant 0$ and all $X \in \mathcal{G}$, $\rho(\lambda X) = \lambda\rho(X)$. This requires that the risk scales with the size of a position. If the size of a position renders it illiquid, then this should be considered when modelling the future net worth.

Monotonicity. For all X and $Y \in \mathcal{G}$ with $X \geqslant Y$, we have $\rho(X) \geqslant \rho(Y)$. If the future net loss X is greater, then X is more risky.

The term *coherent measure of risk* has found its way into the risk management vernacular. It is defined, for example, in the second edition of Philippe Jorion's *Value at Risk* (Jorion, 2001).

Note that the axioms of translation invariance and monotonicity rule out standard deviation as a coherent measure of risk. Indeed, since $\sigma_{X+\alpha r} = \sigma_X$, translation invariance fails, and since σ also penalizes the investor for large profits as well as large losses, monotonicity fails as well. Consider, for example, two portfolios X and Y which are identical except for the free lottery ticket held in Y. We have $X \geqslant Y$, since there is no down-side to the free ticket and therefore the potential losses in Y are smaller than in X. Nevertheless, the standard deviation measure assigns to Y a higher risk, hence monotonicity fails. Markowitz's alternative risk measure semi-variance is not coherent either because it is not subadditive.

4.2. Expected shortfall

VaR is not a coherent measure of risk because it fails to be subadditive in general. One can indeed easily construct scenarios [see Albanese (1997)] where for two positions X and Y it is true that

$$VaR_\alpha(X + Y) > VaR_\alpha(X) + VaR_\alpha(Y).$$

This is contrary to the risk managers feelings, that the overall risk of different trading desks is bounded by the sum of their individual risks. In short, *VaR* fails to aggregate risks in a logical manner. In addition, *VaR* tells us nothing about the size of the loss that exceeds it. Two distributions may have the same *VaR* yet be dramatically different in the tail.

Hence neither the standard deviation nor *VaR* are coherent. On the other hand, the *expected shortfall*, also called *tail conditional expectation*, is a coherent risk measure. Intuitively, the expected shortfall addresses the question: *given that we will have a bad day, how bad do we expect it to be?* It is a more conservative measure than *VaR* and looks at the average of all losses that exceed *VaR*. Formally, the expected shortfall for risk X and high confidence level α is defined as follows:

Definition 4.2. Let X be the random variable whose distribution function F_X describes the negative profit and loss distribution (P&L) of the risky financial position at the specified horizon time τ (thus losses are positive). Then the expected shortfall for X is

$$S_\alpha(X) = \mathbb{E}\big(X | X > VaR_\alpha(X)\big). \tag{22}$$

Suppose, for example, that a portfolio's risk is to be calculated through simulation. If 1000 simulations are run, then for $\alpha = 0.95$, the portfolios *VaR* would be the smallest of the 50 largest losses. The corresponding expected shortfall would be estimated by the numerical average of these 50 largest losses. Expected shortfall, therefore, tells us something about the expected size of a loss exceeding *VaR*. It is subadditive, coherent and puts fewer restrictions on the distribution of X, requiring only a finite first moment to be well defined. Additionally, it may be reconciled with the idea of maximizing expected utility. Levy and Kroll (1978) show that for all utility functions U with the properties described in Section 2.1 and all random variables X and Y (representing losses) that

$$\mathbb{E}U(-X) \geqslant \mathbb{E}U(-Y) \quad \Longleftrightarrow \quad S_\alpha(X) \leqslant S_\alpha(Y) \text{ for all } \alpha \in (0,1).$$

Expected shortfall can be used in portfolio theory as a replacement of the standard deviation if the distribution of $\boldsymbol{X}$ is normal, or more generally, elliptical. As we will see in Section 5.3, in this case any positive homogeneous translation invariant risk measure will yield the same optimal linear portfolio for the same level of expected return.

Unlike standard deviation, expected shortfall, as defined in (22), does not measure deviation from the mean. Bertsimas, Lauprete and Samarov (2000) define shortfall[31] as

$$s_\alpha(X) = \mathbb{E}\big(X | X > VaR_\alpha(X)\big) - \mathbb{E}X. \tag{23}$$

The subtraction of the mean makes it more similar to the standard deviation $\sigma_X = \sqrt{\mathbb{E}(X - \mathbb{E}X)^2}$ and again, as far as portfolio theory is concerned, in the case of elliptical distributions, one obtains the same optimal portfolio for the same level of expected return if one uses s_α to measure risk. In fact, it can be shown that for a linear portfolio $X_p = w_1 X_1 + \cdots + w_n X_n$ of multivariate normally distributed returns $\boldsymbol{X} \sim \mathcal{N}(\boldsymbol{\mu}, \boldsymbol{\Sigma})$, that

$$s_\alpha(X_p) = \frac{\phi(\Phi^{-1}(\alpha))}{1-\alpha}\sigma_p,$$

where $\phi(x)$ and $\Phi(x)$ are respectively, the pdf and cdf of a standard normal random variable evaluated at x. In other words,

$$\arg\min_{\mathbf{A}\mathbf{w}=\boldsymbol{b}} \mathbf{w}^{\mathrm{T}}\boldsymbol{\Sigma}\mathbf{w} = \arg\min_{\mathbf{A}\mathbf{w}=\mathbf{b}} s_\alpha\big(\mathbf{w}^{\mathrm{T}}\boldsymbol{X}\big),$$

[31] We still assume losses are positive. This is at odds with the authors notation.

for all $\alpha \in (0,1)$, where $\mathbf{Aw} = \mathbf{b}$ is *any* set of linear constraints, including constraints that do not require all portfolios to have the same mean. Note, however, that s_α is not coherent since it violates the axioms of translation invariance and monotonicity.

5. Portfolios and dependence

The measure of dependence most popular in the financial community is linear correlation.[32] Its popularity may be traced back to Markowitz' mean variance portfolio theory since, under the assumption of multivariate normality, the correlation is the canonical measure of dependence. Outside of the world of multivariate normal distributions, correlation as a measure of dependence may lead to misleading conclusions (see Section 5.2.1).[33] The linear correlation between two random variables X and Y, defined by

$$\rho(X,Y) = \frac{\operatorname{Cov}(X,Y)}{\sigma_X \sigma_Y}, \tag{24}$$

is a measure of *linear dependence* between X and Y. The word *linear* is used because when variances are finite, $\rho(X,Y) = \pm 1$ if and only if Y is an affine transformation of X *almost surely*, that is if $Y = aX + b$ a.s. for some constants $a \in \mathbb{R} \setminus \{0\}$, and $b \in \mathbb{R}$. When the distribution of returns $\boldsymbol{X}$ is multivariate normal, the dependence structure of the returns is determined completely by the covariance matrix $\boldsymbol{\Sigma}$ or, equivalently, by the correlation matrix $\boldsymbol{\rho}$. One has $\boldsymbol{\Sigma} = [\boldsymbol{\sigma}]\boldsymbol{\rho}[\boldsymbol{\sigma}]$ where $[\boldsymbol{\sigma}]$ is a diagonal matrix with the standard deviations σ_j on the diagonal.

When returns are not multivariate normal, linear correlation may no longer be a meaningful measure of dependence. To deal with potential alternatives, we will introduce the concept of copulas, describe various measures of dependence and focus on elliptical distributions. For additional details and proofs, see Embrechts, McNeil and Straumann (2001), Lindskog (2000b), Nelsen (1999), Joe (1997) and Fang, Kotz and Ng (1990).

5.1. Copulas

When $\mathbf{X} = (X_1, \ldots, X_n) \sim \mathcal{N}(\boldsymbol{\mu}, \boldsymbol{\Sigma})$, the distribution of any linear portfolio of the X_j's is normal with known mean and variance. In the non-normal case, the joint distribution of $\mathbf{X}$,

$$F(x_1, \ldots, x_n) = \mathbb{P}(X_1 \leqslant x_1, \ldots, X_n \leqslant x_n)$$

is not fully described by its mean and covariance. One would like, however, to describe the joint distribution by specifying separately the marginal distributions, that is, the distribution of the components $X_1, \ldots, X_n$, and the dependence structure. One can do this with copulas.

[32] Also known as Pearson's correlation.

[33] Linear correlation is actually the canonical measure of dependence for the class of elliptical distributions. This class will be introduced shortly and may be thought of as an extension of multivariate normal distributions.

Definition 5.1. An n-copula is any function $C:[0,1]^n \to [0,1]$ satisfying the following properties:

(1) For every $\boldsymbol{u} = (u_1, \dots, u_n)$ in $[0,1]^n$ we have that $C(\boldsymbol{u}) = 0$ if at least one component $u_j = 0$ and $C(\boldsymbol{u}) = u_j$ if $\boldsymbol{u} = (1, \dots, 1, u_j, 1, \dots, 1)$.

(2) For every $\boldsymbol{a}, \boldsymbol{b} \in [0,1]^n$ such that $\boldsymbol{a} \leqslant \boldsymbol{b}$

$$\sum_{i_1=1}^{2} \cdots \sum_{i_n=1}^{2} (-1)^{i_1+\cdots+i_n} C(u_{1i_1}, \dots, u_{ni_n}) \geqslant 0, \tag{25}$$

where $u_{j1} = a_j$ and $u_{j2} = b_j$ for $j = 1, \dots, n$.

Corollary 5.1 below provides a concrete way to construct copulas. It is based on the following theorem due to Sklar [see Sklar (1996), Nelsen (1999)], which states that by using copulas one can separate the dependence structure of the multivariate distribution from the marginal behavior.

Theorem 5.1 (Sklar). *Let F be an n-dimensional distribution function with marginals $X_j \sim F_j$ for $j = 1, \dots, n$. Then there exists an n-copula $C:[0,1]^n \to [0,1]$ such that for every $\boldsymbol{x} = (x_1, \dots, x_n) \in \mathbb{R}^n$,*

$$F(x_1, \dots, x_n) = C\big(F_1(x_1), \dots, F_n(x_n)\big). \tag{26}$$

Furthermore, if the F_j are continuous then C is unique. Conversely, if C is an n-copula and F_j are distribution functions, then F in (26) is an n-dimensional distribution function with marginals F_j.

The function C is called the copula of the multivariate distribution of $\mathbf{X}$. Assuming continuity of the marginals F_j, $j = 1, \dots, n$, we see that the copula C of F is the joint distribution of the uniform transformed variables $F_j(X_j)$,

$$C(u_1, \dots, u_n) = F\big(F_1^{-1}(u_1), \dots, F_n^{-1}(u_n)\big). \tag{27}$$

Corollary 5.1. *If the F_j are the cdfs of $U(0,1)$ random variables, then $x_j = F_j(x_j)$, $0 < x_j < 1$, and (26) becomes $F(x_1, \dots, x_n) = C(x_1, \dots, x_n)$. Therefore the copula C may be thought of as the cumulative distribution function (cdf) of a random vector with uniform marginals.*

Copulas allow us to model the joint distribution of $\mathbf{X}$ in two natural steps. First, one models the univariate marginals X_j. Second, one chooses a copula that characterizes the dependence structure of the joint distribution. Any n-dimensional distribution function can serve as a copula. The following examples relate familiar multivariate distributions to their associated copulas and marginals.

Example 5.1. Suppose $X_1, \ldots, X_n$ are independent then

$$\begin{aligned} F(x_1, \ldots, x_n) &= \mathbb{P}(X_1 \leqslant x_1, \ldots, X_n \leqslant x_n) \\ &= \mathbb{P}(X_1 \leqslant x_1) \cdots \mathbb{P}(X_n \leqslant x_n) \\ &= F_1(x_1) \cdots F_n(x_n). \end{aligned}$$

Hence, in the case of independence, $C(u_1, \ldots, u_n) = u_1 \cdots u_n$ for all $(u_1, \ldots, u_n) \in [0, 1]^n$.

Example 5.2. Suppose $(X_1, \ldots, X_n)$ is multivariate standard normal with linear correlation matrix $\boldsymbol{\rho}$. Let $\Phi(z) = \mathbb{P}(Z \leqslant z)$ for $Z \sim \mathcal{N}(0, 1)$. Then

$$\begin{aligned} F(x_1, \ldots, x_n) &= \mathbb{P}(X_1 \leqslant x_1, \ldots, X_n \leqslant x_n) \\ &= \mathbb{P}\big(F_1(X_1) \leqslant F_1(x_1), \ldots, F_n(X_n) \leqslant F_n(x_n)\big) \\ &= C_{\boldsymbol{\rho}}^{\mathrm{Ga}}\big(\Phi(x_1), \ldots, \Phi(x_n)\big), \end{aligned}$$

where

$$C_{\boldsymbol{\rho}}^{\mathrm{Ga}}(u_1, \ldots, u_n) = \frac{1}{\sqrt{|\boldsymbol{\rho}|(2\pi)^n}} \int_{-\infty}^{\Phi^{-1}(u_1)} \cdots \int_{-\infty}^{\Phi^{-1}(u_n)} \mathrm{e}^{-\frac{1}{2}\mathbf{s}^{\mathrm{T}}\boldsymbol{\rho}^{-1}\mathbf{s}}\, \mathrm{d}\mathbf{s} \tag{28}$$

is called the multivariate Gaussian copula.

Example 5.3. Suppose $(X_1, \ldots, X_n)$ is multivariate t with ν degrees of freedom and linear correlation matrix $\boldsymbol{\rho}$.[34] Let $t_\nu(x) = \mathbb{P}(T \leqslant x)$ where $T \sim t_\nu$. Then

$$\begin{aligned} F(x_1, \ldots, x_n) &= \mathbb{P}(X_1 \leqslant x_1, \ldots, X_n \leqslant x_n) \\ &= \mathbb{P}\big(F_1(X_1) \leqslant F_1(x_1), \ldots, F_n(X_n) \leqslant F_n(x_n)\big) \\ &= C_{\boldsymbol{\rho}}^{t_\nu}\big(t_\nu(x_1), \ldots, t_\nu(x_n)\big), \end{aligned}$$

where

$$C_{\boldsymbol{\rho}}^{t_\nu}(u_1, \ldots, u_n) = \frac{\Gamma(\frac{\nu+n}{2})}{\Gamma(\frac{\nu}{2})\sqrt{|\boldsymbol{\rho}|(\nu\pi)^n}} \int_{-\infty}^{t_\nu^{-1}(u_1)} \cdots \int_{-\infty}^{t_\nu^{-1}(u_n)} \left(1 + \frac{\mathbf{s}^{\mathrm{T}}\boldsymbol{\rho}^{-1}\mathbf{s}}{\nu}\right)^{-(\nu+n)/2} \mathrm{d}\mathbf{s} \tag{29}$$

is called the multivariate t_ν copula.

[34] Its cdf is given by (29) where the upper limits $t_\nu^{-1}(u_1), \ldots, t_\nu^{-1}(u_n)$ are replaced by $x_1, \ldots, x_n$ respectively. A multivariate t_ν is easy to generate. Generate a multivariate normal with covariance matrix $\boldsymbol{\Sigma}$ and divide it by $\sqrt{\chi_\nu^2/\nu}$ where χ_ν^2 is an independent chi-squared random variable with ν degrees of freedom.

In Examples 5.2 and 5.3, $|\boldsymbol{\rho}|$ denotes the determinant of the matrix $\boldsymbol{\rho}$. In these examples, the copulas were introduced through the joint distribution, but it is important to remember that the copula characterizes the dependence structure of the multivariate distribution through (26). The Gaussian and t_ν copulas (28) and (29) exist separately from their associated multivariate distributions.

Example 5.4. The bivariate Gumbel copula C_β^{Gu} is given by

$$C_\beta^{\mathrm{Gu}}(u_1,u_2)=\exp\{-[(-\ln u_1)^{1/\beta}+(-\ln u_2)^{1/\beta}]^\beta\}, \tag{30}$$

where $0<\beta\leqslant 1$ is a parameter controlling the dependence, $\beta\to 0^+$ implies perfect dependence (see Section 5.2.3), and $\beta=1$ implies independence.

Example 5.5. The bivariate Clayton copula C_β^{Cl} is given by

$$C_\beta^{\mathrm{Cl}}(u_1,u_2)=\left(u_1^{-\beta}+u_2^{-\beta}-1\right)^{-1/\beta}, \tag{31}$$

where $0<\beta<\infty$ is a parameter controlling the dependence, $\beta\to 0^+$ implies independence, and $\beta\to\infty$ implies perfect dependence. This copula family is sometimes referred to as the Kimeldorf and Sampson family.

Both the Gumbel and Clayton copulas are strict *Archimedean* copulas. Archimedean copulas are defined as follows. Let $\phi:[0,1]\to[0,\infty)$ with $\phi(0)=\infty$ and $\phi(1)=0$ be a continuous, convex, strictly decreasing function. The transformation $\phi^{-1}\phi$ maintains the uniform 1-dimensional distribution since $\phi^{-1}\phi(u)=u$, $u\in[0,1]$. To obtain a 2-dimensional distribution function use instead of $\phi^{-1}\phi(u)$, $u\in[0,1]$, the function $\phi^{-1}(\phi(u)+\phi(v))$, $u,v\in[0,1]$.

Definition 5.2. A strict Archimedean copula with generator ϕ is of the form

$$C(u,v)=\phi^{-1}\big(\phi(u)+\phi(v)\big),\quad u,v\in[0,1]. \tag{32}$$

Example 5.6. The function $\phi(t)=(-\ln t)^{1/\beta}, 0<\beta\leqslant 1$, generates the bivariate Gumbel copula C_β^{Gu} (see Example 5.4).

Example 5.7. The function $\phi(t)=(t^{-\beta}-1)/\beta, \beta>0$, generates the bivariate Clayton copula C_β^{Cl} (see Example 5.5).

Example 5.8. The function $\phi(t)=-\ln((\mathrm{e}^{-\beta t}-1)/(\mathrm{e}^{-\beta}-1))$, $\beta\in\mathbb{R}\setminus\{0\}$, generates the bivariate Frank copula

$$C_\beta^{\mathrm{Fr}}(u,v)=-\frac{1}{\beta}\ln\left(1+\frac{(\mathrm{e}^{-\beta u}-1)(\mathrm{e}^{-\beta v}-1)}{\mathrm{e}^{-\beta}-1}\right)$$

[see Frank (1979)].

If $\phi(0) < \infty$, then the term *strict* in Definition 5.2 is dropped and $\phi^{-1}(s)$ in (32) is replaced by the pseudo-inverse $\phi^{[-1]}(s)$ which equals $\phi^{-1}(s)$ if $0 \leqslant s \leqslant \phi(0)$ and is zero otherwise.

Example 5.9. The function $\phi(t) = 1 - t$, $t \in [0, 1]$, satisfies $\phi(0) = 1$ and hence $\phi^{[-1]}(t) = \max(1 - t, 0)$. It generates the non-strict Archimedean copula

$$C(u, v) = \max(u + v - 1, 0).$$

The class of Archimedean copulas has many nice properties, including various simple multivariate extensions. For more on Archimedean copulas see Lindskog (2000b), Nelsen (1999), Joe (1997) and Embrechts, Lindskog and McNeil (2001).

Figure 7 illustrates how the choice of a copula can affect the joint distribution. Each figure shows contours of constant density of a bivariate distribution (X, Y) with standard normal marginals and linear correlations $\rho \approx 0.7$. The differences in the distributions is due to the choice of the copula. [For an introduction on the choice of a copula, see Frees and Valdez (1998).]

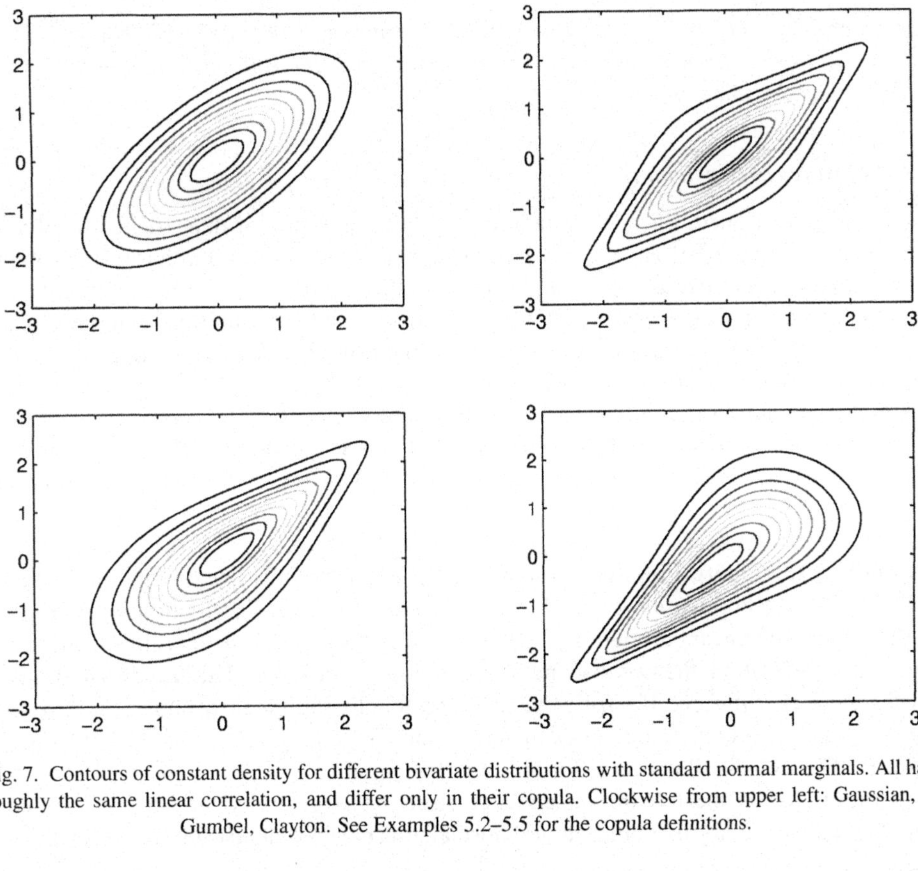

Fig. 7. Contours of constant density for different bivariate distributions with standard normal marginals. All have roughly the same linear correlation, and differ only in their copula. Clockwise from upper left: Gaussian, t_2, Gumbel, Clayton. See Examples 5.2–5.5 for the copula definitions.

The following theorem provides a bound for the joint cdf.

Theorem 5.2 (Fréchet). *Let F be the joint cdf of distribution with univariate marginals $F_1, \dots, F_n$. Then for all $\mathbf{x} \in \mathbb{R}^n$*

$$\underbrace{\max\{0, F_1(x_1) + \dots + F_n(x_n) - (n-1)\}}_{C_L(F_1(x_1),\dots,F_n(x_n))} \leqslant \underbrace{F(x_1, \dots, x_n)}_{C(F_1(x_1),\dots,F_n(x_n))}$$
$$\leqslant \underbrace{\min\{F_1(x_1), \dots, F_n(x_n)\}}_{C_U(F_1(x_1),\dots,F_n(x_n))}.$$

The function $C_U(u_1 \dots, u_n)$ is a copula for all $n \geqslant 2$, but the function $C_L(u_1, \dots, u_n)$ is a copula for $n = 2$ only. If $n = 2$, the copulas C_L and C_U are the bivariate cdf's of the random vectors $(U, 1-U)$ and (U, U) respectively, where $U \sim U(0, 1)$.

Another important property of copulas is their invariance under an increasing transformation of the marginals.

Theorem 5.3. *Let $X_1, \dots, X_n$ be continuous random variables with copula C. Let $\alpha_1, \dots, \alpha_n$ be strictly increasing transformations. Then the random vector $(\alpha_1(X_1), \dots, \alpha_n(X_n))$ has the same copula C as $(X_1, \dots, X_n)$.*

5.2. Measures of dependence

As already mentioned, linear correlation is the only measure of dependence involved in the mean–variance portfolio theory. This theory assumes, either implicitly or explicitly, that returns are multivariate normal. This assumption seems implausible today given the many complex financial products in the marketplace and the empirical evidence against normality. Without the restrictive assumption of normality, is linear correlation still an appropriate measure of dependence?

Linear correlation is often used in the financial community to describe any form of dependence. As illustrated in Embrechts, McNeil and Straumann (2001, 1999), linear correlation is often a very misunderstood measure of dependence. Consider the following example.

Example 5.10. Figure 8 represent 10000 simulations from bivariate distributions $(X, Y)_L$ and $(X, Y)_R$. In both cases X and Y have a standard normal distribution with (approximately) the same linear correlation $\rho \approx 0.7$. Thus, on the basis of the marginal distributions and linear correlation, the two distributions are indistinguishable. The two distributions are however clearly different. If positive values represent losses, the distribution on the right is clearly of greater concern to the risk manager since large losses in X and Y occur simultaneously. The two distributions differ only in their copula.

In the figure on the left the dependence structure is given by the bivariate Gaussian copula. Since the marginals are standard normal, this means that distribution is the bivariate

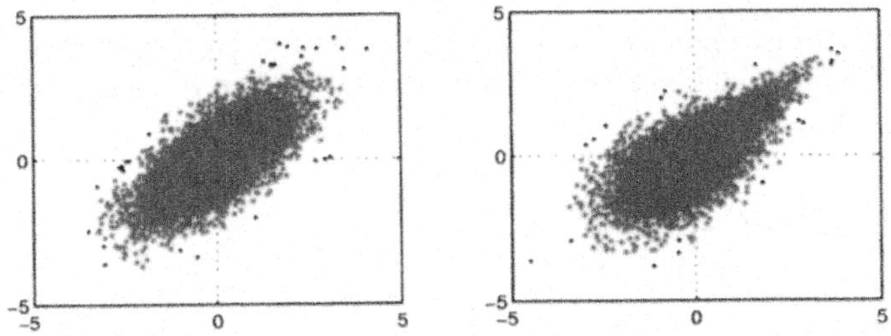

Fig. 8. Simulation of 10000 realizations from bivariate distributions both with standard normal marginals and linear correlation of $\rho \approx 0.7$. The distribution on the left has a Gaussian copula, on the right a Gumbel copula. Compare the shapes with those illustrated in Figure 7, where the population distribution is used.

standard normal distribution with the given correlation coefficient. The copula in the figure on the right the Gumbel copula given in (30) with $\beta = 1/2$. Various values of β were tried until the simulation sample linear correlation was $\rho \approx 0.7$.

We now briefly describe several measures of dependence which may be useful to the risk manager. Again the reader in encouraged to look at the above references, especially Embrechts, McNeil and Straumann (2001) for details.

5.2.1. *Linear correlation*

The linear correlation coefficient ρ, defined in (24), is a commonly misused measure of dependence. To illustrate the confusion involved in interpreting it, consider the following classic example. Let $X \sim \mathcal{N}(\mu, \sigma^2)$ and let $Y = X^2$. Then $\rho(X, Y) = 0$, yet clearly X and Y are dependent. Unless we are willing to make certain assumptions about the multivariate distribution, linear correlation can therefore be a misleading measure of dependence. Since the copula of a multivariate distribution describes its dependence structure we would like to use measures of dependence which are copula-based. Linear correlation is not such a measure.

5.2.2. *Rank correlation*

Two well-known rank correlation measures which are copula based and have better properties than linear correlation are the Kendall's tau and Spearman's rho.

Definition 5.3. Let (X_1, Y_1) and (X_2, Y_2) be two independent copies of (X, Y). Then , denoted ρ_τ, is given by

$$\rho_\tau(X, Y) = \mathbb{P}\big[(X_1 - X_2)(Y_1 - Y_2) > 0\big] - \mathbb{P}\big[(X_1 - X_2)(Y_1 - Y_2) < 0\big].$$

If the marginal distributions F_X and F_Y of X and Y are continuous and if F is the bivariate distribution function of (X, Y) with copula C, then ρ_τ can be expressed in terms of C as follows [see Embrechts, McNeil and Straumann (2001)]:

$$\rho_\tau(X, Y) = 4 \int_0^1 \int_0^1 C(u, v)\, \mathrm{d}C(u, v) - 1.$$

Definition 5.4. Let $X \sim F_X$ and $Y \sim F_Y$. Spearman's correlation, denoted ρ_S, is the linear correlation of $F_X(X)$ and $F_Y(Y)$, that is,

$$\rho_S(X, Y) = \rho\big(F_X(X), F_Y(Y)\big).$$

Spearman's correlation can also be expressed in a form similar to Definition 5.3 [see Lindskog (2000b)]. Let (X_1, Y_1), (X_2, Y_2) and (X_3, Y_3) be three independent copies of (X, Y). Then

$$\rho_S(X, Y) = 3\big(\mathbb{P}[(X_1 - X_2)(Y_1 - Y_3) > 0] - \mathbb{P}[(X_1 - X_2)(Y_1 - Y_3) < 0]\big).$$

If the marginal distributions are continuous, ρ_S is related to the copula of the joint distribution as follows:

$$\rho_S(X, Y) = 12 \int_0^1 \int_0^1 C(u, v)\, \mathrm{d}u\, \mathrm{d}v - 3.$$

Whereas linear correlation is a measure of linear dependence, both Kendall's tau and Spearman's rho are measures of monotonic dependence. Since they are copula based, they are invariant under strictly increasing transformations.[35] Indeed, if α_1, α_2 are strictly increasing transformations, then

$$\begin{aligned}
&\rho_\tau\big(\alpha_1(X_1), \alpha_2(X_2)\big) = \rho_\tau(X_1, X_2), \\
&\rho_S\big(\alpha_1(X_1), \alpha_2(X_2)\big) = \rho_S(X_1, X_2), \quad \text{but} \\
&\rho\big(\alpha_1(X_1), \alpha_2(X_2)\big) \neq \rho(X_1, X_2).
\end{aligned}$$

5.2.3. *Comonotonicity*

An additional important property of these rank correlations is their handling of *perfect dependence*. By perfect dependence we mean intuitively that X and Y are monotone functions of the same source of randomness. Recall that in the bivariate case, the Fréchet bounds C_L and C_U in Theorem 5.2 are themselves copulas. The following theorem shows that if the copula is C_L or C_U then X and Y are perfectly dependent.

[35] Recall that invariance under increasing transformations is a property of copulas.

Theorem 5.4 (Embrechts, McNeil and Straumann, 2001). *Suppose that the copula C of (X, Y) is either C_L or C_U. Then there exist monotone functions α and β and a random variable Z such that*

$$(X, Y) \stackrel{d}{=} \big(\alpha(Z), \beta(Z)\big).$$

If $C = C_L$ then α and β are increasing and decreasing respectively. If $C = C_U$, then both α and β are increasing.

X and Y are said to be *countermonotonic* if they have copula C_L. If they have copula C_U, they are said to be *comonotonic*. In fact, when F_X and F_Y are continuous,

$$C = C_L \quad \Longleftrightarrow \quad Y = T(X) \text{ a.s., } T = F_Y^{-1} \circ (1 - F_X) \searrow,$$

$$C = C_U \quad \Longleftrightarrow \quad Y = T(X) \text{ a.s., } T = F_Y^{-1} \circ F_X \nearrow .$$

Kendall's tau and Spearman's rho handle perfect dependence in a reasonable manner. Indeed,

Theorem 5.5 (Embrechts, McNeil and Straumann, 2001). *Let $(X, Y) \sim F$ with continuous marginals and copula C. Then*

$$\begin{aligned}
\rho_\tau(X, Y) = -1 \quad &\Longleftrightarrow \quad \rho_S(X, Y) = -1 \quad \Longleftrightarrow \quad C = C_L \\
&\Longleftrightarrow \quad X \text{ and } Y \text{ are countermonotonic}, \\
\rho_\tau(X, Y) = 1 \quad &\Longleftrightarrow \quad \rho_S(X, Y) = 1 \quad \Longleftrightarrow \quad C = C_U \\
&\Longleftrightarrow \quad X \text{ and } Y \text{ are comonotonic}.
\end{aligned}$$

The following theorem due to Höffding and Fréchet deals with linear correlation. See Embrechts, McNeil and Straumann (2001) for its proof.

Theorem 5.6. *Let (X, Y) be a random vector with marginals non-degenerate F_X and F_Y and unspecified dependence structure. If X and Y have finite variance, then*

(1) *The set of possible linear correlations is a closed interval $[\rho_{\min}, \rho_{\max}]$ with $\rho_{\min} < 0 < \rho_{\max}$.*
(2) *The extremal linear correlation $\rho = \rho_{\min}$ is attained iff X and Y are countermonotonic; $\rho = \rho_{\max}$ is attained iff X and Y are comonotonic.*
(3) *$\rho_{\min} = -1 \Leftrightarrow X$ and $-Y$ are of the same type;*[36] *$\rho_{\max} = 1 \Leftrightarrow X$ and Y are of the same type.*

The following example shows that linear correlation does not handle perfect dependence in a reasonable manner.

[36] Recall that two random variables are the same type if their distributions are the same up to a change in location and scale.

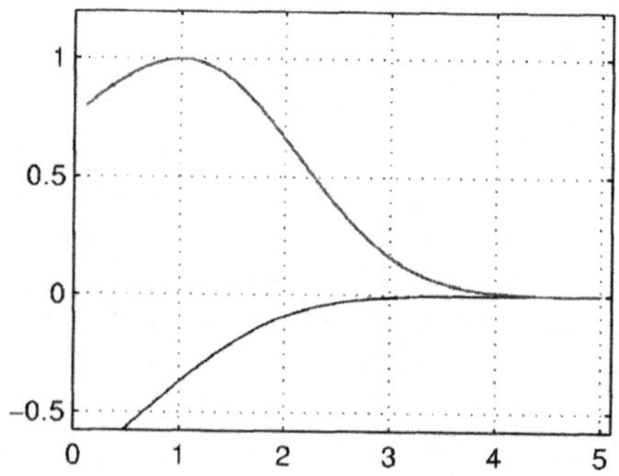

Fig. 9. Range of maximal and minimal linear correlation in Example 5.11. The x-axis is in units of σ. As σ increases, both the maximal and minimal linear correlations tend to zero.

Example 5.11 (Embrechts, McNeil and Straumann, 2001). Let $X \sim \text{Lognormal}(0, 1)$ and $Y \sim \text{Lognormal}(0, \sigma^2)$ with $\sigma > 0$. By Theorem 5.6, $\rho = \rho_{\min}$ and $\rho = \rho_{\max}$ when X and Y are countermonotonic and comonotonic respectively. By Theorem 5.4, $(X, Y) \stackrel{d}{=} (\alpha(Z), \beta(Z))$, and in fact, $(X, Y) \stackrel{d}{=} (\mathrm{e}^Z, \mathrm{e}^{-\sigma Z})$ when X and Y are countermonotonic and $(X, Y) \stackrel{d}{=} (\mathrm{e}^Z, \mathrm{e}^{\sigma Z})$ when X and Y are comonotonic, where $Z \sim \mathcal{N}(0, 1)$. Hence $\rho_{\min} = \rho(\mathrm{e}^Z, \mathrm{e}^{-\sigma Z})$ and $\rho_{\max} = \rho(\mathrm{e}^Z, \mathrm{e}^{\sigma Z})$ where $Z \sim \mathcal{N}(0, 1)$. Using the properties of the lognormal distribution, these maximal and minimal correlations can be evaluated explicitly and one gets

$$\rho_{\min} = \frac{\mathrm{e}^{-\sigma} - 1}{\sqrt{(\mathrm{e} - 1)(\mathrm{e}^{\sigma^2} - 1)}}, \qquad \rho_{\max} = \frac{\mathrm{e}^{\sigma} - 1}{\sqrt{(\mathrm{e} - 1)(\mathrm{e}^{\sigma^2} - 1)}}.$$

As σ increases, the maximal and minimal linear correlation both tend to zero even though X and Y are monotonic functions of the same source of randomness. This is illustrated in Figure 9.

5.2.4. *Tail dependence*

There is a saying in finance that in times of stress all correlations go to one.[37] While it shows that the financial community uses linear correlation to describe any measure of dependence, it can also serve as motivation for the next measure of dependence, known as tail dependence.

Bivariate tail dependence measures the amount of dependence in the upper and lower quadrant tail of the distribution. This is of great interest to the risk manager trying to guard against concurrent bad events in the tails.

[37] See Cizeau, Potters and Bouchaud (2001) for example.

Definition 5.5. Let $X \sim F_X$ and $Y \sim F_Y$ and observe that as $\alpha \to 1^-$, $F_X^{-1}(\alpha) \to \infty$ and $F_Y^{-1}(\alpha) \to \infty$. The coefficient of upper tail dependence λ_U is

$$\lambda_U(X, Y) = \lim_{\alpha \to 1^-} \mathbb{P}\big(Y > F_Y^{-1}(\alpha) | X > F_X^{-1}(\alpha)\big) \tag{33}$$

provided the limit exists. If $\lambda_U = 0$, then X and Y are said to asymptotically independent in the upper tail. If $\lambda_U \in (0, 1]$, then X and Y are asymptotically dependent in the upper tail. The coefficient of lower tail dependence λ_L is similarly defined:

$$\lambda_L(X, Y) = \lim_{\alpha \to 0^+} \mathbb{P}\big(Y < F_Y^{-1}(\alpha) | X < F_X^{-1}(\alpha)\big).$$

Since

$$\begin{aligned}\lambda_U(X, Y) \\ = \lim_{\alpha \to 1^-} \frac{1 - \mathbb{P}(X \leqslant F_X^{-1}(\alpha)) - \mathbb{P}(Y \leqslant F_Y^{-1}(\alpha)) + \mathbb{P}(X \leqslant F_X^{-1}(\alpha), Y \leqslant F_Y^{-1}(\alpha))}{1 - \mathbb{P}(X \leqslant F_X^{-1}(\alpha))},\end{aligned}$$

λ_U, as well as λ_L, can be expressed in terms of copulas. Let (X, Y) have continuous distribution F with copula C. It is easily seen that the coefficient of upper tail dependence λ_U can be expressed as

$$\lambda_U(X, Y) = \lim_{\alpha \to 1^-} \frac{\overline{C}(\alpha, \alpha)}{1 - \alpha}, \tag{34}$$

where $\overline{C}(\alpha, \alpha) = 1 - 2\alpha + C(\alpha, \alpha)$.[38] Similarly,

$$\lambda_L(X, Y) = \lim_{\alpha \to 0^+} \frac{C(\alpha, \alpha)}{\alpha}.$$

Example 5.12. Recall the simulation Example 5.10. In this example, both distributions had the same marginal distributions with the same linear correlation. Yet the distributions were clearly different in the upper tail. This difference came from the choice of copula and may now be quantified by using the notion of upper tail dependence. In Figure 8 on the left, $F(x, y) = C_\rho^{\mathrm{Ga}}(\Phi(x), \Phi(y))$, Φ denotes the standard $\mathcal{N}(0, 1)$ cdf and C_ρ^{Ga} is given by

[38] If $(U_1, U_2)^{\mathrm{T}} \sim C$ then

$$\overline{C}(u_1, u_2) = \mathbb{P}(U_1 > u_1, U_2 > u_2) = 1 - u_1 - u_2 + C(u_1, u_2).$$

(28) that is, the distribution is a bivariate standard normal with linear correlation $\rho = 0.7$. The coefficient of upper tail dependence can be calculated explicitly,[39]

$$\lambda_U(X,Y) = 2 \lim_{x\to\infty} \overline{\Phi}\left(\frac{x\sqrt{1-\rho}}{\sqrt{1+\rho}}\right) = 0,$$

which is a general characteristic of Gaussian copulas. This means that if we go far enough out into the tail then extreme events occur independently in X and Y. In the figure of the right,

$$F(x,y) = C_\beta^{\mathrm{Gu}}(\Phi(x), \Phi(y)),$$

with C_β^{Gu} given by (30), where the dependence parameter β was chosen to give (approximately) the same linear correlation.[40] In the case of the Gumbel copula a simple calculation shows that for all $0 < \beta < 1$, the coefficient of upper tail dependence is

$$\lambda_U(X,Y) = 2 - 2^\beta.$$

Hence, for the Gumbel copula, $\lambda_U \neq 0$ for $0 < \beta < 1$.

Suppose the risk manager tries to account for heavy tails of a distribution by simply modelling the joint distribution as a multivariate t_ν. He will not get $\lambda_U = 0$ as in the case of the multivariate normal distribution.

Example 5.13. If $(X,Y) \sim t_\nu$ with any linear correlation $\rho \in (-1,1)$ then it can be shown (Embrechts, McNeil and Straumann, 2001) that

$$\lambda_U(X,Y) = 2\bar{t}_{\nu+1}\left(\sqrt{\frac{(\nu+1)(1-\rho)}{1+\rho}}\right).$$

Hence for all $\rho \in (-1,1)$ there is upper tail dependence of the bivariate t_ν. The stronger the linear correlation and the lower the degrees of freedom, the stronger the upper tail dependence.

5.3. *Elliptical distributions*

There are distributions other than multivariate normal where linear correlation can be used effectively. These are the spherical, or more generally, the elliptical distributions. Elliptical distributions extend in a natural way the class of multivariate normal distributions. Linear correlation (when it exists) will still be the canonical measure of dependence, yet elliptical distributions can display heavy tails.

39 $\overline{\Phi}(x) = 1 - \Phi(x)$, and, below $\bar{t}_\nu(x) = 1 - t_\nu(x)$.

40 The dependence parameter β of the bivariate Gumbel copula is related to Kendall's tau by $\rho_\tau = 1 - \beta$.

We shall define first the spherical distributions. These extend the class of standard multivariate normal distributions with zero correlations (Fang, Kotz and Ng, 1990; Embrechts, McNeil and Straumann, 2001).

Definition 5.6. The random vector $\mathbf{X} \in \mathbb{R}^n$ is said to be spherically distributed if

$$\boldsymbol{\Gamma}\mathbf{X} \stackrel{d}{=} \mathbf{X} \quad \forall \boldsymbol{\Gamma} \in \mathcal{O}(n),$$

where $\mathcal{O}(n)$ is the group of $n \times n$ orthogonal matrices.

In other words, the distribution of $\mathbf{X}$ is invariant under rotation of the coordinates. Here are further characterizations.

Theorem 5.7. *The random vector $\mathbf{X} \in \mathbb{R}^n$ has a spherical distribution iff its characteristic function $\Psi_{\mathbf{X}}$ satisfies one of the following equivalent conditions:*
(1) $\Psi_{\mathbf{X}}(\boldsymbol{\Gamma}^{\mathrm{T}}\mathbf{t}) = \Psi_{\mathbf{X}}(\mathbf{t})\ \forall \boldsymbol{\Gamma} \in \mathcal{O}(n)$;
(2) *There exists a function $\phi(\cdot) : \mathbb{R}^+ \to \mathbb{R}$ such that $\Psi_{\mathbf{X}}(\mathbf{t}) = \phi(\mathbf{t}^{\mathrm{T}}\mathbf{t})$, that is, $\Psi_X(t) = \phi(\sum_{i=1}^n t_i^2)$, where $t = (t_1, \ldots, t_n)$. Alternatively, spherical distributions admit a stochastic representation, namely, $\mathbf{X} \in \mathbb{R}^n$ has a spherical distribution iff there exists a non-negative random variable R and random vector U independent of R and uniformly distributed over the unit hypersphere $S_n = \{s \in \mathbb{R}^n \mid \|s\| = 1\}$ such that*

$$X \stackrel{d}{=} RU. \tag{35}$$

Example 5.14. Let $\mathbf{X} \sim \mathcal{N}(\mathbf{0}, \mathbf{I}_n)$ then

$$\Psi_{\mathbf{X}}(\mathbf{t}) = \mathrm{e}^{-(1/2)(\mathbf{t}^{\mathrm{T}}\mathbf{t})} = \mathrm{e}^{-(1/2)(\sum_{i=1}^n t_i^2)},$$

and so $\phi(u) = \mathrm{e}^{-u/2}$. Additionally, $R \sim \sqrt{\chi_n^2}$ in the stochastic representation (35).

The function ϕ is called the *characteristic generator* of the spherical distribution. We write

$$\mathbf{X} \sim S_n(\phi)$$

to indicate that $\mathbf{X} \in \mathbb{R}^n$ is spherically distributed with generator ϕ. Note that if $\mathbf{X}$ possesses a density, then Theorem 5.7 requires that it is of the form

$$f(\mathbf{x}) = g(\mathbf{x}^{\mathrm{T}}\mathbf{x}) = g\left(\sum_{i=1}^n x_i^2\right)$$

for some non-negative function g. The curves of constant density are spheroids in $\mathbb{R}^n$.

Table 2
Partial list of spherical distributions used in finance

Type	pdf $f(\mathbf{x})$ or ch.f. $\Psi(\mathbf{t})$
Normal	$f(\mathbf{x}) = c\exp(-\mathbf{x}^{\mathrm{T}}\mathbf{x}/2)$
t_ν	$f(\mathbf{x}) = c(1+\mathbf{x}^{\mathrm{T}}\mathbf{x}/\nu)^{-(\nu+n)/2}$
Logistic	$f(\mathbf{x}) = c\exp(-\mathbf{x}^{\mathrm{T}}\mathbf{x})/[1+\exp(-\mathbf{x}^{\mathrm{T}}\mathbf{x})]^2$
Scale mixture	$f(\mathbf{x}) = c\int_0^\infty t^{-n/2}\exp(-\mathbf{x}^{\mathrm{T}}\mathbf{x}/2t)\,\mathrm{d}G(t)$, $G(t)$ a c.d.f.
Stable laws	$\Psi(\mathbf{t}) = \exp\{r(\mathbf{t}^{\mathrm{T}}\mathbf{t})^{\alpha/2}\}$, $0<\alpha\leqslant 2$ and $r>0$

Example 5.15. If $\mathbf{X} \in \mathbb{R}^n$ has a multivariate t_ν distribution with zero correlation, then

$$f(\mathbf{x}) = \frac{\Gamma(\frac{\nu+n}{2})}{\Gamma(\frac{\nu}{2})(\nu\pi)^{n/2}}\left(1+\frac{\mathbf{x}^{\mathrm{T}}\mathbf{x}}{\nu}\right)^{-(\nu+n)/2}.$$

$\mathbf{X}$ is therefore spherically distributed.

Table 2 gives a partial list of the spherical distributions used in finance.

Recall that if $\mathbf{X} \sim \mathcal{N}(\mathbf{0}, \mathbf{I}_n)$, then $\mathbf{Y} = \boldsymbol{\mu} + \mathbf{AX}$ has a multivariate normal distribution with mean $\boldsymbol{\mu}$ and covariance matrix $\boldsymbol{\Sigma} = \mathbf{AA}^{\mathrm{T}}$. Elliptical distributions are defined from spherical distributions in a similar manner. They are affine transformations of spherical distributions.

Definition 5.7. Let $\mathbf{X} \in \mathbb{R}^n$, $\boldsymbol{\mu} \in \mathbb{R}^n$, and $\boldsymbol{\Sigma} \in \mathbb{R}^{n\times n}$. Then $\mathbf{X}$ has an elliptical distribution with parameters $\boldsymbol{\mu}$ and $\boldsymbol{\Sigma}$ if

$$\mathbf{X} \stackrel{d}{=} \boldsymbol{\mu} + \mathbf{AY},$$

where $\mathbf{Y} \sim S_k(\phi)$, and $\mathbf{A} \in \mathbb{R}^{n\times k}$, $\boldsymbol{\Sigma} = \mathbf{AA}^{\mathrm{T}}$, with $\operatorname{rank}(\boldsymbol{\Sigma}) = k$.

Since the characteristic function of $\mathbf{X}$ may be written

$$\Psi_{\mathbf{X}}(\mathbf{t}) = \mathrm{e}^{\mathrm{i}\mathbf{t}^{\mathrm{T}}\boldsymbol{\mu}}\phi\big(\mathbf{t}^{\mathrm{T}}\boldsymbol{\Sigma}\mathbf{t}\big),$$

we use the notation

$$\mathbf{X} \sim E_n(\boldsymbol{\mu}, \boldsymbol{\Sigma}, \phi).$$

In this representation only $\boldsymbol{\mu}$ is uniquely determined. Since both $\boldsymbol{\Sigma}$ and ϕ are determined up to a positive constant $\boldsymbol{\Sigma}$ may be chosen to be the covariance matrix if variances are finite (which we assume here). An elliptically distributed random variable $\mathbf{X} \sim E_n(\boldsymbol{\mu}, \boldsymbol{\Sigma}, \phi)$ is thus described by its mean, covariance matrix and its characteristic generator. If $\mathbf{X}$ possesses a density, then it is of the form

$$f(x) = |\boldsymbol{\Sigma}|^{-1/2} g\big((\mathbf{x} - \boldsymbol{\mu})^{\mathrm{T}} \boldsymbol{\Sigma}^{-1} (\mathbf{x} - \boldsymbol{\mu})\big) \tag{36}$$

so that contours of constant density are ellipsoids in $\mathbb{R}^n$.[41]

The following theorem describes some properties of linear combinations, marginal distributions and conditional distributions of elliptical distributions.

Theorem 5.8 (Fang, Kotz and Ng, 1990). *Let* $\mathbf{X} \sim E_n(\boldsymbol{\mu}, \boldsymbol{\Sigma}, \phi)$.

(1) *If* $\mathbf{B} \in \mathbb{R}^{m \times n}$ *and* $\boldsymbol{\nu} \in \mathbb{R}^m$, *then*

$$\boldsymbol{\nu} + \mathbf{B}\mathbf{X} \sim E_m\big(\boldsymbol{\nu} + \mathbf{B}\boldsymbol{\mu}, \mathbf{B}\boldsymbol{\Sigma}\mathbf{B}^{\mathrm{T}}, \phi\big).$$

Hence any linear combination of elliptically distributed variates is elliptical with the same characteristic generator.

(2) *Partition* $\mathbf{X}$, $\boldsymbol{\mu}$,*and* $\boldsymbol{\Sigma}$ *into*

$$\mathbf{X} = \begin{pmatrix} \mathbf{X}^{(1)} \\ \mathbf{X}^{(2)} \end{pmatrix}, \qquad \boldsymbol{\mu} = \begin{pmatrix} \boldsymbol{\mu}^{(1)} \\ \boldsymbol{\mu}^{(2)} \end{pmatrix}, \qquad \boldsymbol{\Sigma} = \begin{pmatrix} \boldsymbol{\Sigma}_{11} & \boldsymbol{\Sigma}_{12} \\ \boldsymbol{\Sigma}_{21} & \boldsymbol{\Sigma}_{22} \end{pmatrix},$$

where $\mathbf{X}^{(1)} \in \mathbb{R}^m$, $\boldsymbol{\mu}^{(1)} \in \mathbb{R}^m$ *and* $\boldsymbol{\Sigma}_{11} \in \mathbb{R}^{m \times m}$, $0 < m < n$. *Then*

$$\mathbf{X}^{(1)} \sim E_m\big(\boldsymbol{\mu}^{(1)}, \boldsymbol{\Sigma}_{11}, \phi\big), \qquad \mathbf{X}^{(2)} \sim E_{n-m}\big(\boldsymbol{\mu}^{(2)}, \boldsymbol{\Sigma}_{22}, \phi\big).$$

Hence all marginals of an elliptical distribution are also elliptical with the same generator.

(3) *Partition* $\mathbf{X}$, $\boldsymbol{\mu}$,*and* $\boldsymbol{\Sigma}$ *as above and assume that* $\boldsymbol{\Sigma}$ *is strictly positive definite. Then*

$$\mathbf{X}^{(1)} | \mathbf{X}^{(2)} = \mathbf{x}_0^{(2)} \sim E_m\big(\boldsymbol{\mu}_{1.2}, \boldsymbol{\Sigma}_{11.2}, \tilde{\phi}\big),$$

where

$$\boldsymbol{\mu}_{1.2} = \boldsymbol{\mu}^{(1)} + \boldsymbol{\Sigma}_{12} \boldsymbol{\Sigma}_{22}^{-1} \big(\mathbf{x}_0^{(2)} - \boldsymbol{\mu}^{(2)}\big), \qquad \boldsymbol{\Sigma}_{11.2} = \boldsymbol{\Sigma}_{11} - \boldsymbol{\Sigma}_{12} \boldsymbol{\Sigma}_{22}^{-1} \boldsymbol{\Sigma}_{21}.$$

Hence the conditional distribution of $\mathbf{X}^{(1)}$ *given* $\mathbf{X}^{(2)}$ *is also elliptical, though with different generator.*[42]

[41] For example if $\mathrm{rank}(\boldsymbol{\Sigma}) = n$ and $\mathbf{Y}$ has density of the form $g(\mathbf{y}^{\mathrm{T}}\mathbf{y})$.

[42] The form of the generator $\tilde{\phi}$ can be related to ϕ through the stochastic representation of an elliptically distributed random vector in (35). See Fang, Kotz and Ng (1990) for details.

The importance of the class of elliptical distributions to risk management can be seen in the following theorem. It indicates that the standard approaches to risk management apply to a linear portfolio with elliptically distributed risk factors.

Theorem 5.9 (Embrechts, McNeil and Straumann, 2001). *Suppose* $\mathbf{X} \sim E_n(\boldsymbol{\mu}, \boldsymbol{\Sigma}, \phi)$ *with finite variances for all univariate marginals. Let*

$$\mathcal{P} = \left\{ Z = \sum_{i=1}^{n} w_i X_i \;\middle|\; w_i \in \mathbb{R} \right\}$$

be the set of all linear portfolios. Then:

(1) (Subadditivity of *VaR*.) *For any two portfolios* $Z_1, Z_2 \in \mathcal{P}$ *and* $0.5 \leqslant \alpha < 1$,

$$VaR_\alpha(Z_1 + Z_2) \leqslant VaR_\alpha(Z_1) + VaR_\alpha(Z_2).$$

(2) (Equivalence of variance and any other positive homogeneous risk measure.) *Let* ρ *be any real valued, positive homogeneous risk measure depending only on the distribution of a random variable* X. *Then for* $Z_1, Z_2 \in \mathcal{P}$,

$$\rho(Z_1 - \mathbb{E}Z_1) \leqslant \rho(Z_2 - \mathbb{E}Z_2) \quad \Longleftrightarrow \quad \sigma_{Z_1}^2 \leqslant \sigma_{Z_2}^2.$$

(3) (Markowitz risk minimizing portfolio.) *Let* ρ *be as in* (2), *but also translation invariant, and let*

$$\mathcal{E} = \left\{ Z = \sum_{i=1}^{n} w_i X_i \;\middle|\; w_i \in \mathbb{R},\ \sum_{i=1}^{n} w_i = 1,\ \mathbb{E}Z = r \right\}$$

be the subset of portfolios with the same expected return r. *Then*

$$\arg\min_{Z \in \mathcal{E}} \rho(Z) = \arg\min_{Z \in \mathcal{E}} \sigma_Z^2.$$

The theorem[43] states that:

[43] Because of the importance of Theorem 5.9 and because its proof is illuminating and straightforward we shall sketch it. It is based on the observation that (Z_1, Z_2) is elliptical and so portfolios Z_1, Z_2 and $Z_1 + Z_2$ are all of the same type. Let q_α, $1/2 < \alpha < 1$, denote the α quantile of the corresponding standardized distribution. Then

$$VaR_\alpha(Z_1) = \mathbb{E}Z_1 + \sigma_{Z_1} q_\alpha,$$
$$VaR_\alpha(Z_2) = \mathbb{E}Z_2 + \sigma_{Z_2} q_\alpha,$$
$$VaR_\alpha(Z_1 + Z_2) = \mathbb{E}Z_1 + \mathbb{E}Z_2 + \sigma_{Z_1+Z_2} q_\alpha$$

- For any linear portfolio of elliptical risk factors, *VaR* is a coherent measure of risk.
- If the risk factors are elliptical, the linear correlation is the canonical measure of dependence.
- For elliptical risk factors, the Markowitz mean variance optimal portfolio, for a given level of expected return, will be the same regardless of whether the risk measure is given by the variance, *VaR*, expected shortfall or any other positive homogeneous, translation invariant risk measure. Hence, all the usual techniques of portfolio theory and risk management apply.
- It may be strange at first that the expected shortfall $S_\alpha(X)$, for example, which does not involve subtraction of the mean (see (22)), can be used instead of the variance in Markowitz' risk minimization portfolio theory. This is because one considers a set of portfolios $\mathcal{E}$, all of the same mean. Since $S_\alpha(X - \mathbb{E}X) = S_\alpha(X) - \mathbb{E}X$ and since $\mathbb{E}X$ is the same for all portfolios X in $\mathcal{E}$, the term $\mathbb{E}X$ can be ignored.

Note that elliptical distributions are not required to be thin-tailed. The multivariate normal is but one elliptical distribution. The risk manager may well feel that the risk factors under consideration are better modelled using a heavy-tailed elliptical distribution.[44] The usual techniques then apply, but the risk of a linear portfolio will be greater than if the risk factors were assumed multivariate normal.

6. Univariate extreme value theory

Managing extreme market risk is a goal of any financial institution or individual investor. In an effort to guarantee solvency, financial regulators require most financial institutions to maintain a minimum level of capital in reserve. The recommendation of the Basle Committee (1995b) of a minimum capital reserve requirement based on *VaR* is an attempt to manage extreme market risks. Recall that *VaR* is nothing more that a quantile of a probability

but $\sigma_{Z_1+Z_2} \leqslant \sigma_{Z_1} + \sigma_{Z_2}$ and $q_\alpha > 0$, proving (1). Next, note that there exists $a > 0$ such that $Z_1 - \mathbb{E}Z_1 \stackrel{d}{=} a(Z_2 - \mathbb{E}Z_2)$, so that $a \leqslant 1 \Leftrightarrow \sigma_1^2 \leqslant \sigma_2^2$. Since the risk measure ρ is assumed positive homogeneous and depends only on the distribution of Z,

$$\rho(Z_1 - \mathbb{E}Z_1) = \rho(a(Z_2 - \mathbb{E}Z_2)) = a\rho(Z_2 - \mathbb{E}Z_2)$$

and hence

$$\rho(Z_1 - \mathbb{E}Z_1) \leqslant \rho(Z_2 - \mathbb{E}Z_2) \iff a \leqslant 1 \iff \sigma_{Z_1}^2 \leqslant \sigma_{Z_2}^2 \tag{37}$$

which proves (2). Now consider only portfolios in $\mathcal{E}$. Then (37) holds with $\mathbb{E}Z_1 = \mathbb{E}Z_2 = r$. However, using translation invariance of ρ, $\rho(Z_j - r) = \rho(Z_j) - r$ for $j = 1, 2$. This gives

$$\rho(Z_1) \leqslant \rho(Z_2) \iff \sigma_{Z_1}^2 \leqslant \sigma_{Z_2}^2$$

proving (3).

[44] In a recent paper, Lindskog (2000a) compares estimators for linear correlation showing that the standard covariance estimator (17) performs poorly for heavy-tailed elliptical data. Several alternatives are proposed and compared.

distribution. The minimum capital reserve is then a multiple of this high quantile, usually computed with $\alpha = 0.99$. Therefore it is very important to attempt to model correctly the tail of probability distribution of returns (profit and losses). The primary difficulty is that we are trying to model events about which we know very little. By definition, these events are rare. The model must allow for these rare but very damaging events. Extreme value theory (EVT) approaches the modelling of these rare and damaging events in a statistically sound way. Once the risks have been modelled they may be measured. We will use *VaR* and *Expected Shortfall* to measure them.

Extreme value theory (EVT) has its roots in hydrology, where, for example, one needed to compute how high a sea dyke had to be to guard against a 100 year storm. EVT has recently found its way into the financial community. The reader interested in a solid background may now consult various texts on EVT such as Embrechts, Klüppelberg and Mikosch (1997), Reiss and Thomas (2001) and Beirlant, Teugels and Vynckier (1996). For discussions of the use of EVT in risk management, see Embrechts (2000) and Diebold, Schuermann and Stroughair (2000).

The modelling of extremes may be done in two different ways: modelling the *maximum* of a collection of random variables, and modelling the *largest values* over some high threshold. We start, for historical reasons, with the first method, called block maxima.

6.1. Limit law for maxima

The Fisher–Tippett theorem is one of two fundamental theorems in EVT. It does for the maxima of i.i.d. random variables what the central limit theorem does for sums. It provides the limit law for maxima.

Theorem 6.1 (Fisher–Tippett, 1928). *Let (X_n) be a sequence of i.i.d. random variables with distribution F. Let $M_n = \max(X_1, \ldots, X_n)$. If there exist norming constants $c_n > 0$ and $d_n \in \mathbb{R}$ and some non-degenerate distribution function H such that*

$$\frac{M_n - d_n}{c_n} \xrightarrow{d} H,$$

then H is one of the following three types:

$$\textit{Fréchet} \quad \Phi_\alpha(x) = \begin{cases} 0, & x \leqslant 0, \\ \exp\{-x^{-\alpha}\}, & x > 0, \end{cases} \qquad \alpha > 0,$$

$$\textit{Weibull} \quad \Psi_\alpha(x) = \begin{cases} \exp\{-(-x)^{\alpha}\}, & x \leqslant 0, \\ 1, & x > 0, \end{cases} \qquad \alpha > 0,$$

$$\textit{Gumbel} \quad \Lambda(x) = \exp\{-\mathrm{e}^{-x}\}, \quad x \in \mathbb{R}.$$

The distributions Φ_α, Ψ_α, and Λ are called standard extreme value distributions. The expressions given above are cumulative distribution functions. The Weibull is usually defined

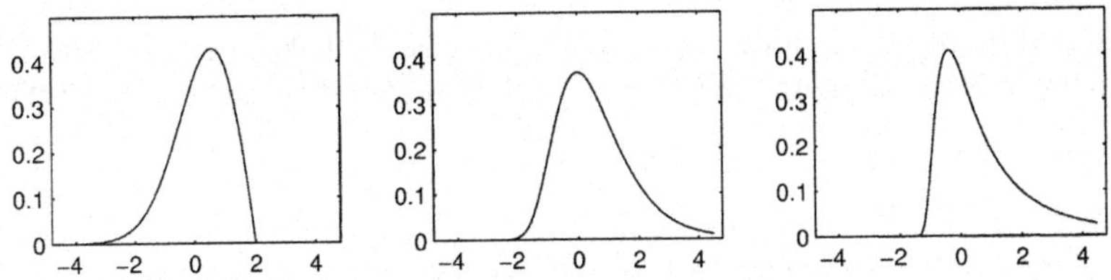

Fig. 10. Densities of the generalized extreme value distribution H_ξ. Left: Weibull with $\xi = -0.5$. Middle: Gumbel with $\xi = 0$. Right: Fréchet with $\xi = 0.5$.

as having support $(0, \infty)$ but, in the context of extreme value theory, it has support on $(-\infty, 0)$, as indicated in the theorem. These distributions are related:

$$X \sim \Phi_\alpha \quad \Longleftrightarrow \quad \ln X^\alpha \sim \Lambda \quad \Longleftrightarrow \quad -\frac{1}{X} \sim \Psi_\alpha.$$

A one-parameter representation of these distributions (due to Jenkinson and von Mises) will be useful. The reparameterized version is called the generalized extreme value (GEV) distribution.

$$H_\xi(x) = \begin{cases} \exp\{-(1+\xi x)^{-1/\xi}\}, & \xi \neq 0, \\ \exp\{-e^{-x}\}, & \xi = 0, \end{cases}$$

where $1 + \xi x > 0$. The standard extreme value distributions Φ_α, Ψ_α, and Λ follow by taking $\xi = \alpha^{-1} > 0$, $\xi = -\alpha^{-1} < 0$, and $\xi = 0$ respectively.[45] There densities are sketched in Figure 10. The parameter ξ is the *shape parameter* of H. Since for any random variable $X \sim F_X$ and constants $\mu \in \mathbb{R}$ and $\sigma > 0$, the distribution function of $\widetilde{X} = \mu + \sigma X$ is given by $F_{\widetilde{X}}(x) = F_X((x - \mu)/\sigma)$, we can add location and scale parameters to the above parameterization, and consider

$$H_{\xi,\mu,\sigma}(x) = H_\xi\left(\frac{x-\mu}{\sigma}\right).$$

If the Fisher–Tippett theorem holds, then we say that F is in the maximum domain of attraction of H and write $F \in MDA(H)$. Most distributions in statistics are in $MDA(H_\xi)$ for some ξ. If $F \in MDA(H_\xi)$ and $\xi = 0$ or $F \in MDA(H_\xi)$ and $\xi < 0$, then F is said to be thin-tailed or short-tailed respectively. Thin-tailed distributions ($\xi = 0$) include the normal, exponential, gamma and lognormal. Short-tailed distributions ($\xi < 0$) have a finite

[45] Consider, for example, the Fréchet distribution where $\xi = \alpha^{-1} > 0$. Since the support of H_ξ is $1 + \xi x > 0$, one has

$$H_{\alpha^{-1}}(x) = \exp\{-(1+\alpha^{-1}x)^{-\alpha}\} = \Phi_\alpha(1+\alpha^{-1}x)$$

for $1 + \alpha^{-1}x > 0$.

right-hand end point and include the uniform and beta distributions. The heavy-tailed distributions, those in the domain of attraction of the Fréchet distribution, $F \in MDA(H_\xi)$, for $\xi > 0$, are of particular interest in finance. They are characterized in the following theorem due to Gnedenko.

Theorem 6.2 (Gnedenko, 1943). *The distribution function $F \in MDA(H_\xi)$ for $\xi > 0$ if and only if $\overline{F}(x) = 1 - F(x) = x^{-1/\xi} L(x)$ for some slowly varying function L.*[46]

Distributions such as the Student-t, α-stable and *Pareto* are in this class. Note that if $X \sim F$ with $F \in MDA(H_\xi)$, $\xi > 0$ then all moments $\mathbb{E}X^\beta$ are infinite for $\beta > 1/\xi$. Note also that $\xi < 1$ corresponds to $\alpha > 1$, where α is as in Theorem 6.1.

6.2. *Block maxima method*

We now explain the *block maxima* method, where one assumes in practice that the maximum is distributed as $H_{\xi,\mu,\sigma}$. The implementation of this method requires a great deal of data. Let $X_1, X_2, \ldots, X_{mn}$ be daily (negative) returns and divide them into m adjacent blocks of size n. Choose the block size n large enough so that our limiting theorem results apply to $M_n^{(j)} = \max(X_{(j-1)n+1}, \ldots, X_{(j-1)n+n})$ for $j = 1, \ldots, m$. Our data set must then be long enough to allow for m blocks of length n. There are three parameters, ξ, μ and σ, which need to be estimated, using for example maximum likelihood based on the extreme value distribution. The value of m must be sufficiently large as well, to allow for a reasonable confidence in the parameter estimation. This is the classic bias-variance trade-off since for a finite data set, increasing the number of blocks m, which reduces the variance, decreases the block size n, which increases the bias. Once the GEV model $H_{\xi,\mu,\sigma}$ is fit using $M_n^{(1)}, \ldots, M_n^{(m)}$, we may estimate quantities of interest.

For example, assuming $n = 261$ trading days per year, we may want to find $R_{261,k}$, the daily loss we expect to be exceeded in one year every k years.[47] If this loss is exceeded in a given day, this day is viewed as an exceedance day and the year to which the day belongs is regarded as an exceedance year. While an exceedance year has at least one exceedance day, we are not concerned here with the total number of exceedance days in that year. This would involve taking into consideration the propensity of extremes to form clusters. Since we want M_{261} to be less than $R_{261,k}$ for $k-1$ of k years, $R_{261,k}$ is the $1 - 1/k$ quantile of M_{261}:

$$R_{261,k} = \inf\left\{ r \;\middle|\; \mathbb{P}(M_{261} \leqslant r) \geqslant 1 - \frac{1}{k} \right\}. \tag{38}$$

[46] The function L is said to be slowly varying (at infinity) if

$$\lim_{x \to \infty} \frac{L(tx)}{L(x)} = 1, \quad \forall t > 0.$$

[47] Note the obvious hydrological analogy: How high to build a sea dyke to guard against a k year storm.

If we assume that M_{261} has approximately the $H_{\xi,\mu,\sigma}$ distribution, the quantile $R_{261,k}$ is given by

$$R_{261,k} = H^{-1}_{\xi,\mu,\sigma}\left(1-\frac{1}{k}\right) \tag{39}$$

$$= \mu + \frac{\sigma}{\xi}\left(\left(-\ln\left(1-\frac{1}{k}\right)\right)^{-\xi} - 1\right), \quad \xi \neq 0, \tag{40}$$

since the inverse function of $y = \exp\{-(1+\xi x)\}^{-1/\xi}$ is $x = (1/\xi)[(-\ln y)^{-\xi} - 1]$. Confidence intervals for $R_{261,k}$ may also be constructed using *profile log-likelihood* functions. The idea is as follows. The GEV distribution $H_{\xi,\mu,\sigma}$ depends on three parameters. Substitute $R_{261,k}$ for μ using (40) and denote the reparameterized H as $H_{\xi,R_{261,k},\sigma}$ after some abuse of notation. Then obtain the log-likelihood $L(\xi, R_{261,k}, \sigma|M_1, \ldots, M_m)$ for our m observations from $H_{\xi,R_{261,k},\sigma}$. Take H_0: $R_{261,k} = r$ as the null hypothesis in an asymptotic likelihood ratio test and let $\Theta_0 = (\xi \in \mathbb{R},\ R_{261,k} = r,\ \sigma \in \mathbb{R}^+)$ and $\Theta = (\xi \in \mathbb{R},\ R_{261,k} \in \mathbb{R},\ \sigma \in \mathbb{R}^+)$ be the constrained and unconstrained parameter spaces respectively. Then under certain regularity conditions we have that

$$-2\left[\sup_{\Theta_0} L(\theta|M_1, \ldots, M_m) - \sup_{\Theta} L(\theta|M_1, \ldots, M_m)\right] \sim \chi^2_1$$

as $m \to \infty$ where $\theta = (\xi, R_{261,k}, \sigma)$ and χ^2_1 is a chi-squared distribution with one degree of freedom. Let $L(\hat{\xi}, r, \hat{\sigma}) = \sup_{\Theta_0} L(\theta|M_1, \ldots, M_m)$ and $L(\hat{\xi}, \widehat{R}_{261,k}, \hat{\sigma}) = \sup_{\Theta} L(\theta|M_1, \ldots, M_m)$ denote the constrained and unconstrained maximum log-likelihood values respectively. The α confidence interval for $R_{261,k}$ is the set

$$\left\{r\colon\ L(\hat{\xi}, r, \hat{\sigma}) \geqslant L(\hat{\xi}, \widehat{R}_{261,k}, \hat{\sigma}) - \frac{1}{2}\chi^2_1(\alpha)\right\},$$

that is, the set r for which the null hypothesis cannot be rejected for level α. See McNeil (1998a) or Këllezi and Gilli (2000) for details.

Example 6.1. We have 7570 data points for the *NASDAQ*, which we subdivided into $m = 31$ blocks of roughly $n = 261$ trading days. (The last block, which corresponds to January 2001, has relatively few trading days, but was included because of the large fluctuations.) Estimating the GEV distribution by maximum likelihood leads to $\hat{\xi} = 0.319$, $\hat{\mu} = 2.80$ and $\hat{\sigma} = 1.38$. The value of $\hat{\xi}$ corresponds to $\hat{\alpha} = 1/\hat{\xi} = 3.14$, which is in the expected range for financial data. The GEV fit is not perfect (see Figure 11). Choosing $k = 20$ yields an estimate of the twenty year return level $\widehat{R}_{261,20} = 9.62\%$. Figure 12, which displays the log-likelihood corresponding to the null-hypothesis that $\widehat{R}_{261,20} = r$, where r is displayed on the abscissa, also provides the corresponding confidence interval.

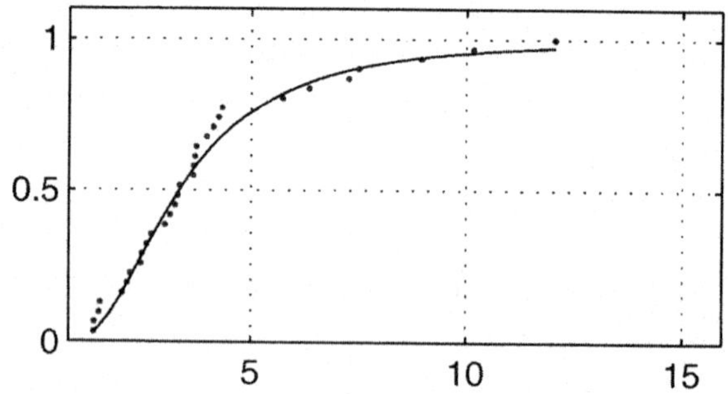

Fig. 11. The GEV distribution $H_{\hat{\xi},\hat{\mu},\hat{\sigma}}$ fitted using the 31 annual maxima of daily (negative, as %) *NASDAQ* returns.

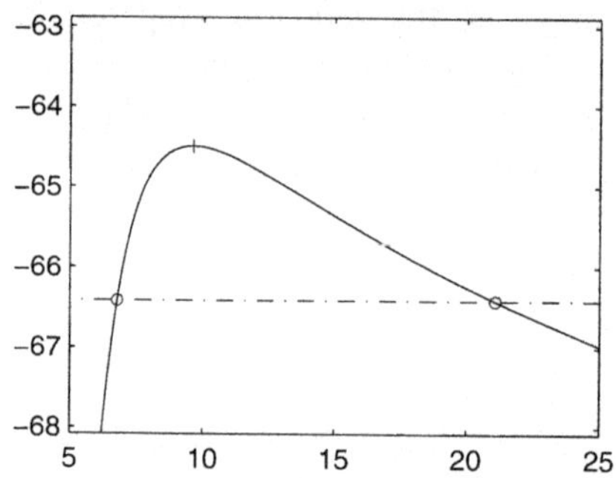

Fig. 12. The profile log-likelihood curve for the 20 year return level $R_{261,20}$ for *NASDAQ*. The abscissa displays return levels (as %) and the ordinate displays log-likelihoods. The point estimate $\widehat{R}_{261,20} = 9.62\%$ corresponds to the location of the maximum and the asymmetric 95% confidence interval, computed using the profile log-likelihood curve, is $(6.79\%, 21.1\%)$.

6.3. *Using the block maxima method for stress testing*

For the purpose of *stress testing* (worst case scenario), it is the high quantiles of the daily return distribution F that we are interested in, not those of M_n. If the $X_i \sim F$ have a continuous distribution, we have

$$\mathbb{P}(M_n \leqslant R_{n,k}) = 1 - \frac{1}{k}.$$

If they are also i.i.d.,

$$\mathbb{P}(M_n \leqslant R_{n,k}) = \big(\mathbb{P}(X \leqslant R_{n,k})\big)^n,$$

where $X \sim F$, and hence

$$\mathbb{P}(X \leqslant R_{n,k}) = \left(1 - \frac{1}{k}\right)^{1/n}. \tag{41}$$

This means that $R_{n,k}$ is the $(1 - 1/k)^{1/n}$ quantile of the marginal distribution F. Suppose we would like to calculate *VaR* at very high quantiles for the purposes of stress testing. The block size n has been fixed for the calibration of the model. This leaves the parameter k for the $R_{n,k}$ return level free. High α quantiles, $x_\alpha = F^{-1}(\alpha)$, of F may then be computed from (41) by choosing $\alpha = (1 - 1/k)^{1/n}$, that is $k = 1/(1 - \alpha^n)$. Hence

$$VaR_\alpha(X) = R_{n,k}, \quad \text{where } k = \frac{1}{1 - \alpha^n}. \tag{42}$$

For the *NASDAQ* data, our choice of $k = 20$, corresponds to

$$\alpha = 0.9998 \quad \text{and} \quad \widehat{VaR}_{\alpha=0.9998}(X) = \widehat{R}_{261,20} = 9.62\%.$$

In practice α is given, and one chooses $k = 1/(1 - \alpha^n)$, then computes $R_{n,k}$ using (40) and thus one obtains $VaR_\alpha(X) = R_{n,k}$.

We assumed independence but, in finance, this assumption is not realistic. At best, the marginal distribution F can be viewed as stationary. For the extension of the Fisher–Tippett theorem to stationary time series see Leadbetter, Lindgren and Rootzén (1983, 1997) and McNeil (1998a). See McNeil (1998b) for a non-technical example pertaining to the block maxima method and the market crash of 1987.

6.4. *Peaks over threshold method*

The more modern approach to modelling extreme events is to attempt to focus not only the largest (maximum) events, but on all events greater than some large preset threshold. This is referred to as *peaks over threshold* (POT) modelling. We will discuss two approaches to POT modelling currently found in the literature. The first is a semi-parametric approach based on a Hill type estimator of the tail index (Beirlant, Teugels and Vynckier, 1996; Danielsson and de Vries, 1997, 2000; Mills, 1999). The second approach is a fully parametric approach based on the generalized Pareto distribution (Embrechts, Klüppelberg and Mikosch, 1997; McNeil and Saladin, 1997; Embrechts, Resnick and Samorodnitsky, 1999).

6.4.1. *Semiparametric approach*

Recall that F_X is in the maximum domain of attraction of the Fréchet distribution if and only if $\overline{F}_X(x) = x^{-\alpha}L(x)$ for some slowly varying function L. Suppose F_X is the distribution function of a loss distribution over some time horizon, where we would like to calculate a quantile based risk measure such as *VaR*. Assume for simplicity that the distribution of large losses is of Pareto type

$$\mathbb{P}(X > x) = cx^{-\alpha}, \quad \alpha > 0,\ x > x_0. \tag{43}$$

The semiparametric approach uses a Hill type estimator for α and order statistics of historical data to invert and solve for *VaR*.

We first focus on *VaR*. Let $X^{(1)} \geqslant X^{(2)} \geqslant \cdots \geqslant X^{(n)}$ be the order statistics of an historical sample of losses of size n, assumed i.i.d. with distribution F_X. If X is of Pareto type in the tail and $X^{(k+1)}$ is a high order statistic then for $x > X^{(k+1)}$,

$$\frac{\overline{F}_X(x)}{\overline{F}_X(X^{(k+1)})} = \left(\frac{x}{X^{(k+1)}}\right)^{-\alpha}.$$

The empirical distribution function estimator $\widehat{\overline{F}}_X(X^{(k+1)}) = k/n$ suggests the following estimator of F_X in the upper tail,

$$\widehat{F}_X(x) = 1 - \frac{k}{n}\left(\frac{x}{X^{(k+1)}}\right)^{-\hat{\alpha}} \quad \text{for } x > X^{(k+1)}.$$

By inverting this relation, one can express x in terms of $\widehat{F}_X(x)$, so that fixing $q = \widehat{F}_X(x)$ one gets[48] $x = \widehat{VaR}_q(X)$. The value of q should be large, namely, $q = \widehat{F}_X(x) > \widehat{F}(X^{(k+1)}) = 1 - k/n$. This yields

$$\widehat{VaR}_q(X) = X^{(k+1)}\left(\frac{n}{k}(1-q)\right)^{-1/\hat{\alpha}}. \tag{44}$$

We obtained an estimator for *VaR* but it depends on k through $X^{(k+1)}$, on the sample size n and $\hat{\alpha}$. To estimate α, Hill (1975) proposed the following estimator $\hat{\alpha}^{(\mathrm{Hill})}$ which is also dependent on the order statistics and sample size:

$$\hat{\alpha}^{(\mathrm{Hill})} = \hat{\alpha}_{k,n}^{(\mathrm{Hill})} = \left(\frac{1}{k}\sum_{i=1}^{k} \ln X^{(i)} - \ln X^{(k+1)}\right)^{-1}. \tag{45}$$

The consistency and asymptotic normality properties of this $\hat{\alpha}^{(\mathrm{Hill})}$ estimator are known in the i.i.d. case and for certain stationary processes. There are however, many issues surrounding Hill-type estimators, see for example Beirlant, Teugels and Vynckier (1996), Embrechts, Klüppelberg and Mikosch (1997) and Drees, de Haan and Resnick (2000).

To obtain $VaR_q(X)$, one also needs to choose the threshold level $X^{(k+1)}$ or, equivalently, k. Danielsson et al. (2001) provide an optimal choice for k by means of a two stage bootstrap method. Even in this case, however, optimal means merely minimizing the *asymptotic* mean squared error, which leaves the user uncertain as to how to proceed in the finite sample case. Traditionally the choice of k is done visually by constructing a Hill plot.

The Hill plot $\{(k, \hat{\alpha}_{k,n}^{(\mathrm{Hill})}): k = 1, \ldots, n-1\}$ is a visual check for the optimal choice of k. The choice of k and therefore of $\hat{\alpha}_{k,n}^{(\mathrm{Hill})}$, is inferred from a stable region of the plot since

[48] We write here VaR_q and not VaR_α since now α represents the heavy-tail exponent.

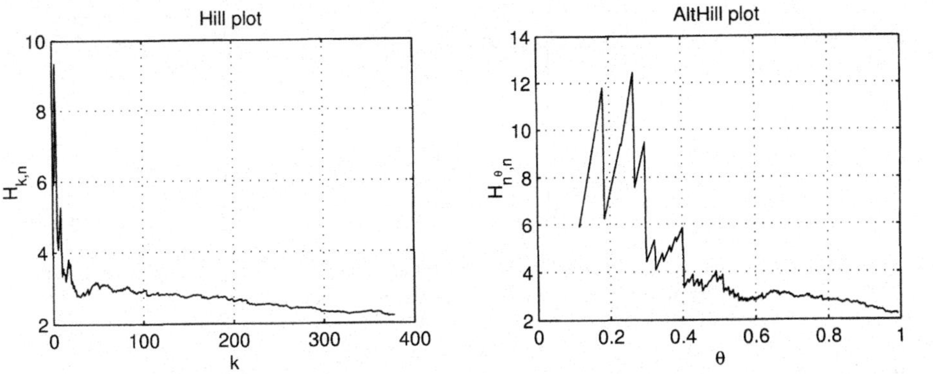

Fig. 13. Hill plots for the *NASDAQ* data set. Left: The Hill plot $\{(k, \hat{\alpha}_{k,n}^{(\mathrm{Hill})}): k = 1, \ldots, n-1\}$. Right: The AltHill plot$\{(\theta, \hat{\alpha}_{\lceil n^\theta \rceil,n}^{(\mathrm{Hill})}): 0 \leqslant \theta < 1\}$. The Hill plot is difficult to read, whereas the AltHill plot gives the user an estimate of $\hat{\alpha}^{\mathrm{AltHill}} \approx 3$.

in the Pareto case, where (43) holds, $\hat{\alpha}_{n-1,n}^{(\mathrm{Hill})}$ is the maximum likelihood estimator for α. In the more general case

$$1 - F(x) \sim x^{-\alpha} L(x), \quad x \to \infty,\ \alpha > 0, \tag{46}$$

where L is a slowly varying function, the traditional Hill plot is often difficult to interpret. Resnick and Stărică (1997) suggest an alternative plot, called an *AltHill* plot by plotting $\{(\theta, \hat{\alpha}_{\lceil n^\theta \rceil,n}^{(\mathrm{Hill})}): 0 \leqslant \theta < 1\}$ where $\lceil n^\theta \rceil$ denotes the smallest integer greater than or equal to n^θ. This plot has the advantage of stretching the left-hand side of the plot, which corresponds to smaller values of k, often making the choice of k easier. See Figure 13 for examples of the Hill and AltHill plots for the ordered negative returns $X^{(j)}$ for the *NASDAQ*.

6.4.2. *Fully parametric approach*

The fully parametric approach uses the generalized Pareto distribution (GPD) and the second fundamental theorem in EVT by Pickands, Balkema and de Haan. The GPD is a two-parameter distribution

$$G_{\xi,\beta}(x) = \begin{cases} 1 - \left(1 + \dfrac{\xi x}{\beta}\right)^{-1/\xi}, & \xi \neq 0, \\ 1 - \exp\left(-\dfrac{x}{\beta}\right), & \xi = 0, \end{cases}$$

where an additional parameter $\beta > 0$ has been introduced. The support of $G_{\xi,\beta}(x)$ is $x \geqslant 0$ for $\xi \geqslant 0$ and $0 \leqslant x \leqslant -\beta/\xi$ for $\xi < 0$. The distribution is heavy-tailed when $\xi > 0$. GPD distributions with $\beta = 1$ are displayed in Figure 14.

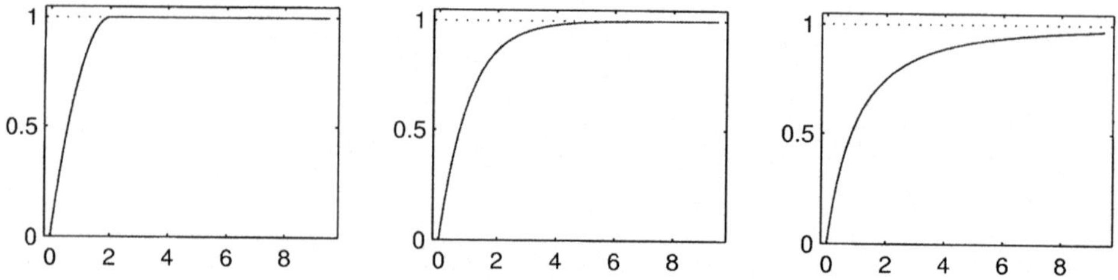

Fig. 14. GPD distribution functions $G_{\xi,\beta}$, all with $\beta = 1$. Left: $\xi = -0.5$, Middle: $\xi = 0$, Right: $\xi = 0.5$, which corresponds to a location adjusted Pareto distribution with $\alpha = 2$.

Definition 6.1. Let $X \sim F$ with right-end-point $x_F = \sup\{x \in \mathbb{R} \mid F(x) < 1\} \leqslant \infty$. For any high threshold $u < x_F$ define the *excess* distribution function

$$F_u(x) = \mathbb{P}(X - u \leqslant x | X > u) \quad \text{for } 0 \leqslant x < x_F - u. \tag{47}$$

The *mean* excess function of X is then

$$\mathrm{e}_X(u) = \mathbb{E}(X - u | X > u). \tag{48}$$

If X has exceeded the high level u, $F_u(x)$ measures the probability that it did not exceed it by more than x. Note that for $0 \leqslant x < x_F - u$, we may express $F_u(x)$ in terms of F,

$$F_u(x) = \frac{F(u + x) - F(u)}{1 - F(u)},$$

and the mean excess function $e_X(u)$ may be expressed as a function of the excess distribution F_u as

$$e_X(u) = \int_0^{x_F - u} x \, \mathrm{d}F_u(x).$$

The following theorem relates F_u to a GPD through the maximum domain of attraction of a GEV distribution. In fact, it completely characterizes the maximum domain of attraction of H_ξ.

Theorem 6.3 (Pickands, 1975, Balkema and de Haan, 1974). *Let $X \sim F$. Then for every $\xi \in \mathbb{R}$, $X \in MDA(H_\xi)$ if and only if*

$$\lim_{u \uparrow x_F} \sup_{0 < x < x_F - u} \left| F_u(x) - G_{\xi, \beta(u)}(x) \right| = 0$$

for some positive function β.

This theorem says that the excess distribution F_u may be replaced by the GPD distribution G when u is very large. To see how it can be used, note that by (47) above, we may write

$$\overline{F}(x) = \overline{F}(u)\overline{F}_u(x-u) \tag{49}$$

for $x > u$. Assuming that u is sufficiently large, we may then approximate F_u by $G_{\xi,\beta(u)}$ and use the empirical estimator, for $\overline{F}(u)$,

$$\widehat{\overline{F}}(u) = \frac{N_u}{n}, \quad \text{where } N_u = \sum_{i=1}^{n} 1_{\{X_i > u\}}$$

and where n is the total number of observations. The upper tail of $F(x)$ may then be estimated by

$$\widehat{F}(x) = 1 - \widehat{\overline{F}} = 1 - \frac{N_u}{n}\left(1 + \hat{\xi}\frac{x-u}{\hat{\beta}}\right)^{-1/\hat{\xi}} \quad \text{for all } x > u. \tag{50}$$

This way of doing things allows us to extrapolate beyond the available data which would not be possible had we chosen an empirical estimator for $F(x)$, $x > u$. We can therefore deal with potentially catastrophic events which have not yet occurred.

The parameters ξ and β of the GPD $G_{\xi,\beta(u)}$ may be estimated by using, for example, maximum likelihood once the threshold u has been chosen. The data points that are used in the maximum likelihood estimation are $X_{i_1} - u, \ldots, X_{i_k} - u$ where $X_{i_1}, \ldots, X_{i_k}$ are the observations that exceed u. Again there is a bias-variance trade-off in the choice of u. To choose a value for u, a graphical tool known as the *mean excess plot* $(u, e_X(u))$ is often used.

The mean excess plot relies on the following theorem for generalized Pareto distributions.

Theorem 6.4 (Embrechts, Klüppelberg and Mikosch, 1997). *Suppose X has GPD distribution with $\xi < 1$ and β. Then, for $u < x_F$,*

$$e_X(u) = \frac{\beta + \xi u}{1 - \xi}, \quad \beta + \xi u > 0.$$

The restriction $\xi < 1$ implies that the heavy-tailed distribution must have at least a finite mean.

If the threshold u is large enough so that F_u is approximately $G_{\xi,\beta}$ then, by Theorem 6.4, the plot $(u, e(u))$ is linear in u. How then is one to pick u? The mean excess plot is a graphical tool for examining the relationship between the possible threshold u and the mean excess function $e_X(u)$ and checking the values of u where there is linearity. In practice it is not $e_X(u)$, but its sample version

$$\hat{e}_X(u) = \frac{\sum_{i=1}^n (X_i - u)^+}{\sum_{i=1}^n 1_{\{X_i > u\}}}$$

which is plotted against u. After using the mean excess plot to pick the upper threshold u one obtains an estimator of the tail of the distribution by applying (50). For the *NASDAQ* data, since linearity seems to start at relatively small values of u (Figure 15), we choose $u = 1.59$ which corresponds to the 95% of the empirical *NASDAQ* return distribution.

To obtain $VaR_\alpha(X)$ for $VaR_\alpha(X) > u$, one simply inverts the tail estimator (50), which yields

$$\widehat{VaR}_\alpha(X) = u + \frac{\hat{\beta}}{\hat{\xi}}\left(\left(\frac{n}{N_u}(1-\alpha)\right)^{-\hat{\xi}} - 1\right). \tag{51}$$

Since *expected shortfall* is a risk measure with better technical properties than *VaR* we would like to find an estimator for it which uses our GPD model of the tail. Recalling the definitions of the expected shortfall (22) and the mean excess function (48) we have that

$$S_\alpha(X) = VaR_\alpha(X) + e_X\big(VaR_\alpha(X)\big).$$

Since the excess distribution F_u is approximated by a GPD $G_{\xi,\beta(u)}$ with $\xi < 1$ then, applying Theorem 6.4, we get for $VaR_\alpha(X) > u$,

$$S_\alpha(X) = VaR_\alpha(X) + \frac{\beta + \xi(VaR_\alpha(X) - u)}{1-\xi} = \frac{\beta + VaR_\alpha(X) - \xi u}{1-\xi}.$$

This suggests the following estimator for expected shortfall,

$$\widehat{S}_\alpha(X) = \frac{\hat{x}_\alpha}{1-\hat{\xi}} + \frac{\hat{\beta} - \hat{\xi} u}{1-\hat{\xi}}, \tag{52}$$

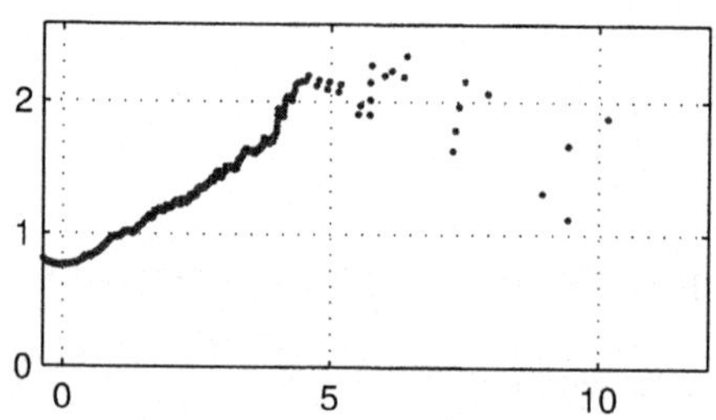

Fig. 15. Sample mean excess plot $(u, \hat{e}_X(u))$ for *NASDAQ*.

where $\hat{x}_\alpha = \widehat{VaR}_\alpha(X)$ may be obtained by using (51). As in the case of block maxima, confidence intervals for $\widehat{VaR}_\alpha$ and $\widehat{S}_\alpha$ may be constructed using profile log-likelihood functions.

6.4.3. Numerical illustration

To illustrate the usefulness of EVT in risk management, we consider the following example. Let $X_1, \ldots, X_n$ represent the daily negative returns of the *NASDAQ* index over most of its history from February 1971 to February 2001, which gives a time series of $n = 7570$ data points.

The price and return series are displayed in Figure 16. Let $X^{(1)} \geqslant \cdots \geqslant X^{(n)}$ be the corresponding order statistics. Suppose the risk manager wants to obtain value at risk and expected shortfall estimates of the returns on the index at some high quantile. Assume that $\{X_i\}_{i=1}^n$ are i.i.d. so that Theorem 6.1 holds. Then, using Theorem 6.3, we model the tail of the excess distribution F_u by a GPD $G_{\xi,\beta}$ and use (49) to model the distribution $F(x)$

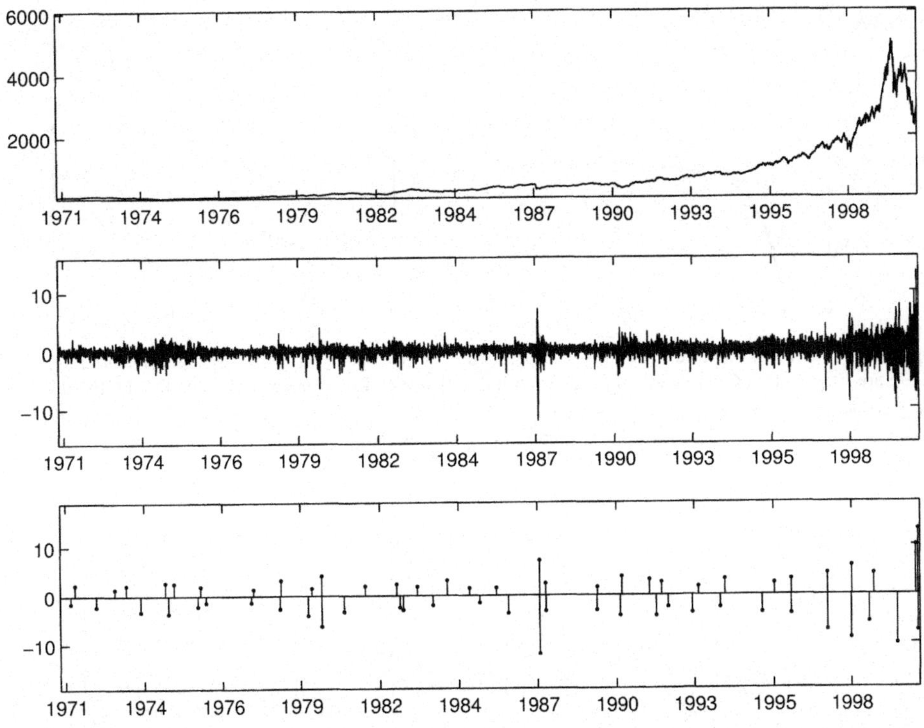

Fig. 16. Time series of *NASDAQ* daily prices, (log) returns and annual maxima and minima daily returns given as a percent for the period February 1971 (when it was created) to February 2001. If P_t is the price (level) at time t, the returns are defined as $100 \ln(P_t/P_{t-1})$ and expressed as %. The crash of 1987 is clearly visible. The *NASDAQ* price level peaked in March of 2000.

of the observations for all $x > u$. We use Theorem 6.4 and the sample mean excess plot, Figure 15, to pick the high threshold $u = 1.59\%$. This leaves us with $k = 379$ observations from which we estimate the parameters of the GPD by maximum likelihood. The estimates give $\xi = 0.189$ and $\beta = 0.915$. The model fit is checked by using a QQ plot displayed in Figure 17. Accepting the model, we go on to calculate the value at risk and expected shortfall for various high quantiles α by using (51) and (52). The results for the *NASDAQ* are plotted in Figure 18 (solid lines). If one had assumed that the observations were normally distributed (dashed lines), both the *VaR* and the expected shortfall would have been significantly underestimated for high quantiles.

For example, at the $\alpha = 0.99$ confidence level, $\widehat{VaR}_\alpha(X) = 6.59\%$ under the normal model versus $\widehat{VaR}_\alpha(X) = 8.19\%$ for the GPD model. For the expected shortfall, the difference is even more dramatic. For the normal model, $\widehat{S}_\alpha(X) = 7.09\%$ versus $\widehat{S}_\alpha(X) = 10.8\%$ for the GPD model. This is to be expected, since under the assumption of normality it may be shown (Embrechts, Klüppelberg and Mikosch, 1997) that

$$\frac{S_\alpha}{VaR_\alpha} \to 1 \quad \text{as } \alpha \to 1^-,$$

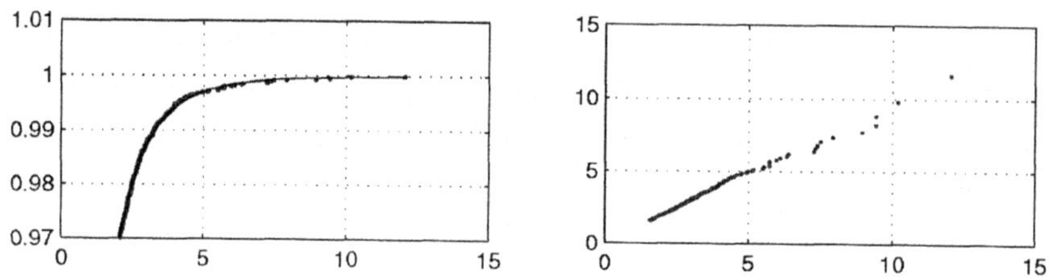

Fig. 17. For the *NASDAQ* return data (as %), there were 379 exceedances above the high threshold $u = 1.59\%$. These are fitted with a GPD distribution $G_{\hat{\xi},\hat{\beta}}$ with $\hat{\xi} = 0.189$ and $\hat{\beta} = 0.915$. Left: The fitted GPD distribution (dark curve) and the empirical one (dotted curve). Right: QQ-plot of sample quantiles versus the quantiles of the fitted $G_{\hat{\xi},\hat{\beta}}$ distribution.

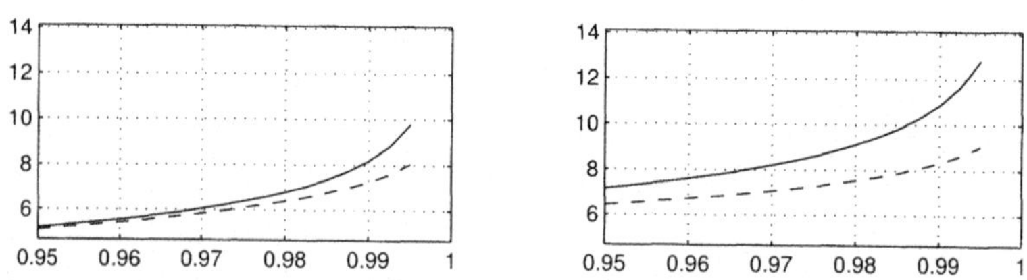

Fig. 18. Risk estimates for *NASDAQ* in percent returns versus α. Left: Value at risk VaR_α, for GPD (solid) and normal (dashed). Right: Expected shortfall S_α, for GDP (solid) and normal (dashed). The parameters of the GPD are fitted by maximum likelihood using 30 years of data. The sample mean and volatility of the normal distribution are computed by (16) using the most recent year of daily observations.

whereas for the GPD model

$$\frac{S_\alpha}{VaR_\alpha} \longrightarrow \frac{1}{1-\xi} \quad \text{as } \alpha \to 1^-.$$

These results indicate that for very high quantiles, the expected shortfall S_α and the value at risk VaR_α are comparable under normality, but for the GPD with $\xi < 1$, S_α tends to be larger than VaR_α.

6.4.4. A GARCH-EVT model for risk

In order to invoke Theorems 6.1 and 6.3 in the numerical illustration above it was necessary to assume that the (negative) returns $\{X_t\}_{t\in\mathbb{Z}}$ were i.i.d. However, from inspection of Figures 16 and 19, it is apparent that this assumption is unrealistic. The time series of returns is characterized by periods of varying volatility, that is, the time series is *heteroscedastic.* The heteroscadicity of the time series may cause problems for the estimation of the parameters of the GPD model since we would expect the high threshold u to be violated more often during periods of high volatility. Smith (2000) suggests using Bayesian techniques to model time-varying GPD parameters. In this section, we review a model proposed by McNeil and Frey (2000) which extends the EVT methodology to models of financial time series that allow for stochastic volatility and apply this model to the *NASDAQ* data set.

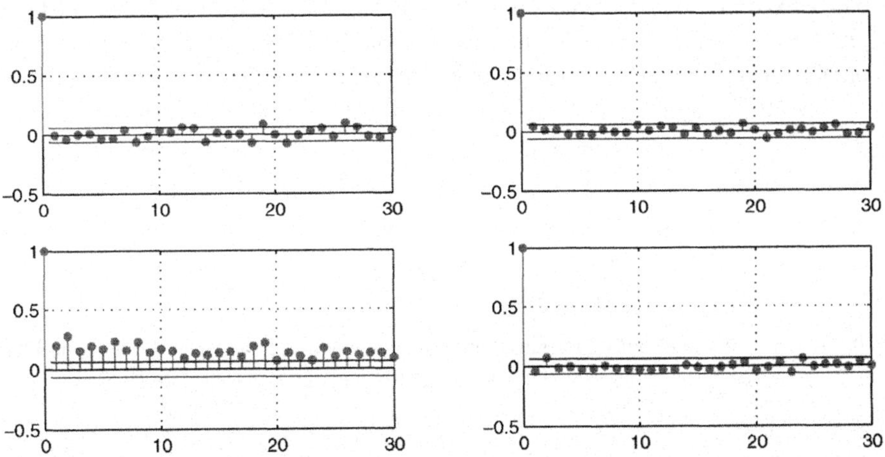

Fig. 19. Sample auto correlation functions with lags on the abscissa and sample autocorrelation on the ordinate: returns (top left), squared returns (bottom left), GARCH innovations (top right), squared GARCH innovations (bottom right). The sample consists of 1000 daily returns for the *NASDAQ* ending February 2001. Horizontal lines indicate the 95% confidence bands ($\pm 1.96/\sqrt{n}$) corresponding to Gaussian white noise.

Recall from Section 3.2.2 that the standard GARCH(1,1) model is given by[49]

$$X_t = \sigma_t Z_t, \quad \text{where } Z_t \sim F_Z \text{ i.i.d.}, \tag{53}$$

$$\sigma_t^2 = \alpha_0 + \alpha_1 X_{t-1}^2 + \beta_1 \sigma_{t-1}^2. \tag{54}$$

Since the time $t+1$ volatility σ_{t+1} is known at time t we have that

$$VaR_\alpha(X_{t+1}|\mathcal{F}_t) := \inf\{x \in \mathbb{R} \mid F_{X_{t+1}|\mathcal{F}_t}(x) \geqslant \alpha\} = \sigma_{t+1} z_\alpha, \tag{55}$$

where $z_\alpha = F_Z^{-1}(\alpha)$. The same argument shows that the conditional expected shortfall

$$S_\alpha(X_{t+1}|\mathcal{F}_t) := \mathbb{E}\big(X_{t+1}|X_{t+1} > VaR_\alpha(X_{t+1}|\mathcal{F}_t), \mathcal{F}_t\big) = \sigma_{t+1}\mathbb{E}(Z|Z > z_\alpha).$$

Traditionally the innovation distribution F_Z is assumed normal. Figures 6 and 20 show that this assumption may still underestimate the tails of the loss portion of the distribution. McNeil and Frey propose a two step procedure to estimate *VaR* and expected shortfall of the conditional distribution. First they use a GARCH(1,1) model for the volatility of the (negative) return series $\{X_t\}$. This gives a series of model implied innovations $Z_t = X_t/\sigma_t$. Second, EVT is used to model the tails of the distribution of these innovations. This approach has the obvious benefit that the resulting innovations Z_t are much closer to satisfying the requirements of Theorems 6.1 and 6.3 than is the original series. We illustrate the methodology with an example using the *NASDAQ* data.

(1) Let $(x_{t-n+1}, \ldots, x_{t-1}, x_t)$ be n daily negative returns of the *NASDAQ*. We take[50] $n = 1000$ and use *pseudo-maximum-likelihood* (PML) to estimate the model parameters $\hat{\boldsymbol{\theta}} = (\hat{\alpha_0}, \hat{\alpha_1}, \hat{\beta_1})$ in (54) under the assumption[51] that F_Z is normal in (53). The parameter vector $\hat{\boldsymbol{\theta}}$ depends on the true distribution of $(X_{t-n+1}, \ldots, X_{t-1}, X_t)$, which

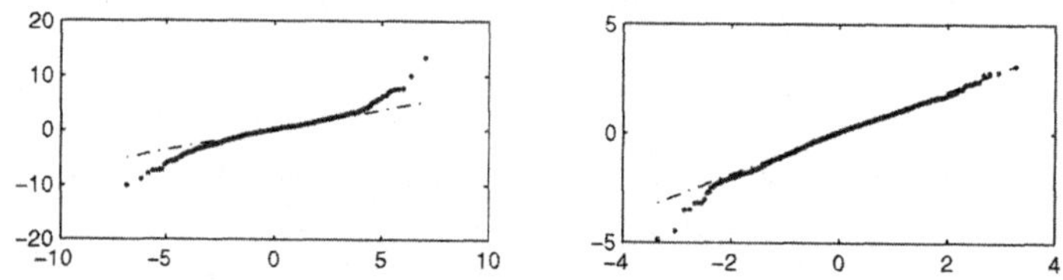

Fig. 20. QQ plots versus the normal for returns (left) and innovations (right) in Figure 19. Notice that the lower (loss) tail of the innovations are still heavier than the normal distribution.

[49] Since the *NASDAQ* series appears to have a zero conditional mean we do not set $X_t = \mu_t + \sigma_t Z_t$ and model the mean μ_t, for example as an AR(1) process $\mu_t = \phi X_{t-1}$.

[50] We keep the sample size moderate in order to avoid the *IGARCH* effect, that is $\alpha_1 + \beta_1 = 1$, corresponding to non-stationarity. See Mikosch and Stărică (2000) for details.

[51] The term *pseudo* refers to the fact that one is not maximizing the true likelihood.

is assumed stationary, and on the distribution F_Z used to compute the likelihood function.[52] When we assume F_Z is normal we fit a model whose distributional assumptions we do not believe. Under standard regularity conditions this is justified since $\hat{\boldsymbol{\theta}}$ is a consistent estimator of $\boldsymbol{\theta}$ (in fact, asymptotically normal) even if F_Z is non-normal. See Gouriéroux (1997) and references therein for details.

(2) The model innovations $(z_{t-n+1}, \ldots, z_{t-1}, z_t) = (x_{t-n+1}/\hat{\sigma}_{t-n+1}, \ldots, x_{t-1}/\hat{\sigma}_{t-1}, x_t/\hat{\sigma}_t)$ are now calculated. If the model is tenable, these innovations should be i.i.d. Figure 19 shows that while the i.i.d. assumption is not realistic for the series of returns, it is defensible for the series of innovations.[53] While the returns appear uncorrelated, their squares clearly are not, and hence the returns are dependent. The GARCH innovations and their squares appear uncorrelated. The i.i.d. assumption is therefore more tenable.

(3) Examination of the QQ plot of the innovations in Figure 20 reveals that the loss tail is heavier than that of the normal. Therefore the EVT tools of Section 6.4.2 are now applied to the innovations $(z_{t-n+1}, \ldots, z_{t-1}, z_t)$. Let $z^{(1)} \leqslant \cdots \leqslant z^{(n)}$ be the order statistics of the innovation sample. We choose the threshold $u = 1.79$, again corresponding to the 95% of the empirical distribution of innovations, which leaves $k = 50$ observations $(z^{(n-k+1)}, \ldots, z^{(n)})$, from which to estimate the GPD parameters by maximum likelihood. The estimates give $\xi = 0.323$ and $\beta = 0.364$.

Observe that $\xi = 0.323$ corresponds to a heavier tail than $\xi = 0.189$ which we found in Section 6.4.3. We are fitting here, however, over a particularly volatile period of 1000 days of the *NASDAQ* ending February 2001, whereas in Section 6.4.3, we considered nearly 30

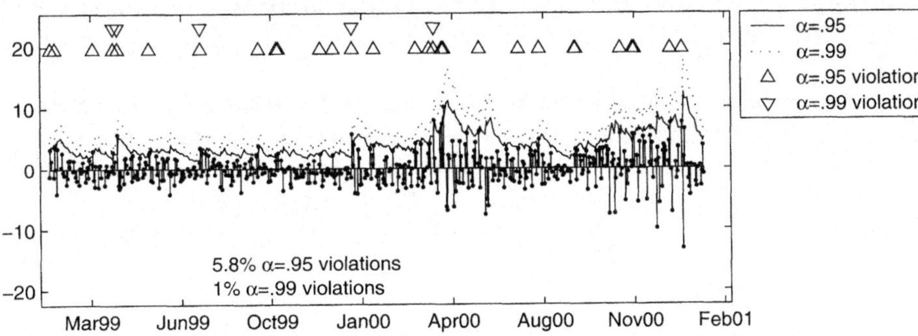

Fig. 21. Backtest results for the GARCH-EVT methodology of McNeil and Frey. Under the assumption that the model correctly estimates the conditional quantiles we expect violations 5% and 1% of the time for $\alpha = 0.95$ and $\alpha = 0.99$ respectively. *VaR* for $\alpha = 0.95$ and $\alpha = 0.99$ are given by the solid and dotted lines respectively. We obtain 5.8% violations of the $\alpha = 0.95$ level and 1% violations of the $\alpha = 0.99$ level.

[52] The condition $\alpha_1 + \beta_1 < 1$ is sufficient for stationarity of the GARCH model. We found $\hat{\alpha}_0 = 0.080$, $\hat{\alpha}_1 = 0.181$ and $\hat{\beta}_1 = 0.811$. However, as indicated in the sequel, the GARCH model is constantly updated, and hence is never used on an infinite horizon.

[53] Ljung-Box tests also found no evidence against the i.i.d. assumption for the innovations.

years worth of returns where for the majority of the time the *NASDAQ* was significantly less volatile (see Figure 16).

Since the model is assumed stationary, we could, in principle, use the estimated GARCH parameters to compute $\hat{\sigma}_{t+1}|\mathcal{F}_t$ using (54) for t beyond February 2001. Using z_α corresponding to the GPD distribution $G_{\xi,\beta}$, we would obtain, by using (55), $\widehat{VaR}_\alpha(X_{t+1}|\mathcal{F}_t)$ for t beyond February 2001. In practice, however, stationarity is not always assured and in any case one wants to use the most recent data available in order to calibrate the model.

In order to backtest the methodology we use the most recent 500 days in our *NASDAQ* data set. For each day, $t+1$, in this data set we use the previous $n = 1000$ days (negative) returns $(X_{t-n+1}, \ldots, X_{t-1}, X_t)$ to calibrate the model and estimate $VaR_\alpha(X_{t+1}|\mathcal{F}_t)$ for $\alpha = 0.95$ and $\alpha = 0.99$ using the steps above. We compare $\widehat{VaR}_\alpha(X_{t+1}|\mathcal{F}_t)$ with the actual loss x_{t+1}. A violation, at the α level, is said to occur whenever $x_{t+1} > \widehat{VaR}_\alpha(X_{t+1}|\mathcal{F}_t)$. Results for the period ending February 2001 are given in Figure 21.

7. Stable Paretian models

The works of Mandelbrot (1963) and Fama (1965) introduced the use of stable distributions to finance. The *excessively peaked* and *heavy-tailed* nature of the return distribution led the authors to reject the standard hypothesis of normally distributed returns in favor of the stable distribution. Since this time, the stable distribution has been used to model both the unconditional, and conditional return distributions. In addition, portfolio theories and market equilibrium models have been constructed using it. For an in depth introduction to the general properties of stable distributions see Samorodnitsky and Taqqu (1994) and the upcoming text Nolan (2001). A major reference for applications in finance is Rachev and Mittnik (2000).

In Definition 3.2, the stable distribution $S_\alpha(\sigma, \beta, \mu)$ is defined as the limiting distribution of the sum of i.i.d. random variables. Like the normal distribution, stable distributions are closed under addition, and are often defined by this property. Recall that if $X_1 \sim \mathcal{N}(\mu_1, \sigma_1^2)$ and $X_2 \sim \mathcal{N}(\mu_2, \sigma_2^2)$ are independent then $X_1 + X_2 \sim \mathcal{N}(\mu_1 + \mu_2, \sigma_1^2 + \sigma_2^2)$. Similarly, for stable random variables, if $X_1 \sim S_\alpha(\sigma_1, \beta_1, \mu_1)$ and $X_2 \sim S_\alpha(\sigma_2, \beta_2, \mu_2)$ are independent, then $X_1 + X_2 \sim S_\alpha(\sigma, \beta, \mu)$ where

$$\sigma = \left(\sigma_1^\alpha + \sigma_2^\alpha\right)^{1/\alpha}, \qquad \beta = \frac{\beta_1\sigma_1^\alpha + \beta_2\sigma_2^\alpha}{\sigma_1^\alpha + \sigma_2^\alpha}, \qquad \mu = \mu_1 + \mu_2.$$

It is in this sense that the stable distribution is a natural heavy-tailed alternative to the normal distribution. However, a common criticism of the stable distribution is that their tails are too heavy. One has $\mathbb{P}(X > x) \sim c_\alpha x^{-\alpha}$ as $x \to \infty$. For $0 < \alpha < 2$, this implies that $\mathbb{E}|X|^p < \infty$ if $0 < p < \alpha$. In particular, $\mathbb{E}X^2 = \infty$, that is, all non-Gaussian stable distributions have infinite variance.

The stable distributions can be defined and parameterized in different ways. One way to specify a stable distribution is through its characteristic function. This is helpful since in general there exists no closed form for the probability density function,[54] which historically, has been an impediment to their widespread use. Today, however, there are efficient computer programs to evaluate their densities using fast Fourier transform methods (Rachev and Mittnik, 2000; Nolan, 2001).

Definition 7.1. A random variable X is said to have a stable distribution if there are parameters $\alpha \in (0, 2]$, $\sigma \in [0, \infty)$, $\beta \in [-1, 1]$ and $\mu \in \mathbb{R}$ such that its characteristic function has the following form:

$$\Psi_X(t) = \begin{cases} \exp\left\{-\sigma^\alpha |t|^\alpha \left(1 - i\beta(\operatorname{sign} t)\tan\frac{\pi\alpha}{2} + i\mu t\right)\right\} & \text{for } \alpha \neq 1, \\ \exp\left\{-\sigma|t|\left(1 + i\beta\frac{2}{\pi}(\operatorname{sign} t)\ln|t|\right) + i\mu t\right\} & \text{for } \alpha = 1. \end{cases} \tag{56}$$

If both the skewness and location parameters β and μ are zero, X is said to be symmetric stable, which is denoted $X \sim S\alpha S$, and its characteristic function takes the simple form

$$\Psi_X(t) = e^{-\sigma^\alpha |t|^\alpha}.$$

If $X \sim S\alpha S$, then it is characterized completely by its index of stability α and its scale parameter σ. If $\alpha = 2$, the Gaussian case, then the scale parameter is $\sigma = \sqrt{\frac{1}{2}\operatorname{Var}(X)}$.

7.1. Stable portfolio theory

In Section 2.2 we introduced the mean–variance portfolio theory of Markowitz. The model assumed that the distribution of asset returns is multivariate normal, and provides efficient portfolios, that is, portfolios with maximum expected return for a given level of risk, where risk is measured by the variance of the portfolio. It is possible to extend the ideas of portfolio theory to the case where asset returns have a multivariate stable distribution, even though, variances are now infinite. We need first to define a stable random vector and specify its characteristic function.

Definition 7.2. The random vector $\boldsymbol{X} = (X_1, \ldots, X_n)$ is said to be a stable random vector in $\mathbb{R}^n$ if for any $a, b > 0$ there exists $c > 0$ and $\boldsymbol{d} \in \mathbb{R}^n$ such that

$$a\boldsymbol{X}_1 + b\boldsymbol{X}_2 \stackrel{d}{=} c\boldsymbol{X} + \boldsymbol{d}, \tag{57}$$

where $\boldsymbol{X}_j$, $j = 1, 2$, are independent copies of $\boldsymbol{X}$.

[54] The exceptions to this rule are the distributions $S_2(\sigma, 0, \mu)$, $S_1(\sigma, 0, \mu)$,and $S_{1/2}(\sigma, 1, \mu)$ which correspond to the Gaussian, Cauchy and Lévy distributions respectively.

The constants in (57) are related by $c^\alpha = a^\alpha + b^\alpha$, where $\alpha \in (0, 2]$ is the index of stability. Setting $n = 1$ in (57) yields one of the alternate definitions of a stable random variable alluded to earlier. In the case of a stable random vector, the scale and skewness parameters σ and β are replaced by a finite measure Γ_X on the unit hypersphere in $\mathbb{R}^n$. For convenience here, let $(\cdot, \cdot)$ denote the inner product so that $(\boldsymbol{t}, \boldsymbol{s}) = \sum_{i=1}^n t_i s_i$.[55]

Theorem 7.1. *Let $0 < \alpha < 2$. Then $\boldsymbol{X} = (X_1, \ldots, X_n)$ is a stable random vector with index of stability α if and only if there exists a finite measure Γ_X on the unit hypersphere $S_n = \{\boldsymbol{s} \in \mathbb{R}^n \mid \|\boldsymbol{s}\| = 1\}$ and a vector $\boldsymbol{\mu} \in \mathbb{R}^n$ such that*

$$\boldsymbol{\Psi}_\alpha(\boldsymbol{t}) = \begin{cases} \exp\left\{ -\int_{S_n} \left|(\boldsymbol{t}, \boldsymbol{s})\right|^\alpha \left(1 - \mathrm{i}\,\mathrm{sign}\big((\boldsymbol{t}, \boldsymbol{s})\big) \tan\frac{\pi\alpha}{2}\right) \Gamma_X(\mathrm{d}\boldsymbol{s}) + \mathrm{i}(\boldsymbol{t}, \boldsymbol{\mu}) \right\}, & \alpha \neq 1, \\ \exp\left\{ -\int_{S_n} \left|(\boldsymbol{t}, \boldsymbol{s})\right| \left(1 + \mathrm{i}\frac{2}{\pi}\,\mathrm{sign}\big((\boldsymbol{t}, \boldsymbol{s})\big) \ln\left|(\boldsymbol{t}, \boldsymbol{s})\right|\right) \Gamma_X(\mathrm{d}\boldsymbol{s}) + \mathrm{i}(\boldsymbol{t}, \boldsymbol{\mu}) \right\}, & \alpha = 1. \end{cases} \tag{58}$$

The pair $(\Gamma_X, \boldsymbol{\mu})$ is unique.

The measure Γ_X is called the *spectral measure* of the stable random vector $\boldsymbol{X}$ and specifies the dependence structure. If $\boldsymbol{X}$ is $S\alpha S$ in $\mathbb{R}^n$, then the characteristic function takes the simple form

$$\boldsymbol{\Psi}_\alpha(\boldsymbol{t}) = \exp\left\{ -\int_{S_n} \left|(\boldsymbol{t}, \boldsymbol{s})\right|^\alpha \Gamma_X(\mathrm{d}\boldsymbol{s}) \right\},$$

where Γ is the unique symmetric spectral measure. The expression in (58) for the characteristic function is also valid for the normal case $\alpha = 2$. When $\alpha = 2$, it reduces to $\boldsymbol{\Psi}_2(\boldsymbol{t}) = \exp\{-\int_{S_n} |(\boldsymbol{t}, \boldsymbol{s})|^2 \Gamma_X(\mathrm{d}\boldsymbol{s})\}$ but in this case Γ_X is no longer unique. To get a feeling for Γ_X, suppose $\boldsymbol{X} = (X_1, X_2)$ and that the distribution is Gaussian. Then

$$\begin{aligned} \int_{S_2} \left|(\boldsymbol{t}, \boldsymbol{s})\right|^2 \Gamma_{(X_1, X_2)}(\mathrm{d}\boldsymbol{s}) &= \int_{S_2} \left|(t_1 s_1 + t_2 s_2)\right|^2 \Gamma_{(X_1, X_2)}(\mathrm{d}\boldsymbol{s}) \\ &= t_1^2 \sigma_1^2 + 2 t_1 t_2 \sigma_{1,2} + t_1^2 \sigma_1^2, \end{aligned}$$

where

$$\sigma_i^2 = \int_{S_2} s_i^2 \Gamma_{(X_1, X_2)}(\mathrm{d}\boldsymbol{s}), \quad i = 1, 2, \quad \text{and} \quad \sigma_{1,2} = \int_{S_2} s_1 s_2 \Gamma_{(X_1, X_2)}(\mathrm{d}\boldsymbol{s}),$$

[55] Previously we wrote $\boldsymbol{t}^{\mathrm{T}} \boldsymbol{s}$ instead of $(\boldsymbol{t}, \boldsymbol{s})$.

and where integration over the circle S_2 means integration on $\{s = (s_1, s_2) \mid s_1^2 + s_2^2 = 1\}$. One recognizes the normal characteristic function with $\operatorname{Var} X_1 = 2\sigma_1^2$, $\operatorname{Var} X_2 = 2\sigma_2^2$ and $\operatorname{Cov}(X_1, X_2) = 2\sigma_{1,2}$. Since different choices of $\Gamma_{(X_1,X_2)}$ can yield the same values for σ_1^2, σ_2^2 and $\sigma_{1,2}$, the choice of Γ_X is not unique in the Gaussian case.

As in the case of a normal random vector, if $\boldsymbol{X}$ is multivariate stable with index of stability $0 < \alpha < 2$, then all linear combinations of the components of $\boldsymbol{X}$ are stable with the same α. So, if $\boldsymbol{X}$ is a stable random vector in $\mathbb{R}^n$, and $\mathbf{w} \in \mathbb{R}^n$, we know that $Y = (\mathbf{w}, \boldsymbol{X}) = \sum_{i=1}^n w_i X_i$ is $S_\alpha(\sigma_Y, \beta_Y, \mu_Y)$. Using the characteristic function (58), it can be shown [see Samorodnitsky and Taqqu (1994), Example 2.3.4], that

$$\sigma_Y = \left(\int_{S_n} |(\mathbf{w}, \boldsymbol{s})|^\alpha \Gamma_X(\mathrm{d}\boldsymbol{s})\right)^{1/\alpha}, \tag{59}$$

$$\beta_Y = \frac{\int_{S_n} |(\mathbf{w}, \boldsymbol{s})|^\alpha \operatorname{sign}(\mathbf{w}, \boldsymbol{s}) \Gamma_X(\mathrm{d}\boldsymbol{s})}{\int_{S_n} |(\mathbf{w}, \boldsymbol{s})|^\alpha \Gamma_X(\mathrm{d}\boldsymbol{s})}, \tag{60}$$

$$\mu_Y = \begin{cases} (\mathbf{w}, \boldsymbol{\mu}) & \text{for } \alpha \neq 1, \\ (\mathbf{w}, \boldsymbol{\mu}) - \dfrac{2}{\pi} \displaystyle\int_{S_n} (\mathbf{w}, \boldsymbol{s}) \ln |(\mathbf{w}, \boldsymbol{s})| \Gamma_X(\mathrm{d}\boldsymbol{s}) & \text{for } \alpha = 1. \end{cases} \tag{61}$$

In the mean–variance portfolio theory, the risk to be minimized for any level of expected return is given by the portfolios' variance. If the asset returns are assumed multivariate stable with index of stability $0 < \alpha < 2$ then the variance is infinite and cannot be used. In the stable portfolio theory, it is assumed that $1 < \alpha < 2$, $\mathbb{E}\boldsymbol{X} = \boldsymbol{\mu}$ and that $\boldsymbol{X} - \boldsymbol{\mu} \sim S\alpha S$. Let $\mathbf{w}$ be the vector of weights for the risky portfolio $X_p = (\mathbf{w}, \mathbf{X})$. Given the relationship between the scale parameter and the variance in the Gaussian case (that is, stable with $\alpha = 2$), it is natural to use the scale parameter σ_{X_p} of the resulting stable distribution instead of the standard deviation. It is given by (59). This brings us to the corresponding stable portfolio problem:

$$\begin{aligned} \min_{\mathbf{w}} \sigma_{X_p} &= \left(\int_{S_n} |(\mathbf{w}, \boldsymbol{s})|^\alpha \Gamma_X(\mathrm{d}\boldsymbol{s})\right)^{1/\alpha} \\ \text{such that } \; & (\mathbf{w}, \boldsymbol{\mu}) \geqslant a, \\ & (\mathbf{w}, \boldsymbol{e}) = 1. \end{aligned} \tag{62}$$

The risk measure $\sigma_{X_p} = \sigma_{(\mathbf{w},\mathbf{X})}$ is a convex function of $\mathbf{w}$ and the problem is generally solved using sequential quadratic programming. See Belkacem (1997) and Rachev and Mittnik (2000) and references therein for details of the procedure and on the estimation of the index of stability, spectral measure and scale parameters. If a risk free asset is included in the asset universe, then we end up with a maximization problem similar to (2) in Section 2.2, but where the risk measure is the scale parameter σ_{X_p} of the risky portfolio.

7.2. Stable asset pricing

Since there exists a portfolio theory under the assumption of a multivariate stable distribution of asset returns ($1 < \alpha < 2$), it is natural to ask whether there exists an analogous CAPM. The answer is positive, and it was first introduced by Fama (1970). For recent descriptions of the stable CAPM see Belkacem, Lévy Véhel and Walter (1996) and, of course, Rachev and Mittnik (2000).

The assumptions behind the stable CAPM are the same as in the Gaussian case in Section 2.3 with the assumption of joint normality of asset returns replaced by that of jointly stable asset returns with index of stability $\alpha \in (1, 2)$. That is, we assume $\mathbb{E}\boldsymbol{X} = \boldsymbol{\mu}$ and that $\boldsymbol{X} - \boldsymbol{\mu} \sim S\alpha S$. Recall from the traditional CAPM and Equations (3) and (4), that the expected premium of holding the risky asset i over the riskless asset is proportional to the expected premium of holding the market portfolio over the riskless asset. The constant of proportionality was the risky assets *beta* given by (4). In the stable CAPM, we require an alternative measure of dependence since covariances do not exist. Naturally, the scale parameter σ replaces the standard deviation.

The *covariation* is a natural alternative to the covariance in the stable case when $1 < \alpha < 2$. This measure possesses many, but not all, of the useful properties of covariance in the Gaussian case. We define and present several of the properties of covariation. Details may be found in Samorodnitsky and Taqqu (1994) and Rachev and Mittnik (2000).

Definition 7.3. Let X_1 and X_2 be jointly $S\alpha S$ with $1 < \alpha \leqslant 2$ and let $\Gamma_{(X_1,X_2)}$ be the spectral measure of the random vector (X_1, X_2). The covariation of X_1 on X_2 is given by

$$[X_1, X_2]_\alpha = \int_{S_2} s_1 s_2^{\langle \alpha - 1 \rangle} \Gamma_{(X_1,X_2)}(\mathrm{d}\boldsymbol{s}) \tag{63}$$

where $s^{\langle p \rangle}$ denotes the *signed power* $s^{\langle p \rangle} = |s|^p (\operatorname{sign} s)$.

In the Gaussian case $\alpha = 2$ it reduces to

$$[X_1, X_2]_2 = \frac{1}{2} \operatorname{Cov}(X_1, X_2). \tag{64}$$

Note, however, that whereas in the Gaussian case the dependence structure is fully characterized by the covariance, in the stable one needs to use $\Gamma_{\boldsymbol{X}}$, and the covariation does *not* fully characterize the dependence structure. We now derive the stable CAPM under the preceding assumptions, following Belkacem, Lévy Véhel and Walter (1996).

Consider a portfolio of a riskless asset with rate of return r and a risky asset X_i with weights w and $1 - w$ respectively. The expected rate of return of the portfolio $X_p = wr +$

$(1-w)X_i$ is then $\mathbb{E}X_p = wr + (1-w)\mathbb{E}X_i$, and its risk, as given by its scale parameter, is $\sigma_p = (1-w)\sigma_i$.[56] The risk-return trade-off is then given by

$$\mathbb{E}X_p = r + \frac{\mathbb{E}X_i - r}{\sigma_i}\sigma_p \tag{65}$$

after setting $w = 1 - \sigma_p/\sigma_i$. Under the assumptions of CAPM, investors have homogeneous beliefs, that is, they all agree on the multivariate stable parameters. This means that all investors hold the market portfolio (as in Section 2.3) as their risky asset and the risk-return trade-off (65) becomes

$$\mathbb{E}X_p = r + \frac{\mathbb{E}X_M - r}{\sigma_M}\sigma_p, \tag{66}$$

where X_M and σ_M are the rate of return and scale parameter respectively of the market.

Now consider the suboptimal portfolio $X_p = wX_i + (1-w)X_M$ obtained by adding to the market portfolio a certain position in asset i (the portfolio is optimal if $w = 0$). Since $\boldsymbol{X} - \boldsymbol{\mu} \sim S\alpha S$ we know that $X_i - \mu_i$ and $X_M - \mu_M$ are jointly $S\alpha S$. By properties of symmetric stable random vectors this means that $X_p \sim S_\alpha(\sigma_p, 0, \mu_p)$, where the scale and location parameters are given by (59) and (61), that is

$$\sigma_p^\alpha = \int_{S_2} \left|ws_1 + (1-w)s_2\right|^\alpha \Gamma_{(X_i,X_p)}(\mathrm{d}s_1, \mathrm{d}s_2), \tag{67}$$

$$\mu_p = \mathbb{E}X_p = w\mu_i + (1-w)\mu_M, \tag{68}$$

respectively. Differentiating with respect to w gives

$$\frac{\partial\mu_p}{\partial w} = \mu_i - \mu_M, \tag{69}$$

$$\begin{aligned}\frac{\partial\sigma_p}{\partial w} &= \frac{1}{\alpha\sigma_p^{\alpha-1}}\frac{\partial\sigma_p^\alpha}{\partial w}\\ &= \frac{1}{\sigma_p^{\alpha-1}}\int_{S_2}(s_1 - s_2)\bigl(ws_1 + (1-w)s_2\bigr)^{\langle\alpha-1\rangle}\Gamma_{(X_i,X_p)}(\mathrm{d}s_1, \mathrm{d}s_2).\end{aligned} \tag{70}$$

So evaluating (69) and (70) at $w = 0$ and using Definition 7.3 we get

$$\left.\frac{\partial\mu_p}{\partial\sigma_p}\right|_{w=0} = \left.\frac{\partial\mu_p}{\partial w}\middle/\frac{\partial\sigma_p}{\partial w}\right|_{w=0} = \frac{\sigma_M^{\alpha-1}(\mu_i - \mu_M)}{[X_i, X_M]_\alpha - \sigma_M^\alpha}, \tag{71}$$

[56] Note that if $X \sim S_\alpha(\sigma, \beta, \mu)$ then $aX + b \sim S_\alpha(|a|\sigma, \mathrm{sign}(a)\beta, a\mu + b)$ if $1 < \alpha < 2$.

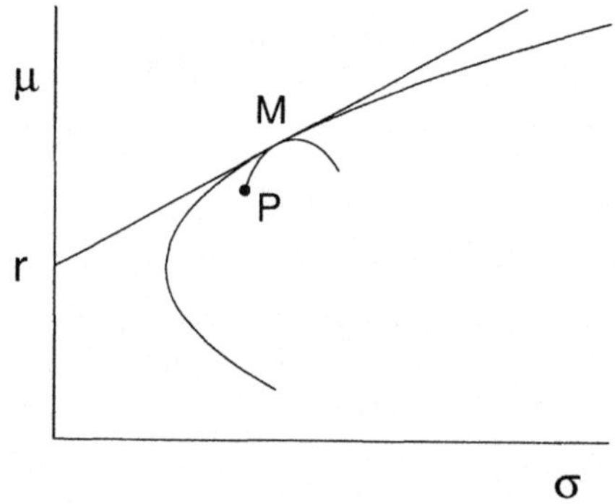

Fig. 22. The stable efficient frontier. The portfolio $X_p = wX_i + (1-w)X_M$ is suboptimal, and hence must be dominated by the efficient frontier.

since at $w=0$ the portfolio X_p becomes X_M and σ_p becomes σ_M. Moreover, in market equilibrium the trade-off between risk and return is given by (66), so that the slope $\partial\mu_p/\partial\sigma_p$ at $w=0$ is given by $(\mu_M - r)/\sigma_M$ (see Figure 22). Hence

$$\frac{\mu_M - r}{\sigma_M} = \frac{\sigma_M^{\alpha-1}(\mu_i - \mu_M)}{[X_i, X_M]_\alpha - \sigma_M^\alpha}. \tag{72}$$

This may be rewritten in the familiar CAPM form (3) as

$$\mathbb{E}(X_i - r) = \beta_i \mathbb{E}(X_M - r),$$

where now, in the stable case,

$$\beta_i = \frac{[X_i, X_M]_\alpha}{\sigma_M^\alpha}. \tag{73}$$

Note that if we assume Gaussian returns, then $\boldsymbol{X} - \boldsymbol{\mu} \sim S\alpha S$ with $\alpha = 2$, and by using (64), we recover

$$\beta_i = \frac{\mathrm{Cov}(X_i, X_M)}{\mathrm{Var}(X_M)},$$

that is, the traditional CAPM result.

Acknowledgments

We would like to thank Paul Embrechts and Filip Lindskog for many valuable comments which led to an improved exposition of this material. This research was partially supported by the NSF Grant ANI-9805623 at Boston University.

References

Albanese, C., 1997. Credit exposure. Diversification risk and coherent VaR. Preprint. Department of Mathematics, University of Toronto.

Artzner, P., Delbaen, F., Eber, J.M., Heath, D., 1997. Thinking coherently. RISK 10 (11).

Artzner, P., Delbaen, F., Eber, J.M., Heath, D., 1999. Coherent measures of risk. Mathematical Finance 9 (3), 203–228.

Barndorff-Nielsen, O.E., 1977. Exponentially decreasing distributions for the logarithm of particle size. Proceeding of the Royal Society London. Series A.

Basle Committee on Banking Supervision, 1995a. An internal model-based approach to market risk capital requirements. Technical report. Basle Committee on Banking Supervision, Basle, Switzerland.

Basle Committee on Banking Supervision, 1995b. Planned supplement to the capital accord to incorporate market risks. Technical report. Basle Committee on Banking Supervision, Basle, Switzerland.

Beirlant, J., Teugels, J.L., Vynckier, P., 1996. Practical Analysis of Extreme Values. Leuven University Press.

Belkacem, L., 1997. How to select optimal portfolio in α-stable markets. Preprint. INRIA.

Belkacem, L., Lévy Véhel, J., Walter, C., 1996. CAPM, risk and portfolio selection in stable markets. Preprint. INRIA.

Bernstein, P.L., 1996. Against the Gods: The Remarkable Story of Risk. Wiley.

Bertsimas, D., Lauprete, G.J., Samarov, A., 2000. Shortfall as a risk measure: Properties, optimization and application. Preprint. MIT.

Black, F., Scholes, M., 1973. The pricing of options and corporate liabilities. Journal of Political Economy 81, 637–654.

Bollerslev, T., 1986. Generalized autoregressive conditional heteroscadicity. Journal of Econometrics 31 (1), 34–105.

Bollerslev, T., Chou, R.Y., Kroner, K.F., 1992. ARCH modelling in finance: A review of the theory and empirical evidence. Journal of Econometrics 52, 307–327.

Broadie, M., Glasserman, P., 1998. Simulation for option pricing and risk management. In: Alexander, C. (Ed.), Risk Management and Analysis. Wiley.

Cizeau, P., Potters, M., Bouchaud, J.-P., 2001. Correlation structure of extreme stock returns. Quantitative Finance 1, 217–222.

Danielsson, J., Hartmann, P., De Vries, C.G., 1998. The cost of conservatism: Extreme returns, value at risk and the Basle multiplication factor. RISK 11, 101–103.

Danielsson, J., De Haan, L., Peng, L., De Vries, C.G., 2001. Using a bootstrap method to choose the sample fraction in the tail index estimation. Journal of Multivariate Analysis 76, 226–248.

Danielsson, J., de Vries, C., 1997. Beyond the sample: Extreme quantile and probability estimation. Preprint, LSE.

Danielsson, J., de Vries, C., 2000. Value at risk and extreme returns. In: Embrechts, P. (Ed.), Extremes and Integrated Risk Management. Risk Books, pp. 85–106.

Diebold, F.X., Schuermann, T., Stroughair, J.D., 2000. Pitfalls and opportunities in the use of extreme value theory in risk management. Journal of Risk Finance 1 (Winter), 30–36.

Dowd, K., 1998. Beyond Value at Risk: The New Science of Risk Management. Wiley.

Drees, H., de Haan, L., Resnick, S., 2000. How to make a Hill plot. The Annals of Statistics 28 (1), 254–274.

Duffie, D., Pan, J., 1997. An overview of value at risk. Journal of Derivatives 4 (3), 7–49.

Eberlein, E., Keller, U., 1995. Hyperbolic distributions in finance. Bernoulli 1, 281–299.

Eberlein, E., Prause, K., 2002. The generalized hyperbolic model: Financial derivatives and risk measures. In: Geman, H. (Ed.), Mathematical Finance – Bachelier Congress 2000. Springer-Verlag, pp. 245–267.

Embrechts, P., 2000. Extreme value theory: Potential and limitations as an integrated risk management tool. Derivatives Use, Trading & Regulation 6, 449–456.

Embrechts, P., Klüppelberg, C., Mikosch, T., 1997. Modelling Extremal Events for Insurance and Finance. Springer, Berlin.

Embrechts, P., Lindskog, F., McNeil, A., 2003. Modelling dependence with copulas and applications to risk management. In: Heavy Tailed Distibutions in Finance. Elsevier. In this volume.

Embrechts, P., McNeil, A.J., Straumann, D., 1999. Correlation: Pitfalls and alternatives. RISK 12 (5), 69–71.

Embrechts, P., McNeil, A.J., Straumann, D., 2001. Correlation and dependence in risk management: Properties and pitfalls. In: Dempster, M., Moffatt, H.K. (Eds.), Risk Management: Value at Risk and Beyond. Cambridge University Press.

Embrechts, P., Resnick, S.I., Samorodnitsky, G., 1999. Extreme value theory as a risk management tool. North American Actuarial Journal 3, 30–41.

Engle, R.F., 1982. Autoregressive conditional heteroscadicity with estimates of the variance of U.K. inflation. Econometrica 50, 987–1008.

Fama, E.F., 1965. The behavior of stock market prices. Journal of Business 38 (1), 34–105.

Fama, E.F., 1970. Risk, return and equilibrium. Journal of Political Economy 79 (1), 30–55.

Fang, K.T., Kotz, S., Ng, K.W., 1990. Symmetric Multivariate and Related Distributions. Chapman & Hall.

Frank, M., 1979. On the simultaneous associativity of $F(x, y)$ and $x + y - F(x, y)$. Aequationes Mathematicae 19, 194–226.

Frees, E.W., Valdez, E.A., 1998. Understanding relationships using copulas. North American Actuarial Journal 2 (1), 1–25.

Gouriéroux, C., 1997. ARCH Models and Financial Applications. Springer-Verlag.

Hill, B.M., 1975. A simple general approach to inference about the tail of a distribution. The Annals of Statistics 3 (5), 1163–1174.

Huang, C., Litzenberger, R., 1988. Foundations of Financial Economics. North-Holland, New York.

Hull, J., White, A., 1987. The pricing of options on assets with stochastic volatilities. Journal of Finance 2, 281–300.

Ingersoll, J.E., 1987. Theory of Financial Decision Making. Rowman & Littlefield.

Joe, H., 1997. Multivariate Models and Dependence Concepts. Chapman & Hall.

Jorion, P., 2001. Value at Risk: The New Benchmark for Controlling Market Risk, 2nd edition. McGraw-Hill.

Këllezi, E., Gilli, M., 2000. Extreme value theory for tail-related risk measures. Preprint. University of Geneva.

Leadbetter, M.R., Lindgren, G., Rootzén, H., 1983. Extremes and Related Properties of Random Sequences and Processes. Springer-Verlag.

Levy, H., Kroll, Y., 1978. Ordering uncertain options with borrowing and lending. Journal of Finance 33, 553–573.

Lindskog, F., 2000a. Linear correlation estimation. Preprint. ETH, Zürich.

Lindskog, F., 2000b. Modelling dependence with copulas. Master's thesis. ETH, Zürich.

Lintner, J., 1965. The valuation of risky assets and the selection of risky investment in stock portfolios and capital budgets. Review of Economics and Statistics 47, 13–37.

Mandelbrot, B.B., 1963. The variation of certain speculative prices. Journal of Business 36, 392–417.

Markowitz, H., 1952. Portfolio selection. Journal of Finance 7, 77–91.

Markowitz, H., 1959. Portfolio Selection: Efficient Diversification of Investments. Wiley.

McNeil, A., 1998a. Calculating quantile risk measures for financial return series using extreme value theory. Preprint. ETH, Zürich.

McNeil, A., 1998b. On extremes and crashes. RISK, January.

McNeil, A., Frey, R., 2000. Estimation of tail-related risk measures for heteroscedastic financial time series: An extreme value approach. Journal of Empirical Finance 7, 271–300.

Mikosch, T., Stărică, C., 2000. Change of structure in financial time series, long range dependence and the GARCH model. Preprint.

McNeil, A., Saladin, T., 1997. The peaks over threshold method for estimating high quantiles of loss distributions. In: Proceedings of the 28th International ASTIN Colloquium.

Mills, T.C., 1999. The Econometric Modelling of Financial Time Series. Cambridge University Press.

Mittnik, S., Paolella, M.S., Rachev, T., 1997. Modelling the persistence of conditional volatilities with GARCH-stable processes. Preprint. University of California, Santa Barbara.

Mittnik, S., Rachev, T., Paolella, M.S., 1998. Stable Paretian modelling in finance: Some empirical and theoretical aspects. In: Adler et al. (Eds.), A Practical Guide to Heavy Tails. Birkhäuser.

Nelsen, R.B., 1999. An Introduction to Copulas. Springer, New York.

Nolan, J., 2001. Stable Distribution: Models for Heavy-Tailed Data. Birkhäuser, forthcoming.

Panorska, A.K., Mittnik, S., Rachev, T., 1995. Stable GARCH models for financial time series. Applied Mathematics Letters 8, 33–37.

Rachev, S., Mittnik, S., 2000. Stable Paretian Models in Finance. Wiley.

Reiss, R.-D., Thomas, M., 2001. Statistical Analysis of Extreme Values, 2nd edition. Birkhäuser.

Resnick, S., Stărică, C., 1997. Smoothing the Hill estimator. Advances in Applied Probability, 29.

RiskMetrics, 1996. Technical document. Technical report. JP Morgan.

Ross, S.A., 1976. The arbitrage theory of capital asset pricing. Journal of Economic Theory 13, 341–360.

Samorodnitsky, G., Taqqu, M., 1994. Stable Non-Gaussian Random Processes. Chapman & Hall.

Sharpe, W.F., 1964. Capital asset prices: A theory of market equilibrium under conditions of risk. Journal of Finance 19, 425–442.

Shiryaev, A.N., 1999. Essentials of Stochastic Finance. World Scientific.

Sklar, A., 1996. Random variables, distribution functions, and copulas – a personal look backward and forward. In: Rüschendorff et al. (Eds.), Distributions with Fixed Marginals and Related Topics. Institute of Mathematical Sciences, Hayward, CA.

Smith, R., 2000. Measuring risk with extreme value theory. In: Embrechts, P. (Ed.), Extremes and Integrated Risk Management. Risk Books, pp. 19–35.

Stahl, G., 1997. Three cheers. RISK 10, 67–69.

Wilson, T.C., 1998. Value at risk. In: Alexander, C. (Ed.), Risk Management and Analysis. Wiley.

Chapter 3

MODELING FINANCIAL DATA WITH STABLE DISTRIBUTIONS

JOHN P. NOLAN

Department of Mathematics and Statistics, American University

Contents

Handbook of Heavy Tailed Distributions in Finance, Edited by S.T. Rachev

Abstract

Stable distributions are a class of probability distributions that allow heavy tails and skewness. In addition to theoretical reasons for using stable laws, they are a rich family that can accurately model different kinds of financial data. We review the basic facts, describe programs that make it practical to use stable distributions, and give examples of these distributions in finance. A non-technical introduction to multivariate stable laws is also given.

1. Basic facts about stable distributions

Stable distributions are a class of probability laws that have intriguing theoretical and practical properties. Their applications to financial modeling comes from the fact that they generalize the normal (Gaussian) distribution and allow heavy tails and skewness, which are frequently seen in financial data. In this chapter, we focus on the basic definition and properties of stable laws, and show how they can be used in practice. We give no proofs; interested readers can find these in Zolotarev (1986), Samorodnitsky and Taqqu (1994), Janicki and Weron (1994), Uchaikin and Zolotarev (1999), Rachev and Mittnik (2000) and Nolan (2003).

The defining characteristic, and reason for the term *stable*, is that they retain their shape (up to scale and shift) under addition: if $X, X_1, X_2, \ldots, X_n$ are independent, identically distributed stable random variables, then for every n

$$X_1 + X_2 + \cdots + X_n \stackrel{d}{=} c_n X + d_n \tag{1}$$

for some constants $c_n > 0$ and d_n. The symbol $\stackrel{d}{=}$ means equality in distribution, i.e., the right- and left-hand sides have the same distribution. The law is called *strictly stable* if $d_n = 0$ for all n. Some authors use the term *sum stable* to emphasize the stability under addition and to distinguish it from other concepts, e.g., max-stable, min-stable, etc. The normal distributions satisfy this property: the sum of normals is normal. Likewise the Cauchy laws and the Lévy laws (see below) satisfy this property. The class of all laws that satisfy (1) is described by four parameters, which we call $(\alpha, \beta, \gamma, \delta)$, see Figure 1 for some density graphs. In general, there are no closed form formulas for stable densities f and cumulative distribution functions F, but there are now reliable computer programs for working with these laws.

The parameter α is called the *index* of the law or the *index of stability* or *characteristic exponent* and must be in the range $0 < \alpha \leqslant 2$. The constant c_n in (1) must be of the form $n^{1/\alpha}$. The parameter β is called the *skewness* of the law, and must be in the range $-1 \leqslant \beta \leqslant 1$. If $\beta = 0$, the distribution is symmetric, if $\beta > 0$ it is skewed toward the right, if $\beta < 0$, it is skewed toward the left. The parameters α and β determine the shape of the distribution. The parameter γ is a scale parameter, it can be any positive number. The parameter δ is a location parameter, it shifts the distribution right if $\delta > 0$, and left if $\delta < 0$.

A confusing issue with stable parameters is that there are multiple definitions of what the parameters mean. There are at least 10 different definitions of stable parameters, see Nolan (2003). The reader should be careful in reading the literature and verify what parameterization is being used. We will describe two different parameterizations, which we denote by $\mathbf{S}(\alpha, \beta, \gamma, \delta_0; 0)$ and $\mathbf{S}(\alpha, \beta, \gamma, \delta_1; 1)$. The first is what we will use in all our applications, because it has better numerical behavior and intuitive meaning. The second parameterization is more commonly used in the literature, so it is important to understand it. The parameters α, β and γ have the same meaning in the two parameterizations, only

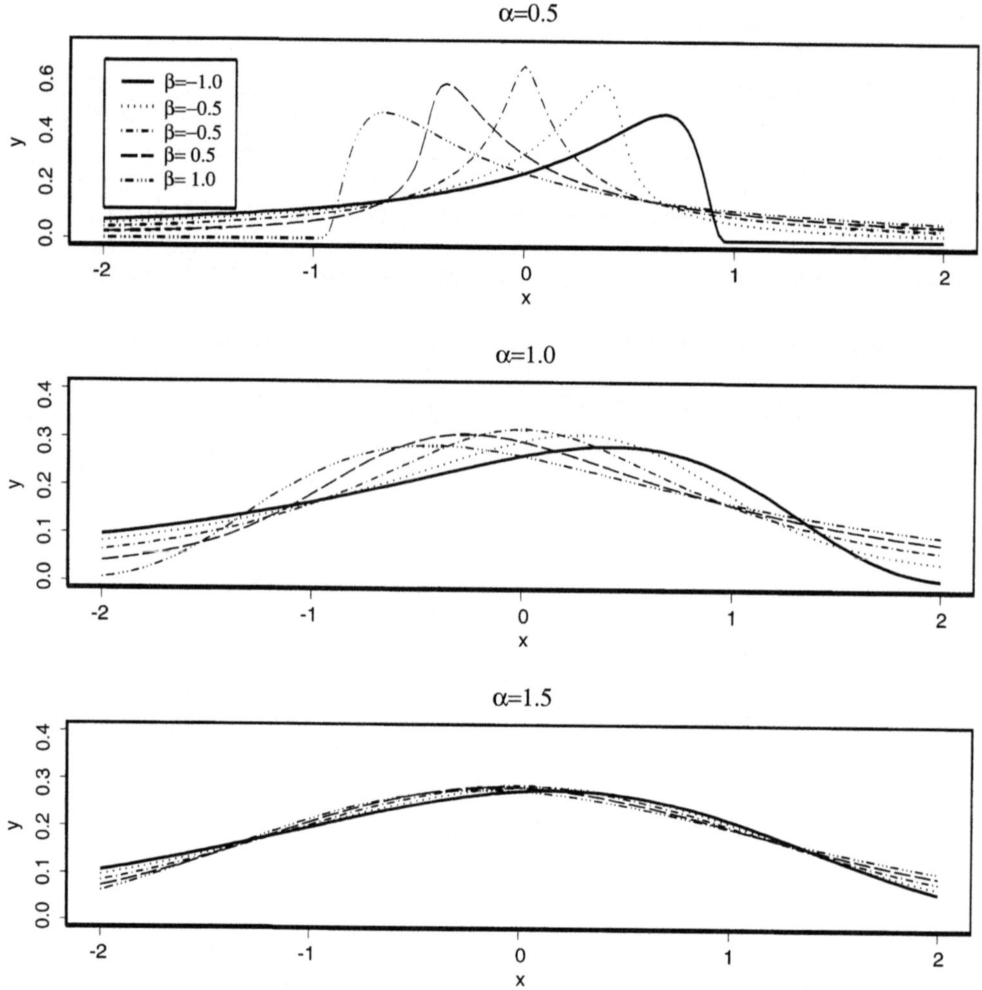

Fig. 1. Standardized stable densities for different α and β in the $S(\alpha, \beta, 1, 0; 0)$ parameterization. The top graph includes a Lévy$(1, -1) = \mathbf{S}(1/2, 1, 1, 0; 0) = \mathbf{S}(1/2, 1, 1, -1; 1)$ graph and the middle graph includes a Cauchy$(1, 0) = \mathbf{S}(1, 0, 1, 0; 0) = \mathbf{S}(1, 0, 1, 0; 1)$ graph.

the location parameter is different. To distinguish between the two, we will sometimes use a subscript to indicate which parameterization is being used: δ_0 for the location parameter in the $\mathbf{S}(\alpha, \beta, \gamma, \delta_0; 0)$ parameterization and δ_1 for the location parameter in the $\mathbf{S}(\alpha, \beta, \gamma, \delta_1; 1)$ parameterization.

Definition 1. A random variable X is $\mathbf{S}(\alpha, \beta, \gamma, \delta_0; 0)$ if it has characteristic function

$$E\exp(\mathrm{i}uX) = \begin{cases} \exp\left(-\gamma^\alpha |u|^\alpha \left[1 + \mathrm{i}\beta\left(\tan\frac{\pi\alpha}{2}\right)(\operatorname{sign} u)\left(|\gamma u|^{1-\alpha} - 1\right)\right] + \mathrm{i}\delta_0 u\right), & \alpha \neq 1, \\ \exp\left(-\gamma|u|\left[1 + \mathrm{i}\beta\frac{2}{\pi}(\operatorname{sign} u)\ln\left(\gamma|u|\right)\right] + \mathrm{i}\delta_0 u\right), & \alpha = 1. \end{cases} \tag{2}$$

Definition 2. A random variable X is $\mathbf{S}(\alpha, \beta, \gamma, \delta_1; 1)$ if it has characteristic function

$$E\exp(\mathrm{i}uX) = \begin{cases} \exp\left(-\gamma^\alpha |u|^\alpha \left[1 - \mathrm{i}\beta\left(\tan\frac{\pi\alpha}{2}\right)(\operatorname{sign} u)\right] + \mathrm{i}\delta_1 u\right), & \alpha \neq 1, \\ \exp\left(-\gamma|u|\left[1 + \mathrm{i}\beta\frac{2}{\pi}(\operatorname{sign} u)\ln|u|\right] + \mathrm{i}\delta_1 u\right), & \alpha = 1. \end{cases} \tag{3}$$

The location parameters are related by

$$\delta_0 = \begin{cases} \delta_1 + \beta\gamma\tan\frac{\pi\alpha}{2}, & \alpha \neq 1, \\ \delta_1 + \beta\frac{2}{\pi}\gamma\ln\gamma, & \alpha = 1, \end{cases} \qquad \delta_1 = \begin{cases} \delta_0 - \beta\gamma\tan\frac{\pi\alpha}{2}, & \alpha \neq 1, \\ \delta_0 - \beta\frac{2}{\pi}\gamma\ln\gamma, & \alpha = 1. \end{cases} \tag{4}$$

Note that if $\beta = 0$, the parameterizations coincide. When $\beta \neq 0$, the parameterizations differ by a shift $\gamma\beta\tan\frac{\pi\alpha}{2}$, which gets infinitely large as $\alpha \to 1$. In particular, the mode of a $\mathbf{S}(\alpha, \beta, \gamma, \delta_1; 1)$ density tends toward ∞ (if $\operatorname{sign}(\alpha - 1)\beta > 0$) or $-\infty$ (otherwise) as $\alpha \to 1$. When α is near 1, computing stable densities and cumulatives in this range is numerically difficult and estimating parameters is unreliable. From the applied point of view, it is preferred to use the $\mathbf{S}(\alpha, \beta, \gamma, \delta_0; 0)$ parameterization, which is jointly continuous in all four parameters. The arguments for using the $\mathbf{S}(\alpha, \beta, \gamma, \delta_1; 1)$ parameterization are historical and algebraic simplicity. It seems unavoidable that both parameterizations will be used, so users of stable distributions should know both and state clearly which they are using.

There are three cases where one can write down closed form expressions for the density and verify directly that they are stable – normal, Cauchy and Lévy distributions.

Example 1 (*Normal or Gaussian distributions*). $X \sim N(\mu, \sigma^2)$ if it has a density

$$f(x) = \frac{1}{\sqrt{2\pi}\sigma}\exp\left(-\frac{(x-\mu)^2}{2\sigma^2}\right), \qquad -\infty < x < \infty.$$

Gaussian laws are stable with $\alpha = 2$ and $\beta = 0$; more precisely $N(\mu, \sigma^2) = \mathbf{S}(2, 0, \sigma/\sqrt{2}, \mu; 0) = \mathbf{S}(2, 0, \sigma/\sqrt{2}, \mu; 1)$.

Example 2 (*Cauchy distributions*). $X \sim \text{Cauchy}(\gamma, \delta)$ if it has density

$$f(x) = \frac{1}{\pi} \frac{\gamma}{\gamma^2 + (x-\delta)^2}, \qquad -\infty < x < \infty.$$

Cauchy laws are stable with $\alpha = 1$ and $\beta = 0$; more precisely, $\text{Cauchy}(\gamma, \delta) = \mathbf{S}(1, 0, \gamma, \delta; 0) = \mathbf{S}(1, 0, \gamma, \delta; 1)$.

Example 3 (*Lévy distributions*). $X \sim \text{Lévy}(\gamma, \delta)$ if it has density

$$f(x) = \sqrt{\frac{\gamma}{2\pi}} \frac{1}{(x-\delta)^{3/2}} \exp\left(-\frac{\gamma}{2(x-\delta)}\right), \qquad \delta < x < \infty.$$

These are stable with $\alpha = 1/2$, $\beta = 1$;

$$\text{Lévy}(\gamma, \delta) = \mathbf{S}\left(\frac{1}{2}, 1, \gamma, \gamma + \delta; 0\right) = \mathbf{S}\left(\frac{1}{2}, 1, \gamma, \delta; 1\right).$$

The graphs in Figure 1 show several qualitative features of stable laws. First, stable distributions have densities and are unimodal. These facts are not obvious: since there is no general formula for stable densities, indirect arguments must be used and it is quite involved to prove unimodality. Second, the $-\beta$ curve is a reflection of the β curve. Third, when α is small, the skewness is significant, when α is large, the skewness parameter matters less and less. The support of a stable density is either all of $(-\infty, \infty)$ or a half-line. The latter case occurs if and only if $0 < \alpha < 1$ and $\beta = +1$ or -1. More precisely, the support of density $f(x|\alpha, \beta, \gamma, \delta; k)$ for a $\mathbf{S}(\alpha, \beta, \delta, \gamma; k)$ law is

$$\begin{cases} \left[\delta - \gamma \tan\frac{\pi\alpha}{2}, \infty\right), & \alpha < 1,\ \beta = +1,\ k = 0, \\ \left(-\infty, \delta + \gamma \tan\frac{\pi\alpha}{2}\right], & \alpha < 1,\ \beta = -1,\ k = 0, \\ [\delta, \infty), & \alpha < 1,\ \beta = +1,\ k = 1, \\ (-\infty, \delta], & \alpha < 1,\ \beta = -1,\ k = 1, \\ (-\infty, \infty), & \text{otherwise.} \end{cases}$$

In particular, to model a positive distribution, a $\mathbf{S}(\alpha, 1, \delta, 0; 1)$ distribution with $\alpha < 1$ is used.

When $\alpha = 2$, the normal law has light tails and all moments exist. Except for the normal law, all stable laws have heavy tails with an asymptotic power law (Pareto) decay. The term *stable Paretian* distributions is used to distinguish the $\alpha < 2$ cases from the normal case. For $X \sim \mathbf{S}(\alpha, \beta, 1, 0; 0)$ with $0 < \alpha < 2$ and $-1 < \beta \leqslant 1$, then as $x \to \infty$,

$$P(X > x) \sim c_\alpha (1+\beta) x^{-\alpha},$$

$$f(x|\alpha, \beta; 0) \sim \alpha c_\alpha (1+\beta) x^{-(\alpha+1)},$$

where $c_\alpha = \Gamma(\alpha)(\sin \frac{\pi\alpha}{2})/\pi$. When $\beta = -1$, the right tail decays faster than any power. The left tail behavior is similar by the symmetry property mentioned above.

One consequence of these heavy tails is that only certain moments exist. This is not a property restricted to stable laws: any distribution with power law decay will not have certain moments. When $\alpha < 2$, it can be shown that the variance does not exist and that when $\alpha \leqslant 1$, the mean does not exist. If we use fractional moments, then the p-th absolute moment $E|X|^p = \int |x|^p f(x)\,dx$ exists if and only if $p < \alpha$. We stress that this is a population moment, and by definition it is finite when the integral just above converges. If the tails are too heavy, the integral will diverge. In contrast, the sample moments of all orders will exist: one can always compute the variance of a sample. The problem is that it does not tell you much about stable laws because the sample variance does not converge to a well-defined population moment (unless $\alpha = 2$).

If X, X_1, X_2 are i.i.d. stable, then for any $a, b > 0$,

$$aX_1 + bX_2 \stackrel{d}{=} cX + d,$$

for some $c > 0$, $-\infty < d < \infty$. This condition is equivalent to (1) and can be taken as a definition of stability. More generally, linear combinations of independent stable laws with the same α are stable: if $X_j \sim \mathbf{S}(\alpha, \beta_j, \gamma_j, \delta_j; k)$ for $j = 1, \dots, n$, then

$$a_1X_1 + a_2X_2 + \cdots + a_nX_n \sim \mathbf{S}(\alpha, \beta, \gamma, \delta; k), \tag{5}$$

where $\beta = (\sum_{j=1}^n \beta_j (\operatorname{sign} a_j)|a_j\gamma_j|^\alpha)/\sum_{j=1}^n |a_j\gamma_j|^\alpha$, $\gamma^\alpha = \sum_{j=1}^n |a_j\gamma_j|^\alpha$, and

$$\delta = \begin{cases} \sum \delta_j + \gamma\beta \tan\frac{\pi\alpha}{2}, & k = 0,\ \alpha \neq 1, \\ \sum \delta_j + \beta\frac{2}{\pi}\gamma \ln\gamma, & k = 0,\ \alpha = 1, \\ \sum \delta_j, & k = 1. \end{cases}$$

This is a generalization of (1): it allows different skewness, scales and locations in the terms. It is essential that all the αs are the same: adding two stable random variables with different αs does not give a stable law.

2. Appropriateness of stable models

Stable distributions have been proposed as a model for many types of physical and economic systems. There are several reasons for using a stable distribution to describe a system. The first is where there are solid theoretical reasons for expecting a non-Gaussian stable model, e.g., reflection off a rotating mirror yielding a Cauchy distribution, hitting times for a Brownian motion yielding a Lévy distribution, the gravitational field of stars yielding the Holtsmark distribution; see Feller (1975) and Uchaikin and Zolotarev (1999) for these and other examples. The second reason is the Generalized Central Limit Theorem, see below, which states that the only possible non-trivial limit of normalized sums of independent identically distributed terms is stable. It is argued that some observed quantities are the sum of many small terms, e.g., the price of a stock, and hence a stable model should be used to describe such systems. The third argument for modeling with stable distributions is empirical: many large data sets exhibit heavy tails and skewness. The strong empirical evidence for these features combined with the Generalized Central Limit Theorem is used to justify the use of stable models. Examples in finance and economics are given in Mandelbrot (1963), Fama (1965), Embrechts, Klüppelberg and Mikosch (1997), and Rachev and Mittnik (2000). Such data sets are poorly described by a Gaussian model, some can be well described by a stable distribution.

The classical Central Limit Theorem says that the normalized sum of independent, identical terms with a finite variance converges to a normal distribution. The *Generalized Central Limit Theorem* shows that if the finite variance assumption is dropped, the only possible resulting limits are stable. Let $X_1, X_2, X_3, \ldots$ be a sequence of independent, identically distributed random variables. There exists constants $a_n > 0$, b_n and a non-degenerate random variable Z with

$$a_n(X_1 + \cdots + X_n) - b_n \xrightarrow{d} Z \tag{6}$$

if and only if Z is stable. A random variable X is in the *domain of attraction* of Z if there exists constants $a_n > 0$, b_n such that (6) holds when $X_1, X_2, X_3, \ldots$ are independent identically distributed copies of X.

The Generalized Central Limit Theorem says that the only possible distributions with a domain of attraction are stable. Characterizations of distributions in the domain of attraction of a stable law are in terms of tail probabilities. The simplest is: if X is a random variable with $x^\alpha P(|X| > x) \to c > 0$ for some $0 < \alpha < 2$ as $x \to \infty$, then X is in the domain of attraction of an α-stable law.

Even if we accept that large data sets have heavy tails, is it ever reasonable to use a stable model? One of the arguments against using stable models is that they have infinite variance, which is inappropriate for real data that have bounded range. However, bounded data are routinely modeled by normal distributions which have infinite support. The only justification for this is that the normal distribution gives a usable description of the shape of the distribution, even though it is clearly inappropriate on the tails for any problem with naturally bounded data. The same justification can be used for stable models: does a stable fit gives an accurate description of the shape of the distribution? The variance is one measure of spread; the scale γ in a stable model is another. Perhaps practioners are so used to using the variance as *the* measure of spread, that they automatically retreat from models without a variance. The parameters δ and γ can play the role of the scale and location usually played by the mean and variance. For the normal distribution, the first and second moment completely specify the distribution; for most distributions they do not.

We propose that the practitioner approach this dispute as an agnostic. The fact is that until recently we have not really been able to compare data sets to a proposed stable model. The next Section shows that estimation of all four stable parameters is feasible and that there are methods to assess whether a stable model accurately describes the data. In some cases there are solid theoretical reasons for believing that a stable model is appropriate; in other cases we will be pragmatic: if a stable distribution describes the data accurately and parsimoniously with four parameters, then we accept it as a model for the observed data.

3. Computation, simulation, estimation and diagnostics

Until recently, it was difficult to use stable laws in practical problems because of computational difficulties. Most of these difficulties have been resolved by the program STABLE,[1] which can compute stable densities, cumulative distribution functions and quantiles. The basic method used in the program are described in Nolan (1997). Later improvements to the program include incorporating the Chambers, Mallows and Stuck (1976) method of simulating stable random variables, improved accuracy in the calculations, and estimation of stable parameters from data sets. Except for α close to 0, it is now possible to quickly and accurately work with stable distributions. We will not discuss details of these programs here, but will focus on the practical problems of estimation and assessing goodness of fit.

The basic estimation problem for stable laws is to estimate the four parameters $(\alpha, \beta, \gamma, \delta)$ from an i.i.d. sample $X_1, X_2, \ldots, X_n$. Because of numerical problems with the 1-parameterization, we will always use the 0-parameterization in estimation. If desired, the parameter δ_1 can be estimated by using (4). There are several methods available for this basic estimation problem: a quantile method of McCulloch (1986), a fractional moment method of Ma and Nikias (1995), sample characteristic function (SCF) method of Kogon and Williams (1998) based on ideas of Koutrouvelis, and maximum likelihood (ML) estimation of DuMouchel (1971) and Nolan (2001). These methods have been compared in a large simulation study, Ojeda (2001), who found that the ML estimates are almost always more accurate, with the SCF estimates next best, followed by the quantile method, and finally the moment method. The ML method has the added advantage that one can give large sample confidence intervals for the parameters, based on numerical computations of the Fisher information matrix.

Perhaps just as important as methods of estimation, are diagnostics for assessing the fit. While a Kolmogorov–Smirnov goodness-of-fit test statistic can be computed, giving a correct significance level to such a test when comparing a data set to a fitted distribution is an involved problem. However, one can adapt standard exploratory data analysis graphical techniques to informally evaluate the closeness of a stable fit. We have found that comparing smoothed data density plots to a proposed fit gives a good sense of how good the fit is near the center of the data. P–P plots allow a comparison over the range of the data.

[1] The program STABLE is available at www.mathstat.american.edu and following the "Faculty" link to the author's homepage.

For technical reasons we recommend the "variance stabilized" P–P plot of Michael (1983). We found Q–Q plots not as satisfactory for comparing heavy tailed data to proposed fit. One reason for this is visual – by definition a heavy tailed data set will have many more extreme values than a typical sample from finite variance population. This forces a Q–Q plot to be visually compressed, with a few extreme values dominating the plot. Also, the heavy tails imply that the extreme order statistics will have a lot of variability, and hence deviations from an ideal straight line Q–Q plot are hard to assess. The next section shows some examples of these techniques on financial data, more examples can be found in Nolan (1999, 2001).

There are methods for more complicated estimation problems involving stable laws. For example, regression models with stable residuals have been described by McCulloch (1998) for the symmetric stable case and Ojeda (2001) for the general case. The problem analyzing time series with stable noise is discussed in Section II of Adler, Feldman and Taqqu (1998), in Nikias and Shao (1995), and in Rachev and Mittnik (2000). McCulloch (1996) and Rachev and Mittnik (2000) give methods of pricing options under stable models.

4. Applications to financial data

The first example we consider is the British Pound vs. German Mark exchange rate. The data set has daily exchange rates for the 16 year period from 2 January 1980 to 21 May 1996. The log of the successive exchange rates was computed as $y_t = \ln(x_{t+1}/x_t)$, yielding 4,274 y_t values. The ML parameter estimates with 95% confidence intervals are 1.495 ± 0.047 for α, -0.182 ± 0.085 for β, 0.00244 ± 0.00008 for γ and 0.00019 ± 0.00013 for δ_0. Figure 2 shows a P–P plot and density for the data vs. the stable fit. The third curve in the density plot is the normal/Gaussian fit to the data.

The next example is another exchange rate one, this time from a developing country. This data set consists of monthly exchange rates between the US Dollar and the Tanzanian Shilling, from January 1975 to September 1997. The log of the successive exchange rates were computed as above for this monthly data, giving a data set with $n = 213$ points. The ML parameter estimates with 95% confidence intervals are 1.088 ± 0.185 for α, 0.112 ± 0.251 for β, 0.0300 ± 0.0055 for γ and 0.00501 ± 0.00621 for δ_0. The more extreme fluctuations of the Tanzanian Shilling exchange rate show up in the smaller estimate of α and in the larger estimate of γ. Figure 3 shows the diagnostics, with the third curve again showing a normal/Gaussian fit.

The third example is from the stock market. McCulloch (1997) analyzed 40 years of monthly stock price data from the Center for Research in Security Prices (CRSP). The data set is 480 values of the CRSP value-weighted stock index, including dividends and adjusted for inflation. The ML estimates with 95% confidence intervals are 1.855 ± 0.110 for α, -0.558 ± 0.615 for β, 2.711 ± 0.213 for γ, and 0.871 ± 0.424 for δ_0. Figure 4 shows the goodness of fit.

Stable distributions may be a useful tool in Value at Risk (VaR) calculations. The goal of VaR calculations is to assess the risk in an asset by estimating population quantiles. Stable

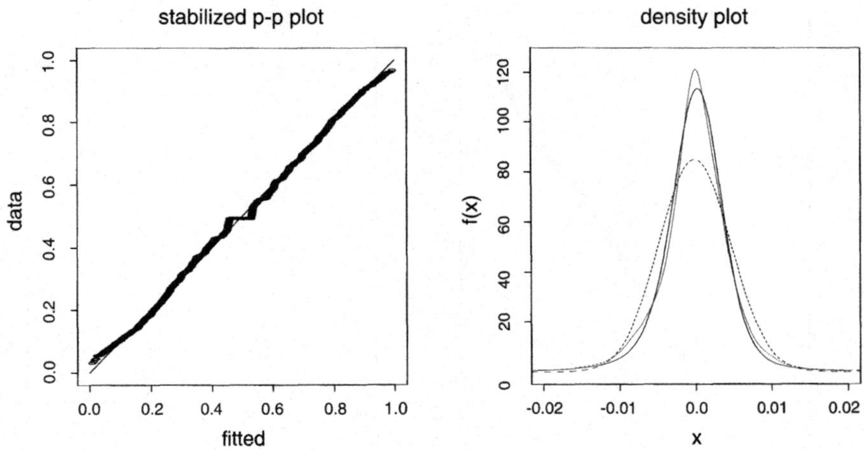

Fig. 2. P–P plot and density plot for Pound vs. Mark exchange rate data. On the density plot, the dotted curve is the smoothed data, the solid curve is the stable fit, the dashed curve is a normal fit.

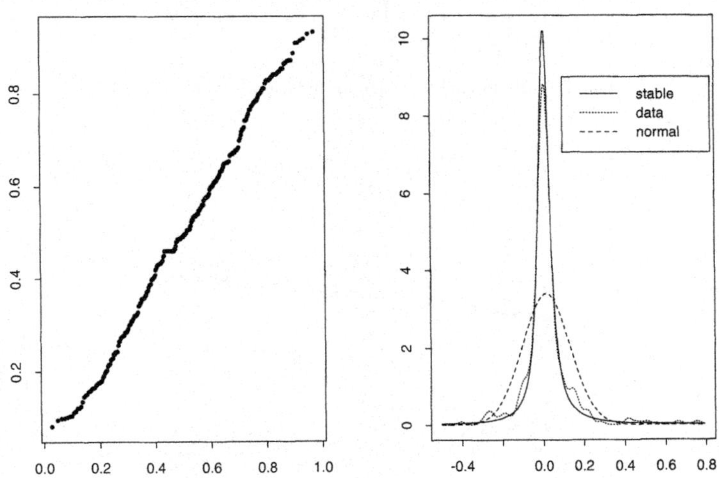

Fig. 3. P–P plot and density plot for the Tanzanian Shilling/US Dollar exchange rate.

distributions have two advantages over normal distributions: they can explicitly model both the heavier tails and asymmetry that are frequently found in financial data. Sometimes the normal distribution can give reasonable VaR estimates, because the sample variance is inflated by the extreme values in the sample. If one is lucky, the poor fitting normal distribution may approximate certain quantiles well, at the cost of poorly approximating other quantiles. Additionally, some practioners compensate for the heavy tail behavior by "adjusting" a normal quantile estimate by some empirical factor. If a stable distribution gives a more accurate fit to the sample, then it is more likely to accurately predict the VaR values. In order to compare different fits, a plot like Figure 5 can be useful. It uses the Deutsch Mark exchange rate data (log ratios of successive values) described above.

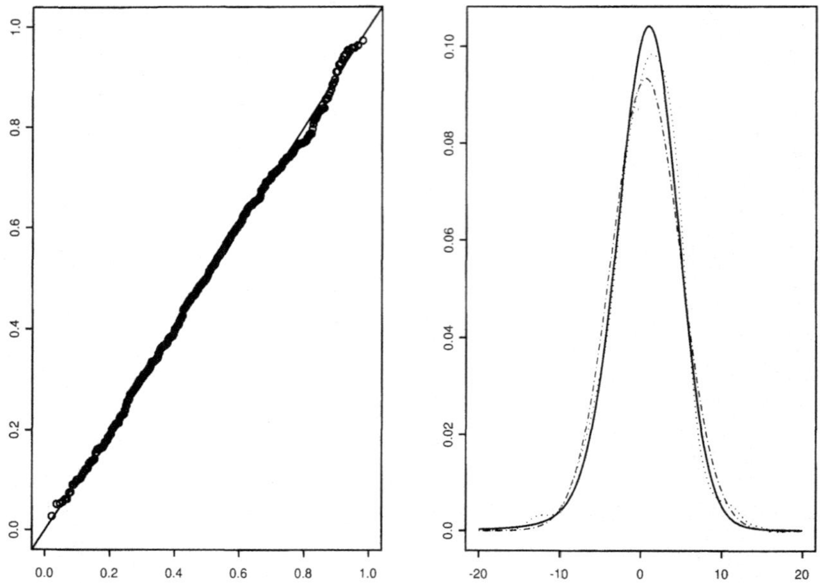

Fig. 4. P–P plot and densities for the CRSP stock price data.

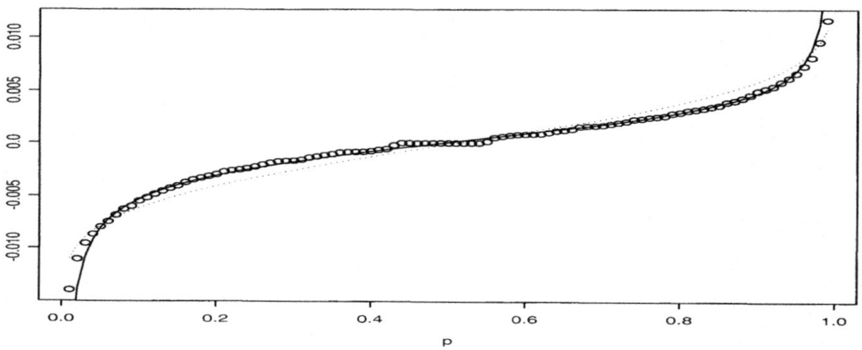

Fig. 5. VaR comparison of quantiles for the Deutsch Mark exchange rate data (circles), quantiles predicted by the stable fit (solid line), and quantiles predicted by the normal distribution (dotted line).

5. Multivariate stable distributions

This section is about d-dimensional stable laws. Such random vectors will be denoted by $\mathbf{X} = (X_1, \ldots, X_d)$. The definition of stability is the same as in (1): for i.i.d. $\mathbf{X}, \mathbf{X}_1, \mathbf{X}_2, \ldots,$

$$\mathbf{X}_1 + \mathbf{X}_2 + \cdots + \mathbf{X}_n \stackrel{d}{=} a_n \mathbf{X} + \mathbf{b}_n, \tag{7}$$

for some $a_n > 0$, and some vector $\mathbf{b}_n \in \mathbf{R}^d$. As in one dimension, an equivalent definition is that $a\mathbf{X}_1 + b\mathbf{X}_2 \stackrel{d}{=} c\mathbf{X} + \mathbf{d}$ for all $a, b > 0$.

If $\mathbf{X}$ is a stable random vector, then every one-dimensional projection $\mathbf{u} \cdot \mathbf{X} = \sum u_i X_i$ is a one-dimensional stable random variable with the same index α for every $\mathbf{u}$. The phrase "jointly stable" is sometimes used to stress the fact that the definition forces all the components X_j to be univariate α-stable with one α. Conversely, suppose $\mathbf{X}$ is a random vector with the property that every one-dimensional projection $\mathbf{u} \cdot \mathbf{X}$ is one-dimensional stable, e.g., $\mathbf{u} \cdot \mathbf{X} \sim \mathbf{S}(\alpha, (\mathbf{u}), \beta(\mathbf{u}), \gamma(\mathbf{u}), \delta(\mathbf{u}); 1)$. Then there is one α that is the index of all projections, i.e., $\alpha(\mathbf{u}) = \alpha$ is constant. If $\alpha \geqslant 1$, then $\mathbf{X}$ is stable. If $\alpha < 1$ and the location parameter function $\delta(\mathbf{u})$ and the vector of location parameters $\boldsymbol{\delta} = (\delta_1, \delta_2, \ldots, \delta_d)$ of the components $X_1, X_2, \ldots, X_d$ (all in the 1 parameterization) are related by

$$\delta(\mathbf{u}) = \mathbf{u} \cdot \boldsymbol{\delta}, \tag{8}$$

then $\mathbf{X}$ is stable. The point here is that we have a way of determining joint stability in terms of univariate stability and, when $\alpha < 1$, Equation (8).

We note that (8) holds automatically when $\alpha > 1$, so the condition is only required when $\alpha < 1$. Furthermore, (8) is necessary when $\alpha \neq 1$, so it cannot be dropped. There are examples, e.g., Section 2.2 of Samorodnitsky and Taqqu (1994), where $\alpha < 1$ and all one-dimensional projections are stable, but (8) fails and $\mathbf{X}$ is not jointly stable.

One way of parameterizing multivariate stable distributions is to use the above results about one dimensional projections. For any vector $\mathbf{u} \in \mathbf{R}^d$,

$$\mathbf{u} \cdot \mathbf{X} \sim \mathbf{S}\big(\alpha, \beta(\mathbf{u}), \gamma(\mathbf{u}), \delta(\mathbf{u}); k\big), \quad k = 0, 1.$$

Thus we know the (univariate) characteristic function of $\mathbf{u} \cdot \mathbf{X}$ for every $\mathbf{u}$, and hence the joint characteristic function of $\mathbf{X}$. Therefore α and the functions $\beta(\cdot)$, $\gamma(\cdot)$ and $\delta(\cdot)$ completely characterize the joint distribution. In fact, knowing these functions on the sphere $\mathbf{S}^d = \{\mathbf{u} \in \mathbf{R}^d \colon |\mathbf{u}| = 1\}$ characterizes the distribution.

The functions $\beta(\cdot)$, $\gamma(\cdot)$ and $\delta(\cdot)$ must satisfy certain regularity conditions. The standard way of describing multivariate stable distributions is in terms of a finite measure Λ on $\mathbf{S}^d$, called the spectral measure. Let $\mathbf{X} = (X_1, \ldots, X_d)$ be jointly stable, say

$$\mathbf{u} \cdot \mathbf{X} \sim \mathbf{S}\big(\alpha, \beta(\mathbf{u}), \gamma(\mathbf{u}), \delta(\mathbf{u}); k\big), \quad k = 0, 1.$$

Then there exists a finite measure Λ on $\mathbf{S}^d$ and a location vector $\boldsymbol{\delta} \in \mathbf{R}^d$ with

$$\begin{aligned} \gamma(\mathbf{u}) &= \left(\int_{\mathbf{S}^d} |\mathbf{u} \cdot \mathbf{s}|^{\alpha} \Lambda(d\mathbf{s})\right)^{1/\alpha}, \\ \beta(\mathbf{u}) &= \frac{\int_{\mathbf{S}^d} |\mathbf{u} \cdot \mathbf{s}|^{\alpha} \operatorname{sign}(\mathbf{u} \cdot \mathbf{s}) \Lambda(d\mathbf{s})}{\int_{\mathbf{S}^d} |\mathbf{u} \cdot \mathbf{s}|^{\alpha} \Lambda(d\mathbf{s})}, \end{aligned} \tag{9}$$

$$\delta(\mathbf{u}) = \begin{cases} \boldsymbol{\delta}\cdot\mathbf{u}, & k=1,\ \alpha\neq 1,\\ \boldsymbol{\delta}\cdot\mathbf{u} - \dfrac{2}{\pi}\displaystyle\int_{\mathbf{S}^d}(\mathbf{u}\cdot\mathbf{s})\ln|\mathbf{u}\cdot\mathbf{s}|\Lambda(d\mathbf{s}), & k=1,\ \alpha=1,\\ \boldsymbol{\delta}\cdot\mathbf{u} + \left(\tan\dfrac{\pi\alpha}{2}\right)\beta(\mathbf{u})\gamma(\mathbf{u}), & k=0,\ \alpha\neq 1,\\ \boldsymbol{\delta}\cdot\mathbf{u} - \dfrac{2}{\pi}\displaystyle\int_{\mathbf{S}^d}(\mathbf{u}\cdot\mathbf{s})\ln(\mathbf{u}\cdot\mathbf{s})\Lambda(d\mathbf{s}) & \\ \quad + \dfrac{2}{\pi}\beta(\mathbf{u})\gamma(\mathbf{u})\ln\gamma(\mathbf{u}), & k=0,\ \alpha=1. \end{cases}$$

Thus another way to parameterize is $\mathbf{X}\sim\mathbf{S}(\alpha,\Lambda,\boldsymbol{\delta};k)$, $k=0,1$. If one knows Λ, then the above equations specify the parameter functions $\beta(\cdot)$, $\gamma(\cdot)$ and $\delta(\cdot)$. Going the other direction is more difficult. If one recognizes a certain form for the parameter functions, then one can specify the spectral measure. In the general case, one can numerically invert the map $\Lambda\to(\beta(\cdot),\gamma(\cdot),\delta(\cdot))$ to get a discrete approximation to Λ.

It is possible for $\mathbf{X}$ to be non-degenerate, but singular. For example, $\mathbf{X}=(X_1,0)$ is formally a two-dimensional stable distribution if X_1 is univariate stable, but it is supported on a one-dimensional subspace. In what follows, we will always assume that $\mathbf{X}$ is non-singular that is, it has a density on $\mathbf{R}^d$. It can be shown that the following are equivalent:

(i) $\mathbf{X}$ is non-singular,

(ii) $\gamma(\mathbf{u})>0$ for all non-zero $\mathbf{u}\in\mathbf{R}^d$, and

(iii) span support$(\Lambda)=\mathbf{R}^d$.

For $\alpha\geqslant 1$, the support of non-singular stable $\mathbf{X}$ is all of $\mathbf{R}^d$. When $\alpha<1$, it can be all of $\mathbf{R}^d$ or a cone, depending on the spectral measure. For A is a subset of $\mathbf{R}^d$, define $\mathrm{CCH}(A)$ = closed convex hull of A = closure of

$$\left\{\mathbf{x}=\mathbf{a}_1b_1+\cdots+\mathbf{a}_nb_n\in\mathbf{R}^d\colon\ \mathbf{a}_1,\ldots,\mathbf{a}_n\in A,\ b_1,\ldots,b_n\geqslant 0\right\}.$$

Note that we only take positive linear combinations of elements of A, so this is not generally the closed span of A. The translate of a cone is denoted by $\mathrm{CCH}(A)+\boldsymbol{\delta}=\{\mathbf{x}+\boldsymbol{\delta}\colon\ \mathbf{x}\in\mathrm{CCH}(A)\}$. Then the support of $\mathbf{X}\sim\mathbf{S}(\alpha,\Lambda,\delta;1)$ is

$$\operatorname{support}\mathbf{X}=\begin{cases}\mathrm{CCH}(\operatorname{support}(\Lambda))+\boldsymbol{\delta}, & \alpha<1,\\ \mathbf{R}^d, & \alpha\geqslant 1.\end{cases}$$

For example, in the two-dimensional case, if the spectral measure is supported in the first quadrant, $\alpha<1$, and $\boldsymbol{\delta}=0$, then the support of the corresponding stable distribution is contained in the first quadrant, i.e., both components are positive.

The tail behavior of $\mathbf{X}$ is easiest to describe in terms of the spectral measure. It is best stated in polar form: let $A\subset\mathbf{S}^d$, then

$$\lim_{r\to\infty}\frac{P(\mathbf{X}\in\mathrm{CCH}(A),\,|\mathbf{X}|>r)}{P(|\mathbf{X}|>r)}=\frac{\Lambda(A)}{\Lambda(\mathbf{S}^d)}.$$

The tail behavior of the densities is more intricate. In the radially symmetric case, $f(\mathbf{x}) \sim c|\mathbf{x}|^{-(d+\alpha)}$ as $|\mathbf{x}| \to \infty$. In other cases, the tail behavior can have very different behavior in different directions. For example, in the bivariate independent case, the joint density factors $f(x_1, x_2) = f_1(x_1) f_2(x_2)$. The one-dimensional results above show $f(x, 0) \sim c_1 x^{-(1+\alpha)}$ along the x-axis, but $f(x, x) \sim c_2 x^{-2(1+\alpha)}$ along the diagonal. The

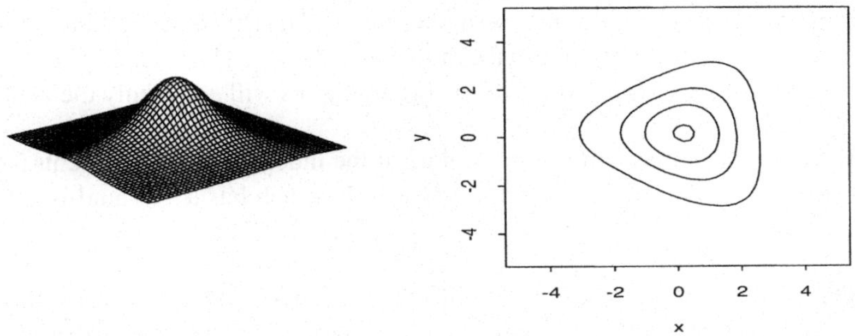

Fig. 6. Density surface and level curves for "triangle" example of a bivariate stable law.

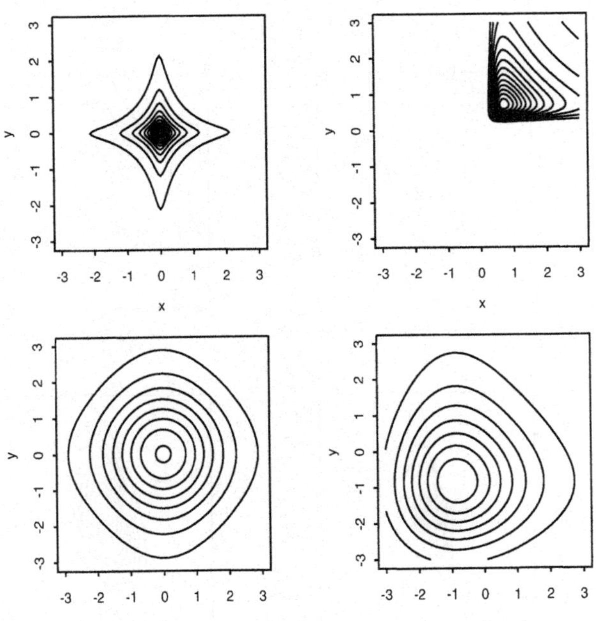

Fig. 7. Contour plots for bivariate stable densities with independent $\mathbf{S}(\alpha, \beta, 1, 0; 1)$ components. The plots show $\alpha = 0.6$, $\beta = 0$ in upper left, $\alpha = 0.6$, $\beta = 1$ in upper right, $\alpha = 1.6$, $\beta = 0$ in lower left, and $\alpha = 1.6$, $\beta = 1$ in lower right.

general case is complicated, depending on the nature (discrete, continuous) and spread of the spectral measure.

We now give some examples of bivariate stable densities, see the next section for information on their computation. In all cases, the shift vector $\boldsymbol{\delta} = 0$.

Example 4. The first example uses $\alpha = 1.2$ and a discrete spectral measure with three unit point masses, distributed on the unit circle at angles $\pi/3$, π and $-\pi/3$. A plot of the density surface and level curves are given in Figure 6. The triangular spread of the spectral measure shows up in the triangular shape of the level curves. The contour plot reveals more about the shape of the surface, so the following examples will show only the contour plots.

Example 5. Figure 7 shows the contour plots of the independent components cases when $\alpha = 0.6, 1.6$ and $\beta = 0, 1$. Note that the upper right graph has $\alpha < 1$ and is supported in the first quadrant.

Example 6. Figure 8 shows a mix of different contours, mostly to show the range of possibilities. The upper left plot shows an elliptically contoured stable distribution with $\alpha = 1.5$ and "covariation matrix"

$$R = \begin{pmatrix} 1.0 & 0.7 \\ 0.7 & 1.0 \end{pmatrix}.$$

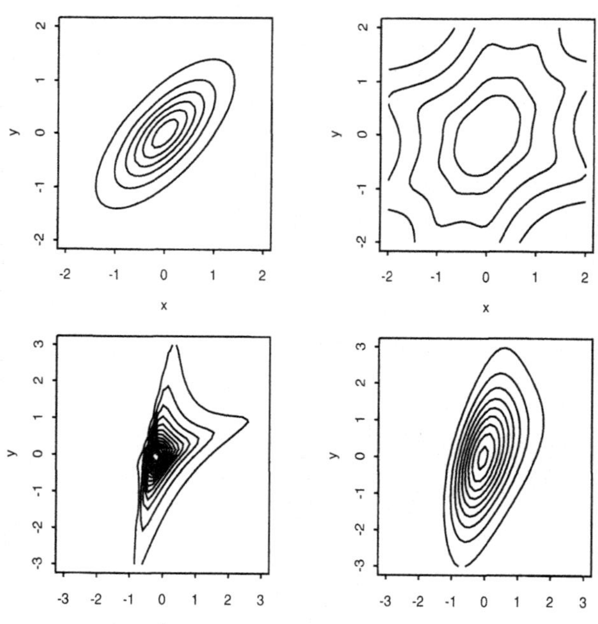

Fig. 8. Contours of miscellaneous bivariate stable distributions.

The upper right plot shows a $\alpha = 0.8$ stable distribution with discrete spectral measure having point masses at angles $-\pi/9$, $\pi/6$, $\pi/3$, $\pi/2$ and uniform weight $\lambda_j = 0.3$. The lower left plot uses $\alpha = 0.7$ with a discrete spectral measure with point masses at angles $\pi/9$, $4\pi/9$, $10\pi/9$, $13\pi/9$ of weight 0.75, 1, 0.25, 1. The lower right plot uses the same discrete spectral measure as the lower left, but with $\alpha = 1.5$.

There are some general statements that can be made about the qualitative behavior of multivariate stable densities. For fixed α, central behavior is determined by overall spread of the spectral measure: if the spectral mass is highly concentrated the density is close to singular, with large values near the center; if the spectral mass is more evenly spread around the sphere, the density is less peaked. On the tails, behavior is determined by the exact distribution of the spectral measure, with the contour lines bulging out in directions where the spectral measure is concentrated. This tail effect is more pronounced for small values of α, where distributions can be highly skewed, and becomes less pronounced as α approaches 2, where contours are all rounded into ellipses.

6. Multivariate computation, simulation, estimation and diagnostics

The computational problems are challenging, and not solved for general multivariate stable distributions. The problems are caused by the both the usual difficulties of working in d dimensions and by the complexity of the possible distributions: spectral measures are an uncountable set of "parameters". The graphs above were computed by the program MVSTABLE (available at the same web-site noted above), which only works in 2 dimensions and has limited accuracy. Density calculations are based on either numerically inverting the characteristic function as described in Nolan and Rajput (1995) or by numerically implementing the symmetric formulas in Abdul-Hamid and Nolan (1998).

One class of accessible models is when the spectral measure is discrete with a finite number of point masses:

$$\Lambda(\cdot) = \sum_{j=1}^{n} \lambda_j 1_{\{\cdot\}}(\mathbf{s}_j). \tag{10}$$

This class is dense in the space of all stable distributions: given an arbitrary spectral measure Λ_1, there is a concrete formula for n and a discrete spectral measure Λ_2 such that the densities of the corresponding stable densities are uniformly close on $\mathbf{R}^d$.

In the case of a discrete spectral measure, the parameter functions $\beta(\cdot)$, $\gamma(\cdot)$ and $\delta(\cdot)$ are computed as finite sums, rather than $(d-1)$-dimensional integrals, which makes all computations easier. It also makes simulation simple in an arbitrary dimension: $\mathbf{X} \sim \mathbf{S}(\alpha, \Lambda, \boldsymbol{\delta}; k)$ where Λ is given by (10) can be simulated by the vector sum

$$\mathbf{X} \stackrel{d}{=} \sum_{j=1}^{n} \lambda_j^{1/\alpha} Z_j \mathbf{s}_j + \boldsymbol{\delta},$$

where $Z_1, \ldots, Z_n$ are i.i.d. univariate $\mathbf{S}(\alpha, 1, 1, 0; k)$ random variables.

Another example where computations are more accessible is the elliptically contoured, or sub-Gaussian, stable distributions described in Section 8. Such densities are easier to compute and simulation is straightforward. Certain sub-stable distributions are also easy to simulate: if $\alpha < \alpha_1$, $\mathbf{X}$ is strictly α_1-stable and A is positive (α/α_1)-stable, then $A^{1/\alpha_1}\mathbf{X}$ is α-stable. Since sums and shifts of multivariate stables are also multivariate stable, one can combine these different classes to simulate a large class of multivariate stable laws.

There are several methods of estimating for multivariate stable distributions. If you know the distribution is isotropic (radially symmetric), then Problem 4, p. 44 of Nikias and Shao (1995) gives a way to estimate α and then the constant scale function/uniform spectral measure from fractional moments. In general one should let the data speak for itself, and see if the spectral measure Λ is constant. The general techniques involve some estimate of α and some estimate of the spectral measure $\hat{\Lambda} = \sum_{k=1}^{m} \lambda_k 1_{\{\cdot\}}(\mathbf{s}_k)$, $\mathbf{s}_k \in \mathbf{S}^d$. Rachev and Xin (1993) and Cheng and Rachev (1995) use the fact that the directional tail behavior of multivariate stable distributions is Pareto, and base an estimate of Λ on this. Nolan, Panorska and McCulloch (2001) define two other estimates of Λ, one based on the joint empirical/sample ch. f. and one based on the one-dimensional projections of the data.

Using the fact that one-dimensional projections are univariate stable gives a way of assessing whether a multivariate data set is stable by looking at just one-dimensional projections of the data. Fit projections in multiple directions using the univariate techniques described above, and see if they are well described by a univariate stable fit. If so, and if the α's are the same for every direction (and if $\alpha < 1$, the location parameters satisfy (8)), then a multivariate stable model is appropriate. We will illustrate this in examples below.

For the purposes of comparing two multivariate stable distributions, the parameter functions $(\alpha, \beta(\mathbf{u}), \gamma(\mathbf{u}), \delta(\mathbf{u}))$ are more useful than Λ itself. This is because the distribution of $\mathbf{X}$ depends more on how Λ distributes mass around the sphere than exactly on the measure. Two spectral measures can be far away in the traditional total variation norm (e.g., one can be discrete and the other continuous), but their corresponding parameter functions and densities can be very close.

The diagnostics suggested for assessing stability of a multivariate data set are:

- Project the data in a variety of directions $\mathbf{u}$ and use the univariate diagnostics described in Section 3 on each of those distributions. Bad fits in any direction indicate that the data is not stable.
- For each direction $\mathbf{u}$, estimate the parameter functions $\alpha(\mathbf{u})$, $\beta(\mathbf{u})$, $\gamma(\mathbf{u})$, $\delta(\mathbf{u})$ by ML estimation. The plot of $\alpha(\mathbf{u})$ should be a constant, significant departures from this indicate that the data has different decay rates in different directions. (Note that $\gamma(\mathbf{t})$ will be a constant iff the distribution is isotropic.)
- Assess the goodness-of-fit by computing a discrete $\hat{\Lambda}$ by one of the methods above. Substitute the discrete $\hat{\Lambda}$ in (9) to compute parameter functions. If it differs from the one obtained above by projection, then either the data is not jointly stable, or not enough points were chosen in the discrete spectral measure approximation.

These techniques are illustrated in the next section.

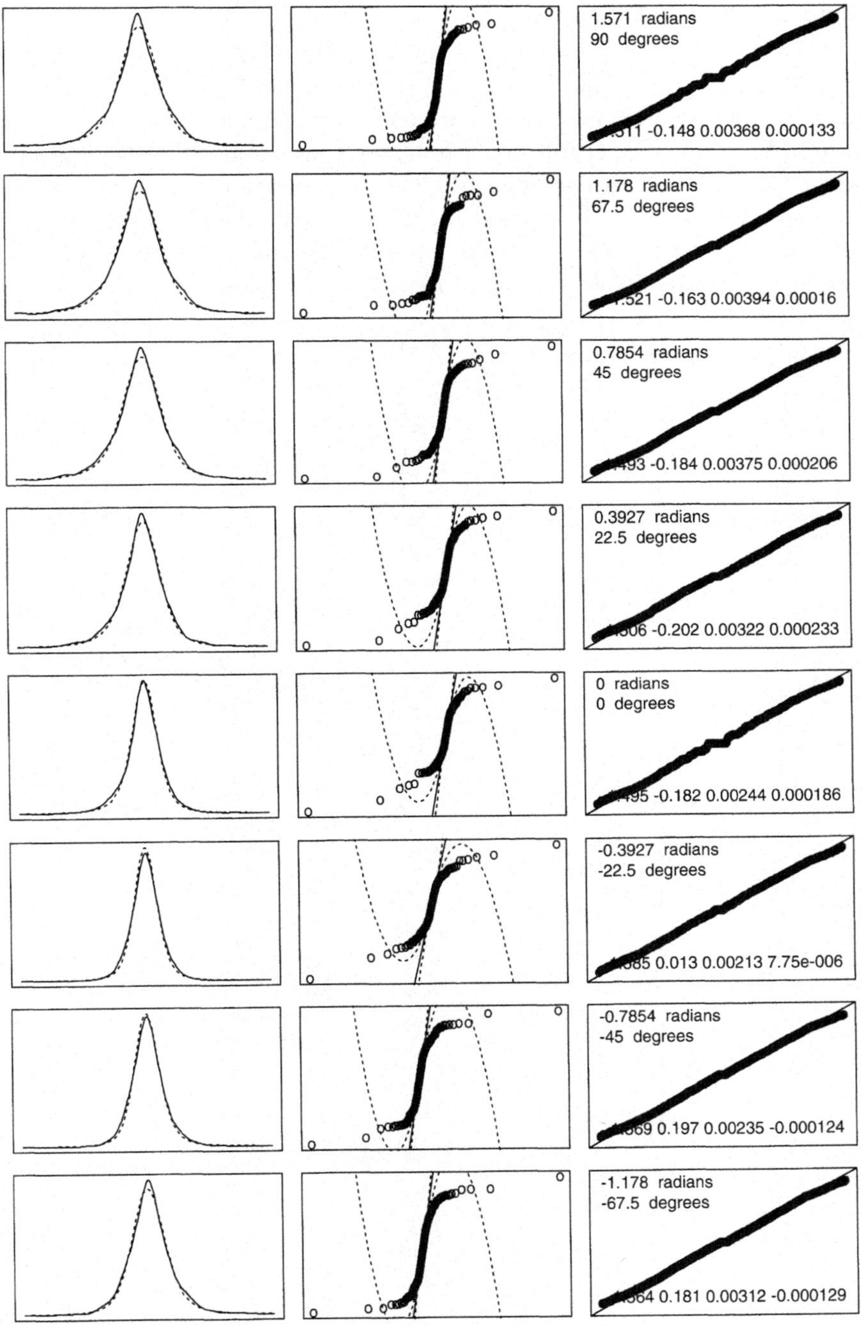

Fig. 9. Projection diagnostics for the German Mark and Japanese Yen exchange rates.

7. Multivariate application

Here we will examine the joint distribution of the German Mark and the Japanese Yen. The data set is the one described above in the univariate example. We are interested in both assessing whether the joint distribution is bivariate stable and in estimating the fit.

Figure 9 shows a sequence of smoothed density, q–q plot and variance stabilized p–p plot for projections in 8 different directions: $\pi/2$, $\pi/3$, $\pi/4$, $\pi/6$, 0, $-\pi/6$, $-\pi/4$, $-\pi/3$. (We restrict to the right half plane because projections in the left half plane are reflections of those in the right half plane.) These projections are similar to Figure 2, in fact the fifth

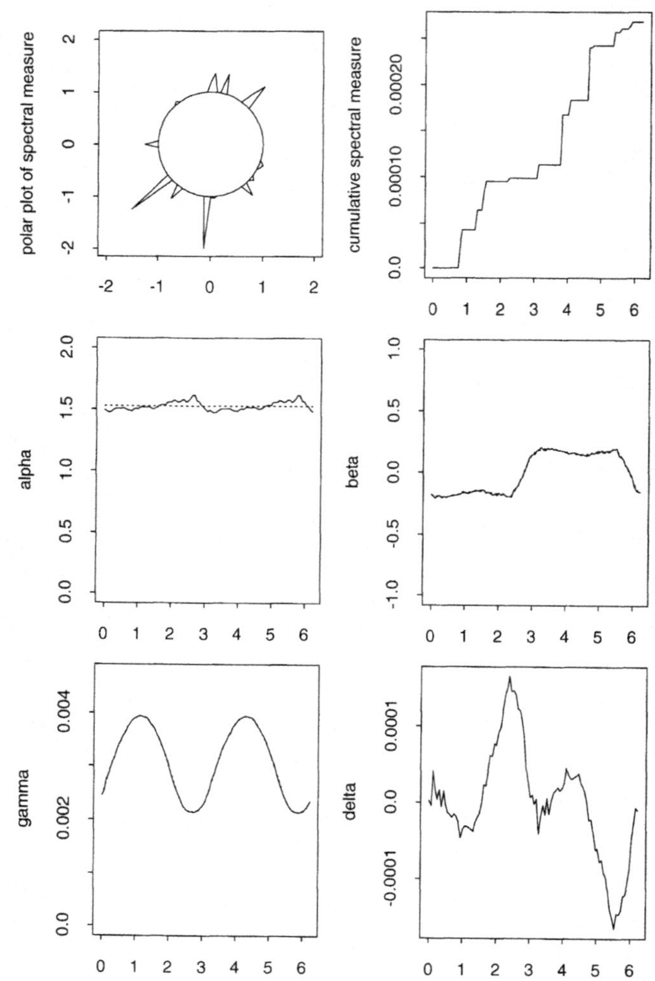

Fig. 10. Estimation results for the German Mark and Japanese Yen exchange rates.

row of Figure 9 is exactly the same as Figure 2. Except on the extreme tails, the stable fit does a good job of describing the data.

The projection functions $\alpha(\mathbf{t})$, $\beta(\mathbf{t})$, $\gamma(\mathbf{t})$, and $\delta(\mathbf{t})$ were estimated and used to compute an estimate of the spectral measure using the projection method. The results are shown in Figure 10. It shows a discrete estimate of the spectral measure (with $m = 100$ evenly spaced point masses) in polar form, a cumulative plot of the spectral measure in rectangular form, and then four plots for the parameter estimates $(\alpha(\mathbf{t}), \beta(\mathbf{t}), \gamma(\mathbf{t}), \delta(\mathbf{t}))$. Also on the $\alpha(\mathbf{t})$ plot is a horizontal line showing the average value of all the estimated indices which is taken as the estimate of the common α that should come from a jointly stable distribution. The plots of $\beta(\mathbf{t})$ and $\gamma(\mathbf{t})$ also show the skewness and scale functions computed from the estimated spectral measure substituted into (9). These curves, which are based on a joint estimate of the spectral measure, are indistinguishable from the direct, separate estimates of the directional parameters.

The fitted spectral measure was used to plot the fitted bivariate density shown in Figure 11. The spread of the spectral measure is spiky, and masks a pattern that is more obvious in the density surface: the approximate elliptical contours of the fitted density. This suggests modeling the data by a sub-Gaussian stable distribution, a topic discussed in the next section.

Some comments on these plots. The polar plots of the spectral measure show a unit circle and lines connecting the points (θ_j, r_j), where $\theta_j = 2\pi(j-1)/m$ and $r_j = 1 + (\lambda_j/\lambda_{\max})$, where $\lambda_{\max} = \max \lambda_j$. The polar plots are spiky, because we are estimating a discrete object. What should be looked at is the overall spread of mass, not specific spikes in the plot. In cases where the spectral measure is really smooth, it may be appropriate to smooth these plots out to better show it's true nature. In cases where the measure is discrete, i.e., the independent case, then one wants to emphasize the spikes. So there is no satisfactory general solution and we just plot the raw data.

Finally, most graphing programs will set vertical scale so that the graph fills the graph. This emphasizes minor fluctuations in the data that are not of practical significance. In the graphs below, the vertical scales for the parameter functions $\alpha(\mathbf{t})$, $\beta(\mathbf{t})$, $\gamma(\mathbf{t})$ are respectively $[0, 2]$, $[-1, 1]$, and $[0, 1.2 \times \max \gamma(\mathbf{t})]$. These bounds show how the functions vary

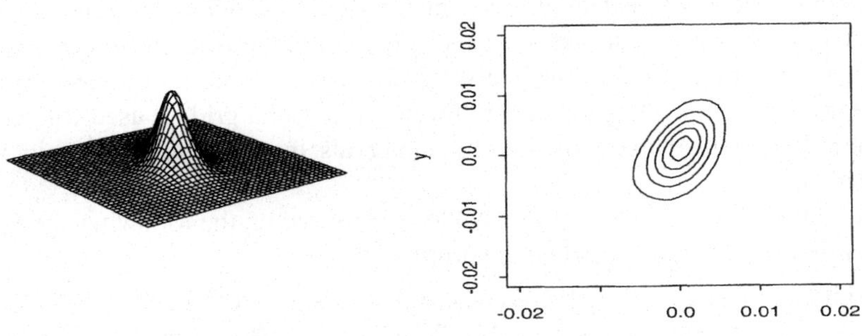

Fig. 11. Estimated density surface and level curves for a bivariate stable fit to the German Mark and Japanese Yen exchange rates.

over their possible range. For $\delta(\mathbf{t})$, we used the bounds $[-1.2 \times \max|\delta(\mathbf{t})|, 1.2 \times \max|\delta(\mathbf{t})|]$, which visually exaggerates the changes in $\delta(\mathbf{t})$. A scale that depends on $\max \gamma(\mathbf{t})$ may be more appropriate.

8. Classes of multivariate stable distributions

There may be cases where we believe that a multivariate sample has certain structure. If so, we can fit a stable model that takes this into account. This may give a more parsimonious fit to the model, especially if the data set is high dimensional. Below we fill focus on elliptically contoured distributions and see that it is computationally accessible. The idea here is to estimate an α and a matrix R so that the scale function is closely approximated by $\gamma(\mathbf{u}) = (\mathbf{u}R\mathbf{u})^{\alpha/2}$. The principle can be generalized to other special classes of distributions. Given some parametric model for the scale function $\gamma(\cdot)$, one can fit parameters, or use a nonparametric model (smoothing or loess) for the scale. Or, one can assume a special form of the spectral measure $\Lambda(\cdot)$, which determines the scale function $\gamma(\cdot)$. The methods of estimation described above do this implicitly, by assuming Λ is discrete as in (10). This can be adapted in many ways. If we assume the components of the data are independent, then we can only allow point masses at "poles", i.e., where the coordinate axes intersect the sphere. If we assume the spectral measure is concentrated on some smaller region, then one can allow point masses only in that region.

If we assume the spectral measure is continuous, then one can use some particular model for its density, say as a sum of terms like $\Lambda(d\mathbf{s}) = \sum_{k=1}^{n} \lambda_k(\mathbf{s})\, d\mathbf{s}$, where the density terms $\lambda_k(\cdot)$ in the sum have some accessible form. If the goal is a computationally accessible model, then an ad hoc approach may be useful. First compute a fit using a discrete spectral measure. If there are clearly defined point masses that are isolated, then include them and try to model the rest as an elliptical model, or using some spectral density.

Since the foreign exchange data seems to be approximately elliptically contoured, there may be interest in categorizing such stable distributions. The main practical advantage to this is that all d-dimensional elliptically contoured stable distributions are parameterized by α and a symmetric, positive definite $d \times d$ matrix. Since the matrix is symmetric, there are a total of $1 + d(d+1)/2$ parameters. This is quite different from the general stable case, which involves an infinite dimensional spectral measure. Even a discrete approximating measure involves a much larger number of terms: if a "polar grid" is used with each of the angle directions divided up evenly with k subintervals, then there are k^{d-1} point masses to be estimated.

For $\mathbf{X}$ an non-singular symmetric α-stable random vector, the following are equivalent:

- $\mathbf{X}$ is elliptically contoured around the origin.
- $\mathbf{X}$ is sub-Gaussian, i.e., $\mathbf{X} \stackrel{d}{=} A^{1/2}\mathbf{G}$, where $A \sim \mathbf{S}(\alpha, 1, \gamma, 0; 1)$ and $\mathbf{G} \sim N(0, R)$.
- The characteristic function is $E \exp(i\mathbf{u} \cdot \mathbf{X}) = \exp(-(\mathbf{u}R\mathbf{u}^{\mathrm{T}})^{\alpha/2})$, for some symmetric, positive definite matrix R.

There is a "random volatility" interpretation of sub-Gaussian distributions. Think of $\mathbf{G}$ as an underlying multivariate normal model for the returns on d assets with random scale

$A^{1/2}$. In general, A can be any positive random variable, but the product will be α-stable only when A is itself a positive $(\alpha/2)$-stable random variable.

Computations with elliptically contoured stable distributions is much simpler than the general stable case. All calculations are essentially reduced to one-dimensional problems: the linear transformation $\mathbf{Y} = R^{-1/2}\mathbf{X}$ gives a radially symmetric distribution. With a radially symmetric density, one only needs to compute it along some one-dimensional ray. In symbols, $f(\mathbf{x}) = \det(R)^{-1/2} f(|R^{-1/2}\mathbf{x}^{\mathrm{T}}|, 0, 0, \ldots, 0) = c(R)g(|R^{-1/2}\mathbf{x}'|)$. The univariate function g can be computed for arbitrary dimension d by numerically evaluating the univariate integral

$$g(x) = (2\pi)^{-d/2} \int_0^\infty \mathrm{e}^{-x^2/(2t)} f\left(t \,\middle|\, \frac{\alpha}{2}, 1, 2\left(\frac{\cos \pi\alpha}{4}\right)^{2/\alpha}, 0; 1\right) \mathrm{d}t.$$

We next describe ways of assessing a d-dimensional data set to see if it is approximately sub-Gaussian and then estimating the parameters of a sub-Gaussian vector.

First perform a one-dimensional stable fit to each coordinate of the data using one of the methods described above, to get estimates $\hat{\boldsymbol{\theta}}_i = (\hat{\alpha}_i, \hat{\beta}_i, \hat{\gamma}_i, \hat{\delta}_i)$. If the α_i's are significantly different, then the data is not jointly α-stable, so it cannot be sub-Gaussian. Likewise, if the β_i's are not all close to 0, then the distribution is not symmetric and it cannot be sub-Gaussian.

If the α_i's are all close, form a pooled estimate of $\alpha = (\sum_{i=1}^{d} \alpha_i)/d =$ average of the indices of each component. Then shift the data by $\hat{\boldsymbol{\delta}} = (\hat{\delta}_1, \hat{\delta}_2, \ldots, \hat{\delta}_d)$ so the distribution is centered at the origin.

Next, test for sub-Gaussian behavior. This can be accomplished by examining two-dimensional projections because of the following result. If $\mathbf{X}$ is a d-dimensional sub-Gaussian α-stable random vector, then every two-dimensional projection

$$\mathbf{Y} = (Y_1, Y_2) = (\mathbf{a}_1 \cdot \mathbf{X}, \mathbf{a}_2 \cdot \mathbf{X}), \tag{11}$$

$(\mathbf{a}_1, \mathbf{a}_2 \in \mathbf{R}^d)$ is a two-dimensional sub-Gaussian α-stable random vector. Conversely, suppose $\mathbf{X}$ is a d-dimensional α-stable random vector with the property that every two-dimensional projection of form (11) is non-singular sub-Gaussian. Then d-dimensional $\mathbf{X}$ is non-singular sub-Gaussian α-stable.

Estimating the $d(d+1)/2$ parameters (upper triangular part) of R can be done in at least two ways. For the first method, set $r_{ii} = \gamma_i^2$, i.e., the square of the scale parameter of the i-th coordinate. Then estimate r_{ij} by analyzing the pair (X_i, X_j) and take $r_{ij} = (\gamma^2(1,1) - r_{ii} - r_{jj})/2$, where $\gamma(1,1)$ is the scale parameter of $(1,1)\cdot(X_i, X_j) = X_i + X_j$. This involves estimating $d + d(d-1)/2 = d(d+1)/2$ one-dimensional scale parameters.

For the second method, note that if $\mathbf{X}$ is α-stable sub-Gaussian, then $E\exp(\mathrm{i}\mathbf{u} \cdot \mathbf{X}) = \exp(-(\mathbf{u}R\mathbf{u}^{\mathrm{T}})^{\alpha/2})$, so

$$\left[-\ln E\exp(\mathrm{i}\mathbf{u} \cdot \mathbf{X})\right]^{2/\alpha} = \mathbf{u}R\mathbf{u}^{\mathrm{T}} = \sum_i u_i^2 r_{ii} + 2\sum_{i<j} u_i u_j r_{ij}.$$

This is a linear function of the r_{ij}'s, so they can be estimated by regression. This method may be more accurate because it uses multiple directions, whereas the first method uses only three directions: (1,0), (0,1) and (1,1). If a two-dimensional fit has already been done, then one has already estimated $\gamma(\mathbf{u})$ on a grid. Note that $\mathbf{u}R\mathbf{u}^{\mathrm{T}} = \gamma^2(\mathbf{u})$ is the square of the scale parameter in the direction $\mathbf{u}$. Sample estimates of $\gamma^2(\mathbf{u})$ on a grid of $\mathbf{u}$ points can be used for the middle term above. In both methods, checks should be made to test that the resulting matrix R is positive definite.

The first method was used to estimate the matrix R for the Deutsche Mark–Japanese Yen data set considered above. The estimated matrix $\widehat{R}$ was

$$\widehat{R} = 10^{-6} \begin{pmatrix} 5.9552 & 4.0783 \\ 4.0783 & 13.9861 \end{pmatrix}.$$

The plot of $\gamma(\mathbf{t})$ shown in the lower left corner of Figure 10 also shows $\sqrt{\mathbf{t}\widehat{R}\mathbf{t}^{\mathrm{T}}}$ as a dashed line. It is virtually indistinguishable from the curve of $\gamma(\mathbf{t})$, supporting the idea that a sub-Gaussian stable fit does a good job of fitting the bivariate data.

9. Operator stable distributions

A brief discussion of operator stable laws is given next. The class of operator stable distributions allows different components of $\mathbf{X}$ to be stable with different indices α_j. It is defined by replacing the real scale term a_n in (7) with a matrix scale term A_n, see Jurek and Mason (1993) or Meerschaert and Schefler (2001). This may be of use in analyzing a portfolio, where different assets have different characteristics, e.g., some have Gaussian behavior and some have heavy tailed behavior, possibly with different tail behavior.

One subclass of the operator stable distributions is obtained by building up from independent groups of α-stable laws: suppose $(X_1, \ldots, X_{d_1})$ has a d_1-dimensional α_1-stable distribution, $(X_{d_1+1}, \ldots, X_{d_1+d_2})$ has a d_2-dimensional α_2-stable distribution, $\ldots$, $(X_{d_1+d_2+\cdots+d_{k-1}+1}, \ldots, X_{d_1+\cdots+d_k})$ has a d_k-dimensional α_k-stable distribution. If all these groups of distributions are independent, then the vector $\mathbf{X} = (X_1, \ldots, X_d)$, $d = d_1 + \cdots + d_k$, has a d-dimensional operator stable law. Also, for any $d \times d$ matrix A, the vector $\mathbf{Y} = A\mathbf{X}$ is an operator stable law. (One usually requires A to be invertible, otherwise the resulting $\mathbf{Y}$ will not be d-dimensional.)

10. Discussion

We have shown that estimation of general stable parameters is now feasible. The diagnostics show that some financial sets with heavy tails are well described by stable distributions. While they do not give a perfect fit, stable models can give a much better fit than Gaussian models.

In practice, the decision to use a stable model should be based on the purpose of the model. In cases where a large data set shows close agreement with a stable fit, confident statements can be made about the population. In other cases where there is a poor fit, one should not use a stable model. These models are not a panacea – not all heavy tailed data sets can be well described by stable distributions. In intermediate cases, one could tentatively use a stable model as a descriptive method of summarizing the general shape of the distribution, but not try to make statements about tail probabilities. In such problems, it may actually be better to use the quantile parameter estimates rather than ML estimates, because the former tries to match the shape of the empirical distribution and ignores the top and bottom 5% of the data.

In multivariate problems where the dimension is large, it will be very difficult to model with a stable distribution unless there is some special structure. If some components are independent, then they should be separated out and analyzed alone. If the dependent components are elliptically contoured or have some other special structure, then Section 8 discusses a way to analyze them. In the general stable case, one may try to group the components into smaller dependent groups, estimate within groups, and then try to characterize dependence between groups. We are not aware of work on this topic.

References

Abdul-Hamid, H., Nolan, J.P., 1998. Multivariate stable densities as functions of their one dimensional projections. Journal of Multivariate Analysis 67, 80–89.

Adler, R., Feldman, R., Taqqu, M. (Eds.), 1998. A Practical Guide to Heavy Tails. Birkhäuser, Boston, MA.

Chambers, J.M., Mallows, C., Stuck, B.W., 1976. A method for simulating stable random variables. JASA 71, 340–344.

Cheng, B.N., Rachev, S.T., 1995. Multivariate stable future prices. Mathematical Finance 5, 133–153.

DuMouchel, W.H., 1971. Stable distributions in statistical inference. Ph.D. Dissertation. Department of Statistics, Yale University.

DuMouchel, W.H., 1973a. On the asymptotic normality of the maximum-likelihood estimate when sampling from a stable distribution. The Annals of Statistics 1, 948–957.

DuMouchel, W.H., 1973b. Stable distributions in statistical inference. 1: Symmetric stable distributions compared to other symmetric long-tailed distributions. JASA 68, 469–477.

DuMouchel, W.H., 1975. Stable distributions in statistical inference. 2: Information from stably distributed samples. JASA 70, 386–393.

DuMouchel, W.H., 1983. Estimating the stable index α in order to measure tail thickness: A critique. The Annals of Statistics 11, 1019–1031.

Embrechts, P., Klüppelberg, C., Mikosch, T., 1997. Modelling Extremal Events for Insurance and Finance. Springer-Verlag, Heidelberg.

Fama, E., 1965. The behavior of stock prices. Journal of Business 38, 34–105.

Feller, W., 1975. An Introduction to Probability Theory and its Applications, Vol. II, 2nd edition. Wiley, New York, NY.

Fofack, H., Nolan, J.P., 1999. Tail behavior, modes and other characteristics of stable distributions. Extremes 2, 39–58.

Janicki, A., Weron, A., 1994. Simulation and Chaotic Behavior of α-Stable Stochastic Processes. Marcel Dekker, New York, NY.

Jurek, Z.J., Mason, J.D., 1993. Operator-Limit Distributions is Probability Theory. Wiley, New York, NY.

Kogon, S.M., Williams, D.B., 1998. Characteristic function based estimation of stable parameters. In: Adler, R., Feldman, R., Taqqu, M. (Eds.), A Practical Guide to Heavy Tailed Data. Birkhäuser, Boston, MA, pp. 311–335.

Koutrouvelis, I.A., 1980. Regression type estimation of the parameters of stable laws. JASA 75, 918–928.
Koutrouvelis, I.A., 1981. An iterative procedure for the estimation of the parameters of stable laws. Communications in Statistics. Simulation and Computation 10, 17–28.
Lévy, P., 1924. Théorie des erreurs la loi de Gauss et les lois exceptionelles. Bulletin de la Société de France 52, 49–85.
Ma, X., Nikias, C.L., 1995. On blind channel identification for impulsive signal environments. In: Proceedings of ICASSP '95. Detroit, MI.
Mandelbrot, B., 1963. The variation of certain speculative prices. Journal of Business 26, 394–419.
McCulloch, J.H., 1986. Simple consistent estimators of stable distribution parameters. Communications in Statistics. Simulation and Computation 15, 1109–1136.
McCulloch, J.H., 1996. Financial applications of stable distributions. In: Maddala, G.S., Rao, C.R. (Eds.), Statistical Methods in Finance, Handbook of Statistics, Vol. 14. North-Holland, New York.
McCulloch, J.H., 1997. Measuring tail thickness to estimate the stable index alpha: a critique, Journal of Business and Economical Statistics 15, 74–81.
McCulloch, J.H., 1998. Linear regression with stable disturbances. In: Adler, R., Feldman, R., Taqqu, M. (Eds.), A Practical Guide to Heavy Tailed Data. Birkhäuser, Boston, MA, pp. 359–376.
Meerschaert, M.M., Scheffler, H.-P., 2001. Limit Distributions for Sums of Independent Random Vectors. Wiley, New York, NY.
Michael, J.R., 1983. The stabilized probability plot. Biometrika 70, 11–17.
Nikias, C.L., Shao, M., 1995. Signal Processing with Alpha-Stable Distributions and Applications. Wiley, New York, NY.
Nolan, J.P., 1997. Numerical computation of stable densities and distribution functions. Communications in Statistics. Stochastic Models 13, 759–774.
Nolan, J.P., 1999. Fitting data and assessing goodness of fit with stable distributions. In: Proceedings of the Conference on Applications of Heavy Tailed Distributions in Economics, Engineering and Statistics. American University, Washington, DC, June 3–5, 1999.
Nolan, J.P., 2001. Maximum likelihood estimation and diagnostics for stable distributions. In: Barndorff-Nielson, Mikosch, Resnick (Eds.), Levý Processes. Birkhäuser, Boston, MA.
Nolan, J.P., 2003. Stable Distributions: Models for Heavy Tailed Data. Birkhäuser, Boston, MA.
Nolan, J.P., Panorska, A.K., 1997. Data analysis for heavy tailed multivariate samples. Communications in Statistics. Stochastic Models 13, 687–702.
Nolan, J.P., Panorska, A.K., McCulloch, J.H., 2001. Estimation of stable spectral measures. Mathematical and Computer Modelling 34, 1113–1122.
Nolan, J.P., Rajput, B., 1995. Calculation of multidimensional stable densities. Communications in Statistics. Simulation and Computation 24, 551–556.
Ojeda, D., 2001. Comparison of stable estimators. Ph.D. Thesis. Department of Mathematics and Statistics, American University.
Paulson, A.S., Holcomb, E.W., Leitch, R., 1975. The estimation of the parameters of the stable laws. Biometrika 62, 163–170.
Rachev, S.T., Mittnik, S., 2000. Stable Paretian Models in Finance. Wiley, New York, NY.
Rachev, S.T., Xin, H., 1993. Test for association of random variables in the domain of attraction of multivariate stable laws. Probability and Mathematical Statistics 14, 125–141.
Samorodnitsky, G., Taqqu, M.S., 1994. Stable Non-Gaussian Random Processes. Chapman and Hall, New York, NY.
Uchaikin, V.V., Zolotarev, V.M., 1999. Chance and Stability: Stable Distributions and their Applications. VSP BV, Zeist.
Zolotarev, V.M., 1986. One-Dimensional Stable Distributions. American Mathematical Society Translations of Mathematical Monographs, Vol. 65. American Mathematical Society, Providence, RI. Translation of the original 1983 (in Russian).

Chapter 4

STATISTICAL ISSUES IN MODELING MULTIVARIATE STABLE PORTFOLIOS

TOMASZ J. KOZUBOWSKI

University of Nevada, Reno

ANNA K. PANORSKA

University of Nevada, Reno

SVETLOZAR T. RACHEV

Department of Statistics and Applied Probability, University of California, Santa Barbara, USA
Institute of Statistics and Mathematical Economics, University of Karlsruhe, Germany
e-mail: rachev@lsoe-4.wiwi.uni-karlsruhe.de

Contents

Handbook of Heavy Tailed Distributions in Finance, Edited by S.T. Rachev

Abstract

Paretian stable distributions have had a relatively successful career in modeling of financial data. We discuss statistical issues common in modeling multivariate portfolios with focus on the estimation of the spectral measure that is important for estimation of the risk and dependence structure of a portfolio. We also briefly discuss alternative multivariate stable models for financial portfolios and estimation of their parameters.

1. Introduction

Statistical analysis of multivariate Paretian stable portfolios presents numerous practical and theoretical challenges. Perhaps the most common practical issues in the modeling of stable portfolios include diagnostics of the stable hypothesis and estimation of the index of stability (or tail index) α and the stable spectral measure $\boldsymbol{\Gamma}$. The index of stability α determines the overall properties of a multivariate stable distribution and the spectral measure $\boldsymbol{\Gamma}$ governs the dependence structure between the components of a stable portfolio. In this work, although we will briefly discuss diagnostics for a stable model and estimation of the index of stability, we will focus on the estimation of the spectral measure.

The fundamental work in the sixties of Mandelbrot [see, e.g., Mandelbrot (1963a, 1963b, 1967)] and Fama (1965a) led to development of a large field of research in the theory and applications of Paretian stable models in finance and economics. For example, the problem of derivative pricing for stable Paretian returns was considered in Dostoglou and Rachev (1999), Janicki et al. (1997), Hurst, Platen and Rachev (1999), Karandikar and Rachev (1995), Rachev and Rüschendorf (1994), Rachev and Samorodnitsky (1993), risk-management issues were treated in Bassi, Embrechts and Kafetzaki (1988), Gamrowski and Rachev (1996), Mittnik, Rachev and Paolella (1998), while the problem of computing optimal portfolios when the returns have Paretian stable distributions was presented in Bawa, Elton and Gruber (1979), Belkacem, Véhel and Walter (2000), Chamberlain, Cheung and Kwan (1990), Gamba (1999), Fama (1965b), Press (1982), Rachev and Han (2000), Ziemba (1974). For an extensive exposition of this subject and further references we invite the reader to peruse a recent volume of Rachev and Mittnik (2000) containing over 1000 references.

The properties of stable distributions that make them attractive for modeling include *domains of attraction* and *stability*. Domains of attraction add robustness to the stable model. We can not expect that the observed data follows *exactly* the distribution specified by the modeler. In fact, any model is an approximation of the underlying distribution of the process generating the data. As an approximation any model has a domain of applicability where its fit to the observations is reasonable and justified. Domain of attraction for a stable model contains many distributions with properties close to the specified stable law. Therefore, a stable distribution provides good approximation for a wide range of observed data. More importantly, decisions will essentially not be affected by using a stable approximation as the model instead of the true distribution. Additionally, it is possible to check whether or not a distribution is in the domain of attraction of a stable model by examining only its tails because the tails completely determine domain of attraction (see Sections 2.1).

Stability implies existence of an *overall parameter*, the *index of stability* α, that remains unchanged across all scales (sampling intervals). This is beneficial because in univariate modeling we can focus on the estimation of only one parameter that controls the main properties of the underlying distribution. In the multivariate case, the index of stability is still crucial, but it is not enough to describe all of the important properties of the stable model. To describe a multivariate portfolio, it is necessary to estimate its spectral measure which carries the information about the dependence structure and risk. The dependence structure and tests for association between the returns on different assets require an estimate of the spectral measure [see Lee, Rachev and Samorodnitcky (1990a, b)]. The risk of a stable portfolio is usually measured by its scale parameter which is defined as a functional of the spectral measure [see Section 2.5 and Rachev and Mittnik (2000)] and α. Here again, estimates of the spectral measure prove necessary for both modeling and decision making processes.

Our chapter is organized as follows. In Section 2 we define stable distributions and review their properties, particularly those relevant to financial modeling and estimation of parameters. In Sections 3, 4, and 5 we discuss estimation of the index of stability (the tail index), spectral measure, and scale parameter, respectively. Then, in Section 6, we present some other heavy tailed multivariate laws related to Paretian stable distributions that have found applications in mathematical finance. We review existing estimation procedures for these laws and introduce new estimators. Finally, in Section 7, we apply some of the procedures discussed in this work to financial data sets and further discuss some practical statistical issues related to stable modeling.

2. Multivariate stable laws

A random vector $\mathbf{X} = (X_1, \ldots, X_d)$ in $\mathbb{R}^d$ is stable (Paretian stable, α-stable) if it obeys the stability property, that is for any $n \geqslant 2$ there is some $\alpha \in (0, 2]$ and a vector $\mathbf{D}_n$ such that

$$\mathbf{X}_1 + \cdots + \mathbf{X}_n \stackrel{d}{=} n^{1/\alpha}\mathbf{X} + \mathbf{D}_n, \tag{1}$$

where the $\mathbf{X}_i$'s are i.i.d. copies of $\mathbf{X}$ [see, e.g., Samorodnitsky and Taqqu (1994)]. Parameter α is called *index of stability*. Stable vectors do not admit densities or distribution functions in a closed form (with a few exceptions) and are usually described in terms of their characteristic functions (Fourier transforms), which are of the form [see, e.g., Samorodnitsky and Taqqu (1994)]:

$$\Phi(\mathbf{t}) = \mathbb{E}\,\mathrm{e}^{\mathrm{i}\langle \mathbf{t}, \mathbf{X}\rangle} = \mathrm{e}^{-I_\alpha(\mathbf{t}) + \mathrm{i}\langle \mathbf{t}, \mathbf{m}\rangle}, \tag{2}$$

where $\mathbf{m} \in \mathbb{R}^d$ is the shift parameter (the mean for $\alpha > 1$) while

$$I_\alpha(\mathbf{t}) = \int_{S_d} \omega_{\alpha,1}\big(\langle \mathbf{t}, \mathbf{s}\rangle\big)\boldsymbol{\Gamma}(\mathrm{d}\mathbf{s}). \tag{3}$$

Here, S_d is the unit sphere in $\mathbb{R}^d$, $\boldsymbol{\Gamma}$ is a finite measure on S_d, called the *spectral measure*, the quantity $\langle \mathbf{t}, \mathbf{s}\rangle = \sum_j t_j s_j$ is the inner product in $\mathbb{R}^d$, and

$$\omega_{\alpha,\beta}(u) = \begin{cases} |u|^\alpha\big(1 - \mathrm{i}\beta\,\mathrm{sign}(u)\tan\frac{\pi\alpha}{2}\big) & \text{for } \alpha \neq 1, \\ |u|\big(1 + \mathrm{i}\beta\frac{2}{\pi}\,\mathrm{sign}(u)\log|u|\big) & \text{for } \alpha = 1. \end{cases} \tag{4}$$

We denote the distribution of a stable r.v. $\mathbf{X}$ with the ch.f. (2) by $S_\alpha(\mathbf{m}, \boldsymbol{\Gamma})$.

The index of stability $\alpha \in (0, 2]$ determines the tail of the stable law and can be thought of as a *shape* parameter. When $\alpha = 2$ we obtain the special case of multivariate normal distribution, while when $\alpha < 2$, the probability $P(X_j > x)$ associated with each component X_j of an α-stable r.v. $\mathbf{X}$ decreases like the power function $x^{-\alpha}$ as x increases to infinity. Spectral measure $\boldsymbol{\Gamma}$ controls the dependence among the components of $\mathbf{X}$. The latter are independent if and only if $\boldsymbol{\Gamma}$ is discrete and concentrated on the intersection of S_d with the coordinate axes.

2.1. Domains of attraction

Stable laws are the only possible limiting distributions of scalar-normalized sums of i.i.d. random vectors. A random vector $\mathbf{X}$ is said to be in the *domain of attraction* of a multivariate stable r.v. $\mathbf{Y}$ if for some $a_n > 0$ and $\mathbf{b}_n \in \mathbb{R}^d$ the following convergence in distribution holds

$$a_n(\mathbf{X}_1 + \cdots + \mathbf{X}_n) + \mathbf{b}_n \xrightarrow{d} \mathbf{Y} \quad \text{as } n \to \infty, \tag{5}$$

where the $\mathbf{X}_i$'s are i.i.d. copies of $\mathbf{X}$. By the stability property (1) it is clear that any stable r.v. belongs to its own domain of attraction. The domain of attraction of a stable law with index $\alpha = 2$ (the normal law) includes all distributions with finite second moments for which the convergence in (5) coincides with the classical Central Limit Theorem. The domain of attraction of a nonnormal stable law admits the following characterization due to Rvačeva[1] (1962), and plays a crucial role in estimating the spectral measure $\boldsymbol{\Gamma}$.

Proposition 2.1. *A random vector* $\mathbf{X}$ *on* $\mathbb{R}^d$ *belongs to the domain of attraction of some full*[2] *stable* $S_\alpha(\mathbf{m}, \boldsymbol{\Gamma})$ *law with* $\alpha < 2$ *if and only if* $V(r) = P(\|\mathbf{X}\| > r)$ *is regularly varying at infinity with index* $-\alpha$ *and*

$$P\left(\frac{\mathbf{X}}{\|\mathbf{X}\|} \in D \text{ given } \|\mathbf{X}\| > r\right) = \frac{P(\mathbf{X}/\|\mathbf{X}\| \in D, \|\mathbf{X}\| > r)}{V(r)} \to \frac{\boldsymbol{\Gamma}(D)}{\boldsymbol{\Gamma}(S_d)} \tag{6}$$

as $r \to \infty$ *for all Borel subsets* D *of the sphere* S_d *with* $\boldsymbol{\Gamma}(\partial D) = 0$.

In other words, the tail behavior of $\mathbf{X}$ in the direction of D is determined by the spectral measure of the set D.

1 The original proof in Rvačeva (1962) seems to contain an error; for a corrected proof and a more modern treatment (in terms of regular variation), see Meerschaert and Scheffler (2001).

2 The probability distribution of a random vector $\mathbf{X}$ on $\mathbb{R}^d$ is *full* if $\langle \mathbf{t}, \mathbf{X} \rangle$ is nondegenerate for every $\mathbf{t} \neq \mathbf{0}$.

2.2. Strictly stable and symmetric stable vectors

A r.v. $\mathbf{X}$ is strictly stable if the relation (1) is valid with $\mathbf{D}_n = \mathbf{0}$. This holds if the shift vector $\mathbf{m}$ is zero for $\alpha \neq 1$ and if

$$\int_{S_d} \mathbf{s}\boldsymbol{\Gamma}(\mathrm{d}\mathbf{s}) = \mathbf{0} \tag{7}$$

if $\alpha = 1$ [see, e.g., Samorodnitsky and Taqqu (1994)]. A r.v. $\mathbf{X}$ is said to be symmetric stable if it is stable and the probabilities $P(\mathbf{X} \in A)$ and $P(-\mathbf{X} \in A)$ are the same for all Borel sets A of $\mathbb{R}^d$. Then, the spectral measure $\boldsymbol{\Gamma}$ of $\mathbf{X}$ is symmetric and the ch.f. (2) reduces to

$$\Phi(\mathbf{t}) = \mathrm{e}^{-\int_{S_d} |\langle \mathbf{t},\mathbf{s}\rangle|^\alpha \boldsymbol{\Gamma}(\mathrm{d}\mathbf{s})}. \tag{8}$$

2.3. One-dimensional case

In one dimension, the unit sphere is the set $\{-1, 1\}$ and the ch.f. (2) reduces to

$$\phi(t) = \mathbb{E}\,\mathrm{e}^{\mathrm{i}tX} = \mathrm{e}^{\mathrm{i}\mu t - \sigma^\alpha \omega_{\alpha,\beta}(t)}, \tag{9}$$

where the parameter α is the index of stability as before, $\beta \in [-1, 1]$ is the skewness parameter, parameters $\mu \in \mathbb{R}$ and $\sigma > 0$ control location and scale, respectively, and $\omega_{\alpha,\beta}$ is given by (4). We shall use the notation $S_\alpha(\sigma, \beta, \mu)$ to denote the stable distribution given by the ch.f. (9). Strictly stable laws in one dimension correspond to $\mu = 0$ for $\alpha \neq 1$ and $\beta = 0$ for $\alpha = 1$. Symmetric univariate stable laws are strictly stable with $\mu = \beta = 0$. Stable distributions are supported on the entire real line, except when $\alpha < 1$ and $|\beta| = 1$, when we obtain totally skewed distributions concentrated on (μ, ∞) for $\beta = 1$ and $(-\infty, \mu)$ for $\beta = -1$.

The following moment formula from Samorodnitsky and Taqqu (1994), is useful in estimating parameters of multivariate stable laws [cf. Nikias and Shao (1995)].

Proposition 2.2. *Let $X \sim S_\alpha(\sigma, \beta, 0)$ with $\alpha \in (0, 2)$ and $\beta = 0$ for $\alpha = 1$. Then for any $p \in (0, \alpha)$ we have*

$$\mathbb{E}|X|^p = \sigma^p C, \tag{10}$$

where

$$C = C(\alpha, \beta, p) = \frac{2^{p-1}\Gamma(1 - p/\alpha)}{p\int_0^\infty u^{-p-1}\sin^2 u\,\mathrm{d}u}\left(1 + \beta^2 \tan^2\frac{\alpha\pi}{2}\right)^{p/(2\alpha)} \times \cos\left(\frac{p}{\alpha}\arctan\left(\beta\tan\frac{\alpha\pi}{2}\right)\right). \tag{11}$$

2.4. Discrete spectral measure

An important special class of stable laws are those with a discrete spectral measure. A measure $\boldsymbol{\Gamma}$ is discrete if

$$\boldsymbol{\Gamma}(A) = \sum_{j=1}^{k} \delta_{\mathbf{s}_j}(A)\gamma_j, \tag{12}$$

where the $\mathbf{s}_j$'s are k points on the unit sphere, $\delta_{\mathbf{s}}$ denotes a point mass at $\mathbf{s}$,

$$\delta_{\mathbf{s}}(A) = \begin{cases} 1 & \text{if } \mathbf{s} \in A, \\ 0 & \text{otherwise,} \end{cases} \tag{13}$$

and $\gamma_j > 0$ for $j = 1, 2, \ldots, k$. If the spectral measure $\boldsymbol{\Gamma}$ has form (12), then the corresponding ch.f. is straightforward to compute, because in this case I_α in (3) takes the form:

$$I_\alpha(\mathbf{t}) = \sum_{j=1}^{k} \omega_{\alpha,1}\big(\langle \mathbf{t}, \mathbf{s}_j \rangle\big)\gamma_j. \tag{14}$$

Because of the simple form of their ch.f.'s, stable laws with discrete spectral measures are much easier to handle in practice than the general ones. In particular, their computer simulation is straightforward, whereas exact algorithms for simulation of general stable vectors are not available. The simulation of stable variates with discrete spectral measure is based on the following representation from Modarres and Nolan (1994) [see also Samorodnitsky and Taqqu (1994), Example 2.3.6].

Proposition 2.3. *Let* $\mathbf{X} \sim S_\alpha(\mathbf{m}, \boldsymbol{\Gamma})$ *with* $\boldsymbol{\Gamma}$ *of the form* (12). *Then*

$$\mathbf{X} \stackrel{d}{=} \begin{cases} \mathbf{m} + \sum_{j=1}^{k} \gamma_j^{1/\alpha} V_j \mathbf{s}_j & \text{if } \alpha \neq 1, \\ \mathbf{m} + \sum_{j=1}^{k} \gamma_j \left(V_j + \frac{2}{\pi} \log \gamma_j \right) \mathbf{s}_j & \text{if } \alpha = 1, \end{cases} \tag{15}$$

where the V_j*'s are i.i.d. totally skewed, one-dimensional standard stable variables* $S_\alpha(1, 1, 0)$.

Since there exist exact algorithms for simulating one-dimensional stable variates [see, e.g., Weron (1996)], representation (15) can be used to generate d-dimensional stable vectors with discrete spectral measure.

Another important aspect of stable laws with discrete spectral measure is their role in approximating general stable distributions. As shown in Byczkowski, Nolan and Rajput (1993) every stable distribution can be approximated by one with a discrete spectral measure.

Proposition 2.4. *Given a stable vector* $\mathbf{X} \sim S_\alpha(\mathbf{m}, \boldsymbol{\Gamma})$ *in* $\mathbb{R}^d$ *with density* p*, for every* $\varepsilon > 0$ *there exists a positive integer* $k = k(\varepsilon, d, \alpha, \boldsymbol{\Gamma})$*, points* $\mathbf{s}_1, \ldots, \mathbf{s}_k$ *on the unit sphere* S_d*, and positive constants* $\gamma_1, \ldots, \gamma_k$ *such that*

$$\sup_{\mathbf{x} \in \mathbb{R}^d} \left| p(x) - p^*(x) \right| < \varepsilon, \tag{16}$$

where p^* *is the density of the stable distribution on* $\mathbb{R}^d$ *with a discrete* $\boldsymbol{\Gamma}$ *given by* (12).

The value of k is given explicitly in Byczkowski, Nolan and Rajput (1993). Because of the above approximation, in practice one usually restricts attention to laws with discrete spectral measure, see Nolan (1998) for further discussion.

2.5. *Linear combinations and risk of a financial portfolio*

Return on a d-asset portfolio can be modeled as a linear combination

$$\langle \mathbf{b}, \mathbf{X} \rangle = b_1 X_1 + \cdots + b_d X_d \tag{17}$$

of the stable vector of returns on individual assets $\mathbf{X}$ and the vector of weights $\mathbf{b}$ indicating the portion with which each asset enters the portfolio. The properties of a portfolio can then be studied via properties of linear combinations of stable random variables.

It is well known that all linear transformations (17), which include marginal distributions of stable vectors, are again stable. In particular, linear combinations of a stable r.v. $\mathbf{X} = (X_1, \ldots, X_d) \sim S_\alpha(\mathbf{m}, \boldsymbol{\Gamma})$ are univariate stable $S_\alpha(\sigma_\mathbf{b}, \beta_\mathbf{b}, \mu_\mathbf{b})$, where

$$\sigma_\mathbf{b} = \left\{ \int_{S_d} \left| \langle \mathbf{b}, \mathbf{s} \rangle \right|^\alpha \boldsymbol{\Gamma}(\mathrm{d}\mathbf{s}) \right\}^{1/\alpha}, \tag{18}$$

$$\beta_\mathbf{b} = \frac{\int_{S_d} |\langle \mathbf{b}, \mathbf{s} \rangle|^\alpha \operatorname{sign}(\langle \mathbf{b}, \mathbf{s} \rangle) \boldsymbol{\Gamma}(\mathrm{d}\mathbf{s})}{\int_{S_d} |\langle \mathbf{b}, \mathbf{s} \rangle|^\alpha \boldsymbol{\Gamma}(\mathrm{d}\mathbf{s})}, \tag{19}$$

$$\mu_\mathbf{b} = \begin{cases} \langle \mathbf{b}, \mathbf{m} \rangle & \text{for } \alpha \neq 1, \\ \langle \mathbf{b}, \mathbf{m} \rangle - \dfrac{2}{\pi} \displaystyle\int_{S_d} \langle \mathbf{b}, \mathbf{s} \rangle \log \left| \langle \mathbf{b}, \mathbf{s} \rangle \right| \boldsymbol{\Gamma}(\mathrm{d}\mathbf{s}) & \text{for } \alpha = 1. \end{cases} \tag{20}$$

Parameter $\sigma_\mathbf{b}^\alpha$ is often called the *risk of a stable portfolio*. We would like to note here, that it is necessary to have information about the spectral measure $\boldsymbol{\Gamma}$ in order to estimate that risk. For the motivation and more discussion of the definition of risk please see Rachev and Mittnik (2000).

2.6. Densities

All fully d-dimensional stable laws are absolutely continuous and admit bounded unimodal densities. In general there are no closed form expressions for stable densities. For numerical stable density computations one can use the following integral representation of stable densities due to Abdul-Hamid and Nolan (1998).

Proposition 2.5. *Let* $\mathbf{X} \sim S_\alpha(\mathbf{m}, \boldsymbol{\Gamma})$ *be a nondegenerate stable random vector in* $\mathbb{R}^d$ *with* $d \geqslant 1$, *and let* $\sigma_{\mathbf{s}}$, $\beta_{\mathbf{s}}$ *and* $\mu_{\mathbf{s}}$ *be given by* (18)–(20). *Then the density of* $\mathbf{X}$ *admits the following form*:

(i) *For* $\alpha \neq 1$,

$$p(\mathbf{x}) = \int_{S_d} g_{\alpha,d}\left(\frac{\langle \mathbf{x} - \mathbf{m}, \mathbf{s}\rangle}{\sigma_{\mathbf{s}}}, \beta_{\mathbf{s}}\right)\sigma_{\mathbf{s}}^{-d}\, \mathrm{d}\mathbf{s}, \tag{21}$$

where

$$g_{\alpha,d}(v, \beta) = \frac{1}{(2\pi)^d}\int_0^\infty \cos\left(vu - \beta u^\alpha \tan\frac{\pi\alpha}{2}\right)u^{d-1}\,\mathrm{e}^{-u^\alpha}\,\mathrm{d}u. \tag{22}$$

(ii) *For* $\alpha = 1$,

$$p(\mathbf{x}) = \int_{S_d} g_{1,d}\left(\frac{\langle \mathbf{x} - \mathbf{m}, \mathbf{s}\rangle - \mu_{\mathbf{s}} + (2/\pi)\beta_{\mathbf{s}}\sigma_{\mathbf{s}}\log\sigma_{\mathbf{s}}}{\sigma_{\mathbf{s}}}, \beta_{\mathbf{s}}\right)\sigma_{\mathbf{s}}^{-d}\, \mathrm{d}\mathbf{s}, \tag{23}$$

where

$$g_{1,d}(v, \beta) = \frac{1}{(2\pi)^d}\int_0^\infty \cos\left(vu - \frac{2}{\pi}\beta u \log u\right)u^{d-1}\,\mathrm{e}^{-u}\,\mathrm{d}u. \tag{24}$$

As remarked by Nolan (1998), this representation is more suitable for approximating multivariate stable densities than the numerical inversion of the stable ch.f. [see Nolan and Rajput (1995)], since $g_{\alpha,d}$ is a function of two variables regardless of the dimension d and it is the same for any stable random vector.

2.7. An alternative parameterization

Note that the spectral measure is not necessarily a probability measure on S_d. An alternative parameterization introduces a scale parameter

$$\sigma = \{\boldsymbol{\Gamma}(S_d)\}^{1/\alpha} \tag{25}$$

and the normalized measure

$$\tilde{\boldsymbol{\Gamma}}(\mathrm{d}\mathbf{s}) = \sigma^{-\alpha}\boldsymbol{\Gamma}(\mathrm{d}\mathbf{s}), \tag{26}$$

so that $\widetilde{\boldsymbol{\Gamma}}(S_d) = 1$ [see, e.g., Davydov and Paulauskas (1999)]. With this new normalized spectral measure, the ch.f. (2) takes the form

$$\Phi(\mathbf{t}) = \mathrm{e}^{-\sigma^\alpha I_\alpha(\mathbf{t}) + \mathrm{i}\langle \mathbf{t}, \mathbf{m}\rangle}, \tag{27}$$

where I_α is as before (with $\widetilde{\boldsymbol{\Gamma}}$ in place of $\boldsymbol{\Gamma}$). We now have four parameters: the stability index $\alpha \in (0, 2]$, the scale parameter $\sigma > 0$, the shift parameter $\mathbf{m} \in \mathbb{R}^d$, and the normalized spectral measure $\widetilde{\boldsymbol{\Gamma}}$. We shall use the notation $S_\alpha^*(\sigma, \mathbf{m}, \widetilde{\boldsymbol{\Gamma}})$ for the distribution corresponding to the ch.f. (27).

2.8. *Association*

A strong form of positive dependence of the components of a d-dimensional r.v. $\mathbf{X} = (X_1, \ldots, X_d)'$ is the *association*, introduced in Esary, Proschan and Walkup (1967). The components of $\mathbf{X}$ are said to be associated if for any functions $f, g : \mathbb{R}^d \to \mathbb{R}$, nondecreasing in each coordinate, we have

$$\mathrm{Cov}\{f(\mathbf{X}), g(\mathbf{Y})\} \geqslant 0 \tag{28}$$

whenever covariance exists. Normal variables are associated if and only if they are nonnegatively correlated (Pitt, 1982). Association of stable variables has been characterized in terms of the spectral measure in Lee, Rachev and Samorodnitsky (1990a).

Proposition 2.6. *Let* $\mathbf{X} = (X_1, \ldots, X_d)' \sim S_\alpha(\mathbf{m}, \boldsymbol{\Gamma})$*, where* $0 < \alpha < 2$*. Then* $X_1, \ldots, X_d$ *are associated if and only if*

$$\boldsymbol{\Gamma}(S_d^-) = 0, \tag{29}$$

where

$$S_d^- = \{\mathbf{s} = (s_1, \ldots, s_d) \in S_d\text{: for some } i, j \in \{1, \ldots, d\},\ s_i > 0 \text{ and } s_j < 0\}. \tag{30}$$

Thus, bivariate stable vectors are associated if and only if their corresponding spectral measure is concentrated on the first and third quadrants.

Remark. Other notions of positive dependence include *positive upper orthant dependence* (PUOD) and *positive lower orthant dependence* (PLOD). The variables $X_1, \ldots, X_d$ are PUOD if

$$P(X_1 > x_1, \ldots, X_d > x_d) \geqslant P(X_1 > x_1) \cdots P(X_d > x_d) \tag{31}$$

for any $x_1, \ldots, x_d$, and they are PLOD if

$$P(X_1 \leqslant x_1, \ldots, X_d \leqslant x_d) \geqslant P(X_1 \leqslant x_1) \cdots P(X_d \leqslant x_d), \tag{32}$$

so that the variables are likely to take on larger or smaller values together. It is well known that association implies both PUOD and PLOD, but one cannot in general reverse these implications. However, as shown in Lee, Rachev and Samorodnitsky (1990a), for stable random vectors association is equivalent to PUOD and also to PLOD, so that all of the above notions of positive dependence are equivalent.

The components of $\mathbf{X} = (X_1, \ldots, X_d)'$ are said to be *negatively associated* if for any $1 \leqslant k < d$ and any functions $f : \mathbb{R}^k \to \mathbb{R}$, $g : \mathbb{R}^{d-k} \to \mathbb{R}$, nondecreasing in each coordinate, we have

$$\operatorname{Cov}\{f(\mathbf{Y}), g(\mathbf{Z})\} \leqslant 0 \tag{33}$$

whenever the covariance exists, where $\mathbf{Y}$ and $\mathbf{Z}$ are any k and $(d-k)$-dimensional subvectors of $\mathbf{X}$ [see Alam and Saxena (1982)]. The negative association of stable random vectors was characterized in Lee, Rachev and Samorodnitsky (1990a).

Proposition 2.7. *Let* $\mathbf{X} = (X_1, \ldots, X_d)' \sim S_\alpha(\mathbf{m}, \boldsymbol{\Gamma})$, *where* $0 < \alpha < 2$. *Then* $X_1, \ldots, X_d$ *are negatively associated if and only if*

$$\boldsymbol{\Gamma}(S_d^+) = 0, \tag{34}$$

where

$$S_d^+ = \{\mathbf{s} = (s_1, \ldots, s_d) \in S_d\colon\ s_i s_j > 0 \text{ for some } i \neq j\}. \tag{35}$$

Thus, a bivariate stable vector has negatively associated components if and only if the corresponding spectral measure is concentrated on the second and forth quadrants.

3. Estimation of the index of stability

In this section we address the issue of estimating the tail index α. We start with the case when the sample comes from a univariate α-stable distribution, and then consider a more general situation where the observations are not necessarily stable, but asymptotically have a stable-Pareto tail with index α, that is

$$P(X_1 > x) = 1 - F(x) \approx x^{-\alpha} L(x), \tag{36}$$

where L is some slowly varying function. Given a multivariate heavy tailed data set $\mathbf{X}_1, \ldots, \mathbf{X}_n$, one can apply the methods of this section to one-dimensional samples corresponding to the norms $||\mathbf{X}_j||$ or the projections $\langle \mathbf{X}_j, \mathbf{b} \rangle$ for some $\mathbf{b} \in \mathbb{R}^d$.

3.1. Estimation of univariate stable parameters

Estimating the parameters of stable distributions is a challenging problem due to the fact that the densities and distributions functions of these laws are not available in closed form. Various estimation methods have been developed over the last 30 years, most of them requiring numerical approximations.

Since the stable characteristic function can be written in a closed form, several estimation techniques are based on fitting the sample characteristic function to its theoretical counterpart. The substantial collection of papers in this area started with Press (1972b), and include Arad (1980), Feuerverger and McDunnough (1977, 1981a, b), Kogon and Williams (1998), Koutrouvelis (1980, 1981), Paulson and Delehanty (1984, 1985), Paulson, Holcomb and Leitch (1975). As noted by McCulloch (1996), these estimation procedures were reported by practitioners to have high efficiency relative to the maximum likelihood approach. However, some of these methods are quite complex and require the practitioner to choose certain arbitrary parameters. A discussion and comparative study of these approaches can be found in Kogon and Williams (1998).

The maximum likelihood (ML) method for the stable case was first proposed by DuMouchel (1971, 1973), who also discussed the asymptotic properties of the estimators. To approximate the loglikelihood function DuMouchel (1971) employed fast Fourier transform (FFT) for the central part of the data and series expansions for the tails. See also DuMouchel (1975, 1983) for numerical approximation of the Fisher information matrix and further comments on this approach. Since this early work, various numerical procedures for approximating stable densities have been developed, which now permit an efficient computation of the likelihood function without the grouping procedure of DuMouchel (1971). For the ML in the symmetric case, see Brorsen and Yang (1990), McCulloch (1979, 1998). Asymmetric stable ML was treated in Brorsen and Preckel (1993), Liu and Brorsen (1995), Mittnik et al. (1999), Nolan (2001), Stuck (1976). As noted in Mittnik et al. (1999), one advantage of the ML approach over most other methods is its ability to handle generalizations to dependent or not identically distributed data arising in financial modeling (for example, regression or various time series models with stable disturbances). An implementation of the ML method for such generalizations can be found in Liu and Brorsen (1995) (stable GARCH), Mittnik, Rachev and Paolella (1998) (ARMA models driven by asymmetric stable distributions), and Brorsen and Preckel (1993), McCulloch (1998) (linear regression). In the last section of our chapter, we utilize the maximum likelihood numerical procedures of Nolan (1998), applicable for the most general i.i.d. stable case (available on the author's web site).

Numerous other methods of estimating stable parameters have been suggested. Perhaps the most commonly used estimators in empirical work are quantile procedures of Fama and Roll (1971) for the symmetric case and their modifications to the general case obtained by McCulloch (1986). Buckle (1995) proposed sampling based Bayesian inference for stable laws, see also Qiou and Ravishanker (1995), Ravishanker and Qiou (1998) for further extensions and discussion of the Bayesian approach. Nikias and Shao (1995) derived moment estimators based on sample fractional moments. Computationally simple estimators based on the modified method of scoring were proposed in

Klebanov, Melamed and Rachev (1994). For further references on estimating stable parameters, see, e.g., McCulloch (1996), Rachev and Mittnik (2000). Comparative studies of various estimators for stable parameters include Akgiray and Lamoureux (1989) and more recent Höpfner and Rüschendorf (1999), Kogon and Williams (1998).

3.2. *Estimation of the tail index α*

Assume that we have a one-dimensional random sample $X_1, \ldots, X_n$ satisfying (36) and belonging to the domain of attraction of an α-stable distribution. There is a large body of literature concerning estimation of the tail index α. Many common estimators of α are based on a subset of the sample order statistics,

$$X_{(1)} \leqslant \cdots \leqslant X_{(n)}. \tag{37}$$

Below we sketch few standard and some recent methods for estimating α and give references for many others.

3.2.1. *The Hill estimator*

The Hill estimator [see Hill (1975)] along with its various modifications is perhaps the most common way of estimating the tail thickness α of a financial data set [see, e.g., Jansen and de Vries (1991), Koedijk, Schafgans and de Vries (1990), Loretan and Phillips (1994), Phillips (1993)]. The estimator uses the k largest order statistics,

$$\hat{\alpha}_{\text{Hill}} = \left(\frac{1}{k} \sum_{j=1}^{k} \log X_{(n+1-j)} - \log X_{(n-k)} \right)^{-1}, \tag{38}$$

and arises as the conditional maximum likelihood estimator for the Pareto distribution $P(X > x) = Cx^{-\alpha}$. With the proper choice of the sequence $k = k(n)$, the estimator is consistent and asymptotically normal, see, e.g., Beirlant and Teugels (1989), Csörgő and Mason (1985), de Haan and Resnick (1998), Deheuvels, Haeusler and Mason (1988), Goldie and Smith (1987), Haeusler and Teugels (1985), Hall (1982), Hall and Welsh (1984, 1985), Mason (1982). For further discussion and extensions, see, e.g., Csörgő, Deheuvels and Mason (1985), Csörgő and Viharos (1995), Dekkers and de Haan (1993), Dekkers, Einmahl and de Haan (1989).

An obvious problem with the Hill estimator and its generalizations discussed below is the practical choice of k. Generally, we must have

$$k \to \infty \quad \text{and} \quad \frac{k}{n} \to 0 \quad \text{as } n \to \infty \tag{39}$$

to achieve strong consistency and asymptotic normality. In practice, one usually plots values of the estimator against the values of k (obtaining the so-called *Hill plot*) and looks for a stabilization (flat spot) in the graph. An alternative, more informative method of doing a Hill plot, is described in Drees, de Haan and Resnick (2000), Resnick and Stărică (1997). We refer the readers to Danielsson, Jansen and de Vries (1996), Embrechts, Klüppelberg and Mikosh (1997), Kratz and Resnick (1996), Mittnik and Paolella (1999), Rachev and Mittnik (2000), Resnick (1998), Resnick and Stărică (1997) and references therein for more details on this and related tail estimators.

3.2.2. A shifted Hill's estimator

Noting that the Hill estimator is scale invariant but not shift invariant, Aban and Meerschaert (2001) proposed the modification that is shift invariant. Their method consists of conditional maximum likelihood estimation for the shifted Pareto distribution $P(X > x) = C(x - s)^{-\alpha}$, and yields the estimators:

$$\hat{\alpha} = \left(\frac{1}{k} \sum_{j=1}^{k} \left[\log(X^*_{(j)} - \hat{s}) - \log(X^*_{(k+1)} - \hat{s}) \right] \right)^{-1}, \tag{40}$$

$$\hat{c} = \frac{k}{n} (X^*_{(k+1)} - \hat{s})^{\hat{\alpha}}, \tag{41}$$

where $\hat{s}$ is obtained by solving the equation

$$\hat{\alpha}(X^*_{(k+1)} - \hat{s})^{-1} = (\hat{\alpha} + 1)k^{-1} \sum_{j=1}^{k} (X^*_{(j)} - \hat{s})^{-1} \tag{42}$$

over the set $\hat{s} < X^*_{(k+1)}$. Here the starred variables indicate the order statistics taken in the decreasing order:

$$X^*_{(1)} \geqslant \cdots \geqslant X^*_{(n)}. \tag{43}$$

Numerical procedures are required to compute the estimators.

3.2.3. The Pickands estimator and its modifications

Pickands (1975) introduced a tail estimator of the form

$$\hat{\alpha}_{\text{Pick}} = \frac{\log 2}{\log(X_{(n-k+1)} - X_{(n-2k+1)}) - \log(X_{(n-2k+1)} - X_{(n-4k+1)})}, \quad 4k < n, \tag{44}$$

see also Drees (1996), Rosen and Weissman (1996). Noting its poor performance on samples from stable distributions, Mittnik and Rachev (1996) introduced a modification of (44) based on Bergström expansion of stable distribution function [see Bergström (1952), and also Janicki and Weron (1994)]. Their *unconditional Pickands estimator* is of the form

$$\hat{\alpha}_{\mathrm{UP}} = \frac{\log 2}{\log X_{(n-k+1)} - \log X_{(n-2k+1)}}. \tag{45}$$

We refer the readers to Rachev and Mittnikl (2000) for further discussion on the practical performance and other modifications of the Pickands estimator.

3.2.4. Least-squares estimators

Taking the logarithm of both sides in relation (36) we observe that for large values of x the points with abscissa $\log x$ and ordinate $\log(1 - F(x))$ should approximately fall on a straight line with slope $-\alpha$. Using the k largest order statistics X_{n+1-j}, $j = 1, \ldots, k$, we can examine the plot of $\log X_{n+1-j}$ versus

$$\log \frac{j}{n} \approx \log\bigl(1 - F(X_{n+1-j})\bigr) \tag{46}$$

and visually estimate the slope of the resulting line. This graphical approach was suggested by Mandelbrot (1963b).

Using these upper order statistics one can estimate the slope by the classical least-squares method [see Kratz and Resnick (1996), Schultze and Steinebach (1996)]. Below we briefly describe the estimators obtained in Schultze and Steinebach (1996). Assuming that in (36) we have $L(x) = \mathrm{e}^c$ (which is the case for stable distributions), Schultze and Steinebach (1996) applied the method of least squares to estimate the intercept c/α and the slope $1/\alpha$ of a straight line fit to

$$\log X_{n+1-j} \approx \frac{c}{\alpha} + \frac{1}{\alpha} \log \frac{n}{j}, \quad j = 1, \ldots, k. \tag{47}$$

This resulted in the following estimator of α:

$$\hat{\alpha}_{\mathrm{LS}}^{(1)} = \left[\frac{1}{k} \sum_{j=1}^{k} \log \frac{n}{j} \log X_{n+1-j} - \frac{1}{k^2} \sum_{j=1}^{k} \log \frac{n}{j} \sum_{j=1}^{k} \log X_{n+1-j} \right]^{-1} \times \left[\frac{1}{k} \sum_{j=1}^{k} \log^2 \frac{n}{j} - \left(\frac{1}{k} \sum_{j=1}^{k} \log \frac{n}{j} \right)^2 \right]. \tag{48}$$

Another estimator was obtained in Schultze and Steinebach (1996) by the least squares method under the assumption of zero intercept in (47):

$$\hat{\alpha}_{\text{LS}}^{(2)} = \sum_{j=1}^{k} \log \frac{n}{j} \log X_{n+1-j} \sum_{j=1}^{k} \log^2 \frac{n}{j}. \tag{49}$$

Finally, Schultze and Steinebach (1996) proposed yet another estimator of α resulting from expressing (47) in the form

$$\alpha \log X_{n+1-j} \approx c + \log \frac{n}{j}, \quad j = 1, \dots, k, \tag{50}$$

and minimizing the sum of squares

$$\sum_{j=1}^{k} \left(\alpha \log X_{n+1-j} - c - \log \frac{n}{j} \right)^2.$$

This produced:

$$\hat{\alpha}_{\text{LS}}^{(3)} = \left[\frac{1}{k} \sum_{j=1}^{k} \log \frac{n}{j} \log X_{n+1-j} - \frac{1}{k^2} \sum_{j=1}^{k} \log \frac{n}{j} \sum_{j=1}^{k} \log X_{n+1-j} \right] \times \left[\frac{1}{k} \sum_{j=1}^{k} \log^2 X_{n+1-j} - \left(\frac{1}{k} \sum_{j=1}^{k} \log X_{n+1-j} \right)^2 \right]^{-1}. \tag{51}$$

Consistency and asymptotic normality of the above estimators are established in Schultze and Steinebach (1996) and Csörgő and Viharos (1997), respectively [see also Kratz and Resnick (1996) for similar results on their QQ estimator].

3.2.5. *The M–S method*

Meerschaert and Scheffler (1998) introduced a simple robust estimator for the tail index α that is based on the asymptotics of the sum and utilizes the entire sample not just the largest order statistics. The estimator is based on the idea that if X_i's are i.i.d. and belong to the domain of attraction of an α-stable law with $0 < \alpha < 2$ (and their distribution function satisfies (36)), then their sample variance,

$$\hat{\sigma}^2 = \frac{1}{n} \sum_{j=1}^{n} (X_j - \overline{X})^2, \tag{52}$$

converges to an $\alpha/2$-stable (totally skewed) r.v. Y:

$$n^{1-2/\alpha}\hat{\sigma}^2 \xrightarrow{d} Y. \tag{53}$$

Taking the logarithm on both sides of (53) we obtain the convergence

$$2\log n\left(\frac{1}{\hat{\alpha}}-\frac{1}{\alpha}\right) \xrightarrow{d} \log Y, \tag{54}$$

where

$$\frac{1}{\hat{\alpha}}=\frac{\log n+\log\hat{\sigma}^2}{2\log n} \tag{55}$$

is the Meerschaert–Scheffler (M–S) estimator of $1/\alpha$. The estimator is consistent and its asymptotic distribution is that of $\log Y$ for some totally skewed $\alpha/2$ positive stable r.v. Y. Moreover, the estimator applies to certain dependent data. Comparing its performance with that of Hill's estimator, Meerschaert and Scheffler (1998) concluded that it works as well as the latter in most cases, and substantially better when applied to stable data, see Meerschaert and Scheffler (1998) for further details.

4. Estimation of the stable spectral measure

4.1. Tail estimators

A method of estimating the spectral measure of a stable r.v. $\mathbf{Y}$ based on a random sample

$$\mathbf{X}_1,\ldots,\mathbf{X}_n \tag{56}$$

from the domain of attraction of $\mathbf{Y}$ was proposed by Rachev and Xin (1993) and Cheng and Rachev (1995). The method, referred to as the Rachev–Xin–Cheng (RXC) method by Nolan and Panorska (1997), is based on the limiting relation in Proposition 2.1. To estimate $\boldsymbol{\Gamma}(D)$, where $\boldsymbol{\Gamma}$ is the (normalized) spectral measure of $\mathbf{Y}$ [cf. parameterization (27)], choose a large value of r and calculate the proportion of the $\mathbf{X}_i$'s with the norm exceeding r that belong to the set D when normalized, that is

$$\widehat{\boldsymbol{\Gamma}}(D)=\frac{\sharp\{\mathbf{X}_i/\|\mathbf{X}_i\|\in D \text{ and } \|\mathbf{X}_i\|>r\}}{\sharp\{\|\mathbf{X}_i\|>r\}}. \tag{57}$$

Equivalently, we can choose an integer $k=k(n)\leqslant n/2$ and consider the set

$$\|\mathbf{X}_{i_1}\|,\ldots,\|\mathbf{X}_{i_k}\| \tag{58}$$

of the k largest order statistics connected with the corresponding sample of the norms:

$$\|\mathbf{X}_1\|, \ldots, \|\mathbf{X}_n\|. \tag{59}$$

Then, the RXC estimator of $\boldsymbol{\Gamma}$ is the discrete measure on S_d that assigns the mass of $1/k$ to each of the unit vectors

$$\frac{\mathbf{X}_{i_1}}{\|\mathbf{X}_{i_1}\|}, \ldots, \frac{\mathbf{X}_{i_k}}{\|\mathbf{X}_{i_k}\|}. \tag{60}$$

The authors suggest taking about 20% of the largest order statistics. Under appropriate technical conditions the estimator is strongly consistent and asymptotically normal.

A similar method was recently proposed by Davydov et al. (2000) and discussed further in Davydov and Paulauskas (1999). We refer to this approach as the Davydov–Paulauskas–Rackauskas (DPR) method. Assuming that the sample (56) is actually from an α-stable distribution with a zero shift vector $\mathbf{m}$ and a symmetric (normalized) spectral measure $\boldsymbol{\Gamma}$, and the sample size n is a perfect square $n = k^2$ for some integer k, the method consists of splitting the data into k groups of k variables each, choosing a vector with the largest norm within each group, leading to a set of k vectors $\mathbf{X}_{i_1}, \ldots, \mathbf{X}_{i_k}$, and again estimating $\boldsymbol{\Gamma}$ by the empirical measure based on the unit vectors (60). The consistency and asymptotic normality of the resulting estimators,

$$\widehat{\boldsymbol{\Gamma}}(D) = \frac{1}{k}\sum_{j=1}^{k} \mathbb{I}_D\left(\frac{\mathbf{X}_{i_j}}{\|\mathbf{X}_{i_j}\|}\right), \tag{61}$$

is established in Davydov and Paulauskas (1999).

Both RXC and DPR methods do not assume any prior knowledge of α and are well suited for the $S_\alpha^*(\mathbf{m}, \sigma, \boldsymbol{\Gamma})$ parameterization, as they provide estimators for the *normalized* spectral measure. Once the spectral measure and the index α are estimated, the scale parameter σ can be estimated by methods described in Section 5.

4.2. *The empirical characteristic function method*

The method described below, proposed in Nolan, Panorska and McCulloch (2001) and investigated in Nolan and Panorska (1997), assumes that the sample comes from an α-stable distribution with shift vector $\mathbf{m}$ equal to zero. First, estimate the index of stability and center the data by the sample mean (if $\alpha > 1$) or sample median (if $\alpha < 1$). In Nolan, Panorska and McCulloch (2001) the value of α was estimated by the average $\frac{1}{d}\sum_{j=1}^{d}\hat{\alpha}_j$, where $\hat{\alpha}_j$ is an estimate of the index obtained from a univariate sample $X_{1j}, \ldots, X_{nj}$ (the quantile method of McCulloch (1986) was used to obtain these). Then, the method uses the sample to estimate the exponent I_α of the stable ch.f. (2) (with $\mathbf{m} = \mathbf{0}$):

$$\hat{I}_\alpha(\mathbf{t}) = -\log \widehat{\boldsymbol{\Phi}}_n(\mathbf{t}), \tag{62}$$

where the quantity $\widehat{\boldsymbol{\Phi}}_n$ is the sample characteristic function,

$$\widehat{\boldsymbol{\Phi}}_n(\mathbf{t}) = \frac{1}{n}\sum_{j=1}^{n} \mathrm{e}^{\mathrm{i}\langle \mathbf{t}, \mathbf{X}_j \rangle}. \tag{63}$$

For some grid $\mathbf{t}_1, \ldots, \mathbf{t}_k \in S_d$, the quantity

$$\hat{I}_{\mathrm{ECF}} = \big(\hat{I}_n(\mathbf{t}_1), \ldots, \hat{I}_n(\mathbf{t}_k)\big)' \tag{64}$$

is the empirical ch.f. (ECF) estimate of I_α. If $\boldsymbol{\Gamma}$ is a discrete measure of the form (12), then the exponent I_α is given by (14), and we can estimate $\gamma = (\gamma_1, \ldots, \gamma_k)'$ by solving the following system of linear equations:

$$I = A\gamma, \tag{65}$$

where $I = \hat{I}_{\mathrm{ECF}}$ is an estimate of I_α given by (64) and A is a $k \times k$ (complex) matrix $[a_{ij}]_{i,j=1,\ldots,k}$ with

$$a_{ij} = \omega_{\alpha,1}\big(\langle \mathbf{t}_i, \mathbf{s}_j \rangle\big). \tag{66}$$

If the grid is chosen so that the inverse of A exists, then the solution of the system (65) is $\gamma = A^{-1}I$.

For a general spectral measure, divide the unit sphere into k non-overlapping patches A_j with some central points $\mathbf{s}_j$, where $j = 1, \ldots, k$, and consider an approximation of $\boldsymbol{\Gamma}$ of the form (12), where $\gamma_i = \boldsymbol{\Gamma}(A_j)$ (which is always possible in view of Proposition 2.4). When $d = 2$, it is convenient to take the arcs

$$A_j = \left(\frac{2\pi(j-3/2)}{k}, \frac{2\pi(j-1/2)}{k}\right], \quad j = 1, \ldots, k, \tag{67}$$

centered at

$$\mathbf{s}_j = \left(\cos\frac{2\pi(j-1)}{k}, \sin\frac{2\pi(j-1)}{k}\right) \in S_d, \quad j = 1, \ldots, k. \tag{68}$$

We would again estimate I_α by (64) and solve the system (65) to obtain the estimates of the weights γ_j.

As reported in Nolan and Panorska (1997), in practice there are some problems with the direct implementation of the above method; the matrix A may be ill-conditioned and the solution of the system (65) may include negative or complex numbers (although the values of γ_j must be real and positive). Thus, in practice one should restate the problem as a constrained quadratic programming problem,

$$\text{minimize } \|I - A\gamma\| = (I - A\gamma)'(I - A\gamma) \text{ subject to } \gamma \geqslant \mathbf{0}, \tag{69}$$

which guarantees a nonnegative solution γ. We refer the readers to Nolan (1998), Nolan and Panorska (1997), Nolan, Panorska and McCulloch (2001) for examples and further discussion of these issues.

4.3. The projection method

The projection (PROJ) method was introduced in McCulloch (1994) and studied in Nolan and Panorska (1997), Nolan, Panorska and McCulloch (2001). As before, assume that the data have been shifted so that the parameter $\mathbf{m}$ is zero. The method is similar to the ECF method, since we estimate the weights γ_j at $\mathbf{s}_j$ of a discrete spectral measure $\boldsymbol{\Gamma}$ of the form (12) by solving the linear system of Equations (65). However, the PROJ method uses a different value of I, the estimate of I_α, obtained from estimators of univariate stable parameters applied to a one-dimensional sample

$$\langle \mathbf{X}_1, \mathbf{t}_j\rangle, \dots, \langle \mathbf{X}_n, \mathbf{t}_j\rangle, \quad j = 1, \dots, k, \tag{70}$$

where $\mathbf{t}_1, \dots, \mathbf{t}_k \in S_d$ is a suitably chosen grid on the unit sphere. More precisely, for each $\mathbf{t} \in \mathbb{R}^d$ the r.v. $\langle \mathbf{X}_1, \mathbf{t}\rangle$ is one-dimensional stable with parameters given by (18)–(20) and ch.f.

$$\psi(u) = \mathbb{E}\,\mathrm{e}^{\mathrm{i}u\langle \mathbf{t},\mathbf{X}\rangle} = \mathbb{E}\,\mathrm{e}^{\mathrm{i}\langle u\mathbf{t},\mathbf{X}\rangle} = \Phi(u\mathbf{t}) = \mathrm{e}^{-I_\alpha(u\mathbf{t})}, \tag{71}$$

where I_α is the characteristic exponent of the $\mathbf{X}_j$'s. Now, we can estimate the scale $\hat{\sigma}(\mathbf{t}_j)$ and skewness $\hat{\beta}(\mathbf{t}_j)$ (and also the shift $\hat{\mu}(\mathbf{t}_j)$ if $\alpha = 1$) of the univariate stable law corresponding to the sample (70), and use them to estimate the ch.f. (9) of this univariate law. Then, we can equate the above estimate with the right-hand side of (71) with $u = 1$ to estimate the quantity I_α on the grid $\mathbf{t}_1, \dots, \mathbf{t}_k$:

$$\hat{I}_n(\mathbf{t}_j) = \begin{cases} \hat{\sigma}^\alpha(\mathbf{t}_j)\left(1 - \mathrm{i}\hat{\beta}(\mathbf{t}_j)\tan\dfrac{\pi\alpha}{2}\right) & \text{for } \alpha \neq 1, \\ \hat{\sigma}(\mathbf{t}_j)\bigl(1 - \mathrm{i}\hat{\mu}(\mathbf{t}_j)\bigr) & \text{for } \alpha = 1. \end{cases} \tag{72}$$

For the index α McCulloch (1994) recommend using the pooled estimate obtained by averaging the univariate estimates obtained for each of the univariate samples (70). Thus, the PROJ estimate of I_α on the grid $\mathbf{t}_1, \dots, \mathbf{t}_k$ is the quantity

$$\hat{I}_{\mathrm{PROJ}} = \bigl(\hat{I}_n(\mathbf{t}_1), \dots, \hat{I}_n(\mathbf{t}_k)\bigr)'. \tag{73}$$

Now, the weights γ_j of the spectral measure are obtained as before by solving the system (65). For examples and further discussion, please see McCulloch (1994), Nolan and Panorska (1997), Nolan, Panorska and McCulloch (2001).

5. Estimation of the scale parameter

Let us consider the problem of estimating the scale parameter σ of the stable $S_\alpha^*(\sigma, \mathbf{m}, \boldsymbol{\Gamma})$ distribution given by the ch.f. (27) (where $\boldsymbol{\Gamma}$ is the normalized spectral measure). As before, we shall assume that the distribution is strictly stable with $\alpha > 1$, so that $\mathbf{m} = 0$. We extend the moment estimators of Davydov and Paulauskas (1999) who considered the case of symmetric spectral measure.

Note that if $\mathbf{X} \sim S_\alpha^*(\sigma, \mathbf{0}, \boldsymbol{\Gamma})$ then $\mathbf{Y} = \sigma^{-1}\mathbf{X} \sim S_\alpha^*(1, \mathbf{0}, \boldsymbol{\Gamma})$, so that for any $0 < p < \alpha$ we have

$$\mathbb{E}\|\mathbf{X}\|^p = \sigma^p C(\alpha, \boldsymbol{\Gamma}, p), \tag{74}$$

where

$$C(\alpha, \boldsymbol{\Gamma}, p) = \mathbb{E}\|\mathbf{Y}\|^p \tag{75}$$

is independent of σ and can be computed for a given values of α and $\boldsymbol{\Gamma}$. Then, approximating $\mathbb{E}\|\mathbf{X}\|^p$ by the corresponding sample moment we obtain

$$\hat{\sigma}_n = \left\{ \frac{1}{nC(\alpha, \boldsymbol{\Gamma}, p)} \sum_{j=1}^{n} \|\mathbf{X}_j\|^p \right\}^{1/p}. \tag{76}$$

Alternatively, we might use moment estimator for univariate stable variables on the i.i.d. observations

$$\langle \mathbf{X}_1, \mathbf{t} \rangle, \ldots, \langle \mathbf{X}_n, \mathbf{t} \rangle \tag{77}$$

for some $\mathbf{t} \in \mathbb{R}^d$. Then, by (18), (19), the above variables are univariate $S_\alpha(\sigma_\mathbf{t}, \beta_\mathbf{t}, 0)$, where

$$\sigma_\mathbf{t} = \sigma \left\{ \int_{S_d} |\langle \mathbf{t}, \mathbf{s} \rangle|^\alpha \boldsymbol{\Gamma}(\mathrm{d}\mathbf{s}) \right\}^{1/\alpha} \tag{78}$$

and

$$\beta_\mathbf{t} = \frac{\int_{S_d} |\langle \mathbf{t}, \mathbf{s} \rangle|^\alpha \operatorname{sign}(\langle \mathbf{t}, \mathbf{s} \rangle) \boldsymbol{\Gamma}(\mathrm{d}\mathbf{s})}{\int_{S_d} |\langle \mathbf{t}, \mathbf{s} \rangle|^\alpha \boldsymbol{\Gamma}(\mathrm{d}\mathbf{s})}. \tag{79}$$

Then, for any $0 < p < \alpha$, we have

$$\mathbb{E}\left|\langle \mathbf{X}, \mathbf{t} \rangle\right|^p = \sigma^p C(\alpha, \beta_\mathbf{t}, p) C_1(\alpha, \boldsymbol{\Gamma}, p), \tag{80}$$

where $C(\alpha, \beta_{\mathbf{t}}, p)$ is given by (11) and

$$C_1(\alpha, \boldsymbol{\Gamma}, p) = \left\{ \int_{S_d} \left| \langle \mathbf{t}, \mathbf{s} \rangle \right|^{\alpha} \boldsymbol{\Gamma}(\mathrm{d}\mathbf{s}) \right\}^{p/\alpha}. \tag{81}$$

Now, approximating (80) with the sample p-moment we obtain the following estimator of σ:

$$\tilde{\sigma}_n = \left\{ \frac{1}{nC(\alpha, \beta_{\mathbf{t}}, p)C_1(\alpha, \boldsymbol{\Gamma}, p)} \sum_{j=1}^{n} \left| \langle \mathbf{X}_j, \mathbf{t} \rangle \right|^{p} \right\}^{1/p}. \tag{82}$$

6. Extensions to other *stable* models

In this section we briefly discuss two generalizations of multivariate stable laws that often compete with them in modeling financial data: the ν-stable laws that arise as limiting distributions in the *random summation* scheme and *operator stable* laws arising as limits in ordinary summation (5) but normalized by linear operators a_n.

6.1. *ν-stable laws*

Let $\mathbf{X}_1, \mathbf{X}_2, \ldots$ be a sequence of i.i.d. random vectors in $\mathbb{R}^d$ and let ν_p, $p \in (0, 1)$, be a family of integer-valued random variables independent of the $\mathbf{X}_i$'s. Assuming that ν_p converges to infinity (in probability) as $p \to 0$, we can study the limiting distributions of the random sums

$$a_p \sum_{j=1}^{\nu_p} (\mathbf{X}_j + \mathbf{b}_p), \tag{83}$$

where $a_p > 0$ and $\mathbf{b}_p \in \mathbb{R}^d$. It follows from transfer theorems [see, e.g., Rosiński (1976)] that if the variables $p\nu_p$ converge in distribution to a positive r.v. Z with the Laplace transform $\lambda(s) = \mathbb{E}\exp(-sZ)$ and the $\mathbf{X}_j$'s are in the domain of attraction of some α-stable distribution with ch.f. $\boldsymbol{\Phi}$, then the random sums (83) will converge to a random variable with the ch.f. of the form

$$\boldsymbol{\Psi}(\mathbf{t}) = \lambda\bigl(-\log \boldsymbol{\Phi}(\mathbf{t})\bigr). \tag{84}$$

The variables with the ch.f. (84), referred to as the ν-stable laws – see, e.g., Klebanov and Rachev (1996), Kozubowski and Panorska (1998, 1999b), can be described by the same parameters as the corresponding stable laws: the tail index α, location vector $\mathbf{m}$, and spectral measure $\boldsymbol{\Gamma}$. *Strictly* ν-stable laws are given by (84) with a strictly stable ch.f. $\boldsymbol{\Phi}$.

We use the notation $\nu_\alpha(\mathbf{m}, \boldsymbol{\Gamma})$ for the distribution corresponding to the ch.f. (84) with $\boldsymbol{\Phi}$ given by (2).

The ν-stable laws are essentially location-scale mixtures of stable laws [see, e.g., Kozubowski and Panorska (1998)] and for a light-tailed r.v. Z have the same tail behavior as the corresponding stable laws. More precisely, the tail behavior of each coordinate of a ν-stable r.v. $\mathbf{X}$ is of the form $P(X_k > x) = \mathrm{O}(x^{-\alpha})$ as $x \to \infty$ under the following conditions [see Kozubowski and Panorska (1996, 1998)]:

- $\mathbb{E}Z < \infty$ if $\mathbf{X}$ is strictly ν-stable,
- $\mathbb{E}Z^{1\vee\alpha} < \infty$ and $\alpha \neq 1$ or $\mathbb{E}|Z \log Z| < \infty$ and $\alpha = 1$, if $\mathbf{X}$ is not strictly ν-stable.

Under the above conditions, the same tail behavior applies to every linear combinations $\langle \mathbf{X}, \mathbf{b} \rangle$ of $\mathbf{X}$, the order statistics of the vector $\mathbf{X}$ (as well as their absolute values), and the norm $\|\mathbf{X}\|$, see Kozubowski and Panorska (1998) for details. Note that these conditions are satisfied, for example, by the geometric stable laws discussed below.

Remark. Although the tails of ν-stable laws are essentially of the same type as those of stable distributions, ν-stable densities may behave very differently *near the mode* than their stable counterparts (may be more peaked, or even infinite) which may lead to an improved fit when modeling financial data.

Kozubowski and Panorska (1999b) showed that if the spectral measure is discrete, then truly d-dimensional ν-stable random vectors admit a representation similar to that of stable laws given in Proposition 2.3:

Proposition 6.1. *Let* $\mathbf{Y} \sim \nu_\alpha(\mathbf{m}, \boldsymbol{\Gamma})$ *with* $\boldsymbol{\Gamma}$ *of the form* (12) *and* $0 < \alpha < 2$. *Then*

$$\mathbf{Y} \stackrel{d}{=} \begin{cases} Z\mathbf{m} + Z^{1/\alpha} \sum_{j=1}^{k} \gamma_j^{1/\alpha} V_j \mathbf{s}_j & \text{if } \alpha \neq 1, \\ Z\mathbf{m} + Z \sum_{j=1}^{k} \left[V_j + \frac{2}{\pi} \log(\gamma_j Z) \right] \gamma_j \mathbf{s}_j & \text{if } \alpha = 1, \end{cases} \tag{85}$$

where the V_j*'s are i.i.d. totally skewed, one-dimensional standard stable variables* $S_\alpha(1, 1, 0)$, *independent of* Z.

Thus, ν-stable random variates are straightforward to simulate if $\boldsymbol{\Gamma}$ is discrete. Distributions with general $\boldsymbol{\Gamma}$ can be approximated by those with discrete spectral measure [see Kozubowski and Panorska (1999b)] as in the stable case, so that in practice we can restrict attention to the case with discrete $\boldsymbol{\Gamma}$.

6.1.1. Geometric stable laws

An important special case are the limiting distributions of (83) when the variables ν_p are geometric with mean $1/p$ in which case the variables $p\nu_p$ converge to a standard exponential variable with the Laplace transform $\lambda(s) = (1+s)^{-1}$. We then obtain the class of *geometric stable law* (GS) laws $GS_\alpha(\mathbf{m}, \boldsymbol{\Gamma})$ with the ch.f.

$$\boldsymbol{\Psi}(\mathbf{t}) = \big(1 + I_\alpha(\mathbf{t}) - \mathrm{i}\langle \mathbf{t}, \mathbf{m}\rangle\big)^{-1}, \tag{86}$$

where $\mathbf{m} \in \mathbb{R}^d$ and I_α is given by (3). In financial applications, where these laws have been successfully applied [see, e.g., Kozubowski and Panorska (1999a), Kozubowski and Rachev (1994), Mittnik and Rachev (1991, 1993a)] the r.v. ν_p represents the moment when the probabilistic structure governing the returns changes, so that the random sum

$$\sum_{j=1}^{\nu_p} \mathbf{X}_j \tag{87}$$

represents the total return up to this random time. In case $\alpha = 2$, we obtain the multivariate Laplace distribution [see, e.g., Kozubowski and Podgórski (2000)], which may be particularly well suited for financial applications due to its simplicity and flexibility [see, e.g., Kozubowski and Podgórski (2001)], although the tails of these laws, being heavier than Gaussian tails, are not as heavy as those of stable and geometric stable laws. More information on theory and applications of GS laws can be found in Kozubowski and Rachev (1999).

6.1.2. Statistical issues

Most estimation procedures for stable laws can be extended to the corresponding ν-stable distributions. For simplicity we consider the problem of estimating α and $\boldsymbol{\Gamma}$ of a strictly geometric stable distribution given by the ch.f. (86) with $\mathbf{m} = \mathbf{0}$ and $\alpha \neq 2$, based on a random sample

$$\mathbf{Y}_1, \ldots, \mathbf{Y}_n. \tag{88}$$

For estimating α, the tail estimators of Section 3.2 can be applied to one-dimensional samples corresponding to (88) by taking the norms of the $\mathbf{Y}_i$'s or their projections $\langle \mathbf{Y}_i, \mathbf{b}\rangle$ for some $\mathbf{b} \in \mathbb{R}^d$. These apply regardless of whether the sample is actually geometric stable or only belongs to a geometric stable domain of attraction. Alternatively, assuming that the $\mathbf{Y}_i$'s are geometric stable, one can use estimators for univariate geometric stable parameters [see, e.g., Kozubowski (1983, 2001), Rachev and Mittnik (2000)] applied to the projections $\langle \mathbf{Y}_i, \mathbf{b}\rangle$.

To estimate the spectral measure $\boldsymbol{\Gamma}$, one can use the RXC tail estimator discussed in Section 4.1 since geometric stable distributions have the same domains of attraction as the corresponding stable laws (that have the same α and $\boldsymbol{\Gamma}$), see, e.g., Klebanov and Rachev (1996). Alternatively, the empirical characteristic function method discussed in Section 4.2

can be modified to accommodate the geometric stable case. Assuming that the sample (88) is from a GS distribution, we estimate the exponent I_α of the GS ch.f. (86) as follows:

$$\hat{I}_\alpha(\mathbf{t}) = \frac{1}{\widehat{\boldsymbol{\Psi}}_n(\mathbf{t})} - 1, \tag{89}$$

where $\widehat{\boldsymbol{\Psi}}_n$ is the sample characteristic function (63) based on the $\mathbf{Y}_i$'s. The rest is the same as in the stable case. For some grid $\mathbf{t}_1, \ldots, \mathbf{t}_k \in S_d$, the quantity (64) is the empirical ch.f. (ECF) estimate of I_α. If $\boldsymbol{\Gamma}$ is a discrete measure of the form (12), then I_α is given by (14), and we can estimate $\gamma = (\gamma_1, \ldots, \gamma_k)'$ by solving the system of linear equations of the form (65), where $I = \hat{I}_{\mathrm{ECF}}$ is an estimate of I_α given by (64) and A is a $k \times k$ (complex) matrix with the entries specified in (66). If the inverse of A exists, then the solution of the system (65) is $\gamma = A^{-1}I$. To avoid the same numerical problems as in the stable case, in practice one should restate the problem as a constrained quadratic programming problem (69). The projection method of Section 4.3 can be extended similarly.

6.2. *Operator stable laws*

If we have a heavy-tail multivariate data with different tail indexes in different directions, then the multivariate stable (as well as the ν-stable) laws are no longer appropriate to model such data. Instead, we can consider the class of multivariate laws with stable marginal distributions, introduced in Resnick and Greenwood (1979), that arise as limiting distributions in the summation scheme (5) where the scaling factors are diagonal matrices, $\mathrm{diag}(a_{n1}, \ldots, a_{nd})$, for some positive a_{ni}'s. The resulting limiting marginally stable random vectors X possess a stability property similar to (1),

$$\mathbf{X}_1 + \cdots + \mathbf{X}_n \stackrel{d}{=} n^E \mathbf{X} + \mathbf{D}_n, \tag{90}$$

where the $\mathbf{X}_i$'s are i.i.d. copies of $\mathbf{X}$, E is a diagonal matrix

$$E = \mathrm{diag}\left(\frac{1}{\alpha_1}, \ldots, \frac{1}{\alpha_d}\right), \quad 0 < \alpha_i \leqslant 2,\ i = 1, \ldots, d, \tag{91}$$

called the *characteristic exponent* of $\mathbf{X}$, and n^E denotes the diagonal matrix

$$n^E = \mathrm{diag}\big(n^{1/\alpha_1}, \ldots, n^{1/\alpha_d}\big). \tag{92}$$

Remark. More general *operator stable* (OS) laws arise as the limits in (5) when the sums are normalized by some linear operators a_n [see Sharpe (1969)]. For a comprehensive review of the theory of OS laws see Jurek and Mason (1993).

Marginally stable OS laws satisfying (90) with the characteristic exponent E of the form (91) can be described in terms of their characteristic function. If all α's are strictly between one and two, we have

$$\Phi(\mathbf{t}) = \exp\left\{C\int_{S_d}\int_0^\infty \left(\mathrm{e}^{\mathrm{i}\langle \mathbf{t}, r^E\mathbf{s}\rangle} - 1 - \mathrm{i}\langle \mathbf{t}, r^E\mathbf{s}\rangle\right)\frac{\mathrm{d}r}{r^2}\boldsymbol{\Gamma}(\mathrm{d}\mathbf{s}) + \mathrm{i}\langle \mathbf{t}, \mathbf{m}\rangle\right\}, \tag{93}$$

where $\mathbf{m} \in \mathbb{R}^d$ is the shift parameter, $C > 0$ controls the scale, and $\boldsymbol{\Gamma}$ is a probability measure on the unit sphere S_d (the normalized *spectral measure*, also called the *mixing measure*). If all the α's of the characteristic exponent E in (91) are equal, then (93) reduces to the stable ch.f. with the same spectral measure. We use the notation $OS(\mathbf{m}, C, E, \boldsymbol{\Gamma})$ to denote the distributions with the ch.f. (93) with E given by (91). Similarly to the stable case, the measure $\boldsymbol{\Gamma}$ determines the dependence among the components of a marginally stable vector. For example, if $X \sim OS(\mathbf{m}, C, E, \boldsymbol{\Gamma})$ is positively or negatively associated, then the spectral measure $\boldsymbol{\Gamma}$ satisfies the condition (29) or (34), respectively [see Mittnik, Rachev and Rüschendorf (1999)].

6.2.1. Statistical issues

Estimating the parameters of an $OS(\mathbf{m}, C, E, \boldsymbol{\Gamma})$ distribution is similar to the stable case. Since all marginal distributions are univariate stable, one can obtain estimates of the α_i's by using the methods for univariate stable laws (see Section 3.1) for each of the d samples

$$X_{1j}, \ldots, X_{nj}, \quad j = 1, \ldots, d. \tag{94}$$

For samples from a domain of attraction of an OS law we can again consider univariate samples (94) and apply the methods of Section 3.2, or use the moment estimator of E based on the sample covariance matrix [see Meerschaert and Scheffler (1999)].

To estimate C and $\boldsymbol{\Gamma}$, one can use a generalization of the tail estimator of the spectral measure for stable laws described in Section 4.1 [see Mittnik, Rachev and Rüschendorf (1999), Scheffler (1999)]. First, write each of the data points (different than zero) in the unique form

$$\mathbf{X}_i = \tau(\mathbf{X}_i)^E\mathbf{s}_i, \tag{95}$$

where $\tau(\mathbf{X}_i) > 0$ is the "radius" of $\mathbf{X}_i$ and $\mathbf{s}_i$ is a point on the unit sphere S_d [these are the so-called *Jurek coordinates*, see Jurek (1984)]. Next, for some integer $k = k(n)$ consider the k largest of the $\tau(\mathbf{X}_i)$'s, that is the k largest order statistics

$$\tau(\mathbf{X}_{i_1}), \ldots, \tau(\mathbf{X}_{i_k}) \tag{96}$$

corresponding to the random sample

$$\tau(\mathbf{X}_1), \ldots, \tau(\mathbf{X}_n). \tag{97}$$

Then, the estimator of $\boldsymbol{\Gamma}$ is the discrete measure on S_d that assigns the mass of $1/k$ to each of the unit vectors

$$\mathbf{s}_{i_1}, \ldots, \mathbf{s}_{i_k} \tag{98}$$

corresponding to these order statistics via (95). Thus, the probability assigned by the estimated spectral measure $\widehat{\boldsymbol{\Gamma}}$ to a set $A \in S_d$ is the fraction of the points (98) falling in the set A.

The corresponding estimator of C is

$$\widehat{C} = \frac{k}{n}\tau(\mathbf{Y}^*), \tag{99}$$

where $\mathbf{Y}^*$ is the k-th largest of the values (97). More details regarding the estimation of $\boldsymbol{\Gamma}$ (including the asymptotic properties of estimators) can be found in Mittnik, Rachev and Rüschendorf (1999), Scheffler (1999).

7. Applications

In this section we present an example of fitting bivariate financial data sets with stable models. We fit a bivariate stable and a bivariate operator stable models to two data sets. Our data consists of 1162 daily DAX30 Index (DAX), FTSE100 Index (UKX), and S&P500 Index (SPX) prices for the period from 1/1/95 to 11/3/99. The raw indexes are first transformed into log-returns by taking natural logarithms of the quotients of their consecutive values. We analyze log-returns (1161 observations) $X_t = \ln(Y_t/Y_{t-1})$, where the Y_t's are the raw daily index prices. The goal is to fit reasonable models to the bivariate vectors (DAX, UKX) and (UKX, SPX). This section is modeled after Nolan and Panorska (1997).

We start with Exploratory Data Analysis (EDA) which focuses on general properties of the data with particular attention to the amount of variability in each data set. We first plot the log-returns of individual indexes as time series (see Figure 1). The plots show relatively large number of high spikes in the returns which points out to high volatility and the possibility that the log-returns' innovations are non-Gaussian.

The next step is to plot density histograms (total area under a density histogram equals one) of the log-returns and check for indications of long tails which again suggest more variability than allowed by a Gaussian distribution. It helps at this time to fit a Gaussian distribution to the data and overlay the histogram with the fitted Gaussian density curve. Fitting a Gaussian model amounts to estimating its mean and standard deviation from the data using the sample mean and sample standard deviation. We also check for unimodality and symmetry of the data. The density histograms of the univariate log-returns overlayed with the Gaussian (and stable) models' densities are presented in Figure 2.

We note that the histograms are much more peaky in the center and have heavier tails than the Gaussian models. Since the histograms are fairly symmetric and unimodal, the

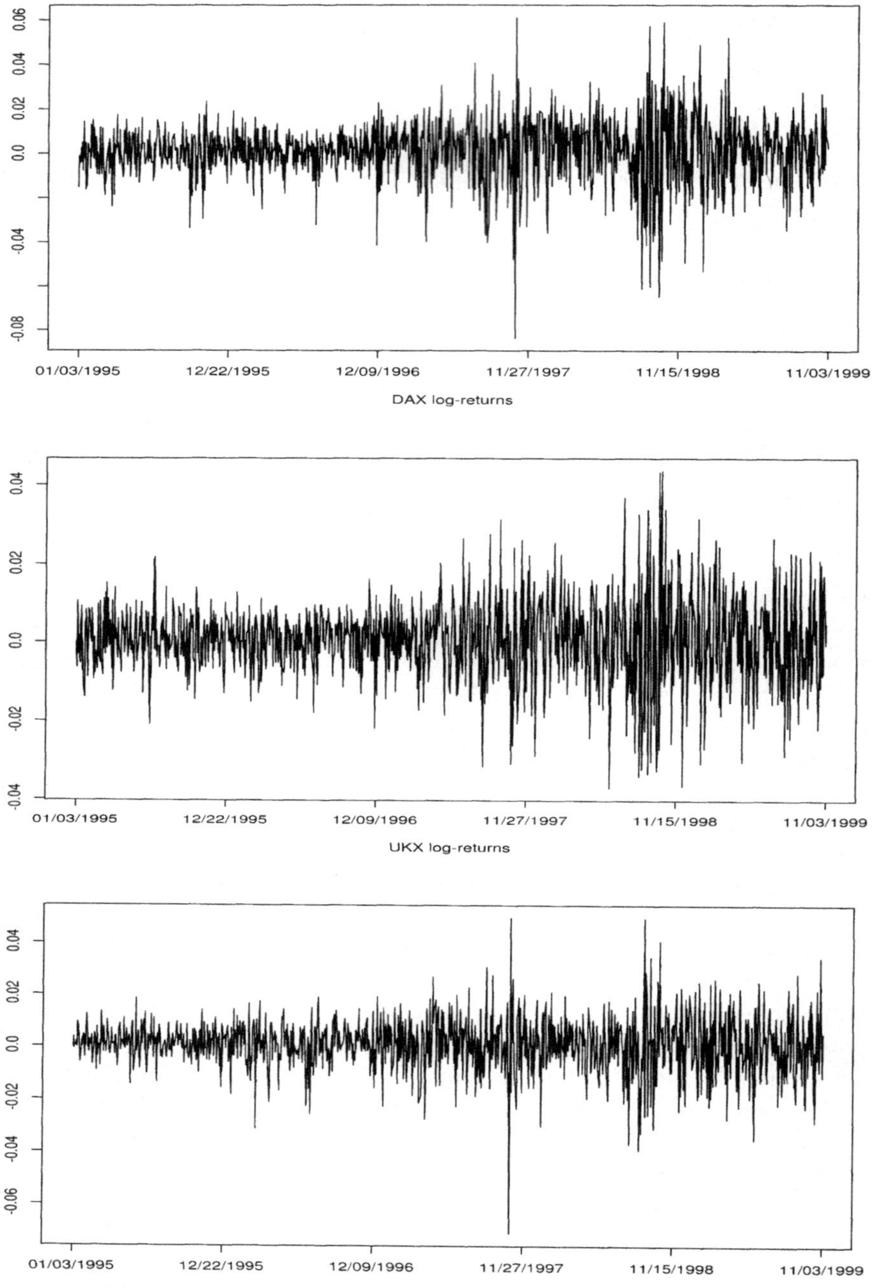

Fig. 1. Time series plots of the daily log-returns for the three indexes (1/1/95–11/3/99). Top panel: DAX log-returns. Middle panel: UKX log-returns. Bottom panel: SPX log-returns.

two distributional problems (sharp center peaks and long tails) with the Gaussian model could be alleviated by the stable approach. The next step in model fitting is deciding if we should use stable or operator stable models. To answer that question we have to estimate the tail indexes α for all three financial indexes' log-returns. If the α's for a pair of indexes are the same, we fit a stable law to their bivariate distribution, otherwise we work with an operator stable model. To fit univariate stable models to the indexes' log-returns we estimated their parameters using maximum likelihood procedure of Nolan (2001) and its numerical implementation (STABLE 2.16) due to Nolan and available on his web page. We report the parameters according to the parametrization used by Samorodnitsky and Taqqu (see (9)). Estimation results are summarized in Table 1.

We used STABLE 2.16 to compute densities of the stable models with the estimated parameters. To evaluate and compare stable and Gaussian fit we overlayed density histograms of the data with stable and Gaussian densities of the models. The results appear in Figure 2.

We note much better fit of the stable models. From now on we will work under the assumption that the individual stock indexes have univariate stable distributions. Since the tail parameters for DAX and UKX and for DAX and SPX are different, we model bivariate distribution of DAX and UKX using an operator stable distribution. The tail parameters for UKX and SPX appear to be the same and thus we will fit a bivariate stable model to UKX and SPX data. To fit these bivariate models we need estimates of the spectral measures for both DAX–UKX and UKX–SPX portfolios. To fit an operator stable distribution, we estimated the spectral measure using the method described in Section 6.2. The estimated cumulative normalized (total mass equal to one) spectral measure in radian coordinates is presented in Figure 3. Since the spectral measure seems to be concentrated on the first and third quadrants, we conclude that these variables are positively associated (see our comments in Section 6.2). Conversion to Jurek coordinates was performed using a Fortran program due to Meerschaert (personal communication), all other numerical and graphical work was done by the authors in Splus2000 Professional.

To estimate the bivariate stable spectral measure for the UKX–SPX portfolio we used all four methods described in Section 4: the tail estimators (RXC and DPR), the projection method (PROJ) and the empirical characteristic function method (ECF). The numerical implementation of the RXC, PROJ and ECF estimation procedures was done with the program MVSTABLE (Version 2.0) of Nolan available on J.P. Nolan's web site[3] with 40 weights, that is using a 40 points estimation grid on the unit circle. The DPR estimator

Table 1

Index	alpha	beta	gamma	delta
UKX	1.28	0	0.0055	0.0004
DAX	1.57	0.31	0.0076	0.0041
SPX	1.28	0	0.0058	0.001

3 See http://academic2.american.edu/~jpnolan/ for Stable 2.16 and MVSTABLE programs.

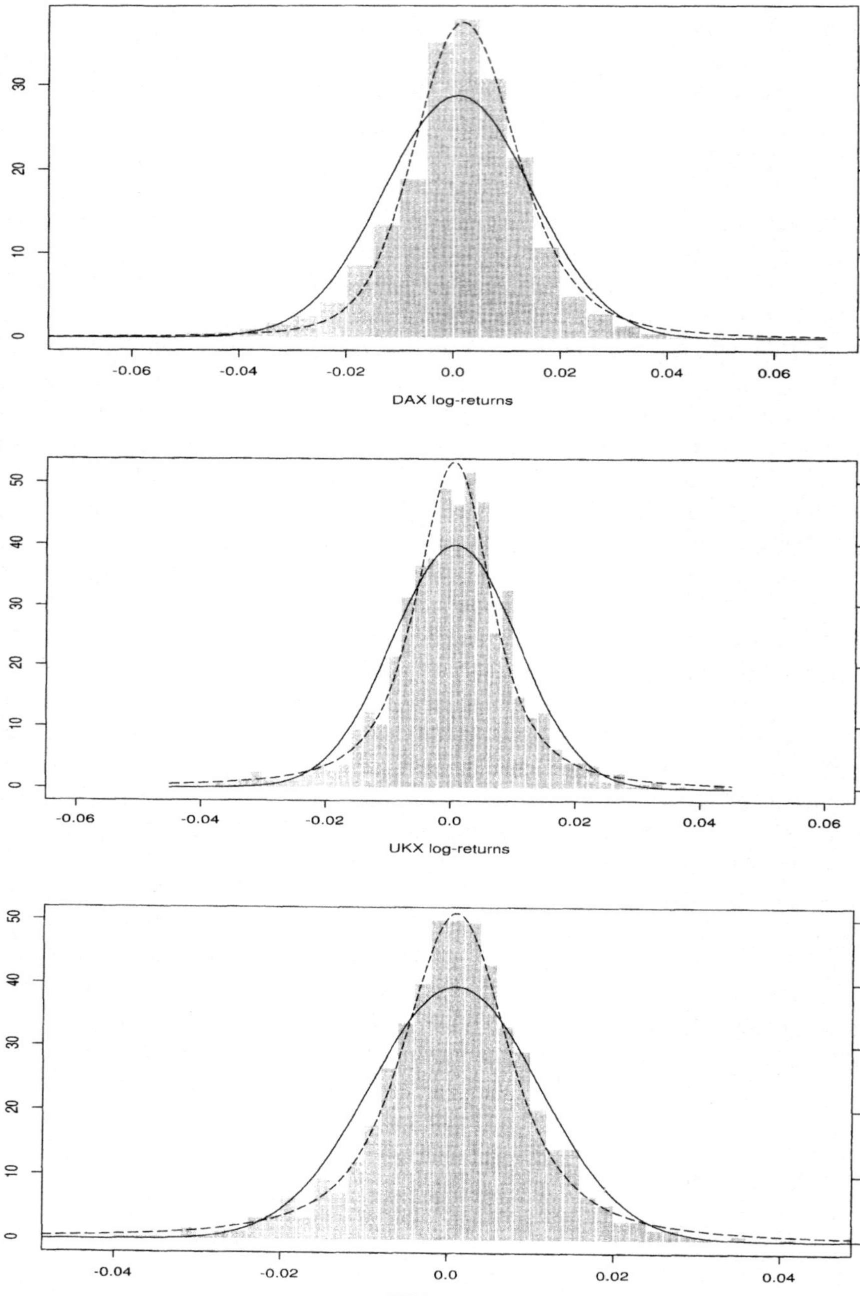

Fig. 2. Density histograms with Gaussian (solid line) and stable (dashed line) fitted densities. Top panel: DAX log-returns. Middle panel: UKX log-returns. Bottom panel: SPX log-returns.

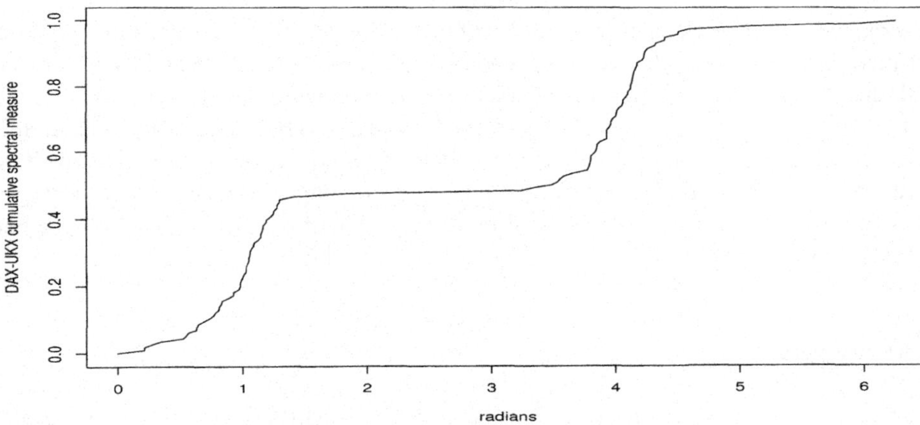

Fig. 3. Estimated cumulative (normalized) operator spectral measure for (DAX, UKX) data.

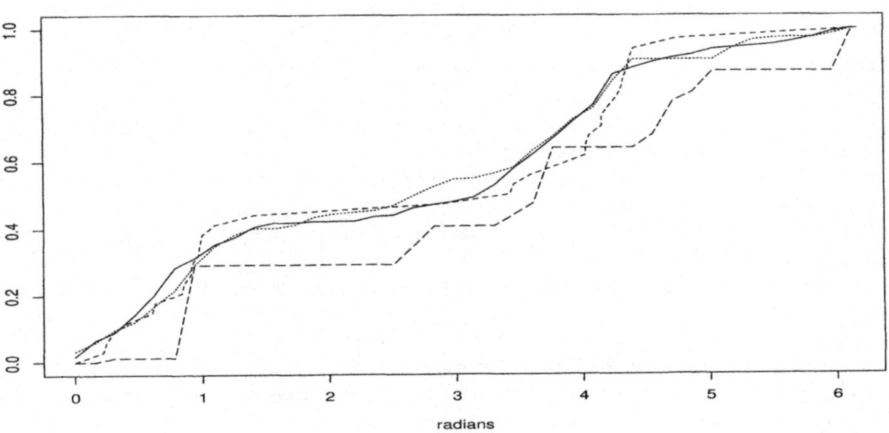

Fig. 4. Estimates of the stable spectral measure for (UKX, SPX) vector: Solid line – RXC estimator, dotted line – ECF estimator, long-dashed line – PROJ estimator, and short-dashed line – DPR estimator.

was computed for the first 1,156 ($= 34^2$) observed vectors of UKX–SPX log-returns (from 1/1/95 to 10/26/99). Numerical work for the DPR estimator was done by the authors. The graph of the estimated cumulative normalized spectral measure in radian coordinates for DAX-SPX is given in Figure 4.

As the spectral measure appears to be concentrated in the first and third quadrants we believe that UKX and SPX are positively associated (see Proposition 2.6).

To check the goodness of fit of our model we suggest methods described in Nolan and Panorska (1997). These include plotting parameters (e.g., scale) of one-dimensional projections of the sample (in several directions) computed first directly from the projected sample and then using the estimate of the spectral measure. A good fit will be indicated by general agreement between the parameters of these projections computed using two different methods. For a more detailed discussion of the choice of gridsize and its relationship with the goodness of fit, we refer the reader to Nolan and Panorska (1997).

To summarize, we performed EDA and fit two data sets (DAX, UKX) and (UKX, SPX) with bivariate operator stable and stable models. The indexes seem to be positively associated, which is an important information in constructing a portfolio. The estimates of the stable spectral measures can be used to estimate risk of a portfolio using the methods described in Section 2.5.

Acknowledgment

We thank Dr. Mark Meerschaert, Department of Mathematics, University of Nevada for helpful comments and for his Fortran routines. The research of the first two authors was partially supported by NSF grant DMS-0139927.

References

Aban, I.B., Meerschaert, M.M., 2001. Shifted Hill's estimator for heavy tails. Communications in Statistics. Simulation and Computation 30 (4), 949–962.

Abdul-Hamid, H., Nolan, J.P., 1998. Multivariate stable densities as functions of one dimensional projections. Journal of Multivariate Analysis 67, 80–89.

Akgiray, V., Lamoureux, C.G., 1989. Estimation of stable law parameters: A comparative study. Journal of Business and Economical Statistics 7, 85–93.

Alam, K., Saxena, K.M.L., 1982. Positive dependence in multivariate distributions. Communications in Statistics A10, 1183–1186.

Arad, R.W., 1980. Parameter estimation for symmetric stable distribution. Econometric Reviews 21, 209–220.

Bassi, F., Embrechts, P., Kafetzaki, M., 1988. Risk management and quantile estimation. In: Adler, R.J., Feldman, R.E., Taqqu, M. (Eds.), A Practical Guide to Heavy Tails: Statistical Techniques and Applications. Birkhäuser, Boston, pp. 111–130.

Bawa, V.C., Elton, E.L., Gruber, M.J., 1979. Simple rules for optimal portfolio selection in stable Paretian markets. Journal of Finance 34, 1041–1047.

Beirlant, J., Teugels, J.L., 1989. Asymptotic normality of Hill's estimator. In: Extreme Value Theory, Oberwolfach, 1987, Lecture Notes in Statistics, Vol. 51. Springer-Verlag, New York, pp. 148–155.

Belkacem, L., Véhel, J., Walter, C., 2000. CAPM, risk and portfolio selection in α-stable markets. Fractals 8 (1), 99–115.

Bergström, H., 1952. On some expansions of stable distributions. Arkiv för Mathematik II 18, 375–378.

Brorsen, B.W., Preckel, P.V., 1993. Linear regression with stably distributed residuals. Communications in Statistics. Theory and Methods 22, 659–667.

Brorsen, W.B., Yang, S.R., 1990. Maximum likelihood estimates of symmetric stable distribution parameters. Communications in Statistics. Simulation and Computation 19 (4), 1459–1464.

Buckle, D.J., 1995. Bayesian inference for stable distributions. Journal of the American Statistical Association 90, 605–613.

Byczkowski, T., Nolan, J.P., Rajput, B., 1993. Approximation of multidimensional stable densities. Journal of Multivariate Analysis 46 (1), 13–31.

Chamberlain, T.W., Cheung, C.S., Kwan, C.C., 1990. Optimal portfolio selection using the general multi-index model: A stable-Paretian framework. Decision Sciences 21, 563–571.

Cheng, B., Rachev, S.T., 1995. Multivariate stable future prices. Mathematical Finance 5 (2), 133–153.

Csörgő, S., Deheuvels, P., Mason, D.M., 1985. Kernel estimates of the tail index of a distribution. The Annals of Statistics 13, 1050–1077.

Csörgő, S., Mason, D.M., 1985. Central limit theorems for sums of extreme values. Mathematical Proceedings of the Cambridge Philosophical Society 98, 547–558.

Csörgő, S., Viharos, L., 1995. On the asymptotic normality of Hill's estimator. Mathematical Proceedings of the Cambridge Philosophical Society 118, 375–382.

Csörgő, S., Viharos, L., 1997. Asymptotic normality of least-squares estimators of tail indices. Bernoulli 3 (3), 351–370.

Danielsson, J., Jansen, D.W., de Vries, C.G., 1996. The method of moments ratio estimator for the tail shape parameter. Communications in Statistics. Theory and Methods 25 (4), 711–720.

Davydov, Yu., Paulauskas, V., 1999. On the estimation of the parameters of multivariate stable distributions. Acta Applicandae Mathematicae 58, 107–124.

Davydov, Yu., Paulauskas, V., Rackauskas, A., 2000. More on p-stable convex sets in Banach spaces. Journal of Theoretical Probability 13 (1), 39–64.

de Haan, L., Resnick, S., 1998. On asymptotic normality of the Hill estimator. Communications in Statistics. Stochastic Models 14 (4), 849–866.

Deheuvels, P., Haeusler, E., Mason, D.M., 1988. Almost sure convergence of the Hill's estimator. Mathematical Proceedings of the Cambridge Philosophical Society 104, 371–381.

Dekkers, A., de Haan, L., 1993. Optimal choice of sample fraction in extreme-value estimation. Journal of Multivariate Analysis 47, 173–195.

Dekkers, A., Einmahl, J., de Haan, L., 1989. A moment estimator for the index of an extreme-value distribution. The Annals of Statistics 17 (4), 1833–1855.

Drees, H., 1996. Refined Pickands estimators with bias correction. Communications in Statistics. Theory and Methods 25 (4), 837–851.

Drees, H., de Haan, L., Resnick, S.I., 2000. How to make a Hill plot. The Annals of Statistics 28 (1), 254–274.

Dostoglou, S.A., Rachev, S.T., 1999. Stable distributions and the term structure of interest rates. Mathematical and Computer Modelling 29, 57–60.

DuMouchel, W., 1971. Stable distributions in statistical inference. Ph.D. Thesis. University of Michigan, Ann Arbor, MI.

DuMouchel, W., 1973. On the asymptotic normality of the maximum-likelihood estimate when sampling from a stable distribution. The Annals of Statistics 1, 948–957.

DuMouchel, W., 1975. Stable distributions in statistical inference 2: Information from stably distributed samples. Journal of the American Statistical Association 70, 386–393.

DuMouchel, W., 1983. Estimating the stable index α in order to measure tail thickness: A critique. The Annals of Statistics 11, 1019–1031.

Embrechts, P., Klüppelberg, C., Mikosh, T., 1997. Modeling Extremal Events for Insurance and Finance. Springer, Berlin.

Esary, J., Proschan, F., Walkup, D., 1967. Association of random variables with applications. The Annals of Mathematical Statistics 38, 1466–1474.

Fama, E.F., 1965a. The behavior of stock market prices. Journal of Business 38, 34–105.

Fama, E.F., 1965b. Portfolio analysis in a stable Paretian market. Management Science 11, 404–419.

Fama, E., Roll, R., 1971. Parameter estimates for symmetric stable distributions. Journal of the American Statistical Association 66, 331–338.

Feuerverger, A., McDunnough, P., 1977. The empirical characteristic function and its applications. The Annals of Statistics 5, 88–97.

Feuerverger, A., McDunnough, P., 1981a. On the efficiency of empirical characteristic function procedures. Journal of the Royal Statistical Society. Series B 43, 20–27.

Feuerverger, A., McDunnough, P., 1981b. On efficient inference in symmetric stable laws and processes. In: Csörgő, M., Dawson, D.A., Rao, N.J.K., Saleh, A.K. (Eds.), Statistics and Related Topics. North-Holland, Amsterdam, pp. 109–122.

Gamba, A., 1999. Portfolio analysis with symmetric stable Paretian returns. In: Canestrelli, E. (Ed.), Current Topics in Quantitative Finance. Springer-Verlag, Heidelberg, pp. 48–69.

Gamrowski, B., Rachev, S.T., 1996. Testing the validity of value-at-risk measures. In: Heyde, C., et al. (Eds.), Applied Probability. Springer-Verlag, Berlin, pp. 307–320.

Goldie, C.M., Smith, R.L., 1987. Slow variation with remainder: Theory and applications. The Quarterly Journal of Mathematics. Oxford. Second Series 38, 45–71.

Haeusler, E., Teugels, J.L., 1985. On the asymptotic normality of Hill's estimator for the exponent of regular variation. The Annals of Statistics 13, 743–756.

Hall, P., 1982. On some simple estimates of an exponent of regular variation. Journal of the Royal Statistical Society. Series B 44, 37–42.

Hall, P., Welsh, A.H., 1984. Best attainable rates of convergence for estimates of regular variation. The Annals of Statistics 12, 1079–1083.

Hall, P., Welsh, A.H., 1985. Adaptive estimates of parameters of regular variation. The Annals of Statistics 13, 331–341.

Heathcote, C.R., Rachev, S.T., Cheng, B., 1995. Testing multivariate symmetry. Journal of Multivariate Analysis 54 (1), 91–112.

Hill, B., 1975. A simple general approach to inference about the tail of a distribution. The Annals of Statistics 3, 1163–1173.

Höpfner, R., Rüschendorf, L., 1999. Comparison of estimators in stable models. Mathematical and Computer Modelling 29, 145–160.

Hurst, S.H., Platen, E., Rachev, S.T., 1999. Option pricing for a logstable asset price model. Mathematical and Computer Modelling 29, 105–119.

Janicki, A., Popova, I., Ritchken, R., Woyczynski, W., 1997. Option pricing bounds in an α-stable security market. Communications in Statistics. Stochastic Models 13, 817–839.

Janicki, A., Weron, A., 1994. Simulation and Chaotic Behavior of α-Stable Stochastic Processes. Marcel Dekker, New York.

Jansen, D.W., de Vries, C.G., 1991. On the frequency of large stock returns: Putting booms and busts into perspective. Review of Economic Statistics 73, 18–24.

Jurek, Z., 1984. Polar coordinates in Banach spaces. Bulletin of the Polish Academy of Sciences. Mathematics 32, 61–66.

Jurek, Z., Mason, J.D., 1993. Operator-Limit Distributions in Probability Theory. Wiley, New York.

Karandikar, R., Rachev, S.T., 1995. A generalized binomial model and option formulae for subordinated stock-price processes. Probability and Mathematical Statistics 15, 427–447.

Klebanov, L.B., Melamed, J.A., Rachev, S.T., 1994. On the joint estimation of stable law parameters. In: Anastassiou, G., Rachev, S.T. (Eds.), Approximation, Probability and Related Fields. Plenum Press, New York, pp. 315–320.

Klebanov, L.B., Rachev, S.T., 1996. Sums of random number of random variables and their approximations with ν-accompanying infinitely divisible laws. Serdica Mathematical Journal 22, 471–496.

Koedijk, K.G., Schafgans, M.M., de Vries, C.G., 1990. The tail index of exchange rate returns. Journal of International Economics 29, 93–116.

Kogon, S.M., Williams, D.B., 1998. Characteristic function based estimation of stable distribution parameters. In: Adler, R.J., Feldman, R.E., Taqqu, M. (Eds.), A Practical Guide to Heavy Tails: Statistical Techniques and Applications. Birkhäuser, Boston, pp. 311–335.

Koutrouvelis, I.A., 1980. Regression-type estimation of the parameters of stable laws. Journal of the American Statistical Association 75, 918–928.

Koutrouvelis, I.A., 1981. An iterative procedure for the estimation of the parameters of the stable law. Communications in Statistics. Simulation and Computation 10, 17–28.

Kozubowski, T.J., 1993. Estimation of the parameters of geometric stable laws. Technical Report No. 253. Department of Statistics and Applied Probability, University of California, Santa Barbara; appeared in: Mathematical and Computer Modelling 29 (10–12), 241–253, 1999.

Kozubowski, T.J., 2001. Fractional moment estimation of Linnik and Mittag-Leffler parameters. Mathematical and Computer Modelling 34, 1023–1035.

Kozubowski, T.J., Panorska, A.K., 1996. On moments and tail behavior of ν-stable random variables. Statistics Probability Letters 29, 307–315.

Kozubowski, T.J., Panorska, A.K., 1998. Weak limits for multivariate random sums. Journal of Multivariate Analysis 67, 398–413.

Kozubowski, T.J., Panorska, A.K., 1999a. Multivariate geometric stable distributions in financial applications. Mathematical and Computer Modelling 29, 83–92.

Kozubowski, T.J., Panorska, A.K., 1999b. Simulation of geometric stable and other limiting multivariate distributions arising in random summation scheme. Mathematical and Computer Modelling 29, 255–262.

Kozubowski, T.J., Podgórski, K., 2000. A multivariate and asymmetric generalization of Laplace distribution. Computational Statistics 15, 531–540.

Kozubowski, T.J., Podgórski, K., 2001. Asymmetric Laplace laws and modeling financial data. Mathematical and Computer Modelling 34, 1003–1021.

Kozubowski, T.J., Rachev, S.T., 1994. The theory of geometric stable distributions and its use in modeling financial data. European Journal of Operational Research 74, 310–324.

Kozubowski, T.J., Rachev, S.T., 1999. Multivariate geometric stable laws. Journal of Computational Analysis and Applications 1 (4), 349–385.

Kratz, M., Resnick, S.I., 1996. The QQ-estimator for heavy tails, Communications Statistics. Stochastic and Models 12 (4), 699–724.

Lee, M.-L.T., Rachev, S.T., Samorodnitsky, G., 1990a. Association of Stable Random Variables. The Annals of Probability 18, 1759–1764.

Lee, M.-L.T., Rachev, S.T., Samorodnitsky, G., 1990b. Dependence of stable random variables. In: Stochastic Inequalities. IMS Lecture Notes–Monograph Series, Vol. 22, pp. 219–234.

Liu, S.M., Brorsen, B.W., 1995. Maximum likelihood estimation of a GARCH-stable model. Journal of Applied Econometrics 10 (3), 273–285.

Loretan, M., Phillips, P.C.B., 1994. Testing the covariance stationarity of heavy-tailed time series. Journal of Empirical Finance 1, 211–248.

Mandelbrot, B.B., 1963a. New methods of statistical economics. Journal of Political Economics 71, 421–440.

Mandelbrot, B.B., 1963b. The variation of certain speculative prices. Journal of Business 36, 394–419.

Mandelbrot, B., 1967. The variation of some other speculative prices. Journal of Business 40, 393–413.

Mason, D.M., 1982. Laws of large numbers for sums of extreme values. The Annals of Probability 10, 754–764.

McCulloch, J.H., 1979. Linear regression with symmetric stable residuals. Working Paper 63. Economics Department, Ohio State University.

McCulloch, J.H., 1986. Simple consistent estimators of stable distribution parameters. Communications of Statistics. Simulation and Computation 15, 1109–1136.

McCulloch, J.H., 1994. Estimation of bivariate stable spectral densities. Unpublished Manuscript. Department of Economics, Ohio State University.

McCulloch, J.H., 1996. Financial applications of stable distributions. In: Maddala, G.S., Rao, C.R. (Eds.), Statistical Methods in Finance, Handbook in Statistics, Vol. 14. Elsevier, Amsterdam, pp. 393–425.

McCulloch, J.H., 1998. Linear regression with stable disturbances. In: Adler, R.J., Feldman, R.E., Taqqu, M. (Eds.), A Practical Guide to Heavy Tails: Statistical Techniques and Applications. Birkhäuser, Boston, pp. 359–376.

Meerschaert, M.M., Scheffler, H.P., 1998. A simple robust estimation method for the thickness of heavy tails. Journal of Statistical Planning and Inference 71, 19–34.

Meerschaert, M.M., Scheffler, H.P., 1999. Moment estimator for random vectors with heavy tails. Journal of Multivariate Analysis 71, 145–159.

Meerschaert, M.M., Scheffler, H.P., 2001. Limit Theorems for Sums of Independent Random Vectors. Wiley, New York.

Mittnik, S., Paolella, M.S., 1999. A simple estimator for the characteristic exponent of the stable Paretian distribution. Mathematical and Computer Modelling 29, 161–176.

Mittnik, S., Rachev, S.T., 1991. Alternative multivariate stable distributions and their applications to financial modelling. In: Cambanis, S., et al. (Eds.), Stable Processes and Related Topics. Birkhäuser, Boston, pp. 107–119.

Mittnik, S., Rachev, S.T., 1993a. Modeling asset returns with alternative stable distributions. Econometric Reviews 12 (3), 261–330.

Mittnik, S., Rachev, S.T., 1993b. Reply to comments on Modeling asset returns with alternative stable distributions, and some extensions. Econometric Reviews 12, 347–389.

Mittnik, S., Rachev, S.T., 1996. Tail estimation of the stable index α. Applied Mathematics Letters 9 (3), 53–56.

Mittnik, S., Rachev, S.T., Doganoglu, T., Chenyao, D., 1999. Maximum likelihood estimation of stable Paretian models. Mathematical and Computer Modelling 29, 275–293.

Mittnik, S., Rachev, S.T., Paolella, M.S., 1998. Stable Paretian modeling in finance: Some empirical and theoretical aspects. In: Adler, R.J., Feldman, R.E., Taqqu, M. (Eds.), A Practical Guide to Heavy Tails: Statistical Techniques and Applications. Birkhäuser, Boston, pp. 79–110.

Mittnik, S., Rachev, S.T., Rüschendorf, L., 1999. Test of association between multivariate stable vectors. Mathematical and Computer Modelling 29, 181–195.

Modarres, R., Nolan, J.P., 1994. A method for simulating stable random vectors. Computers & Statistics 9, 11–19.

Nikias, C.L., Shao, M., 1995. Signal Processing with α-Stable Distributions and Applications. Wiley, New York.

Nolan, J.P., 1998. Multivariate stable distributions: Approximation, estimation, simulation and identification, In: Adler, R.J., Feldman, R.E., Taqqu, M. (Eds.), A Practical Guide to Heavy Tails: Statistical Techniques and Applications. Birkhäuser, Boston, pp. 509–525.

Nolan, J.P., 2001. Maximum likelihood estimation and diagnostics for stable distributions. In: Barndorff-Nielsen, O., Mikosch, T., Resnick, S. (Eds.), Lévy Processes. Birkhäuser, Boston, pp. 379–400.

Nolan, J.P., Panorska, A.K., 1997. Data analysis for heavy tailed multivariate samples. Communications in Statistics. Stochastic Models 13, 687–702.

Nolan, J.P., Panorska, A.K., McCulloch, J.H., 2001. Estimation of stable spectral measures. Mathematical and Computer Modelling 34, 1113–1122.

Nolan, J.P., Rajput, B., 1995. Calculation of multidimensional stable densities. Communications in Statistics. Simulation and Computation 24, 551–556.

Paulson, A.S., Delehanty, T.A., 1984. Some properties of modified integrated squared error estimators for the stable laws. Communications in Statistics. Simulation and Computation 13, 337–365.

Paulson, A.S., Delehanty, T.A., 1985. Modified weighted squared error estimation procedures with special emphasis on the stable laws. Communications in Statistics. Simulation and Computation 14, 927–972.

Paulson, A.S., Holcomb, E.W., Leitch, R., 1975. The estimation of the parameters of the stable law. Biometrika 62, 163–170.

Phillips, P.C.B., 1993. Comments on "Modeling asset returns with alternative stable distributions, and some extensions." Econometric Reviews 12, 331–338.

Pickands, J., 1975. Statistical inference using extreme-order statistics. The Annals of Statistics 3, 1–13.

Pitt, L., 1982. Positively correlated normal variables are associated. The Annals of Probability 10 (2), 496–499.

Press, S.J., 1972a. Multivariate stable distributions. Journal of Multivariate Analysis 2, 444–462.

Press, S.J., 1972b. Estimation of univariate and multivariate stable distributions. Journal of the American Statistical Association 67, 842–846.

Press, S.J., 1982. Applied Multivariate Analysis, 2nd edition, Krieger, Malabar.

Qiou, Z., Ravishanker, N., 1995. Bayesian inference for stable law parameters. Technical Report. University of Connecticut.

Rachev, S.T., Han, S., 2000. Portfolio management with stable distributions. Mathematical Methods of Operations Research 51 (2), 341–352.

Rachev, S., Mittnik, S., 2000. Stable Paretian Models in Finance. Wiley, Chichester.

Rachev, S.T., Rüschendorf, L., 1994. On the Cox–Ross and Rubinstein model for option pricing. Theory of Probability and its Applications 39, 150–190.

Rachev, S.T., Samorodnitsky, G., 1993. Option pricing formula for speculative prices modelled by subordinated stochastic processes. Pliska 19, 175–190.

Rachev, S.T., Xin, H., 1993. Test for association of random variables in the domain of attraction of multivariate stable law. Probability and Mathematical Statistics 14 (1), 125–141.

Ravishanker, N., Qiou, Z., 1998. Bayesian inference for time series with infinite variance stable innovations. In: Adler, R.J., Feldman, R.E., Taqqu, M. (Eds.), A Practical Guide to Heavy Tails: Statistical Techniques and Applications. Birkhäuser, Boston, pp. 259–280.

Resnick, S.I., 1997. Heavy tail modeling and teletraffic data. The Annals of Statistics 25 (5), 1805–1869.

Resnick, S.I., 1998. Why non-linearities can ruin heavy-tailed modeler's day. In: Adler, R.J., Feldman, R.E., Taqqu, M. (Eds.), A Practical Guide to Heavy Tails: Statistical Techniques and Applications. Birkhäuser, Boston, pp. 219–239.

Resnick, S., Greenwood, P., 1979. A bivariate stable characterization and domains of attraction. Journal of Multivariate Analysis 9, 206–221.

Resnick, S.I., Stărică, 1997. Smoothing the Hill estimator. Advances in Applied Probabability 29, 271–293.

Rosen, O., Weissman, I., 1996. Comparison of estimation methods in extreme value theory. Communications in Statistics. Theory and Methods 25 (4), 759–773.

Rosiński, J., 1976. Weak compactness of laws of random sums of identically distributed random vectors in Banach spaces. Colloquium Mathematicum 35, 313–325.

Rvačeva, E., 1962. On domains of attraction of multidimensional distributions. In: Select. Transl. Math. Stat. Prob., Vol. 2. American Mathematical Society, Providence, RI, pp. 183–205.

Samorodnitsky, G., Taqqu, M., 1994. Stable Non-Gaussian Random Processes. Chapman & Hall, New York.

Scheffler, H.P., 1999. On the estimation of the spectral measure of certain nonnormal operator stable laws. Statistics & Probabability Letters 43 (4), 385–392.

Schultze, J., Steinebach, J., 1996. On least squares estimates of an exponential tail coefficient. Statistics & Decisions 14, 353–372.

Sharpe, M., 1969. Operator-stable probability distributions on vector groups. Transactions of the American Mathematical Society 136, 51–65.

Stuck, B.W., 1976. Distinguishing stable probability measures. Part I: Discrete time. Bell System Technical Journal 55, 1125–1182.

Uchaikin, V.V., Zolotarev, V.M., 1999. Chance and Stability: Stable Distributions and their Applications. VSP, Utrecht.

Weron, R., 1996. On the Chambers–Mallows–Stuck method for simulating skewed stable random variables. Statistics & Probability Letters 28, 165–171.

Ziemba, W.T., 1974. Choosing investment portfolios when the returns have stable distributions. In: Hammer, P.L., Zoulendijk, G. (Eds.), Mathematical Programming in Theory and Practice. North Holland, Amsterdam, pp. 443–482.

Chapter 5

JUMP-DIFFUSION MODELS

WOLFGANG J. RUNGGALDIER
Dipartimento di Matematica Pura ed Applicata, Università di Padova, 7 Via Belzoni, I-35131, Padova, Italy
web: http://www.math.unipd.it/runggaldier/index.html

Contents

Handbook of Heavy Tailed Distributions in Finance, Edited by S.T. Rachev

Abstract

We discuss jump-diffusion type models for financial market as well as methods for pricing and hedging of contingent claims in such markets. We consider both, asset price and term structure models, and deal also with situations when there is a stochastic volatility correlated with the jumps and when one has very small time scales, i.e., high frequency data. To make the presentation possibly self-contained, in a preliminary section we recall some basic notions from stochastic analysis for jump-diffusions.

Keywords

jump-diffusions, Poisson point processes, marked point processes, Cox processes, hidden processes, martingale measures, market price of risk, pricing and hedging in incomplete markets, market completion, risk minimization, stochastic volatility, high frequency data, computing expectations of functionals of jump-diffusions

1. Introduction

Most of the standard literature in Finance, in particular for pricing and hedging of contingent claims, is based on the assumption that the prices of the underlying assets follow a diffusion-type process, in particular a geometric Brownian motion (GBM).

Documentation from various empirical studies shows that such models are inadequate, both in relation to their descriptive power, as well as for the mispricing that they might induce. The contributions to the present volume deal with various generalizations of the basic GBM; here we concentrate on the fact that returns of various asset prices and interest rates may exhibit a jumping behaviour. We thus study possible superpositions of jump and diffusion processes, namely what is called *jump-diffusion processes*. Jump-diffusions form a particular class of Levy processes. Our purpose here is not to study the general case of a Levy driving process, but rather to concentrate on the specific aspects of the subclass of jump-diffusions. Jump-diffusion models have also some intuitive appeal in that they let prices and interest rates change continuously most of the time, but they also take into account the fact that from time to time larger jumps may occur that cannot be adequately modeled by pure diffusion-type processes.

Among the earlier empirical studies, documenting a jumping behaviour in prices and interest rates, one may quote Ball and Torous (1985), Jorion (1988). There are also studies, such as Babbs and Webber (1997), putting forward specific sources of jumps in interest rates like moves by central banks. On the other hand, a first approach developing further the basic Black and Scholes (BS) model with the inclusion of jumps appears to be that of Merton (1976). Since the introduction of jumps in the BS model implies that derivative prices are no longer determined by the principle of absence of arbitrage alone, Merton solved the pricing problem by assuming that the jump risk was not systematic. This was later criticized by showing that such an assumption is equivalent to the existence of a market portfolio, that contains the underlying asset and that does not present a jumping behaviour [for a discussion on this point see, e.g., Björk and Näslund (1998)]. Further studies then appeared showing that jumps in stock returns are indeed systematic. Another early approach is that in Cox and Ross (1976), where the market remains however complete since the authors consider just a simple jump-type process with fixed jump amplitude and thus with a single source of randomness. One of the major purposes of this chapter is now to try to give an overview of the state of the art of jump-diffusion modeling in stock and bond markets as well as of the corresponding approaches for pricing and hedging.

It was further documented in empirical studies [see, e.g., Bakshi, Cao and Chen (1997)] that a combination of jumps and stochastic volatility leads to even better fits and allows to avoid implied volatility skews. Stochastic volatility models are treated elsewhere and so in this chapter we limit ourselves to stochastic volatility in conjunction with jump-diffusion modeling, also because empirical documentation gives evidence for a jump-type behaviour in the volatility and of a correlation between jumps in volatility and jumps in prices. In fact, as mentioned, e.g., in Naik (1993), it is natural to expect that, if the volatility jumps, also the price should jump. A further purpose of the present chapter is then to discuss issues related to such more general jump-diffusion-stochastic-volatility modeling.

On very small time scales actual prices do not really change continuously over time, but rather at discrete random points in time in reaction to trade and/or significant new information. We shall show that also such situations can be captured by models featuring a combination of diffusion and jump processes, that is however different from the canonical jump-diffusion processes.

The outline of the chapter is as follows. In Section 2 we recall some preliminary notions from stochastic analysis for jump-diffusion processes, such as a martingale representation result and generalized versions of the Ito formula as well as of the Girsanov measure transformation. We limit ourselves to those notions that will be used in the sequel. In Section 3 we then describe various market models based on jump-diffusion representations. More precisely, in line with the introductory remarks above, we shall consider first canonical jump-diffusion models for stock and bond markets, then consider jump-diffusions correlated with stochastic volatility and, finally, combinations of diffusions and jumps to describe high frequency data. In Section 4 we discuss existence and uniqueness of martingale measures in a jump-diffusion setting, exhibiting also the market price of (jump-diffusion) risk. Some emphasis is given to the notion of *completion of the market* as a tool to obtain a unique martingale measure. In this context it is also pointed out that uniqueness of a martingale measure does not necessarily always imply completeness of the market in the sense of hedging, namely that every claim can be duplicated by a self financing portfolio. In Section 5 we then concentrate on hedging in jump-diffusion market models having two goals in mind: first, investigating whether and when a market, that has been completed to yield a unique martingale measure, is also complete in the sense of hedging. Second, to study the hedging problem when the market cannot be completed or market completion is inappropriate. In such cases there is always some residual risk and so one may want to choose a strategy such as to minimize a criterion related to this risk. Finally, Section 6 is devoted to the problem of pricing in jump-diffusion market models. With jumps and/or stochastic volatility, the market is incomplete. The principle of absence of arbitrage alone is then insufficient to define uniquely a price and so the preference structure of investors has to come into play to determine a *pricing measure*. From the point of view of pure pricing, the problem reduces formally to that of determining a specific martingale measure. In Section 6.1 we mention various approaches to this effect, related to the literature, in particular approaches based on market completion and on the relationship between the choice of a hedging criterion and that of a martingale measure. Once a martingale measure has been chosen, there remains the problem of the actual computation of the expectation of the discounted claim and this is dealt with in Section 6.2.

Unavoidably, this overview of the state of the art may turn out to be incomplete and reflects the specific interests and competences of the author. Among the topics that are not discussed here, we just mention the American-type options in a jump-diffusion setting [for this see, e.g., Mulinacci (1996), Pham (1997)] and Portfolio Optimization [see, e.g., Framstad, Oeksendal and Sulem (2001) and references therein]. The same has to be said about the references to the literature: while we have tried to take into account a good deal of recent papers on the subject, we have only quoted a small selection of previous papers in order to keep the list within a reasonable size. Still, we hope to have succeeded in giving a sufficiently comprehensive account on models and methods related to jump-diffusions in financial markets.

2. Preliminaries

In this section we recall basic definitions and results needed for the study of jump-diffusion models, limiting ourselves to multivariate (univariate) and marked point processes and assuming that the reader is familiar with the corresponding notions concerning diffusion processes. In addition to the basic definitions we recall here a martingale representation result and discuss the Ito formula and Girsanov's measure transformation, generalized to jump-diffusion processes. The main reference for this section is Brémaud (1981), from which most of the contents of the section are taken.

2.1. *Univariate point processes (Poisson jump processes)*

A point process is intended to describe events that occur randomly over time. It can be represented as a sequence of nonnegative random variables

$$0 = T_0 < T_1 < T_2 < \cdots,$$

where the generic T_n is the n-th instant of occurrence of an event. One makes the usual assumption of nonexplosion, according to which

$$T_\infty = \lim \uparrow T_n = +\infty.$$

The process may equivalently be represented via its associated counting process N_t where

$$N_t = n \quad \text{if } t \in [T_n, T_{n+1}),\ n \geqslant 0, \quad \text{or, equivalently,} \quad N_t = \sum_{n \geqslant 1} 1_{\{T_n \leqslant t\}}. \tag{1}$$

It counts the number of events up to and including time t. The nonexplosion condition becomes $N_t < \infty$ for $t \geqslant 0$. Both, T_n and N_t, are defined on some probability space $(\Omega, \mathcal{F}, P)$ with a filtration $\mathcal{F}_t$ to which N_t is adapted.

A point process N_t is called a Poisson point process if

(i) $N_0 = 0$;

(ii) N_t is a process with independent increments;

(iii) $N_t - N_s$ is a Poisson random variable with a given parameter $\Lambda_{s,t}$.

Usually one assumes $\Lambda_{s,t} = \int_s^t \lambda_u \, du$ for a deterministic function λ_t; the latter is called the *intensity* of the Poisson point process N_t. If $\mathcal{F}_t$ is the filtration $\mathcal{F}_t^N$, generated by N_t, and $\lambda_t \equiv 1$, then N_t is called a *standard Poisson process*. It is also easily seen that, if N_t is a Poisson process with intensity $\lambda_t \equiv \lambda$, then $T_{n+1} - T_n$ are i.i.d., exponential random variables with parameter λ. A natural interpretation of the intensity and of this latter property comes from relating the above setup with the usual Poisson model, that is based on the following assumptions:

(a) the probability of one change/jump in an interval of length Δ is $\lambda \cdot \Delta + o(\Delta)$;
(b) the probability of two or more changes/jumps in an interval of length Δ is $o(\Delta)$;
(c) the number of changes/jumps in nonoverlapping intervals are stochastically independent.

In this setup one can in fact consider the two "dually" related random variables:

(1) A discrete random variable X describing the number of changes/jumps in a time interval of given length T and having as distribution the usual Poisson distribution, i.e.,

$$P\{X = k\} = \frac{(\lambda T)^k}{k!} e^{-\lambda T}, \quad k \in \mathbb{N}.$$

(2) A continuous random variable T describing the time that is needed to obtain k successive changes/jumps and for which the distribution is of the Gamma-type with density

$$f_T(t) = \frac{\lambda^k}{\Gamma(k)} t^{k-1} e^{-\lambda t}, \quad T > 0.$$

The parameter λ is the same in both cases and corresponds to the λ in assumption (a) above.

It will be convenient to consider also the case when the intensity of a Poisson process is itself an adapted process being driven by some background process. This can be explained by a two-step randomization procedure: first one draws at random a trajectory of the background process, say Z_t; then one generates a Poisson process with intensity $\lambda_t = \lambda(t, Z_t)$, where the dependence also on t allows to incorporate seasonality effects. We now have a Poisson process N_t conditionally on Z_t and it is called a *doubly stochastic Poisson process*, or a *Cox process* [see Cox (1955)]. Formally, we require that the random intensity λ_t is $\mathcal{F}_0$-measurable, i.e., $\mathcal{F}_\infty^Z \subset \mathcal{F}_0$. For additional details of the intensity of a Poisson process we refer to Brémaud (1981).

Notice that the above characterizations (i)–(iii) of a Poisson process parallel those of a Wiener process: both are processes with independent increments; the increments of a Wiener process are normally distributed, while those of a Poisson process are Poisson distributed. The Wiener process is the basic building block for processes with continuous trajectories, the Poisson process is a basic building block for processes with jumping trajectories. On the other hand, while the Wiener process is itself a martingale, a Poisson process as such is not. It becomes a martingale if one subtracts from N_t the process given by its mean. Indeed,

$$M_t := N_t - \int_0^t \lambda_s \, ds \tag{2}$$

is an $\mathcal{F}_t$-martingale by the $\mathcal{F}_0$-measurability of λ_t, assuming in addition that $E\{\int_0^t \lambda_u \, du\} < \infty$. By (iii) one then has in fact

$$E\{N_t - N_s | \mathcal{F}_s\} = E\left\{\int_s^t \lambda_u \, du | \mathcal{F}_s\right\} \tag{3}$$

which implies that $E\{N_t\} < \infty$ and that M_t in (2) is an $\mathcal{F}_t$-martingale. Equality (3) admits a generalization in the form

$$E\left\{\int_0^\infty C_s \, dN_s\right\} = E\left\{\int_0^\infty C_s \lambda_s \, ds\right\} \tag{4}$$

that has to be valid for all nonnegative, $\mathcal{F}_t$-predictable processes C_t and as such characterizes a doubly stochastic Poisson process with intensity λ_t [see Brémaud (1981)].

2.2. *Multivariate and marked point processes*

Let T_n be a (univariate) point process and Y_n, $n \geqslant 1$, a sequence of random variables with values in $\{1, 2, \ldots, K\}$, all defined on the same $(\Omega, \mathcal{F}, P)$. For each $k = 1, \ldots, K$ we may then consider the counting process

$$N_t(k) := \sum_{n \geqslant 1} 1_{\{T_n \leqslant t\}} 1_{\{Y_n = k\}}.$$

Each $N_t(k)$ is a univariate point process and the various $N_t(k)$'s have no common jumps, i.e., $\Delta N_t(k) \Delta N_t(h) = 0$, $\forall t \geqslant 0$ and all $k \neq h$. Analogously to the case of univariate point processes, here too we have now two equivalent representations, either as the double sequence $(T_n, Y_n)_{n \geqslant 1}$ or as the K-vector process $N_t = (N_t(1), \ldots, N_t(K))$ and this process is called a multivariate, more precisely a K-variate point process. As in the univariate case, here too we shall mainly use the representation as the K-vector process N_t and we have formula (2) with M_t a K-vector and λ_t the K-vector intensity process whose components are the individual intensities of the components $N_t(k)$ of N_t.

Considering the representation (T_n, Y_n), we may interpret T_n as the n-th occurrence of some phenomenon and Y_n as an attribute or *mark* of this phenomenon. We may then speak of (T_n, Y_n) as a *marked point process*, or space-time point process and extend its definition to allow Y_n to take values in a general measurable *mark space* $(E, \mathcal{E})$. We synthesize the foregoing in the following

Definition 2.1. An E-marked point process is a double sequence $(T_n, Y_n)_{n \geqslant 1}$ where
(i) T_n is a (univariate) point process;
(ii) Y_n is a sequence of E-valued random variables.

Obviously, the univariate and multivariate point processes are special cases of a marked point process.

Generalizing the representation of a multivariate point process in the form of the K-vector process N_t, we associate to each $A \in \mathcal{E}$ the counting process

$$N_t(A) := \sum_{n \geqslant 1} 1_{\{T_n \leqslant t\}} 1_{\{Y_n \in A\}}$$

and let simply $N_t := N_t(E)$. Considering the filtration

$$\mathcal{F}_t^N := \sigma\{N_s(A);\ s \leqslant t,\ A \in \mathcal{E}\}$$

define the associated (random) counting measure

$$p\big((0,t], A\big) = N_t(A), \quad t \geqslant 0,\ A \in \mathcal{E}, \tag{5}$$

which is σ-finite under the assumption of nonexplosion of T_n. This measure allows to obtain more concise expressions via integrals of the form

$$\int_0^t \int_E H(s,y) p(\mathrm{d}s, \mathrm{d}y) = \sum_{n \geqslant 1} H(T_n, Y_n) 1_{\{T_n \leqslant t\}} = \sum_{n=1}^{N_t} H(T_n, Y_n). \tag{6}$$

Again, we may represent an E-marked point process equivalently as the double sequence (T_n, Y_n) or as the counting measure $p(\mathrm{d}s, \mathrm{d}y)$.

To introduce now the intensity process in this more general setup, assume that for each $A \in \mathcal{E}$, the point process $N_t(A)$ admits the intensity $\lambda_t(A)$. This then leads to a measure-valued intensity $\lambda_t(\mathrm{d}y)$ so that, generalizing (4), one has

$$E\left\{\int_0^\infty \int_E H(s,y) p(\mathrm{d}s, \mathrm{d}y)\right\} = E\left\{\int_0^\infty \int_E H(s,y) \lambda_s(\mathrm{d}y)\, \mathrm{d}s\right\} \tag{7}$$

that has to be valid for all nonnegative $\mathcal{F}_t$-predictable E-marked processes H (given a filtration $\mathcal{F}_t$ on Ω, $\mathcal{F}_t$*-predictability* here means measurability with respect to $\mathcal{P}(\mathcal{F}_t) \otimes \mathcal{E}$ where $\mathcal{P}(\mathcal{F}_t)$ is the predictable σ-field on $(0,\infty) \times \Omega$). We have also the generalization of (2) in the form

$$q(\mathrm{d}s, \mathrm{d}y) = p(\mathrm{d}s, \mathrm{d}y) - \lambda_s(\mathrm{d}y)\, \mathrm{d}s, \tag{8}$$

where $q(\mathrm{d}s, \mathrm{d}y)$ is a (signed) measure-valued martingale in the sense that

$$\int_0^t \int_E H(s,y) q(\mathrm{d}s, \mathrm{d}y)$$

is a $(P, \mathcal{F}_t)$-martingale (local martingale) for each $\mathcal{F}_t$-predictable E-marked process H, satisfying appropriate integrability conditions. The most common form of intensity is

$$\lambda_t(\mathrm{d}y) = \lambda_t m_t(\mathrm{d}y), \tag{9}$$

where λ_t is nonnegative $\mathcal{F}_t$-predictable and represents the intensity of the Poisson process $N_t(E)$, while $m_t(\mathrm{d}y)$ is a probability measure on E (typically, the Y_n will be i.i.d., independent of $N_t(E)$). The pair $(\lambda_t, m_t(\mathrm{d}y))$ is called the $(P, \mathcal{F}_t)$-*local characteristics* of $p(\mathrm{d}s, \mathrm{d}y)$.

Notice finally that, as in the univariate case, we may let $\lambda_t(\mathrm{d}y)$ depend on some driving $\mathcal{F}_0$-measurable random process Z_t, leading to a doubly stochastic marked point process. If, in the representation (9), λ_t is a deterministic time function, the marked point process is called a *marked Poisson process*.

2.3. Martingale representation

Martingale representation results are widely used in Finance, especially when it comes to solving hedging problems. For pure "Wiener-martingales" we have in fact the well-known result that every square integrable martingale with respect to the filtration generated by a Wiener process is, up to an additive constant, a stochastic integral of the Ito type. We shall now recall a corresponding result for point-process martingales that we formulate in the most general case of a marked point process. We have in fact the following theorem [see Theorem VIII, T8 in Brémaud (1981)]

Theorem 2.2. *Let $(\Omega, \mathcal{F}, \mathcal{F}_t, P)$ be a probability space satisfying the "usual assumptions" where $\mathcal{F}_t = \mathcal{F}_0 \vee \mathcal{F}_t^p$ with $\mathcal{F}_t^p$ the filtration generated by a marked point process, represented by the counting measure $p(\mathrm{d}t, \mathrm{d}y)$. Then any $(P, \mathcal{F}_t)$- martingale M_t admits the representation*

$$M_t = M_0 + \int_0^t \int_E H(s, y) q(\mathrm{d}s, \mathrm{d}y) \tag{10}$$

with $q(\cdot)$ as in (8) and H an integrable (with respect to $\lambda_t(\mathrm{d}y)$) $\mathcal{F}_t$-predictable E-marked process. This representation is essentially unique.

In the case of a multivariate (and therefore also univariate) point process, the representation (10) becomes

$$M_t = M_0 + \sum_{k=1}^{K} \int_0^t H_s(k) \big(\mathrm{d}N_s(k) - \lambda_s(k)\,\mathrm{d}s\big), \tag{11}$$

where $[H_t(1), \ldots, H_t(K)]$ is $\mathcal{F}_t$-predictable with $H_t(k)$ integrable with respect to $\lambda_t(k)$.

This representation result can be generalized according to Jacod and Shiryaev (1987) to include martingales that are simultaneously "Wiener" and point-process martingales and that will have some relevance later on.

Theorem 2.3. *Given a Wiener process w_t and a marked point process $p(\mathrm{d}s, \mathrm{d}y)$, let*

$$\mathcal{F}_t := \sigma\big\{w_s, p\big((0, s], A\big), B;\ 0 \leqslant s \leqslant t,\ A \in \mathcal{E},\ B \in \mathcal{N}\big\}$$

with $\mathcal{N}$ the collection of P-null sets from $\mathcal{F}$. Then any $(P, \mathcal{F}_t)$-local martingale M_t has the representation

$$M_t = M_0 + \int_0^t \phi_s \,\mathrm{d}w_s + \int_0^t \int_E H(s, y)\big(p(\mathrm{d}s, \mathrm{d}y) - \lambda_s(\mathrm{d}y)\,\mathrm{d}s\big), \tag{12}$$

where ϕ_t is predictable and square integrable and H is an $\mathcal{F}_t$-predictable E-marked process, integrable with respect to $\lambda_t(\mathrm{d}y)$.

2.4. Exponential formula; generalized Ito formula

With the definition of a marked point process and of integrals in the form of (6), we may now consider processes of the general type

$$X_t = X_0 + \int_0^t \alpha_s \,\mathrm{d}s + \int_0^t \beta_s \,\mathrm{d}w_s + \int_0^t \int_E \gamma(s, y) p(\mathrm{d}s, \mathrm{d}y) \tag{13}$$

that are called *jump-diffusion processes* and where the coefficients satisfy the implicit integrability conditions, β_t is adapted and $\gamma(t, y)$ is predictable in the sense as defined previously. As usual, we may rewrite (13) in differential form and consider, more specifically, differential equations of the type

$$\mathrm{d}X_t = X_{t-}\left(\alpha_t \,\mathrm{d}t + \beta_t \,\mathrm{d}w_t + \int_E \gamma(t, y) p(\mathrm{d}t, \mathrm{d}y)\right), \tag{14}$$

where we write X_{t-} with $t-$ because of the predictability requirement in the last coefficient and where $\gamma(t, y) > -1$. Notice that [see (6)] the last term in (14) can also be written as

$$\int_E \gamma(t, y) p(\mathrm{d}t, \mathrm{d}y) = \gamma(t, Y_t)\,\mathrm{d}N_t, \tag{15}$$

where $N_t = N_t(E) = p((0, t], E)$ is the total number of jumps and Y_t denotes the piecewise constant, left-continuous time interpolation of the sequence Y_n. Notice also that, in the case of a multivariate (in particular univariate) point process, this last term in (14) takes the form

$$\int_E \gamma(t, y) p(\mathrm{d}s, \mathrm{d}y) = \sum_{k=1}^{K} \gamma_t(k)\,\mathrm{d}N_t(k). \tag{16}$$

We shall not discuss here in detail equations of the form (14), in particular the uniqueness of their solutions, but limit ourselves to show that a solution to (14) is given by the following

Exponential formula

$$X_t = X_0 \exp\left[\int_0^t \left(\alpha_s - \frac{1}{2}\beta_s^2\right) ds + \int_0^t \beta_s \, dw_s + \int_0^t \log\big(1+\gamma(s, Y_s)\big) \, dN_s\right]$$

$$= X_0 \exp\left[\int_0^t \left(\alpha_s - \frac{1}{2}\beta_s^2\right) ds + \int_0^t \beta_s \, dw_s\right] \prod_{n=1}^{N_t} \big(1+\gamma(T_n, Y_n)\big). \tag{17}$$

While the diffusion part in this expression follows from the usual Ito formula, the jump part follows from the so-called *exponential formula* of Stieltjes–Lebesgue Calculus [see Theorem T4 of Appendix A4 in Brémaud (1981)], but it can also be obtained from the generalized Ito formula as we shall show next. For this purpose let a process X_t satisfy the general Equation (13). Given a $C^{1,2}$-function $F(t, X)$, we have the *generalized Ito formula*

$$\begin{aligned} dF(t, X_t) &= F_t(\cdot) \, dt + F_X(\cdot)\alpha_t \, dt + \frac{1}{2} F_{XX}(\cdot)\beta_t^2 \, dt + F_X(\cdot)\beta_t \, dw_t \\ &\quad + \big[F\big(t, X_{t-} + \gamma(t, Y_t)\big) - F(t, X_{t-})\big] dN_t \end{aligned} \tag{18}$$

that, in the specific case of (14), becomes

$$\begin{aligned} dF(t, X_t) &= F_t(\cdot) \, dt + F_X(\cdot)X_t\alpha_t \, dt + \frac{1}{2} F_{XX}(\cdot)X_t^2\beta_t^2 \, dt + F_X(\cdot)X_t\beta_t \, dw_t \\ &\quad + \big[F\big(t, X_{t-}\big(1 + \gamma(t, Y_t)\big)\big) - F(t, X_{t-})\big] dN_t \end{aligned} \tag{19}$$

and where, again, $N_t = N_t(E) = p((0, t], E)$ and $(\cdot)$ stands for (t, X_t); the pedices in F denote partial derivatives. Notice that, if (19) is written in integral form, for the last term on the right we have the two equivalent representations

$$\int_0^t \big[F\big(s, X_{s-}\big(1+\gamma(s, Y_s)\big)\big) - F(s, X_{s-})\big] dN_s = \sum_{n=1}^{N_t} \big[F(T_n, X_{T_n}) - F(T_n, X_{T_n^-})\big],$$

where the right-hand side remains the same also in the more general case of (18).

We shall now use the generalized Ito formula (19) to obtain the solution (17) of Equation (14). Choosing $F(t, X) = \log X$, from (19) and (14) we have

$$dF = \alpha_t \, dt - \frac{1}{2}\beta_t^2 \, dt + \beta_t \, dw_t + \log\big(1+\gamma(t, Y_t)\big) \, dN_t$$

from which

$$\log X_t = \log X_0 + \int_0^t \left(\alpha_s - \frac{1}{2}\beta_s^2\right) ds + \int_0^t \beta_s \, dw_s + \int_0^t \log\big(1+\gamma(s, Y_{s-})\big) \, dN_s, \tag{20}$$

i.e., we obtain (17) by taking the exponential on both sides in (20).

2.5. Absolutely continuous transformation of measures

We recall that in the classical case of Wiener driven diffusion processes, the Girsanov-type measure transformation concerns a translation of the Wiener process that in turn induces a change in the drift of the diffusion equation. In view of its generalization below, we recall here the basic result of Girsanov's transformation, conveniently reformulated for a finite time horizon $t \in [0, T]$.

Theorem 2.4 (Girsanov's measure transformation). *Given a filtered probability space* $(\Omega, \mathcal{F}, \mathcal{F}_t, P)$ *with* $\mathcal{F} = \bigcup_t \mathcal{F}_t$, *let* $t \in [0, T]$ *with* T *given and* θ_t *be a square integrable predictable process. Define* $L = (L_t)$ *by*

$$\mathrm{d}L_t = L_t \theta_t \,\mathrm{d}w_t, \quad L_0 = 1, \tag{21}$$

and suppose that, for all t, $E^P\{L_t\} = 1$. *Then there exists a probability measure* Q *on* $\mathcal{F}$, *equivalent to* P, *with* $\mathrm{d}Q = L_T \,\mathrm{d}P$ *such that*

$$\mathrm{d}w_t = \theta_t \,\mathrm{d}t + \mathrm{d}w_t^Q, \tag{22}$$

where w_t^Q *is a* Q*-Wiener process.*

Conversely, if $\mathcal{F}_t = \mathcal{F}_t^w$, *then every probability measure* Q, *equivalent to* P, *has the above structure.*

Notice that the second statement relies on martingale representation and requires thus the filtration $\mathcal{F}_t$ to be the one generated by the Wiener process.

As mentioned, Girsanov's measure transformation allows to change the drift in a diffusion equation. In fact, suppose that under P we have

$$\mathrm{d}X_t = a_t X_t \,\mathrm{d}t + \sigma_t X_t \,\mathrm{d}w_t$$

and that we would like to change to a measure $Q \sim P$ ($\sim$ meaning *equivalent to*), under which the same X_t satisfies

$$\mathrm{d}X_t = r_t X_t \,\mathrm{d}t + \sigma_t X_t \,\mathrm{d}w_t^Q.$$

In this case just take $\theta_t = \sigma_t^{-1}(r_t - a_t)$.

If, besides a Wiener w_t, we now have also a marked point process represented by a counting measure $p(\mathrm{d}t, \mathrm{d}y)$, a Girsanov-type measure transformation allows, in addition to the translation of the Wiener, to perform also a change in the intensity process of the point process part. We have [see Theorem VIII, T10 in Brémaud (1981), see also Björk, Kabanov and Runggaldier (1997)]

Theorem 2.5. *On the finite time interval* $[0, T]$ *let* $p(\mathrm{d}t, \mathrm{d}y)$ *be an* E*-marked point process with* $(P, \mathcal{F}_t)$*-local characteristics* $(\lambda_t, m_t(\mathrm{d}y))$. *Let* $\psi_t \geqslant 0$ *be* $\mathcal{F}_t$*-predictable and* $h_t(y) \geqslant 0$ *an* $\mathcal{F}_t$*-predictable* E*-indexed process such that,* P*-a.s. and for all* $t \in [0, T]$,

$$\int_0^t \psi_s \lambda_s \,\mathrm{d}s < \infty; \qquad \int_E h_t(y) m_t(\mathrm{d}y) = 1.$$

Define $L_t = L_t^{(1)} \cdot L_t^{(2)}$ *where* $L_t^{(1)}$ *satisfies* (21) *and* $L_t^{(2)}$ *satisfies*

$$\mathrm{d}L_t^{(2)} = \int_E (\psi_t h_t(y) - 1) L_{t-}^{(2)} q(\mathrm{d}t, \mathrm{d}y) \tag{23}$$

with $q(\mathrm{d}t, \mathrm{d}y) = p(\mathrm{d}t, \mathrm{d}y) - \lambda_t m_t(\mathrm{d}y)\,\mathrm{d}t$ – *the martingale measure associated with* $p(\mathrm{d}t, \mathrm{d}y)$. *If* $E^P\{L_t^{(2)}\} = 1$ *for all t, then all the statements of Theorem* 2.4 *hold true in addition to the fact that* $p(\mathrm{d}t, \mathrm{d}y)$ *has the* $(Q, \mathcal{F}_t)$*-local characteristics* $(\psi_t\lambda_t, h_t(y)m_t(\mathrm{d}y))$.

Notice that, using (21) and (23), we have for the Radon–Nikodym derivative L_t

$$\begin{aligned} \mathrm{d}L_t &= \mathrm{d}\big(L_t^{(1)} \cdot L_t^{(2)}\big) = L_{t-}^{(1)}\,\mathrm{d}L_t^{(2)} + L_t^{(2)}\,\mathrm{d}L_t^{(1)} \\ &= L_t\theta_t\,\mathrm{d}w_t + L_{t-}\int_E (\psi_t h_t(y) - 1) q(\mathrm{d}t, \mathrm{d}y), \quad L_0 = 1. \end{aligned} \tag{24}$$

Using the exponential formula (17), we have that a solution of (24) is given by

$$\begin{aligned} L_t = {} & \exp\left\{-\frac{1}{2}\int_0^t \theta_s^2\,\mathrm{d}s + \int_0^t \theta_s\,\mathrm{d}w_s\right\} \\ & \times \exp\left\{\int_0^t\int_E (1 - \psi_s h_s(y))\lambda_s m_s(\mathrm{d}y)\,\mathrm{d}s\right\} \prod_{n=1}^{N_t} (\psi_{T_n} h_{T_n}(Y_n)). \end{aligned} \tag{25}$$

In the case of a multivariate (in particular univariate) point process $(N_t(1), \ldots, N_t(K))$ with $(P, \mathcal{F}_t)$-intensities $(\lambda_t(1), \ldots, \lambda_t(K))$, consider an $\mathcal{F}_t$-predictable process $(\psi_t(1), \ldots, \psi_t(K))$ such that, P-a.s. and for $t \in [0, T]$, $\sum_{k=1}^K \int_0^t \psi_s(k)\lambda_s(k)\,\mathrm{d}s < \infty$. Define then $L_t^{(2)}$ by

$$\mathrm{d}L_t^{(2)} = \sum_{k=1}^K (\psi_t(k) - 1) L_{t-}^{(2)} (\mathrm{d}N_t(k) - \lambda_t(k)\,\mathrm{d}t) \tag{26}$$

instead of by (23) making also corresponding changes in (24) and (25) for the Radon–Nikodym derivative L_t, namely

$$L_t = \exp\left\{-\frac{1}{2}\int_0^t \theta_s^2\,\mathrm{d}s + \int_0^t \theta_s\,\mathrm{d}w_s\right\} \prod_{k=1}^K \left[\exp\left\{\int_0^t (1 - \psi_s(k))\lambda_s(k)\,\mathrm{d}s\right\} \prod_{n=1}^{N_t(k)} \psi_{T_n}(k)\right]. \tag{27}$$

Then, under Q, the intensities become $(\psi_t(1)\lambda_t(1), \ldots, \psi_t(K)\lambda_t(K))$.

Notice, finally, that a condition to have $E^P\{L_t^{(2)}\} = 1$ can be found in Theorem VIII, T11 of Brémaud (1981).

3. Market models with jump-diffusions

In this section we introduce various jump-diffusion type models that were studied in the literature and that we shall be dealing with in the sequel. In the first two subsections we discuss, for asset price and term structure models respectively, the canonical jump-diffusion models in which there are two additive terms: a diffusion term and a jump term. In the last two subsections we then discuss diffusion/jump-diffusion models with stochastic volatility, where the latter is also described in terms of a jumping process. In addition, in the last subsection we model asset price behaviour on very small time scales where actual prices do not change continuously in time but rather at discrete random time points in reaction to trades and significant information. This then leads to a rather peculiar combination of diffusion and jump processes.

3.1. Asset-price and term structure models with additive jumps

As mentioned in the Introduction, the asset price evolution can perhaps be adequately described by a GBM for most of the time, but from time to time a large jump may occur and this cannot be adequately captured by a GBM. It appears thus natural to introduce models, where a jump process can be superimposed on a GBM, e.g., by adding to the diffusion term also a jump term. In a first subsection we discuss this modeling issue in the context of asset prices, while in the second subsection we concentrate on interest rate modeling.

3.1.1. Asset price models with jumps

In this section we adapt the outline of Section 7.2 in Lamberton and Lapeyre (1997). Let the price S_t of a risky asset jump at the random times $T_1, \ldots, T_n, \ldots$ and suppose that the relative/proportional change in its value at a jump time is given by $Y_1, \ldots, Y_n, \ldots$ respectively. We may then assume that, between two jump times, the price S_t follows a Black and Scholes model for a Wiener process w_t, that T_n are the jump times of a Poisson process N_t with intensity λ_t and that Y_n is a sequence of random variables with values in $(-1, \infty)$. This description can be formalized by letting, on the intervals $[T_n, T_{n+1})$,

$$\mathrm{d}S_t = S_t(\mu_t \,\mathrm{d}t + \sigma_t \,\mathrm{d}w_t) \tag{28}$$

while, at $t = T_n$, the jump is given by $\Delta S_n = S_{T_n} - S_{T_n^-} = S_{T_n^-} Y_n$ so that

$$S_{T_n} = S_{T_n^-}(1 + Y_n) \tag{29}$$

which, by the assumption that $Y_n > -1$, leads always to positive values of the prices. Using the standard Ito formula to obtain the solution to (28) as well as a recursive argument based on (29), it is easily seen that, at the generic time t, S_t can be given the following equivalent representations

$$S_t = S_0 \exp\left[\int_0^t \left(\mu_s - \frac{\sigma_s^2}{2}\right) \mathrm{d}s + \int_0^t \sigma_s \,\mathrm{d}w_s\right]\left[\prod_{n=1}^{N_t}(1 + Y_n)\right]$$

$$= S_0 \exp\left[\int_0^t \left(\mu_s - \frac{\sigma_s^2}{2}\right) ds + \int_0^t \sigma_s \, dw_s + \sum_{n=1}^{N_t} \log(1 + Y_n)\right]$$

$$= S_0 \exp\left[\int_0^t \left(\mu_s - \frac{\sigma_s^2}{2}\right) ds + \int_0^t \sigma_s \, dw_s + \int_0^t \log(1 + Y_s) \, dN_s\right], \tag{30}$$

where, as before, Y_t is obtained from Y_n by a piecewise constant and left continuous time interpolation. By the generalized Ito formula (19), the process S_t in (30) is easily seen to be a solution of

$$dS_t = S_{t-}[\mu_t \, dt + \sigma_t \, dw_t + Y_t \, dN_t]. \tag{31}$$

This equation corresponds to (28) with the addition of a jump term and is a particular case of the general jump-diffusion model (14) ((15)) when $\gamma(t, y) = y$. In what follows we shall thus consider the more general version of (31) given by

$$dS_t = S_{t-}\left[\mu_t \, dt + \sigma_t \, dw_t + \gamma(t, Y_t) \, dN_t\right] \tag{32}$$

that corresponds to (14) in the version of (15) and can thus equivalently be represented as

$$dS_t = S_{t-}\left[\mu_t \, dt + \sigma_t \, dw_t + \int_E \gamma(t, y) p(dt, dy)\right]. \tag{33}$$

If the marked point process is in particular a multivariate (or univariate) point process $(N_t(1), \ldots, N_t(K))$, then (32) ((33)) takes the form (see also (16))

$$dS_t = S_{t-}\left[\mu_t \, dt + \sigma_t \, dw_t + \sum_{k=1}^{K} \gamma_t(k) \, dN_t(k)\right]. \tag{34}$$

We finally point out that the marked point process in (32) ((33)) may be doubly stochastic in the sense specified in Sections 2.1 and 2.2 and this allows for further flexibility when it comes to modeling.

Remark 3.1. Occasionally, in the financial literature one finds model (32) ((33)) written in the form

$$dS_t = S_{t-}[\mu_t \, dt + \sigma_t \, dw_t + dJ_t],$$

where, in the specific case when (32) reduces to (31), $J_t := \sum_{n=1}^{N_t} Y_n$, while in the general case $J_t := \sum_{n=1}^{N_t} \gamma(T_n, Y_n)$. Furthermore, in models of the form (31) one may find the last term $Y_t \, dN_t$ written as $(Y_t - 1) \, dN_t$; in this latter case, instead of (29), we would then have $S_{T_n} = S_{T_n^-} Y_n = S_{T_n^-} Y_{T_n}$.

3.1.2. *Term structure models with jumps*

Among the basic objects in term structure models we have the zero-coupon bonds with prices $p(t, T)$ (the price, at t, of a bond maturing at T), forward rates $f(t, T)$ (the rate, contracted at t, for instantaneous borrowing at T), and the short rate $r(t)$. There exist some well-known relationships among these quantities, in particular $f(t,T) = -\partial \log p(t,T)/\partial T$; $r(t) = f(t,t)$. Since interest rates, and therefore also bond prices may indeed jump, one may consider the following jump-diffusion models for the above three quantities

$$dr(t) = a_t\, dt + b_t\, dw_t + \int_E c(t,y)p(dt,dy), \tag{35}$$

$$df(t,T) = \alpha(t,T)\, dt + \sigma(t,T)\, dw_t + \int_E \delta(t,T;y)p(dt,dy), \tag{36}$$

$$dp(t,T) = p(t-,T)\left\{m(t,T)\, dt + v(t,T)\, dw_t + \int_E n(t,T;y)p(dt,dy)\right\}, \tag{37}$$

where the differential is with respect to the time argument t, not the maturity T. Notice that only (37) has the factor $p(t-,T)$ also in the right-hand side. This guarantees (see the exponential formula (17)) positivity of $p(t,T)$ as it should be since $p(t,T)$ is the price of an asset; the interest rates $r(t)$, $f(t,T)$ need not necessarily be positive. Given the well-known relationships between the three quantities in (35)–(37), there obviously has to exist a relationship also between the coefficients in these models. This relationship can be found in Proposition 2.2. of Björk, Kabanov and Runggaldier (1997).

So far we have mentioned only continuously compounded interest rates. In financial markets also discretely compounded or simple rates such as LIBOR rates play an important role. Given a fixed accrual period δ, denote by $L(t,T)$ the forward rate, contracted at $t < T$, for the interval from T to $T+\delta$. Jump-diffusion models for $L(t,T)$ are studied in Glasserman and Kou (1999) under the form

$$dL(t,T) = L(t-,T)\left[\mu(t,T)\, dt + \sigma(t,T)\, dw_t + dJ(t,T)\right], \tag{38}$$

where (see Remark 3.1) $J(t,T) = \sum_{n=1}^{N_t} \gamma(T_n, Y_n)$ for a given marked point process represented by the double sequence (T_n, Y_n) [for a more general setup beyond jump-diffusions see Jamshidian (1999)]. Notice that the relationship

$$L(t,T) = \frac{1}{\delta}\left[\exp\left\{\int_T^{T+\delta} f(t,s)\, ds\right\} - 1\right] \tag{39}$$

between discretely and continuously compounded forward rates induces a relationship between the coefficients of the corresponding dynamic equations (36) and (39).

3.2. *Jump-diffusion models driven by hidden jump processes*

As mentioned in the introduction, empirical studies have led to consider also combinations of jumps and stochastic volatility, where the volatility presents a jump-type behaviour and is possibly also correlated with the jumps in the prices. As pointed out in Naik (1993), it is in fact natural to expect that, when the volatility jumps, also the price should jump. One can capture these aspects by a jump-diffusion model, where the coefficients depend on a hidden/latent jump process Z_t that affects also the intensity of the marked point process in the jump term (doubly stochastic marked point process). Formally, and limiting ourselves to asset price models of the form of (33) (that are equivalent to (32) and include (34)), we then have

$$\mathrm{d}S_t = S_{t-}\left[\mu_t(Z_t)\,\mathrm{d}t + \sigma_t(Z_t)\,\mathrm{d}w_t + \int_E \gamma(t, y; Z_{t-})p(\mathrm{d}t, \mathrm{d}y)\right], \tag{40}$$

where Z_t is any jump process with non-predictable jumps (could also be a Markov jump process) and $p(\mathrm{d}t, \mathrm{d}y)$ is the counting measure of a doubly stochastic marked point process with intensity $\lambda_t(Z_{t-}, \mathrm{d}y)$. Notice that Z_t affects the jump part both through the intensity as well as through the proportional jump sizes and it affects them in a predictable way.

3.3. Asset prices as diffusions sampled at the jump times of a jump process

As was mentioned in the Introduction, on very small time scales the real asset prices do not change continuously over time, but rather only at discrete random points in time in reaction to trades and/or significant new information. This makes jump processes attractive also for modeling high frequency data and here we give a description of such a modeling approach according to Frey and Runggaldier (2001, 1999). Marked point processes as models for high frequency data were also studied independently by various authors in the recent literature [see, e.g., Geman, Madan and Yor (1999), Rogers and Zane (1998), Rydberg and Shephard (1999)]. The models in Frey and Runggaldier (2001, 1999) are more in the spirit of jump-diffusions in that they consider a combination, although not an additive one, of a diffusion and a jump process as follows: given is a *background price process* of the diffusion type and this process is then sampled according to the random jump times of a jump process. This setup allows also to incorporate a possible correlation between (stochastic) volatility and price jumps in the way mentioned in the previous section, by letting again Z_t be a hidden process that drives the volatility of the background diffusion process and at the same time also the intensity of the (doubly stochastic) jump process that determines the random sampling times. In more formal terms, the logarithm Λ_t of the *background price process* is supposed to satisfy

$$\mathrm{d}\Lambda_t = \sqrt{v_t(Z_t)}\,\mathrm{d}w_t \tag{41}$$

with w_t a Wiener process independent of Z_t. The process Z_t is the hidden or latent state variable process that can be interpreted as modeling the rate at which new information is absorbed by the market. It may be given as a diffusion or as a finite state Markov process. Next consider a univariate doubly stochastic Poisson process (a Cox process) N_t with intensity $\lambda_t = \lambda_t(Z_{t-})$. The time dependence of this λ as well as of v in (41) is introduced to incorporate systematic patterns in trading activity. The actual price process is now such that its logarithm L_t satisfies

$$L_t = \Lambda_{T_{n-1}} \quad \text{for } t \in [T_{n-1}, T_n) \tag{42}$$

with T_n the jump times of N_t. The given model can thus be interpreted as a stochastic volatility model, evaluated at random times T_n. It is easily seen that the process L_t in (42) satisfies

$$\mathrm{d}L_t = (\Lambda_t - \Lambda_{T_{N_t-}})\,\mathrm{d}N_t, \tag{43}$$

where T_{N_t-} is the time of the last jump strictly prior to t and it is thus a marked point process with local characteristics $(\lambda_t(Z_t), \mathcal{N}(0, \int_{T_{N_t-}}^t v_s\,\mathrm{d}s))$ where $\mathcal{N}(m, \sigma^2)$ denotes a Gaussian r.v. with mean m and variance σ^2.

Notice that we may choose an intensity of the form

$$\lambda_t(Z_t) = \lambda_t^{(1)} + \lambda_t^{(2)} Z_t \tag{44}$$

so that N_t can be seen as the sum $N_t = N_t^{(1)} + N_t^{(2)}$ of two independent jump processes: $N_t^{(1)}$ with deterministic intensity $\lambda_t^{(1)}$ corresponding to *noise trading* and $N_t^{(2)}$ corresponding to *informed trading*.

One interesting aspect of the above model is that it makes it clear how sample path properties matter when it comes to volatility estimation: the volatility in a diffusion model, i.e., its quadratic variation, can be approximated arbitrarily well by the sum of the observed squared increments. For the given piecewise constant processes the empirical quadratic variation is useless for volatility estimation, even if computed over very small time intervals.

We finally point out that the definition, that was given in Section 2 concerning a doubly stochastic Poisson process, in particular that λ_t is $\mathcal{F}_0$-measurable, has as consequence the fact that N_t and Z_t cannot have common jumps and that the actual trading activity, namely the realization of the point process N_t, does not affect the law of Z_t. In economic terms this means that, in the given model, trading is caused purely by exogenous factors such as fundamental information, and not by the observed past trading activity.

4. Martingale measures: Existence and uniqueness (Market price of risk and market completion)

In each of the models discussed in Section 3, individual asset prices are driven by at least two independent sources of randomness so that the corresponding market models are incomplete. Based on the extended Girsanov-type measure transformation recalled in Section 2.5, in this section we shall discuss existence and, where applicable, uniqueness of martingale measures, thereby exhibiting also the market price of (jump-diffusion) risk. Uniqueness of the martingale measure will be mainly related to *completion of the market*. We want to point out that, as will be shown in more detail in the next Section 5 on hedging, it is not necessarily true that, if a market is *completed* to yield a unique martingale measure, then it is also genuinely complete in the sense that every contingent claim can be duplicated by a self financing portfolio. In fact, for marked point process with an infinite mark space, i.e., with an infinite number of sources of randomness, it will be shown in

Section 5.1.2 that uniqueness of the martingale measure implies only some form of *approximate completeness*. In this Section 4 we shall limit ourselves to the jump-diffusion asset price and term structure models of Section 3.1. In Section 4.1 below we treat the case of jump-diffusion models for asset prices and show that the market can relatively easily be completed to yield a unique martingale measure if the jump part corresponds to a marked point process with a finite number of marks (multivariate point processes). For an infinite number of marks the situation is studied in more detail in Section 4.2 below in the context of term structure models.

4.1. The case of jump-diffusion asset price models

We start with a jump-diffusion model, where the jump part corresponds to a univariate Poisson point process with P-intensity λ_t, namely (see (34) for $K = 1$)

$$\begin{aligned} dS_t &= S_{t-}[\mu_t\, dt + \sigma_t\, dw_t + \gamma_t\, dN_t] \\ &= S_{t-}\big[(\mu_t + \gamma_t\lambda_t)\, dt + \sigma_t\, dw_t + \gamma_t\, dM_t\big] \end{aligned} \tag{45}$$

with (see (2)) $M_t = N_t - \int_0^t \lambda_s\, ds$ the P-martingale corresponding to N_t. The Radon–Nikodym derivative for an absolutely continuous change of measure from P to Q, that implies a translation of the Wiener by θ_t and a change of the Poisson intensity from λ_t to $\psi_t\lambda_t$, is (see (27) for $K = 1$)

$$L_t = \exp\left\{\int_0^t \left[(1-\psi_s)\lambda_s - \frac{1}{2}\theta_s^2\right] ds + \int_0^t \theta_s\, dw_s + \int_0^t \log\psi_s\, dN_s\right\}. \tag{46}$$

Defining the Wiener and Poisson martingales w_t^Q and M_t^Q by (see (22) and (2))

$$\begin{cases} dw_t^Q = dw_t - \theta_t\, dt, \\ dM_t^Q = dN_t - \psi_t\lambda_t\, dt \end{cases} \tag{47}$$

the dynamics of S_t under Q become

$$dS_t = S_{t-}\big[(\mu_t + \sigma_t\theta_t + \gamma_t\psi_t\lambda_t)\, dt + \sigma_t\, dw_t^Q + \gamma_t\, dM_t^Q\big]. \tag{48}$$

Taking as numeraire the usual money market account B_t, where $dB_t = r_t B_t\, dt$, we immediately see that Q is a martingale measure, i.e., a measure under which $\widetilde{S}_t := B_t^{-1} S_t$ is a martingale, if θ_t and $\psi_t \geqslant 0$ are chosen such that $\mu_t + \sigma_t\theta_t + \gamma_t\psi_t\lambda_t = r_t$. From here we see that, for each pair (θ_t, ψ_t) with $\psi_t \geqslant 0$ arbitrary and

$$\theta_t = \sigma_t^{-1}(r_t - \mu_t - \gamma_t\psi_t\lambda_t), \tag{49}$$

we obtain a martingale measure, i.e., we can obtain infinitely many martingale measures, one for each choice of ψ_t.

Concerning the market price of risk ρ_t, from (45) and (49) we have

$$\rho_t := \mu_t + \gamma_t\lambda_t - r_t = \gamma_t\lambda_t - \sigma_t\theta_t - \gamma_t\psi_t\lambda_t = -\sigma_t\theta_t - \gamma_t\lambda_t(\psi_t - 1) \tag{50}$$

from where we see that $(-\theta_t)$ can be interpreted as *risk premium per unit of diffusion volatility*, whereas $-\lambda_t(\psi_t - 1)$ can be interpreted as *risk premium per unit of jump volatility*. On an arbitrage-free market all assets have, at a given time t, the same diffusion- and jump-risk premia and they determine, via the Girsanov transformation, i.e., via (46), the equivalent martingale measure Q.

We obtained infinitely many martingale measures because, for a single risky asset, we had two independent sources of randomness. One may thus expect that, by adding a further asset, one can *complete* the market to obtain a unique martingale measure. Consider then, in addition to S_t in (45), an asset with price $\overline{S}_t$ satisfying

$$\mathrm{d}\overline{S}_t = \overline{S}_{t-}[\bar{\mu}_t\,\mathrm{d}t + \bar{\sigma}_t\,\mathrm{d}w_t + \bar{\gamma}_t\,\mathrm{d}N_t]. \tag{51}$$

Notice that $\overline{S}_t$ could correspond to the price of a derivative asset with underlying S_t. In fact, if one is given the explicit expression of this derivative price in terms of S_t, i.e., $\overline{S}_t = F(t, S_t)$, then (51) is straightforwardly obtained from (45) by use of the generalized Ito formula (19). Since the two risk premia θ_t and $\lambda_t(\psi_t - 1)$ have to be the same for all assets, we may impose (49) on both assets with prices S_t and $\overline{S}_t$ respectively, namely

$$\theta_t = \sigma_t^{-1}(r_t - \mu_t - \gamma_t\psi_t\lambda_t) = \bar{\sigma}_t^{-1}(r_t - \bar{\mu}_t - \bar{\gamma}_t\psi_t\lambda_t) \tag{52}$$

from where one immediately gets

$$\psi_t\lambda_t = \frac{r_t(\sigma_t - \bar{\sigma}_t) + (\mu_t\bar{\sigma}_t - \sigma_t\bar{\mu}_t)}{\sigma_t\bar{\gamma}_t - \gamma_t\bar{\sigma}_t}. \tag{53}$$

Inserting this expression in (49) it follows that

$$\theta_t = \frac{\gamma_t(\bar{\mu}_t - r_t) - \bar{\gamma}_t(\mu_t - r_t)}{\sigma_t\bar{\gamma}_t - \gamma_t\bar{\sigma}_t}. \tag{54}$$

We have thus obtained unique risk premia and, consequently, a unique martingale measure provided the coefficients in (45) and (51) are such that $\sigma_t\bar{\gamma}_t - \gamma_t\bar{\sigma}_t \neq 0$ and that $\psi_t\lambda_t$ in (53) is positive.

With the unique martingale measure we may expect to have also obtained a *complete* market in the sense that, by investing in a self financing way in the two assets with prices S_t and $\overline{S}_t$, one can duplicate any claim. In Section 5.1.1 we shall show that, for the given market model, this is indeed the case.

It is easily seen that, if the jump part in the jump-diffusion model corresponds to a multivariate Poisson process, i.e., if instead of (45) we have (see (34))

$$\begin{aligned} \mathrm{d}S_t &= S_{t-}\left[\mu_t\,\mathrm{d}t + \sigma_t\,\mathrm{d}w_t + \sum_{k=1}^{K}\gamma_t(k)\,\mathrm{d}N_t(k)\right] \\ &= S_{t-}\left[\left(\mu_t + \sum_{k=1}^{K}\gamma_t(k)\lambda_t(k)\right)\mathrm{d}t + \sum_{k=1}^{K}\gamma_t(k)\,\mathrm{d}M_t(k)\right] \end{aligned} \tag{55}$$

with $M_t(k) = N_t(k) - \int_0^t \lambda_s(k)\,\mathrm{d}s$, then the previous results admit a straightforward extension. In particular, (49) becomes

$$\theta_t = \sigma_t^{-1}\left(r_t - \mu_t - \sum_{k=1}^{K} \gamma_t(k)\psi_t(k)\lambda_t(k)\right) \tag{56}$$

and the market price of risk is

$$\rho_t = -\sigma_t\theta_t - \sum_{k=1}^{K} \gamma_t(k)\lambda_t(k)\big(\psi_t(k) - 1\big). \tag{57}$$

This time the generic k-th term $\gamma_t(k)\lambda_t(k)(\psi_t(k) - 1)$ on the right can be interpreted as *risk premium per unit of jump volatility of type k*.

Again we obtain infinitely many martingale measures by choosing freely $\psi_t(k) \geqslant 0$, ($k = 1, \ldots, K$), and θ_t according to (56). Having now $K + 1$ independent sources of randomness, we may expect that one can *complete* the market by adding K further assets to obtain a unique equivalent martingale measure. This can be done along the lines of (52)–(54) although this time the calculations are more complicated and the conditions on the coefficients more cumbersome.

Finally, we consider the more general model (33) (or, equivalently, (32)) with a possibly infinite number of marks. Using the P-martingale measure $q(\cdot)$ in (8), by analogy to (45) and (55) we may rewrite (33) as

$$\begin{aligned} \mathrm{d}S_t &= S_{t-}\left[\mu_t\,\mathrm{d}t + \sigma_t\,\mathrm{d}w_t + \int_E \gamma(t, y)p(\mathrm{d}t, \mathrm{d}y)\right] \\ &= S_{t-}\left[\left(\mu_t + \int_E \gamma(t, y)\lambda_t(\mathrm{d}y)\right)\mathrm{d}t + \sigma_t\,\mathrm{d}w_t + \int_E \gamma(t, y)q(\mathrm{d}t, \mathrm{d}y)\right]. \end{aligned} \tag{58}$$

Using the particular form of the intensity given in (9), we also have

$$\int_E \gamma(t, y)\lambda_t(\mathrm{d}y) = \bar{\gamma}_t\lambda_t \quad \text{with } \bar{\gamma}_t = \int_E \gamma(t, y)m_t(\mathrm{d}y) \tag{59}$$

and so (58) becomes, quite analogously to (45),

$$\mathrm{d}S_t = S_{t-}\left[(\mu_t + \bar{\gamma}_t\lambda_t)\,\mathrm{d}t + \sigma_t\,\mathrm{d}w_t + \int_E \gamma(t, y)q(\mathrm{d}t, \mathrm{d}y)\right]. \tag{60}$$

Consider then, instead of (46), the more general Radon–Nikodym derivative (25) that we rewrite here in the form analogous to (46) as

$$L_t = \exp\left\{\int_0^t \left[(1 - \psi_s\bar{h}_s)\lambda_s - \frac{1}{2}\theta_s^2\right]\mathrm{d}s + \int_0^t \theta_s\,\mathrm{d}w_s + \int_0^t \log\big(\psi_s h_s(Y_s)\big)\,\mathrm{d}N_s\right\}, \tag{61}$$

where $\bar{h}_s = \int_E h_s(y) m_s(\mathrm{d}y)$. Define next the Wiener and jump martingales w_t^Q and $q^Q(\mathrm{d}t, \mathrm{d}y)$ by (see (22) and (8), (9) as well as (47))

$$\begin{cases} \mathrm{d}w_t^Q = \mathrm{d}w_t - \theta_t \,\mathrm{d}t, \\ q^Q(\mathrm{d}t, \mathrm{d}y) = p(\mathrm{d}t, \mathrm{d}y) - \psi_t \lambda_t h_t(y) m_t(\mathrm{d}y)\,\mathrm{d}t. \end{cases} \tag{62}$$

The dynamics of S_t under Q then become

$$\mathrm{d}S_t = S_{t-}\left[(\mu_t + \sigma_t\theta_t + \overline{\Gamma}_t\psi_t\lambda_t)\,\mathrm{d}t + \sigma_t\,\mathrm{d}w_t^Q + \int_E \gamma(t, y) q^Q(\mathrm{d}t, \mathrm{d}y)\right], \tag{63}$$

where $\overline{\Gamma}_t = \int_E \gamma(t, y) h_t(y) m_t(\mathrm{d}y)$. The measure Q is now a martingale measure if θ_t and $\psi_t \geqslant 0$ as well as $h_t(y) \geqslant 0$ are chosen so that $\mu_t + \sigma_t\theta_t + \overline{\Gamma}_t\psi_t\lambda_t = r_t$, which leads to the following relation corresponding to (49)

$$\theta_t = \sigma_t^{-1}(r_t - \mu_t - \overline{\Gamma}_t\psi_t\lambda_t). \tag{64}$$

Again, this leads to infinitely many martingale measures but, unless the mark space is finite, to *complete* the market in order to obtain a unique equivalent martingale measure one needs infinitely many assets. We shall discuss this situation in more detail in the context of bond markets in the next subsection.

To complete the analogy with the previous cases, notice that this time the market price of risk becomes (by (60) and (64))

$$\begin{aligned} \rho_t &:= \mu_t + \bar{\gamma}_t\lambda_t - r_t = \bar{\gamma}_t\lambda_t - \sigma_t\theta_t - \overline{\Gamma}_t\psi_t\lambda_t = -\sigma_t\theta_t - \lambda_t(\overline{\Gamma}_t\psi_t - \bar{\gamma}_t) \\ &= -\sigma_t\theta_t - \lambda_t \int_E \gamma(t, y)\big[\psi_t h_t(y) - 1\big] m_t(\mathrm{d}y). \end{aligned} \tag{65}$$

This time one may interpret $[\psi_t h_t(y) - 1] m_t(\mathrm{d}y)$ as *risk premium per unit of jump volatility of type* y.

In this latter context of a more general model of type (45) we want to point out that a methodology to obtain all equivalent martingale measures has also been worked out in Prigent (2001).

We close this subsection by mentioning that, depending on the purpose, one can single out some specific martingale measures among the various possible ones in a jump-diffusion model, where the market has not been completed. As an example, the construction of the so-called *minimal martingale measure* in a univariate Poisson jump diffusion model can be found in Runggaldier and Schweizer (1995). From a more practical point of view, an obvious possibility is always that of calibrating the model to market data.

4.2. *The case of jump-diffusion term structure models*

Consider first a term structure model where, under a given measure P, the (continuously compounded) forward rates $f(t, T)$ and the (zero coupon) bond prices $p(t, T)$ satisfy (36) and (37) respectively, namely

$$\mathrm{d}f(t, T) = \alpha(t, T)\,\mathrm{d}t + \sigma(t, T)\,\mathrm{d}w_t + \int_E \delta(t, T; y) p(\mathrm{d}t, \mathrm{d}y), \tag{66}$$

$$\mathrm{d}p(t,T)=p(t-,T)\left\{m(t,T)\,\mathrm{d}t+v(t,T)\,\mathrm{d}w_t+\int_E n(t,T;y)p(\mathrm{d}t,\mathrm{d}y)\right\}. \tag{67}$$

We shall also make the ad hoc assumptions that all objects are specified in a way to guarantee the validity of the various operations that will have to be performed, such as differentiation under the integral sign and interchange of the order of integration.

For later use we recall from Björk, Kabanov and Runggaldier (1997) the relationship between the coefficients in (66) and (67): if $f(t,T)$ satisfies (66), then $p(t,T)$ satisfies (67) with

$$\begin{cases} m(t,T)=r(t)+A(t,T)+\frac{1}{2}\|S(t,T)\|^2, \\ v(t,T)=S(t,T), \\ n(t,T;y)=\mathrm{e}^{D(t,T;y)}-1, \end{cases} \tag{68}$$

where $r(t)=f(t,t)$ is the short rate and

$$\begin{cases} A(t,T)=-\int_t^T \alpha(t,s)\,\mathrm{d}s, \\ S(t,T)=-\int_t^T \sigma(t,s)\,\mathrm{d}s, \\ D(t,T;y)=-\int_t^T \delta(t,s;y)\,\mathrm{d}s. \end{cases} \tag{69}$$

In the given bond market there are, at least theoretically, infinitely many assets, namely the bonds for all possible maturities $T>t$. A martingale measure Q is now a measure under which all these bond prices, discounted with respect to the money market account, are (local) martingales. We are therefore not even sure whether in such a given market model there exists a martingale measure and so our first purpose is to investigate the existence of such a measure.

Following essentially Björk, Kabanov and Runggaldier (1997) and considering general marked point processes, we also take the general form of the Radon–Nikodym derivative L_t, namely (see (24) where, for simplicity, we put $\psi_t\equiv 1$)

$$\mathrm{d}L_t=L_t\theta_t\,\mathrm{d}w_t+L_{t-}\int_E \big(h_s(y)-1\big)q(\mathrm{d}s,\mathrm{d}y), \tag{70}$$

where (see (8) and (9))

$$q(\mathrm{d}s,\mathrm{d}y)=p(\mathrm{d}s,\mathrm{d}y)-\lambda_s m_s(\mathrm{d}y), \tag{71}$$

i.e., we assume that, under P, the local characteristics of the marked point process $p(\mathrm{d}s,\mathrm{d}y)$ are $(\lambda_t,m_t(\mathrm{d}y))$. By Theorem 2.5 we know that, under the measure Q that corresponds to L_t in (70), the local characteristics become $(\lambda_t,h_t(y)m_t(\mathrm{d}y))$ so that, defining (see also (62))

$$\begin{cases} \mathrm{d}w_t^Q=\mathrm{d}w_t-\theta_t\,\mathrm{d}t, \\ q^Q(\mathrm{d}t,\mathrm{d}y)=p(\mathrm{d}t,\mathrm{d}y)-\lambda_t h_t(y)m_t(\mathrm{d}y)\,\mathrm{d}s \end{cases} \tag{72}$$

the bond prices $p(t,T)$ satisfy, under Q, the dynamics

$$\mathrm{d}p(t,T) = p(t-,T)\Big\{\Big[m(t,T) + v(t,T)\theta_t + \lambda_t \int_E n(t,T;y)h_t(y)m_t(\mathrm{d}y)\Big]\mathrm{d}t + v(t,T)\,\mathrm{d}w_t^Q + \int_E n(t,T;y)q^Q(\mathrm{d}t,\mathrm{d}y)\Big\}. \tag{73}$$

A necessary condition for the existence of martingale measure Q is then that there exist a predictable process θ_t and a predictable E-indexed process $h_t(y) \geqslant 0$ such that the conditions of Theorem 2.5 hold and

$$m(t,T) + v(t,T)\theta_t + \lambda_t \int_E n(t,T;y)h_t(y)m_t(\mathrm{d}y) = r(t). \tag{74}$$

Notice that this implies for the market price of risk a relation analogous to (65), namely

$$\begin{aligned}\rho_t &:= m(t,T) + \lambda_t \int_E n(t,T;y)m_t(\mathrm{d}y) - r(t)\\ &= -v(t,T)\theta_t - \lambda_t \int_E n(t,T;y)\big[h_t(y) - 1\big]m_t(\mathrm{d}y).\end{aligned} \tag{75}$$

We shall now translate condition (74), involving the coefficients of (67), into a condition involving the coefficients of (66), namely of the forward rates. Using (68), condition (74) becomes

$$A(t,T) + \frac{1}{2}\big\|S(t,T)\big\|^2 + S(t,T)\theta_t + \int_E h_s(y)\nu(t,T;\mathrm{d}y) = 0 \tag{76}$$

with $\nu(t,T;\mathrm{d}y) := (\mathrm{e}^{D(t,T;y)} - 1)\lambda_t m_t(\mathrm{d}y)$ and with $A,\ S$ and D as in (69).

When building a term structure model it is often convenient to specify all objects directly under a martingale measure Q and this obviously imposes some restrictions on the coefficients in the models. Concentrating on forward rates, assume that we want model (66) to be valid under a martingale measure Q, i.e., we are postulating that $P = Q$ and so we have to choose $\theta_t \equiv 0,\ h_t(y) \equiv 1$. Notice now that (76) has to hold for all maturities so that, inserting the above choices of θ_t and $h_t(y)$ and differentiating with respect to T, we obtain (using also (69)) the following necessary condition

$$\alpha(t,T) = \sigma(t,T)\int_t^T \sigma(t,s)\,\mathrm{d}s - \int_E \delta(t,T;y)\,\mathrm{e}^{D(t,T;y)}\lambda_t h_t(y)m_t(\mathrm{d}y) \tag{77}$$

which is a clear extension of the classical *Heath–Jarrow–Morton drift condition* for the pure diffusion case.

Having investigated the existence of a martingale measure, we may next look for conditions implying its uniqueness. Concentrating again on forward rates, a necessary condition for the existence of a martingale measure has been seen to be the existence of a predictable θ_t and a predictable E-indexed $h_t(y) \geqslant 0$ such that relation (76) holds. Quite obviously then, if (76) admits a unique solution in θ_t and $h_t(y)$, the martingale measure is unique.

To formalize this fact, consider the following linear operator [for technical details, that for simplicity we neglect here, we refer to Björk, Kabanov and Runggaldier (1997)]

$$\mathcal{K}_t : (\theta, h(y)) \to S(t, \cdot)\theta + \int_E h(y)\big(e^{D(t,\cdot;y)} - 1\big)\lambda_t m_t(dy). \tag{78}$$

The operator $\mathcal{K}_t$ is an integral operator of the first kind and we refer to it as *martingale operator*. The martingale measure is then unique if and only if, dP, dt-*a.e.*, we have

$$\operatorname{Ker} \mathcal{K}_t = 0. \tag{79}$$

We may now wonder whether, in the present context of infinitely many sources of randomness, the uniqueness of the martingale measure implies completeness in the sense that every contingent claim can be replicated by a self financing portfolio. The answer is no; in fact, as we shall mention in Section 5.1.2 below, we obtain only a form of *approximate completeness*.

We finally remark that the relationship (39) between discretely and continuously compounded forward rates has allowed Glasserman and Kou (1999) to carry over the just mentioned results for continuously compounded forward rates also to the case when one has simple forwards instead. In fact, a model of the term structure of simple forwards $L(t, T)$ (see (38)) is defined in Glasserman and Kou (1999) to be arbitrage-free, if it can be embedded in an arbitrage-free model of instantaneous forwards $f(t, T)$ via (39).

5. Hedging in jump-diffusion market models

In the previous section we have seen that, as a consequence of its incompleteness, in a jump-diffusion market model we have in general infinitely many martingale measures. We have then investigated the method of *market completion* as a tool to obtain a unique martingale measure. On the other hand, from the *second fundamental theorem of asset pricing* one has that, in general, if a market admits a unique equivalent martingale measure, then it is also complete in the sense that every contingent claim can be hedged by a self financing portfolio.

We shall investigate the hedging problem in a jump-diffusion market model having in mind two goals: for the first goal, in the context of asset price models, we shall show in Section 5.1.1 that completed market models with a unique martingale measure are complete also in the sense of hedging if there are only a finite number of marks for the jumping component (there is a finite number of sources of randomness). If however there are an infinite number of marks (an infinite number of sources of randomness) then, in the context of bond markets, in Section 5.1.2 we shall show that the completed market models with a unique martingale measure are only *approximately complete* in the sense of hedging.

In the context of the first goal we also want to add here that Jensen (1999) approximates a given jump-diffusion market model, having an infinite number of marks, by a sequence of jump-diffusion models with a finite number of marks that are therefore complete also in the sense of hedging.

For the second goal, in Section 5.2 we shall consider the case when one cannot have a complete market or when it is not appropriate to complete it. In such a case one has to determine the hedging strategy according to some specific hedging criterion. We shall consider the (local) risk minimization and the related minimum variance criteria and show that they lead to hedging strategies that are quite natural extensions of those in complete markets. While so far only the models of Section 3.1 have been further investigated, the discussion in Section 5.2 will center mainly around the model of Section 3.3.

In part, this section can also be seen as preliminary to the next Section 6 on pricing. In fact, if a market is complete in the sense of hedging, then by the criterion of absence of arbitrage the initial value of the self financing and hedging strategy has to correspond to the arbitrage-free price of the contingent claim. If the market cannot be completed, the criterion of absence of arbitrage alone is not sufficient to define a price and the preference structure of the investors has to come into play. Since, typically, the initial value of a hedging portfolio satisfying a specific hedging criterion can be expressed as expectation of the discounted claim under a specific martingale measure, the choice of a hedging criterion implies also the choice of a martingale measure and thus of a pricing kernel. We shall discuss these issues in more detail in Section 6.1 below.

5.1. Hedging when the market is completed

5.1.1. Asset-price models

In this subsection we consider the univariate jump-diffusion model of Section 4.1. We had seen that, considering in addition to the asset with price S_t satisfying (45), also the asset with price $\overline{S}_t$ satisfying (51) with coefficients such that $\psi_t\lambda_t$ in (53) is positive and $\sigma_t\bar{\gamma}_t - \gamma_t\bar{\sigma}_t \neq 0$, then there exists a unique martingale measure Q corresponding to the choice of ψ_t and θ_t according to (53) and (54). Basing ourselves on Jeanblanc-Piqué and Pontier (1990), we show now that in this situation any claim can be hedged with a self financing portfolio.

Given a maturity T, consider as *claim* a (square-integrable) random variable H_T, measurable with respect to $\mathcal{F}_T$, where $\mathcal{F}_t := \sigma\{S_0, \overline{S}_0, w_s, N_s, s \leqslant t\}$, completed with the null sets. In addition to the two risky assets with prices S_t and $\overline{S}_t$, we suppose given also a nonrisky asset, whose price we take for simplicity identically equal to 1 (equivalent to assuming all prices discounted with respect to the nonrisky asset). An investment strategy is then a triple $\Phi_t = [\phi_t, \bar{\phi}_t, \eta_t]$, where η_t denotes the number of units of the nonrisky asset held in the portfolio at time t and ϕ_t, $\bar{\phi}_t$ are the number of shares of the two risky assets respectively. Let ϕ_t, $\bar{\phi}_t$ be predictable and η_t be adapted. The value, at time t, of a portfolio corresponding to the strategy Φ is then

$$V_\Phi(t) = \phi_t S_t + \bar{\phi}_t \overline{S}_t + \eta_t. \tag{80}$$

We want Φ to be such that the corresponding portfolio is self financing and duplicates the claim, i.e., that it satisfies

$$\begin{cases} \mathrm{d}V_\Phi(t) = \phi_t \,\mathrm{d}S_t + \bar{\phi}_t \,\mathrm{d}\overline{S}_t, \\ V_\Phi(T) = H_T. \end{cases} \tag{81}$$

It follows from Section 4.1 that, under the unique martingale measure Q, the discounted prices of the two risky assets, that for simplicity we continue denoting by S_t and $\overline{S}_t$, are the martingales satisfying

$$\begin{cases} \mathrm{d}S_t = S_{t-}\big[\sigma_t \,\mathrm{d}w_t^Q + \gamma_t \,\mathrm{d}M_t^Q\big], \\ \mathrm{d}\overline{S}_t = \overline{S}_{t-}\big[\bar{\sigma}_t \,\mathrm{d}w_t^Q + \bar{\gamma}_t \,\mathrm{d}M_t^Q\big], \end{cases} \tag{82}$$

where w_t^Q and M_t^Q are as in (47) with $\psi_t\lambda_t$ and θ_t according to (53) and (54). Replacing $\mathrm{d}S_t$ and $\mathrm{d}\overline{S}_t$ from (82) in (81), it follows that also $V_\Phi(t)$ is a $(Q, \mathcal{F}_t)$-martingale satisfying

$$V_\Phi(t) = V_\Phi(0) + \int_0^t [\phi_s S_s \sigma_s + \bar{\phi}_s \overline{S}_s \bar{\sigma}_s] \,\mathrm{d}w_s^Q + \int_0^t [\phi_s S_{s-} \gamma_s + \bar{\phi}_s \overline{S}_{s-} \bar{\gamma}_s] \,\mathrm{d}M_t^Q. \tag{83}$$

Consider next the $(Q, \mathcal{F}_t)$-martingale

$$M(t) := E^Q\{H_T | \mathcal{F}_t\}. \tag{84}$$

By the martingale representation theorem (see Theorem 2.3 applied here to the particular case of a univariate Poisson point process) there exist two $\mathcal{F}_t$-predictable processes $\xi_t^{(1)}$ and $\xi_t^{(2)}$ such that

$$M(t) = M(0) + \int_0^t \xi_s^{(1)} \,\mathrm{d}w_s^Q + \int_0^t \xi_s^{(2)} \,\mathrm{d}M_s^Q. \tag{85}$$

Comparing (83) and (85), one sees immediately that, by putting

$$V_\Phi(0) = M(0) = E^Q\{H_T | \mathcal{F}_0\} \tag{86}$$

and choosing ϕ_t, $\bar{\phi}_t$ such that (integrating with respect to a Wiener process one may change S_t into S_{t-})

$$\begin{cases} \phi_t S_{t-} \sigma_t + \bar{\phi}_t \overline{S}_{t-} \bar{\sigma}_t = \xi_t^{(1)}, \\ \phi_t S_{t-} \gamma_t + \bar{\phi}_t \overline{S}_{t-} \bar{\gamma}_t = \xi_t^{(2)} \end{cases} \tag{87}$$

we have $V_\Phi(t) = M(t)$. Since $M(T) = H_T$ by definition, with the choices (86) and (87) we obtain a self financing and hedging strategy (the value of η_t follows from (80)). Notice that, in order to obtain a unique solution of (87), we have to require that $\sigma_t \bar{\gamma}_t - \gamma_t \bar{\sigma}_t \neq 0$, which is exactly one of the conditions required after (53) and (54) to obtain a unique equivalent martingale measure.

What we have just shown is an existence result leading to the completeness (in the sense of hedging) of the given market when the martingale measure is unique. To actually determine the hedging strategy, we need an explicit expression for the processes $\xi_t^{(1)}$ and $\xi_t^{(2)}$ that, in the case of a simple claim of the form $H_T = H(S_T, \overline{S}_T)$, can be obtained by analogy to the pure diffusion case using the generalized Ito formula (19). Due to the Markov property of $(S_t, \overline{S}_t)$, we may in fact put

$$M(t) = M(t; S_t, \overline{S}_t) = E^Q\big\{H(S_T, \overline{S}_T) | \mathcal{F}_t\big\}. \tag{88}$$

Formula (19) then leads to

$$\begin{aligned} \mathrm{d}M(t) = \Big[& M_t(\cdot) + \frac{1}{2} M_{SS}(\cdot) S_{t-}^2 \sigma_t^2 + \frac{1}{2} M_{\bar{S}\bar{S}}(\cdot) \overline{S}_{t-}^2 \bar{\sigma}_t^2 + M_{S\bar{S}} S_{t-} \overline{S}_{t-} \sigma_t \bar{\sigma}_t \\ & + \big(M\big(t; S_{t-}(1+\gamma_t), \overline{S}_{t-}(1+\bar{\gamma}_t)\big) - M(t; S_{t-}, \overline{S}_{t-}) \\ & - M_S(\cdot)\gamma_t - M_{\bar{S}}(\cdot)\bar{\gamma}_t\big)\psi_t \lambda_t \Big] \mathrm{d}t + \big[M_S(\cdot) S_t \sigma_t + M_{\bar{S}}(\cdot) \overline{S}_t \bar{\sigma}_t\big] \mathrm{d}w_t^Q \\ & + \big[M\big(t; S_{t-}(1+\gamma_t), \overline{S}_{t-}(1+\bar{\gamma}_t)\big) - M(t; S_{t-}, \overline{S}_{t-})\big] \mathrm{d}M_t^Q. \end{aligned} \tag{89}$$

Since $M(t)$ is a Q-martingale, the drift (finite variation) term in (89) has to vanish and so it follows from (89) and (85) that

$$\begin{cases} \xi_t^{(1)} = M_S(t; S_t, \overline{S}_t) S_t \sigma_t + M_{\bar{S}}(t; S_t, \overline{S}_t) \overline{S}_t \bar{\sigma}_t, \\ \xi_t^{(2)} = M\big(t; S_{t-}(1+\gamma_t), \overline{S}_{t-}(1+\bar{\gamma}_t)\big) - M(t; S_{t-}, \overline{S}_{t-}). \end{cases} \tag{90}$$

For a related result see also Shirakawa (1990). We conclude this subsection by pointing out that, analogously to Section 4.1, the procedure that we have described here for the case of a univariate point process can quite naturally be extended to the case of multivariate point processes, provided the market is completed with the addition of an appropriate number of further assets.

5.1.2. Term structure models

We consider the term structure model discussed in Section 4.2 assuming that the condition for uniqueness of the martingale measure given by the injectivity (see (79)) of the integral operator $\mathcal{K}_t$ in (78) is satisfied. This subsection is mainly based on Björk, Kabanov and Runggaldier (1997) [see also Jarrow and Madan (1999) for a related approach].

In this market, where the basic assets are zero-coupon bonds with prices $p(t, T)$ for any maturity $T > t$ in addition to a nonrisky asset (money market account B_t), we have first to define a portfolio.

Definition 5.1. On the given bond market a portfolio is a pair $(\eta_t, \xi_t(\mathrm{d}T))$ where
(i) η_t is predictable;
(ii) $\forall t$, $\xi_t(\cdot)$ is a signed finite measure on $[t, \infty)$.

Intuitively, η_t is the number of units of the riskfree asset held in the portfolio at time t, $\xi_t(\mathrm{d}T)$ is the "number" of bonds, with maturities in $[T, T + \mathrm{d}T)$, held at time t. Some integrability assumptions are also required, but we leave them here as implicit. The *value process* of the portfolio (η, ξ), discounted with respect to B_t, is

$$V_t(\eta, \xi) = \eta_t + \int_t^\infty p(t, T)\xi_t(\mathrm{d}T) \tag{91}$$

where, with some abuse of notation, we denote by $p(t, T)$ also the discounted value of a T-bond.

Definition 5.2. The portfolio (η, ξ) is self-financing if

$$\mathrm{d}V_t(\eta, \xi) = \int_t^\infty \xi_t(\mathrm{d}T)\,\mathrm{d}p(t, T). \tag{92}$$

The integral in the right-hand side in (92) needs an appropriate definition. Justified by the development in Björk et al. (1997), we shall simply replace here $\mathrm{d}p(t, T)$ in (92) by its expression under the (unique) martingale measure. To obtain this expression, recall the condition (77) (or, equivalently, (76) with $\theta_t = 0,\ h_t(y) = 1$) on the coefficients of the forward rate dynamics in order that these dynamics hold under a martingale measure. Translating, via (68), these conditions back to the bond price dynamics and taking also into account the definition of $q^Q(\mathrm{d}t, \mathrm{d}y)$ in (72), one has

$$\mathrm{d}p(t, T) = p(t-, T)\left[S(t, T)\,\mathrm{d}w_t^Q + \int_E \left(\mathrm{e}^{D(t,T;y)} - 1\right)q^Q(\mathrm{d}t, \mathrm{d}y)\right] \tag{93}$$

(recall that we take here for $p(t, T)$ the discounted values). Given a contingent claim $H_T \in \mathcal{F}_T$, that we assume here to be bounded, the conditions for self financing and perfect hedging can be expresses as (combining (92) with (93))

$$\begin{cases} V_t(\eta, \xi) = V_0(\eta, \xi) + \displaystyle\int_0^t \int_s^\infty \xi_s(\mathrm{d}T)p(s, T)S(s, T)\,\mathrm{d}w_s^Q \\ \qquad + \displaystyle\int_0^t \int_E \int_s^\infty \xi_s(\mathrm{d}T)p(s-, T)\left(\mathrm{e}^{D(s,T;y)} - 1\right)q^Q(\mathrm{d}s, \mathrm{d}y), \\ V_T(\eta, \xi) = H_T, \end{cases} \tag{94}$$

where the inner integral is with respect to T and the outer with respect to s.

Paralleling the development in the previous Section 5.1.1, consider next the $(Q, \mathcal{F}_t)$-martingale

$$M(t) := E^Q\{H_T \mid \mathcal{F}_t\} \tag{95}$$

which, by the martingale representation Theorem 2.3, admits the representation (see (12) under the measure Q)

$$M(t) = M(0) + \int_0^t \phi_s\,\mathrm{d}w_s^Q + \int_0^t \int_E H(s, y)q^Q(\mathrm{d}s, \mathrm{d}y) \tag{96}$$

for predictable (and appropriately integrable) ϕ and H. Comparing (94) with (96) one sees that, by putting

$$V_0(\eta, \phi) = M(0) = E^Q\{H_T \mid \mathcal{F}_0\} \tag{97}$$

and choosing $\xi_t(\mathrm{d}T)$ such that

$$\begin{cases} \displaystyle\int_t^\infty \xi_t(\mathrm{d}T)p(t, T)S(t, T) = \phi_t, \\ \displaystyle\int_t^\infty \xi_t(\mathrm{d}T)p(t-, T)\left(\mathrm{e}^{D(t,T;y)} - 1\right) = H(t, y) \end{cases} \tag{98}$$

we have $V_t(\eta,\xi) = M(t)$ and, in particular, $V_T(\eta,\xi) = H_T$, i.e., we have obtained a self financing and hedging strategy (the value of η_t follows from (91)). Everything now hinges upon the (unique) solvability of (98). To this effect consider the integral operator $\mathcal{K}_t^*$ implicit in the left-hand side of (98), namely

$$\mathcal{K}_t^* : \xi \to \begin{bmatrix} \int_t^\infty p(t,T)S(t,T)\xi(\mathrm{d}T) \\ \int_t^\infty p(t-,T)\big(\mathrm{e}^{D(t,T;\cdot)} - 1\big)\xi(\mathrm{d}T) \end{bmatrix} \tag{99}$$

so that the conditions (98) become

$$\mathcal{K}_t^* \xi = \begin{bmatrix} \phi_t \\ H(t,\cdot) \end{bmatrix}. \tag{100}$$

The integral operator $\mathcal{K}_t^*$ will be called *hedging operator* and the market is complete if $\mathcal{K}_t^*$ is surjective. Combining this result with that of Section 4.2 on the uniqueness of the martingale measure, namely (79), we may synthesize them into

Proposition 5.3. *For the given term structure model* (66), (67) *we have that*
(i) *the martingale measure is unique, if the martingale operators $\mathcal{K}_t$ in* (78) *are injective*;
(ii) *the market is complete if the hedging operators $\mathcal{K}_t^*$ in* (99) *are surjective.*

It turns out that the operators $\mathcal{K}_t^*$ are adjoint to $\mathcal{K}_t$. If the spaces, on which they act, are finite-dimensional, then the injectivity of $\mathcal{K}_t$ implies surjectivity of $\mathcal{K}_t^*$ and thus that uniqueness of the martingale measure implies completeness. Unfortunately, our spaces here are infinite-dimensional and so, due to the duality relationship $(\mathrm{Ker}\,\mathcal{K})^\perp = \mathrm{cl}(\mathrm{Im}\,\mathcal{K}^*)$ between bounded linear operators, the injectivity of $\mathcal{K}_t$ implies denseness of $\mathcal{K}_t^*$. In other words, the uniqueness of the martingale measure implies only an *approximate completeness*. For details we refer to Björk, Kabanov and Runggaldier (1997).

For the case when the mark space E is infinite, Björk, Kabanov and Runggaldier (1997) also give a characterization of the hedgeable claims, based on a Laplace-transform technique and under assumptions that hold, e.g., in the case of an affine term structure. When the mark space E is finite, in Björk, Kabanov and Runggaldier (1997) it is furthermore shown that, under appropriate assumptions, any claim can be hedged with a finite number of bonds, whose maturities can be chosen in an essentially arbitrary way and such that they remain fixed as the running time t varies.

5.2. *Hedging when the market is not complete*

If one cannot have a complete market or market completion is not appropriate, one has to accept some residual risk, due either to non-self-financing or nonperfect hedging, and choose an investment strategy that minimizes the unhedgeable risk. For this purpose various criteria have been proposed and here we describe one such criterion for the case of a slight variant of the market model described in Section 3.3.

We assume here that the actual price S_t of the risky asset satisfies a model of the form of (41), namely

$$\mathrm{d}S_t = S_t\sqrt{v_t(Z_t)}\,\mathrm{d}w_t, \tag{101}$$

where Z_t is supposed to be a diffusion-type process of the form

$$\mathrm{d}Z_t = \alpha_t(Z_t)\,\mathrm{d}t + \beta_t(Z_t)\,\mathrm{d}\overline{w}_t \tag{102}$$

for a Wiener $\overline{w}_t$, independent of w_t. Given a univariate, doubly stochastic Poisson process N_t with intensity $\lambda_t = \lambda_t(Z_t)$, suppose that the prices of the risky asset can only be observed at the jump times T_n of N_t, i.e., the observation process Y_t is given by (see (43))

$$\mathrm{d}Y_t = (S_t - S_{T_{N_{t-}}})\,\mathrm{d}N_t \tag{103}$$

so that the information of the hedger can be modeled by the filtration

$$\mathcal{F}_t^Y = \sigma\{N_s, Y_s;\ s \leqslant t\} \subset \mathcal{F}_t = \sigma\{S_0, Z_0, w_s, \overline{w}_s, N_s;\ s \leqslant t\}.$$

Notice that the only difference with respect to the model described in Section 3.3 is that here the actual price process S_t varies continuously in time according to (101), but is observed only at the discrete time points T_n; there, the process according to (101) is only a *background process* and the actual price process is given by the values of the background process, sampled at the time points T_n according to (103). Notice also that, according to (101), the process S_t is implicitly assumed to be a $(P, \mathcal{F}_t)$-martingale. On one hand, this will make our hedging procedure below applicable; on the other hand it can be justified by assuming that [see, e.g., Becherer (2001)] S_t is discounted with respect to a P-numeraire portfolio, which is a tradable numeraire such that the discounted assets become martingales with respect to the original measure P.

Our hedging criterion will be that of (local) risk minimization according to Föllmer and Sondermann (1986), Föllmer and Schweizer (1991), that keeps the requirement of perfect hedging and relaxes the self financing requirement into mean self financing. More precisely, considering as strategy a pair (η_t, ξ_t) of $\mathcal{F}_t^Y$-predictable processes with η_t and ξ_t denoting the number of units of the numeraire and the given asset respectively, that are held in the portfolio at time t, we give the following

Definition 5.4. Assuming prices are discounted with respect to the numeraire, define

$$V_t = V_t(\eta, \xi) := \xi_t S_t + \eta_t \quad \text{as value process,}$$
$$C_t = C_t(\eta, \xi) := V_t - \int_0^t \xi_s\,\mathrm{d}S_s \quad \text{as cost process.}$$

Notice that, if $C_t(\eta, \xi) = const.$, the strategy (η, ξ) is self financing. We shall now relax this assumption by allowing $C_t(\eta, \xi)$ to be a $(P, \mathcal{F}_t^Y)$-martingale and, given a (square-integrable) claim $H(S_T)$ (already discounted with respect to the numerarire), determine a hedging strategy (η^*, ξ^*) that, for all $t = T_n$ $(n = 1, 2, \ldots)$, minimizes

$$R_t^Y(\eta, \xi) := E\{(C_T(\eta, \xi) - C_t(\eta, \xi))^2 | \mathcal{F}_t^Y\} \tag{104}$$

with respect to the hedging strategies (η, ξ) for which $C_t(\eta, \xi)$ is a $(P, \mathcal{F}_t^Y)$-martingale. The strategy (η^*, ξ^*) will be called an $\mathcal{F}_t^Y$*-risk minimizing strategy*.

Notice that there is a close relationship between risk minimizing strategies in the just specified sense and variance-minimizing strategies that are self financing and minimize the variance of the residual hedging error.

To compute an $\mathcal{F}_t^Y$-risk minimizing strategy we shall proceed in two steps following Frey and Runggaldier (1999) [see also Fischer, Platen and Runggaldier (1999) and Frey (2000)]. In the first step we determine an $\mathcal{F}_t$-risk minimizing strategy, namely a risk minimizing strategy where the (hypothetical) information of the hedger corresponds to the full filtration $\mathcal{F}_t$, instead of the subfiltration $\mathcal{F}_t^Y$. For this purpose define the P-martingale

$$g(t, S_t, Z_t) := E\{H(S_T)|\mathcal{F}_t\}, \tag{105}$$

where the notation is justified by the Markov property of (S_t, Z_t). Assuming sufficient regularity of $g(\cdot)$, we proceed analogously to the last part of Section 5.1.1 applying Ito's formula to $g(t, S_t, Z_t)$ thereby obtaining

$$\begin{aligned} H(S_T) = g(0, S_0, Z_0) &+ \int_0^T \left[g_t(\cdot) + g_Z(\cdot)\alpha_t(\cdot)\right] \mathrm{d}t \\ &+ \int_0^T \left[\frac{1}{2} g_{SS}(\cdot)v_t(\cdot)S_t^2 + \frac{1}{2} g_{ZZ}(\cdot)\beta_t^2(\cdot)\right] \mathrm{d}t \\ &+ \int_0^T g_S(\cdot)\,\mathrm{d}S_t + \int_0^T g_Z(\cdot)\beta_t(\cdot)\,\mathrm{d}\overline{w}_t. \end{aligned} \tag{106}$$

Since $g(t, S_t, Z_t)$ is a P-martingale, the finite variation terms in (106) vanish, leading to

$$H(S_T) = g(0, S_0, Z_0) + \int_0^T g_S(t, S_t, Z_t)\,\mathrm{d}S_t + M_T^H \tag{107}$$

which is of the form of a *Kunita–Watanabe decomposition* of $H(S_T)$, namely a decomposition of the form

$$H(S_T) = H_0 + \int_0^T \xi_t^H\,\mathrm{d}S_t + M_T^H, \tag{108}$$

where M^H is a P-martingale that, due to the independence of $\overline{w}_t$ and w_t, is orthogonal to the P-martingale S. It then follows from Föllmer and Sondermann (1986) and Föllmer and Schweizer (1991) that the $\mathcal{F}_t$-risk minimizing strategy is given by

$$\begin{cases} \xi_t^{\mathcal{F}} = \xi_t^H = g_S(t, S_t, Z_t), \\ \eta_t^{\mathcal{F}} = g(t, S_t, Z_t) - \xi_t^{\mathcal{F}} S_t \end{cases} \tag{109}$$

so that $V_t(\eta^{\mathcal{F}}, \xi^{\mathcal{F}}) = g(t, S_t, Z_t)$. This strategy appears as a very natural extension of the classical Black Scholes strategy in the pure diffusion case. Notice that, to actually determine $(\eta_t^{\mathcal{F}}, \xi_t^{\mathcal{F}})$ and its value, one needs to compute $g(t, S_t, Z_t)$, which can be achieved either by computing the expectation in (105) (numerical simulations may be used) or by solving the PDE that results from (106) by setting equal to zero the finite variation terms. Details can be found in Frey and Runggaldier (1999).

Coming to the second step, it follows from a general result in Schweizer (1994) [see also Di Masi, Platen and Runggaldier (1995)] that the $\mathcal{F}_t^Y$-risk minimizing strategy is obtained by projecting the $\mathcal{F}_t$-risk minimizing strategy onto the subfiltration $\mathcal{F}_t^Y$. This *projection property*, which is due to the quadratic nature of the risk minimizing criterion, makes this latter criterion very attractive every time one has to deal with partial information. More precisely, the $\mathcal{F}_t^Y$-risk minimizing strategy (η^*, ξ^*) is given by

$$\begin{cases} \xi_t^* = E\{v_t(Z_t)S_t^2\xi_t^{\mathcal{F}}(S_t, Z_t)|\mathcal{F}_{t-}^Y\}/E\{v_t(Z_t)S_t^2|\mathcal{F}_{t-}^Y\}, \\ \eta_t^* = E\{H(S_T) - \xi_t^* S_t|\mathcal{F}_t^Y\}. \end{cases} \tag{110}$$

Notice that, according to the model, the hedger will compute the strategy (η^*, ξ^*) only at the jump times T_n of N_t, when he receives new information [for details and a stochastic filtering-type algorithm to compute the projection in (110) see again Frey and Runggaldier (1999)].

We close the section mentioning that, for a standard jump-diffusion model of the type of Section 3.1.1 with a marked point process, a self financing strategy that minimizes the variance of the residual hedging error can be found in Chapter 7 of Lamberton and Lapeyre (1997).

6. Pricing in jump-diffusion models

6.1. General aspects

With the introduction of jumps and/or stochastic volatility the market becomes incomplete. Consequently, the principle of absence of arbitrage does not lead to a uniquely defined price. One obtains actually an entire range of prices [see Eberlein and Jacod (1997), Bellamy and Jeanblanc (2000)] and the preference structure of the investors has to come into play to determine the *pricing measure*. From the point of view of pure pricing, the problem then reduces to determining a specific martingale measure or, equivalently, the market price of risk. To this effect there are various possibilities and in this section we mention some of them, the last two of which will be discussed in more detail.

(i) Historically it appears that a first approach to pricing in markets that are incomplete due to jumps in the prices and to a jumping volatility has been based on *general equilibrium* with a representative agent [see, e.g., Ahn and Thompson (1988), Naik and Lee (1990), Ahn (1992)].

(ii) A somewhat related and rather recent approach is that of *pricing by utility maximization*, in which the density of the martingale measure (the *pricing kernel*) is related to the marginal utility of terminal wealth [see, e.g., Frittelli (2000) and the references therein; for a specific jump-diffusion setting see Miyahara (1998)].

(iii) An alternative possibility is given by more econometric-type approaches based on *estimating/filtering the market price of risk* on the basis of market data. Related to such an approach is the approach described in Herzel (1998) for a diffusion model with a volatility that may jump at a random time and where the price of a European call turns out to be a monotone function of a parameter λ characterizing the martingale measures. There exists then a unique λ^* consistent with the option price thus allowing to price all the other derivatives consistently with this option. This corresponds basically to completing the market with the given option.

(iv) Approaches based on *market completion*. In the previous Section 4.1 we have discussed various ways to complete both stock as well as bond markets of the jump-diffusion type. As we have seen, this completion leads always to a unique martingale measure, but it does not necessarily imply also completeness in the sense that every claim can be hedged with a self financing portfolio. On the other hand, the uniqueness alone of an equivalent martingale measure is already sufficient to obtain a unique arbitrage-free price of a claim as the expectation of its discounted value under this measure. In all cases where one achieves also completeness in the sense of hedging (essentially all cases except when there are an infinite number of sources of randomness) then, always by absence of arbitrage, the (unique) initial value of the self financing and hedging portfolio has to coincide with the price computed as expectation under the unique martingale measure. The approach based on market completion has been widely used an implemented in various economic setups and here we mention just Shirakawa (1990, 1991), Jeanblanc-Piqué and Pontier (1990), Naik (1993), Mercurio and Runggaldier (1993), Jarrow and Madan (1995, 1999). It has the advantage to lead to a unique price on the basis of the principle of absence of arbitrage alone, without having to make assumptions on a non-priced jump risk and without the need to introduce a general equilibrium model. On the other hand it requires that the stochastic evolution of more than just the underlying asset has to be specified and, without specific criteria, the completion may occasionally be rather arbitrary.

(v) In the previous Section 5, in the context of hedging it was mentioned that, if the market cannot be completed, then one has to accept some residual risk and it becomes natural to determine the hedging strategy on the basis of a risk minimization criterion. On the other hand, in the previous point (iv) we recalled the fact that, in a complete/completed market the initial value of a self financing and hedging portfolio has to coincide with the arbitrage-free price of the claim. By analogy, it appears then natural to define as price of a claim in a noncomplete market the *initial value of a portfolio minimizing a given hedging criterion*. Quite typically, the initial value of such a portfolio turns out to be the expectation of the discounted value of the given claim under a specific martingale measure. In other words, there is a correspondence between hedging criteria and martingale measures and the choice of a specific pricing measure can be based on the choice of a specific hedging criterion. An approach along these lines appears thus related to the pricing approach by utility maximization, mentioned in point (ii) above. As an example, let us point out that the criterion of risk minimization discussed in Section 5.2 leads to the so-called *minimal martingale measure* that was already mentioned at the end of Section 4.1. It has been further shown in Runggaldier and Schweizer (1995) that, if in a jump-diffusion model claims are priced according to the minimal martingale measure, then convergence of asset prices implies convergence of option prices. This stability result for prices computed according to the minimal martingale measure makes the risk minimization criterion discussed in Section 5.2 an attractive criterion for hedging. [For further extensions of this stability property see Prigent (1999), Hubalek and Schachermayer (1998).]

6.2. Computational aspects

Assume that for a jump-diffusion model we have selected a specific martingale measure according to one of the approaches mentioned in the previous Section 6.1. We have then to compute the expectation of the (discounted value of the) claim under this martingale measure. In this section we shall mention some of the possible methods to accomplish this.

We consider first the univariate jump-diffusion model (45) under a generic martingale measure Q with intensity of the Poisson process N_t given by $\psi_t \lambda_t$. If Q corresponds to the unique martingale measure obtained from a market completion as in Section 4.1, then $\psi_t \lambda_t$ has to be taken according to (53). For simplicity we assume that all the prices are already discounted and so we can put $r_t \equiv 0$. The dynamics of S_t under Q are given by (see (48), (49), (47))

$$\mathrm{d}S_t = S_{t-}\big[-\gamma_t \psi_t \lambda_t \,\mathrm{d}t + \sigma_t \,\mathrm{d}w_t^Q + \gamma_t \,\mathrm{d}N_t\big]. \tag{111}$$

We want to compute the value of a European call option, namely $E^Q\{(S_T - K)^+\}$. For this purpose we adapt an approach from Mercurio and Runggaldier (1993), assuming first that in (111) we have $\gamma_t \equiv \gamma$, i.e., the jump coefficient is constant [for this case see also Aase (1988)]. We have

$$E^Q\big\{(S_T - K)^+\big\} = E^Q\big\{E^Q\big\{(S_T - K)^+ | N_T\big\}\big\}. \tag{112}$$

For a fixed k, i.e., when $N_T = k$ ($k = 0, 1, \ldots$), using the exponential formula (17) for the specific case when (14) is given by (111), we have

$$S_T^{(k)} = S_0 \,\mathrm{e}^{k \log(1+\gamma)} \exp\left[-\int_0^T \left(\gamma \psi_s \lambda_s + \frac{1}{2}\sigma_s^2\right) \mathrm{d}s + \int_0^T \sigma_s \,\mathrm{d}w_s^Q\right] \tag{113}$$

namely

$$\log S_T^{(k)} \sim \mathcal{N}\big(\cdot; m_T, \sigma_T^2\big) \tag{114}$$

with

$$\begin{cases} m_T = \log S_0 + k \log(1+\gamma) - \displaystyle\int_0^T \left(\gamma \psi_s \lambda_s + \frac{1}{2}\sigma_s^2\right) \mathrm{d}s, \\ \sigma_T^2 = \displaystyle\int_0^T \sigma_s^2 \,\mathrm{d}s, \end{cases} \tag{115}$$

i.e., $S_T^{(k)}$ is lognormal with mean and variance given by m_T and σ_T respectively. Next compute (with $\Phi(\cdot)$ the cumulative standard Gaussian distribution function)

$$V_0^{(k)} := E_{S_0}^Q\big\{\big(S_T^{(k)} - K\big)^+\big\} = \int_{\log k}^{+\infty} (\mathrm{e}^x - K)\,\mathrm{d}\mathcal{N}\big(x; m_T, \sigma_T^2\big)\,\mathrm{d}x$$

$$= \frac{1}{\sqrt{2\pi \sigma_T^2}} \int_{\log k}^{+\infty} \mathrm{e}^x \mathrm{e}^{-\frac{1}{2\sigma_T^2}(x-m_T)^2} \,\mathrm{d}x - \frac{K}{2\sigma_T^2} \int_{\log k}^{+\infty} \mathrm{e}^{-\frac{1}{2\sigma_T^2}(x-m_T)^2} \,\mathrm{d}x$$

$$
\begin{aligned}
&= \mathrm{e}^{m_T - \frac{1}{2}\sigma_T^2} \Phi\left(\frac{m_T + \sigma_T^2 - \log K}{\sigma_T}\right) - K\Phi\left(\frac{m_T - \log K}{\sigma_T}\right) \\
&:= (1+\gamma)^k G(k, S_0)
\end{aligned} \tag{116}
$$

with

$$
G(k, S_0) = S_0 \exp\left(-\int_0^T \gamma \psi_s \lambda_s \, \mathrm{d}s\right) \Phi(x) - \frac{K}{(1+\gamma)^k} \Phi(y), \tag{117}
$$

where

$$
\begin{cases}
x = \dfrac{\log(S_0(1+\gamma)^k/K) \int_0^T (-\gamma\psi_s\lambda_s + \frac{1}{2}\sigma_s^2)\, \mathrm{d}s}{\sqrt{\int_0^T \sigma_s^2 \, \mathrm{d}s}}, \\
y = x - \sqrt{\int_0^T \sigma_s^2 \, \mathrm{d}s}.
\end{cases} \tag{118}
$$

Coming back to (112) we then have

$$
E^Q\{(S_T - K)^+\} = E^Q\{V_0^{(N_T)}\} = \sum_{k=0}^{\infty} (1+\gamma)^k G(k, S_0)\left(\frac{H^k}{k!} \mathrm{e}^{-H}\right) \tag{119}
$$

with $H = \int_0^T \psi_s \lambda_s \, \mathrm{d}s$. Notice that, for actual computations, the infinite sum in the right in (119) has to be truncated at a sufficiently large positive integer.

The result for $\gamma_t \equiv \gamma$ can be easily extended to the case when γ_t is a piecewise constant deterministic time function. To this effect, given a positive integer m and a subdivision $0 = t_0^m < t_1^m < \cdots < t_m^m = T$, let

$$
\gamma_t^{(m)} = \gamma_0 1_{\{0\}}(t) + \sum_{j=1}^{m} \gamma_j 1_{(t_{j-1}^m, t_j^m]}(t); \quad \gamma_j > -1. \tag{120}
$$

Furthermore, let $\mathcal{P}_j$ $(j = 1, \ldots, m)$, be independent Poisson random variables with parameters $H_j = \int_{t_{j-1}^m}^{t_j^m} \psi_s \lambda_s \, \mathrm{d}s$. The generalization of formula (119) is then

$$
\begin{aligned}
&E^Q\{(S_T - K)^+\} \\
&= \sum_{k_1,\ldots,k_m=0}^{\infty} \exp\left[\sum_{j=1}^{m} k_j \log(1+\gamma_j)\right] G(k_1, \ldots, k_m, S_0) \prod_{j=1}^{m} \left[\frac{(H_j)^{k_j}}{(k_j)!} \mathrm{e}^{-H_j}\right]
\end{aligned} \tag{121}
$$

with

$$
G(k_1, \ldots, k_m, S_0) = S_0 \exp\left(-\int_0^T \gamma_s^{(m)} \psi_s \lambda_s \, \mathrm{d}s\right) \Phi(x) - \frac{K}{\prod_{j=1}^m (1+\gamma_j)^{k_j}} \Phi(y), \tag{122}
$$

where

$$\begin{cases} x = \dfrac{\log(S_0 \prod_{j=1}^{m}(1+\gamma_j)^{k_j}/K) \int_0^T (-\gamma_t^{(m)} \psi_s \lambda_s + \frac{1}{2}\sigma_s^2)\,\mathrm{d}s}{\sqrt{\int_0^T \sigma_s^2\,\mathrm{d}s}}, \\ y = x - \sqrt{\int_0^T \sigma_s^2\,\mathrm{d}s}. \end{cases} \tag{123}$$

Coming finally to the case of a more general deterministic time function γ_t for the jump coefficient, we assume that there exist piecewise constant deterministic time functions $\gamma_t^{(m)}$ and $\sigma_t^{(m)}$ such that

$$\gamma_t^{(m)} \uparrow \gamma_t, \quad \sigma_t^{(m)} \uparrow \sigma_t \quad \text{as } m \to \infty. \tag{124}$$

Consider then a sequence of fictitious risky assets, whose (discounted) values $S_t^{(m)}$ are martingales with respect to the same martingale measure Q as is S_t in (111), namely they satisfy

$$\mathrm{d}S_t^{(m)} = S_{t-}^{(m)}\big[-\gamma_t^{(m)} \psi_t \lambda_t\,\mathrm{d}t + \sigma_t^{(m)}\,\mathrm{d}w_t^Q + \gamma_t^{(m)}\,\mathrm{d}N_t\big]. \tag{125}$$

For each of the processes $S_t^{(m)}$ we can compute

$$v_0^{(m)} = E^Q\{(S_T^{(m)} - K)^+\} \tag{126}$$

according to (121)–(123). In Mercurio and Runggaldier (1993) it is now shown that

$$\lim_{m\to\infty} v_0^{(m)} = v_0 = E^Q\{(S_T - K)^+\}, \tag{127}$$

i.e., if γ_t is a generic time function, that can be approximated from below by a sequence of piecewise constant time functions, then the corresponding option value can be approximated arbitrarily closely by computable expressions. In Mercurio and Runggaldier (1993) it is also shown that, for given m, $v_0^{(m)}$ can be interpreted as initial value of a mean self financing and risk minimizing portfolio in the sense of Section 5.2 when the asset price evolves in discrete time according to the process $S_t^{(m)}$ of (125), evaluated at the discrete time points t_j. In line with the last part of point (v) of the previous Section 6.1, we may thus consider the approximating values $v_0^{(m)}$ as option values themselves, computed according to the minimal martingale measure.

After having discussed the univariate jump-diffusion model (45), we turn now to the general jump-diffusion model with a marked point process and which can equivalently be represented either by (32) or (33). We opt here for the representation (32). i.e.,

$$\mathrm{d}S_t = S_{t-}\big[\mu_t\,\mathrm{d}t + \sigma_t\,\mathrm{d}w_t + \gamma(t, Y_t)\,\mathrm{d}N_t\big]. \tag{128}$$

In what follows we shall make the further

Assumption 6.1.
(i) $\gamma(t, Y_t) \equiv \gamma(Y_t)$, i.e., γ is independent of the current time;

(ii) considering the representation of the marked point process as double sequence (T_n, Y_n), assume that T_n is independent of Y_n and the Y_n form a sequence of independent random variables, the generic one Y_n having law $m(\mathrm{d}y)$.

The driving marked point process has thus local P-characteristics $(\lambda_t, m(\mathrm{d}y))$.

Suppose that we have chosen a specific martingale measure Q and that we want to compute $v_0 = E^Q\{H(S_T)\}$ where, typically, we may have $H(S) = (S - K)^+$. For this purpose, in what follows we adapt a procedure from Chapter 7 in Lamberton and Lapeyre (1997).

Recall first from Theorem 2.5 that a general absolutely continuous measure transformation from P to Q transforms the P-local characteristics into Q-local characteristics of the form $(\psi_t\lambda_t, h_t(y)m(\mathrm{d}y))$. Recalling furthermore (63) with (62) and (64), it is easily seen that, under the measure Q corresponding to the above local characteristics, the discounted value of S_t satisfies

$$\mathrm{d}S_t = S_{t-}\left[-\overline{\Gamma}_t\bar{\lambda}_t\,\mathrm{d}t + \sigma_t\,\mathrm{d}w_t^Q + \gamma(Y_t)\,\mathrm{d}N_t\right], \tag{129}$$

where we have put $\overline{\Gamma}_t = \int_E \gamma(y)h_t(y)m(\mathrm{d}y)$ and $\bar{\lambda}_t = \psi_t\lambda_t$. Using the exponential formula (17) to integrate (129), that is of the form of (14) with the representation (15), one immediately finds that, for a given initial asset price S_0, the value $v_0(S_0)$ of the claim $H(S_T)$ is given by

$$v_0(S_0) = E^Q\left\{H\left(S_0\exp\left[-\int_0^T\left(\overline{\Gamma}_t\bar{\lambda}_t + \frac{\sigma_t^2}{2}\right)\mathrm{d}t + \int_0^T\sigma_t\,\mathrm{d}w_t^Q\right]\prod_{n=1}^{N_T}(1+\gamma(Y_n))\right)\right\}. \tag{130}$$

Next let

$$V(S_0) := E^Q\left\{H\left(S_0\exp\left[-\int_0^T\frac{\sigma_t^2}{2}\,\mathrm{d}t + \int_0^T\sigma_t\,\mathrm{d}w_t^Q\right]\right)\right\} \tag{131}$$

so that, for $H(S) = (S - K)^+$, the $V(S_0)$ is given by the Black–Scholes formula, i.e., $V(S_0) = BS(S_0)$. With the use of $V(S_0)$ we can now write

$$\begin{aligned} v_0(S_0) &= E^Q\left\{V\left(S_0\exp\left[-\int_0^T\overline{\Gamma}_t\bar{\lambda}_t\,\mathrm{d}t\right]\prod_{n=1}^{N_T}(1+\gamma(Y_n))\right)\right\} \\ &= \sum_{k=0}^{\infty}E^Q\left\{V\left(S_0\exp\left[-\int_0^T\overline{\Gamma}_t\bar{\lambda}_t\,\mathrm{d}t\right]\prod_{n=1}^{k}(1+\gamma(Y_n))\right)\right\}\left(\frac{H^k}{k!}\mathrm{e}^{-H}\right), \end{aligned} \tag{132}$$

where, due to the local characteristics under Q, we have $H = \int_0^T\psi_s\lambda_s\,\mathrm{d}s$ and where the expectation is with respect to the joint distribution of the Y_n that in Assumption 6.1 were supposed to be independent. This latter expectation can be explicitly computed in special cases, in more complicated cases one has to use simulations. Again, for the actual computations, the infinite sum has to be truncated at a sufficiently large positive integer.

We close this section by mentioning that in Glasserman and Kou (1999), for the term structure models of simple forwards in the jump-diffusion setup described therein, the authors study the pricing of some derivative securities after having characterized arbitrage-free dynamics. The derivative prices are also used to investigate what types of patterns in implied volatilities are produced through jumps.

References

Aase, K.K., 1988. Contingent claim valuation when the security price is a combination of an Ito process and a random point process. Stochastic Processes and their Applications 28, 185–220.

Ahn, C., 1992. Option pricing when jump risk is systematic. Mathematical Finance 2, 299–308.

Ahn, C.M., Thompson, H.E., 1988. Jump-diffusion processes and the term structure of interest rates. Journal of Finance 43, 155–174.

Babbs, S., Webber, N., 1997. Term structure modelling under alternative official regimes. In: Dempster, M.H.A., Pliska, S.R. (Eds.), Mathematics of Derivative Securities. Cambridge University Press.

Bakshi, G., Cao, C., Chen, Z., 1997. Empirical performance of alternative option pricing models. Journal of Finance 52 (5), 2003–2049.

Ball, C.A., Torous, W.N., 1985. On jumps in common stock prices and their impact on call option pricing. Journal of Finance 40 (1), 155–173.

Becherer, D., 2001. The numeraire portfolio for unbounded semimartingales. Finance and Stochastics 5, 327–341.

Bellamy, N., Jeanblanc, M., 2000. Incompleteness of markets driven by a mixed diffusion. Finance and Stochastics 4, 201–222.

Björk, T., Kabanov, Yu., Runggaldier, W.J., 1997. Bond market structure in the presence of marked point processes. Mathematical Finance 40 (1), 211–239.

Björk, T., Di Masi, G.B., Kabanov, Yu., Runggaldier, W.J., 1997. Towards a general theory of bond markets. Finance and Stochastics 1, 141–174.

Björk, T., Näslund, B., 1998. Diversified portfolios in continuous time. European Financial Review 1, 361–387.

Brémaud, P., 1981. Point Processes and Queues: Martingale Dynamics. Springer-Verlag, New York.

Cox, D., 1955. Some statistical methods connected with series of events. Journal of the Royal Statistical Society. Series B 17, 129–164.

Cox, J.C., Ross, S.A., 1976. The valuation of options for alternative stochastic processes. Journal of Financial Economics 3, 145–166.

Di Masi, G.B., Platen, E., Runggaldier, W.J., 1995. Hedging of options under discrete observation on assets with stochastic volatility. In: Seminar on Stochastic Analysis, Random Fields and Applications. In: Progress in Probability, Vol. 36. Birkhäuser, pp. 359–364.

Eberlein, E., Jacod, J., 1997. On the range of options prices. Finance and Stochastics 2, 131–140.

Fischer, P., Platen, E., Runggaldier, W.J., 1999. Risk-minimizing hedging strategies under partial information. In: Seminar on Stochastic Analysis, Random Fields and Applications. In: Progress in Probability, Vol. 45. Birkhäuser, pp. 173–186.

Föllmer, H., Schweizer, M., 1991. Hedging of contingent claims under incomplete information. In: Davis, M.H.A., Elliott, R.J. (Eds.), Applied Stochastic Analysis. In: Stochastic Monographs, Vol. 5. Gordon and Breach, London, pp. 389–414.

Föllmer, H., Sondermann, D., 1986. Hedging of non-redundant contingent claims. In: Hildenbrand, W., Mas-Colell, A. (Eds.), Contributions to Mathematical Economics, North-Holland, pp. 205–223.

Framstad, N.C., Oeksendal, B., Sulem, A., 2001. Optimal consumption and portfolio in a jump-diffusion market with proportional transaction costs. Journal of Mathematical Economics 35, 233–257.

Frey, R., 2000. Risk minimization with incomplete information in a model for high frequency data. Mathematical Finance 10, 215–225.

Frey, R., Runggaldier, W.J., 1999. Risk-minimizing hedging strategies under restricted information: The case of stochastic volatility models observable only at discrete random times. Mathematical Methods Operations Research 50, 339–350.

Frey, R., Runggaldier, W.J., 2001. A nonlinear filtering approach to volatility estimation with a view towards high frequency data. International Journal of Theoretical and Applied Finance 4 (2), 199–210.

Frittelli, M., 2000. The minimal entropy martingale measure and the valuation problem in incomplete markets. Mathematical Finance 10, 39–52.

Geman, H., Madan, D., Yor, M., 1999. Asset prices are Brownian motions: only in business time. In: Avellaneda, M. (Ed.), Quantitative Analysis in Financial Markets. World Scientific, Singapore.

Glasserman, P., Kou, S.G., 1999. The term structure of simple forward rates with jump risk. Preprint. Columbia University.

Herzel, S., 1998. A simple model for option pricing with jumping stochastic volatility. International Journal of Theoretical and Applied Finance 1, 487–505.

Hubalek, F., Schachermayer, W., 1998. When does convergence of asset price processes imply convergence of option prices? Mathematical Finance 5, 385–403.

Jacod, J., Shiryaev, A.N., 1987. Limit Theorems for Stochastic Processes. Springer-Verlag, Berlin.

Jamshidian, F., 1999. LIBOR market model with semimartingales. Working Paper. Net Analytic Ltd., London.

Jarrow, R., Madan, D., 1995. Option pricing using the term structure of interest rates to hedge systematic discontinuous asset returns. Mathematical Finance 5, 311–336.

Jarrow, R., Madan, D., 1999. Hedging contingent claims on semimartingales. Finance and Stochastics 3, 111–134.

Jeanblanc-Piqué, M., Pontier, M., 1990. Optimal portfolio for a small investor in a market model with discontinuous prices. Applied Mathematical Optimization 22, 287–310.

Jensen, B., 1999. Option pricing in the jump-diffusion model with a random jump amplitude: A complete market approach. Centre for Analytical Finance. WP Series No. 42, Aarhus.

Jorion, P., 1988. On jump processes in the foreign exchange and stock markets. Review of Financial Studies 1 (4), 427–445.

Lamberton, D., Lapeyre, B., 1997. Introduction au Calcul Stochastique Appliqué à la Finance. Ellipses. English translation by Chapman and Hall.

Mercurio, F., Runggaldier, W.J., 1993. Option pricing for jump-diffusions: Approximations and their interpretation. Mathematical Finance 3, 191–200.

Merton, R., 1976. Option pricing when the underlying stock returns are discontinuous. Journal of Financial Economics 5, 125–144.

Miyahara, Y., 1998. Minimal entropy martingale measures of jump type processes in incomplete asset markets. WP Nagoya City University.

Mulinacci, S., 1996. An approximation of American option prices in a jump-diffusion model. Stochastic Processes and their Applications 62, 1–17.

Naik, V., 1993. Option valuation and hedging strategies with jumps in the volatility of asset returns. Journal of Finance 48 (5), 1969–1984.

Naik, V., Lee, M., 1990. General equilibrium pricing of options on the market portfolio with discontinuous returns. Review of Financial Studies 3, 493–521.

Pham, H., 1997. Optimal stopping, free boundary and American option in a jump diffusion model. Applied Mathematical Optimization 35, 145–164.

Prigent, J.L., 1999. Incomplete markets: convergence of option values under the minimal martingale measure. Advances in Applied Probabilities 31, 1058–1077.

Prigent, J.L., 2001. Option pricing with a general marked point process. Mathematics of Operations Research 26 (1), 50–66.

Rogers, L.C.G., Zane, O., 1998. Designing models for high frequency data. Preprint. University of Bath.

Runggaldier, W.J., Schweizer, M., 1995. Convergence of option values under incompleteness. In: Seminar on Stochastic Analysis, Random Fields and Applications. In: Progress in Probability, Vol. 36. Birkhäuser, pp. 365–384.

Rydberg, T., Shephard, N., 1999. A modelling framework for prices and trades made at the New York stock exchange. Nuffield College working paper series 1999-W14.

Schweizer, M., 1994. Risk minimizing hedging strategies under restricted information. Mathematical Finance 4, 327–342.

Shirakawa, H., 1990. Security market model with Poisson and diffusion type return process. Institute of Human and Social Sciences, Tokyo Institute of Technology.

Shirakawa, H., 1991. Interest rate option pricing with Poisson–Gaussian forward rate curve processes. Mathematical Finance 1, 77–94.

Chapter 6

HYPERBOLIC PROCESSES IN FINANCE

BO MARTIN BIBBY

Department of Mathematics and Physics, The Royal Veterinary and Agricultural University, Thorvaldsensvej 40, DK-1871 Frederiksberg C, Denmark

MICHAEL SØRENSEN

Department of Statistics and Operations Research, Institute of Mathematical Sciences, University of Copenhagen, Universitetsparken 5, DK-2100 København Ø, Denmark

Contents

Handbook of Heavy Tailed Distributions in Finance, Edited by S.T. Rachev

Abstract

Distributions that have tails heavier than the normal distribution are ubiquitous in finance. For purposes such as risk management and derivative pricing it is important to use relatively simple models that can capture the heavy tails and other relevant features of financial data. A class of distributions that is very often able to fit the distributions of financial data is the class of generalized hyperbolic distributions. This has been established in numerous investigations, see, e.g., Eberlein ad Keller (1995), Bibby and Sørensen (1997), Hurst (1997), Eberlein, Keller and Prause (1998), Rydberg (1999), Küchler et al. (1999), Jiang (2000), and Barndorff-Nielsen and Shephard (2001c). The class of generalized hyperbolic distributions includes the standard hyperbolic distributions, the normal inverse Gaussian distributions, the scaled t-distributions and the variance-gamma distributions. The use of scaled t-distributions in finance was studied by Praetz (1972) and Blattberg and Gonedes (1974), while Madan and Seneta (1990) introduced the variance-gamma distributions in the financial literature. The normal distribution appears as a limit of generalized hyperbolic distributions. The tail behaviour of the generalized hyperbolic distributions thus span a range from Gaussian tails via exponential tails to the power tails of the t-distributions.

In Section 1 we present the generalized hyperbolic distributions and their most important properties. We also discuss the generalized inverse Gaussian distributions which play an important role in the theory of generalized hyperbolic distributions and processes. This class of distributions is also of interest in its own right as a model of positive quantities in finance. Its right-hand tail behaviour spans a range from exponential decrease to a Pareto tail. In the following sections we present a number of stochastic process models for which the marginal distributions or the distributions of increments (or both) are generalized hyperbolic. The models are increasingly complex. They are thus able to fit an increasing number of the stylized features of financial data. The well established features of financial data are for instance reviewed in Barndorff-Nielsen (1998) and Rydberg (2000). In Section 2 we discuss Lévy process models, while in Section 3 we discuss models defined by stochastic differential equations. These include classical diffusion models and Ornstein–Uhlenbeck models driven by Lévy processes as well as superpositions of such models. In the final Section 4 we present generalized hyperbolic stochastic volatility models.

1. Hyperbolic and related distributions

In this section we present the *generalized hyperbolic distributions* and describe their most important properties. We will also discuss the *generalized inverse Gaussian distributions* which play an important role in the theory of generalized hyperbolic distributions and processes. As mentioned earlier, this class of distributions is also of independent interest as a model of positive quantities in finance. We will present a few examples of how well these distributions fit financial data.

1.1. The generalized hyperbolic distribution

The generalized hyperbolic distributions were introduced by Barndorff-Nielsen (1977) and include, among others, the *hyperbolic distributions*, the *normal-inverse Gaussian* (*NIG*) *distributions*, the *scaled t-distributions* and the *variance-gamma distributions*. We shall discuss these sub-classes in more detail later. First we present the generalized hyperbolic distributions and their properties.

A generalized hyperbolic distribution has five parameters. If X follows a generalized hyperbolic distribution we write

$$X \sim H(\lambda, \alpha, \beta, \delta, \mu).$$

The probability density function of a generalized hyperbolic distribution is given by

$$\frac{(\gamma/\delta)^{\lambda}}{\sqrt{2\pi}K_{\lambda}(\delta\gamma)} \cdot \frac{K_{\lambda-1/2}(\alpha\sqrt{\delta^2+(x-\mu)^2})}{(\sqrt{\delta^2+(x-\mu)^2}/\alpha)^{1/2-\lambda}} \cdot \mathrm{e}^{\beta(x-\mu)}, \quad x \in \mathbb{R}, \tag{1}$$

where $\gamma^2 = \alpha^2 - \beta^2$, and K_λ is the modified Bessel function of the third kind with index λ. Definitions and results concerning Bessel functions are collected in an appendix.

The parameter domain for the class of generalized hyperbolic distributions is given by

$$\begin{aligned}
&\delta \geqslant 0, \quad \alpha > 0, \quad \alpha^2 > \beta^2, \quad \text{if } \lambda > 0,\\
&\delta > 0, \quad \alpha > 0, \quad \alpha^2 > \beta^2, \quad \text{if } \lambda = 0,\\
&\delta > 0, \quad \alpha \geqslant 0, \quad \alpha^2 \geqslant \beta^2, \quad \text{if } \lambda < 0.
\end{aligned}$$

In all cases $\mu \in \mathbb{R}$. If $\delta = 0$ or $\alpha^2 = \beta^2$ the generalized hyperbolic density in (1) is defined as the limit expression obtained by using (A.5). Note that if β is equal to zero, the distribution is symmetric.

The class of generalized hyperbolic distributions is closed under affine transformation. That is, if $X \sim H(\lambda, \alpha, \beta, \delta, \mu)$ and Y is defined as $Y = aX + b$, for some positive a, then we have that

$$Y \sim H\left(\lambda, \frac{\alpha}{a}, \frac{\beta}{a}, a\delta, a\mu + b\right). \tag{2}$$

From (2) we also see that the parameter λ is invariant under affine transformations of a generalized hyperbolic random variable.

From (A.3) it follows that the mode points for the generalized hyperbolic distribution are solutions to the equation

$$\frac{x-\mu}{\sqrt{\delta^2+(x-\mu)^2}} \cdot \frac{K_{\lambda-3/2}(\alpha\sqrt{\delta^2+(x-\mu)^2})}{K_{\lambda-1/2}(\alpha\sqrt{\delta^2+(x-\mu)^2})} = \frac{\beta}{\alpha}. \tag{3}$$

If $\beta = 0$, it follows immediately that the distribution is unimodal with mode point μ. If $\lambda \geqslant \frac{3}{2}$, the ratio of the modified Bessel functions in (3) increases monotonically from 0 to 1, and therefore the distribution is unimodal. See Blæsild (1978) for further discussion of features of the generalized hyperbolic density function.

The Laplace transform of the generalized hyperbolic distribution is given by

$$L(z) = e^{\mu z} \cdot \frac{\gamma^\lambda \cdot K_\lambda(\delta\gamma_z)}{\gamma_z^\lambda \cdot K_\lambda(\delta\gamma)}, \quad |\beta + z| < \alpha, \tag{4}$$

where $\gamma_z^2 = \alpha^2 - (\beta+z)^2$. From (A.3) we get that

$$\mathrm{E}X = \mu + \frac{\delta\beta K_{\lambda+1}(\delta\gamma)}{\gamma K_\lambda(\delta\gamma)}, \tag{5}$$

and

$$\operatorname{Var} X = \frac{\delta K_{\lambda+1}(\delta\gamma)}{\gamma K_\lambda(\delta\gamma)} + \frac{\beta^2\delta^2}{\gamma^2}\left(\frac{K_{\lambda+2}(\delta\gamma)}{K_\lambda(\delta\gamma)} - \frac{K_{\lambda+1}^2(\delta\gamma)}{K_\lambda^2(\delta\gamma)}\right). \tag{6}$$

Expressions for the skewness and kurtosis involve modified Bessel functions in a rather complicated way and can be found in Barndorff-Nielsen and Blæsild (1980).

Sometimes it is useful to reparametrize the generalized hyperbolic density in terms of the parameters λ, τ, ζ, δ, and μ, where $\tau = \beta/\gamma$ and $\zeta = \delta\gamma$. Using this parametrization, the generalized hyperbolic density has the form,

$$\frac{\sqrt{\zeta}}{\sqrt{2\pi}\delta K_\lambda(\zeta)} \cdot \frac{K_{\lambda-1/2}(\zeta\sqrt{1+\tau^2}\sqrt{1+((x-\mu)/\delta)^2})}{(\sqrt{1+((x-\mu)/\delta)^2}/\sqrt{1+\tau^2})^{1/2-\lambda}} \cdot e^{\tau\zeta(x-\mu)/\delta}, \quad x \in \mathbb{R}. \tag{7}$$

The parameters τ, ζ, and λ are invariant under affine transformations of a random variable following the generalized hyperbolic distribution. More precisely, the result equivalent to (2) is that $Y \sim H(\lambda, \tau, \zeta, a\delta, a\mu + b)$. From this result we see that δ is a scaling parameter and μ is a location parameter. In Figure 1 generalized hyperbolic densities are drawn for different values of λ, τ, and ζ. In all cases the mean value is 0 and the variance is 1. The tail behaviour of the distributions is more easily seen in Figure 2, where the logarithm of the same densities are plotted.

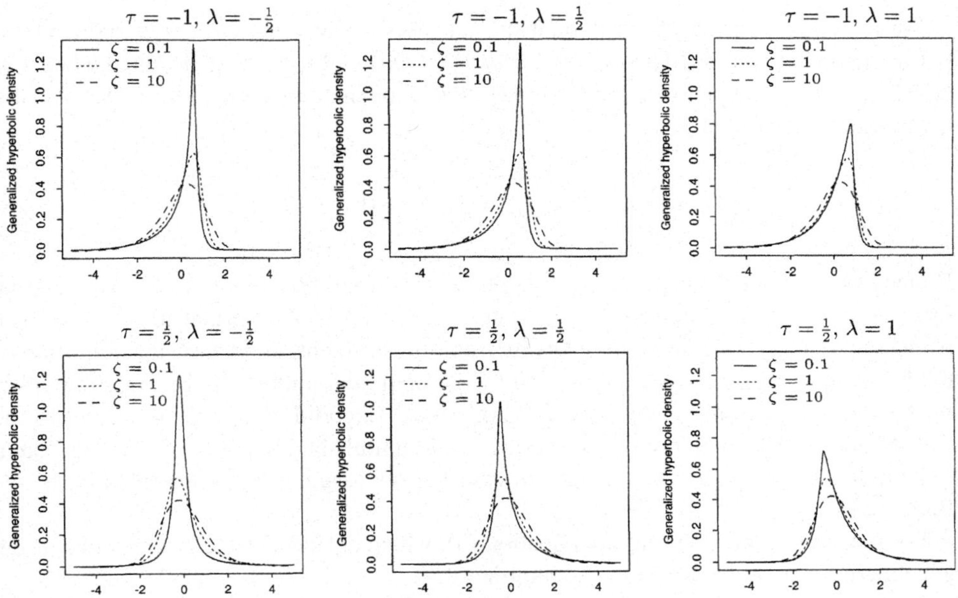

Fig. 1. Generalized hyperbolic densities with mean 0 and variance 1 for different values of the parameters λ, τ, and ζ.

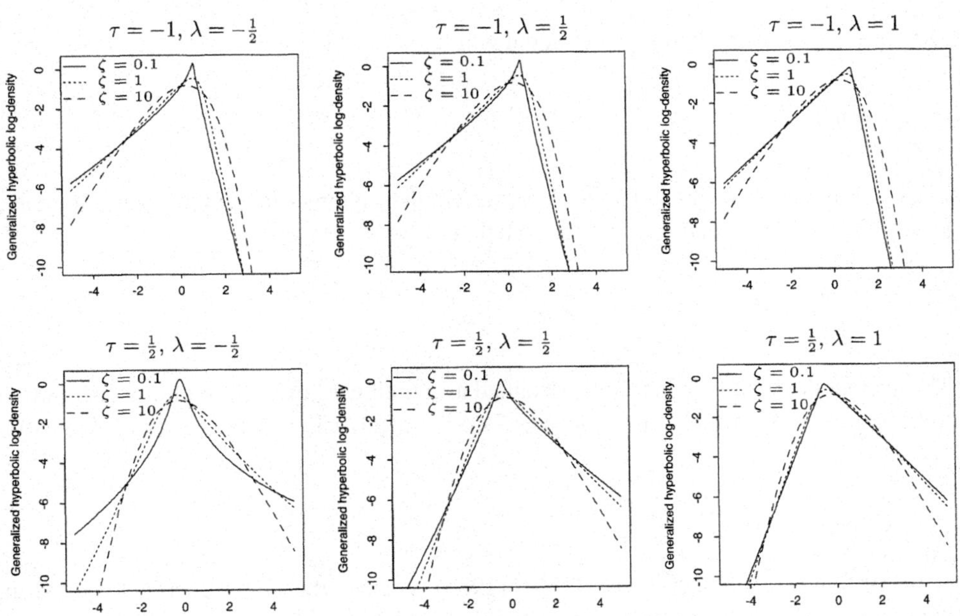

Fig. 2. The logarithm of generalized hyperbolic densities with mean 0 and variance 1 for different values of the parameters λ, τ, and ζ.

We shall now consider the important special cases of the generalized hyperbolic distribution mentioned earlier. The *hyperbolic distributions* is the subclass obtained when λ is equal to 1. With λ equal to 1 in (1), we get the following expression for the density of a hyperbolic distribution,

$$\frac{\gamma}{2\alpha\delta K_1(\delta\gamma)} \exp\left\{-\alpha\sqrt{\delta^2+(x-\mu)^2}+\beta(x-\mu)\right\}, \quad x \in \mathbb{R}. \tag{8}$$

From (8) we see that the logarithm of the density of a hyperbolic distribution is a hyperbola, which should be compared to the parabolic log-density of the normal distribution. The name of the hyperbolic distribution stems from this observation. In fact, the definition of the hyperbolic distributions was inspired by the empirical finding by the founding father of the physics of wind blown sand, Brigadier R.A. Bagnold, that the log-density of the distribution of the logarithm of the grain size of natural sand deposits looks more like a hyperbola than like a parabola, as had previously been assumed by geomorphologists, see Bagnold (1941).

For the hyperbolic distributions Equation (3), which determines the mode points of the generalized hyperbolic distribution, simplifies to

$$\frac{x-\mu}{\sqrt{\delta^2+(x-\mu)^2}} = \frac{\beta}{\alpha},$$

implying that the distribution is unimodal with mode point

$$x = \mu + \frac{\delta\beta}{\gamma}.$$

Letting δ tend to zero and using (A.5), we get the asymmetric Laplace distribution as a special case of the hyperbolic distribution, that is,

$$\frac{\alpha^2-\beta^2}{2\alpha} e^{\beta(x-\mu)-\alpha|x-\mu|}, \quad x \in \mathbb{R}.$$

The normal distribution can also be obtained as a limit case of the hyperbolic distribution. Letting $\alpha, \delta \to \infty$ in such a way that $\delta/\alpha \to \sigma^2$, we get, using (A.6), the normal density:

$$\frac{1}{\sqrt{2\pi\sigma^2}} e^{-\frac{1}{2\sigma^2}(x-\mu)^2}, \quad x \in \mathbb{R}.$$

According to Barndorff-Nielsen et al. (1985) we have that the skewness (γ_1) and the kurtosis (γ_2) for large values of ζ and small values of β/α satisfy that

$$(\gamma_1, \gamma_2) \sim \left(3\chi, 3\xi^2\right),$$

where

$$\chi = \frac{\beta/\alpha}{\sqrt{1+\zeta}} \quad \text{and} \quad \xi = \frac{1}{\sqrt{1+\zeta}}.$$

Based on this observation Barndorff-Nielsen et al. (1985) suggested that the parameters χ and ξ are natural measures of asymmetry and "kurtosis" for the hyperbolic distribution. Note that they are invariant under location-scale transformations. The parameters χ and ξ vary in the so-called *shape triangle* defined by

$$\{(\chi, \xi) \in \mathbb{R}^2 \mid 0 \leqslant |\chi| < \xi < 1\}. \tag{9}$$

Note that the normal and the (possibly skew) Laplace distributions are obtained as limit distributions when $\xi \to 1$ and $\xi \to 0$, respectively. In Figure 3 hyperbolic log density functions are plotted for different values of χ and ξ in the shape triangle.

In Figure 4 a histogram based on 2666 observations of the daily returns of IBM-stocks (returns are increments on a logarithmic scale of the stock prices) in the period from 1 January 1990 to 20 March 2000 is given. Each point indicates the mid-point of the top of a column in the histogram. The best generalized hyperbolic, hyperbolic, and normal densities are superimposed on the histogram. The parameter values corresponding to the generalized

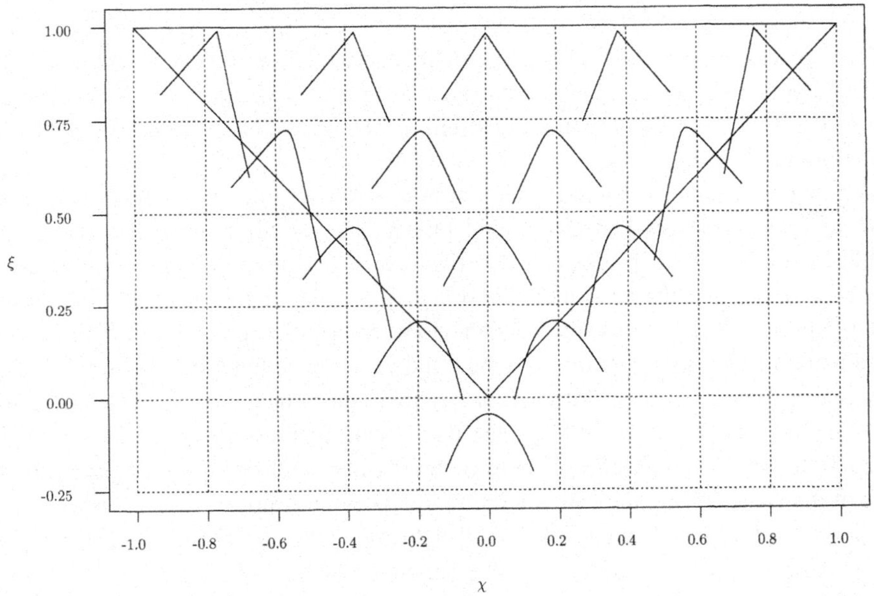

Fig. 3. Hyperbolic log densities with mean 0 and variance 1 for different values of the parameters χ and ξ ($-0.8, -0.6, \ldots, 0.8$ for χ and $0.0, 0.25, \ldots, 1.0$ for ξ). The log densities are placed at the corresponding values of χ and ξ.

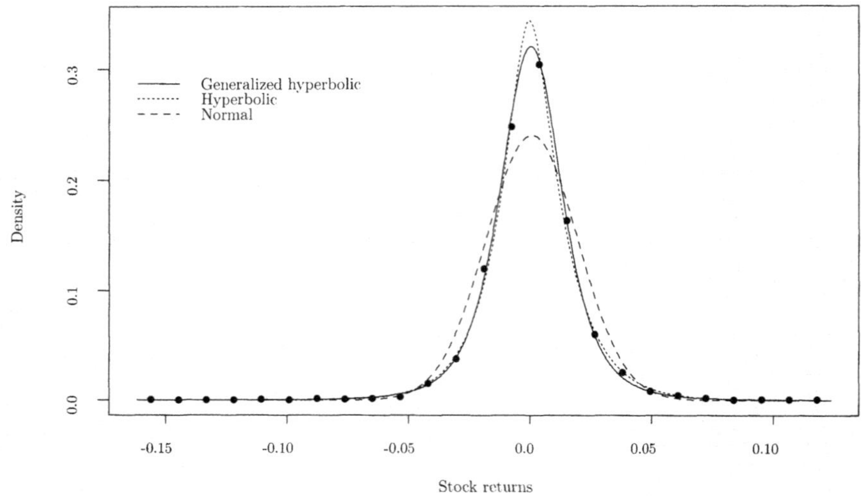

Fig. 4. A histogram of 2666 daily IBM-stock returns. Superimposed are the best fitting generalized hyperbolic, hyperbolic, and normal densities. The parameter values corresponding to the generalized hyperbolic density are $\alpha = 5.174$, $\beta = 0.0048$, $\delta = 0.0262$, $\mu = 0.0002$, and $\lambda = -1.933$. The parameter values corresponding to the hyperbolic density are $\alpha = 82.26$, $\beta = 3.725$, $\delta = 0.0060$, and $\mu = -0.0007$.

hyperbolic density are $\alpha = 5.174$, $\beta = 0.0048$, $\delta = 0.0262$, $\mu = 0.0002$, and $\lambda = -1.933$. For the hyperbolic density the parameter values are $\alpha = 82.26$, $\beta = 3.725$, $\delta = 0.0060$, and $\mu = -0.0007$. In Figure 5 the logarithms of the same histogram points and the same densities are plotted.

Log-histograms and log-densities are very useful when the interest is focussed on tail behaviour. From Figures 4 and 5 it is evident that a heavy-tailed distribution such as a generalized hyperbolic or hyperbolic distribution provides a good fit to the data, and certainly a much better fit than the normal distribution, in particular in the tails. A plot like Figure 5, which emphasizes differences in tail behaviour, reveals that the extreme tails of the histogram are a bit heavier than those of the fitted generalized hyperbolic distribution. There is no reason to be overly concerned about this minor discrepancy, because, first, it should be remembered that it is measured on a logarithmic scale, and secondly, the two log-histogram points in the extreme left tail are based on only 1 and 2 observations, respectively, while each of the two points in the extreme right tail represents 2 observations.

The *normal-inverse Gaussian (NIG) distributions* is the subclass obtained for λ equal to $-\frac{1}{2}$. The density of the normal-inverse Gaussian distribution is given by

$$\frac{\alpha\delta}{\pi} e^{\delta\gamma} \cdot \frac{K_1(\alpha\sqrt{\delta^2 + (x-\mu)^2})}{\sqrt{\delta^2 + (x-\mu)^2}} \cdot e^{\beta(x-\mu)}, \quad x \in \mathbb{R}. \tag{10}$$

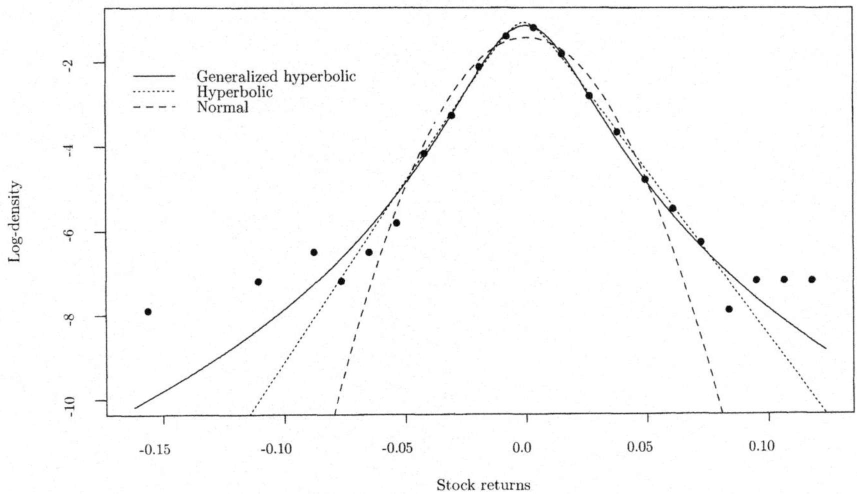

Fig. 5. The logarithm of the histogram in Figure 4 of 2666 daily IBM-stock returns. Superimposed are the logarithms of the best fitting generalized hyperbolic, hyperbolic, and normal densities. The parameter values are as in Figure 4.

If the distribution of X has density function (10), we write

$$X \sim NIG(\alpha, \beta, \delta, \mu).$$

If we let α tend to zero, it follows from (A.5) that the *NIG*-distribution converges to the Cauchy distribution with location parameter μ and scale parameter δ.

The Laplace transform of a *NIG*-distribution is especially simple:

$$L(z) = \mathrm{e}^{\mu z + \delta(\gamma - \gamma_z)}, \quad |\beta + z| < \alpha, \tag{11}$$

where $\gamma_z^2 = \alpha^2 - (\beta + z)^2$. Expressions for the mean and variance are also simple in the case of a *NIG*-distribution:

$$\mathrm{E}X = \mu + \frac{\delta\beta}{\gamma}, \qquad \mathrm{Var}\, X = \frac{\delta\alpha^2}{\gamma^3}.$$

The skewness is $3\delta\alpha^2\beta\gamma^{-5}$ and the kurtosis is $3\delta\alpha^2(\alpha^2 + 4\beta^2)\gamma^{-7}$. Although these expressions are quite simple, it is also for the *NIG*-distributions informative to use the shape triangle, which can be defined in complete analogy with that for the hyperbolic distributions, see, e.g., Rydberg (1997). In Figure 6 *NIG* log-density functions are drawn for different values of χ and ξ in the shape triangle defined in the same way as for the hyperbolic distribution.

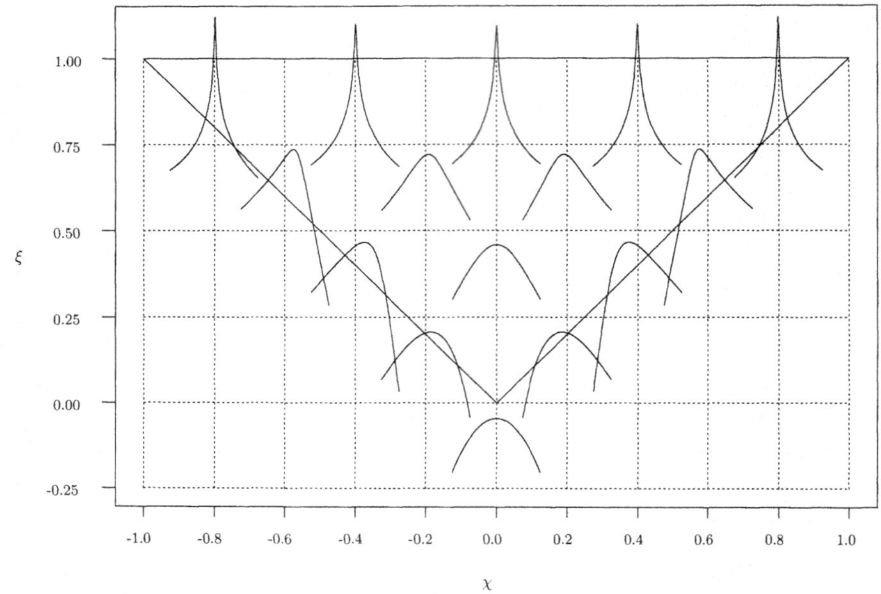

Fig. 6. Normal-inverse Gaussian log densities with mean 0 and variance 1 for different values of the parameters χ and ξ ($-0.8, -0.6, \ldots, 0.8$ for χ and $0.0, 0.25, \ldots, 1.0$ for ξ). The log-densities are placed in the shape triangle at the corresponding values of χ and ξ.

Finally, but not least, the class of normal-inverse Gaussian distributions is closed under convolution when the parameters α and β are fixed, that is if X_1 and X_2 are independent so that $X_i \sim NIG(\alpha, \beta, \delta_i, \mu_i)$, $i = 1, 2$, then we have that

$$X_1 + X_2 \sim NIG(\alpha, \beta, \delta_1 + \delta_2, \mu_1 + \mu_2). \tag{12}$$

Only two subclasses of the generalized hyperbolic distributions are closed under convolution. The other class with this important property is the class of *variance-gamma* (*VG*) distributions, which is obtained when δ is equal to 0. This is only possible when $\lambda > 0$ and $\alpha > |\beta|$. The variance-gamma distributions (with $\beta = 0$) were introduced in the financial literature by Madan and Seneta (1990). Another and perhaps more natural name for the full class is the *normal-gamma* (*NG*) distributions. The density function is given by

$$\frac{\gamma^{2\lambda}}{\sqrt{\pi}\,\Gamma(\lambda)(2\alpha)^{\lambda-1/2}} |x-\mu|^{\lambda-1/2} K_{\lambda-1/2}\big(\alpha|x-\mu|\big)\, \mathrm{e}^{\beta(x-\mu)}, \quad x \in \mathbb{R}, \tag{13}$$

where Γ denotes the gamma-function. If X follows a variance-gamma distribution, we write

$$X \sim VG(\lambda, \alpha, \beta, \mu).$$

The reader is reminded that the parameter domain is $\lambda > 0$, $\alpha > |\beta| \geqslant 0$ and $\mu \in \mathbb{R}$. The Laplace transform of a *VG*-distribution is simple:

$$L(z) = \mathrm{e}^{\mu z}\left(\frac{\gamma}{\gamma_z}\right)^{2\lambda}, \quad |\beta + z| < \alpha, \tag{14}$$

where again $\gamma_z^2 = \alpha^2 - (\beta + z)^2$. From (14) (or from (5) and (6)) it easily follows that

$$\mathrm{E}X = \mu + \frac{2\beta\lambda}{\gamma^2}, \qquad \mathrm{Var}\, X = \frac{2\lambda}{\gamma^2}\left(1 + 2\left(\frac{\beta}{\gamma}\right)^2\right).$$

The class of variance-gamma distributions is closed under convolution when α and β are fixed. If X_1 and X_2 are independent random variables such that $X_i \sim VG(\lambda_i, \alpha, \beta, \mu_i)$, $i = 1, 2$, then we have that

$$X_1 + X_2 \sim VG(\lambda_1 + \lambda_2, \alpha, \beta, \mu_1 + \mu_2). \tag{15}$$

This convolution property follows from (14).

By (A.6), the tails of a *VG*-distribution decrease as $|x|^{\lambda-1}\,\mathrm{e}^{-\alpha|x|+\beta x}$ when $x \to \pm\infty$. The logarithm of the densities of variance-gamma distributions are plotted for different values of λ in Figure 7. In all cases $\beta = 0$, the mean is zero, and the variance is one. From this figure appears a disadvantage of the class of *VG*-distributions. The probability density is very peaked at the centre for $\lambda < 1$, while for $\lambda \geqslant 1$ the tail-behaviour does not fit the tails found in typical financial data like those in Figure 5 as well as other generalized hyperbolic distributions like for instance the *NIG*-distribution.

We will finally consider the subclass of the generalized hyperbolic distributions that is obtained when $\alpha = |\beta|$, or equivalently $\gamma = 0$. This is only possible when $\lambda < 0$ and $\delta > 0$. It is convenient to introduce the reparametrization $\nu = -2\lambda$. For $\gamma = 0$ we obtain the density function

$$\frac{\delta^\nu}{\sqrt{\pi}\,2^{(\nu-1)/2}\Gamma(\nu/2)} \cdot \frac{K_{(\nu+1)/2}(|\beta|\sqrt{\delta^2 + (x-\mu)^2})}{(\sqrt{\delta^2 + (x-\mu)^2}/|\beta|)^{(\nu+1)/2}} \cdot \mathrm{e}^{\beta(x-\mu)}, \quad x \in \mathbb{R}, \tag{16}$$

where $\nu > 0, \delta > 0, \beta \in \mathbb{R}$ and $\mu \in \mathbb{R}$. A natural name for this distribution is the *asymmetric scaled t-distribution,* as will soon be clear. From (A.6) it follows that when β is positive, the left-hand tail decreases as $|x|^{-(\nu/2+1)}\,\mathrm{e}^{2\beta x}$, while the right-hand tail decreases as $x^{-(\nu/2+1)}$. When β is negative, the behaviour of the two tails is interchanged. The expectation exists provided $\nu > 2$, and the variance exists when $\nu > 4$. More generally, the n-th moment exists when $\nu > 2n$. The Laplace transform of the distribution given by (16) is

$$\mathrm{e}^{\mu z}\frac{(-\delta z(z + 2\beta))^{\nu/2} K_{\nu/2}(-\delta z(z + 2\beta))}{\Gamma(\nu/2)2^{\nu/2-1}} \tag{17}$$

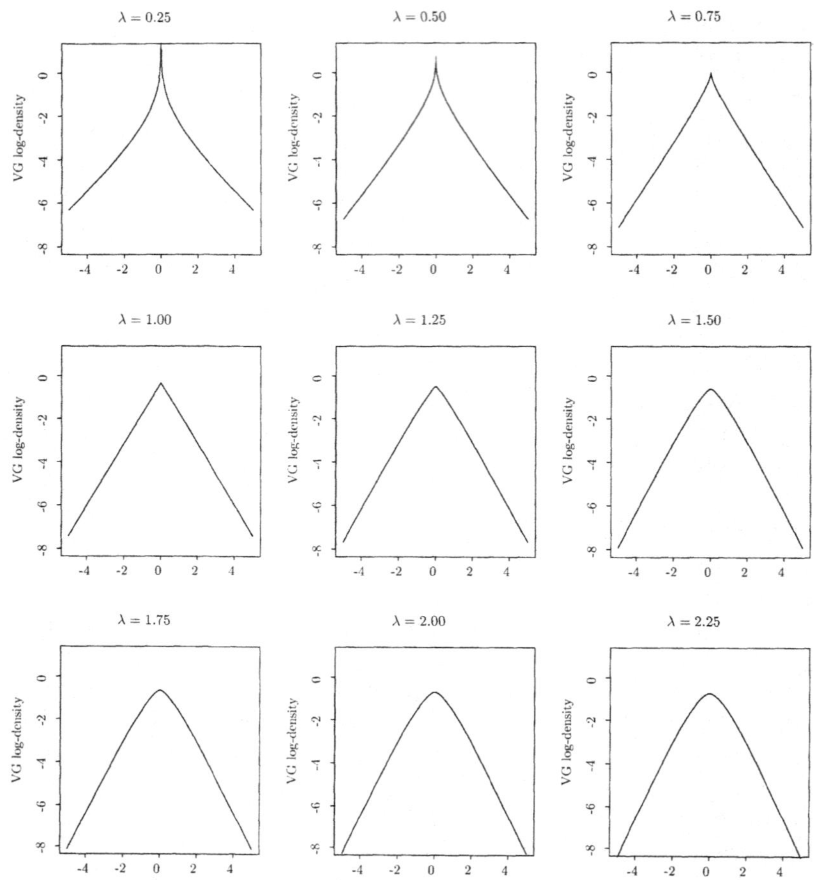

Fig. 7. The logarithm of the densities of variance-gamma distributions with $\beta = 0$, mean 0, and variance 1 for different values of the parameter λ.

with domain $-2\beta < z \leqslant 0$ when $\beta > 0$ and $0 \leqslant z < -2\beta$ when $\beta < 0$. When $\beta = 0$, the domain is the set $\{0\}$, and we obtain the density function

$$\frac{\Gamma((\nu+1)/2)}{\delta\sqrt{\pi}\Gamma(\nu/2)(1+((x-\mu)/\delta)^2)^{(\nu+1)/2}}, \quad x \in \mathbb{R},$$

which is the well-known density of the scaled t-distribution with ν degrees of freedom.

1.2. The generalized inverse Gaussian distribution

The second class of distributions, that we consider in this section, is the class of *generalized inverse Gaussian* (*GIG*) distributions. The *GIG*-distributions are described by three parameters and defined on the positive half axis. The generalized inverse Gaussian density is of the form

$$\frac{(\gamma/\delta)^{\lambda}}{2K_{\lambda}(\delta\gamma)} \cdot x^{\lambda-1} \cdot \exp\left\{-\frac{1}{2}\left(\delta^2 x^{-1} + \gamma^2 x\right)\right\}, \quad x > 0. \tag{18}$$

The parameter domain is given by

$$\begin{aligned} &\delta > 0, \quad \gamma \geqslant 0, \quad \text{if } \lambda < 0, \\ &\delta > 0, \quad \gamma > 0, \quad \text{if } \lambda = 0, \\ &\delta \geqslant 0, \quad \gamma > 0, \quad \text{if } \lambda > 0. \end{aligned}$$

The class of generalized inverse Gaussian distributions was first proposed in 1946 by Étienne Halphen, who used it to model the distribution of the monthly flow of water in hydroelectric stations, see Seshardi (1997). The class was rediscovered by Sichel (1973) who used it to construct mixtures of Poisson distributions and by Barndorff-Nielsen (1977) who used it to construct the class of generalized hyperbolic distributions, but also realized its broad usefulness and initiated an in depth study of the class. We shall return to the relation to the generalized hyperbolic distributions later. The generalized inverse Gaussian distributions were briefly mentioned by Goog (1953) as an intermediate between Pearson's curves of Type III and V. The class of generalized inverse Gaussian distributions was investigated extensively in Jørgensen (1982).

Using (A.5) we see that for $\lambda > 0$ and $\gamma > 0$ the gamma distribution emerges as limit distribution when δ tends to zero, that is we get the following density for positive λ and γ,

$$\frac{(\gamma^2/2)^{\lambda}}{\Gamma(\lambda)} \cdot x^{\lambda-1} \cdot e^{\gamma^2 x/2}, \quad x > 0.$$

Similarly, the inverse gamma distribution with density given by

$$\frac{(2/\delta^2)^{\lambda}}{\Gamma(-\lambda)} \cdot x^{\lambda-1} \cdot e^{(\delta^2/2)/x}, \quad x > 0,$$

is obtained when γ tends to zero for $\lambda < 0$ and $\delta > 0$. This distribution has a tail of the Pareto type. Finally, for $\lambda = -\frac{1}{2}$ we get the inverse Gaussian distribution with density function given by

$$\frac{\delta}{\sqrt{2\pi x^3}} \cdot e^{-\gamma(x-\delta/\gamma)^2/(2x)}, \quad x > 0.$$

The generalized inverse Gaussian distributions are unimodal with mode point given by

$$\begin{cases} \dfrac{\lambda - 1 + \sqrt{(\lambda-1)^2 + \delta^2\gamma^2}}{\gamma^2} & \text{if } \gamma > 0, \\ \dfrac{\delta^2}{2(1-\lambda)} & \text{if } \gamma = 0. \end{cases}$$

If X has a generalized inverse Gaussian distribution, we write $X \sim GIG(\lambda, \delta, \gamma)$. In Figure 8 generalized inverse Gaussian densities are plotted for different values of λ and $\omega = \delta\gamma$. In all cases the variance is 1.

The Laplace transform of the $GIG(\lambda, \delta, \gamma)$-distribution is

$$L(z) = \frac{K_\lambda(\omega\sqrt{1 - 2z/\gamma^2})}{K_\lambda(\omega)(1 - 2z/\gamma^2)^{\lambda/2}} \tag{19}$$

for $\delta > 0$ and $\gamma > 0$. The domain of L is $z < \gamma^2/2$ when $\lambda \geqslant 0$ and $z \leqslant \gamma^2/2$ when $\lambda < 0$. In the cases $\delta = 0$ or $\gamma = 0$, the Laplace transform is obtained from (19) by (A.5). For $\delta = 0$,

$$L(z) = \left(1 - \frac{2z}{\gamma^2}\right)^{-\lambda}, \quad z < \frac{\gamma^2}{2},$$

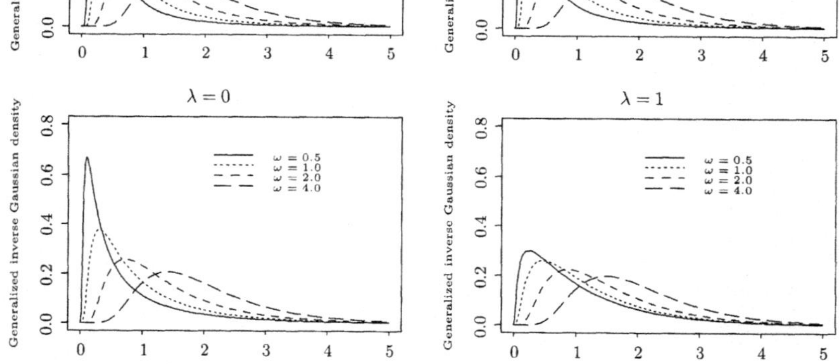

Fig. 8. Generalized inverse Gaussian densities with variance 1 for different values of the parameters λ and $\omega = \delta\gamma$.

which is the well-known Laplace transform of the gamma-distribution. For $\gamma = 0$ we obtain

$$L(z) = \frac{2K_\lambda(\sqrt{-2\delta^2 z})}{\Gamma(-\lambda)(-\delta^2 z/2)^{\lambda/2}}, \quad z \leqslant 0.$$

For positive values of δ and γ the moments of X are given by

$$\mathrm{E}X^j = \left(\frac{\delta}{\gamma}\right)^j \frac{K_{\lambda+j}(\omega)}{K_\lambda(\omega)}, \quad j = 1, 2, \ldots. \tag{20}$$

When either δ or γ is zero, the moments of X are also known and are obtained as limits of (20). The variance of X is given by

$$\operatorname{Var} X = \left(\frac{\delta}{\gamma}\right)^2 \left(\frac{K_{\lambda+2}(\omega)}{K_\lambda(\omega)} - \frac{K_{\lambda+1}^2(\omega)}{K_\lambda^2(\omega)}\right). \tag{21}$$

In Figure 9 a histogram of 307 monthly observations of interest rates in the period from June 1964 to December 1989 is given along with a fitted generalized inverse Gaussian density corresponding to the parameter values $\delta = 0.2693$, $\gamma = 11.23$, and $\lambda = -7.0707$. More precisely, the data are annualized monthly yields of U.S. one-month Treasury bills. The same data set was studied in Chan et al. (1992).

There is the following important relationship between the generalized hyperbolic distribution and the generalized inverse Gaussian distribution, which was, in fact, how the gen-

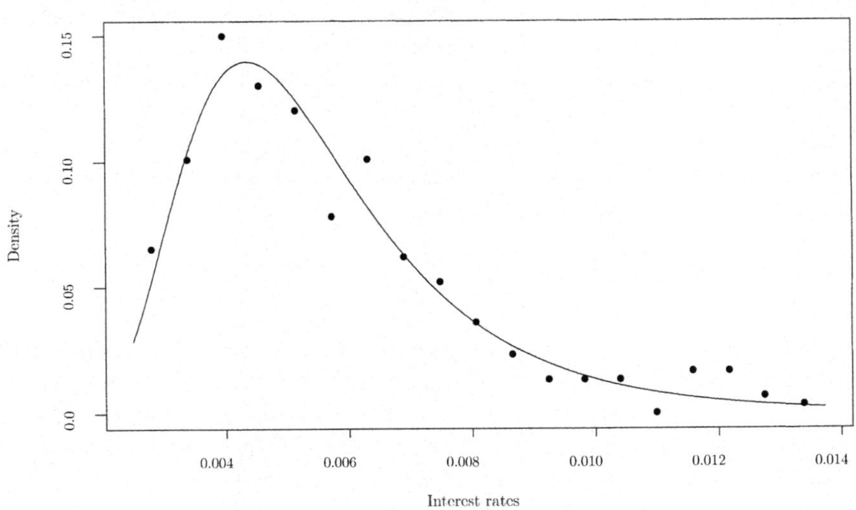

Fig. 9. A histogram of 307 monthly interest rates. The generalized inverse Gaussian density with parameters $\delta = 0.2693$, $\gamma = 11.23$, and $\lambda = -7.0707$ is superimposed.

eralized hyperbolic distribution was originally derived in Barndorff-Nielsen (1977). The generalized hyperbolic distribution is a normal variance–mean mixture where the mixing distribution is generalized inverse Gaussian. What is meant by this is that if

$$X|W = w \sim N(\mu + \beta w, w),$$

and $W \sim GIG(\lambda, \delta, \gamma)$, then the marginal distribution of X will be generalized hyperbolic, $X \sim H(\lambda, \alpha, \beta, \delta, \mu)$, where $\alpha^2 = \beta^2 + \gamma^2$. This property provides a possible interpretation of non-Gaussian stochastic variation described by a generalized hyperbolic distribution.

As special cases we have that the normal-inverse Gaussian distribution appears when the mixing distribution is an inverse Gaussian distribution, and the variance-gamma distribution emerges as a normal variance–mean mixture where the mixing distribution is a gamma distribution. This explains the names of the distributions. The asymmetric scaled t-distribution is a normal variance–mean mixture with an inverse gamma mixing distribution. As a special case we get the well-known result that the t-distribution is a normal variance mixture ($\beta = 0$) with an inverse gamma mixing distribution.

The mixing result implies that there is the following simple relationship between the Laplace transform, L_X, of the generalized hyperbolic distribution $H(\lambda, \alpha, \beta, \delta, \mu)$ and that of the $GIG(\lambda, \delta, \sqrt{\alpha^2 - \beta^2})$-distribution, L_W:

$$L_X(z) = e^{\mu z} \cdot L_W\left(\beta z + \frac{1}{2}z^2\right).$$

Barndorff-Nielsen and Halgreen (1977) showed that generalized inverse Gaussian distributions are *infinitely divisible*. Using that the generalized hyperbolic distributions are normal variance-mean mixtures with generalized inverse Gaussian mixing distributions, they also proved that generalized hyperbolic distributions are infinitely divisible. Halgreen (1979) showed that generalized hyperbolic distributions and generalized inverse Gaussian distribution are even *self-decomposable*. In the following section, the properties of infinite divisibility and self-decomposability will turn out to be important because they allow the construction of certain hyperbolic stochastic process models.

1.3. Statistical inference

Inference for the parameters when dealing with independent and identically generalized hyperbolic or generalized inverse Gaussian distributed observations should be based on the likelihood function. The C-program HYP described in Blæsild and Sørensen (1992) can be used for maximum likelihood estimation in the situation where independent and identically (possibly multi-dimensional) hyperbolic distributed observations are considered. The program HYP also has the facility of basing the inference on the multinomial likelihood function obtained by only observing the number of observations in given intervals. More precisely, if $I_1, \ldots, I_k$ are disjoint intervals with union the entire real line and y_j denotes the number of observations in I_j, $j = 1, \ldots, k$, then the multinomial log-likelihood function is given by

$$\ell(\alpha, \beta, \delta, \mu) = \sum_{j=1}^{k} y_j \log p_j, \tag{22}$$

where p_j is the probability that a hyperbolic distributed random variable takes a value in I_j, that is,

$$p_j = \int_{I_j} \frac{\gamma}{2\alpha\delta K_1(\delta\gamma)} \exp\left\{-\alpha\sqrt{\delta^2 + (x-\mu)^2} + \beta(x-\mu)\right\} \mathrm{d}x, \quad j = 1, \ldots, k. \tag{23}$$

Inference based on grouped observations from other distributions can of course be carried out in a similar way using (22) and the equivalent of (23). Küchler et al. (1999) note that if the observations are not independent then inference based on the multinomial likelihood function for grouped observations will be more robust to effects of the dependence than inference based on the original likelihood function for independent observations.

2. Lévy processes

A homogeneous Lévy process X is a stochastic process with $X_0 = 0$ and with the property that its increments over non-overlapping time intervals are independent. Moreover, the increment, $X_{t+s} - X_s$, over any time interval of length t has the same distributions as X_t. The homogeneous Lévy processes are also called processes with independent, stationary increments or additive processes. The mathematical theory of Lévy processes can be found in Bertoin (1996) or Sato (1999). An example of a Lévy process that is well-known from, for instance, the Black–Scholes–Merton option pricing theory is the Brownian motion (or Wiener process), where the increments are normally distributed.

For every generalized hyperbolic distribution there exists a homogeneous Lévy process X such that the probability distribution of the value of the process, X_t, at a fixed time point t is that particular generalized hyperbolic distribution. A thorough review of the theory of these generalized hyperbolic Lévy processes and their application in finance can be found in Eberlein (2001), see also Prause (1999) and Eberlein and Raible (2001). The distributions that can appear as the distribution of the instantaneous value of a homogeneous Lévy process are exactly those that have the property called infinite divisibility. As mentioned in Section 1 the generalized hyperbolic distributions are infinitely divisible. Usually, the distribution of the value X_s at a time point s different from t will not be generalized hyperbolic. However, in the case of the *NIG* and *VG* distributions, the convolution properties (12) and (15) imply that the value of the Lévy process will be *NIG*-distributed, respectively *VG*-distributed, at all time points. This makes the *NIG* and *VG* Lévy processes more natural generalized hyperbolic Lévy processes than the other generalized hyperbolic Lévy processes. Simulation of the *NIG* Lévy process was studied in Rydberg (1997).

A generalized hyperbolic Lévy processes can be written in the form

$$X_t = \lambda t + Z_t,$$

where Z_t is a pure jump martingale with infinitely many small jumps in every finite time interval, however small. The behaviour of Z_t is reflected in the so-called Lévy measure, see (27) and the discussion following this formula. The Lévy measure of the generalized hyperbolic distribution is

$$q(x) = \begin{cases} \dfrac{e^{\beta x}}{|x|}\left(\displaystyle\int_0^\infty \frac{\exp(-|x|\sqrt{2y+\alpha^2})}{\pi^2 y(J_\lambda^2(\delta\sqrt{2y}) + Y_\lambda^2(\delta\sqrt{2y}))}\,dy + \lambda\, e^{-\alpha|x|}\right) & \text{if } \lambda \geqslant 0, \\ \dfrac{e^{\beta x}}{|x|}\displaystyle\int_0^\infty \frac{\exp(-|x|\sqrt{2y+\alpha^2})}{\pi^2 y(J_{-\lambda}^2(\delta\sqrt{2y}) + Y_{-\lambda}^2(\delta\sqrt{2y}))}\,dy & \text{if } \lambda < 0. \end{cases} \tag{24}$$

Here J_λ and Y_λ denote Bessel functions of the first and second kind, respectively, see the appendix. The Lévy measure was essentially found by Halgreen (1979), see also Prause (1999). For the *NIG*-distribution this expression simplifies to

$$q(x) = \pi^{-1}\delta\alpha|x|^{-1}K_1\big(\alpha|x|\big)\, e^{\beta x}, \tag{25}$$

where K_1 is a modified Bessel function of the third kind. The behaviour near zero is particularly important, so the following expansion for generalized hyperbolic distributions (Raible, 2000) is useful:

$$x^2 q(x) = \frac{\delta}{\pi} + \frac{\lambda + 1/2}{2}|x| + \frac{\delta\beta}{\pi}x + o\big(|x|\big) \tag{26}$$

as $x \to 0$. We see that for every generalized hyperbolic distribution the Lévy measure has infinite mass in every neighbourhood of the origin. The process Z_t is given by

$$Z_t = \int_0^t \int_{\mathbb{R}\setminus\{0\}} x\big(\mu^X(du, dx) - q(x)\, du\, dx\big), \tag{27}$$

where the integer-valued random measure μ^X is defined by

$$\mu^X(dt, dx) = \sum_{s>0} 1_{\{\Delta X_s \neq 0\}} \varepsilon_{(s, \Delta X_s)}(dt, dx).$$

Here ε_a denotes the Dirac measure at a, and $\Delta X_s = X_s - X_{s-}$ is the jump of the process X at time s (for most time points $\Delta X_s = 0$). Integrals of the type (27) are treated in, e.g., Jacod and Shiryaev (1987) or Protter (1990). The random measure μ^X is Poissonian with intensity measure $q(x)\,\mathrm{d}x\,\mathrm{d}t$. This implies that for any closed interval A that does not contain the origin, the number of jumps in the time interval $[0, t]$ with a size that belongs to A, i.e.,

$$N_t^A = \mu^X\big([0, t], A\big),$$

is a Poisson process with intensity $\int_A q(x)\,\mathrm{d}x$, which is a finite number. In particular, N_t^A is Poisson distributed with mean value $t\int_A q(x)\,\mathrm{d}x$. As the boundary of the interval A tends to zero, the mean value goes to infinity, cf. (26). It is interesting to note that a generalized hyperbolic Lévy process has no continuous Brownian motion component and has infinitely many jumps on every time interval.

The generalized hyperbolic Lévy processes do, however, have a nice relation to the Brownian motion. Let B be a standard Brownian motion, and let $\tau(t)$ be a Lévy process for which the distribution of $\tau(1)$ is a generalized inverse Gaussian distribution. Then the process

$$X_t = \mu t + \beta\tau(t) + B_{\tau(t)} \tag{28}$$

is a generalized hyperbolic Lévy process. Because the increments of τ are generalized inverse Gaussian distributed and hence can only be positive, the process τ is increasing and can thus be interpreted as a time that increases with a randomly varying speed. A process τ with this property is called a subordinator, and the construction (28) is called *subordination*. The randomly increasing time τ has been interpreted as an *operational time* or a *business time* reflecting, for instance, the volume of trade at an exchange. Some times a lot is happening at the exchange and the business time increases rapidly. At other times the exchange is tranquil and the business time goes only slowly. That the distribution of X_1 is generalized hyperbolic follows because this distribution is a variance-mean mixture of normal distributions where the mixing distribution is the generalized inverse Gaussian distribution, see Section 1.2. The fact that a Lévy process τ exists such that $\tau(1)$ is generalized inverse Gaussian distributed follows because these distributions are infinitely divisible, as mentioned in Section 1. In the case of a *NIG*-distribution, the construction by subordination can be done in the following simple way (Barndorff-Nielsen, 1998). Let (U_t, V_t) be a two-dimensional standard Brownian motion starting at $(0, 0)$ and with drift vector (β, γ), where $\gamma > 0$. Let $\tau(t)$ denote the first time the second component V attains the value $\delta t > 0$ with $\delta > 0$. Then $\{\tau(t)\colon t > 0\}$ is an inverse Gaussian Lévy process, and

$$X_t = \mu t + U_{\tau(t)}$$

is a *NIG*-Lévy process. Specifically, X_t is $NIG(\alpha, \beta, \delta t, \mu t)$ distributed, where $\alpha = \sqrt{\beta^2 + \gamma^2}$.

Construction of financial models by subordination was first proposed by Praetz (1972) who used a scaled t-distribution to model stock returns and obtained a good fit to weekly returns from the Sydney Stock Exchange. This is a particular example of a generalized hyperbolic distribution where the mixing distribution is an inverse gamma distribution, see Section 1.1. Praetz attributed the mixing of normals to the change in activity at the exchange. Clark (1973) and Epps and Epps (1976) found that there is a dependency between trading volume and the variance of returns, but did not suggest generalized hyperbolic models. These finding have been confirmed by Ané and Geman (2000). In Madan and Seneta (1990), Madan and Lime (1991) and Madan and Chang (1996) the so-called variance gamma model is introduced and studied as a model for share market returns. This model is the generalized hyperbolic Lévy process with a gamma mixing distribution. For a discussion of the subordination approach in finance, see, e.g., Hurst, Platen and Rachev (1997).

The use of generalized hyperbolic Lévy processes to model the prices of stocks and other assets and the corresponding theory of option pricing has been thoroughly investigated by Eberlein and Keller (1995), Keller (1997), Eberlein, Keller and Prause (1998) and Eberlein and Prause (2002). Eberlein and Jacod (1997) proved that the set of equivalent martingale measures is large and that the corresponding price range is the entire non-arbitrage interval. A theory of the term structure of interest rates based on the hyperbolic Lévy process was developed in Eberlein and Raible (1999). A useful review can be found in Eberlein (2001).

For the processes discussed in this section, estimation based on observations at equidistant discrete time points is as easy as estimation for independent generalized hyperbolic distributions, because the increments of the process between the observation times are independent. Usually one would use a Lévy process for which the increments are generalized hyperbolic and then estimate the parameters, for instance by means of the computer program mentioned in Section 1.3. A simple check of the fit of the model to the data can be made as follows. If, for instance, the data are daily observations, then it should be checked that the distributions calculated from the estimated model of the increments over a number of suitably chosen longer time spans fit the corresponding increments calculated from the data. For the *NIG* and *VG* Lévy processes these distributions are simply given by the formulae (12) and (15). For an example of this procedure, see Eberlein (2001).

3. Stochastic differential equations

In this section we present various methods for constructing diffusion processes with generalized hyperbolic and generalized inverse Gaussian marginal distributions. A diffusion process is the solution of a stochastic differential equation driven by a Wiener process. Estimation of parameters based on discrete-time observations of a diffusion process is considered too. Furthermore, we consider Ornstein–Uhlenbeck type processes driven by Lévy processes and models given as sums of processes defined by stochastic differential equations.

3.1. Diffusion models

We consider a one-dimensional diffusion process $\{X_t\}$ and suppose that it is the unique weak solution to the stochastic differential equation

$$dX_t = b(X_t;\theta)\,dt + \sigma(X_t;\theta)\,dW_t, \tag{29}$$

where $\sigma(x;\theta)$ is positive for all x in the state space (l, r) $(-\infty \leqslant l < r \leqslant \infty)$ and all θ in some p-dimensional parameter space Θ. We will focus on ergodic diffusions and denote the density of the corresponding invariant probability measure by μ_θ.

Diffusion processes with a specific marginal distribution are typically constructed by determining drift b and diffusion coefficient σ so that the invariant distribution is of the required type. This method will result in the appropriate marginal distribution for large values of t or for all t provided that the initial distribution is equal to the invariant distribution (i.e., $X_0 \sim \mu_\theta$). Under mild conditions we have the following relationship between the drift, diffusion coefficient, and the density of the invariant distribution,

$$2b(x;\theta) - v'(x;\theta) = v(x;\theta)\frac{\mu'_\theta(x)}{\mu_\theta(x)}, \quad l < x < r,\ \theta \in \Theta, \tag{30}$$

where v denotes the squared diffusion coefficient, $v(x;\theta) = \sigma^2(x;\theta)$.

Using (30), Bibby and Sørensen (2001) discussed a method for constructing diffusion processes with a prescribed marginal (invariant) distribution. Letting the drift be given by

$$b(x;\theta) = \frac{1}{2}v(x;\theta)\frac{d}{dx}\log\bigl[v(x;\theta)f(x)\bigr],$$

where f is a function that is integrable on the interval (l, r), it was shown under some regularity conditions that the diffusion process given by (29) has invariant density μ_θ proportional to f, irrespective of the choice of the function v. Bibby and Sørensen (2001) also considered the special case where

$$v(x;\theta) = \sigma^2 f(x)^{-\kappa}, \quad \sigma^2 > 0,\ \kappa \in [0, 1],$$

in particular the situation where the invariant density was hyperbolic. This led to the following stochastic differential equation,

$$dX_t = \frac{1}{2}\sigma^2(1-\kappa)f(X_t)^{-\kappa}\left[\beta - \frac{\alpha(X_t-\mu)}{\sqrt{\delta^2+(X_t-\mu)^2}}\right]dt + \sigma f(X_t)^{-\kappa/2}\,dW_t, \tag{31}$$

where f is proportional to the hyperbolic density function given by (8), that is

$$f(x) = \exp\bigl[-\alpha\sqrt{\delta^2+(x-\mu)^2} + \beta(x-\mu)\bigr].$$

Note that the drift is towards the mode point of the hyperbolic distribution, $\mu + \beta\delta/\gamma$. The diffusion process given by (31) was successfully used to describe the logarithm of the price of *VW*-stocks after a linear trend had been subtracted.

In Bibby and Sørensen (1997) the special case where $\kappa = 1$ was considered in the situation of a hyperbolic invariant density. Note that this results in a diffusion process with no drift, that is the solution to the stochastic differential equation given by

$$\mathrm{d}X_t = \sigma \exp\left\{\frac{1}{2}\alpha\sqrt{\delta^2 + (X_t - \mu)^2} - \frac{1}{2}\beta(X_t - \mu)\right\} \mathrm{d}W_t. \tag{32}$$

It turns out that this is an example of a local martingale which is not a martingale. Also the hyperbolic diffusion process given as the solution of (32) was fitted successfully to the logarithm of stock-prices (minus a linear trend) in Bibby and Sørensen (1997). The construction leading to the hyperbolic diffusion (31) can obviously be made similarly for any generalized hyperbolic distribution. In the special case $\kappa = 1$, this was done in Rydberg (1999), where the corresponding *NIG*-diffusion was fitted successfully to stock prices (minus a linear trend).

In Küchler et al. (1999) a hyperbolic diffusion process with constant diffusion coefficient was discussed. This corresponds to letting the function v be equal to a constant σ^2, or to $\kappa = 0$ in (31), and gives the following stochastic differential equation,

$$\mathrm{d}X_t = \frac{1}{2}\sigma^2\left[\beta - \alpha\frac{X_t - \mu}{\sqrt{\delta^2 + (X_t - \mu)^2}}\right]\mathrm{d}t + \sigma\,\mathrm{d}W_t. \tag{33}$$

The hyperbolic diffusion process given by (33) was first proposed in Barndorff-Nielsen (1978).

For values of κ between the two extremes 0, corresponding to stationarity being obtained by pure reversion, and 1, where stationarity is obtained by pure diffusion, both these effects are present to varying degrees.

Sørensen (1997b) considers the construction of diffusion processes with a generalized inverse Gaussian invariant distribution. If v is a positive function, then the solution to the stochastic differential equation

$$\begin{aligned}\mathrm{d}X_t = {} & \left(v(X_t)v'(X_t) + \frac{1}{2}v(X_t)^2\left[(\lambda - 1)X_t^{-1} - \frac{\gamma^2}{2} + \frac{1}{2}\delta^2 X_t^{-2}\right]\right)\mathrm{d}t \\ & + v(X_t)\,\mathrm{d}W_t\end{aligned} \tag{34}$$

will have a generalized inverse Gaussian invariant density given by (18) under suitable regularity conditions on v. The focus in Sørensen (1997b) is on the special case where $v(x) = \kappa x^\alpha$ for constants $\alpha \geqslant 0$ and $\kappa > 0$. With this choice of diffusion coefficient, the diffusion process is the solution to the stochastic differential equation given by

$$\mathrm{d}X_t = \left(\beta_1 X_t^{2\alpha-1} - \beta_2 X_t^{2\alpha} + \beta_3 X_t^{2(\alpha-1)}\right)\mathrm{d}t + \kappa X_t^\alpha\,\mathrm{d}W_t, \tag{35}$$

where

$$\beta_1 = \frac{1}{2}\kappa^2(\lambda - 1) + \kappa^2\alpha, \qquad \beta_2 = \frac{1}{4}(\kappa\gamma)^2, \qquad \beta_3 = \frac{1}{4}(\kappa\delta)^2.$$

Note that if $\alpha = \frac{1}{2}$ and $\beta_3 = 0$, then the diffusion process is the solution to

$$\mathrm{d}X_t = (\beta_1 - \beta_2 X_t)\,\mathrm{d}t + \kappa\sqrt{X_t}\,\mathrm{d}W_t, \tag{36}$$

that is the Cox–Ingersoll–Ross process (CIR-process) used in finance to model short term interest rates, see Cox, Ingersoll Jr. and Ross (1985).

A completely different way of constructing hyperbolic diffusion models was proposed in Jensen and Pedersen (1999). These authors consider processes given by $X_t = h(Y_t)$, where Y is a stationary Ornstein–Uhlenbeck process:

$$\mathrm{d}Y_t = -\alpha Y_t\,\mathrm{d}t + \tau\,\mathrm{d}W_t$$

with $\alpha > 0$ and $\tau > 0$. Suppose F is the distribution function of a given probability distribution, and let Φ denote the distribution function of the standard normal distribution. If $\tau^2 = 2\alpha$ and $h(y) = F^{-1}(\Phi(y))$, then the distribution of X_t will have the distribution function F. If, in particular, F is the distribution function of a generalized hyperbolic distribution, we obtain a generalized hyperbolic diffusion process. Unfortunately, there is no explicit expression for the distribution function of a generalized hyperbolic distribution. An advantage of this approach is that there is an expression for the transition density involving the function h. Since the distribution function of a generalized hyperbolic distribution, and hence h, can be calculated numerically, it is relatively easy to calculate the likelihood function, which is usually not the case for diffusion models. A disadvantage is that the drift and diffusion coefficients of the diffusion process X are not explicit functions.

3.2. Statistical inference for diffusion processes

Inference for discretely observed diffusion processes is made difficult by the fact that the likelihood function is generally not tractable. In recent years many different methods have been proposed to overcome this obstacle. We will here briefly discuss the methods most commonly used in connection with financial data. For an excellent overview of a wide variety of procedures for estimating parameters based on discretely observed diffusions, see H. Sørensen (2000).

Approximate likelihood methods are considered by Pedersen (1995), Aït-Sahalia (2002), and Poulsen (1999). In Pedersen (1995) it is shown that the likelihood function can be calculated to any given precision using simulations and the Euler approximation in a clever way. Unfortunately, the method is very computer intensive. Honoré (1997) successfully applied the Pedersen method to the CKLS-model for interest rates (proposed by Chan et al. (1992)). In Aït-Sahalia (2002) an analytical approximation to the likelihood function based on a truncated Hermite expansion is developed. Poulsen (1999) obtained an approximation to the likelihood function by numerically solving the Chapman–Kolmogorov forward equations. He used his method to fit the CKLS-model to interest rate data. Asymptotic results for the maximum likelihood estimator based on discrete time observations of a diffusion model were derived in Dacunha-Catelle and Florens-Zmirou (1986).

Inference for diffusion processes based on martingale estimating functions is considered in Bibby and Sørensen (1995, 1996, 1997). For observations $X_{t_1}, X_{t_2}, \ldots, X_{t_n}$ the martingale estimating functions introduced in Bibby and Sørensen (1995, 1996) are of the form

$$G_n(\theta) = \sum_{i=1}^{n} g_i(X_{t_{i-1}};\theta)\big[X_{t_i} - \mathrm{E}_\theta(X_{t_i}|X_{t_{i-1}})\big] + \sum_{i=1}^{n} h_i(X_{t_{i-1}};\theta)\big[\big(X_{t_i} - \mathrm{E}_\theta(X_{t_i}|X_{t_{i-1}})\big)^2 - \mathrm{Var}_\theta(X_{t_i}|X_{t_{i-1}})\big]. \tag{37}$$

Note that in analogy with the unknown score function, G_n is a sum of functions of consecutive pairs of observations, and G_n is a martingale with respect to the natural filtration. The conditional expectations in (37) can easily be calculated using simulations, and an estimator for the parameter θ is then obtained by solving the equation $G_n(\theta) = 0$. In Bibby and Sørensen (1995) the resulting estimator is shown to be consistent and asymptotically normal as the number of observations tends to infinity. An optimal choice of the functions g_i and h_i as well as simpler approximately optimal functions that are useful in practice are given in Bibby and Sørensen (1995, 1996). As mentioned earlier the hyperbolic diffusion process given by (32) was fitted to the log-prices of stocks after a linear trend had been subtracted in Bibby and Sørensen (1997). The parameters in this hyperbolic diffusion model were estimated using the martingale estimating function

$$K_n(\theta) = \sum_{i=1}^{n} \frac{\dot{v}(X_{t_{i-1}};\theta)}{(t_i - t_{i-1})v(X_{t_{i-1}};\theta)^3}\big[(X_{t_i} - X_{t_{i-1}})^2 - \mathrm{E}_\theta\big((X_{t_i} - X_{t_{i-1}})^2|X_{t_{i-1}}\big)\big],$$

where v is the squared diffusion coefficient and a dot denotes differentiation with respect to the parameter θ. This is an approximately optimal modification of (37) taking into account that the diffusion has no drift.

Kessler and Sørensen (1999) considered martingale estimating functions based on eigenfunctions of the infinetisimal generator of the diffusion process. The advantage of such martingale estimating functions is that they are adapted to concrete models and are easy to calculate in cases where the eigenfunctions are explicitly known. Unfortunately this is not often the case.

It is usually easy to obtain an estimator from a simple estimating function of the form

$$F_n(\theta) = \sum_{i=1}^{n} f(X_{t_i};\theta),$$

where the function f satisfies that $\int_l^r f(x,\theta)\mu_\theta(x)\,\mathrm{d}x = 0$ with μ_θ denoting the density of the invariant probability measure. Such simple estimating functions were studied by Hansen and Scheinkman (1995), Kessler (2000), and Jacobsen (2001). The advantage of these estimating functions is that they are indeed simple and fast to work with because it is straightforward to explicitly find functions f with the property needed. The main disadvantages are that only parameters appearing in the invariant density can be estimated using simple estimating functions and that the estimators may be far from efficient because the dependence structure in the data is ignored. An improved version of the simple estimating function where each term in the sum depends on a pair of consecutive observations was considered by Hansen and Scheinkman (1995) and Jacobsen (2001). Optimality questions were treated in Kessler (2000) and Jacobsen (2001). For the improved version it is also not possible to estimate all parameters, see the discussion in Hansen and Scheinkman (1995). A review of estimating function inference for diffusion models can be found in Sørensen (1997a) and Bibby, Jacobsen and Sørensen (2002).

Indirect inference procedures based on auxiliary models and extensive simulations were proposed by Gouriéroux, Monfort and Renault (1993) and Gallant and Tauchen (1996). These procedures have gained some popularity in the finance literature under the name of the efficient method of moments. However, the quality of the estimators depend on the choice of the auxiliary model, which is not a straightforward matter.

Finally, Bayesian MCMC-methods have been applied to diffusion models by Eraker (2001) and Elerian, Chib and Shepard (2001). In these methods, the likelihood function is calculated in a way similar to that in Pedersen (1995).

3.3. *Ornstein–Uhlenbeck processes*

A stochastic process X is called a process of the Ornstein–Uhlenbeck type, if it satisfies a stochastic differential equation of the form

$$\mathrm{d}X_t = -\lambda X_t\,\mathrm{d}t + \mathrm{d}Z_t, \tag{38}$$

where $\lambda > 0$ and where the driving process Z is a homogeneous Lévy process. It is not difficult to see that

$$X_t = \mathrm{e}^{-\lambda t} X_0 + \int_0^t \mathrm{e}^{-\lambda(t-s)}\,\mathrm{d}Z_s. \tag{39}$$

If X is stationary and square integrable, the autocorrelation function of X is

$$\rho(u) = \exp(-\lambda u). \tag{40}$$

When the process Z is the standard Wiener process, the solution X is the usual Ornstein–Uhlenbeck process. Ornstein–Uhlenbeck type processes have been studied by Wolfe (1982), Sato and Yamazato (1982, 1984) and Sato, Watanabe and Yamazato (1994); see also Jurek and Vervaat (1983), Jurek and Mason (1993), and Barndorff-Nielsen, Jensen and Sørensen (1998). A necessary and sufficient condition for (38) to have a stationary solution is that $\mathrm{E}(\log(1+|Z(1)|)) < \infty$.

For every generalized hyperbolic distribution there exists a stationary Ornstein–Uhlenbeck type process such that for all $t \geqslant 0$ the distribution of X_t is the given generalized hyperbolic distribution. The same is true for all generalized inverse Gaussian distributions. This is because these distributions have the property called self-decomposability, as discussed in Section 1. The Lévy process driving the *NIG* Ornstein–Uhlenbeck type process was studied by Barndorf-Nielsen (1998), while the process driving the symmetric variance-gamma Ornstein–Uhlenbeck type process, was found by Jiang (2000). For symmetric distributions, the driving Lévy process is, in the case of the *NIG* Ornstein–Uhlenbeck process, the sum of a *NIG* Lévy process and a compound Poisson process, while for the variance-gamma Ornstein–Uhlenbeck process, it is simply a compound Poisson process.

As for most ordinary diffusion processes, the likelihood function is usually not explicitly available for processes of the Ornstein–Uhlenbeck type. Since these processes are Markov processes, a simple and natural approach to statistical inference goes via estimating functions based on conditional moments defined in analogy with those discussed in Section 3.2.

3.4. *Compound processes*

Quite often, the exponentially decreasing autocorrelation function (40) is too simple to fit financial data. However, models with a much more flexible covariance structure are easily obtained by summing independent Ornstein–Uhlenbeck type processes, as was proposed by Barndorff-Nielsen, Jensen and Sørensen (1998). The process

$$X_t = X_t^{(1)} + \cdots + X_t^{(m)}, \tag{41}$$

where the processes $X_t^{(i)}$, $i = 1, \dots, m$, are independent Ornstein–Uhlenbeck type processes given by

$$\mathrm{d}X_t^{(i)} = -\lambda_i X_t^{(i)}\,\mathrm{d}t + \mathrm{d}Z_t^{(i)} \tag{42}$$

for independent Lévy processes $Z_t^{(i)}$, $i = 1, \dots, m$, has an autocorrelation function of the form

$$\rho(u) = \Phi_1 \exp(-\lambda_1 u) + \cdots + \Phi_m \exp(-\lambda_m u), \tag{43}$$

where Φ_i is proportional to the variance of $X_t^{(i)}$, and $\Phi_1 + \cdots + \Phi_m = 1$. A much better fit to financial data than that obtained by (40) can often obtained even for $m = 2$. Examples can be found in Barndorff-Nielsen, Jensen and Sørensen (1998) and Barndorff-Nielsen and Shephard (2001c).

For every generalized hyperbolic distribution and for every generalized inverse Gaussian distribution there exists a stationary process X of the form (41), (42) such that for all $t \geqslant 0$

the distribution of X_t is that particular distribution. Again this is because these distributions are self-decomposable, see Barndorff-Nielsen, Jensen and Sørensen (1998). More complex types of superpositions of Ornstein–Uhlenbeck type processes were investigated in Barndorff-Nielsen (2001).

The construction (41) can be made for diffusion models with linear drift and non-linear diffusion coefficient too, see Bibby, Skovgaard and Sørensen (2002). As an example, suppose we want a stationary stochastic process with autocorrelation function (43) for given values of $\lambda_1, \ldots, \lambda_m$ and $\Phi_1, \ldots, \Phi_m$, and such that the marginal distribution of X_t is a gamma distribution with shape parameter α and scale parameter β. This can be obtained by defining m independent processes as the stationary solutions to

$$\mathrm{d}X_t^{(i)} = -\lambda_i \left(X_t^{(i)} - \Phi_i \alpha\beta\right) \mathrm{d}t + \sqrt{2\beta\lambda_i X_t^{(i)}}\, \mathrm{d}W_t^{(i)}, \tag{44}$$

$i = 1, \ldots, m$. Each of the processes, $X_t^{(i)}$, is a CIR-process, (36), which is a particular example of the generalized inverse Gaussian diffusions given by (35). Since $X_t^{(i)}$ is gamma distributed with shape parameter $\alpha\Phi_i$ and scale parameter β, it follows that X_t defined by (41) has the required gamma distribution, and since the autocorrelation function of $X_t^{(i)}$ is $\exp(-\lambda_i u)$, the autocorrelation function of the sum X_t is given by (43). This construction will come in handy in Section 4, where processes of the type (41) will be used as models for stochastic volatility.

Empirical autocorrelations that might be interpreted as an indication of long range dependence, may often alternatively be approximated very well by autocorrelation functions of the type (43). However, if a model with genuine long range dependence is desirable, a *NIG*-process of this type can be constructed as follows.

Let $X^{(i)}$, $i = 1, 2, \ldots$, be a sequence of independent *NIG* Ornstein–Uhlenbeck processes with *NIG*-parameters $(\alpha, \beta, 0, \delta_i)$, where

$$\delta_i \sim i^{-1-2(1-H)},$$

for some $H \in (0, 1)$, and all with the same value of the drift parameter λ. Barndorff-Nielsen (1998) showed that the process

$$X_t = \sum_{i=1}^{\infty} X_{t/i}^{(i)}, \tag{45}$$

which is stationary and well-defined as a mean-square limit, has as its marginal distribution the *NIG* distribution with parameters $(\alpha, \beta, 0, \delta)$, where $\delta = \sum_{i=1}^{\infty} \delta_i$. Moreover, its autocorrelation function $r(u)$ satisfies

$$r(u) \sim L(u) u^{-2(1-H)},$$

for some slowly varying function L. Thus if $\frac{1}{2} < H < 1$, the process X exhibits long range dependence with exponent H. The construction of long range dependent processes by a sum of the type (45) is similar to a construction proposed by Cox (1984). Almost the same construction was used in Barndorff-Nielsen, Jensen and Sørensen (1990). The construction (45) can also be applied to a sequence of independent stationary *NIG*-diffusions given as solutions of stochastic differential equations defined in analogy to (31).

Likelihood inference for the various compound processes considered here is complicated by the fact that the likelihood function is not explicitly available. A feasible alternative is provided by prediction-based estimating functions, see M. Sørensen (2000).

4. Stochastic volatility models

A generalization of the Black–Scholes model for the logarithm of an asset price

$$\mathrm{d}X_t = \left(\mu + \beta\sigma^2\right)\mathrm{d}t + \sigma\,\mathrm{d}W_t, \tag{46}$$

that takes into account the empirical finding that the volatility σ^2 varies randomly over time is a stochastic volatility process:

$$\mathrm{d}X_t = (\mu + \beta v_t)\,\mathrm{d}t + \sqrt{v_t}\,\mathrm{d}W_t. \tag{47}$$

Here the volatility v_t is a stochastic process that cannot be observed directly. If the data are observations at the time points Δi, $i = 0, 1, 2, \ldots, n$, then the returns $Y_i = X_{i\Delta} - X_{(i-1)\Delta}$ can be written in the form

$$Y_i = \mu\Delta + \beta S_i + \sqrt{S_i}\,A_i, \tag{48}$$

where

$$S_i = \int_{(i-1)\Delta}^{i\Delta} v_t\,\mathrm{d}t, \tag{49}$$

and where the A_is are independent, standard normal distributed random variables. If the integrated volatility S_i is independent of A_i, and if it is generalized inverse Gaussian distributed, then the distribution of the return Y_i is generalized hyperbolic. This follows from the representation of the generalized hyperbolic distributions as variance–mean mixtures of normal distributions mentioned in Section 1.2. Unfortunately, no continuous time process v with the property that the integrated volatility (49) is exactly generalized inverse Gaussian distributed is presently known. Therefore we will instead consider models where the volatility process v is stationary with v_t generalized inverse Gaussian distributed. For small values of Δ, the distribution of S_i will then be close to a generalized inverse Gaussian distribution, and hence the distribution of Y_i will be close to a generalized hyperbolic distribution. Thus we obtain models that are not exactly generalized hyperbolic, but which have marginal distribution with much the same tail properties when Δ is not too large.

When Δ tends to infinity, the distribution of $\Delta^{-1/2}(Y_i - \mu\Delta - \beta S_i) = \sqrt{S_i/\Delta}A_i$ tends to a normal distribution with mean zero and variance equal to the mean volatility, $\mathrm{E}(v_t)$, provided that the process v is ergodic. This is in accordance with the empirical finding that the distribution of returns over short periods have heavy tails and are well approximated by generalized hyperbolic distributions, whereas the distribution of returns over long periods is close to a normal distribution. Limit theorems relating, for small Δ, the distribution of Y_i to the generalized hyperbolic distribution obtained by assuming that S_i is exactly generalized inverse Gaussian distributed are given in Genon-Catalot, Jeantheau and Larédo (1998). A rather different type of discrete time stochastic volatility models with exactly generalized hyperbolic distributed returns was proposed in Banrdorff-Nielsen (1997). It should be noted that stochastic volatility models can be interpreted as being obtained by *subordination*. Here the operational time or business time is the integral of the volatility process $\tau(t) = \int_0^t v_s \, \mathrm{d}s$, which can be interpreted as discussed in Section 2.

A simple specification of the volatility process v is to assume that it is one of the stationary and ergodic generalized inverse Gaussian diffusions defined in Section 3 as the solution of (35). A particularly simple choice is to assume that v is the stationary CIR-model given by (36), for which v_t is gamma-distributed so that a variance-gamma stochastic volatility model is obtained. This model was proposed by Hull and White (1988) and was considered further by Heston (1993). Its advantage is that analytically it is relatively tractable. For instance, all moments and mixed moments can be found explicitly, see, e.g., M. Sørensen (2000). A problem is that because of the linear drift, the autocorrelation function is an exponential function, whereas it is a well-established empirical fact that the autocorrelation function of the volatility process decreases more slowly than a single exponential function. Under relatively weak regularity conditions a diffusion model has an exponentially decreasing autocorrelation function. A sufficient condition is that it is ρ-mixing, for which simple conditions are given in Jeantheau and Larédo (2000). For this reason, stochastic volatility models with a diffusion volatility process can usually not fit the autocorrelation of the volatility process well.

In applications where the autocorrelation of the volatility process is important, a solution is to use the construction in Section 3.4, i.e., to define the volatility process as the sum

$$v_t = v_t^{(1)} + \cdots + v_t^{(m)}, \tag{50}$$

where $v_t^{(1)}, \ldots, v_t^{(m)}$ are independent CIR-processes, with $v_t^{(i)}$ defined like the process $X^{(i)}$ given by (44). Also in this case a variance-gamma model is obtained, which is exactly as analytically tractable as the variance-gamma model just discussed, but the autocorrelation structure of the volatility process (50) is given by (43) and is thus very flexible. This approach is studied for more general diffusion models in Bibby and Sørensen (2002).

It has been found empirically that for equities a fall in the price is associated with an increase in the future volatility. This phenomenon is referred to as leverage, Black (1976) and Nelson (1991). Stochastic volatility models of the form (47), where the Wiener process driving the price process is independent of the volatility process, as we have so far assumed, cannot deal with leverage, because for such a model the future fluctuations of the volatility are independent of the present price. We can, however, easily generalize the model to allow for the leverage phenomenon. Again we let the volatility process v be given by (50), and denote the Wiener process driving the kth CIR-process $v_t^{(k)}$ by $B^{(k)}$. Then we define the log-price process by

$$\mathrm{d}X_t = (\mu + \beta v_t)\,\mathrm{d}t + \sqrt{v_t}\,\mathrm{d}\widetilde{W}_t, \tag{51}$$

where $\widetilde{W}$ is the standard Wiener process

$$\widetilde{W}_t = \frac{W_t + \rho \overline{B}_t}{\sqrt{1+\rho^2}}$$

with $\rho \in \mathbb{R}$ and

$$\overline{B}_t = \frac{B_t^{(1)} + \cdots + B_t^{(m)}}{\sqrt{m}}. \tag{52}$$

A lengthy calculation shows that for $\beta = 0$ the covariance between Y_i and Y_{i+j}^2 $(j \geqslant 1)$ is

$$\frac{\rho}{\sqrt{1+\rho^2}} \frac{1}{\sqrt{m}} \sum_{k=1}^{m} b_k\, \mathrm{e}^{-\lambda_k \Delta j}.$$

Here

$$b_k = \sqrt{2\beta}\, \mathrm{e}^{\lambda_k \Delta}\left(1 - \mathrm{e}^{-\lambda_k \Delta}\right)^2 \lambda_k^{-3/2} \mathrm{E}\left(\sqrt{v_1^{(k)} v_1}\right) > 0,$$

where β is the shape parameter of the gamma distribution of the volatility, and λ_k is the speed of reversion of the kth volatility component. We see that the correlation between Y_i and Y_{i+j}^2 is negative if $\rho < 0$, which is exactly what we wanted. For $\rho = 0$ there is no leverage effect as expected. Note that the effect decreases as j tends to infinity. The decrease is of the same type as that of the autocorrelation function (43), but with different weights. It is thus very flexible and can in particular be slow.

Barndorff-Nielsen and Shephard (2001b, c) proposed to model the volatility process v as an Ornstein–Uhlenbeck type process, i.e., a solution to the stochastic differential equation (38). Such a process can be chosen stationary with a generalized inverse Gaussian marginal distribution, as discussed in Section 3.3. Processes of this type have the advantage that the drift is linear and the coefficient in front of the driving Lévy process is constant, which, analogous to the situation for the classical Wiener-driven Ornstein–Uhlenbeck process, implies an unusual analytic tractability. For instance the integrated volatility, which is a key quantity in finance, has the simple structure

$$\int_s^t v_s\, \mathrm{d}s = \lambda^{-1}\big((Z_t - Z_s) - (v_t - v_s)\big),$$

where $s < t$, and where Z is the driving Lévy process. This relation implies, for instance, that stochastic volatility processes of this type can be simulated as accurately as the volatility process can be simulated. This is because the random variables S_i, given by (49), are simple functions of the processes Z and v. An efficient method of simulating Ornstein–Uhlenbeck type processes is based on results by Rosiński (1991) and Rosiński (2001), see the exposition in Barndorff-Nielsen and Shephard (2001b).

Barndorff-Nielsen and Shephard (2001a) have studied the distributional properties of integrated Ornstein–Uhlenbeck type processes in detail. For the Ornstein–Uhlenbeck type volatility process with inverse Gaussian marginal distributions they found that while the integrated volatility process is not distributed exactly as the inverse Gaussian distribution, its tails have the same behaviour as this distribution. This implies that the returns will have the expected *NIG* tail behaviour.

For an Ornstein–Uhlenbeck type volatility process v, the autocorrelations of the discrete time processes S_i and Y_i^2 have the following simple form. Here S_i is given by (49), while Y_i denotes the return given by (48).

$$\operatorname{cor}(S_i, S_{i+j}) = d\exp\bigl(-\lambda\Delta(j-1)\bigr), \tag{53}$$

and

$$\operatorname{cor}\bigl(Y_i^2, Y_{i+j}^2\bigr) = c\exp\bigl(-\lambda\Delta(j-1)\bigr), \tag{54}$$

where

$$\begin{aligned} 1 \geqslant d &= \frac{[1-\exp(-\lambda\Delta)]^2}{2[\exp(-\lambda\Delta)-1+\lambda\Delta]} \\ \geqslant c &= \frac{[1-\exp(-\lambda\Delta)]^2}{6[\exp(-\lambda\Delta)-1+\lambda\Delta]+2(\lambda\Delta)^2(\xi/\omega)^2} \geqslant 0, \end{aligned}$$

with ξ and ω denoting the mean and variance of the volatility v_t. Therefore, as discussed in Barndorff-Nielsen and Shephard (2001c), S and Y^2 are constrained ARMA(1, 1) processes with common autoregressive parameter, and with the moving average root being stronger for S than for Y^2. The ARMA structure implies that the return process Y is weak GARCH(1, 1) in the sense of Drost and Nijman (1993). Note that the formulae (53) and (54) also hold for the stochastic volatility model discussed above, where the volatility process is a CIR-diffusion. Hence for this model, the processes S and Y^2 have the same ARMA structure.

Barndorff-Nielsen and Shephard (2001c) also proposed a model with a Lévy-driven Ornstein–Uhlenbeck volatility process that allows for the leverage phenomenon. The log-price is modelled by

$$\mathrm{d}X_t = (\mu + \beta v_t)\,\mathrm{d}t + \sqrt{v_t}\,\mathrm{d}W_t + \rho\,\mathrm{d}\overline{Z}_t, \tag{55}$$

where $\overline{Z}_t = Z_t - \mathrm{E}(Z_t)$ is the centered version of the Lévy process Z that drives the volatility process. This model has properties similar to those of the model with leverage discussed above (when $m = 1$). It is not a generalized hyperbolic model in the sense of the other stochastic volatility models in this section because of the term $\rho\,\mathrm{d}\overline{Z}_t$. It is not clear to what extend the model is approximately hyperbolic. A complication is that the log-price process is a diffusion with jumps rather than a classical diffusion process driven by a Wiener process.

As already mentioned in Section 3.3, the autocorrelation function of an Ornstein–Uhlenbeck type process decreases exponentially, which, as also mentioned earlier, is faster than what is typically found in financial data. Volatility processes of the form (50), where $v_t^{(1)}, \ldots, v_t^{(m)}$ are independent, stationary Ornstein–Uhlenbeck type processes such that the marginal distribution of v is a generalized inverse Gaussian distribution, have a much more flexible autocorrelation structure. That such a volatility process exists was discussed in Section 3.4. Stochastic volatility models of this type often provide a much better fit to financial data. An example of this is given in Barndorff-Nielsen and Shephard (2001c). Also models where the volatility process is a sum of independent Ornstein–Uhlenbeck processes are analytically tractable.

Statistical inference for stochastic volatility models cannot easily be based on the likelihood function as it is not explicitly available and quite hard to simulate. Harvey, Ruiz and Shephard (1994) proposed a pseudo-likelihood method based on a Gaussian approximation that allowed them to apply the Kalman filter. More recently, likelihood based methods for stochastic volatility models have been proposed by Kim, Shephard and Chib (1998), and simulation based Bayesian methods using Markov chain Monte Carlo have been developed by Elerian, Chib and Shephard (2001) and Eraker (2001). A new and quite simple way of obtaining an approximate likelihood function for stochastic volatility models, which seems very promising, has been proposed by H. Sørensen (2001). The method takes advantage of the fact that lag-k conditional densities are relatively easy to obtain by simulation for stochastic volatility models. Other methods are the indirect inference methods of Gouriéroux, Monfort and Renault (1993), Galant and Tauchen (1996), and Gallant and Long (1997). The prediction-based estimating functions of M. Sørensen (2000) can be applied to all models discussed in this section, while the estimators proposed by Genon-Catalot, Jeantheau and Larédo (1999) based on limit results (where the time between observations goes to zero) in Genon-Catalot, Jeantheau and Larédo (1998) are developed for volatility processes of the diffusion type. Recently methods based on realized volatility have been proposed, see Gloter (1999) and Banrdorff-Nielsen and Shephard (2002). Surveys that discuss the literature on stochastic volatility models up to 1995 can be found in Ghyseles, Harvey and Renault (1996) and Shephard (1996).

Acknowledgment

We are grateful to Ole E. Barndorff-Nielsen for his several useful comments on an earlier version of this chapter. The research of Michael Sørensen was supported by MaPhySto, The

Centre for Mathematical Physics and Stochastics, funded by a grant from The Danish National Research Foundation, and by the European Commission through the DYNSTOCH Network under the Human Potential Programme. Michael Sørensen was also supported by the Centre for Analytical Finance and the Danish Mathematical Finance Network, both financed by the Danish Social Science Research Council. The data were put at our disposal by the Centre for Analytical Finance.

Appendix

In this appendix a few definitions and results concerning Bessel functions are collected.

The *modified Bessel function of the third kind* with index $\lambda \in \mathbb{R}$ can be defined by the following integral representation,

$$K_\lambda(x) = \frac{1}{2}\int_0^\infty u^{\lambda-1} e^{-x(u+u^{-1})/2}\, du, \quad x > 0.$$

The modified Bessel function has the following properties:

$$K_{-\lambda}(x) = K_\lambda(x), \tag{A.1}$$

$$K_{\lambda+1}(x) = \frac{2\lambda}{x} K_\lambda(x) + K_{\lambda-1}(x), \tag{A.2}$$

$$K'_\lambda(x) = -\frac{\lambda}{x} K_\lambda(x) - K_{\lambda-1}(x). \tag{A.3}$$

For $\lambda = n + 1/2$, $n = 0, 1, 2, \ldots$, we have that

$$K_{n+1/2}(x) = \sqrt{\frac{\pi}{2x}}\, e^{-x}\left\{1 + \sum_{i=1}^{n} \frac{(n+i)!}{(n-i)!i!}(2x)^{-i}\right\}. \tag{A.4}$$

For small values of the argument it holds that

$$K_\lambda(x) \sim \Gamma(\lambda) 2^{\lambda-1} x^{-\lambda}, \quad x \downarrow 0, \text{ if } \lambda > 0. \tag{A.5}$$

Similarly, we have for large values of the argument that

$$K_\lambda(x) = \sqrt{\frac{\pi}{2x}} \cdot e^{-x} \cdot \left\{1 + \frac{4\lambda^2 - 1}{8x} + \frac{(4\lambda^2-1)(4\lambda^2-9)}{2!(8x)^2} + \frac{(4\lambda^2-1)(4\lambda^2-9)(4\lambda^2-25)}{3!(8x)^3} + \cdots\right\}. \tag{A.6}$$

The *Bessel function of the first kind* with index $\lambda \in \mathbb{R}$ can be defined for $x > 0$ by

$$J_\lambda(x) = \frac{1}{\pi}\int_0^\pi \cos\bigl(x\sin(u) - \lambda u\bigr)\,du - \frac{\sin(\lambda\pi)}{\pi}\int_0^\infty e^{-x\sinh(u)-\lambda u}\,du.$$

For $\lambda > -\frac{1}{2}$ we have

$$J_\lambda(x) = \frac{2(x/2)^\lambda}{\sqrt{\pi}\Gamma(\lambda+1/2)}\int_0^1 (1-u^2)^{\lambda-1/2}\cos(xu)\,du, \quad x \in \mathbb{R},$$

where Γ denotes the gamma function.

The *Bessel function of the second kind* with index $\lambda \in \mathbb{R}$ can be defined for $x > 0$ by

$$Y_\lambda(x) = \frac{1}{\pi}\int_0^\pi \sin\bigl(x\sin(u) - \lambda u\bigr)\,du - \frac{1}{\pi}\int_0^\infty \bigl[e^{\lambda u} + e^{-\lambda u}\cos(\lambda\pi)\bigr]e^{-x\sinh(u)}\,du.$$

The function $Y_\lambda(x)$ is often alternatively denoted $N_\lambda(x)$ and is sometimes called Weber's function. The relationship between $J_\lambda(x)$ and $Y_\lambda(x)$ is

$$Y_\lambda(x) = \frac{J_\lambda(x)\cos(\lambda\pi) - J_{-\lambda}(x)}{\sin(\lambda\pi)}.$$

In connection with the *NIG*-distribution, it is useful to know that

$$J_{1/2}(x) = \sqrt{\frac{2}{\pi x}}\sin(x) \quad \text{and} \quad Y_{1/2}(x) = -\sqrt{\frac{2}{\pi x}}\cos(x).$$

References

Aït-Sahalia, Y., 2002. Maximum likelihood estimation of discretely sampled diffusions: A closed-form approximation approach. Econometrica 70, 223–262.

Ané, T., Geman, H., 2000. Order flow, transaction clock and normality of asset returns. Journal of Finance 55, 2259–2284.

Bagnold, R.A., 1941. The Physics of Blown Sand and Desert Dunes. Methuen, London.

Barndorff-Nielsen, O.E., 1977. Exponentially decreasing distributions for the logarithm of particle size. Proceedings of the Royal Society London. Series A 353, 401–419.

Barndorff-Nielsen, O.E., 1978. Hyperbolic distributions and distributions on hyperbolae. Scandinavian Journal of Statistics 5, 151–157.

Barndorff-Nielsen, O.E., 1997. Normal inverse Gaussian distributions and stochastic volatility modelling. Scandinavian Journal of Statistics 24, 1–13.

Barndorff-Nielsen, O.E., 1998. Processes of normal inverse Gaussian type. Finance and Stochastics 2, 41–68.

Barndorff-Nielsen, O.E., 2001. Superposition of Ornstein–Uhlenbeck type processes. Theory of Probability and its Applications, 45.

Barndorff-Nielsen, O.E., Blæsild, P., 1980. Hyperbolic distributions and ramifications: Contributions to theory and application. Research Report 68. Department of Theoretical Statistics, Institute of Mathematics, University of Aarhus.

Barndorff-Nielsen, O.E., Blæsild, P., Jensen, J.L., Sørensen, M., 1985. The fascination of sand. In: Atkinson, A.C., Fienberg, S.E., (Eds.), A Celebration of Statistics, Springer, New York, pp. 57–87.

Barndorff-Nielsen, O.E., Halgreen, C., 1977. Infinite divisibility of the hyperbolic and generalized inverse Gaussian distributions. Zeitschrift für Wahrscheinlichkeitstheorie und verwandte Gebiete 38, 309–312.

Barndorff-Nielsen, O.E., Jensen, J.L., Sørensen, M., 1990. Parametric modelling of turbulence. Philosophical Transactions of the Royal Society of London. Series A 332, 439–455.

Barndorff-Nielsen, O.E., Jensen, J.L., Sørensen, M., 1998. Some stationary processes in discrete and continuous time. Advances in Applied Probability 30, 989–1007.

Barndorff-Nielsen, O.E., Shephard, N., 2001a. Integrated OU processes and non-Gaussian OU-based stochastic volatility models. Research report. MaPhySto, University of Aarhus. Scandinavian Journal of Statistics, forthcoming.

Barndorff-Nielsen, O.E., Shephard, N., 2001b. Modelling by Lévy processes for financial econometrics. In: Barndorff-Nielsen, O.E., Mikosch, T., Resnick, S. (Eds.), Lévy Processes: Theory and Applications. Birkhäuser, Boston.

Barndorff-Nielsen, O.E., Shephard, N., 2001c. Non-Gaussian Ornstein–Uhlenbeck-based models and some of their uses in financial econometrics (with discussion). Journal of the Royal Statistical Society. Series B 63, 167–241.

Barndorff-Nielsen, O.E., Shephard, N., 2002. Econometric analysis of realised volatility and its use in estimating stochastic volatility models. Journal of the Royal Statistical Society. Series B 64, 253–280.

Bertoin, J., 1996. Lévy Processes. Cambridge University Press.

Bibby, B.M., Jacobsen, M., Sørensen, M., 2002. Estimating functions for discretely sampled diffusion-type models. In: Aït-Sahalia, Y., Hansen, L.P. (Eds.), Handbook of Financial Econometrics. North-Holland, Amsterdam, forthcoming.

Bibby, B.M., Skovgaard, I.M., Sørensen, M., 2002. Diffusion-type models with given marginals and autocorrelation function. Preprint. Department of Theoretical Statistics, University of Copenhagen, submitted.

Bibby, B.M., Sørensen, M., 1995. Martingale estimation functions for discretely observed diffusion processes. Bernoulli 1, 17–39.

Bibby, B.M., Sørensen, M., 1996. On estimation for discretely observed diffusions: A review. Theory of Stochastic Processes 2, 49–56.

Bibby, B.M., Sørensen, M., 1997. A hyperbolic diffusion model for stock prices. Finance and Stochastics 1, 25–41.

Bibby, B.M., Sørensen, M., 2001. Simplified estimating functions for diffusion models with a high-dimensional parameter. Scandinavian Journal of Statistics 28 (1), 99–112.

Bibby, B.M., Sørensen, M., 2002. Flexible stochastic volatility models of the diffusion type. In preparation.

Black, F., 1976. Studies of stock price volatility changes. Proceedings of Business and Economics Statistics Section American Statistical Association 177–181.

Blæsild, P., 1978. The shape of the generalized inverse Gaussian and hyperbolic distributions. Research Report 37. Department of Theoretical Statistics, Institute of Mathematics, University of Aarhus.

Blæsild, P., Sørensen, M.K., 1992. HYP – A computer program for analyzing data by means of the hyperbolic distribution. Research Report No. 248. Department of Theoretical Statistics, Institute of Mathematics, University of Aarhus.

Blattberg, R.C., Gonedes, N., 1974. A comparison of the stable and student distributions as models for stock prices. Journal of Business 47, 244–280.

Chan, K.C., Karolyi, G.A., Longstaff, F.A., Sanders, A.B., 1992. An empirical comparison of alternative models of the short-term interest rate. Journal of Finance 47, 1209–1227.

Clark, P.K., 1973. A subordinated stochastic process with finite variance for speculative prices. Econometrica, 41.

Cox, D.R., 1984. Long-range dependence: A review. In: David, H.A., David, H.T. (Eds.), Statistics: An Appraisal. Iowa State University Press.

Cox, J.C., Ingersoll, J.E., Jr., Ross, S.A., 1985. A theory of the term structure of interest rates. Econometrica 53 (2), 385–407.

Dacunha-Castelle, D., Florens-Zmirou, D., 1986. Estimation of the coefficients of a diffusion from discrete observations. Stochastics 19, 263–284.

Drost, F.C., Nijman, T.E., 1993. Temporal aggregation of GARCH processes. Econometrica 61, 909–927.

Eberlein, E., 2001. Application of generalized hyperbolic Lévy motions to finance. In: Barndorff-Nielsen, O.E., Mikosch, T., Resnick, S. (Eds.), Lévy Processes – Theory and Applications. Birkhäuser, Boston, pp. 319–337.

Eberlein, E., Jacod, J., 1997. On the range of option prices. Finance and Stochastics 1, 131–140.

Eberlein, E., Keller, U., 1995. Hyperbolic distributions in finance. Bernoulli 1, 281–299.

Eberlein, E., Keller, U., Prause, K., 1998. New insights into smile, mispricing and value at risk: The hyperbolic model. Journal of Business 71, 371–406.

Eberlein, E., Prause, K., 2002. The generalized hyperbolic model: Financial derivatives and risk measures. In: Geman, H., Madan, D., Pliska, S., Vorst, T. (Eds.), Mathematical Finance – Bachelier Congress 2000. Springer, Heidelberg, pp. 245–267.

Eberlein, E., Raible, S., 1999. Term structure models driven by general Lévy processes. Mathematical Finance 9, 31–53.

Eberlein, E., Raible, S., 2001. Some analytic facts on the generalized hyperbolic model. In: Casacuberta, C., et al. (Eds.), Proceedings of the Third European Meeting of Mathematicians, Vol. II. In: Progress in Mathematics, Vol. 202. Birkhäuser, Boston, pp. 367–378.

Elerian, O., Chib, S., Shepard, N., 2001. Likelihood inference for discretely observed non-linear diffusions. Econometrica 69, 959–993.

Epps, T.W., Epps, M.L., 1976. The stochastic dependence of security price changes and transaction volumes: Implications for the mixture-of-distributions hypothesis. Econometrica 44, 305–321.

Eraker, B., 2001. MCMC analysis of diffusion models with application to finance. Journal of Business and Economic Statistics 19, 177–191.

Gallant, A.R., Long, J.R., 1997. Estimating stochastic differential equations efficiently by minimum chi-square. Biometrika 84, 125–141.

Gallant, A.R., Tauchen, G., 1996. Which moments to match? Econometric Theory 12, 657–681.

Genon-Catalot, V., Jeantheau, T., Larédo, C., 1998. Limit theorems for discretely observed stochastic volatility models. Bernoulli 4, 283–303.

Genon-Catalot, V., Jeantheau, T., Larédo, C., 1999. Parameter estimation for discretely observed stochastic volatility models. Bernoulli 5, 855–872.

Genon-Catalot, V., Jeantheau, T., Larédo, C., 2000. Stochastic volatility models as hidden Markov models and statistical applications. Bernoulli 6, 1051–1079.

Ghysels, E., Harvey, A.C., Renault, E., 1996. Stochastic volatility. In: Rao, C.R., Maddala, G.S. (Eds.), Statistical Methods in Finance. North-Holland, Amsterdam, pp. 119–191.

Gloter, A., 1999. Parameter estimation for a hidden diffusion. Preprint 20/99. University of Marne-la-Vallée.

Good, I.J., 1953. The population frequencies of species and the estimation of population parameters. Biometrika 40, 237–264.

Gouriéroux, C., Monfort, A., Renault, E., 1993. Indirect inference. Journal of Applied Econometrics 8, S85–S118.

Halgreen, C., 1979. Self-decomposability of generalized inverse Gaussian and hyperbolic distributions. Zeitschrift für Wahrscheinlichkeitstheorie und verwandte Gebiete 47, 13–17.

Hansen, L.P., Scheinkman, J.A., 1995. Back to the future: Generating moment implications for continuous-time Markov processes. Econometrica 63, 767–804.

Harvey, A.C., Ruiz, E., Shephard, N., 1994. Multivariate stochastic variance models. Review of Economic Studies 61, 247–264.

Heston, S.L., 1993. A closed-form solution for options with stochastic volatility with applications to bond and currency options. Review of Financial Studies 6, 327–343.

Honoré, P., 1997. Maximum likelihood estimation of non-linear continuous-time term-structure models. Working Paper 1997-7. Department of Finance, Aarhus School of Business.

Hull, J., White, A., 1988. An analysis of the bias in option pricing caused by a stochastic volatility. Advances in Futures and Options Research 3, 29–61.

Hurst, S.R., 1997. On the stochastic dynamics of stock market volatility. Ph.D. Thesis. Australian National University.

Hurst, S.R., Platen, E., Rachev, S.T., 1997. Subordinated market index models: A comparison. Financial Engineering and the Japanese Markets 4, 97–124.

Jacobsen, M., 2001. Discretely observed diffusions; classes of estimating functions and small Δ-optimality. Scandinavian Journal of Statistics 28 (1), 123–150.

Jacod, J., Shiryaev, A.N., 1987. Limit Theorems for Stochastic Processes. Springer, New York.

Jensen, J.L., Pedersen, J., 1999. Ornstein–Uhlenbeck type processes with non-normal distribution. Journal of Applied Probability 36, 389–402.

Jiang, W., 2000. Some simulation-based models towards mathematical finance. Ph.D. Thesis. University of Aarhus.

Jørgensen, B., 1982. Statistical Properties of the Generalized Inverse Gaussian Distribution. Lecture Notes in Statistics, Vol. 9. Springer-Verlag, New York.

Jurek, Z.J., Mason, J.D., 1993. Operator-Limit Distributions in Probability Theory. Wiley, New York.

Jurek, Z.J., Vervaat, W., 1983. An integral representation for self-decomposable Banach space valued random variables. Zeitschrift für Wahrscheinlichkeitstheorie und verwandte Gebiete 62, 247–262.

Keller, U., 1997. Realistic modelling of financial derivatives. Ph.D. Thesis. Universität Freiburg.

Kessler, M., 2000. Simple and explicit estimating functions for a discretely observed diffusion process. Scandinavian Journal of Statistics 27, 65–82.

Kessler, M., Sørensen, M., 1999. Estimating equations based on eigenfunctions for a discretely observed diffusion process. Bernoulli 5, 299–314.

Kim, S., Shephard, N., Chib, S., 1998. Stochastic volatility: Likelihood inference and comparison with ARCH models. Review of Economic Studies 65, 361–393.

Küchler, U., Neumann, K., Sørensen, M., Streller, A., 1999. Stock returns and hyperbolic distributions. Mathematical and Computer Modelling 29, 1–15.

Madan, D.B., Chang, E.C., 1996. Volatility smiles, skewness premia and risk metrics: Application of a four parameter closed form generalization of geometric Brownian motion to the pricing of options. Presented at a Conference on Mathematical Finance, University of Aarhus, Denmark.

Madan, D.B., Milne, F., 1991. Option pricing with V.G. martingale components. Mathematical Finance 1, 39–55.

Madan, D.B., Seneta, E., 1990. The variance gamma (V.G.) model for share market returns. Journal of Business 63, 511–524.

Nelson, D.B., 1991. Conditional heteroskedasticity in asset pricing: A new approach. Econometrica 59, 347–370.

Pedersen, A.R., 1995. A new approach to maximum likelihood estimation for stochastic differential equations based on discrete observations. Scandinavian Journal of Statistics 22, 55–71.

Poulsen, R., 1999. Approximate maximum likelihood estimation of discretely observed diffusion processes. Working Paper 29. Centre for Analytical Finance, Aarhus.

Praetz, P.D., 1972. The distribution of share prices. Journal of Business 45, 49–55.

Prause, K., 1999. The generalized hyperbolic model: Estimation, financial derivatives, and risk measure. Ph.D. Thesis. Universität Freiburg.

Protter, P., 1990. Stochastic Integration and Differential Equations – A New Approach. Springer, New York.

Raible, S., 2000. Lévy processes in finance: Theory, numerics, and empirical facts. Ph.D. Thesis. Universität Freiburg.

Rosiński, J., 1991. On a class of infinitely divisible processes represented as mixtures of Gaussian processes. In: Cambanis, S., Samorodnitsky, G., Taqqu, M.S. (Eds.), Stable Processes and Related Topics. Birkhäuser, Boston, pp. 27–41.

Rosiński, J., 2001. Series representations of Lévy processes from the perspective of point processes. In: Barndorff-Nielsen, O.E., Mikosch, T., Resnick, S. (Eds.), Lévy Processes – Theory and Applications. Birkhäuser, Boston.

Rydberg, T.H., 1997. The normal inverse Gaussian Lévy process: Simulation and approximation. Communications in Statistics. Stochastic Models 13, 887–910.

Rydberg, T.H., 1999. Generalized hyperbolic diffusion processes with applications in finance. Mathematical Finance 9, 183–201.

Rydberg, T.H., 2000. Realistic statistical modelling of financial data. International Statistical Review 68, 233–258.
Sato, K., 1999. Lévy Processes and Infinitely Divisible Distributions. Cambridge University Press.
Sato, K., Watanabe, T., Yamazato, M., 1994. Recurrence conditions for multivariate processes of Ornstein–Uhlenbeck type. Journal of the Mathematical Society of Japan 46, 245–265.
Sato, K., Yamazato, M., 1982. Stationary processes of the Ornstein–Uhlenbeck type. In: Ito, K., Prohorov, J.V. (Eds.), Probability Theory and Mathematical Statistics. In: Lecture Notes in Mathematics, Vol. 1021. Springer-Verlag, Berlin.
Sato, K., Yamazato, M., 1984. Operator-selfdecomposable distributions as limit distributions of processes of the Ornstein–Uhlenbeck type. Stochastic Processes and their Applications 17, 73–100.
Seshadri, V., 1997. Halphen's laws. In: Kotz, S., Read, C.B., Banks, D.L. (Eds.), Encyclopedia of Statistical Sciences, Update Volume 1. Wiley, New York, pp. 302–306.
Shephard, N., 1996. Statistical aspects of ARCH and stochastic volatility. In: Cox, D.R., Hinkley, D.V., Barndorff-Nielsen, O.E. (Eds.), Time Series Models in Econometrics, Finance and other Fields. Chapman and Hall, London, pp. 1–67.
Sichel, H.S., 1973. Statistical evaluation of diamondiferous deposits. Journal of South African Institut Min. Metall. 76, 235–243.
Sørensen, H., 2000. Inference for diffusion processes and stochastic volatility models. Ph.D. Thesis. Department of Statistics and Operations Research, University of Copenhagen.
Sørensen, H., 2001. Simulated likelihood approximations for stochastic volatility models. Preprint 1. Department of Theoretical Statistics, University of Copenhagen. Forthcoming in Scandinavian Journal of Statistics.
Sørensen, M., 1997a. Estimating functions for discretely observed diffusions: A review. In: Basawa, I.V., Godambe, V.P., Taylor, R.L. (Eds.), Selected Proceedings of the Symposium on Estimating Functions. In: IMS Lecture Notes – Monograph Series, Vol. 32. Institute of Mathematical Statistics, Hayward.
Sørensen, M., 1997b. Exponential family inference for diffusion models. Research Report No. 383. Department of Theoretical Statistics, Institute of Mathematics, University of Aarhus.
Sørensen, M., 2000. Prediction-based estimating functions. Econometrics Journal 3, 123–147.
Wolfe, S.J., 1982. On a continuous analogue of the stochastic difference equation $X_n = \rho X_{n-1} + B_n$. Stochastic Processes and their Applications 12, 301–312.

Chapter 7

STABLE MODELING OF MARKET AND CREDIT VALUE AT RISK

SVETLOZAR T. RACHEV

Department of Statistics and Applied Probability, University of California, Santa Barbara, USA
Institute of Statistics and Mathematical Economics, University of Karlsruhe, Germany
e-mail: rachev@lsoe-4.wiwi.uni-karlsruhe.de

EDUARDO S. SCHWARTZ

Anderson School of Management, University of California, Los Angeles, USA

IRINA KHINDANOVA

Colorado School of Mines

Contents

Handbook of Heavy Tailed Distributions in Finance, Edited by S.T. Rachev

Abstract

The chapter examines the use of stable Paretian distributions in modeling market and credit Value at Risk (VaR). The in-sample- and forecast-evaluations show that stable market VaR modeling outperforms the "normal" modeling for high values of the VaR confidence level. The chapter also develops a new technique for estimating correlation, constructs a new method for simulating portfolio values, and assesses portfolio VaR in various cases of credit instruments' distributions: independent, symmetric dependent, and skewed dependent.

1. Introduction

One of the most important tasks of financial institutions is evaluating the exposure to market and credit risks. Market risks arise from variations in prices of equities, commodities, exchange rates, and interest rates. Credit risks refer to potential losses that might occur because of a change in the counterparty's credit quality such as a rating migration or a default. The dependence on market and credit risks can be measured by changes in the portfolio value, or profits and losses. A commonly used methodology for estimation of risks is the *Value at Risk* (VaR). In the text below, *the market VaR* implies the VaR measurements associated with market risks and *the credit VaR* means the VaR linked to credit risks.

A *VaR measure* is the highest possible loss over a certain period of time at a given confidence level. For example, if the daily VaR for a given portfolio of assets is reported to be \$2 million at the 95 percent confidence level, it means that, without abrupt changes in the market conditions, one-day losses will exceed \$2 million 5 percent of the time.

Formally, a $\mathrm{VaR} = \mathrm{VaR}_{t,\tau}$ is defined as the upper bound of the one-sided confidence interval:

$$\Pr\left[\Delta P(\tau) < -\mathrm{VaR}\right] = 1 - c, \tag{1}$$

where c is the confidence level and $\Delta P(\tau) = \Delta P_t(\tau)$ is the *relative change* (*return*) in the portfolio value over the time horizon τ.

$$\Delta P_t(\tau) = P(t+\tau) - P(t),$$

where $P(t) = \log S(t)$, $S(t)$ is the portfolio value at t, the time period is $[t, T]$, with $T - t = \tau$, and t is the current time.

The essence of the VaR computations is estimation of low quantiles in the portfolio return distributions. The VaR techniques suggest different ways of constructing the portfolio return distributions. The traditional methods are the parametric method, historical simulation, Monte Carlo simulation, and stress-testing. One of the parametric approaches, the variance–covariance method, is based on the normal assumption for the distribution of financial returns. However, financial data often violate the normality assumption. The empirical observations exhibit "fat" tails and excess kurtosis. The historical method does not impose distributional assumptions but it is not reliable in estimating low quantiles of ΔP with a small number of observations in the tails. The performance of the Monte Carlo method depends on the quality of distributional assumptions on the underlying risk factors. A well-known methodology of constructing credit portfolio return distributions is the CreditMetrics™ product of J.P. Morgan.[1] It is based on the rating transition model of Jarrow, Lando and Turnbull (1997) and assumptions that joint credit quality changes are driven by joint movements of firms' assets values.

[1] See Gupton, Finger and Bhatia (1997).

The existing methods do not provide satisfactory evaluation of VaR. The main drawback is inadequate approximation of distributional forms of portfolio returns. Given the nature (heavy tails, excess kurtosis, and skewness[2]) of empirical financial data, the stable Paretian distributions seem to be the most appropriate distributional models.[3]

The chapter examines the use of stable Paretian distributions in modeling market and credit VaR. The stable distributions are described by four parameters: α tail index, β skewness, μ location, and σ scale. Modeling with such parameters will depict fat tails and skewness of distributions. Empirical analysis reported here confirms that, indeed, stable modeling captures heavy-tailedness and asymmetry of financial returns, and, therefore, produces more accurate risk estimates. The in-sample- and forecast-evaluations show that stable market VaR modeling outperforms the "normal" modeling for high values of the VaR confidence level. The stable distributions possess the additivity property: a linear combination of independent stable (or jointly stable) random variables with stability index α is again a stable random variable with the same α. The additivity property provides analytic formulas for parameters of portfolio returns. In the case of independent instruments, the formulas are simple and can be used for estimating portfolio risk without simulations. An analyst can employ "independent" risk measurements as lower bounds of portfolio risk estimates. A symmetric stable random variable can be interpreted as a transformation of a normal random variable. Based on this property, a new technique is developed here for estimating correlation. A stable random variable can be decomposed into the "symmetry" and "skewness" parts. Building on this feature, we construct a new method for simulating a distribution of portfolio values. We apply this method for portfolio risk evaluation in various cases of credit instruments' distributions: independent, symmetric dependent, and skewed dependent.

The remainder of the chapter is organized as follows. In Section 2 we discuss computation of VaR using the variance–covariance method, which is based on the normality assumption for the distribution of financial returns. Section 3 provides a finance-oriented description of stable distributions. In Section 4 we estimate the market VaR measurements employing normal and stable modeling of financial returns.[4] Section 5 investigates stable modeling of credit returns and discusses risk assessment for individual credit instruments. Section 6 considers portfolio risk estimation for independent portfolio assets and derives lower bounds for risk measurements. Sections 7 and 8 present, respectively, evaluation of portfolio risk in two cases of dependent portfolio instruments': symmetric and skewed. Section 9 describes a main framework of the one-factor model. Section 10 discusses credit risk evaluation for portfolio assets. Section 11 explains portfolio credit risk estimation. Section 12 states conclusions.

[2] Skewness is most pronounced in distributions of value changes of credit instruments. For references, see Gupton, Finger and Bhatia (1997), Federal Reserve System Task Force on Internal Credit Risk Models (1998), Basle Committee on Banking Supervision (1999).

[3] Cheng and Rachev (1995), Chobanov et al. (1996), Fama (1965), Gamrowski and Rachev (1994, 1995a, b), Mandelbrot (1962, 1963a, b, 1967), McCulloch (1996), Mittnik and Rachev (1991, 1993a, b), Mittnik, Rachev and Chenyao (1996), Mittnik, Rachev and Paolella (1998).

[4] See also Gamrowski and Rachev (1996).

2. "Normal" modeling of VaR

From the definition of VaR = $\mathrm{VaR}_{t,\tau}$ in Equation (1), the VaR values are obtained from the probability distribution of portfolio value returns:

$$1 - c = F_{\Delta P}(-\mathrm{VaR}) = \int_{-\infty}^{-\mathrm{VaR}} f_{\Delta P}(x)\,\mathrm{d}x,$$

where $F_{\Delta P}(x) = \Pr(\Delta P \leqslant x)$ is the cumulative distribution function (cdf) of portfolio returns in one period, and $f_{\Delta P}(x)$ is the probability density function (pdf) of ΔP.[5]

If the changes in the portfolio value are characterized by a parametric distribution, VaR can be computed using the distribution parameters. In this section we review "normal" modeling – a parametric method based on the normal distribution assumption. It is often called the variance–covariance method. We describe applications of the methodology for computing VaR of a single asset and portfolio VaR.

2.1. *VaR for a single asset*

Assume that a portfolio consists of a single asset, which depends only on one risk factor. Traditionally, in this setting, the distribution of asset returns is assumed to be *the univariate normal distribution*, identified by two parameters: the mean, μ, and the standard deviation, σ. The problem of calculating VaR is then reduced to finding the $(1-c)$-th percentile of the standard normal distribution z_{1-c}:

$$1 - c = \int_{-\infty}^{X^*} g(x)\,\mathrm{d}x = \int_{-\infty}^{z_{1-c}} \phi(z)\,\mathrm{d}z = N(z_{1-c}), \qquad \text{with } X^* = z_{1-c}\sigma + \mu,$$

where $\phi(z)$ is the standard normal density function, $N(z)$ is the cumulative normal distribution function, X is the portfolio return, $g(x)$ is the normal distribution function for returns with mean μ and standard deviation σ, and X^* is the lowest return at a given confidence level c.

In many applications investors assume that the expected return μ equals 0. This assumption is based on the conjecture that the magnitude of μ is substantially smaller than the magnitude of the standard deviation σ and, therefore, can be ignored. Then we have

$$X^* = z_{1-c}\sigma \quad \text{and, therefore,} \quad \mathrm{VaR} = -Y_0 X^* = -Y_0 z_{1-c}\sigma,$$

where Y_0 is the initial portfolio value.

[5] If $f_{\Delta P}(x)$ does not exist, then VaR can be obtained from the cdf $F_{\Delta P}$.

2.2. Portfolio VaR

If a portfolio consists of many assets, the computation of VaR is performed in several steps. Portfolio assets are decomposed into "building blocks", which depend on a finite number of risk factors. Exposures of the portfolio securities are combined into risk categories. Then, the total portfolio risk is obtained by aggregating risk factors and their correlations. We denote:

- X_p is the portfolio return in one period,
- N is the number of assets in the portfolio,
- X_i is the i-th asset return in one period ($\tau = 1$), $X_i = \Delta P(1) = P_i(1) - P_i(0)$, where P_i is the log-spot price of asset i, $i = 1, \dots, N$. More generally, X_i can be the risk factor that enters linearly[6] in the portfolio return.
- w_i is the i-th asset's weight in the portfolio, $i = 1, \dots, N$.

The portfolio return is

$$X_p = \sum_{i=1}^{N} w_i X_i.$$

In matrix notation,

$$X_p = w^{\mathrm{T}} X,$$

where

$$w = (w_1, w_2, \dots, w_N)^{\mathrm{T}}, \quad X = (X_1, X_2, \dots, X_N)^{\mathrm{T}}.$$

Then the portfolio variance is

$$V(X_p) = w^{\mathrm{T}} \underline{\Sigma} w = \sum_{i=1}^{N} w_i^2 \sigma_{ii} + \sum_{i=1}^{N} \sum_{\substack{j=1 \\ i \neq j}}^{N} w_i w_j \rho_{ij} \sigma_i \sigma_j,$$

where σ_{ii} is the variance of returns on the i-th asset, σ_i is the standard deviation of returns on the i-th asset, ρ_{ij} is the correlation between the returns on the i-th and the j-th assets, $\underline{\Sigma}$ is the covariance matrix, $\underline{\Sigma} = [\sigma_{ij}]$, $1 \leqslant i \leqslant N$, $1 \leqslant j \leqslant N$.

If all portfolio returns are *jointly normally distributed*, the portfolio return, as a linear combination of normal variables, is also *normally distributed*. The portfolio VaR based on the normal distribution assumption is

$$\mathrm{VaR} = -Y_0 z_{1-c} \sigma(X_P),$$

6 If the risk factor does not enter linearly (as in a case of an option), then a linear approximation is used.

where $\sigma(X_p)$ is the portfolio standard deviation (the *portfolio volatility*),

$$\sigma(X_p) = \sqrt{V(X_p)}.$$

Thus, risk can be represented by a combination of linear exposures to normally distributed factors. Hence, estimation of risk reduces to evaluation of the covariance matrix of portfolio risk factors (in the simplest case, individual asset returns).

The simplicity of normal modeling explains its common use for VaR computation despite the fact that financial data often violate the normality assumption. We conjecture that stable distributions are more adequate distributional models. In the following sections we analyze the stable modeling of market and credit VaR. We begin the analysis with providing a finance-oriented description of stable distributions.

3. A finance-oriented description of stable distributions

In this part we describe parameters and some finance-oriented properties of stable distributions. We also examine methods of estimating parameters of stable laws.

3.1. Parameters and properties of stable distributions

A random variable R is said to be *stable*[7] if for any $a > 0$ and $b > 0$ there exist constants $c > 0$ and $d \in \mathbb{R}$ such that

$$aR_1 + bR_2 \stackrel{d}{=} cR + d,$$

where R_1 and R_2 are independent copies of R and $\stackrel{d}{=}$ denotes the equality in distribution.

In general, stable distributions do not have closed form expressions for the density and distribution functions. Stable random variables (R) are commonly described by their characteristic functions:

$$\Phi_R(\theta) = E\big(\exp(\mathrm{i}R\theta)\big) = \exp\left\{-\sigma^\alpha|\theta|^\alpha\left(1 - \mathrm{i}\beta\,\mathrm{sign}(\theta)\tan\frac{\pi\alpha}{2}\right) + \mathrm{i}\mu\theta\right\}, \quad \text{if } \alpha \neq 1,$$

$$\Phi_R(\theta) = E\big(\exp(\mathrm{i}R\theta)\big) = \exp\left\{-\sigma|\theta|\left(1 + \mathrm{i}\beta\frac{2}{\pi}\,\mathrm{sign}(\theta)\ln\theta\right) + \mathrm{i}\mu\theta\right\}, \quad \text{if } \alpha = 1,$$

where α is the *index of stability*, $0 < \alpha \leqslant 2$, β is the *skewness parameter*, $-1 \leqslant \beta \leqslant 1$, σ is the *scale parameter*, $\sigma \geqslant 0$, and μ is the *location parameter*, $\mu \in \mathbb{R}$. To indicate the dependence of a stable random variable R on its parameters, we write $R \sim S_\alpha(\beta, \sigma, \mu)$. If

[7] Often R is called *α-stable* or *Pareto stable* or *Pareto–Lévy-stable* (for $\alpha < 2$).

the *index of stability* $\alpha = 2$, then the stable distribution reduces to the Gaussian distribution. In empirical studies, the modeling of financial return data is done typically with stable distributions having $1 < \alpha < 2$.[8] Stable distributions are unimodal and the smaller α is, the stronger the leptokurtic feature of the distribution (the peak of the density becomes higher and the tails are heavier). Thus, the index of stability can be interpreted as a *measure of kurtosis*. When $\alpha > 1$, the *location parameter* μ measures the mean of the distribution. If the *skewness parameter* $\beta = 0$, the distribution of R is symmetric and the characteristic function is

$$\Phi_R(\theta) = E\big(\exp(\mathrm{i}R\theta)\big) = \exp\big\{-\sigma^\alpha|\theta|^\alpha + \mathrm{i}\mu\theta\big\}.$$

If $\beta > 0$, the distribution is skewed to the right. If $\beta < 0$, the distribution is skewed to the left. Larger magnitudes of β indicate stronger skewness. If $\beta = 0$ and $\mu = 0$, then the stable random variable R is called *symmetric α-stable* ($s\alpha s$). The *scale parameter* (the volatility) σ allows any stable random variable R to be expressed as $R = \sigma R_0$, where R_0 has a unit scale parameter, and the same index of stability α and skewness parameter β as R. The scale parameter generalizes the definition of standard deviation. The stable analog of variance is the *variation*: $\nu_\alpha = \sigma^\alpha$.

In VaR estimations we are interested in investigating the behavior of the distributions in the tails. The *tails* of the stable (non-Gaussian) distributions have a power decay and are characterized by the following properties:

$$\lim_{\lambda\to+\infty} \lambda^\alpha P(R > \lambda) = k_\alpha \frac{1+\beta}{2}\sigma^\alpha$$

and

$$\lim_{\lambda\to+\infty} \lambda^\alpha P(R < -\lambda) = k_\alpha \frac{1-\beta}{2}\sigma^\alpha,$$

where

$$k_\alpha = \frac{1-\alpha}{\Gamma(2-\alpha)\cos(\pi\alpha/2)}, \quad \text{if } \alpha \neq 1, \qquad k_\alpha = \frac{2}{\pi}, \quad \text{if } \alpha = 1.$$[9]

The p-th absolute moment, $E|R|^p = \int_0^\infty P(|R|^p > x)\,\mathrm{d}x$, is

- *finite* if $p < \alpha$ or $\alpha = 2$, and
- *infinite* otherwise.

[8] The financial returns modeled with α-stable laws exhibit finite means but infinite variances.

[9] Note that, in contrast to the normal case, the tails of the non-Gaussian (Pareto) stable distributions are much fatter, which will be an important issue in estimating VaR.

Thus, the second moment of any non-Gaussian stable distribution is infinite.

Stable distributions possess the *additivity property*: a linear combination of independent stable random variables with stability index α is again a stable random variable with the same α.[10]

Example. If $R_1, R_2, \ldots, R_n$ are independent stable random variables with stability index α, $R_i \sim S_\alpha(\beta_i, \sigma_i, \mu_i)$, then $R = \sum_{i=1}^n w_i R_i$ is a stable random variable with the same α and parameters:

(a) if $\alpha \neq 1$,

$$\sigma = \left(\left(|w_1|\sigma_1\right)^\alpha + \cdots + \left(|w_n|\sigma_n\right)^\alpha\right)^{1/\alpha},$$

$$\beta = \frac{\operatorname{sign}(w_i)\beta_1(|w_1|\sigma_1)^\alpha + \cdots + \operatorname{sign}(w_n)\beta_n(|w_n|\sigma_n)^\alpha}{(|w_1|\sigma_1)^\alpha + \cdots + (|w_n|\sigma_n)^\alpha},$$

$$\mu = w_1\mu_1 + \cdots + w_n\mu_n;$$

(b) if $\alpha = 1$,

$$\sigma = |w_1|\sigma_1 + \cdots + |w_n|\sigma_n,$$

$$\beta = \frac{\operatorname{sign}(w_1)\beta_1|w_1|\sigma_1 + \cdots + \operatorname{sign}(w_n)\beta_n|w_n|\sigma_n}{|w_1|\sigma_1 + \cdots + |w_n|\sigma_n},$$

$$\mu = w_1\mu_1 + \cdots + w_n\mu_n - \frac{2}{\pi}\left(w_1 \ln|w_1|\sigma_1\beta_1 + \cdots + w_n \ln|w_n|\sigma_n\beta_n\right).$$

Since the Pareto-stable distributions have infinite variances, one cannot estimate risk by variance and dependence by correlations. We shall introduce variance- and covariance-similar notions for stable laws. These notions are based on the multivariate assumptions of stable distributions.

A random vector R of dimension d is *stable* if for any $a > 0$ and $b > 0$ there exist $c > 0$ and a d-dimensional vector D such that

$$aR_1 + bR_2 \stackrel{d}{=} cR + D,$$

where R_1 and R_2 are independent copies of R.

If a random vector is stable with $\alpha > 1$, then it means that all components of the vector are stable with the same index of stability and any linear combination (for example, portfolio returns) is again stable.[11]

[10] This property is shared only by normal and stable laws, and is the main advantage of the use of stable laws for portfolio returns.

[11] We shall model the dependence structure of the vector of returns $(R_1, \ldots, R_d)$ of a portfolio by assuming that $(R_1, \ldots, R_d)$ is an α-stable vector.

The characteristic function of a d-dimensional vector is given by:
(a) if $\alpha \neq 1$,

$$\begin{aligned}\Phi_R(\theta) &= \Phi_R(\theta_1, \theta_2, \ldots, \theta_d)\\ &= E\exp(\mathrm{i}\theta^{\mathrm{T}} R)\\ &= \exp\left\{-\int_{S_d} |\theta^{\mathrm{T}} s|^{\alpha}\left(1 - \mathrm{i}\,\mathrm{sign}(\theta^{\mathrm{T}} s)\tan\frac{\pi\alpha}{2}\right)\Gamma(\mathrm{d}s) + \mathrm{i}\theta^{\mathrm{T}}\mu\right\},\end{aligned}$$

(b) if $\alpha = 1$,

$$\Phi_R(\theta) = \exp\left\{-\int_{S_d} |\theta^{\mathrm{T}} s|\left(1 + \mathrm{i}\frac{2}{\pi}\,\mathrm{sign}(\theta^{\mathrm{T}} s)\ln|\theta^{\mathrm{T}} s|\right)\Gamma(\mathrm{d}s) + \mathrm{i}\theta^{\mathrm{T}}\mu\right\},$$

where Γ is a bounded nonnegative measure on the unit sphere S_d, s is the integrand unit vector ($s \in S_d$) and μ is the shift vector. The measure Γ is named a *spectral measure*. Let H be the distribution function of Γ. Then, the characteristic function in polar coordinates is as follows
(a) if $\alpha \neq 1$,

$$\begin{aligned}\Phi_R(\theta) = \exp\biggl\{-|\theta|^{\alpha}\int_0^{\pi}\int_0^{\pi}\cdots\int_0^{2\pi}&|\cos(\theta,\psi)|^{\alpha}\\ &\times\bigl(1 - \mathrm{sgn}(\cos(\theta,\psi))\bigr)\tan\frac{\pi\alpha}{2}\,\mathrm{d}H(\psi) + \mathrm{i}\theta^{\mathrm{T}}\mu\biggr\},\end{aligned}$$

(b) if $\alpha = 1$,

$$\begin{aligned}\Phi_R(\theta) = \exp\biggl\{-|\theta|^{\alpha}\int_0^{\pi}\int_0^{\pi}\cdots\int_0^{2\pi}&|\cos(\theta,\psi)|^{\alpha}\\ &\times\bigl(1 - \mathrm{sgn}(\cos(\theta,\psi))\bigr)\frac{2}{\pi}\ln\bigl(\rho|\cos(\theta,\psi)|\bigr)\,\mathrm{d}H(\psi) + \mathrm{i}\theta^{\mathrm{T}}\mu\biggr\},\end{aligned}$$

where for θ given by its polar coordinates, $\theta(\rho\sin\phi_1\cdots\sin\phi_{d-1}, \rho\sin\phi_1\cdots\sin\phi_{d-2}\times\cos\phi_{d-1}, \rho\sin\phi_1\cdots\sin\phi_{d-3}\cos\phi_{d-2}, \ldots, \rho\cos\phi_1)$, we denote

$$\begin{aligned}\cos(\theta,\psi) = {}&\left(\prod_{i=1}^{d-1}\sin\phi_i\sin\psi_i\right) + \left(\prod_{i=1}^{d-2}\sin\phi_i\sin\psi_i\right)\cos\phi_{d-1}\cos\psi_{d-1}\\ &+\cdots+\cos\phi_1\cos\psi_1.\end{aligned}$$

If $\alpha > 1$, then μ is the mean vector, $\mu = ER$. The scale parameter of a linear combination of the components of a stable vector R satisfies the relation:

$$\sigma^{\alpha}(w^{\mathrm{T}}R) = \sigma^{\alpha}(w_1 R_1 + \cdots + w_d R_d) = \int_{S_d} |w^{\mathrm{T}}s|^{\alpha}\,\Gamma(\mathrm{d}s).$$

Viewing $R = (R_1, \ldots, R_d)$ as the vector of individual returns in a portfolio with weights $w_1, \ldots, w_d$, $\sigma^{\alpha}(w^{\mathrm{T}}R)$ will be the portfolio risk-measure. As we defined above, $\nu_{\alpha} = \sigma^{\alpha}$ is the *variation*, the stable equivalent of variance. Similarly to the traditional interpretation of covariance as an indicator of *dependence*, one can use the *covariation* to estimate the dependence between two $s\alpha s$ distributions:

$$[R_1; R_2]_{\alpha} = \frac{1}{\alpha}\frac{\partial\sigma^{\alpha}(w_1 R_1 + w_2 R_2)}{\partial w_1}\bigg|_{w_1=0;\ w_2=1} = \int_{S_n} s_1 s_2^{\langle\alpha-1\rangle}\,\Gamma(\mathrm{d}s),$$

where (R_1, R_2) is a $s\alpha s$ vector $(1 < \alpha \leqslant)$ and $x^{\langle k\rangle} = |x|^k \operatorname{sgn}(x)$ (signed power). The matrix of covariations $[R_i; R_j]_{\alpha}$, $1 \leqslant i \leqslant d$, $1 \leqslant j \leqslant d$, determines the dependence structure among the individual returns in the portfolio.

3.2. *Estimation of parameters of stable distributions*[12]

We shall examine the methods of estimating the stable parameters and their applicability in VaR computations, where the primary concern is the tail behavior of distributions. It has been proposed that it is more useful to evaluate directly the tail index (the index of stability) instead of fitting the whole distribution. The latter method is claimed to negatively affect the estimation of the tail behavior by its use of "center" observations. We shall describe both approaches: tail estimation and entire-distribution modeling. We suggest a method, which combines the two techniques: it is designed for fitting the overall distribution with greater emphasis on the tails.

3.2.1. *Tail estimation*

Tail estimators for the index of stability α are based on the asymptotic Pareto tail behavior of stable distributions.[13] We shall consider the following estimators of tail thickness: the Hill, the Pickands, and the modified unconditional Pickands.[14]

[12] For additional references on estimation of the four parameters of stable univariate laws, see Chobanov et al. (1996), Gamrowski and Rachev (1994, 1995a, b), Klebanov, Melamed and Rachev (1994), Kozubowski and Rachev (1994), McCulloch (1996), Mittnik and Rachev (1991), Rachev and SenGupta (1993). For the multivariate case estimation of: the spectral measure, the index of stability, the covariation and tests for dependence of stable distributed returns, see Cheng and Rachev (1995), Gamrowski and Rachev (1994, 1995a, b, 1996), Heathcote, Cheng and Rachev (1995), Mittnik and Rachev (b), Rachev and Xin (1993).

[13] See Section 3.1.

[14] For details on the Hill, Pickands, and the modified unconditional Pickands estimators, see Mittnik, Paolella and Rachev (1998c) and references therein.

The Hill estimator[15] is described by

$$\hat{\alpha}_{\text{Hill}} = \frac{1}{\frac{1}{k}\sum_{j=1}^{k}\ln(X_{n+1-j:n}) - \ln X_{n-k:n}},$$

where $X_{j:n}$ denotes the j-th order statistic of sample $X_1, \ldots, X_n$;[16] the integer k points where the tail area "starts". The selection of k is complicated by a tradeoff: it must be adequately small so that $X_{n-k:n}$ is in the tail of the distribution; but if it is too small, the estimator is not accurate. The disadvantage of the estimator is the condition to explicitly determine the order statistic $X_{n-k:n}$. It is proved that, for stable Paretian distributions, the Hill estimator is consistent and asymptotically normal. Mittnik, Paolella and Rachev (1998c) found that, the small sample performance of $\hat{\alpha}_{\text{Hill}}$ does not resemble its asymptotic behavior, even for $n > 10\,000$ (see Figure 1[17]). It is necessary to have enormous data series in order to obtain unbiased estimates of α, for example, with $\alpha = 1.9$, reasonable estimates are produced only for $n > 100\,000$ (see Figure 2[18]). Alternatives to the Hill estimator are the Pickands and the modified unconditional Pickands estimators.

The "original" Pickands estimator[19] takes the form

$$\hat{\alpha}_{\text{Pick}} = \frac{\ln 2}{\ln(X_{n-k+1:n} - X_{n-2k+1:n}) - \ln(X_{n-2k+1:n} - X_{n-4k+1:n})}, \qquad 4k < n.$$

The Pickands estimator requires choice of the optimal k, which depends on the true unknown α. Mittnik and Rachev (1996) proposed a new tail estimator named "the modified unconditional Pickands (MUP) estimator", $\hat{\alpha}_{\text{MUP}}$. An estimate of α is obtained by applying the nonlinear least squares method to the following system:

$$k_2 = X_2, X_1^{-1} k_1 + \varepsilon,$$

where

$$X_1 = \begin{bmatrix} X_{n-k+1:n}^{-\alpha} & X_{n-k+1:n}^{-2\alpha} \\ X_{n-2k+1:n}^{-\alpha} & X_{n-2k+1:n}^{-2\alpha} \end{bmatrix}, \qquad X_2 = \begin{bmatrix} X_{n-3k+1:n}^{-\alpha} & X_{n-3k+1:n}^{-2\alpha} \\ X_{n-4k+1:n}^{-\alpha} & X_{n-4k+1:n}^{-2\alpha} \end{bmatrix},$$

[15] Hill (1975).

[16] Given a sample of observations $X_1, \ldots, X_n$, we rearrange the sample in increasing order $X_{1:n} \leqslant \cdots \leqslant X_{n:n}$, then the j-th order statistic is equal to $X_{j:n}$.

[17] In Figure 1, the true value of α is 1.9, the sample size is $n = 10000$; the x-axis shows values of k from 1 to $n/2 = 5000$. Notice that the estimator for $\hat{\alpha} = \hat{\alpha}(k(n), n)$ is unbiased when $\lim_{n\to\infty}(k(n)/n) \to 0$. So, unbiasedness of the estimator requires very small values of k. However, for a small value of k, the variance of the estimator is large. A close look at the estimator $\hat{\alpha}(k, n)$ suggests value of $\hat{\alpha}$ around 2.2, whereas $\alpha = 1.9$.

[18] In Figure 2, the true α is again 1.9, the sample size is $n = 500{,}000$, $k = 1, \ldots, n/2 = 250{,}000$. One can see that, for very small values of k, $\alpha \approx 1.9$.

[19] Pickands (1975).

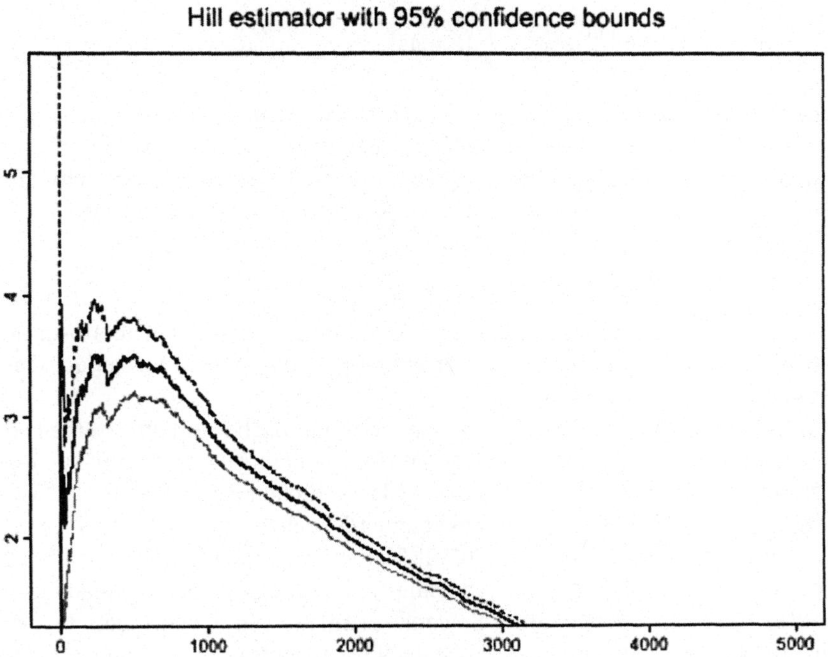

Fig. 1. Hill estimator for 10 000 standard stable observations with index $\alpha = 1.9$.

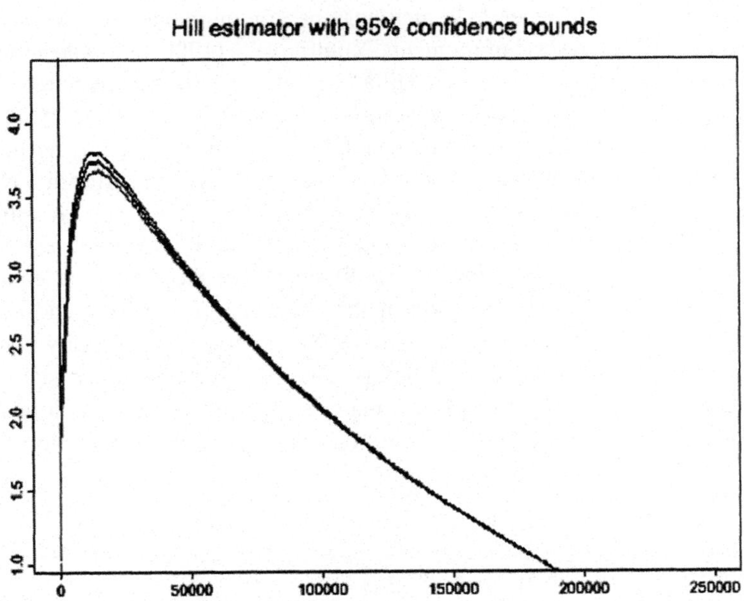

Fig. 2. Hill estimator for 500 000 standard stable observations with index $\alpha = 1.9$.

$$k_1 = \left[\frac{k-1}{2k-1}\right], \quad \text{and} \quad k_2 = \left[\frac{3k-1}{4k-1}\right].$$

Mittnik, Paolella and Rachev (1998c) found that the optimal k for $\hat{\alpha}_{\text{MUP}}$ is far less dependent on α than in the case of either the Hill or Pickands estimators. Studies demonstrated that $\hat{\alpha}_{\text{MUP}}$ is approximately unbiased for $\alpha \in [1.00, 1.95)$ and nearly normally distributed for large sample sizes. The MUP estimator appears to be useful in empirical analysis.

3.2.2. Entire-distribution modeling

We shall describe the following methods of estimating stable parameters with fitting the entire distribution: quantile approaches, characteristic function (CF) techniques, and maximum likelihood (ML) methods.

Fama and Roll (1971) suggested the first quantile approach based on observed properties of stable quantiles. Their method was designed for evaluating parameters of symmetric stable distributions with index of stability $\alpha > 1$. The estimators exhibited a small asymptotic bias. McCulloch (1986) offered a modified quantile technique, which provided consistent and asymptotically normal estimators of all four stable parameters, for $\alpha \in [0.6, 2.0]$ and $\beta \in [-1, 1]$. The estimators are derived using functions of five sample quantiles: the 5%, 25%, 50%, 75%, and 95% quantiles. Since the estimators do not consider observations in the tails (below the 5% quantile and above the 95% quantile), the McCulloch method does not appear to be suitable for estimating parameters in VAR modeling.

Characteristic function techniques are built on fitting the sample CF to the theoretical CF. Press (1972a, b) proposed several CF methods: the minimum distance, the minimum r-th mean distance, and the method of moments. Koutrouvelis (1980, 1981) developed the iterative regression procedure. Kogon and Williams (1998) modified the Koutrouvelis method by eliminating iterations and limiting the estimation to a common frequency interval.[20] CF estimators are consistent and under certain conditions are asymptotically normal.[21]

Maximum likelihood methods for estimating stable parameters differ in a way of computing the stable density. DuMouchel (1971) evaluated the density by grouping data and applying the fast Fourier transform to "center" values and asymptotic expansions – in the tails. Mittnik, Rachev and Paolella (1998) calculated the density at equally spaced grid points via a fast Fourier transform of the characteristic function and at intermediate points – by linear interpolation. Nolan (1998a) computed the density using numerical approximation of integrals in the Zolotarev integral formulas for the stable density.[22] DuMouchel (1973) proved that the ML estimator is consistent and asymptotically normal. In Section 4 we analyze applicability of the ML method in VAR estimations.

[20] For additional references, see Arad (1980), Feuerverger and McDunnough (1981), Mittnik, Rachev and Paolella (1998), Paulson, Holcomb and Leitch (1975).

[21] Heathcote, Cheng and Rachev (1995).

[22] For additional references, see Mittnik et al. (1997).

3.2.3. Tail estimation: Fast Fourier transform method

Tail estimation using the Fourier Transform (FT) method is based on fitting the characteristic function in a neighborhood of the origin $t = 0$. Here we use the classical tail estimate:

$$P\left(X \leqslant -\frac{1}{a}\right) \leqslant P\left(|X| \geqslant \frac{1}{a}\right) \leqslant \frac{K}{a} \int_0^a \left(1 - \mathrm{Re}\{f_X(t)\}\right) \mathrm{d}t, \quad \text{for all } a > 0,$$

where $\mathrm{Re}\{f_X(t)\}$ is the real part of the characteristic function $f_X(t)$ and the constant $K = 1/(1 - \sin 1) < 1/7$. Precise estimation of the characteristic function guarantees accurate tail estimation, which leads to an adequate evaluation of VaR.

Suppose that the distribution of returns r is symmetric-α-stable,[23] that is: the *characteristic function* of r is given by

$$f_r(t) = E\,\mathrm{e}^{\mathrm{i}rt} = \mathrm{e}^{\mathrm{i}t\mu - |ct|^\alpha}.$$

If $\alpha > 1$,[24] then, given observations $r_1, \ldots, r_n$, we estimate μ by the sample mean $\bar{\mu} = \bar{r} = \frac{1}{n}\sum_{i=1}^n r_i$. For large values of n, the characteristic function of observations $R_i = r_i - \bar{r}$ approaches $f_R(t) = \mathrm{e}^{-|ct|^\alpha}$. Consider the *empirical characteristic function* of the centered observations: $\hat{f}_{R,n}(t) = \frac{1}{n}\sum_{k=1}^n \mathrm{e}^{\mathrm{i}R_k t}$. Because the theoretical characteristic function, $f_R(t)$, is real and positive, we have that

$$\hat{f}_{R,n}(t) = \mathrm{Re}\left(\frac{1}{n}\sum_{k=1}^n \mathrm{e}^{\mathrm{i}R_k t}\right) = \frac{1}{n}\sum_{k=1}^n \cos(R_k t).$$

Now the problem of estimating α and c is reduced to determining $\hat{\alpha}$ and $\hat{c}$ such that

$$\int_0^M \left|\hat{f}_{R,n} - f_R(t, \hat{\alpha}, \hat{c})\right| = \int_0^M \left|\frac{1}{n}\sum_{k=1}^n \cos(R_k t) - \mathrm{e}^{-(\hat{c}t)^{\hat{\alpha}}}\right| \mathrm{d}t$$

is minimal, where M is a sufficiently large value.

The realization of the FT method is performed in the following steps:

Step 1. Given the asset returns $r_1, \ldots, r_n$, compute the centered returns $R_i = r_i - \bar{r}$, $i = 1, \ldots, n$, where $\bar{r} = \frac{1}{n}\sum_{i=1}^n r_i$.

Step 2. Construct the sample characteristic function

$$\hat{f}(t_j) = \frac{1}{n}\sum_{k=1}^n \cos(R_k t_j),$$

[23] Empirical evidence suggests that β does not play a significant role for VAR estimation.

[24] As we have already observed, in all financial return data, fitting an α-stable model results in $\alpha > 1$, which implies existence of the first moment.

where $t_j = j\frac{\kappa\pi}{\tau}$, $j = 1, \ldots, \tau$, $\kappa\pi$ is the maximal value of t, τ is the number of grid points on $(0, \kappa\pi]$.[25]

Step 3. Do the search for best $\hat{\alpha}$ and $\hat{c}$ such that

$$\sum_{j=1}^{\tau}\left|\frac{1}{n}\sum_{k=1}^{n}\cos(R_k t_j) - e^{-(\hat{c}t_j)^{\hat{\alpha}}}\right|$$

is minimal.

4. VaR estimates for stable distributed financial returns

In this section we consider a stable VaR model, which assumes that the portfolio return distribution follows a stable law. We derive "stable" VaR estimates and analyze their properties applying in-sample and forecast evaluations. We use "normal" VaR measurements as benchmarks for investigating characteristics of "stable" VaR measurements.

We conduct analysis for various financial data sets:

- the Yen/British Pound (BP) exchange rate,
- the BP/US$ exchange rate,
- the Deutsche Mark (DM)/BP exchange rate,
- the S&P 500 index,
- the DAX30 index,
- the CAC40 index,
- the Nikkei 225 index,
- the Dow Jones Commodities Price Index (DJCPI).

A short description of the data is given in Table 1.

4.1. *In-sample evaluation of VaR estimates*

In this part we evaluate stable and normal VaR models by examining distances between the VaR estimates and the empirical VaR measures.

By a formal definition of VaR in Equation (1), VaR estimates, $\overline{\mathrm{VaR}}_{t,\tau}$, are such that

$$\Pr\left[\Delta P_t(\tau) < -\overline{\mathrm{VaR}}_{t,\tau}\right] \approx 1 - c, \tag{2}$$

where c is the confidence level, $\Delta P_t(\tau)$ is the relative change in the portfolio value over the time horizon τ, i.e., $\Delta P_t(\tau) = R_{t,\tau}$ is the portfolio return at moment t over the time horizon τ and t is the current time.

[25] For computation purposes, we have chosen $\kappa = 20$ and $\tau = 10000$. In the realization of the FT method we selected the following grid steps ht: if $0 \leqslant t \leqslant 1$, $ht = 20\pi/50000$: if $t > 1$, $ht = 20\pi/1000$. In order to emphasize the tail behavior, we refined the mesh near $t = 0$ and named that approach *FT-Tail* (FTT): if $0 \leqslant t \leqslant 0.1$, $ht = 20\pi/100000$; if $0.1 \leqslant t \leqslant 1.0$, $ht = 20\pi/10000$; if $t > 1$, $ht = 20\pi/1000$. The numerical results are reported in Section 4.

Table 1
Financial data series

Series	Source	Number of observations	Time period	Frequency
Yen/BP	Datastream	6285	1.02.74–1.30.98	Daily (D)
BP/US$	D. Hindanov	6157	1.03.74–1.30.98	D
DM/BP	Datastream	6285	1.02.74–1.30.98	D
S&P 500	Datastream	7327	1.01.70–1.30.98	D
DAX30	Datastream	8630	1.04.65–1.30.98	D
CAC40	Datastream	2756	7.10.87–1.30.98	D
Nikkei 225	Datastream	4718	1.02.80–1.30.98	D
DJCPI	Datastream	5761	1.02.76–1.30.98	D

For the purpose of testing VaR models financial regulators advise to choose a time horizon of one day, so we take $\tau = 1$. In the text below, if the time horizon is not stated explicitly, it is assumed to equal one day. At each time t, an estimate $\overline{\mathrm{VaR}}_t$ is obtained using lw recent observations of portfolio returns $R_{t-1}, R_{t-2}, \ldots, R_{t-lw}$:

$$\overline{\mathrm{VaR}}_t = \mathrm{VaR}(R_{t-1}, R_{t-2}, \ldots, R_{t-lw}). \tag{3}$$

The lw parameter is called the *window length*. In this subsection, VaR is estimated employing the entire sample of observations, i.e., $lw = N$, where N is the sample size. Hence, we do not point out the present time t.

We obtain "stable" ("normal") VaR measurements at the confidence level c in two steps:

(i) fitting empirical data by a stable (normal) distribution,

(ii) calculating a VaR as the negative of the $(1-c)$-th quantile of a fitted stable (normal) distribution.

"Stable" fitting is implemented using three methods: maximum likelihood (ML), Fourier Transform (FT), and Fourier Transform-Tail (FTT).[26] Estimated parameters of densities and corresponding confidence intervals are presented in Table 2. In the FT and FTT fitting we assume that distributions of returns are symmetric, i.e., the skewness parameter β is equal to zero. Since the index of stability $\alpha > 1$ for our data series, the location parameter μ is approximated by the sample mean. The ML estimates were computed applying the STABLE program by J.P. Nolan.[27] The confidence intervals (CI) for the FT and FTT parameter estimates were derived using a bootstrap method with 1000 replications.[28] Empirical analysis showed that a set of 1000 replications is:

(i) satisfactory for constructing 95% CI;

(ii) insufficient for obtaining reliable 99% CI.

[26] Evaluation of parameters of stable distributions is provided in Section 3.2.

[27] The STABLE program is described in Nolan (1997).

[28] For references on bootstrapping, see Heathcote, Cheng and Rachev (1995); for discussion on CI based on ML parameter estimates, see Nolan (1998a).

Table 2
Parameters of stable and normal densities[a]

Series	Normal		Stable				
	Mean	Standard deviation	Method	α	β	μ	σ
Yen/BP	−0.012	0.649	ML	1.647	−0.170	−0.023	0.361
			FT	1.61		−0.018	0.34
				[1.57,1.66]		[−0.095,0.015]	[0.33,0.36]
				[1.55,1.68]		[−0.178,0.025]	[0.33,0.37]
			FTT	1.50		−0.018	0.32
				[1.46,1.55]		[−0.131,0.034]	[0.31,0.34]
				[1.44,1.64]		[−0.261,0.070]	[0.31,0.39]
BP/US$	0.006	0.658	ML	1.582	0.038	0.007	0.349
			FT	1.57		0.006	0.33
				[1.53,1.65]		[−0.096,0.045]	[0.32,0.36]
				[1.51,1.75]		[−0.393,0.065]	[0.32,0.47]
			FTT	1.45		0.006	0.31
				[1.41,1.51]		[−0.134,0.070]	[0.30,0.33]
				[1.40,1.62]		[−0.388,0.097]	[0.30,0.47]
DM/BP	−0.012	0.489	ML	1.590	−0.195	−0.018	0.256
			FT	1.60		−0.012	0.24
				[1.54,1.75]		[−0.064,0.013]	[0.23,0.26]
				[1.53,1.75]		[−0.165,0.022]	[0.23,0.27]
			FTT	1.45		−0.012	0.23
				[1.41,1.55]		[−0.114,0.038]	[0.22,0.26]
				[1.40,1.77]		[−0.402,0.061]	[0.22,0.40]
S&P 500	0.032	0.930	ML	1.708	0.004	0.036	0.512
			FT	1.82		0.032	0.54
				[1.78,1.84]		[−0.013,0.057]	[0.53,0.54]
				[1.77,1.84]		[−0.062,0.067]	[0.53,0.55]
			FTT	1.60		0.032	0.48
				[1.56,1.65]		[−0.066,0.078]	[0.47,0.49]
				[1.54,1.66]		[−0.120,0.095]	[0.46,0.50]
DAX30	0.026	1.002	ML	1.823	−0.084	0.027	0.592
			FT	1.84		0.026	0.60
				[1.81,1.88]		[−0.015,0.050]	[0.59,0.60]
				[1.80,1.89]		[−0.050,0.057]	[0.58,0.62]
			FTT	1.73		0.026	0.57
				[1.69,1.77]		[−0.031,0.061]	[0.56,0.58]
				[1.68,1.79]		[−0.124,0.073]	[0.56,0.59]
CAC40	0.028	1.198	ML	1.784	−0.153	0.027	0.698
			FT	1.79		0.028	0.70
				[1.73,1.85]		[−0.050,0.088]	[0.68,0.73]
				[1.71,1.87]		[−0.174,0.103]	[0.67,0.74]

[a] The CIs right below the estimates are the 95% CIs, the next CIs are the 99% CIs.

Table 2
(Continued)

Series	Normal		Stable				
	Mean	Standard deviation	Method	α	β	μ	σ
			FTT	1.76		0.028	0.69
				[1.71,1.84]		[−0.053,0.091]	[0.67,0.72]
				[1.69,1.87]		[−0.394,0.101]	[0.66,0.77]
Nikkei 225	0.020	1.185	ML	1.444	−0.093	−0.002	0.524
			FT	1.58		0.02	0.59
				[1.53,1.64]		[−0.127,0.102]	[0.57,0.62]
				[1.52,1.67]		[−0.421,0.130]	[0.57,0.69]
			FTT	1.30		0.02	0.49
				[1.26,1.47]		[−0.451,0.316]	[0.47,0.69]
				[1.05,1.67]		[−1.448,0.860]	[0.47,1.10]
DJCPI	0.006	0.778	ML	1.569	−0.060	0.003	0.355
			FT	1.58		0.006	0.35
				[1.53,1.66]		[−0.026,0.100]	[0.34,0.37]
				[1.52,1.67]		[−0.140,0.120]	[0.33,0.39]
			FTT	1.49		0.006	0.33
				[1.44,1.55]		[−0.160,0.062]	[0.32,0.36]
				[1.44,1.69]		[−0.396,0.100]	[0.32,0.46]

In our experiments, sets of 1000 replications generated:

(i) 95% CI for α and σ whose bounds coincided up to two decimal points; 95% CI for μ with slightly varying bounds;

(ii) varying 99% CI, with insignificant variation of left limits.

VaR measurements were calculated at confidence levels $c = 99\%$ and $c = 95\%$. The 99% (95%) VaR was determined as the negative of the 1% (5%) quantile. For calculating stable quantiles we used our program, built on the Zolotarev integral representation form of the cumulative distribution function. The 99% and 95% VaR estimates are reported in Tables 3 and 4, respectively. Biases of stable and normal VaR measurements are provided in Table 5.[29]

We accompany our computations with plots of:

- daily price levels,
- daily returns,
- fitted empirical, normal, and stable densities with the ML, FT, and FTT estimated parameters,
- daily empirical, normal, and stable VAR* estimates at the 99% and 95% confidence levels.[30]

[29] Biases are computed by subtracting the empirical VAR from the model VAR estimates.

[30] The VAR* numbers are the negative values of the VAR estimates, VAR* = −VAR.

Combined plots of densities and VaR estimation are displayed in Figures 3–10. In order to illustrate that confidence intervals for the FT parameter estimates are sufficiently narrow, we show stable densities and VaR measures at boundary values of confidence intervals for $\hat{\alpha}_{\text{Yen,FT}}$ and $\hat{\sigma}_{\text{Yen,FT}}$ in Figures 11–14.

As Figures 3–10 demonstrate, the VaR estimates obtained at confidence level $c = 95\%$ seem to belong to the area between the "tail" and the "center". The VaR at level $c = 99\%$ is really in the tail area. Hence, we compare performance of stable and normal models separately for the cases $c = 95\%$ and $c = 99\%$.

In general, the stable modeling (ML, FT, and FTT) provided evaluations of the 99% VaR greater than the empirical 99% VaR (see Figures 3–10, Tables 3 and 4). It underestimated the sample 99% VaR in the applications of two methods: FT – for the CAC40, S&P 500, and DAX30 indices, and ML – for the DAX30 index. Biased downwards stable VaR estimates were closer to the true VaR than the normal estimates (see Table 5). Among the methods of stable approximation, the FT method provided more accurate VaR estimates for 7 data sets (see Table 5). For all analyzed data sets, the normal modeling underestimated the empirical 99% VaR. Stable modeling provided more accurate 99% VaR estimates: mean absolute bias[31] under the stable (FT) method is 42% smaller than under the normal method.

At 95% confidence level, the stable VaR estimates were lower than the empirical VaR for all data sets. The normal VaR measurements exceeded the true VaR, except the Yen/BP exchange rate series (see Table 6). For the exchange rate series (Yen/BP, BP/US$, and DM/BP), the normal method resulted in more exact VaR estimates. For the S&P 500, DAX30, CAC40, and DJCPI indices, stable methods underestimated VaR, though the estimates were closer to the true VaR than the normal estimates. Mean absolute biases under stable and normal modeling are of comparable magnitudes.

In-sample examination of VaR models showed:

- the stable modeling generally results in conservative and accurate 99% VaR estimates, which is preferred by financial institutions and regulators,[32]
- the normal approach leads to overly optimistic forecasts of losses in the 99% VaR estimation,
- from a conservative point of view, the normal modeling is acceptable for the 95% VaR estimation,
- the stable models underestimate the 95% VaR. In fact, the stable 95% VaR measurements are closer to the empirical VaR than the normal 95% VaR measurements.

The next step in evaluating VaR models is analysis of their forecasting characteristics.

[31] Let $b_{m,s}$ be a bias of a VaR estimate: $b_{m,s} = \text{VaR}_{m,s} - \text{VaR}_{\text{Empirical},s}$. The mean absolute bias equals $MAB_m = (\sum_{s=1}^{8} |b_{m,s}|)/8$, where m denotes normal, stable-ML, stable-FT, and stable-FTT methods, and s – a series.

[32] In the 99% VaR estimation for data series from Table 1, mean absolute bias under the stable modeling was 42% smaller than under the normal modelling.

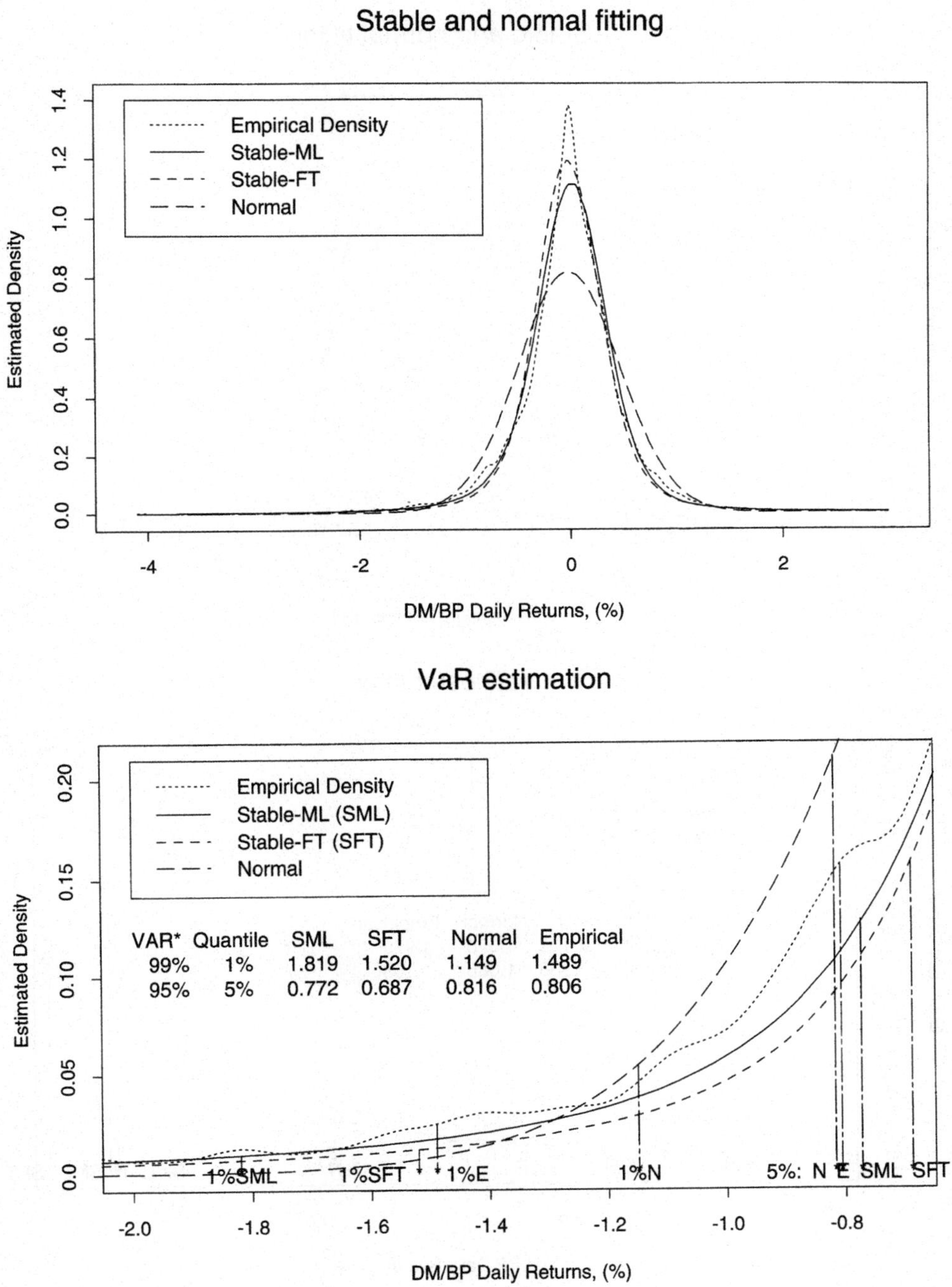

Fig. 3. VAR estimation for the DM/BP exchange rate.

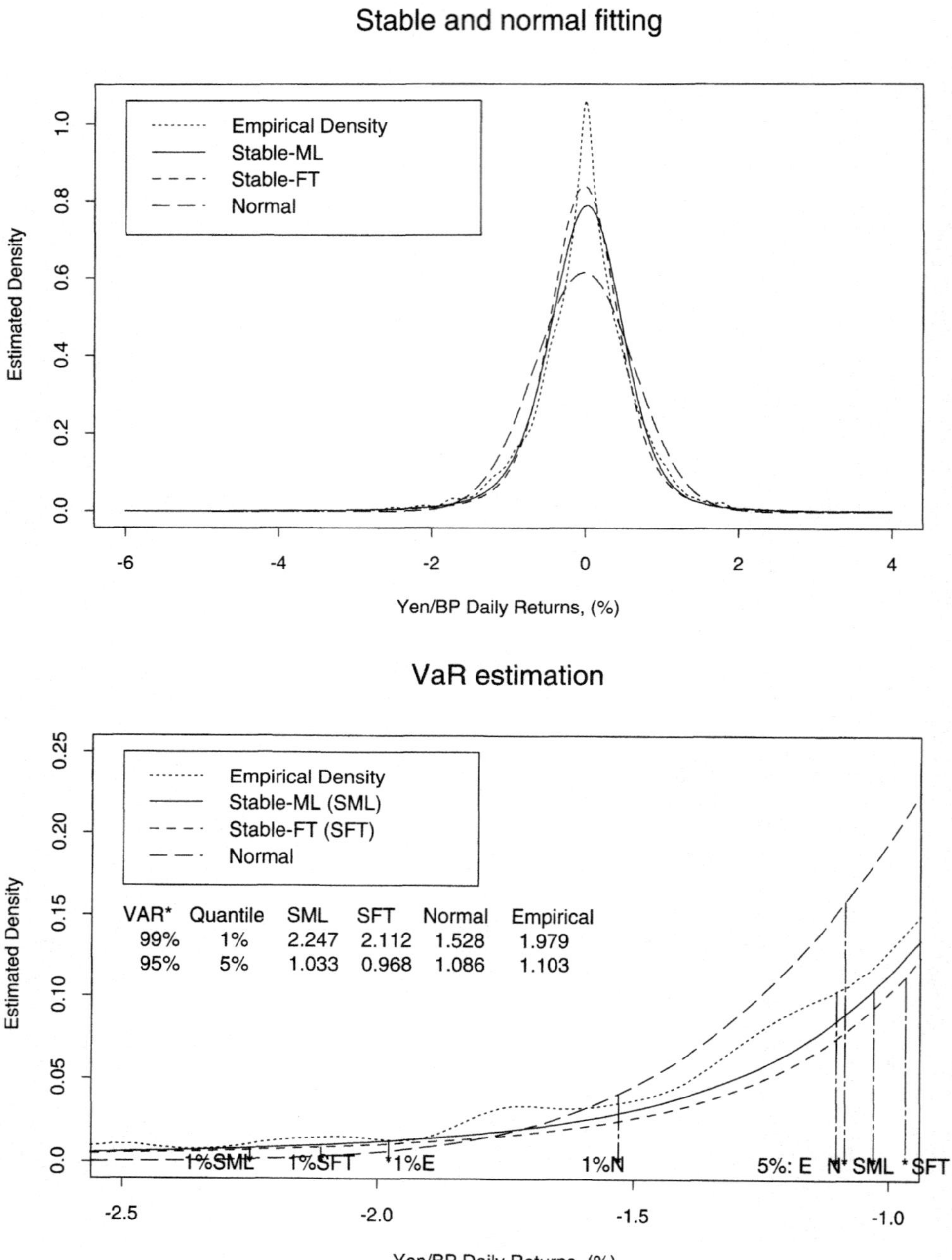

Fig. 4. VAR estimation for the Yen/BP exchange rate.

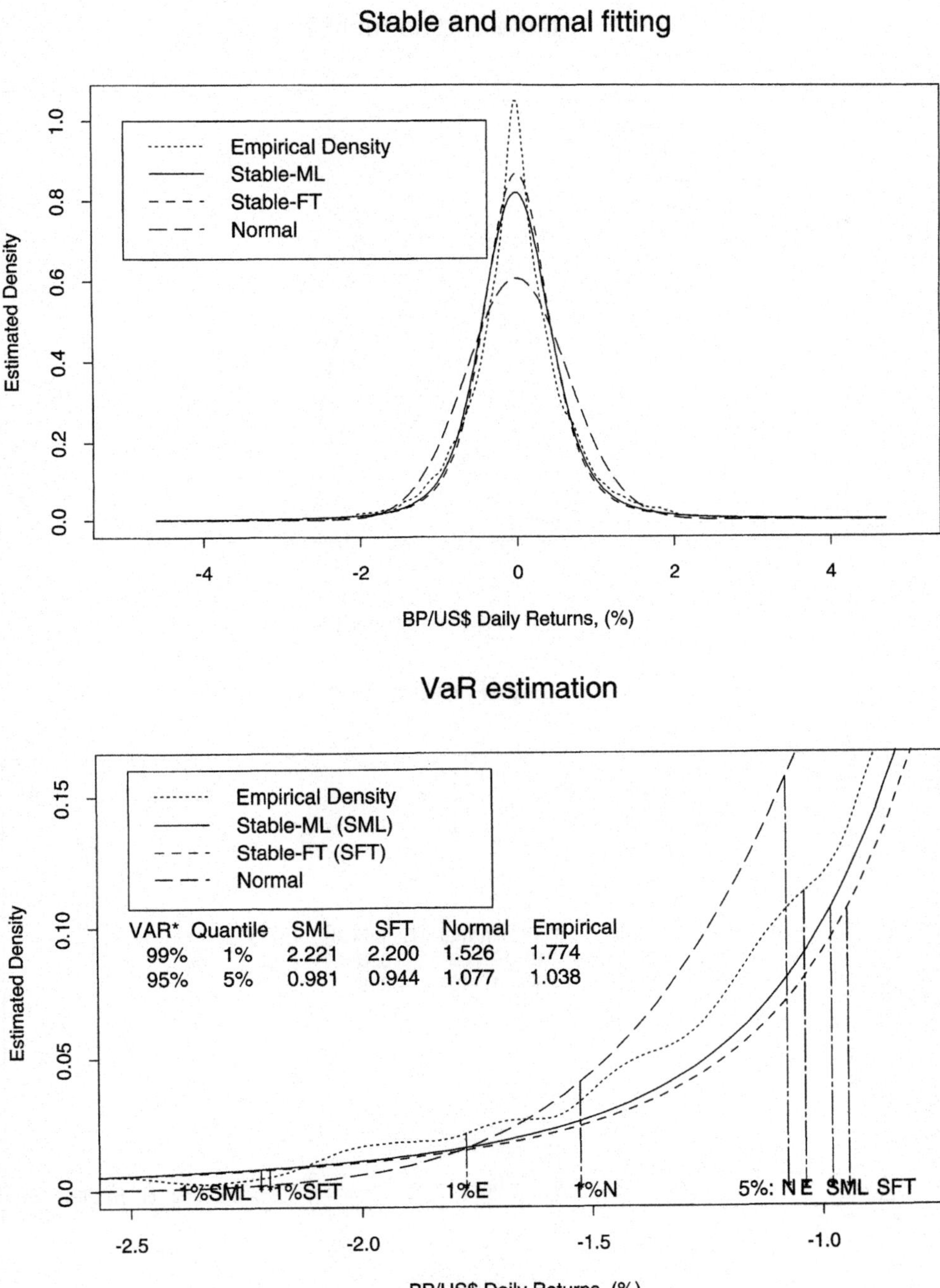

Fig. 5. VAR estimation for the BP/US$ exchange rate.

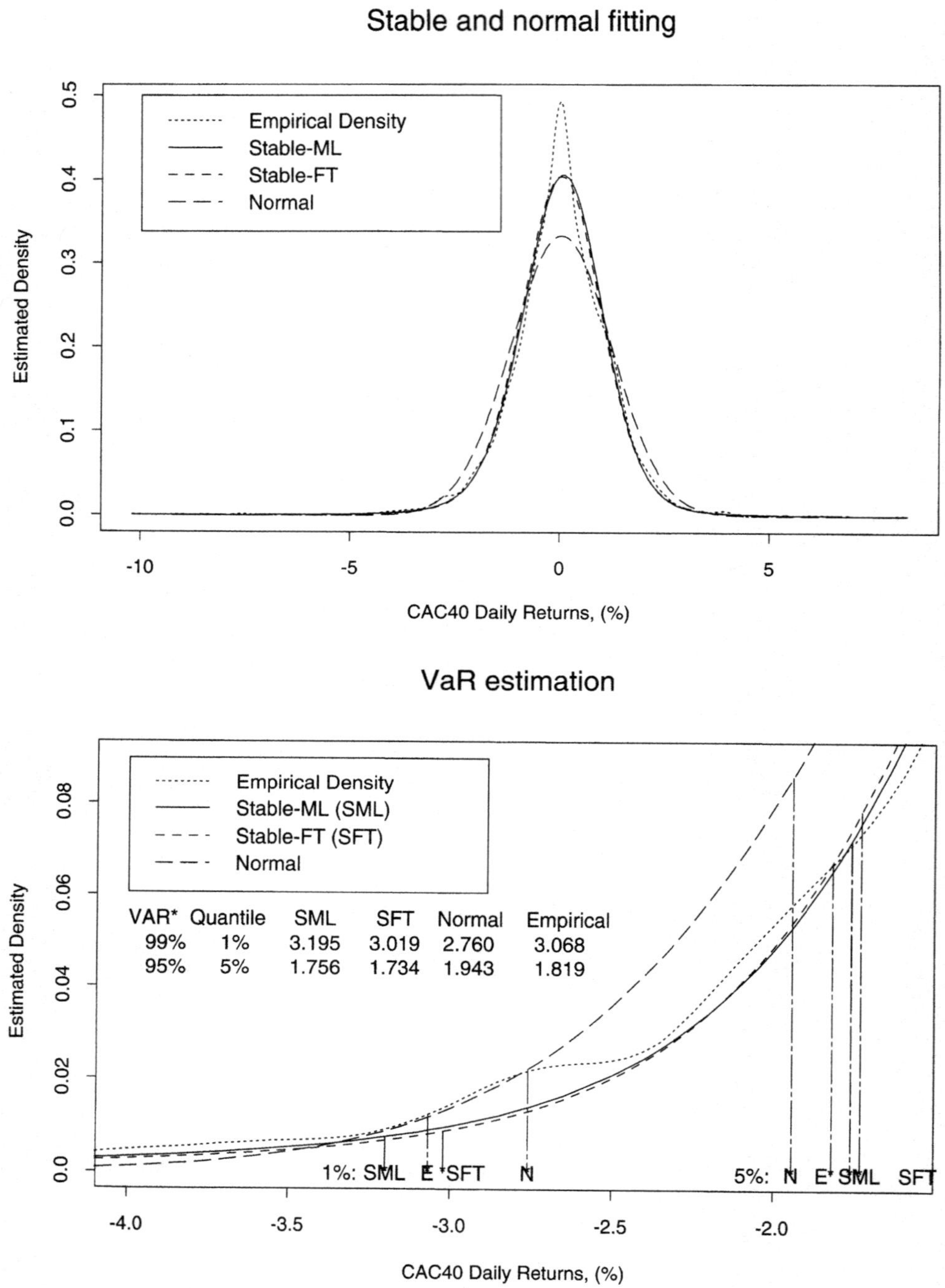

Fig. 6. VAR estimation for the CAC40 index.

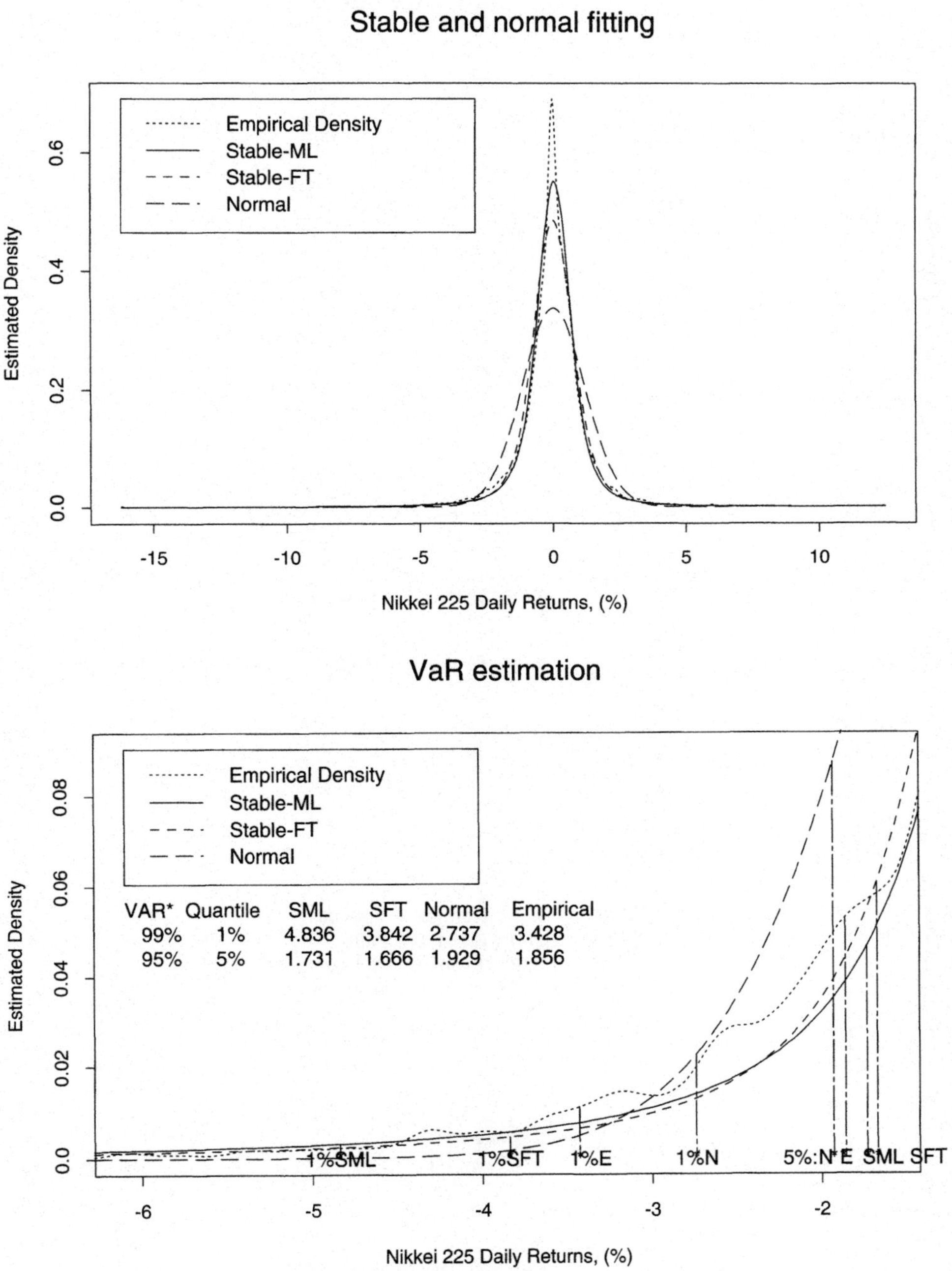

Fig. 7. VAR estimation for the Nikkei 225 index.

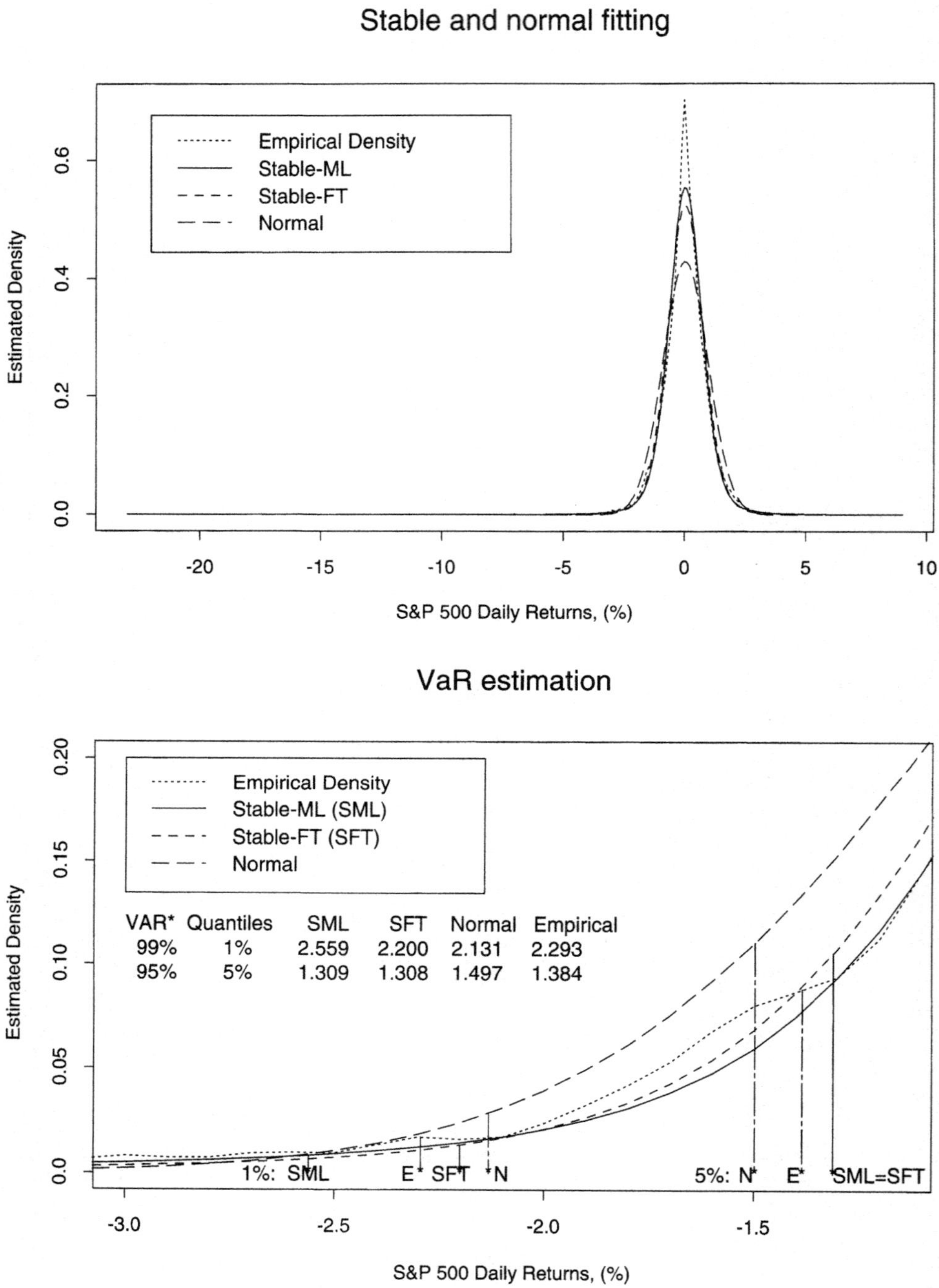

Fig. 8. VAR estimation for the S&P 500 index.

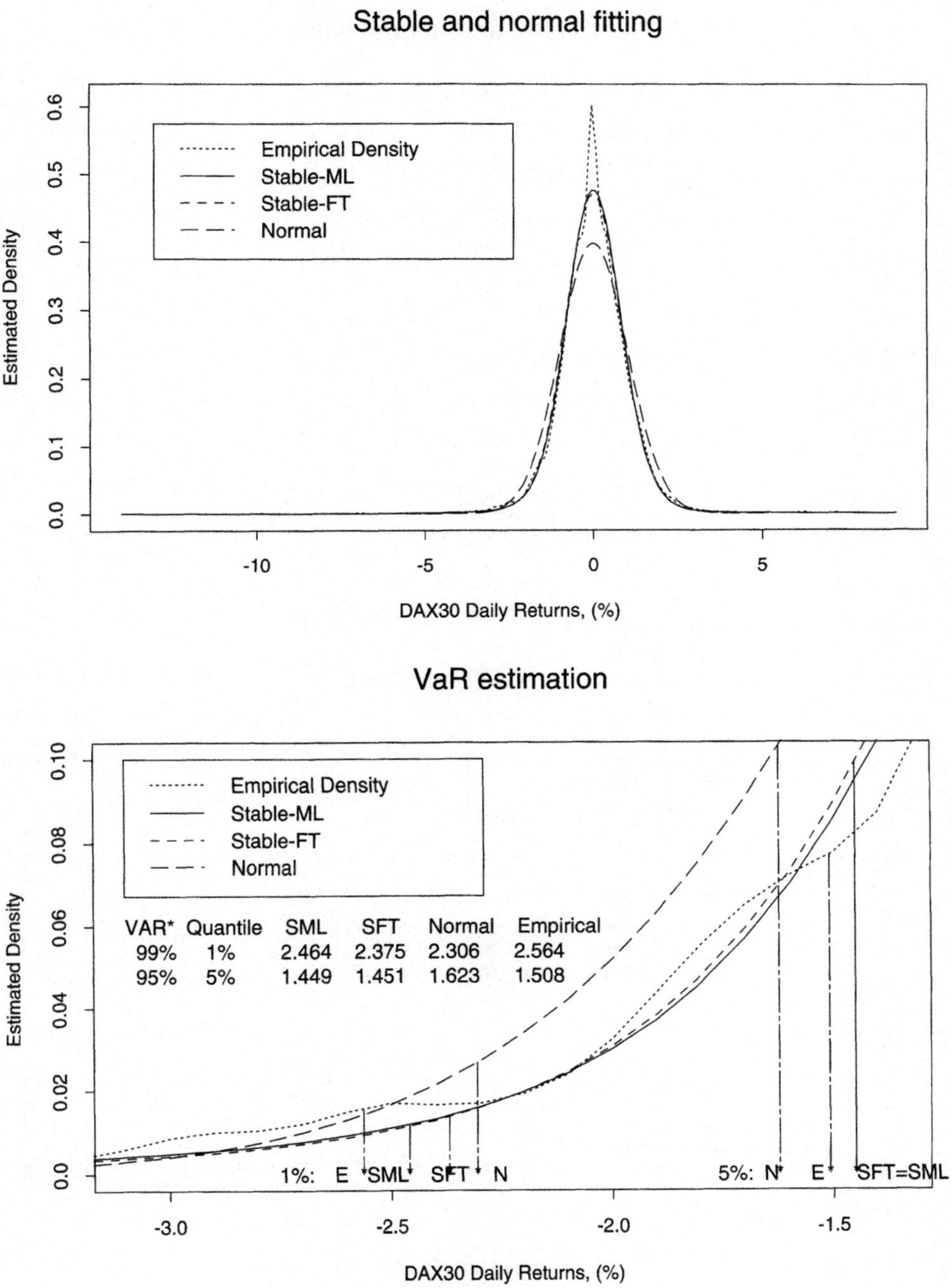

Fig. 9. VAR estimation for the DAX30 index.

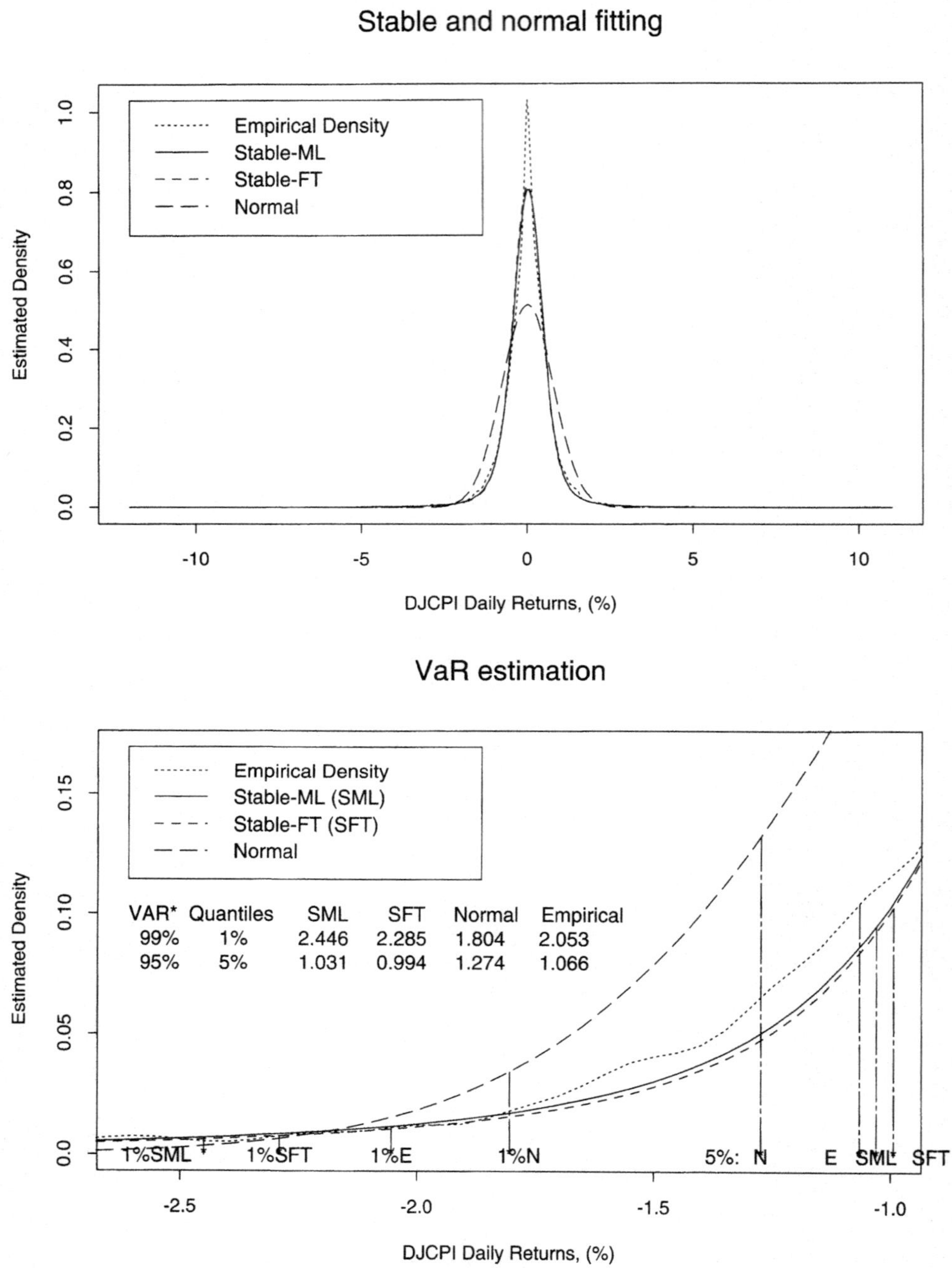

Fig. 10. VAR estimation for the DJCPI index.

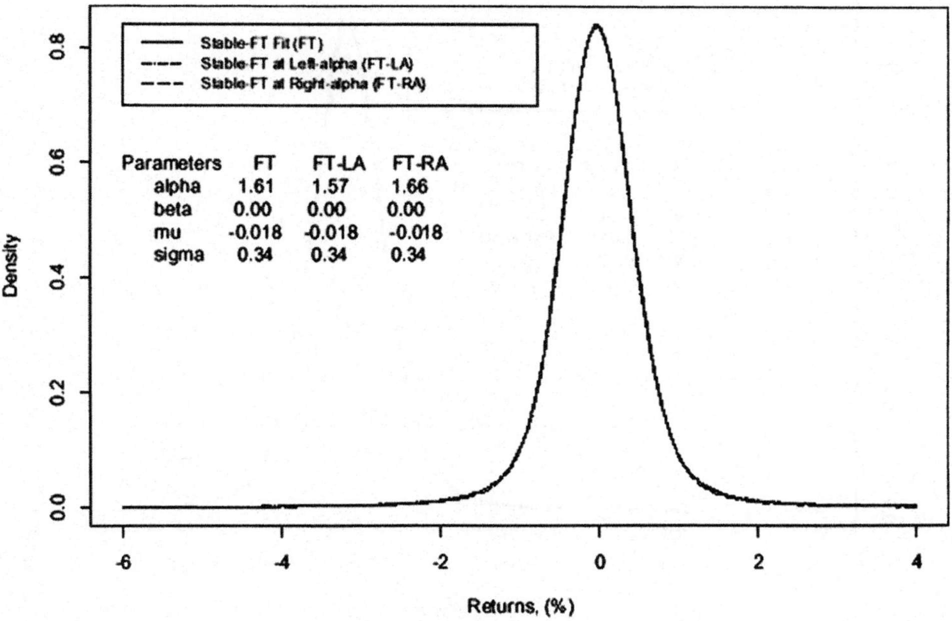

Fig. 11. Stable fitting at limiting values of a confidence interval for alpha.

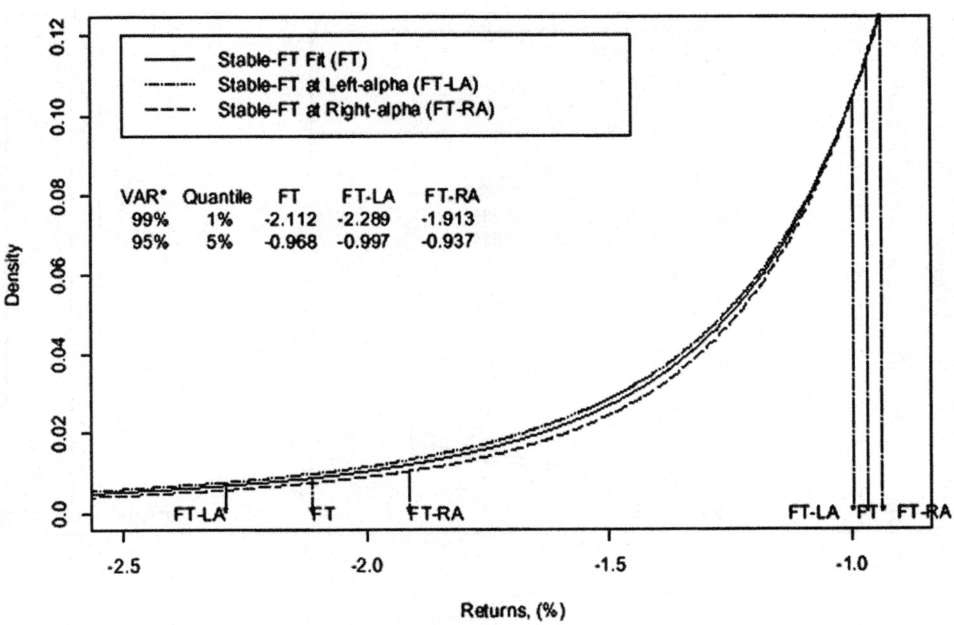

Fig. 12. VAR estimation at limiting values of a confidence interval for alpha.

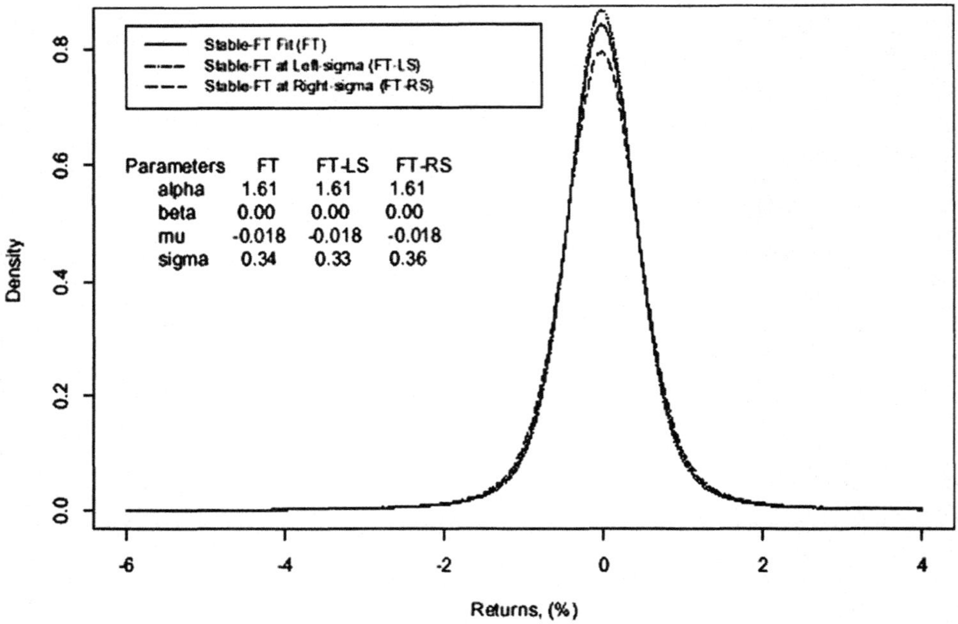

Fig. 13. Stable fitting at limiting values of a confidence interval for sigma.

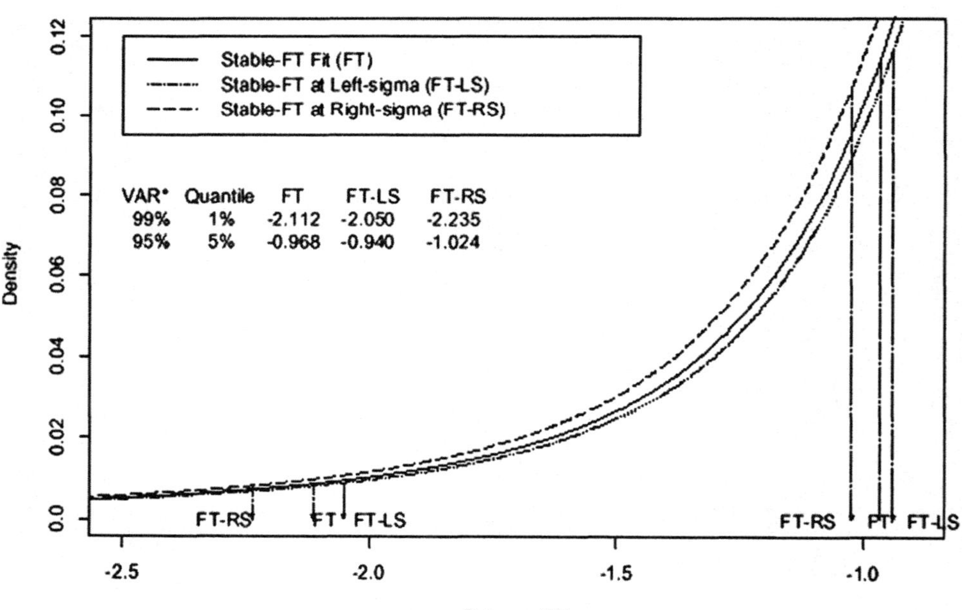

Fig. 14. VAR estimation at limiting values of a confidence interval for sigma.

Table 3
Empirical, normal, and stable 99% VaR estimates[a]

Series	99% VaR				
	Empirical	Normal	Stable		
			ML	FT	FTT
Yen/BP	1.979	1.528	2.247	2.112 [1.968, 2.252] [1.919, 2.415]	2.494 [2.276, 2.736] [2.230, 2.836]
BP/US$	1.774	1.526	2.221	2.200 [2.014, 2.412] [1.956, 2.593]	2.668 [2.436, 2.925] [2.358, 3.029]
DM/BP	1.489	1.149	1.819	1.520 [1.190, 1.712] [1.179, 1.742]	1.996 [1.792, 2.211] [1.700, 2.329]
S&P 500	2.293	2.131	2.559	2.200 [2.117, 2.358] [2.106, 2.470]	2.984 [2.757, 3.243] [2.700, 3.336]
DAX30	2.564	2.306	2.464	2.375 [2.260, 2.502] [2.240, 2.569]	2.746 [2.557, 2.949] [2.523, 2.997]
CAC40	3.068	2.760	3.195	3.019 [2.753, 3.364] [2.682, 3.520]	3.144 [2.788, 3.504] [2.700, 3.841]
Nikkei 225	3.428	2.737	4.836	3.842 [3.477, 4.254] [3.367, 4.453]	6.013 [5.190, 6.701] [4.658, 19.950]
DJCPI	2.053	1.804	2.446	2.285 [1.955, 2.423] [1.916, 2.474]	2.603 [2.382, 2.870] [2.288, 3.035]

[a] The CIs right below the estimates are the 95% CIs, the next CIs are the 99% CIs.

4.2. *Forecast-evaluation of VaR estimates*

In this section we investigate the forecasting properties of stable and normal VaR modeling by comparing predicted VaR with observed returns.

We test the null hypothesis that Equation (1) for a time horizon of 1 day ($\tau = 1$) holds at any time t:

$$\Pr[\Delta P_t < -\mathrm{VaR}_t] = 1 - c, \tag{4}$$

where ΔP_t is the relative change (return) in the portfolio value, i.e., $\Delta P_t = R_t$ is the portfolio return at moment t, VaR_t is the VaR measure at time t, c is the VaR confidence level,

Table 4
Empirical, normal, and stable 95% VaR estimates[a]

Series	95% VaR				
	Empirical	Normal	Stable		
			ML	FT	FTT
Yen/BP	1.103	1.086	1.033	0.968 [0.926,1.047] [0.911,1.186]	0.995 [0.937,1.132] [0.911,1.329]
BP/US$	1.038	1.077	0.981	0.944 [0.898,1.072] [0.876,1.599]	0.986 [0.917,1.158] [0.895,1.588]
DM/BP	0.806	0.816	0.772	0.687 [0.652,0.749] [0.641,0.894]	0.748 [0.695,0.894] [0.678,1.418]
S&P 500	1.384	1.497	1.309	1.308 [1.275,1.361] [1.265,1.411]	1.319 [1.265,1.423] [1.246,1.503]
DAX30	1.508	1.623	1.449	1.451 [1.415,1.500] [1.402,1.533]	1.452 [1.405,1.521] [1.395,1.650]
CAC40	1.819	1.943	1.756	1.734 [1.653,1.837] [1.621,1.944]	1.734 [1.647,1.845] [1.616,2.288]
Nikkei 225	1.856	1.929	1.731	1.666 [1.570,1.839] [1.558,2.280]	1.840 [1.582,2.512] [1.500,5.022]
DJCPI	1.066	1.274	1.031	0.994 [0.888,1.047] [0.870,1.200]	1.011 [0.944,1.188] [0.915,1.615]

[a] The CIs right below the estimates are the 95% CIs, the next CIs are the 99% CIs.

t is the current time, $t \in [1, T]$, and T is the length of the testing interval. The test is performed by checking whether $\Pr[R_t < -\overline{\mathrm{VaR}}_t]$ is reasonably close to $1 - c$, where $\overline{\mathrm{VaR}}_t$ is the estimate of VaR_t. Recall that $\overline{\mathrm{VaR}}_t$ is computed using the last lw observations.[33]

Let b_t be the indicator function $\mathbf{1}\{R_t < -\mathrm{VaR}_t\}$, $1 \leqslant t \leqslant T$. If Equation (4) holds, then

$$b_t = 1\{R_t < -\overline{\mathrm{VaR}}_t\} = \begin{cases} 1, & \text{probability} = 1 - c, \\ 0, & \text{probability} = c. \end{cases}$$

[33] See Equation (3).

Table 5
Biases of normal and stable 99% VaR estimates

Series	99% VaR_m − 99% $\text{VaR}_{\text{Empirical}}$			
	Normal	Stable		
		ML	FT	FTT
Yen/BP	−0.451	0.268	0.133	0.515
BP/US$	−0.248	0.447	0.426	0.894
DM/BP	−0.340	0.330	0.031	0.507
S&P 500	−0.162	0.266	−0.093	0.691
DAX30	−0.258	−0.100	−0.189	0.182
CAC40	−0.308	0.127	−0.049	0.076
Nikkei 225	−0.691	1.408	0.414	2.585
DJCPI	−0.249	0.393	0.232	0.550
Mean absolute bias	0.338	0.416	0.196	0.750

Table 6
Biases of normal and stable 95% VaR estimates

Series	95% VaR_m − 95% $\text{VaR}_{\text{Empirical}}$[a]			
	Normal	Stable		
		ML	FT	FTT
Yen/BP	−0.017	−0.070	−0.135	−0.108
BP/US$	0.039	−0.057	−0.094	−0.052
DM/BP	0.010	−0.034	−0.119	−0.058
S&P 500	0.113	−0.075	−0.076	−0.065
DAX30	0.115	−0.059	−0.057	−0.056
CAC40	0.124	−0.063	−0.085	−0.085
Nikkei 225	0.073	−0.125	−0.190	−0.016
DJCPI	0.208	−0.035	−0.072	−0.055
Mean absolute bias	0.087	0.065	0.104	0.070

[a] m denotes normal, stable-ML, stable-FT, and stable-FTT methods.

Let us denote by E the number of exceedings $(R_t < -\overline{\text{VaR}}_t)$[34] over the testing interval $[1, T]$. If Equation (4) is valid, then the variable $E = \sum_{t=1}^{T} b_t$ has a binomial distribution. We can formulate a testing rule: reject the null hypothesis at level of significance x if

$$\sum_{t=0}^{E} \binom{T}{t} (1-c)^t c^{T-t} \leqslant \frac{x}{2} \quad \text{or} \quad \sum_{t=0}^{E} \binom{T}{t} (1-c)^t c^{T-t} \geqslant 1 - \frac{x}{2}.$$

[34] In nominal levels, an exceeding implies a case when actual losses exceeded the predicted losses.

Table 7
Admissible VaR exceedings and exceeding frequencies

VaR confidence level, c	Length of a testing interval, T	Admissible VaR exceedings, E		Admissible VaR frequencies, E/T (%)	
		Significance level, x		Significance level, x	
		5%	1%	5%	1%
95%	500	[17,33]	[14,36]	[3.40,6.60]	[2.80,7.20]
	1500	[61,89]	[56,94]	[4.07,5.93]	[3.73,6.27]
99%	500	[2,8]	[0,10]	[0.40,1.60]	[0.00,2.00]
	1500	[9,21]	[6,23]	[0.60,1.40]	[0.40,1.53]

For large T and sufficiently high VaR confidence levels, the binomial distribution can be approximated by the normal distribution. Hence, the testing rule for large T is: reject the null hypothesis at level of significance x if

$$E < T(1-c) - z_{1-x/2}\sqrt{T(1-c)c} \quad \text{or} \quad E > T(1-c) + z_{1-x/2}\sqrt{T(1-c)c},$$

where z_p is the $p\%$ standard normal quantile. The bounds of admissible VaR exceedings E and exceedings frequencies, E/T, for testing at level of significance 5% and 1% are provided in Table 7.

We examined forecasting properties of stable and VaR models for data series described in Table 1. In testing procedures we considered the following parameters:

- window lengths $lw = 260$ observations (data over 1year) and $lw = 1560$ observations (data over 6 years),
- lengths of testing intervals $T = 500$ days and $T = 1500$ days.

Evaluation results are reported in Tables 8 and 9. We indicate by the bold font the numbers, which are outside of acceptable ranges.

From Table 8 we can see that normal models for the 99% VaR computations commonly produce numbers of exceedings above the acceptable range, which implies that normal modeling significantly underestimates VaR (losses). At window length of 260 observations, stable modeling is not satisfactory. It provided permissible number of exceptions only for the BP/US$ and DJCPI series. At sample size of 1560 and testing interval of 500 observations, exceedings by the stable-FT method are outside of the admissible interval for the S&P 500, DAX30, and CAC40 indices. Testing on the longer interval with $T = 1500$ showed that numbers of "stable" exceptions are within permissible range. Table 8 demonstrates that increasing the window length from 260 observations to 1560 observations reduces the number of stable-FT exceedings. In contrast, extending the window length for normal models does not decrease E, in some cases, even elevates it. Results illustrate that stable modeling outperforms normal modeling in the 99% VaR estimations.

The 95% VaR normal estimates (except the DAX30 series), obtained using 260 observations, are within the permissible range. Increasing the window length generally worsens the

Table 8
99% VaR exceedings

Series	Length of a testing interval, T	99% VaR exceedings							
		Window length = 260 obs.				Window length = 1560 obs.			
		Normal		FT		Normal		FT	
		E	E/T (%)	E	E/T (%)	E	E/T (%)	E	E/T (%)
Yen/BP	500	**15**	**3.00**	**13**	**2.60**	10	2.00	2	0.40
	1500	**40**	**1.67**	**34**	**2.27**	**45**	**3.00**	21	1.40
BP/US$	500	10	2.00	5	1.00	1	0.20	0	0.00
	1500	**26**	**1.73**	13	0.86	17	1.33	**5**	**0.33**
DM/BP	500	**18**	**3.60**	**14**	**2.80**	**17**	**3.40**	8	1.60
	1500	**45**	**3.00**	**33**	**2.20**	**50**	**3.33**	19	1.27
S&P 500	500	**17**	**3.40**	**13**	**2.60**	**25**	**5.00**	**13**	**2.60**
	1500	**35**	**2.33**	**27**	**1.80**	**28**	**1.87**	14	0.93
DAX30	500	**21**	**4.20**	**14**	**2.80**	**19**	**3.80**	**18**	**3.60**
	1500	**41**	**2.73**	**29**	**1.93**	**25**	**1.67**	20	1.33
CAC40	500	**16**	**3.20**	**14**	**2.80**	**14**	**2.80**	**13**	**2.60**
	1500	**34**	**2.27**	**29**	**1.93**	17	1.63	19	1.27
Nikkei 225	500	**15**	**3.00**	**14**	**2.80**	**13**	**2.60**	7	1.40
	1500	**31**	**2.07**	23	1.53	**26**	**1.73**	10	0.67
DJCPI	500	**12**	**2.40**	7	1.40	**15**	**3.00**	10	2.00
	1500	**29**	**1.93**	15	1.00	**28**	**1.87**	17	1.13

normal VaR measurements. The stable-FT method provided sufficient 95% VaR estimates for the Yen/BP and BP/US$ exchange rates and the CAC40 and Nikkei 225 indices.

A study of the predictive power of VaR models suggests that:

- the normal modeling significantly underestimates 99% VaR,
- the stable method results in reasonable 99% VaR estimates,
- 95% normal measurements are in the admissible range for the window length of 260 observations. Increasing *lw* to 1560 observations might deteriorate the precision of the estimates.

5. Stable modeling and risk assessment for individual credit returns

Recall that the stable distributions are characterized by four parameters: α-tail index, β-skewness, μ-location, and σ-scale. Modeling with such parameters allows for heavy tails and skewness of the distributions. Our empirical analysis confirms that, indeed: (i) credit returns exhibit asymmetry and heavy-tails; (ii) stable modeling captures these features of the returns.

Table 9
95% VaR exceedings

Series	Length of a testing interval, T	95% VaR exceedings							
		Window length = 260 obs.				Window length = 1560 obs.			
		Normal		FT		Normal		FT	
		E	E/T (%)	E	E/T (%)	E	E/T (%)	E	E/T (%)
Yen/BP	500	35	7.00	**38**	**7.60**	27	5.40	31	6.2
	1500	94	6.27	**104**	**6.93**	**109**	**7.27**	**122**	**8.13**
BP/US$	500	33	6.60	**45**	**9.00**	**10**	**2.00**	17	3.40
	1500	73	4.87	**96**	**6.40**	**46**	**3.07**	57	3.80
DM/BP	500	32	6.40	**38**	**7.60**	29	5.80	**37**	**7.40**
	1500	89	5.93	**114**	**7.60**	**105**	**7.00**	**139**	**9.27**
S&P 500	500	34	6.80	**39**	**7.80**	**43**	**8.60**	**47**	**9.40**
	1500	79	5.27	**98**	**6.53**	62	4.13	69	4.60
DAX30	500	**47**	**9.40**	**50**	**10**	**42**	**8.40**	**45**	**9.00**
	1500	**98**	**6.53**	**109**	**7.27**	62	4.13	79	5.27
CAC40	500	32	6.40	34	6.80	31	6.20	32	6.40
	1500	81	5.40	87	5.80	**51**	**4.90**	82	5.47
Nikkei 225	500	**37**	**7.40**	**40**	**8.00**	28	5.60	33	6.60
	1500	85	5.67	90	6.00	68	4.53	87	5.80
DJCPI	500	29	5.80	35	7.00	**37**	**7.40**	**46**	**9.20**
	1500	70	4.67	93	6.20	77	5.13	**108**	**7.20**

The "assets" used in the study are the Merrill Lynch indices of the US government and corporate bonds with maturities from one to 10 years and credit ratings from "BB" to "AAA". Returns on indices are modeled as stable-distributed: $R_i \sim S_{\alpha_i}(\sigma_i, \beta_i, \mu_i)$, where $i = 1, \ldots, 21$. Some analysis of the indices is provided in Table 10. Daily returns series are illustrated on Figure 15 and in Appendix A. The benchmark for assessment of the stable model properties is the "normal" model, i.e., approximation of returns by normal distributions. By categorization of stable distributions, a normal distribution has a tail index $\alpha = 2$ and a symmetric distribution has a skewness parameter $\beta = 0$. Values of $\alpha < 2$ indicate thicker tails than the tails of the normal distribution. In general, as α is smaller, the tails are heavier and the peak of the density is higher. If $\beta < 0$, the distribution is skewed to the left. If $\beta > 0$, the distribution is skewed to the right. Larger absolute magnitudes of β point to stronger skewness. The stable and normal parameter estimates for the bond indices are presented in Table 10. For all 17 considered indices, the tail index α is less than two, which reveals heavy-tailedness, and the skewness parameter β is below zero, which implies skewness to the left. The fitted empirical, stable, and normal densities of indices are displayed in Figure 16 and in Appendix A.

Table 10
Normal and stable parameter estimates of bond indices

Index[a]	Rating or issuer	Maturity (year)	Normal		Stable			
			Mean	St. dev.	Tail index α	Skewness β	Location μ	Scale σ
G102	US gov-t	1–3	0.026	0.096	1.696	−0.160	0.029	0.055
G202	US gov-t	3–5	0.030	0.204	1.739	−0.134	0.036	0.122
G302	US gov-t	5–7	0.032	0.275	1.781	−0.134	0.032	0.169
G402	US gov-t	7–10	0.033	0.352	1.808	−0.172	0.033	0.218
C1A1	AAA	1–3	0.027	0.096	1.654	−0.080	0.053	0.027
C2A1	AAA	3–5	0.029	0.175	1.695	−0.112	0.029	0.099
C3A1	AAA	5–7	0.032	0.249	1.710	−0.116	0.031	0.145
C4A1	AAA	7–10	0.032	0.319	1.739	−0.155	0.031	0.190
C1A2	AA	1–3	0.028	0.099	1.686	−0.105	0.027	0.056
C2A2	AA	3–5	0.029	0.177	1.722	−0.111	0.029	0.104
C3A2	AA	5–7	0.032	0.250	1.757	−0.121	0.032	0.150
C4A2	AA	7–10	0.033	0.325	1.778	−0.148	0.033	0.198
C1A3	A	1–3	0.028	0.098	1.688	−0.135	0.027	0.056
C2A3	A	3–5	0.030	0.180	1.702	−0.122	0.029	0.104
C3A3	A	5–7	0.032	0.255	1.743	−0.133	0.033	0.151
C4A3	A	7–10	0.033	0.333	1.753	−0.167	0.033	0.199
C1A4	BBB	1–3	0.029	0.112	1.653	−0.113	0.029	0.054
C2A4	BBB	3–5	0.032	0.183	1.662	−0.042	0.033	0.096
C3A4	BBB	5–7	0.034	0.249	1.690	−0.125	0.035	0.140
C4A4	BBB	7–10	0.035	0.316	1.694	−0.136	0.035	0.180
H0A1	BB	1–3	0.027	0.185	1.686	−0.252	0.042	0.104

[a] Each index set, except H0A1, includes 2418 daily observations from 3.13.90 to 7.29.99. Source of index series: Merrill Lynch, used with permission.

In order to assess riskiness of the individual credit series and properties of stable modeling in the credit risk evaluation, the 99% and 95% Value at Risk (VaR) measurements were computed. The stable and normal VaR estimates are reported in Table 11. Normal VaR measurements are given for comparison purposes. The differences between empirical and modeled VaR are given in Appendix B, Table B.1. The VaR evaluation for the bond indices is illustrated on Figures 17 and in Appendix A. Results of VaR estimations lead to the following conclusions:[35]

Since credit returns have skewed and heavy-tailed distributions, VaR measurements provide more adequate indication of risk than symmetric measurements (standard deviation or, in case of stable distributions, scale parameter) do.

[35] This section computes "in-sample" VaR. Hence, the conclusions discuss in-sample VaR properties.

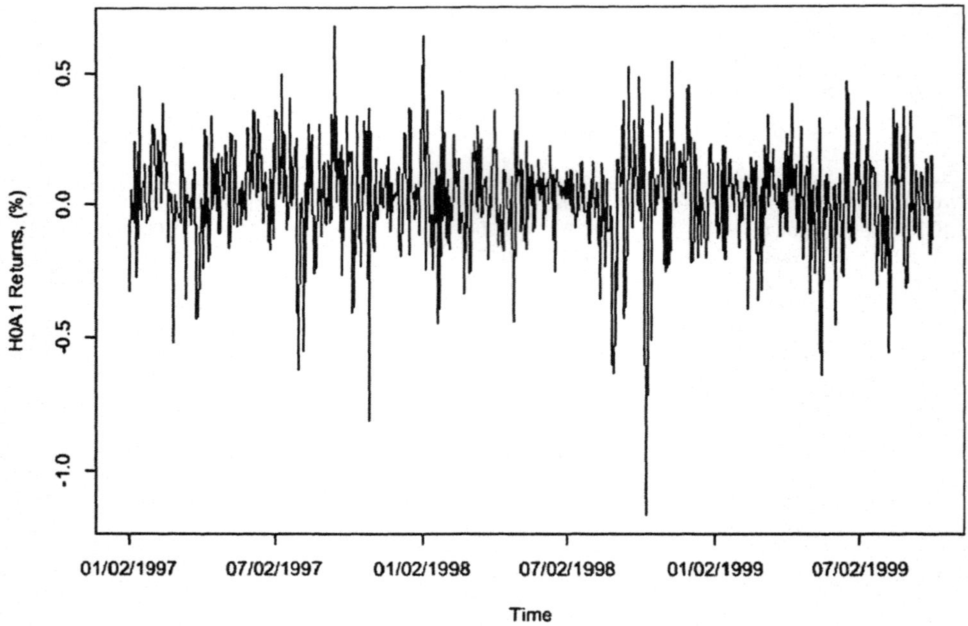

Fig. 15. H0A1 daily returns.

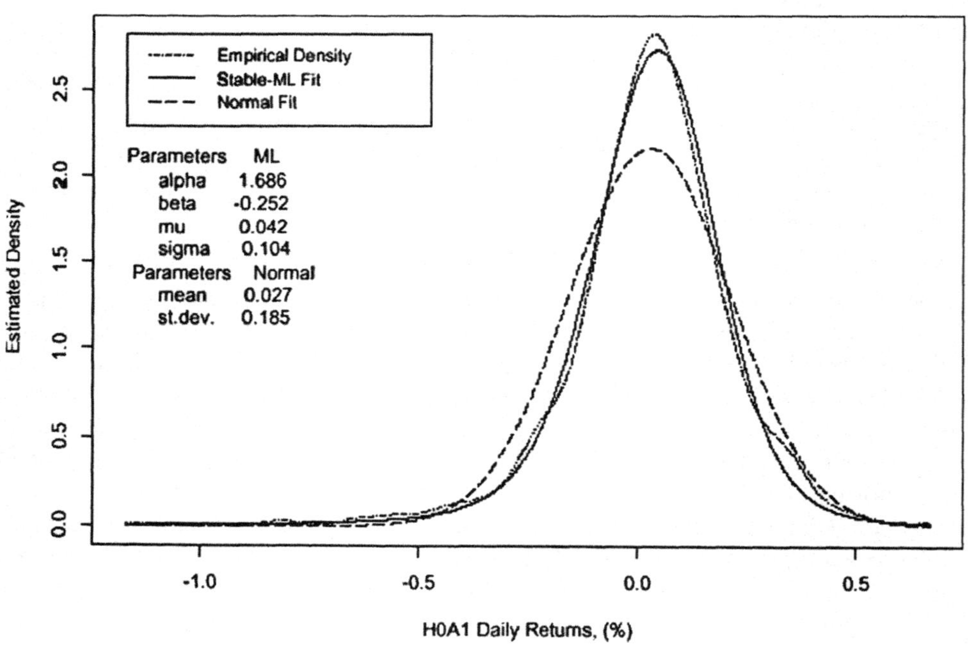

Fig. 16. Stable and normal fitting of the HOA1 index.

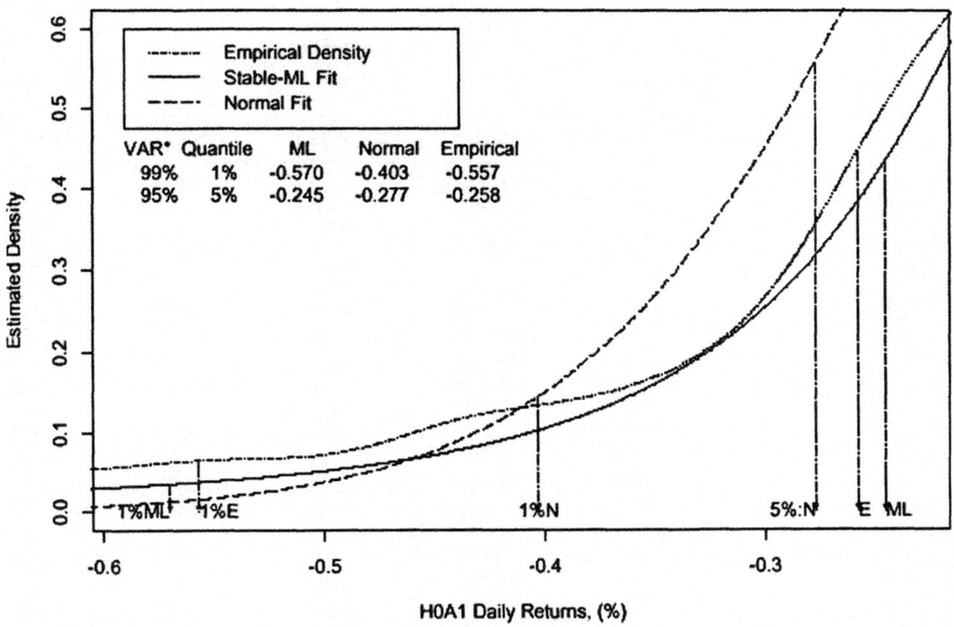

Fig. 17. VAR estimation for the HOA1 index.

- the stable modeling produces conservative and accurate 99% VaR estimates, which is preferred by financial institutions and regulators. "Conservative" VaR estimates exceed empirical VaR, which implies that forecasts of losses were greater than observed losses,
- the stable modeling underestimates the 95% VaR,
- the normal modeling gives overly optimistic forecasts of losses in the 99% VaR estimation,
- the normal modeling is acceptable for the 95% VaR estimation.

The stable modeling for high values of the VaR confidence level is superior because it adequately describes heavy tails and skewness in the data. Our empirical analysis demonstrates advantages of stable modeling in evaluation of riskiness of single credit returns series. The next step is to examine properties of stable modeling in evaluation of portfolio risk.

6. Portfolio credit risk for independent credit returns

Suppose that a portfolio includes n credit assets. Then, the portfolio return is given by $R_P = \sum_{i=1}^{n} w_i R_i$, where R_i is the return on the i-th asset, w_i is the weight of the i-th asset, $i = 1, \ldots, n$, $\sum_{i=1}^{n} w_i = 1$. The modeling in this section assumes that distributions

Table 11
Empirical, normal, and stable VaR estimates for bond indices

Index	99% VaR estimates			95% VaR estimates		
	Empirical	Normal	Stable	Empirical	Normal	Stable
G102	0.242	0.198	0.275	0.127	0.132	0.119
G202	0.518	0.446	0.576	0.303	0.306	0.283
G302	0.739	0.609	0.747	0.412	0.421	0.399
G402	0.928	0.785	0.932	0.545	0.545	0.518
C1A1	0.238	0.196	0.284	0.129	0.130	0.119
C2A1	0.437	0.377	0.509	0.244	0.258	0.236
C3A1	0.687	0.548	0.734	0.369	0.378	0.353
C4A1	0.883	0.712	0.931	0.480	0.494	0.467
C1A2	0.237	0.201	0.285	0.132	0.134	0.125
C2A2	0.443	0.382	0.505	0.254	0.261	0.244
C3A2	0.663	0.550	0.689	0.373	0.380	0.355
C4A2	0.870	0.722	0.890	0.482	0.501	0.474
C1A3	0.237	0.207	0.286	0.135	0.134	0.125
C2A3	0.469	0.390	0.530	0.260	0.267	0.248
C3A3	0.705	0.560	0.719	0.376	0.386	0.361
C4A3	0.893	0.741	0.949	0.487	0.514	0.485
C1A4	0.262	0.231	0.290	0.124	0.155	0.119
C2A4	0.478	0.392	0.511	0.243	0.268	0.228
C3A4	0.711	0.545	0.741	0.361	0.375	0.343
C4A4	0.862	0.702	0.960	0.467	0.486	0.451
H0A1	0.557	0.403	0.570	0.258	0.277	0.245

of R_i are: (i) independent α-stable and (ii) characterized by the same index of stability, $R_i \sim S_\alpha(\sigma_{R_i}, \beta_{R_i}, 0)$,[36] $i = 1, \ldots, n$. The additivity property of independent stable random variables provides analytic formulas for parameters of portfolio returns R_P. The formulas lead to estimates of portfolio parameters and risk without simulations. In practice, the "independent" risk measurements are lower bounds of portfolio risk.

By the additivity property of stable distributions, a linear combination of independent stable random variables is again a stable random variable. Therefore, $R_P = \sum_{i=1}^{n} w_i R_i$ follows a stable law:

$$R_P \sim S_\alpha(\sigma_{R_P}, \beta_{R_P}, 0),$$

[36] We assume that $a > 1$ (this assumption is always supported by the empirical studies) and the mean $\mu_{R_i} = 0$. If $\mu_{R_i} \neq 0$, we "center" the R_i observations: $R_i^* = R_i - \mu_{R_i}$.

where α is the tail index, σ_{R_P} is the scale parameter β_{R_P} is the skewness parameter,

$$\sigma_{R_P} = \left[\sum_{i=1}^{n} (|w_i|\sigma_{R_i})^{\alpha} \right]^{1/\alpha},$$

$$\beta_{R_P} = \frac{\sum_{i=1}^{n} [\text{sign}(w_i)\beta_{R_i}(|w_i|\sigma_{R_i})^{\alpha}]}{\sum_{i=1}^{n} (|w_i|\sigma_{R_i})^{\alpha}}.$$

Thus, the distribution of the portfolio returns is characterized by three parameters: tail index (index of stability) α, skewness β_{R_P}, and scale σ_{R_P}. The parameter α is exogenous. Estimation of β_{R_P} and σ_{R_P} can be implemented in three steps:

Step 1: Find estimates of σ_{R_i} and β_{R_i} by stable fitting sets of R_{it}, $t = 1, \ldots, T$, $i = 1, \ldots, n$.

Step 2: Evaluate portfolio parameters σ_{R_P} and β_{R_P}:

$$\hat{\sigma}_{R_P} = \left[\sum_{i=1}^{n} (|w_i|\hat{\sigma}_{R_i})^{\alpha} \right]^{1/\alpha}, \tag{5}$$

$$\hat{\beta}_{R_P} = \frac{\sum_{i=1}^{n} [\text{sign}(w_i)\hat{\beta}_{R_i}(|w_i|\hat{\sigma}_{R_i})^{\alpha}]}{\sum_{i=1}^{n} (|w_i|\hat{\sigma}_{R_i})^{\alpha}}. \tag{6}$$

Having estimates of parameters of the portfolio credit risk, $\hat{\sigma}_{R_P}$ and $\hat{\beta}_{R_P}$, the *portfolio VaR* is the negative of the appropriate quantile of the $\widehat{R}_P$-distribution.

As an illustration of the method, portfolio risk is estimated for equally weighted returns on indices of the investment grade corporate bonds: C1A1, C2A1, C3A1, C4A1, C1A2, C2A2, C3A2, C4A2, C1A3,C2A3, C3A3, C4A3, C1A4, C2A4, C3A4, and C4A4.[37] Description of indices is given in Table 10 of Section 5. By assumption, the indices are (i) characterized by the same tail index α and (ii) independent. Fix α at 1.708, the average of the α values for the bond return series (see Table 10), and recalculate other stable parameters: β_{R_i}, μ_{R_i}, and σ_{R_i}. New estimates are reported in Table 12. The condition requiring the same tail index α for all analyzed series does not appear to be very restrictive: new parameter estimates (given in Table 12) do not differ much from the previous parameter estimates (reported in Table 10); the new stable VaR estimates (see Table B.2 in Appendix B) are close to the initial stable VaR measurements (Table 11).

The μ estimates are all small. Further on, we shall assume $\mu = 0$. Portfolio parameters following formulas (1), (2) are $\hat{\sigma}_{U_P} = 0.659$, $\hat{\beta}_{U_P} = -0.125$. Thus, $\widehat{R}_P \sim$

[37] A digit after letter "C" denotes the maturity band: 1 – from 1 to 3 years, 2 – from 3 to 5 years, 3 – from 5 to 7 years, 4 – from 7 to 10 years; a digit after letter "A" denotes credit rating: 1 – "AAA", 2 – "AA", 3 – "A", 4 – "BBB".

Table 12
Stable fitting of the bond indices with fixed α

Bond indices	Maturity (years)	Stable parameters at $\alpha = 1.708$		
		β	μ	σ
C1A1	1–3	−0.084	0.027	0.054
C2A1	3–5	−0.111	0.029	0.099
C3A1	5–7	−0.116	0.031	0.144
C4A1	7–10	−0.146	0.031	0.188
C1A2	1–3	−0.107	0.027	0.057
C2A2	3–5	−0.105	0.029	0.103
C3A2	5–7	−0.098	0.033	0.148
C4A2	7–10	−0.128	0.032	0.194
C1A3	1–3	−0.144	0.027	0.057
C2A3	3–5	−0.120	0.030	0.104
C3A3	5–7	−0.125	0.032	0.149
C4A3	7–10	−0.151	0.032	0.196
C1A4	1–3	−0.118	0.029	0.054
C2A4	3–5	−0.045	0.033	0.098
C3A4	5–7	−0.128	0.035	0.141
C4A4	7–10	−0.143	0.035	0.180

$S_{1.708}(0.659, -0.125, 0)$. The *portfolio c% VaR* is calculated as the negative of the $(1-c)$-th quantile of the $\widehat{R}_P$-distribution. For the analyzed portfolio, the 99% VaR equals 3.518 and the 95% VaR equals 1.757. As credit returns typically have the non-negative dependence structure, the assumption of independence for single credit returns results in the lowest VaR measurement, the lower bound for portfolio VaR estimates. The upper bound of the portfolio VaR measurements is given by the *non-diversified VaR*, the sum of the stand-alone VaR values.[38] For our portfolio, the non-diversified stable 99% VaR is 9.813 and the non-diversified stable 95% VaR is 4.733. Analysis in Section 5 showed the 99% stable VaR estimates slightly exceed the empirical 99% VaR, whereas the 95% stable VaR evaluation underestimates the empirical 95% VaR. Therefore, 9.813 is a biased upwards estimate of the portfolio non-diversified 99% VaR and 4.733 is a biased downwards measurement of the portfolio non-diversified 95% VaR.

7. Stable modeling of portfolio risk for symmetric dependent credit returns

In this section we suppose that distributions of credit returns are symmetric α-stable and dependent. We interpret a symmetric random variable as a transformation of a normal ran-

[38] The stand-alone VaR is the VaR for the individual asset.

dom variable. Based on this interpretation, we develop a new methodology for correlation estimation. We apply the methodology for portfolio risk assessment.

We evaluate portfolio risk by determining portfolio VaR: (i) simulating a distribution of the $R_P = \sum_{i=1}^{n} w_i R_i$ values; (ii) finding a certain quantile of the R_P distribution, say, the 1% quantile, which corresponds to the 99% VaR confidence level. The aim of simulations is to project possible portfolio return values R_P at time $T+1$ given: (i) observations of individual returns over time: $R_{i1}, R_{i2}, \ldots, R_{iT}$, $i = 1, \ldots, n$; (ii) weights of portfolio assets $w_1, \ldots, w_n$. The simulations must account for dependence among individual credit returns R_i, $i = 1, \ldots, n$. A traditional approach of quantifying dependence is to calculate the covariance matrix. Under the α-stable assumption for distributions of R_i, computation of the covariance matrix is impossible.

We suggest a new method for deriving the dependence (association) structure. The method assumes that R_i are symmetric strictly stable: $R_i \sim S_{\alpha_{R_i}}(\sigma_{R_i}, 0, 0)$. A symmetric α-stable ($S\alpha S$) random variable can be interpreted as a random rescaling transformation of a normal random variable (see Property 1 below). If a collection of $S\alpha S$ variables is obtained by applying a similar transformation to dependent normal variables, the dependence structure among variables will remain. Thus, the dependence among $S\alpha S$ random variables can be explained by the dependence among underlying normal random variables.

Property 1.[39] Assume that:

(i) G is a normal random variable with a zero mean:

$$G \sim S_2(\sigma_G, 0, 0) = N\left(0, 2\sigma_G^2\right),$$

(ii) Y is a symmetric α-stable random variable, $\alpha < 2$:

$$Y \sim S_\alpha(\sigma_Y, 0, 0),$$

(iii) S is a positive $\frac{\alpha}{2}$-stable random variable:

$$S \sim S_{\alpha/2}\left(\frac{\sigma_Y^2}{\sigma_G^2}\left(\cos\left(\frac{\pi\alpha}{4}\right)\right)^{2/\alpha}, 1, 0\right),$$

(iv) S and G are independent.

Then, the symmetric α-stable random variable Y can be represented as a random rescaling transformation of the normal random variable G:

$$Y = S^{1/2} G.$$

Simulations of the portfolio return values R_P can be divided into two fragments:

39 Property 1 is a slightly modified version of Proposition 1.3.1 in Samorodnitsky and Taqqu (1994).

(i) generating individual returns $\widetilde{R}_i$ with the same dependence structure as the R_i's. We derive the dependence among R_i supposing that $R_i \sim S_{\alpha_{R_i}}(\sigma_{R_i}, 0, 0)$. Based on Property 1, R_i can be expressed as a transformation of a normal random variable:

$$R_i = S_i^{1/2} G_i, \tag{7}$$

where

$$G_i \sim S_2(\sigma_{G_i}, 0, 0) = N(0, 2\sigma_{G_i}^2), \tag{8}$$

$$S_i \sim S_{\alpha_{R_i}/2}\left(\frac{\sigma_{R_i}^2}{\sigma_{G_i}^2}\left(\cos\left(\frac{\pi\alpha}{4}\right)\right)^{2/\alpha_{R_i}}, 1, 0\right), \tag{9}$$

S_i is independent of G_i, $i = 1, \ldots, n$.
Random rescaling transformations of normal variables G_i into R_i preserve the dependence structure. Hence, the dependence among R_i can be explained by the dependence among G_i, $i = 1, \ldots, n$. Based on this property, we generate dependent normal variables $\widetilde{G}_i$, maintaining the initial dependence,[40] then, we generate $\widetilde{R}_i = \widetilde{S}_i^{1/2}\widetilde{G}_i$, where $\widetilde{S}_i$ is a simulated value of S_i;

(ii) computing $\widetilde{R}_P = \sum_{i=1}^n w_i \widetilde{R}_i$.

The simulations are performed according to the following algorithm:[41]

Step 1: Estimate stable parameters of R_i: $\alpha_{R_i}, \sigma_{R_i}, \mu_{R_i}$.[42]

Step 2: "Center" the R_i observations: $R_i^* = R_i - \mu_{R_i}$. Further on, we shall assume $\mu_{R_i} = 0$ and consider R_i^* as R_i : $R_i \sim S_{\alpha_{R_i}}(\sigma_{R_i}, 0, 0)$, $i = 1, \ldots, n$.

Step 3: Assume: (i) R_i can be decomposed according to expressions (7)–(9); (ii) the covariance matrix of $(G_i)_{1\leqslant i\leqslant n}$ is equal to the covariance matrix of truncated $(R_i)_{1\leqslant i\leqslant n}$. Evaluate the covariance matrix of $(G_i)_{1\leqslant i\leqslant n}$ at time $T+1$, $\Sigma_{T+1} = \{c_{ij,T+1|T}\}$, $i = 1, \ldots, n$, $j = 1, \ldots, n$, using exponential weighting:

$$c_{i,T+1|T}^2 = (1-\theta)\sum_{k=0}^K \theta^k R_{i,T-k}^2, \tag{10}$$

$$c_{ij,T+1|T}^2 = (1-\theta)\sum_{k=0}^K \theta^k R_{i,T-k} R_{j,T-k}, \tag{11}$$

[40] Variables G_i, which enter formulas (1) and (8), are not observable. We suppose the dependence structure of Gaussian variables $(G_i)_{1\leqslant i\leqslant n}$ is "inherited" from the dependence structure of truncated values of stable variables $(R_i)_{1\leqslant i\leqslant n}$. Because we believe that the "outliers" are very important for the description of the dependence structure, we take the truncation value for R_i sufficiently large.

[41] The algorithm is implemented in the Mercury Software Package for Market Risk (VaR). See Rachev et al. (1999).

[42] This section assumes $\beta_{R_i} = 0$.

where $T+1|T$ denotes a forecast for time $T+1$ conditional on information up to time T; θ is a decay factor, $0<\theta<1$; K is a number of observations' lags. Exponential weighting (6), (7) allows to account for volatility and correlation clustering (GARCH effects).[43] Formulas (6), (7) can be expressed in recursive (GARCH-type) form:[44]

$$c_{i,T+1|T}^2=\theta c_{i,T|T-1}^2+(1-\theta)R_{i,T}^2,$$

$$c_{ij,T+1|T}^2=\theta c_{ij,T|T-1}^2+(1-\theta)R_{i,T}R_{j,T}.$$

Step 4: Generate a value of the multivariate normal random variable $G=(G_1,G_2,\ldots,G_n)$ with the covariance matrix Σ_{T+1}.
Step 5: Simulate values of stable random variables

$$S_i\sim S_{\alpha_{R_i}/2}\left(\frac{2\sigma_{R_i}^2}{c_i^2}\left(\cos\left(\frac{\pi\alpha}{4}\right)\right)^{2/\alpha_{R_i}},1,0\right),\quad i=1,\ldots,n.$$

Step 6: Compute $\widetilde{R}_i=S_i^{1/2}G_i$, $i=1,\ldots,n$.
Step 7: Calculate $\widetilde{R}_P=\sum_{i=1}^n w_i\widetilde{R}_i$.
Step 8: Repeat Steps 4–7 a large number of times to form an $\widetilde{R}_P$-distribution.

Obtain a portfolio VaR measurement as the negative of a specified quantile of the $\widetilde{R}_P$-distribution.

We evaluate portfolio risk for equally weighted returns on indices of the investment grade corporate bonds: C1A1, C2A1, C3A1, C4A1, C1A2, C2A2, C3A2, C4A2, C1A3, C2A3, C3A3, C4A3, C1A4, C2A4, C3A4, and C4A4. Description of indices is given in Table 10 of Section 5. We impose an assumption that returns on these indices are symmetric-α-stable. We compute the 99% and 95% VaR measurements in two procedures: (i) simulation of portfolio returns following the above described algorithm; (ii) calculation of the 99% (95%) VaR as the negative of the 1% (5%) quantile. In step 3 of the portfolio returns simulations, derivation of the covariance matrix Σ_{T+1}, we used different truncation points and decay factor values. In order to estimate accuracy of simulations, we calculate the Kolmogorov Distance (KD) and Anderson–Darling (AD) statistics:

$$KD=\sup_x\left|F_e(x)-F_s(x)\right|,$$

$$AD=\sup_x\left\{\frac{|F_e(x)-F_s(x)|}{\sqrt{F_e(x)(1-F_e(x))}}\right\},$$

where $F_e(x)$ is the empirical cumulative density function (cdf) and $F_s(x)$ is the simulated cdf. The computation results are summarized in Table 13.

43 An exponential weighting methodology follows the RiskMetrics' exponentially weighted moving average model. See Longerstaey and Zangari (1996).

44 Formulas are adapted from Longerstaey and Zangari (1996).

Table 13
Portfolio VaR for symmetric dependent credit returns

Decay factor θ	Truncation points (%)	Portfolio VaR		Kolmogorov distance	Anderson–Darling
		99% VaR	95% VaR		
0.85	10–90	7.508	4.886	3.880	0.086
	5–95	7.777	5.153	3.736	0.093
	No	8.286	5.346	4.859	0.111
0.94	10–90	7.793	5.147	3.556	0.081
	5–95	8.076	5.248	4.362	0.104
	1–99	8.389	5.434	5.650	0.128
	No	8.114	5.252	5.212	0.117
0.975	10–90	8.028	5.036	3.452	0.077
	5–95	8.166	5.318	9.085	0.234
	1–99	8.469	5.493	5.805	0.130
	No	8.516	5.470	7.274	0.167

The 99% VaR estimates in Table 13 are within the 99% VaR range $(3.518, 9.813)$ derived in Section 6. At each truncation band, increasing the decay factor leads to higher values of the 99% VaR. Thus, as the decay factor grows, the 99% VaR generally rises. At each value of the decay factor, in general, reduction of truncated observations produced higher VaR numbers. We explain the latter observation by positive correlation in tails (concurrent extreme events). Consideration of a larger number of tail observations results in higher VaR. The KD and AD statistics, in general, decline with smaller decay factors. We examine how selection of the decay factor and the truncation method affects estimation of *marginal risks*. The marginal risk is a risk added by an asset to the portfolio risk. It is computed as the difference between the portfolio risk with an analyzed asset and the portfolio risk without the asset. We report the examination results in Table 14.

The decay factor of 0.85 does not produce cases "Marginal VaR > Stand-alone VaR" and "Within one maturity band, higher ratings contribute more risk". In sum, the decay factor = 0.85 results in the lower KD and AD statistics and does not lead to irregular cases; the no-truncation method better accounts for correlation in tails. Hence, we would recommend the choice of the decay factor = 0.85 and the no-truncation method.

In Table 15 we report marginal 99% VaR, stand-alone 99% VaR, and diversification effects at the decay factor of 0.85 and the no-truncation method. Marginal VaR estimates of Table 15 are consistent with the expectation that, for a given credit rating, bonds with longer maturities contribute more risk. Having marginal VaR numbers, we can identify concentration risks. We find that the C4A3 bond index makes the highest addition to the portfolio 99% VaR: the C4A3 marginal VaR of 0.920 exceeds all other marginal VaR. Marginal risks for all bond indices are smaller than stand-alone risks, which indicates that, indeed, diversification reduces risk. From Table 15, we notice that the C4A1 and C3A4 bond indices have highest diversification effects.

Table 14
Marginal risk for symmetric dependent credit returns

Decay factor	Truncation (%)	Cases: Marginal VaR > Stand-alone VaR	Cases: Higher ratings assets contribute more risk
0.85	10–90	0	0
	5–95	0	0
	No	0	0
0.94	10–90	0	0
	5–95	0	0
	1–99	3	2
	No	0	2
0.975	10–90	0	0
	5–95	0	4
	1–99	2	4
	No	3	4

Table 15
Marginal VaR, stand-alone 99% VaR, and diversification effects for bond indices (decay factor = 0.85, no truncation)

Bond indices	Marginal VaR	Stand-alone VaR	Diversification effect
C1A1	0.199	0.284	0.085
C2A1	0.338	0.509	0.171
C3A1	0.572	0.734	0.162
C4A1	0.713	0.931	0.218
C1A2	0.245	0.285	0.040
C2A2	0.494	0.505	0.011
C3A2	0.575	0.689	0.114
C4A2	0.788	0.890	0.102
C1A3	0.190	0.286	0.096
C2A3	0.403	0.530	0.127
C3A3	0.592	0.719	0.127
C4A3	0.920	0.949	0.029
C1A4	0.185	0.290	0.105
C2A4	0.338	0.511	0.173
C3A4	0.522	0.741	0.219
C4A4	0.803	0.960	0.157

We studied stable modeling of portfolio risk under the assumptions of the independent and symmetric dependent instruments. In the next section we consider portfolio risk evaluation in the most general case – skewed dependent instruments.

8. Stable modeling of portfolio risk for skewed dependent credit returns

We quantify portfolio risk R_P by generating a distribution of its possible values and deriving a portfolio VaR from the constructed distribution of R_P. In a case of portfolio assets with skewed dependent credit returns, simulations of the R_P values should reflect the "cumulative" skewness and maintain the dependence (association) among them. In order to do that, we decompose single credit returns R_i into two independent parts: the first part accounts for dependence and the second – for skewness. Then, we obtain the portfolio dependence and skewness components separately aggregating the dependence and skewness parts of individual credit returns. Simulations of the portfolio credit returns values R_P can be divided into three portions: (i) generation of the portfolio dependence component maintaining the dependence structure among individual credit returns, (ii) generation of the portfolio skewness component, and (iii) computation of R_P as a sum of the two generated components. Explanations of our methodology are provided below.

A stable random variable $R \sim S_\alpha(\sigma, \beta, 0)$ can be decomposed (in distribution) into two independent stable random variables $R^{(1)}$ and $R^{(2)}$:

$$R \overset{d}{=} R^{(1)} + R^{(2)},$$

where

$$R^{(1)} \sim S_\alpha(\sigma_1, \beta_1, 0), \qquad R^{(2)} \sim S_\alpha(\sigma_2, \beta_2, 0),$$

$$\sigma = \left(\sigma_1^\alpha + \sigma_2^\alpha\right)^{1/\alpha}, \tag{12}$$

$$\beta = \frac{\beta_1\sigma_1^\alpha + \beta_2\sigma_2^\alpha}{\sigma_1^\alpha + \sigma_2^\alpha}. \tag{13}$$

Suppose that: (i) $R^{(1)}$ is a symmetric stable variable: $\beta_1 = 0$; (ii) $\sigma_1 = \sigma_2 = \sigma^*$. Then, formulas (12) and (13) can be reduced to the following expressions:

$$\sigma = 2^{1/\alpha}\sigma^*, \tag{14}$$

$$\beta = \frac{1}{2}\beta_2. \tag{15}$$

From Equations (14) and (15), we have

$$\sigma^* = 2^{-1/\alpha}\sigma, \qquad \beta_2 = 2\beta.$$

In sum, a stable random variable $R \sim S_\alpha(\sigma, \beta, 0)$ can be decomposed (in distribution) into two independent stable random variables: symmetric $R^{(1)}$ and skewed $R^{(2)}$:

$$R \overset{d}{=} R^{(1)} + R^{(2)}, \tag{16}$$

where

$$R^{(1)} \sim S_\alpha\big(2^{-1/\alpha}\sigma, 0, 0\big), \tag{17}$$

$$R^{(2)} \sim S_\alpha\big(2^{-1/\alpha}\sigma, 2\beta, 0\big). \tag{18}$$

Using methodology (16)–(18), we can divide individual credit returns $R_i \sim S_{\alpha_{R_i}}(\sigma_{R_i}, \beta_{R_i}, 0)$ into the "dependence" and "skewness" parts. First, we partition R_i into the "symmetry" and "skewness" fragments:

$$R_i \stackrel{d}{=} R_i^{(1)} + R_i^{(2)},$$

where

$$R_i^{(1)} \sim S_{\alpha_{R_i}}\big(2^{-1/\alpha_{R_i}}\sigma_{R_i}, 0, 0\big), \qquad R_i^{(2)} \sim S_{\alpha_{R_i}}\big(2^{-1/\alpha_{R_i}}\sigma_{R_i}, 2\beta_{R_i}, 0\big),$$

parts $R_i^{(1)}$ and $R_i^{(2)}$ are independent, $i = 1, \ldots, n$. Second, we suppose: (i) $R_i^{(1)}$, $i = 1, \ldots, n$, are dependent and (ii) $R_i^{(2)}$, $i = 1, \ldots, n$, are independent. Consequently, symmetric terms $R_i^{(1)}$ explain dependence (association) among R_i's and terms $R_i^{(2)}$ account for skewness of R_i's.

Based on Property 1 (see Section 7), $R_i^{(1)} \sim S_{\alpha_{R_i}}(2^{-1/\alpha_{R_i}}\sigma_{R_i}, 0, 0)$ can be written as a transformation of a normal random variable:

$$R_i^{(1)} = S_i^{1/2} G_i,$$

where

$$G_i \sim S_2(\sigma_{G_i}, 0, 0) = N\big(0, 2\sigma_{G_i}^2\big),$$

$$S_i \sim S_{\alpha_{R_i}/2}\left(\frac{2^{-2/\alpha_{R_i}}\sigma_{R_i}^2}{\sigma_{G_i}^2}\left(\cos\left(\frac{\pi\alpha}{4}\right)\right)^{2/\alpha_{R_i}}, 1, 0\right),$$

S_i is independent of G_i, $i = 1, \ldots, n$.

Random rescaling transformations of normal variables G_i into $R_i^{(1)}$ maintain the dependence structure. Therefore, from the dependence among G_i's we can determine the dependence among $R_i^{(1)}$, or the dependence among R_i.

Adding separately the dependence and skewness terms of R_i's, we obtain the two components of the portfolio returns R_P:

$$R_P = R_P^{(1)} + R_P^{(2)}, \tag{19}$$

where $R_P^{(1)} = \sum_{i=1}^n w_i R_i^{(1)} = \sum_{i=1}^n w_i S_i^{1/2} G_i$ is the "dependence" component and $R_P^{(2)} = \sum_{i=1}^n w_i R_i^{(2)}$ is the "skewness" component.

We simulate the R_P values based on decomposition (19): $\widetilde{R}_P = \widetilde{R}_P^{(1)} + \widetilde{R}_P^{(2)}$. The simulations are executed according to the next algorithm:[45]

Step 1: Estimate stable parameters of R_i: $\alpha_{R_i}, \beta_{R_i}, \sigma_{R_i}, \mu_{Ri}$.
Step 2: "Center" the R_i observations: $R_i^* = R_i - \mu_{R_i}$. Further on, we shall assume $\mu_{R_i} = 0$ and consider R_i^* as R_i: $R_i \sim S_{\alpha_{R_i}}(\sigma_{R_i}, \beta_{R_i}, 0)$, $i = 1, \ldots, n$.
Step 3: Evaluate the covariance matrix of normal random variables $(G_i)_{1\leqslant i\leqslant n}$ at time $T+1$, $\Sigma_{T+1} = \{c_{ij,T+1|T}\}$, $i = 1, \ldots, n$, $j = 1, \ldots, n$, using exponential weighting:

$$c_{i,T+1|T}^2 = (1-\theta)\sum_{k=0}^{K} \theta^k R_{i,T-k}^2,$$

$$c_{ij,T+1|T}^2 = (1-\theta)\sum_{k=0}^{K} \theta^k R_{i,T-k} R_{j,T-k},$$

where $T+1|T$ denotes a forecast for time $T+1$ conditional on information up to time T; θ is a decay factor, $0 < \theta < 1$; K is a number of observations' lags.
Step 4: Generate a value of the multivariate normal random variable $G = (G_1, G_2, \ldots, G_n)$ with the covariance matrix Σ_{T+1}.
Step 5: Simulate values of stable random variables

$$S_i \sim S_{\alpha_{R_i}/2}\left(\frac{2^{1-2/\alpha_{R_i}}\sigma_{R_i}^2}{c_i^2}\left(\cos\left(\frac{\pi\alpha}{4}\right)\right)^{2/\alpha_{R_i}}, 1, 0\right), \quad i = 1, \ldots, n.$$

Step 6: Compute $\widetilde{R}_i^{(1)} = S_i^{1/2} G_i$, $i = 1, \ldots, n$.
Step 7: Generate $\widetilde{R}_i^{(2)} \sim S_{\alpha_{R_i}}(2^{-1/\alpha_{R_i}}\sigma_{R_i}, 2\beta_{R_i}, 0)$, $i = 1, \ldots, n$.
Step 8: Calculate $\widetilde{R}_P = \sum_{i=1}^n w_i \widetilde{R}_i^{(1)} + \sum_{i=1}^n w_i \widetilde{R}_i^{(2)}$.
Step 9: Repeat Steps 4–8 a large number of times to form an $\widetilde{R}_P$-distribution.

Derive a portfolio VaR estimate as the negative of a chosen quantile of the $\widetilde{R}_P$-distribution.

We implement the suggested procedure (Step 1–Step 9) for the risk assessment of the same portfolio of indices as in Section 7. We suppose that returns on indices are dependent skewed-α-stable. The portfolio VaR estimates are presented in Table 16.

The 99% portfolio VaR estimates fall within the 99% VaR range (3.518, 9.813) of Section 6. From Table 16, the VaR magnitude generally: (i) increases when the decay factor θ increases from 0.85 to 0.94; (ii) declines when θ changes from 0.94 to 0.975. Thus, the decay factor $\theta = 0.94$ leads to more conservative VaR estimates. The 1%–99% truncation band appears to produce the lowest KD and AD statistics. Based on our observations, we

[45] This algorithm is an extended version of the algorithm in Section 7.

Table 16
Portfolio VaR for skewed dependent credit returns

Decay factor θ	Truncation points (%)	Portfolio VaR		Kolmogorov distance	Anderson–Darling
		99% VaR	95% VaR		
0.85	10–90	4.939	2.904	7.22	0.20
	5–95	5.380	3.162	5.64	0.18
	No	5.449	3.236	5.43	0.17
0.94	10–90	5.101	3.009	6.53	0.19
	5–95	5.456	3.248	5.24	0.17
	1–99	5.596	3.363	4.70	0.14
	No	5.455	3.231	5.13	0.17
0.975	10–90	5.112	3.021	6.54	0.19
	5–95	5.416	3.238	5.34	0.17
	1–99	5.471	3.307	4.37	0.14
	No	5.298	3.238	5.43	0.15

would recommend to employ $\theta = 0.94$ and the 1%–99% truncation band in VaR derivations under the assumption of skewed dependent credit returns. We computed marginal VaRs for the same combinations of the decay factor and the truncation band as in Table 16. The marginal VaR estimates were smaller than the corresponding stand-alone VaR measurements, which supports feasibility of suggested procedure for simulating portfolio returns.

We have applied stable modeling to the total risk assessment of credit returns. Below we analyze stable modeling of isolated credit risk.

9. One-factor model of portfolio credit risk

In this section we outline a *one-factor model* for quantifying portfolio credit risk. The model is built on two postulations: (i) constituent parts of the credit returns are the credit-risk-free part and the credit risk premium; (ii) the credit risk spread follows a stable law. Applying the one-factor model, in the following sections we quantify credit risk for single instruments and then estimate portfolio credit risk as a cumulative result of stable distributed individual credit risks.

Similarly to the previous sections, we assume that a portfolio includes n assets. Then, the portfolio return is given by $R_P = \sum_{i=1}^{n} w_i R_i$, where R_i is the return on the i-th asset, w_i is the weight of the i-th asset, $i = 1, \ldots, n$, $\sum_{i=1}^{n} w_i = 1$. We conjecture that individual returns R_i depend on one credit-risk-free factor Y_i:

$$R_i = a_i + b_i Y_i + U_i, \tag{20}$$

where a_i and b_i are constants, U_i is the residual representing compensation for credit risk and random noise,[46] $i = 1, \ldots, n$.

Suppose the i-th portfolio instrument is a corporate bond of maturity τ with returns R_i. There are two possible choices for an underlying credit-risk-free factor Y_i: (i) returns on a Treasury bond of the same maturity τ; (ii) returns on a τ-year bond with a credit rating AAA. Then, the spread $U_i = R_i - a_i - b_i Y_i$ reflects charges for credit risk. If the j-th portfolio asset is a swap with a counterparty that has a low credit rating, say BBB, we can choose, as an underlying factor Y_j, returns on a similar swap with a company that has a credit rating AAA: $R_j = a_j + b_j Y_j + U_j$, the term U_j accounts for the credit risk of the BBB-swap.

We impose the following assumptions on the components of model (20):

(i) Credit risk spreads U_i are strictly stable, $U_i \sim S_{\alpha_{U_i}}(\sigma_{U_i}, \beta_{U_i}, 0)$,[47] $\alpha_{U_i} > 1$.

(ii) Default-free factors Y_i are strictly stable, $Y_i \sim S_{\alpha_{Y_i}}(\sigma_{Y_i}, \beta_{Y_i}, 0)$,[48] $\alpha_{Y_i} > 1$.

(iii) U_i and Y_i are independent of each other, $i = 1, \ldots, n$.

Then, the portfolio return R_P can be decomposed into three components:

$$R_P = A + Y_P + U_P,$$

where Y_P expresses aggregate effect of underlying factors, U_P represents *portfolio credit risk*,

$$A = \sum_{i=1}^{n} w_i a_i, \qquad Y_P = \sum_{i=1}^{n} w_i b_i Y_i, \qquad U_P = \sum_{i=1}^{n} w_i U_i.$$

We evaluate the portfolio credit risk U_P in two steps: (i) quantifying credit risk of each asset U_i; (ii) estimating U_P as a cumulative result of individual U_i, $i = 1, \ldots, n$. Section 10 discusses credit risk evaluation for single portfolio assets. Section 11 examines portfolio credit risk estimation under the assumptions of independent, symmetric dependent, and skewed dependent credit risks.

10. Credit risk evaluation for portfolio assets

Approximations of the credit risk premium values U_i for portfolio assets can be obtained using model (20):

$$\widehat{U}_{it} = R_{ti} - \hat{a}_i - \hat{b}_i Y_{ti}, \tag{21}$$

[46] We interpret the yield spread as the credit risk premium and include the noise factor into the credit risk part. The noise factor could incorporate taxability, liquidity, and other premiums.

[47] The shift of U_i is, in fact, incorporated in a_i.

[48] Y_i is the centered return. If the returns of portfolio instruments, Z_i, are non-centered, then we take $Y_{it} = Z_{it} - \overline{Z}_i$, $t = 1, \ldots, T$.

where $\hat{a}_i$ and $\hat{b}_i$ are the OLS estimates,

$$\hat{a}_i = \frac{\sum_{t=1}^{T} Y_{it}^2 \sum_{t=1}^{T} R_{it} - \sum_{t=1}^{T} Y_{it} \sum_{t=1}^{T} R_{it} Y_{it}}{T \sum_{t=1}^{T} Y_{it}^2 - (\sum_{t=1}^{T} Y_{it})^2}, \tag{22}$$

$$\hat{b}_i = \frac{T \sum_{t=1}^{T} R_{it} Y_{it} - \sum_{t=1}^{T} Y_{it} \sum_{t=1}^{T} R_{it}}{T \sum_{t=1}^{T} Y_{it}^2 - (\sum_{t=1}^{T} Y_{it})^2}, \tag{23}$$

$$i = 1, \ldots, n; \quad t = 1, \ldots, T.$$

Estimators $\hat{a}_i$ and $\hat{b}_i$, given by expressions (22) and (23), are unbiased.[49]

We analyze credit risk of corporate bonds applying one-factor model (20). Assume that returns on an index of the US corporate bonds, R_i, are described by returns on a credit-risk-free factor, Y_i, and a credit spread, U_i:

$$R_i = a_i + b_i Y_i + U_i,$$

where a_i and b_i are constants, $i = 1, \ldots, 16$. We examine returns on the same 16 indices as in Section 5 (see Table 10): $R_i \in \{R_{\text{C1A1}}, R_{\text{C2A1}}, R_{\text{C3A1}}, R_{\text{C4A1}}, R_{\text{C1A2}}, R_{\text{C2A2}}, R_{\text{C3A2}}, R_{\text{C4A2}}, R_{\text{C1A3}}, R_{\text{C2A3}}, R_{\text{C3A3}}, R_{\text{C4A3}}, R_{\text{C1A4}}, R_{\text{C2A4}}, R_{\text{C3A4}}$, and $R_{\text{C4A4}}\}$. We choose, as corresponding credit-risk-free factors, returns on the indices of US government bonds in the same maturity band: $Y_i \in \{R_{\text{G1O2}}, R_{\text{G2O2}}, R_{\text{G3O2}}, R_{\text{G4O2}}\}$.[50] For example, if we consider returns on the index of bonds with maturity from one to three years, R_{C1A1}, then the returns on the index of the government bonds with maturity from one to three years, R_{G1O2}, serve as the underlying credit-risk-free factor. We approximate the percentage return values of the individual credit risks U_i, following approach (21): (i) run OLS regressions of model (20), (ii) compute the residuals' series $\widehat{U}_i$. Coefficients of the OLS regressions are given in Appendix B, Table B.3. Obtained sets of OLS credit risk premiums $\widehat{U}_i$ are plotted in Figure 18 and in figures of Appendix C. Empirical densities of $\widehat{U}_i$ are shown in Figure 19 and in Appendix C. We observe that the credit risk spread series $\widehat{U}_i$ exhibit volatility clusters and heavy tails. Such behavior of the individual returns sets can be captured by stable and GARCH models.

Stable modeling of the credit risk premiums $\widehat{U}_i$, entailed values of $\alpha < 1.6$, $\beta \approx 0$, and $\mu \approx 0$ (see Table 17). These values of parameter estimates indicate that credit risk spreads of the corporate bonds' indices are fat-tailed and almost symmetric. Table 17 presents the following α and β values of the credit risks of the bond indices with a maturity band from one to three years: AAA bonds: $\alpha = 1.333$ and $\beta = 0.011$; AA bonds: $\alpha = 1.379$ and $\beta = 0.030$; A bonds: $\alpha = 1.393$ and $\beta = -0.021$; BBB bonds: $\alpha = 1.412$ and $\beta = 0.004$.

[49] For analysis of asymptotic properties of OLS estimators (22) and (23) under the stable distribution assumption for the disturbance term, see Götzenberger, Rachev and Schwartz (1999).

[50] A digit after letter "G" denotes the maturity band: 1 – from 1 to 3 years, 2 – from 3 to 5 years, 3 – from 5 to 7 years, 4 – from 7 to 10 years.

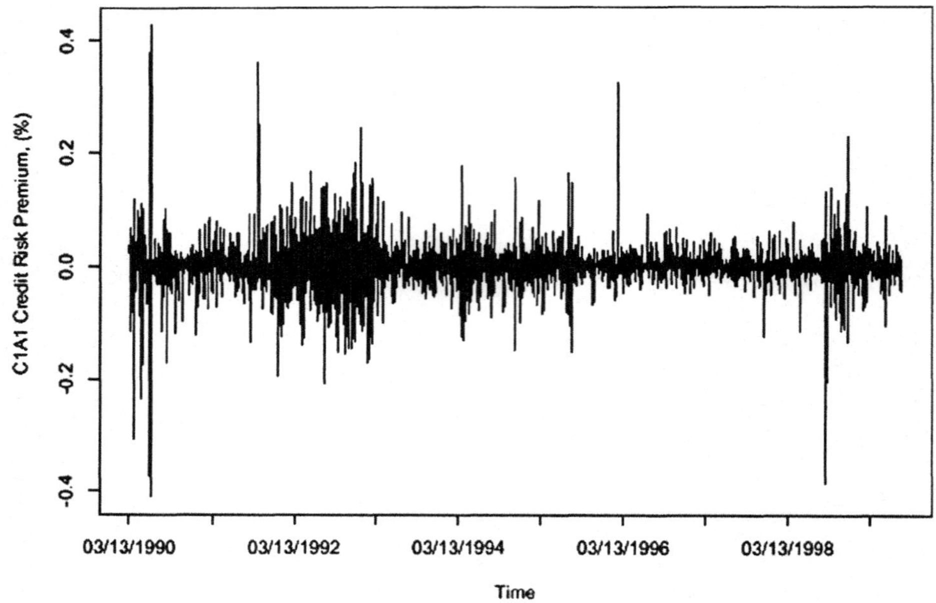

Fig. 18. OLS credit risk premium of the C1A1 bond index.

Table 17
Stable and normal fitting of the OLS credit risk premiums of bond indices

OLS credit risk of bond indices	Maturity (years)	Normal		Stable			
		Mean	Standard deviation	α	β	μ	σ
C1A1	1–3	0.0	0.045	1.333	0.011	0.000	0.017
C2A1	3–5	0.0	0.075	1.528	−0.089	−0.001	0.033
C3A1	5–7	0.0	0.096	1.590	−0.023	0.000	0.047
C4A1	7–10	0.0	0.116	1.456	−0.026	0.000	0.051
C1A2	1–3	0.0	0.037	1.379	0.030	0.001	0.015
C2A2	3–5	0.0	0.064	1.523	−0.074	0.000	0.029
C3A2	5–7	0.0	0.086	1.591	−0.060	0.000	0.044
C4A2	7–10	0.0	0.110	1.426	0.005	0.001	0.050
C1A3	1–3	0.0	0.038	1.393	−0.021	0.000	0.015
C2A3	3–5	0.0	0.069	1.483	−0.084	0.000	0.029
C3A3	5–7	0.0	0.098	1.519	−0.073	0.000	0.042
C4A3	7–10	0.0	0.124	1.366	−0.017	0.001	0.048
C1A4	1–3	0.0	0.074	1.412	0.004	0.001	0.018
C2A4	3–5	0.0	0.096	1.527	−0.024	0.001	0.033
C3A4	5–7	0.0	0.113	1.552	−0.077	0.000	0.048
C4A4	7–10	0.0	0.1424	1.480	−0.055	0.001	0.055

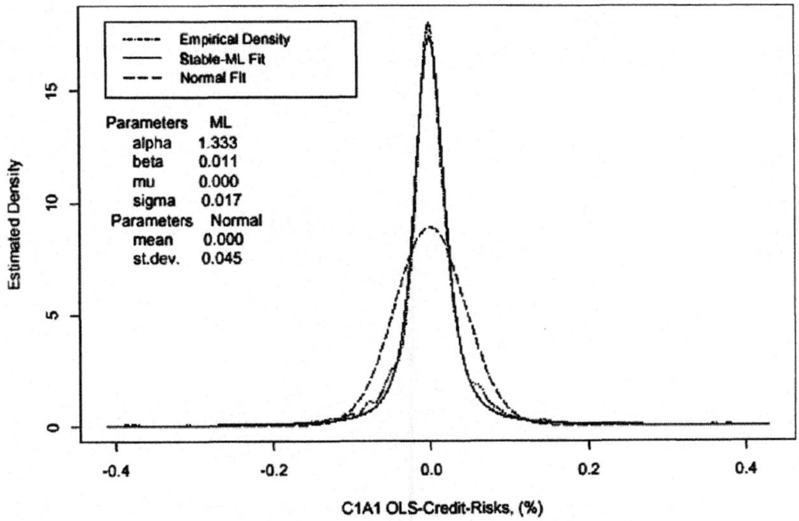

Fig. 19. Stable and normal fitting of C1A1 OLS-credit-risks.

Plots of the stable and normal fitting of the OLS credit risk spreads $\widehat{U}_i$ are shown on Figure 19 and in Appendix C. Figures demonstrate that stable modeling well captures excess kurtosis and heavy tails of the credit risks $\widehat{U}_i$.

The GARCH approach models clustering of volatilities and fat tails, by expressing the conditional variance as an explicit function of past information:

$$R_{i,t} = a_i + b_i Y_{i,t} + U_{i,t}, \tag{24}$$

where

$$U_{i,t} = \sigma_{i,t}\varepsilon_{i,t}, \tag{25}$$

$$\varepsilon_{i,t} \sim N(0,1), \tag{26}$$

$$\sigma_{i,t}^2 = c_i + \sum_{j=1}^{p} \gamma_{i,j}\sigma_{i,t-j}^2 + \sum_{j=1}^{q} \eta_{i,j}U_{i,t-j}^2, \tag{27}$$

$$i = 1,\dots,n; \quad t = 1,\dots,T.$$

We shall name model (24)–(27) as a GARCH(p,q)-*normal* model because it is based on the normality assumption for the disturbance term. In order to detect GARCH-dependencies, we examine sample autocorrelation and partial autocorrelation functions of the squared residuals $\widehat{U}_i$. Visual inspection of the correlograms suggests values of p and q.

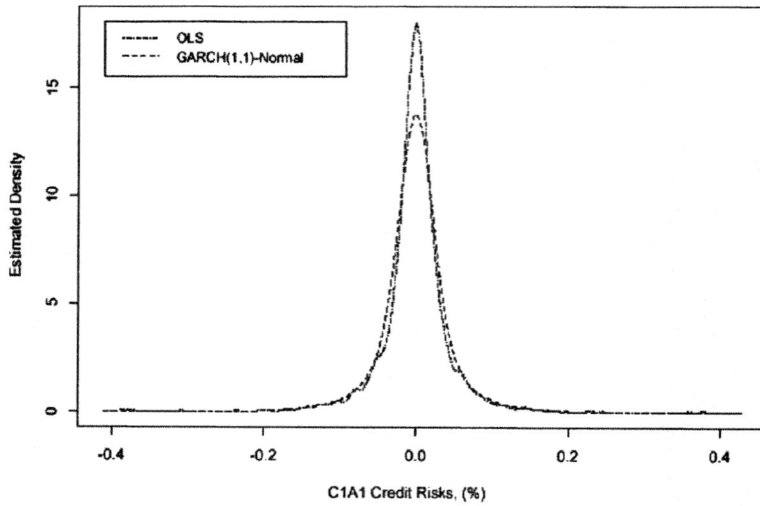

Fig. 20. Credit risks: OLS and GARCH.

Applying the Box–Jenkins methodology, we find that $p = q = 1$ is adequate to capture temporal dependence of volatilities:

$$\sigma_{i,t}^2 = c_i + \gamma_i \sigma_{i,t-1}^2 + \eta_i U_{i,t-1}^2. \tag{28}$$

Coefficients of model (24)–(26) and (28) with $R_i \in \{R_{\mathrm{C1A1}}, R_{\mathrm{C2A1}}, R_{\mathrm{C3A1}}, R_{\mathrm{C4A1}}, R_{\mathrm{C1A2}}, R_{\mathrm{C2A2}}, R_{\mathrm{C3A2}}, R_{\mathrm{C4A2}}, R_{\mathrm{C1A3}}, R_{\mathrm{C2A3}}, R_{\mathrm{C3A3}}, R_{\mathrm{C4A3}}, R_{\mathrm{C1A4}}, R_{\mathrm{C2A4}}, R_{\mathrm{C3A4}}$, and $R_{\mathrm{C4A4}}\}$ and $Y_i \in \{R_{\mathrm{G1O2}}, R_{\mathrm{G2O2}}, R_{\mathrm{G3O2}}, R_{\mathrm{G4O2}}\}$ are reported in Appendix B, Table B.4. Densities of the GARCH(1,1)-normal residuals $U_{i,t} = \sqrt{c_i + \gamma_i \sigma_{i,t-1}^2 + \eta_i U_{i,t-1}^2} \times \varepsilon_{i,t}$ are displayed in Figures 20 and in Appendix D. Graphs demonstrate that the GARCH credit risk series have lower peaks.

In the portfolio context, implementation of the GARCH models is computationally complex because a number of parameters rapidly increases as the portfolio expands.[51] Hence, we evaluate portfolio credit risk U_P based on stable modeling of individual credit risks with accounting for GARCH effects by exponential weighting of observations.[52] In estimation of U_P, we separately investigate cases of independent, symmetric dependent, skewed dependent credit risks of portfolio instruments.

[51] For references on the multivariate GARCH, see Engle and Kroner (1995).

[52] An approach of modeling time-varying volatilities by exponential weighting follows the RiskMetrics' exponentially weighted moving average model described in Morgan (1995).

11. Portfolio credit risk

In this section we follow the one-factor model of Section 9 and evaluate portfolio credit risk as a cumulative effect of stable distributed individual credit risks. We impose different assumptions on their distributions: independent, symmetric dependent, and skewed dependent. We show implementation of the approach on a portfolio of equally weighted OLS-credit-risk premiums from Section 10.

11.1. Independent credit risks

Suppose credit-risk-premiums are: (i) characterized by the same tail index α; (ii) independent. Then, by the additivity property of stable variables (see Section 3), the portfolio credit risk $U_P = \sum_{i=1}^{n} w_i U_i$ is stably distributed:

$$U_P \sim S_\alpha(\sigma_{U_P}, \beta_{U_P}, 0),$$

where α is the tail index, σ_{U_P} is the scale parameter, β_{U_P} is the skewness parameter,

$$\sigma_{U_P} = \left[\sum_{i=1}^{n} (|w_i|\sigma_{U_i})^\alpha\right]^{1/\alpha}, \tag{29}$$

$$\beta_{U_P} = \frac{\sum_{i=1}^{n}[\operatorname{sign}(w_i)\beta_{U_i}(|w_i|\sigma_{U_i})^\alpha]}{\sum_{i=1}^{n}(|w_i|\sigma_{U_i})^\alpha}. \tag{30}$$

Consider a portfolio of equally weighted OLS-credit-risk premiums from Section 10. Assume credit-risk-premiums are independent and have the same tail index α. We take $\alpha = 1.472$, the average of the α values for the credit-risk-premium series (see Table 17), and recompute other stable parameters: β_{U_i}, μ_{U_i}, and σ_{U_i}. New estimates are reported in Table 18. Similarly to returns on bond indices, a condition of the same tail index α for all analyzed credit risk series does not seem to be very restraining: new parameter estimates (Table 18) do not deviate much from the previous parameter estimates (Table 17).

Since obtained estimates of μ are very small, we assume $\mu = 0$. We evaluate portfolio parameters applying formulas (29), (30): $\hat{\sigma}_{U_P} = 0.015$, $\hat{\beta}_{U_P} = -0.038$. Thus, $\widehat{U}_P \sim S_{1.472}(0.015, -0.038, 0)$. The 99% (95%) *credit VaR* is derived as the negative of the 1% (5%) quantile of the $\widehat{U}_P$- distribution: the 99% (95%) VaR equals 0.125 (0.046). Having analytic formulas for the U_P parameters, we obtained estimates of portfolio credit risk without simulations.

11.2. Symmetric dependent credit risks

In order to assess portfolio credit risk, we obtain portfolio credit VaR. It is computed in two steps: (i) simulating a distribution of the $U_P = \sum_{i=1}^{n} w_i U_i$ values; (ii) inferring port-

Table 18
Stable fitting of the OLS credit risk premiums with fixed α

OLS credit risk of bond indices	Maturity (years)	Stable parameters at $\alpha = 1.472$		
		β	μ	σ
C1A1	1–3	0.000	0.000	0.018
C2A1	3–5	−0.090	−0.001	0.032
C3A1	5–7	−0.019	0.000	0.045
C4A1	7–10	−0.019	0.001	0.052
C1A2	1–3	0.023	0.001	0.015
C2A2	3–5	−0.072	−0.001	0.029
C3A2	5–7	−0.039	0.000	0.042
C4A2	7–10	−0.004	0.000	0.051
C1A3	1–3	−0.040	0.000	0.015
C2A3	3–5	−0.084	0.000	0.029
C3A3	5–7	−0.067	0.000	0.041
C4A3	7–10	−0.032	0.001	0.049
C1A4	1–3	−0.010	0.001	0.019
C2A4	3–5	0.011	0.001	0.033
C3A4	5–7	−0.071	−0.001	0.046
C4A4	7–10	−0.053	0.001	0.055

folio credit VaR from the simulated U_P distribution. This section examines the case of symmetric individual credit risks U_i: $U_i \sim S_{\alpha_{U_i}}(\sigma_{U_i}, 0, 0)$, $i = 1, \ldots, n$.

We simulate U_P applying the methodology from Section 7:

(i) generate individual credit risks $\widetilde{U}_i$ with the same dependence structure as the U_i's. We express U_i as a transformation of a normal random variable:

$$U_i = S_i^{1/2} G_i,$$

where

$$G_i \sim S_2(\sigma_{G_i}, 0, 0) = N\big(0, 2\sigma_{G_i}^2\big),$$

$$S_i \sim S_{\alpha_{U_i}/2}\left(\frac{\sigma_{U_i}^2}{\sigma_{G_i}^2}\left(\cos\left(\frac{\pi\alpha}{4}\right)\right)^{2/\alpha_{U_i}}, 1, 0\right),$$

S_i is independent of G_i, $i = 1, \ldots, n$.

The dependence among U_i can be explained by the dependence among G_i, $i = 1, \ldots, n$. We form dependent normal variables $\widetilde{G}_i$, preserving the initial dependence. Next, we generate $\widetilde{U}_i = \widetilde{S}_i^{1/2} \widetilde{G}_i$, where $\widetilde{S}_i$ is a simulated value of S_i;

(ii) calculate $\widetilde{U}_P = \sum_{i=1}^n w_i \widetilde{U}_i$.

Table 19
Portfolio credit VaR for symmetric credit risks

Decay factor θ	Truncation points (%)	Portfolio VaR		Kolmogorov distance	Anderson–Darling
		99% VaR	95% VaR		
0.85	10–90	3.502	1.918	8.071	0.210
	5–95	3.710	1.896	8.898	0.228
	No	3.396	1.856	7.692	0.199
0.94	10–90	3.594	1.963	7.680	0.200
	5–95	3.643	1.941	8.162	0.209
	1–99	3.476	1.975	8.847	0.227
	No	3.321	1.792	6.736	0.164
0.975	10–90	3.623	1.877	7.578	0.194
	5–95	3.435	1.943	9.085	0.234
	1–99	3.578	2.004	9.665	0.254
	No	3.293	1.739	7.174	0.167

A portfolio credit VaR can be measured from the $\widetilde{U}_P$-distribution.

As an illustration of the approach, we estimate credit risk for a portfolio of equally weighted OLS-credit-risk premiums of bond indices (see Section 10) assuming they are symmetric.[53] The estimation results are presented in Table 19. The portfolio credit VaR does not demonstrate a certain pattern of dependence on the decay factor. For each decay factor, reduction of the truncated observations does not seem to affect the portfolio credit VaR in a particular fashion. The no-truncation method approach led to the smallest VaR measurements. Possibly, the credit risk residuals of the investment grade indices have negative correlations in far tails. Taking into account more observations with negative correlations reduces the VaR estimates. Since the decay factor does not influence the VaR results in a specific way and the KD and AD statistics are smaller at the no-truncation approach, in further analysis, we consider the no-truncation method and arbitrarily select the decay factor of 0.85. Computation of the marginal VaR, stand-alone VaR, diversification effects for the no-truncation approach and the decay factor = 0.85 is summarized in Table 20.

From Table 20, highest contributions to portfolio credit risk are made by the C4A4, C4A3, and C4A2 bond indices: their marginal 99% VaR equal 0.366, 0.295, and 0.296. The credit risk premium of the C4A1 index displays the largest diversification effect.

11.3. Skewed dependent credit risks

For estimation of portfolio risk for the skewed dependent credit risks, we propose to employ the approach of Section 8: (i) split individual credit risks U_i into the dependence and

[53] The symmetry proposition is plausible: the skewness parameters of credit risks premiums of bond indices are small (see Table 16).

Table 20
Marginal VaR, stand-alone 99% VaR, and diversification effects for credit risk premiums of bond indices (decay factor = 0.85, no truncation)

Bond indices	Marginal VaR	Stand-alone VaR	Diversification effect
C1A1	0.175	0.191	0.016
C2A1	0.203	0.251	0.048
C3A1	0.162	0.305	0.143
C4A1	0.145	0.441	0.296
C1A2	0.024	0.148	0.124
C2A2	0.153	0.222	0.069
C3A2	0.180	0.290	0.110
C4A2	0.296	0.453	0.157
C1A3	0.013	0.149	0.136
C2A3	0.097	0.244	0.147
C3A3	0.203	0.325	0.122
C4A3	0.295	0.507	0.212
C1A4	0.079	0.168	0.089
C2A4	0.091	0.243	0.152
C3A4	0.142	0.346	0.204
C4A4	0.366	0.457	0.091

skewness parts; (ii) find the portfolio dependence and skewness components by combining the dependence and skewness parts of single credit risks; (iii) evaluate the portfolio credit risk as a sum of the dependence and skewness fragments. Details are given below.

We divide individual credit risks $U_i \sim S_{\alpha_{U_i}}(\sigma_{U_i}, \beta_{U_i}, 0)$ into the "dependence" and "skewness" parts, applying methodology (16)–(18) (see Section 8):

$$U_i \stackrel{d}{=} U_i^{(1)} + U_i^{(2)},$$

where

$$U_i^{(1)} \sim S_{\alpha_{U_i}}\left(2^{-1/\alpha_{U_i}}\sigma_{U_i}, 0, 0\right), \qquad U_i^{(2)} \sim S_{\alpha_{U_i}}\left(2^{-1/\alpha_{U_i}}\sigma_{U_i}, 2\beta_{U_i}, 0\right),$$

parts $U_i^{(1)}$ and $U_i^{(2)}$ are independent, $i = 1, \ldots, n$. We assume: (i) $U_i^{(1)}$, $i = 1, \ldots, n$, are dependent and (ii) $U_i^{(2)}$, $i = 1, \ldots, n$, are independent. Then, symmetric components $U_i^{(1)}$ explain dependence (association) among U_i's and components $U_i^{(2)}$ depict skewness of U_i's.

By Property 1 (see Section 7), $U_i^{(1)} \sim S_{\alpha_{U_i}}(2^{-1/\alpha_{U_i}}\sigma_{U_i}, 0, 0)$ can be interpreted as a transformation of a normal random variable:

$$U_i^{(1)} = S_i^{1/2} G_i,$$

where

$$G_i \sim S_2(\sigma_{G_i}, 0, 0) = N\big(0, 2\sigma_{G_i}^2\big),$$

$$S_i \sim S_{\alpha_{U_i}/2}\left(\frac{2^{-2/\alpha_{U_i}}\sigma_{U_i}^2}{\sigma_{G_i}^2}\left(\cos\left(\frac{\pi\alpha}{4}\right)\right)^{2/\alpha_{U_i}}, 1, 0\right),$$

S_i is independent of G_i, $i = 1, \ldots, n$.

Random rescaling transformations of normal variables G_i into $U_i^{(1)}$ maintain the dependence structure. Hence, we can derive the dependence among $U_i^{(1)}$, or the dependence among U_i, from the dependence among G_i's.

Combining separately the dependence and skewness terms of U_i's, we obtain the two components of the portfolio credit risk U_P:

$$U_P = U_P^{(1)} + U_P^{(2)},$$

$$U_P^{(1)} = \sum_{i=1}^{n} w_i U_i^{(1)} = \sum_{i=1}^{n} w_i S_i^{1/2} G_i,$$

$$U_P^{(2)} = \sum_{i=1}^{n} w_i U_i^{(2)},$$

where $U_P^{(1)}$ is the "dependence" component and $U_P^{(2)}$ is the "skewness" component. The portfolio credit risk can be evaluated as a sum of the dependence and skewness fragments.

We suggested methodologies for portfolio credit risk assessment and demonstrated their applications on analysis of returns on bond indices. The methodologies can be employed for risk evaluation of any financial instruments if they have fat-tailed and/or skewed distributions.

12. Conclusions

The Value-at-Risk (VaR) measurements are widely applied to estimate the exposure to market and credit risks. The traditional approaches to VaR computations – the delta method, historical simulation, Monte Carlo simulation, and stress-testing – do not provide satisfactory evaluation of possible losses. The delta-normal methods do not describe well financial data with heavy tails. Hence, they underestimate VaR measurements in the tails. The historical simulation does not produce robust VaR estimates since it is not reliable in approximating low quantiles with a small number of observations in the tails. The stress-testing VaR estimates are subjective. The Monte Carlo VaR numbers might be affected by model misspecification.

This work proposes the application of stable distributions in market and credit VaR estimation. Our empirical analysis verifies that stable modeling well captures skewness and heavy-tails of market and credit returns and isolated credit risks. The superior fit allows to derive accurate risk estimates. The in-sample- and forecast-evaluation shows that stable VaR modeling outperforms the normal modeling for high values of the VaR confidence level:

- the stable modeling generally produces conservative and accurate 99% VaR estimates, which is preferred by financial institutions and regulators,
- the normal method leads to overly optimistic forecasts of losses in the 99% VaR estimation,
- the normal modeling is acceptable for the 95% VaR estimation.

Based on the properties of stable distributions, we design new methods for the correlation estimation and simulating portfolio values. We employ the methods in evaluation of portfolio and marginal VaR for three cases of the credit returns: independent, symmetric dependent, and skewed dependent. We suggest a one-factor model of credit risks. Applying the one-factor model, we quantify credit risk for individual assets and then assess portfolio credit risk as an aggregate effect of stable distributed individual credit risks.

The stable Paretian model, while sharing the main properties of the normal distribution leading to the CLT (Central Limit Theorem), provides at the same time superior fit in modeling market and credit VaR. However, additional research is needed. Future work is this direction will be construction of models that capture the features of financial empirical data such as heavy tails, time-varying volatility, and short and long range dependence.[54] In order to describe thick tails, one can employ the conditional heteroskedastic models based on the stable hypothesis.[55] ARMA-stable-GARCH models can incorporate both heavy tails and time-varying volatility.[56] The fractional-stable GARCH model can capture all observed phenomena in financial data: heavy tails, time-varying volatility, and short- and long-range dependence. An analysis of VaR estimation with ARMA-α-stable, ARMA-stable-GARCH, and fractional-stable GARCH models will be provided elsewhere.

[54] For some preliminary results see Liu and Brorsen (1995), Mittnik, Rachev and Paolella (1998), Mittnik, Paolella and Rachev (1997, 1998a, b), Panorska, Mittnik and Rachev (1995).

[55] These models are named as ARMA-α-stable models.

[56] For discussion of stable-GARCH models see Panorska, Mittnik and Rachev (1995) and Mittnik, Paolella and Rachev (1997).

Appendix A. Stable modeling of credit returns in figures

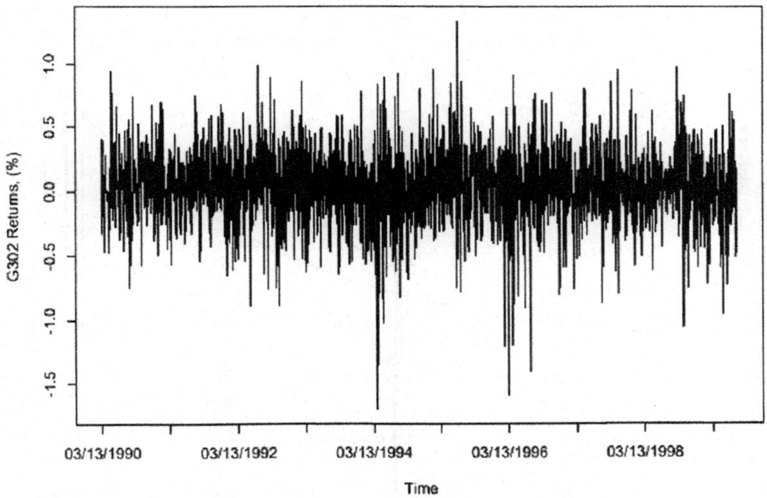

Fig. A.1. G302 daily returns.

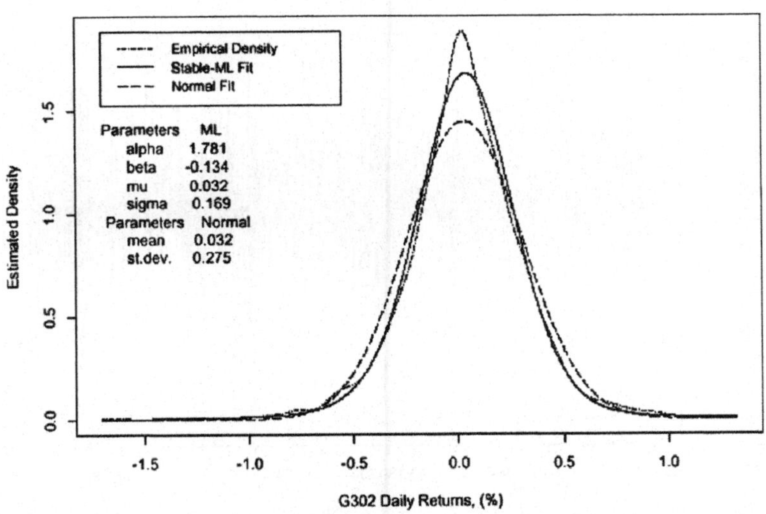

Fig. A.2. Stable and normal fitting of the G302 index.

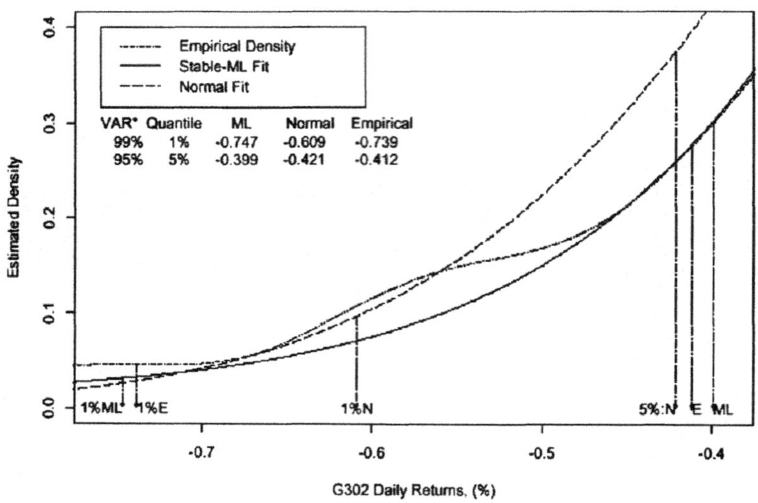

Fig. A.3. VaR estimation for the G302 index.

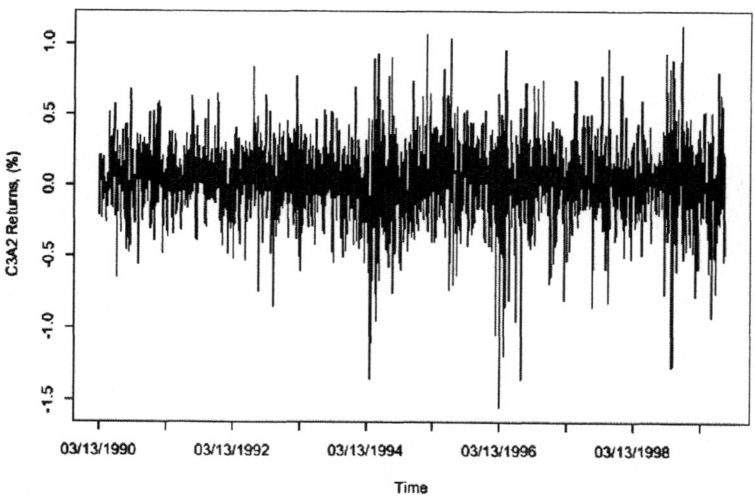

Fig. A.4. C3A2 daily returns.

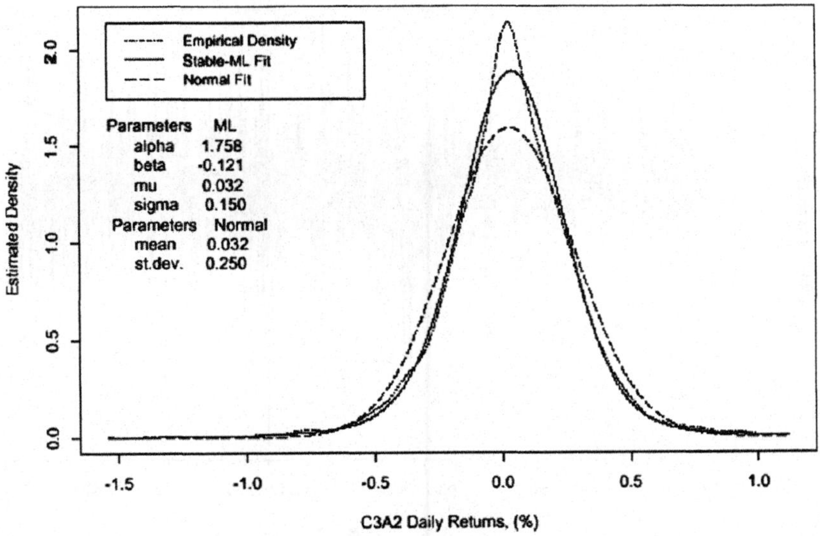

Fig. A.5. Stable and normal fitting of the C3A2 index.

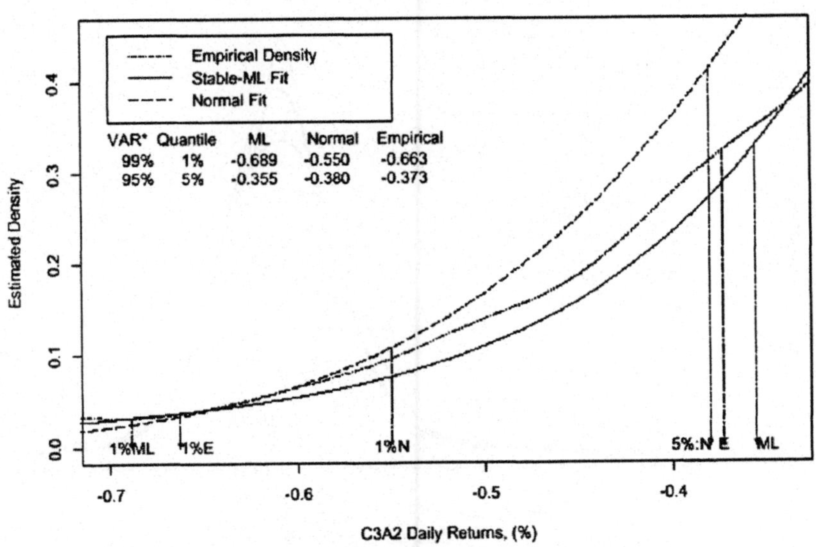

Fig. A.6. VaR estimation for the C3A2 index.

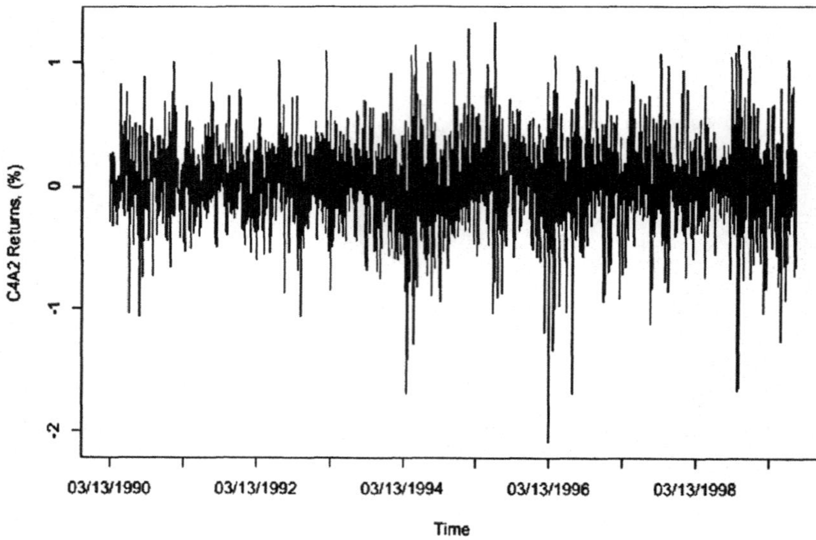

Fig. A.7. C4A2 daily returns.

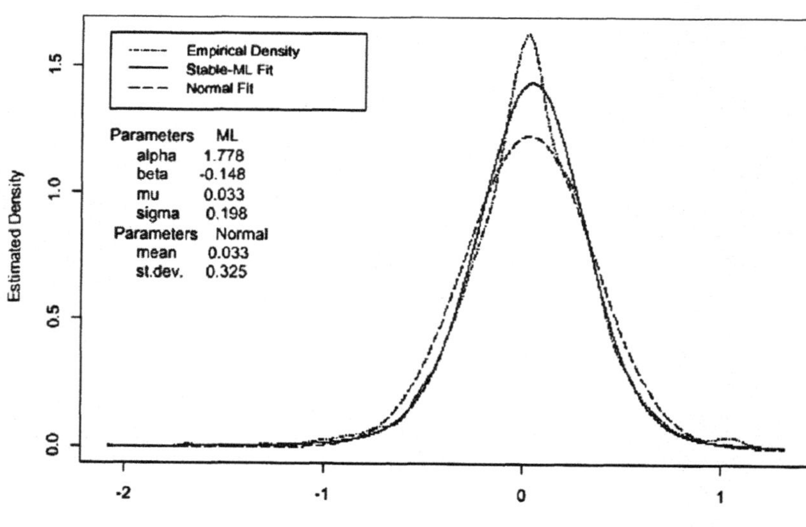

Fig. A.8. Stable and normal fitting of the C4A2 index.

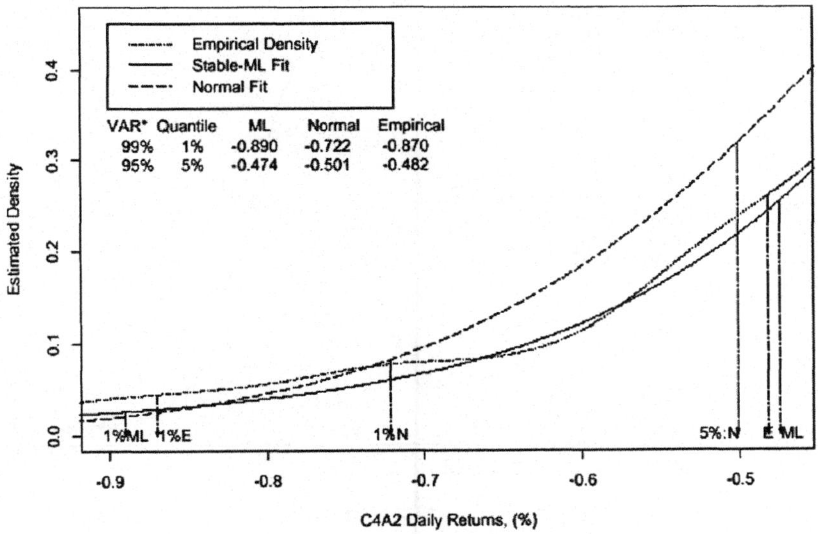

Fig. A.9. VaR estimation for the C4A2 index.

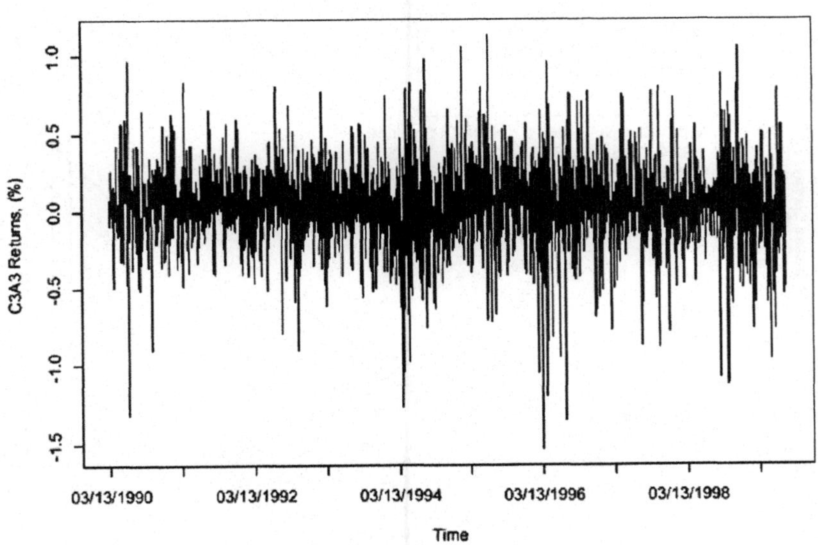

Fig. A.10. C3A3 daily returns.

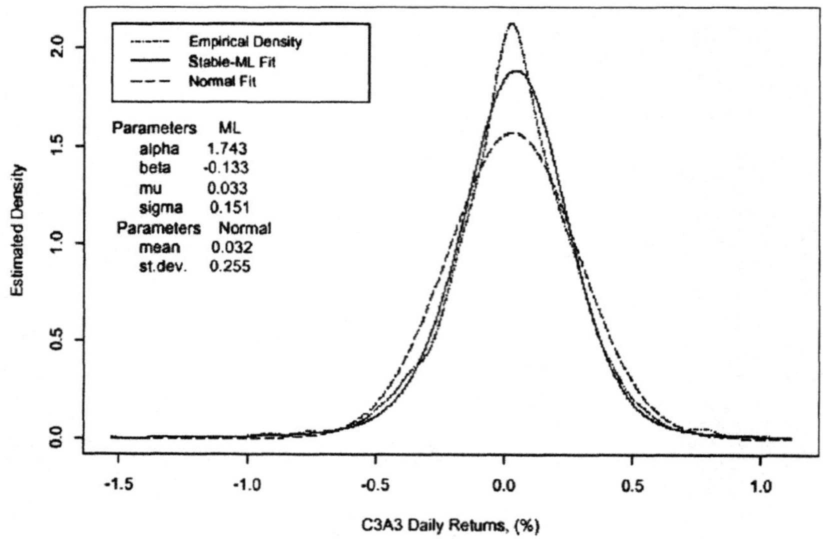

Fig. A.11. Stable and normal fitting of the C3A3 index.

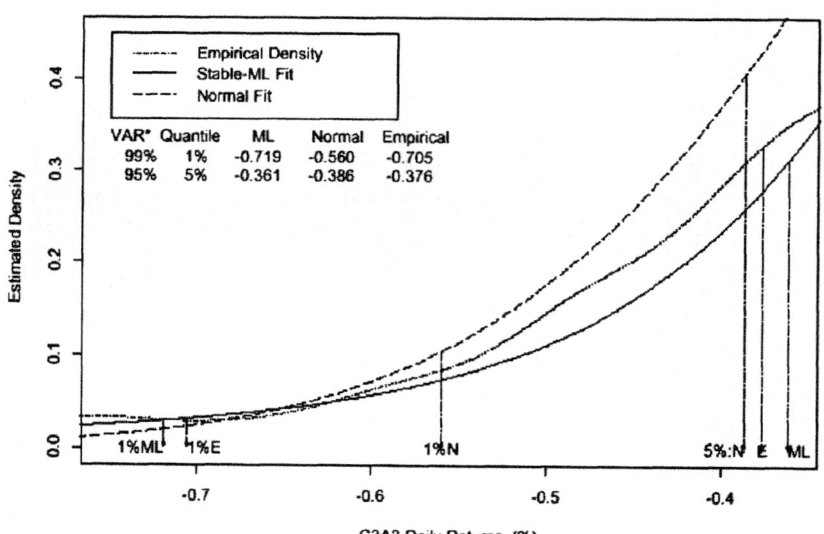

Fig. A.12. VaR estimation for the C3A3 index.

Appendix B. Tables

Table B.1
Deviations of VaR estimates for bond indices

Index	99% VaR_{model} − 99% $VaR_{empirical}$		95% VaR_{model} − 95% $VaR_{empirical}$	
	Normal	Stable	Normal	Stable
G102	−0.044	0.033	0.005	−0.008
G202	−0.072	0.058	0.003	−0.020
G302	−0.130	0.008	0.009	−0.013
G402	−0.143	0.004	0.000	−0.027
C1A1	−0.042	0.046	0.001	−0.010
C2A1	−0.060	0.072	0.014	−0.008
C3A1	−0.139	0.047	0.009	−0.016
C4A1	−0.171	0.048	0.014	−0.013
C1A2	−0.036	0.048	0.002	−0.007
C2A2	−0.061	0.062	0.007	−0.010
C3A2	−0.113	0.026	0.007	−0.018
C4A2	−0.148	0.020	0.019	−0.008
C1A3	−0.030	0.049	−0.001	−0.010
C2A3	−0.079	0.061	0.007	−0.012
C3A3	−0.145	0.014	0.010	−0.015
C4A3	−0.152	0.056	0.027	−0.002
C1A4	−0.031	0.028	0.031	−0.005
C2A4	−0.086	0.033	0.025	−0.015
C3A4	−0.166	0.030	0.014	−0.018
C4A4	−0.160	0.098	0.019	−0.016
H0A1	−0.154	0.013	0.019	−0.013

Table B.2
Stable VaR estimates for bond indices with fixed α

Bond index	99% VaR		95%	
	Different α	Fixed $\alpha = 1.708$	Different α	Fixed $\alpha = 1.708$
C1A1	0.284	0.257	0.119	0.116
C2A1	0.509	0.494	0.236	0.233
C3A1	0.734	0.732	0.353	0.351
C4A1	0.931	0.979	0.467	0.471
C1A2	0.285	0.273	0.125	0.123
C2A2	0.505	0.517	0.244	0.245
C3A2	0.689	0.747	0.355	0.360
C4A2	0.890	1.003	0.474	0.485
C1A3	0.286	0.277	0.125	0.124
C2A3	0.530	0.523	0.248	0.247
C3A3	0.719	0.763	0.361	0.365
C4A3	0.949	1.022	0.485	0.491
C1A4	0.290	0.260	0.119	0.116
C2A4	0.511	0.471	0.228	0.224
C3A4	0.741	0.716	0.343	0.340
C4A4	0.960	0.934	0.451	0.447

Table B.3
Coefficients of OLS regressions

Dependent variable	Variables	Coeff.	Dependent variable	Variables	Coeff.
R_{C1A1}	C	0.004723	R_{C1A3}	C	0.003887
	R_{G102}	0.882424		R_{G102}	0.946025
R_{C2A1}	C	0.006183	R_{C2A3}	C	0.005709
	R_{G202}	0.770132		R_{G202}	0.816271
R_{C3A1}	C	0.005550	R_{C3A3}	C	0.005051
	R_{G302}	0.835640		R_{G302}	0.853295
R_{C4A1}	C	0.003735	R_{C4A3}	C	0.003806
	R_{G402}	0.847226		R_{G402}	0.877039
R_{C1A2}	C	0.003357	R_{C1A4}	C	0.006401
	R_{G102}	0.951165		R_{G102}	0.874032
R_{C2A2}	C	0.005733	R_{C2A4}	C	0.009603
	R_{G202}	0.808308		R_{G202}	0.760162
R_{C3A2}	C	0.004730	R_{C3A4}	C	0.008296
	R_{G302}	0.853315		R_{G302}	0.804311
R_{C4A2}	C	0.004118	R_{C4A4}	C	0.007725
	R_{G402}	0.868154		R_{G402}	0.803091

Table B.4
GARCH-normal coefficients

Dependent variable	Variables	Coeff.	Std. errors	Variance equation		
				Variables	Coeff.	Std. errors
R_{C1A1}	C	0.003581	0.000525	C	2.50E−05	2.02E−06
	R_{G102}	0.937996	0.004895	ARCH(1)	0.116681	0.003541
				GARCH(1)	0.885367	0.002486
R_{C2A1}	C	0.004948	0.000981	C	7.67E−05	5.76E−06
	R_{G202}	0.838944	0.003466	ARCH(1)	0.130119	0.005720
				GARCH(1)	0.870004	0.004714
R_{C3A1}	C	0.004199	0.001331	C	0.000152	1.66E−05
	R_{G302}	0.893949	0.003650	ARCH(1)	0.130479	0.003970
				GARCH(1)	0.866746	0.002716
R_{C4A1}	C	0.004014	0.001539	C	0.000355	3.22E−05
	R_{G402}	0.887583	0.003538	ARCH(1)	0.153744	0.008756
				GARCH(1)	0.830941	0.008452
R_{C1A2}	C	0.002746	0.000411	C	4.77E−06	7.97E−07
	R_{G102}	0.946016	0.003830	ARCH(1)	0.096428	0.002594
				GARCH(1)	0.914737	0.002586
R_{C2A2}	C	0.004229	0.000885	C	1.34E−05	2.38E−06
	R_{G202}	0.890123	0.003547	ARCH(1)	0.056718	0.002501
				GARCH(1)	0.943510	0.001441
R_{C3A2}	C	0.002970	0.001078	C	0.000609	4.66E−05
	R_{G302}	0.894861	0.003899	ARCH(1)	0.289805	0.017996
				GARCH(1)	0.669240	0.015835
R_{C4A2}	C	0.003420	0.001329	C	0.000302	2.42E−05
	R_{G402}	0.918195	0.003240	ARCH(1)	0.180168	0.009135
				GARCH(1)	0.817444	0.007086
R_{C1A3}	C	0.002271	0.000421	C	7.06E−06	9.92E−07
	R_{G102}	1.003079	0.003215	ARCH(1)	0.137045	0.003494
				GARCH(1)	0.887812	0.002061
R_{C2A3}	C	0.005204	0.000664	C	2.01E−05	3.21E−06
	R_{G202}	0.903683	0.002247	ARCH(1)	0.124285	0.004417
				GARCH(1)	0.905287	0.002271
R_{C3A3}	C	0.005840	0.001114	C	0.000223	2.20E−05
	R_{G302}	0.915408	0.003059	ARCH(1)	0.253670	0.007199
				GARCH(1)	0.777935	0.004480
R_{C4A3}	C	0.004076	0.001308	C	0.000792	3.16E−05
	R_{G402}	0.942102	0.002728	ARCH(1)	0.401945	0.015612
				GARCH(1)	0.639974	0.007830
R_{C1A4}	C	0.002450	0.00570	C	−3.27E−06	5.58E−07
	R_{G102}	1.036861	0.003468	ARCH(1)	0.101918	0.001997
				GARCH(1)	0.945209	0.000666
R_{C2A4}	C	0.007017	0.000839	C	5.77E−05	3.97E−06
	R_{G202}	0.879618	0.003199	ARCH(1)	0.231563	0.006013
				GARCH(1)	0.841086	0.002770
R_{C3A4}	C	0.007452	0.001276	C	3.99E−05	7.17E−06
	R_{G302}	0.893645	0.003645	ARCH(1)	0.101316	0.003132
				GARCH(1)	0.907304	0.002295
R_{C4A4}	C	0.007402	0.001393	C	0.000194	1.72E−05
	R_{G402}	0.887104	0.002892	ARCH(1)	0.179030	0.005716
				GARCH(1)	0.840838	0.003809

Appendix C. OLS credit risk evaluation for portfolio assets in figures

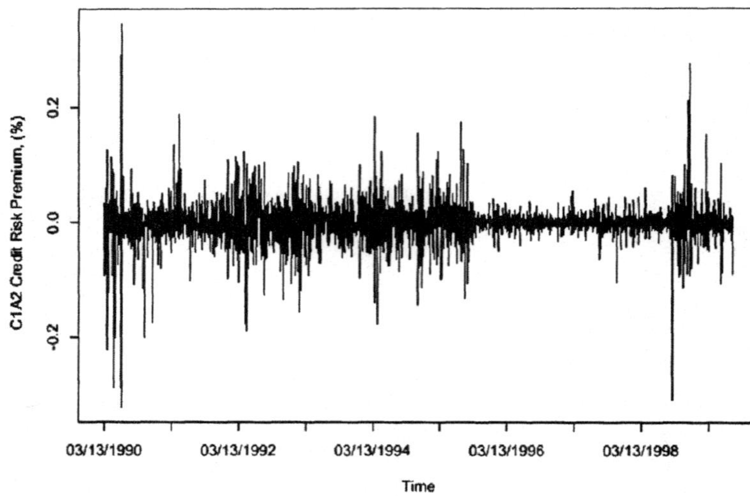

Fig. C.1. OLS credit risk premium for the C1A2 bond index.

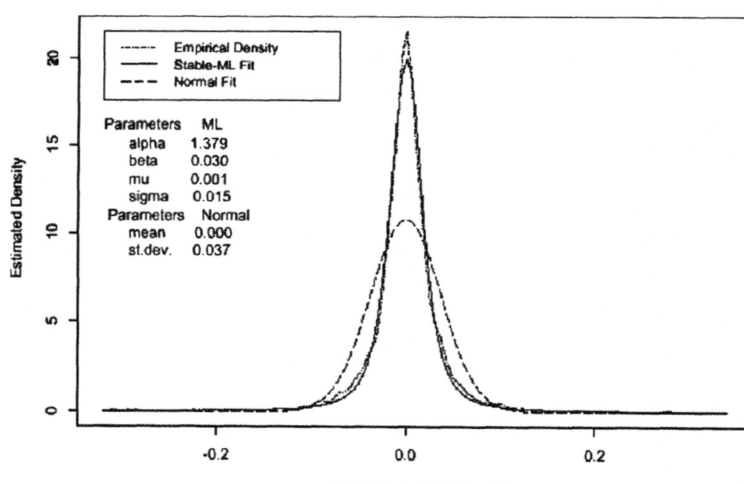

Fig. C.2. Stable and normal fitting of C1A2 OLS-credit-risks.

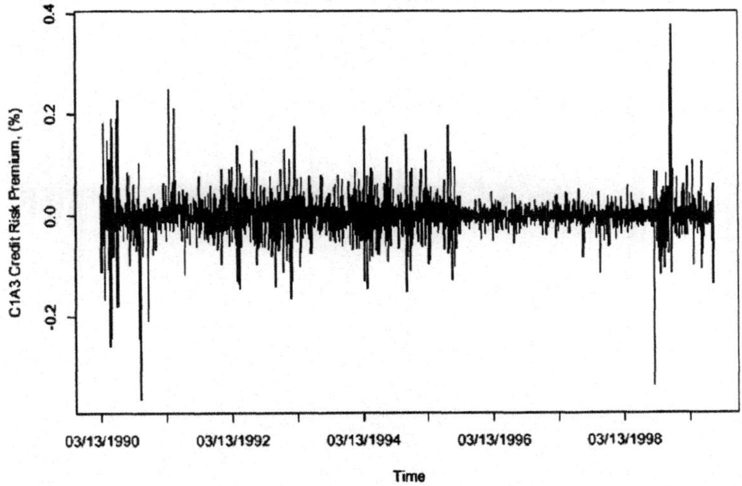

Fig. C.3. OLS credit risk premium for the C1A3 bond index.

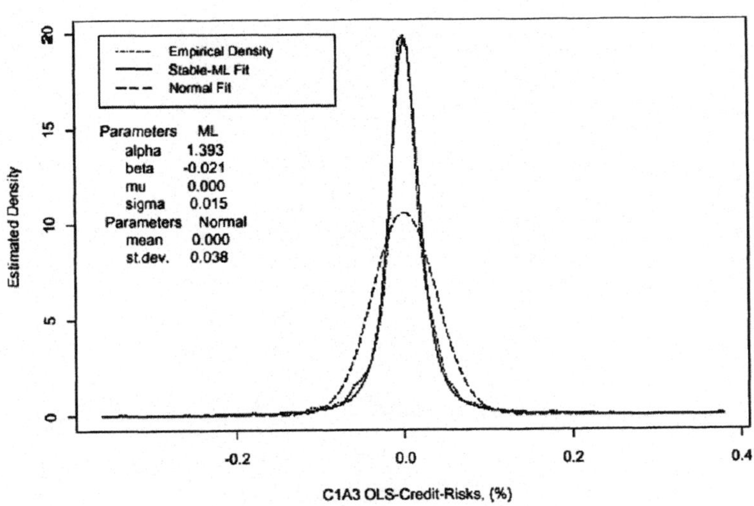

Fig. C.4. Stable and normal fitting of C1A3 OLS-credit-risks.

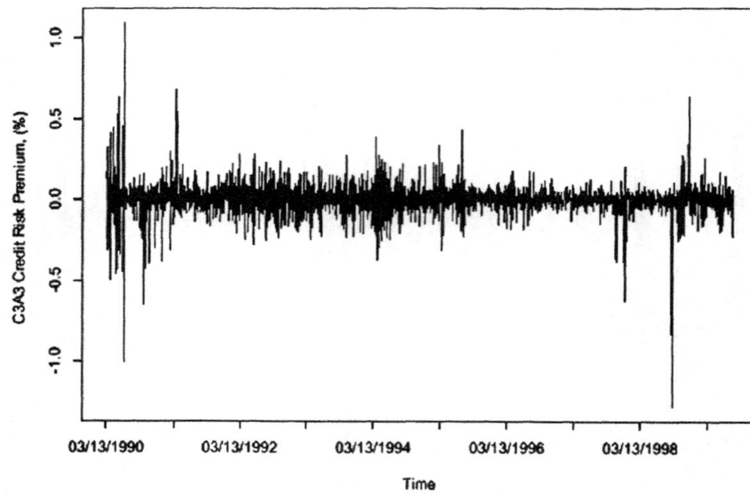

Fig. C.5. OLS credit risk premium for the C3A3 bond index.

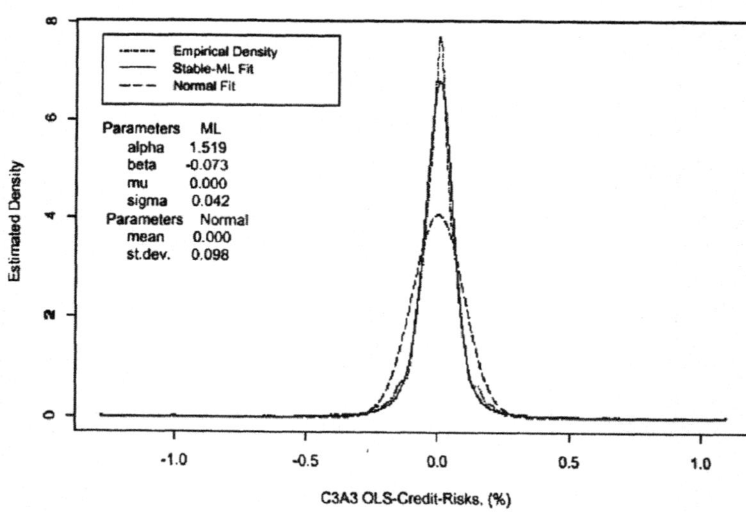

Fig. C.6. Stable and normal fitting of C3A3 OLS-credit-risks.

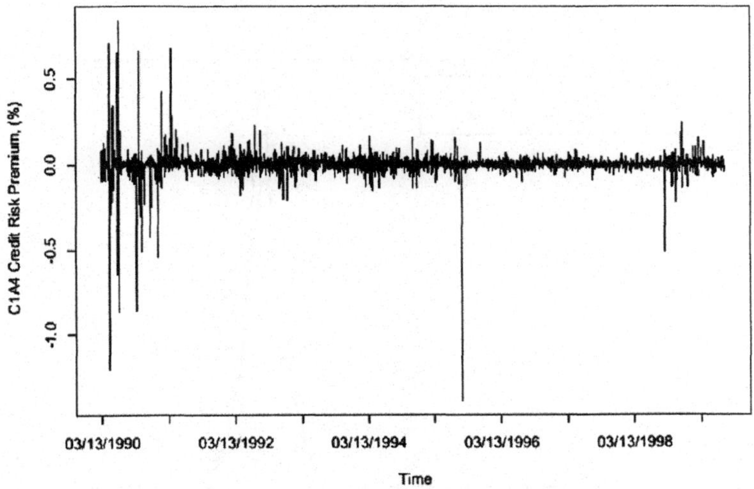

Fig. C.7. OLS credit risk premium for the C1A4 bond index.

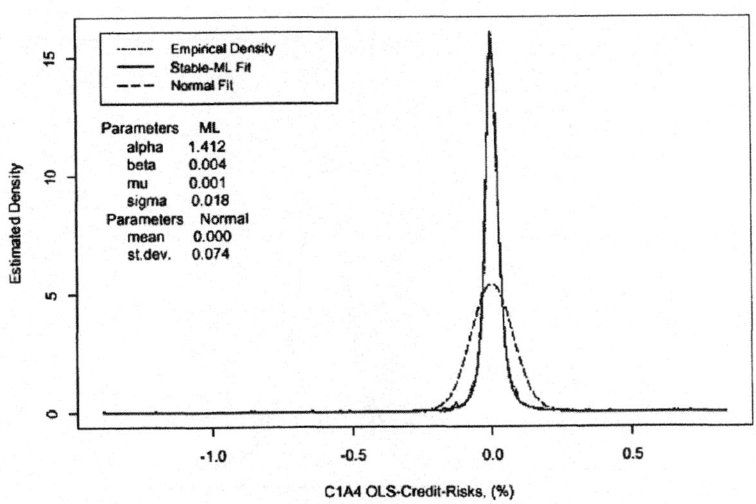

Fig. C.8. Stable and normal fitting of C1A4 OLS-credit-risks.

Appendix D. GARCH credit risk evaluation for portfolio assets in figures

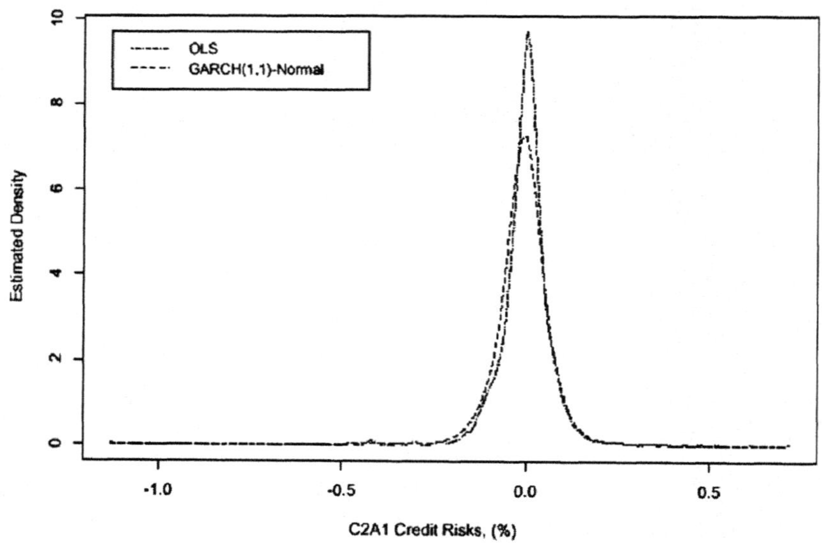

Fig. D.1. C2A1 credit risks: OLS and GARCH.

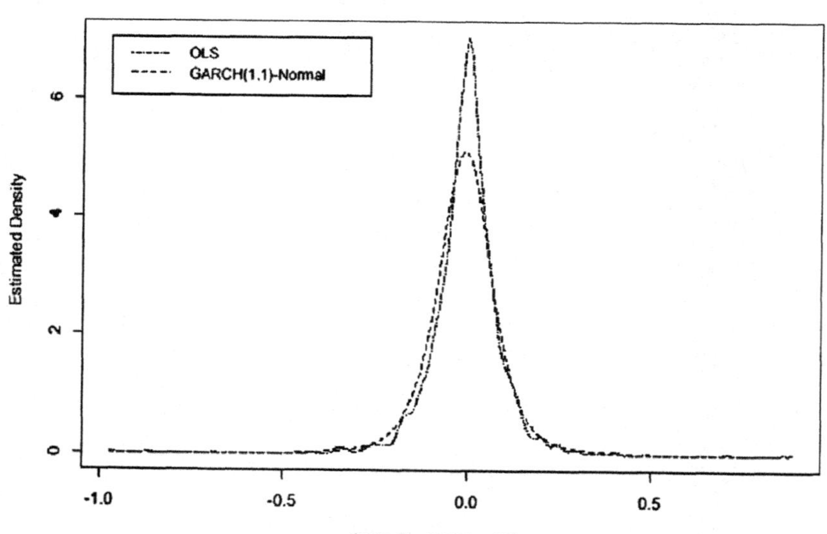

Fig. D.2. C3A1 credit risks: OLS and GARCH.

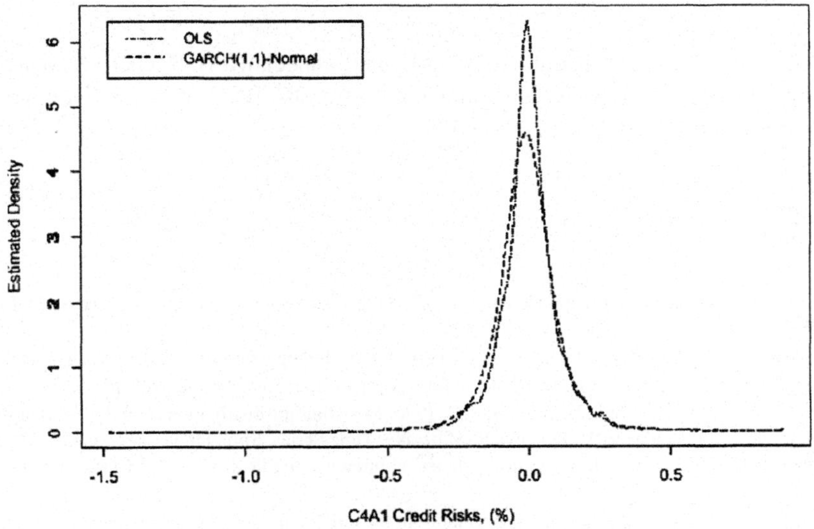

Fig. D.3. C4A1 credit risks: OLS and GARCH.

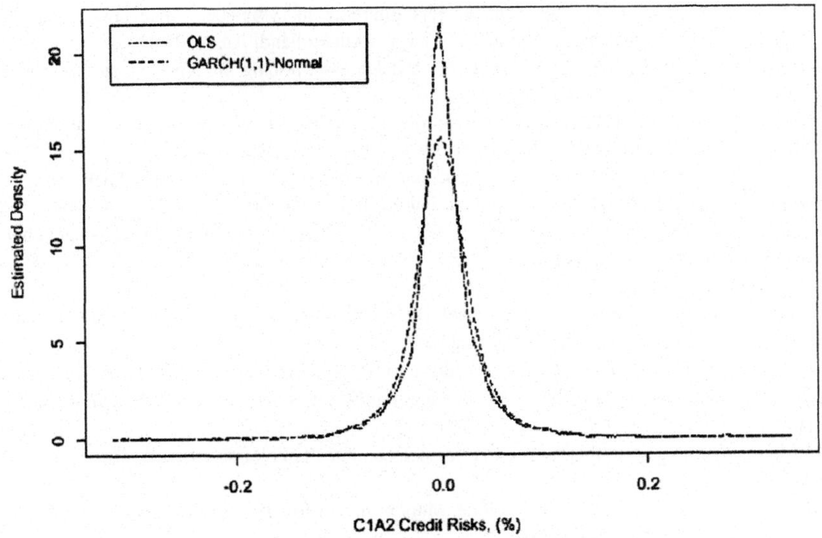

Fig. D.4. C1A2 credit risks: OLS and GARCH.

Acknowledgments

We thank C. Marinelli of Columbia University and, especially, B. Racheva-Iotova of Bravo Consulting Group for computational assistance. We also thank Kristina Tetereva of University of Karlsruhe, Germany, for providing data.

References

Arad, R.W., 1980. Parameter estimation for symmetric stable distributions. International Economic Review 21, 209–220.

Basle Committee on Banking Supervision, 1999. Credit risk modelling: Current practices and applications.

Cheng, B., Rachev, S.T., 1995. Multivariate stable futures prices. Journal of Mathematical Finance 5, 133–153.

Chobanov, G., Mateev, P., Mittnik, S., Rachev, S.T., 1996. Modeling the distribution of highly volatile exchange-rate time series. In: Robinson, P., Rosenblatt, M. (Eds.), Time Series. Springer-Verlag, pp. 130–144.

DuMouchel, W.H., 1971. Stable distributions in statistical inference. Ph.D. dissertation. Department of Statistics, Yale University.

DuMouchel, W.H., 1973. On the asymptotic normality of the maximum-likelihood estimate when sampling from a stable distribution. The Annals of Statistics 1, 948–957.

Engle, R.F., Kroner, K., 1995. Multivariate simultaneous generalized GARCH. Econometric Theory, 122–150.

Fama, E., 1965. The behavior of stock market prices. Journal of Business 38, 34–105.

Fama, E., Roll, R., 1971. Parameter estimates for symmetric stable distributions. Journal of the American Statistical Association 66, 331–338.

Federal Reserve System Task Force on Internal Credit Risk Models, 1998. Credit risk models at major U.S. banking institutions: Current state of the art and implications for assessment of capital adequacy.

Feuerverger, A., McDunnough, P., 1981. On efficient inference in symmetric stable laws and processes. In: Csörgö, M., et al. (Eds.), Statistics and Related Topics. North-Holland, Amsterdam.

Gamrowski, B., Rachev, S.T., 1994. Stable models in testable asset pricing. In: Approximation, Probability, Related Fields. Plenum Press, New York, pp. 223–236.

Gamrowski, B., Rachev, S.T., 1995a. A testable version of the Pareto-stable CAPM. Technical Report 292. Department of Statistics, Applied Probability, University of California, Santa Barbara.

Gamrowski, B., Rachev, S.T., 1995b. Financial models using stable laws. In: Prohorov, Yu.V. (Ed.), Probability Theory, its Applications. In: Surveys in Applied, Industrial Mathematics, Vol. 2, pp., 556–604.

Gamrowski, B., Rachev, S.T., 1996. Testing the validity of value at risk measures. In: Heyde, Prohorov, Pyke, Rachev (Eds.), Athens Conference on Applied Probability and Time Series, Vol. I: Applied Probability. Springer-Verlag, pp. 307–320.

Götzenberger, G., Rachev, S.T., Schwartz, E., 1999. Performance measurements: The stable Paretian approach. Working Paper. University of Karlsruhe, Germany.

Gupton, G.M., Finger, C.C., Bhatia, M., 1997. CreditMetrics™ – Technical Document. JP Morgan, New York.

Heathcote, C.R., Cheng, B., Rachev, S.T., 1995. Testing multivariate symmetry. Journal of Multivariate Analysis 54, 91–112.

Hill, B.M., 1975. A simple general approach to inference about the tail of a distribution. The Annals of Statistics 3 (5), 1163–1174.

Jarrow, R.A., Lando, D., Turnbull, S.M., 1997. A Markov model for the term structure of credit risk spreads. Review of Financial Studies 10 (2), 481–523.

JP Morgan, 1995. RiskMetrics, 3rd edition. JP Morgan.

Klebanov, L.B., Melamed, J.A., Rachev, S.T., 1994. On the joint estimation of stable law parameters. In: Approximation, Probability, and Related Fields. Plenum Press, New York, pp. 315–320.

Kogon, S.M., Williams, D.B., 1998. Characteristic function based estimation of stable distribution parameters. In: Adler, R., et al. (Eds.), A Practical Guide to Heavy Tails: Statistical Techniques, Applications. Birkhäuser, Boston, pp. 311–335.

Koutrouvelis, I.A., 1980. Regression-type estimation of the parameters of stable laws. Journal of American Statistical Association 75, 918–928.

Koutrouvelis, I.A., 1981. An iterative procedure for the estimation of the parameters of stable laws. Communications in Statistics. Simulation and Computation B10, 17–28.

Kozubowski, T.J., Rachev, S.T., 1994. The theory of geometric stable distributions and its use in modeling financial data. European Journal of Operations Research: Financial Modeling 74, 310–324.

Liu, S.-M., Brorsen, B.W., 1995. Maximum likelihood estimation of a GARCH-stable model. Journal of Applied Econometrics 10, 273–285.

Longerstaey, J., Zangari, P., 1996. RiskMetrics™ Technical Document, 4th edition. JP Morgan, New York.

Mandelbrot, B.B., 1962. Sur certain prix spéculatifs: Faits empiriques et modéle basé sur les processes stables additifs de Paul Lévy. Comptes Rendus 254, 3968–3970.

Mandelbrot, B.B., 1963a. New methods in statistical economics. Journal of Political Economy 71, 421–440.

Mandelbrot, B.B., 1963b. The variation of certain speculative prices. Journal of Business 26, 394–419.

Mandelbrot, B.B., 1967. The valuation of some other speculative prices. Journal of Business 40, 393–413.

McCulloch, J.H., 1986. Simple consistent estimators of stable distribution parameters. Communications in Statistics. Simulation and Computation 15, 1109–1136.

McCulloch, J.H., 1996. Financial applications of stable distributions. In: Maddala, G.S., Rao, C.R. (Eds.), Handbook of Statistics – Statistical Methods in Finance. Elsevier, Amsterdam, Vol. 14, pp. 393–425.

Mittnik, S., Paolella, M.S., Rachev, S.T., 1997. Modeling the persistence of conditional volatilities with GARCH-stable processes. Manuscript. Institute of Statistics, Econometrics, University of Kiel, Germany.

Mittnik, S., Paolella, M.S., Rachev, S.T., 1998a. Unconditional and conditional distributional models for the Nikkei's index. Asia-Pacific Financial Markets 5, 99–128.

Mittnik, S., Paolella, M.S., Rachev, S.T., 1998b. The prediction of down-side risk with GARCH-stable models. Technical Report. Institute of Statistics, Econometrics, University of Kiel, Germany.

Mittnik, S., Paolella, M.S., Rachev, S.T., 1998c. A tail estimator for the index of the stable Paretian distribution. Communications in Statistics. Theory and Methods 27, 1239–1262.

Mittnik, S., Rachev, S.T., 1991. Alternate multivariate stable distributions, their applications to financial modeling. In: Cambanis, S., et al. (Eds.), Stable Processes and Related Topics. Birkhäuser, Boston, pp. 107–119.

Mittnik, S., Rachev, S.T., 1993a. Modeling asset returns with alternative stable distributions. Econometric Reviews 12 (3), 261–330.

Mittnik, S., Rachev, S.T., 1993b. Reply to comments on "Modeling asset returns with alternate stable laws" and some extensions. Econometric Reviews 12 (3), 347–389.

Mittnik, S., Rachev, S.T., 1996. Tail estimation of the stable index α. Applied Mathematics Letters 9, 53–56.

Mittnik, S., Rachev, S.T., Chenyao, D., 1996. Distribution of exchange rates: A geometric summation-stable model. In: Proceedings of the Seminar on Data Analysis. Sozopol, Bulgaria, September 12–17, 1996.

Mittnik, S., Rachev, S.T., Doganoglu, T., Chenyao, D., 1997. Maximum likelihood estimation of stable Paretian models. Manuscript. Institute of Statistics, Econometrics, University of Kiel, Germany.

Mittnik, S., Rachev, S.T., Paolella, M.S., 1998. Stable Paretian modeling in finance: Some empirical and theoretical aspects. In: Adler, R., et al. (Eds.), A Practical Guide to Heavy Tails: Statistical Techniques and Applications. Birkhäuser, Boston, pp. 79–110.

Nolan, J.P., 1997. Numerical computation of stable densities, distribution functions. Communications in Statistics. Stochastic Models 13 (4), 759–774.

Nolan, J.P., 1998a. Maximum likelihood estimation and diagnostics for stable distributions. Working Paper. Department of Mathematics, Statistics, American University.

Nolan, J.P., 1998b. Parameterizations of stable distributions. Statistics & Probability Letters 38, 187–195.

Panorska, A.K., Mittnik, S., Rachev, S.T., 1995. Stable GARCH models for financial time series. Applied Mathematics Letters 815, 33–37.

Paulson, A.S., Holcomb, E.W., Leitch, R.A., 1975. The estimation of the parameters of the stable laws. Biometrica 62, 163–170.

Pickands, J., 1975. Statistical inference using extreme order statistics. The Annals of Statistics 3 (1), 119–131.

Press, J.S., 1972a. Estimation of univariate and multivariate stable distributions. Journal of American Statistical Association 67 (340), 842–846.

Press, J.S., 1972b. Applied Multivariate Analysis. Holt, Rinehart, Winston, New York.

Rachev, S.T., Racheva-Jotova, B., Hristov, B., Mandev, I., 1999. Technical documentation of Mercury 1.0, Software package for market risk (VaR) modeling of stable distributed financial returns.

Rachev, S.T., SenGupta, A., 1993. Laplace–Weibull mixtures for modeling price changes. Management Science 1029–1038.

Rachev, S.T., Xin, H., 1993. Test on association of random variables in the domain of attraction of multivariate stable law. Probability and Mathematical Statistics 14 (1), 125–141.

Samorodnitsky, G., Taqqu, M.S., 1994. Stable Non-Gaussian Random Processes: Stochastic Models with Infinite Variance. Chapman & Hall, New York.

Chapter 8

MODELLING DEPENDENCE WITH COPULAS AND APPLICATIONS TO RISK MANAGEMENT

PAUL EMBRECHTS, FILIP LINDSKOG* and ALEXANDER MCNEIL†

Department of Mathematics, ETHZ, CH-8092 Zürich, Switzerland
web: www.math.ethz.ch/finance

Contents

* Research of the second author was supported by Credit Suisse Group, Swiss Re and UBS AG through RiskLab, Switzerland.

† The third author acknowledges financial support from Swiss Re.

Handbook of Heavy Tailed Distributions in Finance, Edited by S.T. Rachev

1. Introduction

Integrated Risk Management (IRM) is concerned with the quantitative description of risks to a financial business. Whereas the qualitative aspects of IRM are extremely important, in the present contribution we only concentrate on the quantitative ones. Since the emergence of Value-at-Risk (VaR) in the early nineties and its various generalisations and refinements more recently, regulators and banking and insurance professionals have build up a huge system aimed at making the global financial system safer. Whereas the steps taken no doubt have been very important towards increasing the overall risk awareness, continuously questions have been asked concerning the quality of the safeguards as constructed.

All quantitative models are based on assumptions vis-a-vis the markets on which they are to be applied. Standard hedging techniques require a high level of liquidity of the underlying instruments, prices quoted for many financial products are often based on "normal" conditions. The latter may be interpreted in a more economic sense, or more specifically referring to the distributional (i.e., normal, Gaussian) behaviour of some underlying data. Especially for IRM, deviations from the "normal" would constitute a prime source of investigation. Hence the classical literature is full of deviations from the so-called random walk (Brownian motion) model and heavy tails appear prominently. The latter has for instance resulted in the firm establishment of Extreme Value Theory (EVT) as a standard tool within IRM. Within market risk management, the so-called stylised facts of econometrics summarise this situation: market data returns tend to be uncorrelated, but dependent, they are heavy tailed, extremes appear in clusters and volatility is random.

Our contribution aims at providing tools for going one step further: what would be the stylised facts of dependence in financial data? Is there a way of understanding so-called normal (i.e., Gaussian) dependence and how can we construct models which allow to go beyond normal dependence? Other problems we would like to understand better are spillover, the behaviour of correlations under extreme market movements, the pros and contras of linear correlation as a measure of dependence, the construction of risk measures for functions of dependent risks. One example concerning the latter is the following: suppose we have two VaR numbers corresponding to two different lines of business. In order to cover the joint position, can we just add the VaR? Under which conditions is this always the upper bound? What can go wrong if these conditions are not fulfilled? A further type of risk where dependence play a crucial role is credit risk: how to define, stress test and model default correlation. The present chapter is not solving the above problem, it presents however tools which are crucial towards the construction of solutions.

The notion we concentrate on is that of copula, well known for some time within the statistics literature. The word copula first appeared in the statistics literature 1959 (Sklar, 1959), although similar ideas and results can be traced back to Hoeffding (1940). Copulas allow us to construct models which go beyond the standard ones at the level of dependence. They yield an ideal tool to stress test a wide variety of portfolios and products in insurance and finance for extreme moves in correlation and more general measures of dependence. As such, they gradually are becoming an extra, but crucial, element of best practice IRM.

After Section 2 in which we define the concept of copula in full generality, we turn in Section 3 to an overview of the most important notions of dependence used in IRM. Sections 4, 5 and 6 introduces the most important families of copulas, their properties both methodological as well as with respect to simulation. Throughout these sections, we stress the importance of the techniques introduced within an IRM framework. Finally in Section 7 we discuss some specific examples.

We would like to stress that the present chapter only gives a first introduction aimed at bringing together from the extensive copula world those results which are immediately usable in IRM. Topics not included are statistical estimation of copulas and the modelling of dependence, through copulas, in a dynamic environment. As such, the topics listed correspond to a one-period point of view. Various extensions are possible; the interested reader is referred to the bibliography for further reading.

2. Copulas

The standard "operational" definition of a copula is a multivariate distribution function defined on the unit cube $[0, 1]^n$, with uniformly distributed marginals. This definition is very natural if one considers how a copula is derived from a continuous multivariate distribution function; indeed in this case the copula is simply the original multivariate distribution function with transformed univariate marginals. This definition however masks some of the problems one faces when constructing copulas using other techniques, i.e., it does not say what is meant by a multivariate distribution function. For that reason, we start with a slightly more abstract definition, returning to the "operational" one later. Below, we follow Nelsen (1999) in concentrating on general multivariate distributions at first and then studying the special properties of the copula subset. For further details we refer to Nelsen (1999).

Throughout this chapter, for a function H, we denote by $\operatorname{Dom} H$ and $\operatorname{Ran} H$ the domain and range respectively of H. Furthermore, a function f will be called increasing whenever $x \leqslant y$ implies that $f(x) \leqslant f(y)$. We may also refer to this as f is nondecreasing. A statement about points of a set $S \subset \mathbb{R}^n$, where S is typically the real line or the unit cube $[0, 1]^n$, is said to hold almost everywhere if the set of points of S where the statement fails to hold has Lebesgue measure zero.

2.1. *Mathematical introduction*

Definition 2.1. Let $S_1, \ldots, S_n$ be nonempty subsets of $\overline{\mathbb{R}}$, where $\overline{\mathbb{R}}$ denotes the extended real line $[-\infty, \infty]$. Let H be a real function of n variables such that $\operatorname{Dom} H = S_1 \times \cdots \times S_n$ and for $\mathbf{a} \leqslant \mathbf{b}$ ($a_k \leqslant b_k$ for all k) let $B = [\mathbf{a}, \mathbf{b}]$ $(= [a_1, b_1] \times \cdots \times [a_n, b_n])$ be an n-box whose vertices are in $\operatorname{Dom} H$. Then the H-volume of B is given by

$$V_H(B) = \sum \operatorname{sgn}(\mathbf{c}) H(\mathbf{c}),$$

where the sum is taken over all vertices $\mathbf{c}$ of B, and $\operatorname{sgn}(\mathbf{c})$ is given by

$$\operatorname{sgn}(\mathbf{c}) = \begin{cases} 1, & \text{if } c_k = a_k \text{ for an even number of } k\text{'s}, \\ -1, & \text{if } c_k = a_k \text{ for an odd number of } k\text{'s}. \end{cases}$$

Equivalently, the H-volume of an n-box $B = [\mathbf{a}, \mathbf{b}]$ is the n-th order difference of H on B

$$V_H(B) = \Delta_{\mathbf{a}}^{\mathbf{b}} H(\mathbf{t}) = \Delta_{a_n}^{b_n} \cdots \Delta_{a_1}^{b_1} H(\mathbf{t}),$$

where the n first order differences are defined as

$$\Delta_{a_k}^{b_k} H(\mathbf{t}) = H(t_1, \ldots, t_{k-1}, b_k, t_{k+1}, \ldots, t_n) - H(t_1, \ldots, t_{k-1}, a_k, t_{k+1}, \ldots, t_n).$$

Definition 2.2. A real function H of n variables is n-increasing if $V_H(B) \geqslant 0$ for all n-boxes B whose vertices lie in $\operatorname{Dom} H$.

Suppose that the domain of a real function H of n variables is given by $\operatorname{Dom} H = S_1 \times \cdots \times S_n$ where each S_k has a smallest element a_k. We say that H is grounded if $H(\mathbf{t}) = 0$ for all $\mathbf{t}$ in $\operatorname{Dom} H$ such that $t_k = a_k$ for at least one k. If each S_k is nonempty and has a greatest element b_k, then H has marginals, and the one-dimensional marginals of H are the functions H_k with $\operatorname{Dom} H_k = S_k$ and with $H_k(x) = H(b_1, \ldots, b_{k-1}, x, b_{k+1}, \ldots, b_n)$ for all x in S_k. Higher-dimensional marginals are defined in an obvious way. One-dimensional marginals are just called marginals.

Lemma 2.1. *Let $S_1, \ldots, S_n$ be nonempty subsets of $\overline{\mathbb{R}}$, and let H be a grounded n-increasing function with domain $S_1 \times \cdots \times S_n$. Then H is increasing in each argument.*

Lemma 2.2. *Let $S_1, \ldots, S_n$ be nonempty subsets of $\overline{\mathbb{R}}$, and let H be a grounded n-increasing function with marginals and domain $S_1 \times \cdots \times S_n$. Then, if $\mathbf{x} = (x_1, \ldots, x_n)$ and $\mathbf{y} = (y_1, \ldots, y_n)$ are any points in $S_1 \times \cdots \times S_n$,*

$$\left|H(\mathbf{x}) - H(\mathbf{y})\right| \leqslant \sum_{k=1}^{n} \left|H_k(x_k) - H_k(y_k)\right|.$$

For the proof, see Schweizer and Sklar (1983).

Definition 2.3. An n-dimensional distribution function is a function H with domain $\overline{\mathbb{R}}^n$ such that H is grounded, n-increasing and $H(\infty, \ldots, \infty) = 1$.

It follows from Lemma 2.1 that the marginals of an n-dimensional distribution function are distribution functions, which we denote $F_1, \ldots, F_n$.

Definition 2.4. An n-dimensional copula is a function C with domain $[0,1]^n$ such that
(1) C is grounded and n-increasing.
(2) C has marginals C_k, $k=1,2,\ldots,n$, which satisfy $C_k(u)=u$ for all u in $[0,1]$.

Note that for any n-copula C, $n \geqslant 3$, each k-dimensional marginal of C is a k-copula. Equivalently, an n-copula is a function C from $[0,1]^n$ to $[0,1]$ with the following properties:
(1) For every $\mathbf{u}$ in $[0,1]^n$, $C(\mathbf{u})=0$ if at least one coordinate of $\mathbf{u}$ is 0, and $C(\mathbf{u})=u_k$ if all coordinates of $\mathbf{u}$ equal 1 except u_k.
(2) For every $\mathbf{a}$ and $\mathbf{b}$ in $[0,1]^n$ such that $a_i \leqslant b_i$ for all i, $V_C([\mathbf{a},\mathbf{b}]) \geqslant 0$.

Since copulas are joint distribution functions (on $[0,1]^n$), a copula C induces a probability measure on $[0,1]^n$ via

$$V_C\big([0,u_1]\times\cdots\times[0,u_n]\big)=C(u_1,\ldots,u_n)$$

and a standard extension to arbitrary (not necessarily n-boxes) Borel subsets of $[0,1]^n$. A standard result from measure theory says that there is a unique probability measure on the Borel subsets of $[0,1]^n$ which coincides with V_C on the set of n-boxes of $[0,1]^n$. This probability measure will also be denoted V_C.

From Definition 2.4 it follows that a copula C is a distribution function on $[0,1]^n$ with uniformly distributed (on $[0,1]$) marginals. The following theorem follows directly from Lemma 2.2.

Theorem 2.1. *Let C be an n-copula. Then for every* $\mathbf{u}$ *and* $\mathbf{v}$ *in* $[0,1]^n$,

$$\big|C(\mathbf{v})-C(\mathbf{u})\big| \leqslant \sum_{k=1}^{n} |v_k-u_k|.$$

Hence C is uniformly continuous on $[0,1]^n$.

2.2. *Sklar's Theorem*

The following theorem is known as Sklar's Theorem. It is perhaps the most important result regarding copulas, and is used in essentially all applications of copulas.

Theorem 2.2. *Let H be an n-dimensional distribution function with marginals $F_1,\ldots,F_n$. Then there exists an n-copula C such that for all* $\mathbf{x}$ *in* $\overline{\mathbb{R}}^n$,

$$H(x_1,\ldots,x_n)=C\big(F_1(x_1),\ldots,F_n(x_n)\big). \tag{2.1}$$

If $F_1,\ldots,F_n$ are all continuous, then C is unique; otherwise C is uniquely determined on $\operatorname{Ran} F_1\times\cdots\times\operatorname{Ran} F_n$. *Conversely, if C is an n-copula and $F_1,\ldots,F_n$ are distribution functions, then the function H defined above is an n-dimensional distribution function with marginals $F_1,\ldots,F_n$.*

For the proof, see Sklar (1996).

From Sklar's Theorem we see that for continuous multivariate distribution functions, the univariate marginals and the multivariate dependence structure can be separated, and the dependence structure can be represented by a copula.

Let F be a univariate distribution function. We define the generalized inverse of F as $F^{-1}(t) = \inf\{x \in \mathbb{R} \mid F(x) \geqslant t\}$ for all t in $[0, 1]$, using the convention $\inf \emptyset = -\infty$.

Corollary 2.1. *Let H be an n-dimensional distribution function with continuous marginals $F_1, \ldots, F_n$ and copula C (where C satisfies* (2.1)*). Then for any $\mathbf{u}$ in* $[0, 1]^n$,

$$C(u_1, \ldots, u_n) = H\big(F_1^{-1}(u_1), \ldots, F_n^{-1}(u_n)\big).$$

Without the continuity assumption, care has to be taken; see Nelsen (1999) or Marshall (1996).

Example 2.1. Let Φ denote the standard univariate normal distribution function and let Φ_R^n denote the standard multivariate normal distribution function with linear correlation matrix R. Then

$$C(u_1, \ldots, u_n) = \Phi_R^n\big(\Phi^{-1}(u_1), \ldots, \Phi^{-1}(u_n)\big)$$

is the Gaussian or normal n-copula.

2.3. The Fréchet–Hoeffding bounds for joint distribution functions

Consider the functions M^n, Π^n and W^n defined on $[0, 1]^n$ as follows:

$$M^n(\mathbf{u}) = \min(u_1, \ldots, u_n),$$
$$\Pi^n(\mathbf{u}) = u_1 \cdots u_n,$$
$$W^n(\mathbf{u}) = \max(u_1 + \cdots + u_n - n + 1, 0).$$

The functions M^n and Π^n are n-copulas for all $n \geqslant 2$ whereas the function W^n is not a copula for any $n \geqslant 3$ as shown in the following example.

Example 2.2. Consider the n-cube $[1/2, 1]^n \subset [0, 1]^n$.

$$\begin{aligned} V_{W^n}\big(\big[\tfrac{1}{2}, 1\big]^n\big) &= \max(1 + \cdots + 1 - n + 1, 0) - n \max\big(\tfrac{1}{2} + 1 + \cdots + 1 - n + 1, 0\big) \\ &\quad + \tbinom{n}{2} \max\big(\tfrac{1}{2} + \tfrac{1}{2} + 1 + \cdots + 1 - n + 1, 0\big) + \cdots \\ &\quad + \max\big(\tfrac{1}{2} + \cdots + \tfrac{1}{2} - n + 1, 0\big) \\ &= 1 - \tfrac{n}{2} + 0 + \cdots + 0. \end{aligned}$$

Hence W^n is not a copula for $n \geqslant 3$.

The following theorem is called the Fréchet–Hoeffding bounds inequality (Fréchet, 1957).

Theorem 2.3. *If C is any n-copula, then for every* $\mathbf{u}$ *in* $[0, 1]^n$,

$$W^n(\mathbf{u}) \leqslant C(\mathbf{u}) \leqslant M^n(\mathbf{u}).$$

For more details, including geometrical interpretations, see Mikusinski, Sherwood and Taylor (1992). Although the Fréchet–Hoeffding lower bound W^n is never a copula for $n \geqslant 3$, it is the best possible lower bound in the following sense.

Theorem 2.4. *For any $n \geqslant 3$ and any* $\mathbf{u}$ *in* $[0, 1]^n$, *there is an n-copula C (which depends on* $\mathbf{u}$*) such that*

$$C(\mathbf{u}) = W^n(\mathbf{u}).$$

For the proof, see Nelsen (1999), p. 42.

We denote by $\overline{C}$ the joint survival function for n random variables with joint distribution function C, i.e., if $(U_1, \ldots, U_n)^{\mathrm{T}}$ has distribution function C, then $\overline{C}(u_1, \ldots, u_n) = \mathbb{P}\{U_1 > u_1, \ldots, U_n > u_n\}$.

Definition 2.5. If C_1 and C_2 are copulas, C_1 is smaller than C_2 (written $C_1 \prec C_2$) if

$$C_1(\mathbf{u}) \leqslant C_2(\mathbf{u}) \quad \text{and} \quad \overline{C}_1(\mathbf{u}) \leqslant \overline{C}_2(\mathbf{u}),$$

for all $\mathbf{u}$ in $[0, 1]^n$.

Note that in the bivariate case, $\overline{C}(u_1, u_2) = 1 - u_1 - u_2 + C(u_1, u_2)$ and hence, $\overline{C}_1(u_1, u_2) \leqslant \overline{C}_2(u_1, u_2)$ if and only if $C_1(u_1, u_2) \leqslant C_2(u_1, u_2)$.

The Fréchet–Hoeffding lower bound W^2 is smaller than every 2-copula, and every n-copula is smaller than the Fréchet–Hoeffding upper bound M^n. This partial ordering of the set of copulas is called a concordance ordering. It is a partial ordering since not every pair of copulas is comparable in this order. However many important parametric families of copulas are totally ordered. We call a one-parameter family $\{C_\theta\}$ positively ordered if $C_{\theta_1} \prec C_{\theta_2}$ whenever $\theta_1 \leqslant \theta_2$. Examples of such one-parameter families will be given later.

2.4. *Copulas and random variables*

Let $X_1, \ldots, X_n$ be random variables with continuous distribution functions $F_1, \ldots, F_n$, and joint distribution function H. Then $(X_1, \ldots, X_n)^{\mathrm{T}}$ has a unique copula C, where C is given by (2.1). The standard copula representation of the distribution of the random vector $(X_1, \ldots, X_n)^{\mathrm{T}}$ then becomes:

$$H(x_1, \ldots, x_n) = \mathbb{P}\{X_1 \leqslant x_1, \ldots, X_n \leqslant x_n\} = C\big(F_1(x_1), \ldots, F_n(x_n)\big).$$

The transformations $X_i \mapsto F_i(X_i)$ used in the above representation are usually referred to as the probability-integral transformations (to uniformity) and form a standard tool in simulation methodology. Since $X_1, \ldots, X_n$ are independent if and only if $H(x_1, \ldots, x_n) = F_1(x_1) \cdots F_n(x_n)$ for all $x_1, \ldots, x_n$ in $\overline{\mathbb{R}}$, the following result follows from Theorem 2.2.

Theorem 2.5. *Let $(X_1, \ldots, X_n)^{\mathrm{T}}$ be a vector of continuous random variables with copula C, then $X_1, \ldots, X_n$ are independent if and only if $C = \Pi^n$.*

One nice property of copulas is that for strictly monotone transformations of the random variables, copulas are either invariant, or change in certain simple ways. Note that if the distribution function of a random variable X is continuous, and if α is a strictly monotone function whose domain contains $\operatorname{Ran} X$, then the distribution function of the random variable $\alpha(X)$ is also continuous.

Theorem 2.6. *Let $(X_1, \ldots, X_n)^{\mathrm{T}}$ be a vector of continuous random variables with copula C. If, for $k = 1, \ldots, n$, α_k is strictly increasing on* $\operatorname{Ran} X_k$, *then also $(\alpha_1(X_1), \ldots, \alpha_n(X_n))^{\mathrm{T}}$ has copula C.*

Proof: Let $F_1, \ldots, F_n$ denote the distribution functions of $X_1, \ldots, X_n$ and let $G_1, \ldots, G_n$ denote the distribution functions of $\alpha_1(X_1), \ldots, \alpha_n(X_n)$, respectively. Let $(X_1, \ldots, X_n)^{\mathrm{T}}$ have copula C, and let $(\alpha_1(X_1), \ldots, \alpha_n(X_n))^{\mathrm{T}}$ have copula C_α. Since α_k is strictly increasing,

$$G_k(x) = \mathbb{P}\{\alpha_k(X_k) \leqslant x\} = \mathbb{P}\{X_k \leqslant \alpha_k^{-1}(x)\} = F_k(\alpha_k^{-1}(x))$$

for any x in $\overline{\mathbb{R}}$, hence

$$\begin{aligned} C_\alpha\big(G_1(x_1), \ldots, G_n(x_n)\big) &= \mathbb{P}\{\alpha_1(X_1) \leqslant x_1, \ldots, \alpha_n(X_n) \leqslant x_n\} \\ &= \mathbb{P}\{X_1 \leqslant \alpha_1^{-1}(x_1), \ldots, X_n \leqslant \alpha_n^{-1}(x_n)\} \\ &= C\big(F_1(\alpha_1^{-1}(x_1)), \ldots, F_n(\alpha_n^{-1}(x_n))\big) \\ &= C\big(G_1(x_1), \ldots, G_n(x_n)\big). \end{aligned}$$

Since $X_1, \ldots, X_n$ are continuous, $\operatorname{Ran} G_1 = \cdots = \operatorname{Ran} G_n = [0, 1]$. Hence it follows that $C_\alpha = C$ on $[0, 1]^n$. □

From Theorem 2.2 we know that the copula function C "separates" an n-dimensional distribution function from its univariate marginals. The next theorem will show that there is also a function, $\widehat{C}$, that separates an n-dimensional survival function from its univariate survival marginals. Furthermore this function can be shown to be a copula, and this survival copula can rather easily be expressed in terms of C and its k-dimensional marginals.

Theorem 2.7. *Let $(X_1, \ldots, X_n)^{\mathrm{T}}$ be a vector of continuous random variables with copula $C_{X_1,\ldots,X_n}$. For $i = 1, \ldots, n$, let α_i be strictly monotone on* $\operatorname{Ran} X_i$, *and let $(\alpha_1(X_1), \ldots, \alpha_n(X_n))^{\mathrm{T}}$ have copula $C_{\alpha_1(X_1),\ldots,\alpha_n(X_n)}$. Furthermore let α_k be strictly decreasing for some k. Without loss of generality let $k = 1$. Then*

$$\begin{aligned}&C_{\alpha_1(X_1),\ldots,\alpha_n(X_n)}(u_1, u_2, \ldots, u_n)\\&\quad= C_{\alpha_2(X_2),\ldots,\alpha_n(X_n)}(u_2, \ldots, u_n) - C_{X_1,\alpha_2(X_2),\ldots,\alpha_n(X_n)}(1 - u_1, u_2, \ldots, u_n).\end{aligned}$$

Proof: For $i = 1, \ldots, n$, let X_i have distribution function F_i and let $\alpha_1(X_i)$ have distribution function G_i. Then

$$\begin{aligned}&C_{\alpha_1(X_1),\alpha_2(X_2),\ldots,\alpha_n(X_n)}\big(G_1(x_1), \ldots, G_n(x_n)\big)\\&\quad= \mathbb{P}\big\{\alpha_1(X_1) \leqslant x_1, \ldots, \alpha_n(X_n) \leqslant x_n\big\}\\&\quad= \mathbb{P}\big\{X_1 > \alpha_1^{-1}(x_1), \alpha_2(X_2) \leqslant x_2, \ldots, \alpha_n(X_n) \leqslant x_n\big\}\\&\quad= \mathbb{P}\big\{\alpha_2(X_2) \leqslant x_2, \ldots, \alpha_n(X_n) \leqslant x_n\big\}\\&\qquad- \mathbb{P}\big\{X_1 \leqslant \alpha_1^{-1}(x_1), \alpha_2(X_2) \leqslant x_2, \ldots, \alpha_n(X_n) \leqslant x_n\big\}\\&\quad= C_{\alpha_2(X_2),\ldots,\alpha_n(X_n)}\big(G_2(x_2), \ldots, G_n(x_n)\big)\\&\qquad- C_{X_1,\alpha_2(X_2),\ldots,\alpha_n(X_n)}\big(F_1\big(\alpha_1^{-1}(x_1)\big), G_2(x_2), \ldots, G_n(x_n)\big)\\&\quad= C_{\alpha_2(X_2),\ldots,\alpha_n(X_n)}\big(G_2(x_2), \ldots, G_n(x_n)\big)\\&\qquad- C_{X_1,\alpha_2(X_2),\ldots,\alpha_n(X_n)}\big(1 - G_1(x_1), G_2(x_2), \ldots, G_n(x_n)\big),\end{aligned}$$

from which the conclusion follows directly. □

By using the two theorems above recursively it is clear that the copula $C_{\alpha_1(X_1),\ldots,\alpha_n(X_n)}$ can be expressed in terms of the copula $C_{X_1,\ldots,X_n}$ and its lower-dimensional marginals. This is exemplified below.

Example 2.3. Consider the bivariate case. Let α_1 be strictly decreasing and let α_2 be strictly increasing. Then

$$\begin{aligned}C_{\alpha_1(X_1),\alpha_2(X_2)}(u_1, u_2) &= u_2 - C_{X_1,\alpha_2(X_2)}(1 - u_1, u_2)\\&= u_2 - C_{X_1,X_2}(1 - u_1, u_2).\end{aligned}$$

Let α_1 and α_2 be strictly decreasing. Then

$$\begin{aligned}C_{\alpha_1(X_1),\alpha_2(X_2)}(u_1, u_2) &= u_2 - C_{X_1,\alpha_2(X_2)}(1 - u_1, u_2)\\&= u_2 - \big(1 - u_1 - C_{X_1,X_2}(1 - u_1, 1 - u_2)\big)\\&= u_1 + u_2 - 1 + C_{X_1,X_2}(1 - u_1, 1 - u_2).\end{aligned}$$

Here $C_{\alpha_1(X_1),\alpha_2(X_2)}$ is the survival copula, $\widehat{C}$, of $(X_1, X_2)^{\mathrm{T}}$, i.e.,

$$\overline{H}(x_1, x_2) = \mathbb{P}\{X_1 > x_1, X_2 > x_2\} = \widehat{C}(\overline{F}_1(x_1), \overline{F}_2(x_2)).$$

Note also that the joint survival function of n $U(0, 1)$ random variables whose joint distribution function is the copula C is $\overline{C}(u_1, \ldots, u_n) = \widehat{C}(1 - u_1, \ldots, 1 - u_n)$.

The mixed k-th order partial derivatives of a copula C, $\partial^k C(\mathbf{u})/\partial u_1 \cdots \partial u_k$, exist for almost all $\mathbf{u}$ in $[0, 1]^n$. For such $\mathbf{u}$, $0 \leqslant \partial^k C(\mathbf{u})/\partial u_1 \cdots \partial u_k \leqslant 1$. For details, see Nelsen (1999, p. 11). With this in mind, let

$$C(u_1, \ldots, u_n) = A_C(u_1, \ldots, u_n) + S_C(u_1, \ldots, u_n),$$

where

$$A_C(u_1, \ldots, u_n) = \int_0^{u_1} \cdots \int_0^{u_n} \frac{\partial^n}{\partial s_1 \cdots \partial s_n} C(s_1, \ldots, s_n)\, \mathrm{d}s_1 \cdots \mathrm{d}s_n,$$
$$S_C(u_1, \ldots, u_n) = C(u_1, \ldots, u_n) - A_C(u_1, \ldots, u_n).$$

Unlike multivariate distributions in general, the marginals of a copula are continuous, hence a copula has no individual points $\mathbf{u}$ in $[0, 1]^n$ for which $V_C(\mathbf{u}) > 0$. If $C = A_C$ on $[0, 1]^n$, then C is said to be absolutely continuous. In this case C has density $\frac{\partial^n}{\partial u_1 \cdots \partial u_n} C(u_1, \ldots, u_n)$. If $C = S_C$ on $[0, 1]^n$, then C is said to be singular, and $\frac{\partial^n}{\partial u_1 \cdots \partial u_n} C(u_1, \ldots, u_n) = 0$ almost everywhere in $[0, 1]^n$. The support of a copula is the complement of the union of all open subsets A of $[0, 1]^n$ with $V_C(A) = 0$. When C is singular its support has Lebesgue measure zero and conversely. However a copula can have full support without being absolutely continuous. Examples of such copulas are so-called Marshall–Olkin copulas which are presented later.

Example 2.4. Consider the bivariate Fréchet–Hoeffding upper bound M given by $M(u, v) = \min(u, v)$ on $[0, 1]^2$. It follows that $\frac{\partial^2}{\partial u \partial v} M(u, v) = 0$ everywhere on $[0, 1]^2$ except on the main diagonal (which has Lebesgue measure zero), and $V_M(B) = 0$ for every rectangle B in $[0, 1]^2$ entirely above or below the main diagonal. Hence M is singular.

One of the main aims of this chapter is to present effective algorithms for random variate generation from the various copula families studied. The properties of the specific copula family is often essential for the efficiency of the corresponding algorithm. We now present a general algorithm for random variate generation from copulas. Note however that in most cases it is not an efficient one to use.

Consider the general situation of random variate generation from the n-copula C. Let

$$C_k(u_1, \ldots, u_k) = C(u_1, \ldots, u_k, 1, \ldots, 1), \quad k = 2, \ldots, n - 1,$$

denote k-dimensional marginals of C, with

$$C_1(u_1) = u_1 \quad \text{and} \quad C_n(u_1, \ldots, u_n) = C(u_1, \ldots, u_n).$$

Let $U_1, \ldots, U_n$ have joint distribution function C. Then the conditional distribution of U_k given the values of $U_1, \ldots, U_{k-1}$, is given by

$$\begin{aligned} C_k(u_k | u_1, \ldots, u_{k-1}) &= \mathbb{P}\{U_k \leqslant u_k | U_1 = u_1, \ldots, U_{k-1} = u_{k-1}\} \\ &= \frac{\partial^{k-1} C_k(u_1, \ldots, u_k)}{\partial u_1 \cdots \partial u_{k-1}} \Big/ \frac{\partial^{k-1} C_{k-1}(u_1, \ldots, u_{k-1})}{\partial u_1 \cdots \partial u_{k-1}}, \end{aligned}$$

given that the numerator and denominator exist and that the denominator is not zero. The following algorithm generates a random variate $(u_1, \ldots, u_n)^{\mathrm{T}}$ from C. As usual, let $U(0, 1)$ denote the uniform distribution on $[0, 1]$.

Algorithm 2.1.

- Simulate a random variate u_1 from $U(0, 1)$.
- Simulate a random variate u_2 from $C_2(\cdot \mid u_1)$.

  ⋮
- Simulate a random variate u_n from $C_n(\cdot \mid u_1, \ldots, u_{n-1})$.

This algorithm is in fact a particular case of what is called "the standard construction". The correctness of the algorithm can be seen from the fact that for independent $U(0, 1)$ random variables $Q_1, \ldots, Q_n$,

$$\left(Q_1, C_2^{-1}(Q_2 | Q_1), \ldots, C_n^{-1}(Q_n | Q_1, C_2^{-1}(Q_2 | Q_1), \ldots)\right)^{\mathrm{T}}$$

has distribution function C. To simulate a value u_k from $C_k(\cdot | u_1, \ldots, u_{k-1})$ in general means simulating q from $U(0, 1)$ from which u_k can be obtained from the equation $q = C_k(u_k | u_1, \ldots, u_{k-1})$ by numerical rootfinding. When $C_k^{-1}(q | u_1, \ldots, u_{k-1})$ has a closed form (and hence there is no need for numerical rootfinding) this algorithm can be recommended.

Example 2.5. Let the copula C be given by $C(u, v) = (u^{-\theta} + v^{-\theta} - 1)^{-1/\theta}$, for $\theta > 0$. Then

$$\begin{aligned} C_{2|1}(v|u) &= \frac{\partial C}{\partial u}(u, v) = -\frac{1}{\theta}\left(u^{-\theta} + v^{-\theta} - 1\right)^{-1/\theta - 1}\left(-\theta u^{-\theta - 1}\right) \\ &= \left(u^{\theta}\right)^{(-1-\theta)/\theta}\left(u^{-\theta} + v^{-\theta} - 1\right)^{-1/\theta - 1} \\ &= \left(1 + u^{\theta}\left(v^{-\theta} - 1\right)\right)^{(-1-\theta)/\theta}. \end{aligned}$$

Solving the equation $q = C_{2|1}(v|u)$ for v yields

$$C_{2|1}^{-1}(q \mid u) = v = \left(\left(q^{-\theta/(1+\theta)} - 1\right)u^{-\theta} + 1\right)^{-1/\theta}.$$

The following algorithm generates a random variate $(u, v)^{\mathrm{T}}$ from the above copula C.

- Simulate two independent random variates u and q from $U(0, 1)$.
- Set $v = ((q^{-\theta/(1+\theta)} - 1)u^{-\theta} + 1)^{-1/\theta}$.

3. Dependence concepts

Copulas provide a natural way to study and measure dependence between random variables. As a direct consequence of Theorem 2.6, copula properties are invariant under strictly increasing transformations of the underlying random variables. Linear correlation (or Pearson's correlation) is most frequently used in practice as a measure of dependence. However, since linear correlation is not a copula-based measure of dependence, it can often be quite misleading and should not be taken as the canonical dependence measure. Below we recall the basic properties of linear correlation, and then continue with some copula based measures of dependence.

3.1. Linear correlation

Definition 3.1. Let $(X, Y)^{\mathrm{T}}$ be a vector of random variables with nonzero finite variances. The linear correlation coefficient for $(X, Y)^{\mathrm{T}}$ is

$$\rho(X, Y) = \frac{\mathrm{Cov}(X, Y)}{\sqrt{\mathrm{Var}(X)}\sqrt{\mathrm{Var}(Y)}}, \tag{3.1}$$

where $\mathrm{Cov}(X, Y) = \mathbb{E}(XY) - \mathbb{E}(X)\mathbb{E}(Y)$ is the covariance of $(X, Y)^{\mathrm{T}}$, and $\mathrm{Var}(X)$ and $\mathrm{Var}(Y)$ are the variances of X and Y.

Linear correlation is a measure of linear dependence. In the case of perfect linear dependence, i.e., $Y = aX + b$ almost surely for $a \in \mathbb{R} \setminus \{0\}$, $b \in \mathbb{R}$, we have $|\rho(X, Y)| = 1$. More important is that the converse also holds. Otherwise, $-1 < \rho(X, Y) < 1$. Furthermore linear correlation has the property that

$$\rho(\alpha X + \beta, \gamma Y + \delta) = \mathrm{sign}(\alpha\gamma)\rho(X, Y),$$

for $\alpha, \gamma \in \mathbb{R} \setminus \{0\}$, $\beta, \delta \in \mathbb{R}$. Hence linear correlation is invariant under strictly increasing *linear* transformations. Linear correlation is easily manipulated under linear operations. Let A, B be $m \times n$ matrices; $a, b \in \mathbb{R}^m$ and let $\mathbf{X}$, $\mathbf{Y}$ be random n-vectors. Then

$$\mathrm{Cov}(A\mathbf{X} + a, B\mathbf{Y} + b) = A\,\mathrm{Cov}(\mathbf{X}, \mathbf{Y})B^{\mathrm{T}}.$$

From this it follows that for $\alpha \in \mathbb{R}^n$,

$$\operatorname{Var}\left(\alpha^{\mathrm{T}}\mathbf{X}\right) = \alpha^{\mathrm{T}} \operatorname{Cov}(\mathbf{X})\alpha,$$

where $\operatorname{Cov}(\mathbf{X}) := \operatorname{Cov}(\mathbf{X}, \mathbf{X})$. Hence the variance of a linear combination is fully determined by pairwise covariances between the components, a property which is crucial in portfolio theory.

Linear correlation is a popular but also often misunderstood measure of dependence. The popularity of linear correlation stems from the ease with which it can be calculated and it is a natural scalar measure of dependence in elliptical distributions (with well known members such as the multivariate normal and the multivariate t-distribution). However most random variables are not jointly elliptically distributed, and using linear correlation as a measure of dependence in such situations might prove very misleading. Even for jointly elliptically distributed random variables there are situations where using linear correlation, as defined by (3.1), does not make sense. We might choose to model some scenario using heavy-tailed distributions such as t_2-distributions. In such cases the linear correlation coefficient is not even defined because of infinite second moments.

3.2. *Perfect dependence*

For every n-copula C we know from the Fréchet–Hoeffding inequality (Theorem 2.3) that

$$W^n(u_1, \ldots, u_n) \leqslant C(u_1, \ldots, u_n) \leqslant M^n(u_1, \ldots, u_n).$$

Furthermore, for $n = 2$ the upper and lower bounds are themselves copulas and we have seen that W and M are the bivariate distributions functions of the random vectors $(U, 1 - U)^{\mathrm{T}}$ and $(U, U)^{\mathrm{T}}$, respectively, where $U \sim U(0, 1)$ (i.e., U is uniformly distributed on $[0, 1]$). In this case we say that W describes perfect negative dependence and M describes perfect positive dependence.

Theorem 3.1. *Let $(X, Y)^{\mathrm{T}}$ have one of the copulas W or M. Then there exist two monotone functions $\alpha, \beta : \mathbb{R} \to \mathbb{R}$ and a random variable Z so that*

$$(X, Y) =_d \left(\alpha(Z), \beta(Z)\right),$$

with α increasing and β decreasing in the former case (W) and both α and β increasing in the latter case (M). The converse of this result is also true.

For a proof, see Embrechts, McNeil and Straumann (2002). In a different form this result was already in Fréchet (1951).

Definition 3.2. If $(X, Y)^{\mathrm{T}}$ has the copula M then X and Y are said to be comonotonic; if it has the copula W they are said to be countermonotonic.

Note that if any of F and G (the distribution functions of X and Y, respectively) have discontinuities, so that the copula is not unique, then W and M are possible copulas. In the case of F and G being continuous, a stronger version of the result can be stated:

$$C = W \quad \Leftrightarrow \quad Y = T(X) \text{ a.s.}, \ T = G^{-1} \circ (1 - F) \text{ decreasing},$$

$$C = M \quad \Leftrightarrow \quad Y = T(X) \text{ a.s.}, \ T = G^{-1} \circ F \text{ increasing}.$$

Other characterizations of comonotonicity can be found in Denneberg (1994).

3.3. Concordance

Let $(x, y)^{\mathrm{T}}$ and $(\tilde{x}, \tilde{y})^{\mathrm{T}}$ be two observations from a vector $(X, Y)^{\mathrm{T}}$ of continuous random variables. Then $(x, y)^{\mathrm{T}}$ and $(\tilde{x}, \tilde{y})^{\mathrm{T}}$ are said to be concordant if $(x - \tilde{x})(y - \tilde{y}) > 0$, and discordant if $(x - \tilde{x})(y - \tilde{y}) < 0$.

The following theorem can be found in Nelsen (1999, p. 127). Many of the results in this section are direct consequences of this theorem.

Theorem 3.2. *Let $(X, Y)^{\mathrm{T}}$ and $(\widetilde{X}, \widetilde{Y})^{\mathrm{T}}$ be independent vectors of continuous random variables with joint distribution functions H and $\widetilde{H}$, respectively, with common marginals F (of X and $\widetilde{X}$) and G (of Y and $\widetilde{Y}$). Let C and $\widetilde{C}$ denote the copulas of $(X, Y)^{\mathrm{T}}$ and $(\widetilde{X}, \widetilde{Y})^{\mathrm{T}}$, respectively, so that $H(x, y) = C(F(x), G(y))$ and $\widetilde{H}(x, y) = \widetilde{C}(F(x), G(y))$. Let Q denote the difference between the probability of concordance and discordance of $(X, Y)^{\mathrm{T}}$ and $(\widetilde{X}, \widetilde{Y})^{\mathrm{T}}$, i.e., let*

$$Q = \mathbb{P}\{(X - \widetilde{X})(Y - \widetilde{Y}) > 0\} - \mathbb{P}\{(X - \widetilde{X})(Y - \widetilde{Y}) < 0\}.$$

Then

$$Q = Q(C, \widetilde{C}) = 4 \iint_{[0,1]^2} \widetilde{C}(u, v)\, \mathrm{d}C(u, v) - 1.$$

Proof: Since the random variables are all continuous,

$$\mathbb{P}\{(X - \widetilde{X})(Y - \widetilde{Y}) < 0\} = 1 - \mathbb{P}\{(X - \widetilde{X})(Y - \widetilde{Y}) > 0\}$$

and hence $Q = 2\mathbb{P}\{(X - \widetilde{X})(Y - \widetilde{Y}) > 0\} - 1$. But

$$\mathbb{P}\{(X - \widetilde{X})(Y - \widetilde{Y}) > 0\} = \mathbb{P}\{X > \widetilde{X}, Y > \widetilde{Y}\} + \mathbb{P}\{X < \widetilde{X}, Y < \widetilde{Y}\},$$

and these probabilities can be evaluated by integrating over the distribution of one of the vectors $(X, Y)^{\mathrm{T}}$ or $(\widetilde{X}, \widetilde{Y})^{\mathrm{T}}$. Hence

$$\mathbb{P}\{X > \widetilde{X}, Y > \widetilde{Y}\} = \mathbb{P}\{\widetilde{X} < X, \widetilde{Y} < Y\}$$

$$= \iint_{\mathbb{R}^2} \mathbb{P}\{\widetilde{X} < x, \widetilde{Y} < y\} \,\mathrm{d}C\big(F(x), G(y)\big)$$
$$= \iint_{\mathbb{R}^2} \widetilde{C}\big(F(x), G(y)\big) \,\mathrm{d}C\big(F(x), G(y)\big).$$

Employing the probability-integral transforms $u = F(x)$ and $v = G(y)$ then yields

$$\mathbb{P}\{X > \widetilde{X}, Y > \widetilde{Y}\} = \iint_{[0,1]^2} \widetilde{C}(u, v) \,\mathrm{d}C(u, v).$$

Similarly,

$$\mathbb{P}\{X < \widetilde{X}, Y < \widetilde{Y}\} = \iint_{\mathbb{R}^2} \mathbb{P}\{\widetilde{X} > x, \widetilde{Y} > y\} \,\mathrm{d}C\big(F(x), G(y)\big)$$
$$= \iint_{\mathbb{R}^2} \big\{1 - F(x) - G(y) + \widetilde{C}\big(F(x), G(y)\big)\big\} \,\mathrm{d}C\big(F(x), G(y)\big)$$
$$= \iint_{[0,1]^2} \big\{1 - u - v + \widetilde{C}(u, v)\big\} \,\mathrm{d}C(u, v).$$

But since C is the joint distribution function of a vector $(U, V)^{\mathrm{T}}$ of $U(0, 1)$ random variables, $\mathbb{E}(U) = \mathbb{E}(V) = 1/2$, and hence

$$\mathbb{P}\{X < \widetilde{X}, Y < \widetilde{Y}\} = 1 - \frac{1}{2} - \frac{1}{2} + \iint_{[0,1]^2} \widetilde{C}(u, v) \,\mathrm{d}C(u, v)$$
$$= \iint_{[0,1]^2} \widetilde{C}(u, v) \,\mathrm{d}C(u, v).$$

Thus

$$\mathbb{P}\{(X - \widetilde{X})(Y - \widetilde{Y}) > 0\} = 2 \iint_{[0,1]^2} \widetilde{C}(u, v) \,\mathrm{d}C(u, v),$$

and the conclusion follows. □

Corollary 3.1. *Let C, $\widetilde{C}$, and Q be as given in Theorem* 3.2. *Then*
(1) *Q is symmetric in its arguments*: $Q(C, \widetilde{C}) = Q(\widetilde{C}, C)$.
(2) *Q is nondecreasing in each argument: if $C \prec C'$, then $Q(C, \widetilde{C}) \leqslant Q(C', \widetilde{C})$.*
(3) *Copulas can be replaced by survival copulas in Q, i.e., $Q(C, \widetilde{C}) = Q(\widehat{C}, \widehat{\widetilde{C}})$.*

The following definition can be found in Scarsini (1984).

Definition 3.3. A real valued measure κ of dependence between two continuous random variables X and Y whose copula is C is a measure of concordance if it satisfies the following properties:
(1) κ is defined for every pair X, Y of continuous random variables.
(2) $-1 \leqslant \kappa_{X,Y} \leqslant 1$, $\kappa_{X,X} = 1$ and $\kappa_{X,-X} = -1$.
(3) $\kappa_{X,Y} = \kappa_{Y,X}$.
(4) If X and Y are independent, then $\kappa_{X,Y} = \kappa_{\Pi} = 0$.
(5) $\kappa_{-X,Y} = \kappa_{X,-Y} = -\kappa_{X,Y}$.
(6) If C and $\widetilde{C}$ are copulas such that $C \prec \widetilde{C}$, then $\kappa_C \leqslant \kappa_{\widetilde{C}}$.
(7) If $\{(X_n, Y_n)\}$ is a sequence of continuous random variables with copulas C_n, and if $\{C_n\}$ converges pointwise to C, then $\lim_{n\to\infty} \kappa_{C_n} = \kappa_C$.

Let κ be a measure of concordance for continuous random variables X and Y. As a consequence of Definition 3.3, if If Y is almost surely an increasing function of X, then $\kappa_{X,Y} = \kappa_M = 1$, and if Y is almost surely a decreasing function of X, then $\kappa_{X,Y} = \kappa_W = -1$. Moreover, if α and β are almost surely strictly increasing functions on $\operatorname{Ran} X$ and $\operatorname{Ran} Y$ respectively, then $\kappa_{\alpha(X),\beta(Y)} = \kappa_{X,Y}$.

3.4. Kendall's tau and Spearman's rho

In this section we discuss two important measures of dependence (concordance) known as Kendall's tau and Spearman's rho. They provide the perhaps best alternatives to the linear correlation coefficient as a measure of dependence for nonelliptical distributions, for which the linear correlation coefficient is inappropriate and often misleading. For more details about Kendall's tau and Spearman's rho and their estimators (sample versions) we refer to Kendall and Stuart (1979), Kruskal (1958), Lehmann (1975), Capéraà and Genest (1993). For other interesting scalar measures of dependence see Schweizer and Wolff (1981).

Definition 3.4. Kendall's tau for the random vector $(X, Y)^{\mathrm{T}}$ is defined as

$$\tau(X, Y) = \mathbb{P}\{(X - \widetilde{X})(Y - \widetilde{Y}) > 0\} - \mathbb{P}\{(X - \widetilde{X})(Y - \widetilde{Y}) < 0\},$$

where $(\widetilde{X}, \widetilde{Y})^{\mathrm{T}}$ is an independent copy of $(X, Y)^{\mathrm{T}}$.

Hence Kendall's tau for $(X, Y)^{\mathrm{T}}$ is simply the probability of concordance minus the probability of discordance.

Theorem 3.3. *Let $(X, Y)^{\mathrm{T}}$ be a vector of continuous random variables with copula C. Then Kendall's tau for $(X, Y)^{\mathrm{T}}$ is given by*

$$\tau(X, Y) = Q(C, C) = 4 \iint_{[0,1]^2} C(u, v)\, \mathrm{d}C(u, v) - 1.$$

Note that the integral above is the expected value of the random variable $C(U, V)$, where $U, V \sim U(0,1)$ with joint distribution function C, i.e., $\tau(X, Y) = 4\mathbb{E}(C(U, V)) - 1$.

Definition 3.5. Spearman's rho for the random vector $(X, Y)^{\mathrm{T}}$ is defined as

$$\rho_S(X, Y) = 3\big(\mathbb{P}\{(X - \widetilde{X})(Y - Y') > 0\} - \mathbb{P}\{(X - \widetilde{X})(Y - Y') < 0\}\big),$$

where $(X, Y)^{\mathrm{T}}$, $(\widetilde{X}, \widetilde{Y})^{\mathrm{T}}$ and $(X', Y')^{\mathrm{T}}$ are independent copies.

Note that $\widetilde{X}$ and Y' are independent. Using Theorem 3.2 and the first part of Corollary 3.1 we obtain the following result.

Theorem 3.4. *Let $(X, Y)^{\mathrm{T}}$ be a vector of continuous random variables with copula C. Then Spearman's rho for $(X, Y)^{\mathrm{T}}$ is given by*

$$\begin{aligned}\rho_S(X, Y) &= 3Q(C, \Pi) = 12 \iint_{[0,1]^2} uv \,\mathrm{d}C(u, v) - 3 \\ &= 12 \iint_{[0,1]^2} C(u, v)\,\mathrm{d}u\,\mathrm{d}v - 3.\end{aligned}$$

Hence, if $X \sim F$ and $Y \sim G$, and we let $U = F(X)$ and $V = G(Y)$, then

$$\begin{aligned}\rho_S(X, Y) &= 12 \iint_{[0,1]^2} uv \,\mathrm{d}C(u, v) - 3 = 12\mathbb{E}(UV) - 3 \\ &= \frac{\mathbb{E}(UV) - 1/4}{1/12} = \frac{\mathrm{Cov}(U, V)}{\sqrt{\mathrm{Var}(U)}\sqrt{\mathrm{Var}(V)}} \\ &= \rho\big(F(X), G(Y)\big).\end{aligned}$$

In the next theorem we will see that Kendall's tau and Spearman's rho are concordance measures according to Definition 3.3.

Theorem 3.5. *If X and Y are continuous random variables whose copula is C, then Kendall's tau and Spearman's rho satisfy the properties in Definition* 3.3 *for a measure of concordance.*

For a proof, see Nelsen (1999, p. 137).

Example 3.1. Kendall's tau and Spearman's rho for the random vector $(X, Y)^{\mathrm{T}}$ are invariant under strictly increasing componentwise transformations. This property does not hold for linear correlation. It is not difficult to construct examples, the following construction is instructive in its own right. Let X and Y be standard exponential random variables with copula C, where C is a member of the Farlie–Gumbel–Morgenstern family, i.e., C is given by

$$C(u,v) = uv + \theta uv(1-u)(1-v),$$

for some θ in $[-1, 1]$. The joint distribution function H of X and Y is given by

$$H(x,y) = C\big(1-\mathrm{e}^{-x}, 1-\mathrm{e}^{-y}\big).$$

Let ρ denote the linear correlation coefficient. Then

$$\rho(X,Y) = \frac{\mathbb{E}(XY) - \mathbb{E}(X)\mathbb{E}(Y)}{\sqrt{\mathrm{Var}(X)}\sqrt{\mathrm{Var}(Y)}} = \mathbb{E}(XY) - 1,$$

where

$$\begin{aligned}\mathbb{E}(XY) &= \int_0^\infty\int_0^\infty xy\,\mathrm{d}H(x,y)\\ &= \int_0^\infty\int_0^\infty xy\big((1+\theta)\,\mathrm{e}^{-x-y} - 2\theta\,\mathrm{e}^{-2x-y} - 2\theta\,\mathrm{e}^{-x-2y} + 4\theta\,\mathrm{e}^{-2x-2y}\big)\,\mathrm{d}x\,\mathrm{d}y\\ &= 1 + \frac{\theta}{4}.\end{aligned}$$

Hence $\rho(X,Y) = \theta/4$. But

$$\begin{aligned}\rho\big(1-\mathrm{e}^{-X}, 1-\mathrm{e}^{-Y}\big) = \rho_S(X,Y) &= 12\iint_{[0,1]^2} C(u,v)\,\mathrm{d}u\,\mathrm{d}v - 3\\ &= 12\iint_{[0,1]^2} \big(uv + \theta uv(1-u)(1-v)\big)\,\mathrm{d}u\,\mathrm{d}v - 3\\ &= 12\left(\frac{1}{4} + \frac{\theta}{36}\right) - 3 = \frac{\theta}{3}.\end{aligned}$$

Hence $\rho(X,Y)$ is not invariant under strictly increasing transformations of X and Y and therefore linear correlation is not a measure of concordance.

Although the properties listed under Definition 3.3 are useful, there are some additional properties that would make a measure of concordance even more useful. Recall that for a random vector $(X,Y)^{\mathrm{T}}$ with copula C,

$$\begin{aligned} C = M &\implies \tau_C = \rho_C = 1,\\ C = W &\implies \tau_C = \rho_C = -1.\end{aligned}$$

The following theorem states that the converse is also true.

Theorem 3.6. *Let X and Y be continuous random variables with copula C, and let κ denote Kendall's tau or Spearman's rho. Then the following are true:*
(1) $\kappa(X,Y)=1 \Leftrightarrow C=M$.
(2) $\kappa(X,Y)=-1 \Leftrightarrow C=W$.

For a proof, see Embrechts, McNeil and Straumann (2002).

From the definitions of Kendall's tau and Spearman's rho it follows that both are increasing functions of the value of the copula under consideration. Thus they are increasing with respect to the concordance ordering given in Definition 2.5. Moreover, for continuous random variables all values in the interval $[-1,1]$ can be obtained for Kendall's tau or Spearman's rho by a suitable choice of the underlying copula. This is however not the case with linear correlation as is shown in the following example from Embrechts, McNeil and Straumann (2002).

Example 3.2. Let $X \sim \mathrm{LN}(0,1)$ (Lognormal) and $Y \sim \mathrm{LN}(0,\sigma^2)$, $\sigma > 0$. Then $\rho_{\min} = \rho(\mathrm{e}^Z, \mathrm{e}^{-\sigma Z})$ and $\rho_{\max} = \rho(\mathrm{e}^Z, \mathrm{e}^{\sigma Z})$, where $Z \sim \mathcal{N}(0,1)$. $\rho_{\min}$ and $\rho_{\max}$ can be calculated, yielding:

$$\rho_{\min} = \frac{\mathrm{e}^{-\sigma}-1}{\sqrt{\mathrm{e}-1}\sqrt{\mathrm{e}^{\sigma^2}-1}}, \qquad \rho_{\max} = \frac{\mathrm{e}^{\sigma}-1}{\sqrt{\mathrm{e}-1}\sqrt{\mathrm{e}^{\sigma^2}-1}},$$

from which follows that $\lim_{\sigma\to\infty}\rho_{\min} = \lim_{\sigma\to\infty}\rho_{\max} = 0$. Hence the linear correlation coefficient can be almost zero, even if X and Y are comonotonic or countermonotonic.

Kendall's tau and Spearman's rho are measures of dependence between two random variables. However the extension to higher dimensions is obvious, we simply write pairwise correlations in an $n \times n$ matrix in the same way as is done for linear correlation.

3.5. Tail dependence

The concept of tail dependence relates to the amount of dependence in the upper-right-quadrant tail or lower-left-quadrant tail of a bivariate distribution. It is a concept that is relevant for the study of dependence between extreme values. It turns out that tail dependence between two continuous random variables X and Y is a copula property and hence the amount of tail dependence is invariant under strictly increasing transformations of X and Y.

Definition 3.6. Let $(X,Y)^{\mathrm{T}}$ be a vector of continuous random variables with marginal distribution functions F and G. The coefficient of upper tail dependence of $(X,Y)^{\mathrm{T}}$ is

$$\lim_{u\nearrow 1} \mathbb{P}\{Y > G^{-1}(u) \mid X > F^{-1}(u)\} = \lambda_U$$

provided that the limit $\lambda_U \in [0, 1]$ exists. If $\lambda_U \in (0, 1]$, X and Y are said to be asymptotically dependent in the upper tail; if $\lambda_U = 0$, X and Y are said to be asymptotically independent in the upper tail.

Since $\mathbb{P}\{Y > G^{-1}(u) \mid X > F^{-1}(u)\}$ can be written as

$$\frac{1 - \mathbb{P}\{X \leqslant F^{-1}(u)\} - \mathbb{P}\{Y \leqslant G^{-1}(u)\} + \mathbb{P}\{X \leqslant F^{-1}(u), Y \leqslant G^{-1}(u)\}}{1 - \mathbb{P}\{X \leqslant F^{-1}(u)\}},$$

an alternative and equivalent definition (for continuous random variables), from which it is seen that the concept of tail dependence is indeed a copula property, is the following which can be found in Joe (1997, p. 33).

Definition 3.7. If a bivariate copula C is such that

$$\lim_{u \nearrow 1} \frac{1 - 2u + C(u, u)}{1 - u} = \lambda_U$$

exists, then C has upper tail dependence if $\lambda_U \in (0, 1]$, and upper tail independence if $\lambda_U = 0$.

Example 3.3. Consider the bivariate Gumbel family of copulas given by

$$C_\theta(u, v) = \exp\left(-\left[(-\ln u)^\theta + (-\ln v)^\theta\right]^{1/\theta}\right),$$

for $\theta \geqslant 1$. Then

$$\frac{1 - 2u + C(u, u)}{1 - u} = \frac{1 - 2u + \exp(2^{1/\theta} \ln u)}{1 - u} = \frac{1 - 2u + u^{2^{1/\theta}}}{1 - u},$$

and hence

$$\lim_{u \nearrow 1} \frac{1 - 2u + C(u, u)}{1 - u} = 2 - \lim_{u \nearrow 1} 2^{1/\theta} u^{2^{1/\theta} - 1} = 2 - 2^{1/\theta}.$$

Thus for $\theta > 1$, C_θ has upper tail dependence.

For copulas without a simple closed form an alternative formula for λ_U is more useful. An example is given in the case of the Gaussian copula

$$C_R(u, v) = \int_{-\infty}^{\Phi^{-1}(u)} \int_{-\infty}^{\Phi^{-1}(v)} \frac{1}{2\pi\sqrt{1 - R_{12}^2}} \exp\left\{-\frac{s^2 - 2R_{12}st + t^2}{2(1 - R_{12}^2)}\right\} \mathrm{d}s\, \mathrm{d}t,$$

where $-1 < R_{12} < 1$ and Φ is the univariate standard normal distribution function. Consider a pair of $U(0,1)$ random variables (U, V) with copula C. First note that $\mathbb{P}\{V \leqslant v | U = u\} = \partial C(u,v)/\partial u$ and $\mathbb{P}\{V > v | U = u\} = 1 - \partial C(u,v)/\partial u$, and similarly when conditioning on V. Then

$$\begin{aligned}
\lambda_U &= \lim_{u \nearrow 1} \frac{\overline{C}(u,u)}{1-u} = -\lim_{u \nearrow 1} \frac{\mathrm{d}\overline{C}(u,u)}{\mathrm{d}u} \\
&= -\lim_{u \nearrow 1} \left(-2 + \frac{\partial}{\partial s} C(s,t) \Big|_{s=t=u} + \frac{\partial}{\partial t} C(s,t) \Big|_{s=t=u} \right) \\
&= \lim_{u \nearrow 1} \big(\mathbb{P}\{V > u | U = u\} + \mathbb{P}\{U > u | V = u\} \big).
\end{aligned}$$

Furthermore, if C is an exchangeable copula, i.e., $C(u,v) = C(v,u)$, then the expression for λ_U simplifies to

$$\lambda_U = 2 \lim_{u \nearrow 1} \mathbb{P}\{V > u | U = u\}.$$

Example 3.4. Let $(X, Y)^{\mathrm{T}}$ have the bivariate standard normal distribution function with linear correlation coefficient ρ. That is $(X, Y)^{\mathrm{T}} \sim C(\Phi(x), \Phi(y))$, where C is a member of the Gaussian family given above with $R_{12} = \rho$. Since copulas in this family are exchangeable,

$$\lambda_U = 2 \lim_{u \nearrow 1} \mathbb{P}\{V > u | U = u\},$$

and because Φ is a distribution function with infinite right endpoint,

$$\begin{aligned}
\lim_{u \nearrow 1} \mathbb{P}\{V > u | U = u\} &= \lim_{x \to \infty} \mathbb{P}\{\Phi^{-1}(V) > x | \Phi^{-1}(U) = x\} \\
&= \lim_{x \to \infty} \mathbb{P}\{X > x | Y = x\}.
\end{aligned}$$

Using the well known fact that $Y | X = x \sim \mathcal{N}(\rho x, 1 - \rho^2)$ we obtain

$$\lambda_U = 2 \lim_{x \to \infty} \overline{\Phi}\left(\frac{x - \rho x}{\sqrt{1 - \rho^2}} \right) = 2 \lim_{x \to \infty} \overline{\Phi}\left(\frac{x\sqrt{1-\rho}}{\sqrt{1+\rho}} \right),$$

from which it follows that $\lambda_U = 0$ for $R_{12} < 1$. Hence the Gaussian copula C with $\rho < 1$ does not have upper tail dependence.

The concept of lower tail dependence can be defined in a similar way. If the limit $\lim_{u \searrow 0} C(u,u)/u = \lambda_L$ exists, then C has lower tail dependence if $\lambda_L \in (0, 1]$, and lower tail independence if $\lambda_L = 0$. For copulas without a simple closed form an alternative formula for λ_L is more useful. Consider a random vector $(U, V)^{\mathrm{T}}$ with copula C. Then

$$\begin{aligned}
\lambda_L &= \lim_{u \searrow 0} \frac{C(u,u)}{u} = \lim_{u \searrow 0} \frac{\mathrm{d}C(u,u)}{\mathrm{d}u} \\
&= \lim_{u \searrow 0} \left(\frac{\partial}{\partial s} C(s,t)\Big|_{s=t=u} + \frac{\partial}{\partial t} C(s,t)\Big|_{s=t=u} \right) \\
&= \lim_{u \searrow 0} \big(\mathbb{P}\{V < u | U = u\} + \mathbb{P}\{U < u | V = u\} \big).
\end{aligned}$$

Furthermore if C is an exchangeable copula, i.e., $C(u, v) = C(v, u)$, then the expression for λ_L simplifies to

$$\lambda_L = 2 \lim_{u \searrow 0} \mathbb{P}\{V < u | U = u\}.$$

Recall that the survival copula of two random variables with copula C is given by

$$\widehat{C}(u, v) = u + v - 1 + C(1 - u, 1 - v),$$

and the joint survival function for two $U(0, 1)$ random variables whose joint distribution function is C is given by

$$\overline{C}(u, v) = 1 - u - v + C(u, v) = \widehat{C}(1 - u, 1 - v).$$

Hence it follows that

$$\lim_{u \nearrow 1} \frac{\overline{C}(u,u)}{1-u} = \lim_{u \nearrow 1} \frac{\widehat{C}(1-u, 1-u)}{1-u} = \lim_{u \searrow 0} \frac{\widehat{C}(u,u)}{u},$$

so the coefficient of upper tail dependence of C is the coefficient of lower tail dependence of $\widehat{C}$. Similarly the coefficient of lower tail dependence of C is the coefficient of upper tail dependence of $\widehat{C}$.

4. Marshall–Olkin copulas

In this section we discuss a class of copulas called Marshall–Olkin copulas. To be able to derive these copulas and present explicit expressions for rank correlation and tail dependence coefficients without tedious calculations, we begin with bivariate Marshall–Olkin copulas. We then continue with the general n-dimensional case and suggest applications of Marshall–Olkin copulas to the modelling of dependent risks. For further details about Marshall–Olkin distributions we refer to Marshall and Olkin (1967). Similar ideas are contained in Muliere and Scarsini (1987).

4.1. Bivariate Marshall–Olkin copulas

Consider a two-component system where the components are subject to shocks, which are fatal to one or both components. Let X_1 and X_2 denote the lifetimes of the two components. Furthermore assume that the shocks follow three independent Poisson processes with parameters $\lambda_1, \lambda_2, \lambda_{12} \geqslant 0$, where the index indicates whether the shocks effect only component 1, only component 2 or both. Then the times Z_1, Z_2 and Z_{12} of occurrence of these shocks are independent exponential random variables with parameters λ_1, λ_2 and λ_{12} respectively. Hence

$$\begin{aligned}\overline{H}(x_1, x_2) &= \mathbb{P}\{X_1 > x_1, X_2 > x_2\} \\ &= \mathbb{P}\{Z_1 > x_1\}\mathbb{P}\{Z_2 > x_2\}\mathbb{P}\{Z_{12} > \max(x_1, x_2)\}.\end{aligned}$$

The univariate survival functions for X_1 and X_2 are $\overline{F}_1(x_1) = \exp(-(\lambda_1 + \lambda_{12})x_1)$ and $\overline{F}_2(x_2) = \exp(-(\lambda_2 + \lambda_{12})x_2)$. Furthermore, since $\max(x_1, x_2) = x_1 + x_2 - \min(x_1, x_2)$,

$$\begin{aligned}\overline{H}(x_1, x_2) &= \exp\bigl(-(\lambda_1 + \lambda_{12})x_1 - (\lambda_2 + \lambda_{12})x_2 + \lambda_{12}\min(x_1, x_2)\bigr) \\ &= \overline{F}_1(x_1)\overline{F}_2(x_2)\min\bigl(\exp(\lambda_{12}x_1), \exp(\lambda_{12}x_2)\bigr).\end{aligned}$$

Let $\alpha_1 = \lambda_{12}/(\lambda_1 + \lambda_{12})$ and $\alpha_2 = \lambda_{12}/(\lambda_2 + \lambda_{12})$. Then $\exp(\lambda_{12}x_1) = \overline{F}_1(x_1)^{-\alpha_1}$ and $\exp(\lambda_{12}x_2) = \overline{F}_2(x_2)^{-\alpha_2}$, and hence the survival copula of $(X_1, X_2)^{\mathrm{T}}$ is given by

$$\widehat{C}(u_1, u_2) = u_1 u_2 \min\bigl(u_1^{-\alpha_1}, u_2^{-\alpha_2}\bigr) = \min\bigl(u_1^{1-\alpha_1}u_2, u_1 u_2^{1-\alpha_2}\bigr).$$

This construction leads to a copula family given by

$$C_{\alpha_1,\alpha_2}(u_1, u_2) = \min\bigl(u_1^{1-\alpha_1}u_2, u_1 u_2^{1-\alpha_2}\bigr) = \begin{cases} u_1^{1-\alpha_1}u_2, & u_1^{\alpha_1} \geqslant u_2^{\alpha_2}, \\ u_1 u_2^{1-\alpha_2}, & u_1^{\alpha_1} \leqslant u_2^{\alpha_2}. \end{cases}$$

This family is known as the Marshall–Olkin family. Marshall–Olkin copulas have both an absolutely continuous and a singular component. Since

$$\frac{\partial^2}{\partial u_1 \partial u_2} C_{\alpha_1,\alpha_2}(u_1, u_2) = \begin{cases} u_1^{-\alpha_1}, & u_1^{\alpha_1} > u_2^{\alpha_2}, \\ u_2^{-\alpha_2}, & u_1^{\alpha_1} < u_2^{\alpha_2}, \end{cases}$$

the mass of the singular component is concentrated on the curve $u_1^{\alpha_1} = u_2^{\alpha_2}$ in $[0, 1]^2$ as seen in Figure 1.

Kendall's tau and Spearman's rho are quite easily evaluated for this copula family. For Spearman's rho, applying Theorem 3.4 yields:

$$\rho_S(C_{\alpha_1,\alpha_2}) = 12 \iint_{[0,1]^2} C_{\alpha_1,\alpha_2}(u, v)\,\mathrm{d}u\,\mathrm{d}v - 3$$

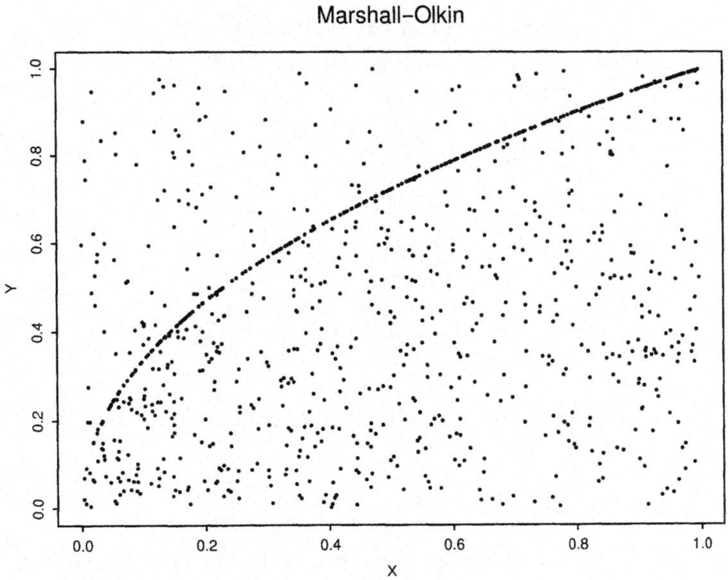

Fig. 1. A simulation from the Marshall–Olkin copula with $\lambda_1 = 1.1$, $\lambda_2 = 0.2$ and $\lambda_{12} = 0.6$.

$$= 12\int_0^1 \left(\int_0^{u^{\alpha_1/\alpha_2}} u^{1-\alpha_1} v \, dv + \int_{u^{\alpha_1/\alpha_2}}^1 u v^{1-\alpha_2} \, dv \right) du - 3$$

$$= \frac{3\alpha_1\alpha_2}{2\alpha_1 + 2\alpha_2 - \alpha_1\alpha_2}.$$

To evaluate Kendall's tau we use the following theorem, a proof of which is found in Nelsen (1999, p. 131).

Theorem 4.1. *Let C be a copula such that the product $(\partial C/\partial u)(\partial C/\partial v)$ is integrable on $[0, 1]^2$. Then*

$$\iint_{[0,1]^2} C(u, v) \, dC(u, v) = \frac{1}{2} - \iint_{[0,1]^2} \frac{\partial}{\partial u} C(u, v) \frac{\partial}{\partial u} C(u, v) \, du \, dv.$$

Using Theorems 3.3 and 4.1 we obtain

$$\tau(C_{\alpha_1,\alpha_2}) = 4 \iint_{[0,1]^2} C_{\alpha_1,\alpha_2}(u, v) \, dC_{\alpha_1,\alpha_2}(u, v) - 1$$

$$= 4\left(\frac{1}{2} - \iint_{[0,1]^2} \frac{\partial}{\partial u} C_{\alpha_1,\alpha_2}(u, v) \frac{\partial}{\partial u} C_{\alpha_1,\alpha_2}(u, v) \, du \, dv \right) - 1$$

$$= \frac{\alpha_1\alpha_2}{\alpha_1 + \alpha_2 - \alpha_1\alpha_2}.$$

Thus, all values in the interval [0, 1] can be obtained for $\rho_S(C_{\alpha_1,\alpha_2})$ and $\tau(C_{\alpha_1,\alpha_2})$. The Marshall–Olkin copulas have upper tail dependence. Without loss of generality assume that $\alpha_1 > \alpha_2$, then

$$\begin{aligned}
\lim_{u \nearrow 1} \frac{\overline{C}(u,u)}{1-u} &= \lim_{u \nearrow 1} \frac{1 - 2u + u^2 \min(u^{-\alpha_1}, u^{-\alpha_2})}{1-u} \\
&= \lim_{u \nearrow 1} \frac{1 - 2u + u^2 u^{-\alpha_2}}{1-u} \\
&= \lim_{u \nearrow 1} \left(2 - 2u^{1-\alpha_2} + \alpha_2 u^{1-\alpha_2}\right) = \alpha_2,
\end{aligned}$$

and hence $\lambda_U = \min(\alpha_1, \alpha_2)$ is the coefficient of upper tail dependence.

4.2. *A multivariate extension*

We now present the natural multivariate extension of the bivariate Marshall–Olkin family. Consider an n-component system, where each nonempty subset of components is assigned a shock which is fatal to all components of that subset. Let $\mathcal{S}$ denote the set of nonempty subsets of $\{1, \ldots, n\}$. Let $X_1, \ldots, X_n$ denote the lifetimes of the components, and assume that shocks assigned to different subsets s, $s \in \mathcal{S}$, follow independent Poisson processes with intensities λ_s. Let Z_s, $s \in \mathcal{S}$, denote the time of first occurrence of a shock event for the shock process assigned to subset s. Then the occurrence times Z_s are independent exponential random variables with parameters λ_s, and $X_j = \min_{s: j \in s} Z_s$ for $j = 1, \ldots, n$.

There are in total $2^n - 1$ shock processes, each in one-to-one correspondence with a nonempty subset of $\{1, \ldots, n\}$.

Example 4.1. Let $n = 4$. Then

$$\begin{aligned}
X_1 &= \min(Z_1, Z_{12}, Z_{13}, Z_{14}, Z_{123}, Z_{124}, Z_{134}, Z_{1234}), \\
X_2 &= \min(Z_2, Z_{12}, Z_{23}, Z_{24}, Z_{123}, Z_{124}, Z_{234}, Z_{1234}), \\
X_3 &= \min(Z_3, Z_{13}, Z_{23}, Z_{34}, Z_{123}, Z_{134}, Z_{234}, Z_{1234}), \\
X_4 &= \min(Z_4, Z_{14}, Z_{24}, Z_{34}, Z_{124}, Z_{134}, Z_{234}, Z_{1234}).
\end{aligned}$$

If for example $\lambda_{13} = 0$, then $Z_{13} = \infty$ almost surely.

We now turn to the question of random variate generation from Marshall–Olkin n-copulas. Order the $l := |\mathcal{S}| = 2^n - 1$ nonempty subsets of $\{1, \ldots, n\}$ in some arbitrary way, $s_1, \ldots, s_l$, and set $\lambda_k := \lambda_{s_k}$ (the parameter of Z_{s_k}) for $k = 1, \ldots, l$. The following algorithm generates random variates from the Marshall–Olkin n-copula.

Algorithm 4.1.

- Simulate l independent random variates $v_1, \ldots, v_l$ from $U(0, 1)$.
- Set $x_i = \min_{1 \leqslant k \leqslant l,\, i \in s_k,\, \lambda_k \neq 0}(-\ln v_k/\lambda_k)$, $i = 1, \ldots, n$.
- Set $\Lambda_i = \sum_{k=1}^{l} 1\{i \in s_k\}\lambda_k$, $i = 1, \ldots, n$.
- Set $u_i = \exp(-\Lambda_i x_i)$, $i = 1, \ldots, n$.

Then $(x_1, \ldots, x_n)^{\mathrm{T}}$ is an n-variate from the n-dimensional Marshall–Olkin distribution and $(u_1, \ldots, u_n)^{\mathrm{T}}$ is an n-variate from the corresponding Marshall–Olkin n-copula. Furthermore, Λ_i is the shock intensity "felt" by component i.

Since the (i, j)-bivariate marginal of a Marshall–Olkin n-copula is a Marshall–Olkin copula with parameters

$$\alpha_i = \Big(\sum_{s:\, i \in s,\, j \in s} \lambda_s\Big) \Big/ \Big(\sum_{s:\, i \in s} \lambda_s\Big) \quad \text{and} \quad \alpha_j = \Big(\sum_{s:\, i \in s,\, j \in s} \lambda_s\Big) \Big/ \Big(\sum_{s:\, j \in s} \lambda_s\Big),$$

the Kendall's tau and Spearman's rho rank correlation matrices are easily evaluated. The (i, j) entries are given by

$$\frac{\alpha_i \alpha_j}{\alpha_i + \alpha_j - \alpha_i \alpha_j} \quad \text{and} \quad \frac{3\alpha_i \alpha_j}{2\alpha_i + 2\alpha_j - \alpha_i \alpha_j},$$

respectively. As seen above, evaluating the rank correlation matrix given the full parameterization of the Marshall–Olkin n-copula is straightforward. However given a (Kendall's tau or Spearman's rho) rank correlation matrix we cannot in general obtain a unique parameterization of the copula. By setting the shock intensities for subgroups with more then two elements to zero, we obtain the perhaps most natural parameterization of the copula in this situation. However this also means that the copula only has bivariate dependence.

4.3. *A useful modelling framework*

In general the huge number of parameters for high-dimensional Marshall–Olkin copulas make them unattractive for high-dimensional risk modelling. However, we now give an example of how an intuitively appealing and easier parameterized model for modelling dependent loss frequencies can be set up, for which the survival copula of times to first losses is a Marshall–Olkin copula.

Suppose we are interested in insurance losses occurring in several different lines of business or several different countries. In credit-risk modelling we might be interested in losses related to the default of various different counterparties or types of counterparty. A natural approach to modelling this dependence is to assume that all losses can be related to a series of underlying and independent shock processes. In insurance these shocks might be natural catastrophes; in credit-risk modelling they might be a variety of underlying economic events. When a shock occurs this may cause losses of several different types; the common shock causes the numbers of losses of each type to be dependent. It is commonly assumed that the different varieties of shocks arrive as independent Poisson processes, in which case the counting processes of the losses are also Poisson and can be handled easily analytically. In reliability such models are known as *fatal shock models*, when the shock always destroys the component, and *nonfatal shock models*, when components have a chance of surviving the shock. A good basic reference on such models is Barlow and Proschan (1975).

Suppose there are m different types of shocks and for $e = 1, \dots, m$, let $\{N^{(e)}(t),\ t \geqslant 0\}$ be a Poisson process with intensity $\lambda^{(e)}$ recording the number of events of type e occurring in $(0, t]$. Assume further that these shock counting processes are independent. Consider losses of n different types and for $j = 1, \dots, n$, let $\{N_j(t),\ t \geqslant 0\}$ be a counting process that records the frequency of losses of the jth type occurring in $(0, t]$. At the rth occurrence of an event of type e the Bernoulli variable $I_{j,r}^{(e)}$ indicates whether a loss of type j occurs. The vectors

$$\mathbf{I}_r^{(e)} = \left(I_{1,r}^{(e)}, \dots, I_{n,r}^{(e)}\right)^{\mathrm{T}}$$

for $r = 1, \dots, N^{(e)}(t)$ are considered to be independent and identically distributed with a multivariate Bernoulli distribution. In other words, each new event represents a new independent opportunity to incur a loss but, for a fixed event, the loss trigger variables for losses of different types may be dependent. The form of the dependence depends on the specification of the multivariate Bernoulli distribution with independence as a special case. We use the following notation for p-dimensional marginal probabilities of this distribution (the subscript r is dropped for simplicity):

$$P\left(I_{j_1}^{(e)} = i_{j_1}, \dots, I_{j_p}^{(e)} = i_{j_p}\right) = p_{j_1,\dots,j_p}^{(e)}(i_{j_1}, \dots, i_{j_p}), \quad i_{j_1}, \dots, i_{j_p} \in \{0, 1\}.$$

We also write $p_j^{(e)}(1) = p_j^{(e)}$ for one-dimensional marginal probabilities, so that in the special case of conditional independence we have $p_{j_1,\dots,j_p}^{(e)}(1, \dots, 1) = \prod_{k=1}^{p} p_{j_k}^{(e)}$. The counting processes for events and losses are thus linked by

$$N_j(t) = \sum_{e=1}^{m} \sum_{r=1}^{N^{(e)}(t)} I_{j,r}^{(e)}.$$

Under the Poisson assumption for the event processes and the Bernoulli assumption for the loss indicators, the loss processes $\{N_j(t),\ t \geqslant 0\}$ are clearly Poisson themselves, since they are obtained by superpositioning m independent (possibly thinned) Poisson processes generated by the m underlying event processes. The random vector $(N_1(t), \dots, N_n(t))^{\mathrm{T}}$ can be thought of as having a *multivariate Poisson* distribution.

The presented nonfatal shock model has an equivalent fatal shock model representation, i.e., of the type presented in Section 4.2. Hence the random vector $(X_1, \dots, X_n)^{\mathrm{T}}$ of times to first losses of different types, where $X_j = \inf\{t \geqslant 0 \mid N_j(t) > 0\}$, has an n-dimensional Marshall–Olkin distribution whose survival copula is a Marshall–Olkin n-copula. From this it follows that Kendall's tau, Spearman's rho and coefficients of tail dependence for $(X_i, X_j)^{\mathrm{T}}$ can be easily calculated. For more details on this model, see Lindskog and McNeil (2001).

5. Elliptical copulas

The class of elliptical distributions provides a rich source of multivariate distributions which share many of the tractable properties of the multivariate normal distribution and enables modelling of multivariate extremes and other forms of nonnormal dependences. Elliptical copulas are simply the copulas of elliptical distributions. Simulation from elliptical distributions is easy, and as a consequence of Sklar's Theorem so is simulation from elliptical copulas. Furthermore, we will show that rank correlation and tail dependence coefficients can be easily calculated. For further details on elliptical distributions we refer to Fang, Kotz and Ng (1987) and Cambanis, Huang and Simons (1981).

5.1. Elliptical distributions

Definition 5.1. If $\mathbf{X}$ is a n-dimensional random vector and, for some $\mu \in \mathbb{R}^n$ and some $n \times n$ nonnegative definite, symmetric matrix Σ, the characteristic function $\varphi_{\mathbf{X}-\mu}(\mathbf{t})$ of $\mathbf{X} - \mu$ is a function of the quadratic form $\mathbf{t}^{\mathrm{T}}\Sigma\mathbf{t}$, $\varphi_{\mathbf{X}-\mu}(\mathbf{t}) = \phi(\mathbf{t}^{\mathrm{T}}\Sigma\mathbf{t})$, we say that $\mathbf{X}$ has an elliptical distribution with parameters μ, Σ and ϕ, and we write $\mathbf{X} \sim E_n(\mu, \Sigma, \phi)$.

When $n = 1$, the class of elliptical distributions coincides with the class of one-dimensional symmetric distributions. A function ϕ as in Definition 5.1 is called a characteristic generator.

Theorem 5.1. $\mathbf{X} \sim E_n(\mu, \Sigma, \phi)$ *with* $\operatorname{rank}(\Sigma) = k$ *if and only if there exist a random variable* $R \geqslant 0$ *independent of* $\mathbf{U}$, *a* k*-dimensional random vector uniformly distributed on the unit hypersphere* $\{\mathbf{z} \in \mathbb{R}^k \mid \mathbf{z}^{\mathrm{T}}\mathbf{z} = 1\}$, *and an* $n \times k$ *matrix* A *with* $AA^{\mathrm{T}} = \Sigma$, *such that*

$$\mathbf{X} =_d \mu + RA\mathbf{U}.$$

For the proof of Theorem 5.1 and the relation between R and ϕ see Fang, Kotz and Ng (1987) or Cambanis, Huang and Simons (1981).

Example 5.1. Let $\mathbf{X} \sim \mathcal{N}_n(\mathbf{0}, \mathbf{I}_n)$. Since the components $X_i \sim \mathcal{N}(0, 1)$, $i = 1, \ldots, n$, are independent and the characteristic function of X_i is $\exp(-t_i^2/2)$, the characteristic function of $\mathbf{X}$ is

$$\exp\left\{-\frac{1}{2}(t_1^2 + \cdots + t_n^2)\right\} = \exp\left\{-\frac{1}{2}\mathbf{t}^{\mathrm{T}}\mathbf{t}\right\}.$$

From Theorem 5.1 it then follows that $\mathbf{X} \sim E_n(\mathbf{0}, \mathbf{I}_n, \phi)$, where $\phi(u) = \exp(-u/2)$.

If $\mathbf{X} \sim E_n(\mu, \Sigma, \phi)$, where Σ is a diagonal matrix, then $\mathbf{X}$ has uncorrelated components (if $0 < \mathrm{Var}(X_i) < \infty$). If $\mathbf{X}$ has independent components, then $\mathbf{X} \sim \mathcal{N}_n(\mu, \Sigma)$. Note that the multivariate normal distribution is the only one among the elliptical distributions where uncorrelated components imply independent components. A random vector $\mathbf{X} \sim E_n(\mu, \Sigma, \phi)$ does not necessarily have a density. If $\mathbf{X}$ has a density it must be of the form $|\Sigma|^{-1/2} g((\mathbf{X}-\mu)^{\mathrm{T}} \Sigma^{-1} (\mathbf{X}-\mu))$ for some nonnegative function g of one scalar variable. Hence the contours of equal density form ellipsoids in $\mathbb{R}^n$. Given the distribution of $\mathbf{X}$, the representation $E_n(\mu, \Sigma, \phi)$ is not unique. It uniquely determines μ but Σ and ϕ are only determined up to a positive constant. More precisely, if $\mathbf{X} \sim E_n(\mu, \Sigma, \phi)$ and $\mathbf{X} \sim E_n(\mu^*, \Sigma^*, \phi^*)$, then

$$\mu^* = \mu, \qquad \Sigma^* = c\Sigma, \qquad \phi^*(\cdot) = \phi\left(\frac{\cdot}{c}\right),$$

for some constant $c > 0$.

In order to find a representation such that $\mathrm{Cov}(\mathbf{X}) = \Sigma$, we use Theorem 5.1 to obtain

$$\mathrm{Cov}(\mathbf{X}) = \mathrm{Cov}(\mu + RA\mathbf{U}) = A\mathbb{E}\big(R^2\big)\,\mathrm{Cov}(\mathbf{U})A^{\mathrm{T}},$$

provided that $\mathbb{E}(R^2) < \infty$. Let $\mathbf{Y} \sim \mathcal{N}_n(\mathbf{0}, \mathbf{I}_n)$. Then $\mathbf{Y} =_d \|\mathbf{Y}\|\mathbf{U}$, where $\|\mathbf{Y}\|$ is independent of $\mathbf{U}$. Furthermore $\|\mathbf{Y}\|^2 \sim \chi_n^2$, so $\mathbb{E}(\|\mathbf{Y}\|^2) = n$. Since $\mathrm{Cov}(\mathbf{Y}) = \mathbf{I}_n$ we see that if $\mathbf{U}$ is uniformly distributed on the unit hypersphere in $\mathbb{R}^n$, then $\mathrm{Cov}(\mathbf{U}) = \mathbf{I}_n/n$. Thus $\mathrm{Cov}(\mathbf{X}) = AA^{\mathrm{T}}\mathbb{E}(R^2)/n$. By choosing the characteristic generator $\phi^*(s) = \phi(s/c)$, where $c = \mathbb{E}(R^2)/n$, we get $\mathrm{Cov}(X) = \Sigma$. Hence an elliptical distribution is fully described by μ, Σ and ϕ, where ϕ can be chosen so that $\mathrm{Cov}(\mathbf{X}) = \Sigma$ (if $\mathrm{Cov}(\mathbf{X})$ is defined). If $\mathrm{Cov}(\mathbf{X})$ is obtained as above, then the distribution of $\mathbf{X}$ is uniquely determined by $\mathbb{E}(\mathbf{X})$, $\mathrm{Cov}(\mathbf{X})$ and the type of its univariate marginals, e.g., normal or t_4, say.

Theorem 5.2. *Let $\mathbf{X} \sim E_n(\mu, \Sigma, \phi)$, let B be a $q \times n$ matrix and $\mathbf{b} \in \mathbb{R}^q$. Then*

$$\mathbf{b} + B\mathbf{X} \sim E_q\big(\mathbf{b} + B\mu, B\Sigma B^{\mathrm{T}}, \phi\big).$$

Proof: By Theorem 5.1, $\mathbf{b} + B\mathbf{X}$ has the stochastic representation

$$\mathbf{b} + B\mathbf{X} =_d \mathbf{b} + B\mu + RBA\mathbf{U}. \ \square$$

Partition $\mathbf{X}$, μ and Σ into

$$\mathbf{X} = \begin{pmatrix} \mathbf{X}_1 \\ \mathbf{X}_2 \end{pmatrix}, \qquad \mu = \begin{pmatrix} \mu_1 \\ \mu_2 \end{pmatrix}, \qquad \Sigma = \begin{pmatrix} \Sigma_{11} & \Sigma_{12} \\ \Sigma_{21} & \Sigma_{22} \end{pmatrix},$$

where $\mathbf{X}_1$ and μ_1 are $r \times 1$ vectors and Σ_{11} is a $r \times r$ matrix.

Corollary 5.1. *Let* $\mathbf{X} \sim E_n(\mu, \Sigma, \phi)$. *Then*

$$\mathbf{X}_1 \sim E_r(\mu_1, \Sigma_{11}, \phi), \qquad \mathbf{X}_2 \sim E_{n-r}(\mu_2, \Sigma_{22}, \phi).$$

Hence marginal distributions of elliptical distributions are elliptical and of the same type (with the same characteristic generator). The next result states that the conditional distribution of $\mathbf{X}_1$ given the value of $\mathbf{X}_2$ is also elliptical, but in general not of the same type as $\mathbf{X}_1$.

Theorem 5.3. *Let* $\mathbf{X} \sim E_n(\mu, \Sigma, \phi)$ *with* Σ *strictly positive definite. Then*

$$\mathbf{X}_1 | \mathbf{X}_2 = \mathbf{x} \sim E_r(\tilde{\mu}, \tilde{\Sigma}, \tilde{\phi}),$$

where $\tilde{\mu} = \mu_1 + \Sigma_{12}\Sigma_{22}^{-1}(\mathbf{x} - \mu_2)$ *and* $\tilde{\Sigma} = \Sigma_{11} - \Sigma_{12}\Sigma_{22}^{-1}\Sigma_{21}$. *Moreover,* $\tilde{\phi} = \phi$ *if and only if* $\mathbf{X} \sim \mathcal{N}_n(\mu, \Sigma)$.

For the proof and details about $\tilde{\phi}$, see Fang, Kotz and Ng (1987). For the extension to the case where rank$(\Sigma) < n$, see Cambanis, Huang and Simons (1981).

The following lemma states that linear combinations of independent, elliptically distributed random vectors with the same dispersion matrix Σ (up to a positive constant) remain elliptical.

Lemma 5.1. *Let* $\mathbf{X} \sim E_n(\mu, \Sigma, \phi)$ *and* $\tilde{\mathbf{X}} \sim E_n(\tilde{\mu}, c\Sigma, \tilde{\phi})$ *for* $c > 0$ *be independent. Then for* $a, b \in \mathbb{R}$, $a\mathbf{X} + b\tilde{\mathbf{X}} \sim E_n(a\mu + b\tilde{\mu}, \Sigma, \phi^*)$ *with* $\phi^*(u) = \phi(a^2u)\tilde{\phi}(b^2cu)$.

Proof: By Definition 5.1, it is sufficient to show that for all $\mathbf{t} \in \mathbb{R}^n$

$$\begin{aligned}
\varphi_{a\mathbf{X}+b\tilde{\mathbf{X}}-a\mu-b\tilde{\mu}}(\mathbf{t}) &= \varphi_{a(\mathbf{X}-\mu)}(\mathbf{t})\varphi_{b(\tilde{\mathbf{X}}-\tilde{\mu})}(\mathbf{t}) \\
&= \phi\big((a\mathbf{t})^{\mathrm{T}}\Sigma(a\mathbf{t})\big)\tilde{\phi}\big((b\mathbf{t})^{\mathrm{T}}(c\Sigma)(b\mathbf{t})\big) \\
&= \phi\big(a^2\mathbf{t}^{\mathrm{T}}\Sigma\mathbf{t}\big)\tilde{\phi}\big(b^2c\mathbf{t}^{\mathrm{T}}\Sigma\mathbf{t}\big). \; \square
\end{aligned}$$

As usual, let $\mathbf{X} \sim E_n(\mu, \Sigma, \phi)$. Whenever $0 < \mathrm{Var}(X_i), \mathrm{Var}(X_j) < \infty$,

$$\rho(X_i, X_j) := \frac{\mathrm{Cov}(X_i, X_j)}{\sqrt{\mathrm{Var}(X_i)\,\mathrm{Var}(X_j)}} = \frac{\Sigma_{ij}}{\sqrt{\Sigma_{ii}\Sigma_{jj}}}.$$

This explains why linear correlation is a natural measure of dependence between random variables with a joint nondegenerate ($\Sigma_{ii} > 0$ for all i) elliptical distribution. Throughout this section we call the matrix R, with $R_{ij} = \Sigma_{ij}/\sqrt{\Sigma_{ii}\Sigma_{jj}}$, the linear correlation matrix of $\mathbf{X}$. Note that this definition is more general than the usual one and in this situation (elliptical distributions) makes more sense. Since an elliptical distribution is uniquely determined by μ, Σ and ϕ, the copula of a nondegenerate elliptically distributed random vector is uniquely determined by R and ϕ.

One practical problem with elliptical distributions in multivariate risk modelling is that all marginals are of the same type. To construct a realistic multivariate distribution for some given risks, it may be reasonable to choose a copula of an elliptical distribution but different types of marginals (not necessarily elliptical). One big drawback with such a model seems to be that the copula parameter R can no longer be estimated directly from data. Recall that for nondegenerate elliptical distributions with finite variances, R is just the usual linear correlation matrix. In such cases, R can be estimated using (robust) linear correlation estimators. One such robust estimator is provided by the next theorem. For nondegenerate nonelliptical distributions with finite variances and elliptical copulas, R does not correspond to the linear correlation matrix. However, since the Kendall's tau rank correlation matrix for a random vector is invariant under strictly increasing transformations of the vector components, and the next theorem provides a relation between the Kendall's tau rank correlation matrix and R for nondegenerate elliptical distributions, R can in fact easily be estimated from data.

Theorem 5.4. *Let* $\mathbf{X} \sim E_n(\mu, \Sigma, \phi)$ *with* $\mathbb{P}\{X_i = \mu_i\} < 1$ *and* $\mathbb{P}\{X_j = \mu_j\} < 1$. *Then*

$$\tau(X_i, X_j) = \left(1 - \sum_{x \in \mathbb{R}} \left(\mathbb{P}\{X_i = x\}\right)^2\right) \frac{2}{\pi} \arcsin(R_{ij}), \tag{5.1}$$

where the sum extends over all atoms of the distribution of X_i. *If* $\operatorname{rank}(\Sigma) \geqslant 2$, *then* (5.1) *simplifies to*

$$\tau(X_i, X_j) = \left(1 - \left(\mathbb{P}\{X_i = \mu_i\}\right)^2\right) \frac{2}{\pi} \arcsin(R_{ij}).$$

For a proof, see Lindskog, McNeil and Schmock (2001). Note that if $\mathbb{P}\{X_i = \mu_i\} = 0$ for all i, which is true for, e.g., multivariate t-distribution or normal distributions with strictly positive definite dispersion matrices Σ, then

$$\tau(X_i, X_j) = \frac{2}{\pi} \arcsin(R_{ij})$$

for all i and j.

The nonparametric estimator of R, $\sin(\pi\widehat{\tau}/2)$ (dropping the subscript for simplicity), provided by the above theorem, inherits the robustness properties of the Kendall's tau estimator and is an efficient (low variance) estimator of R for both elliptical distributions and nonelliptical distributions with elliptical copulas.

5.2. *Gaussian copulas*

The copula of the n-variate normal distribution with linear correlation matrix R is

$$C_R^{\mathrm{Ga}}(\mathbf{u}) = \Phi_R^n\left(\Phi^{-1}(u_1), \ldots, \Phi^{-1}(u_n)\right),$$

where Φ_R^n denotes the joint distribution function of the n-variate standard normal distribution function with linear correlation matrix R, and Φ^{-1} denotes the inverse of the distribution function of the univariate standard normal distribution. Copulas of the above form are called Gaussian copulas. In the bivariate case the copula expression can be written as

$$C_R^{\mathrm{Ga}}(u,v) = \int_{-\infty}^{\Phi^{-1}(u)} \int_{-\infty}^{\Phi^{-1}(v)} \frac{1}{2\pi\sqrt{1-R_{12}^2}} \exp\left\{-\frac{s^2 - 2R_{12}st + t^2}{2(1-R_{12}^2)}\right\} \mathrm{d}s\,\mathrm{d}t.$$

Note that R_{12} is simply the usual linear correlation coefficient of the corresponding bivariate normal distribution. Example 3.4 shows that Gaussian copulas do not have upper tail dependence. Since elliptical distributions are radially symmetric, the coefficient of upper and lower tail dependence are equal. Hence Gaussian copulas do not have lower tail dependence.

We now address the question of random variate generation from the Gaussian copula C_R^{Ga}. For our purpose, it is sufficient to consider only strictly positive definite matrices R. Write $R = AA^{\mathrm{T}}$ for some $n \times n$ matrix A, and if $Z_1, \ldots, Z_n \sim \mathcal{N}(0,1)$ are independent, then

$$\mu + A\mathbf{Z} \sim \mathcal{N}_n(\mu, R).$$

One natural choice of A is the Cholesky decomposition of R. The Cholesky decomposition of R is the unique lower-triangular matrix L with $LL^{\mathrm{T}} = R$. Furthermore Cholesky decomposition routines are implemented in most mathematical software. This provides an easy algorithm for random variate generation from the Gaussian n-copula C_R^{Ga}.

Algorithm 5.1.
- Find the Cholesky decomposition A of R.
- Simulate n independent random variates $z_1, \ldots, z_n$ from $\mathcal{N}(0,1)$.
- Set $\mathbf{x} = A\mathbf{z}$.
- Set $u_i = \Phi(x_i)$, $i = 1, \ldots, n$.
- $(u_1, \ldots, u_n)^{\mathrm{T}} \sim C_R^{\mathrm{Ga}}$.

As usual Φ denotes the univariate standard normal distribution function.

5.3. *t-copulas*

If $\mathbf{X}$ has the stochastic representation

$$\mathbf{X} =_d \mu + \frac{\sqrt{\nu}}{\sqrt{S}}\mathbf{Z}, \tag{5.2}$$

where $\mu \in \mathbb{R}^n$, $S \sim \chi_\nu^2$ and $\mathbf{Z} \sim \mathcal{N}_n(\mathbf{0}, \Sigma)$ are independent, then $\mathbf{X}$ has an n-variate t_ν-distribution with mean μ (for $\nu > 1$) and covariance matrix $\frac{\nu}{\nu-2}\Sigma$ (for $\nu > 2$). If $\nu \leqslant 2$

then $\mathrm{Cov}(\mathbf{X})$ is not defined. In this case we just interpret Σ as being the shape parameter of the distribution of $\mathbf{X}$.

The copula of $\mathbf{X}$ given by (5.2) can be written as

$$C^t_{\nu,R}(\mathbf{u}) = t^n_{\nu,R}\big(t_\nu^{-1}(u_1), \ldots, t_\nu^{-1}(u_n)\big),$$

where $R_{ij} = \Sigma_{ij}/\sqrt{\Sigma_{ii}\Sigma_{jj}}$ for $i, j \in \{1, \ldots, n\}$ and where $t^n_{\nu,R}$ denotes the distribution function of $\sqrt{\nu}\mathbf{Y}/\sqrt{S}$, where $S \sim \chi^2_\nu$ and $\mathbf{Y} \sim \mathcal{N}_n(\mathbf{0}, R)$ are independent. Here t_ν denotes the (equal) marginals of $t^n_{\nu,R}$, i.e., the distribution function of $\sqrt{\nu}Y_1/\sqrt{S}$. In the bivariate case the copula expression can be written as

$$C^t_{\nu,R}(u, v) = \int_{-\infty}^{t_\nu^{-1}(u)} \int_{-\infty}^{t_\nu^{-1}(v)} \frac{1}{2\pi\sqrt{1-R_{12}^2}} \left\{1 + \frac{s^2 - 2R_{12}st + t^2}{\nu(1-R_{12}^2)}\right\}^{-(\nu+2)/2} \mathrm{d}s\, \mathrm{d}t.$$

Note that R_{12} is simply the usual linear correlation coefficient of the corresponding bivariate t_ν-distribution if $\nu > 2$.

If $(X_1, X_2)^{\mathrm{T}}$ has a standard bivariate t-distribution with ν degrees of freedom and linear correlation matrix R, then $X_2|X_1 = x$ is t-distributed with $\upsilon + 1$ degrees of freedom and

$$\mathbb{E}(X_2|X_1 = x) = R_{12}x, \qquad \mathrm{Var}(X_2|X_1 = x) = \left(\frac{\nu + x^2}{\nu + 1}\right)\left(1 - R_{12}^2\right).$$

This can be used to show that the t-copula has upper (and because of radial symmetry) equal lower tail dependence:

$$\begin{aligned}
\lambda_U &= 2 \lim_{x\to\infty} \mathbb{P}(X_2 > x | X_1 = x) \\
&= 2 \lim_{x\to\infty} \bar{t}_{\nu+1}\left(\left(\frac{\nu+1}{\nu+x^2}\right)^{1/2} \frac{x - R_{12}x}{\sqrt{1-\rho_l^2}}\right) \\
&= 2 \lim_{x\to\infty} \bar{t}_{\nu+1}\left(\left(\frac{\nu+1}{\nu/x^2+1}\right)^{1/2} \frac{\sqrt{1-R_{12}}}{\sqrt{1+R_{12}}}\right) \\
&= 2\bar{t}_{\nu+1}\left(\frac{\sqrt{\nu+1}\sqrt{1-R_{12}}}{\sqrt{1+R_{12}}}\right).
\end{aligned}$$

From this it is also seen that the coefficient of upper tail dependence is increasing in R_{12} and decreasing in ν, as one would expect. Furthermore, the coefficient of upper (lower) tail dependence tends to zero as the number of degrees of freedom tends to infinity for $R_{12} < 1$.

Coefficients of upper tail dependence for the bivariate t-copula are given in Table 1. The last row represents the Gaussian copula, i.e., no tail dependence.

Table 1

$\nu\backslash R_{12}$	-0.5	0	0.5	0.9	1
2	0.06	0.18	0.39	0.72	1
4	0.01	0.08	0.25	0.63	1
10	0.00	0.01	0.08	0.46	1
∞	0	0	0	0	1

It should be mentioned that the expression given above is just a special case of a general formula for the coefficient(s) of tail dependence for elliptical distributions with tail dependence. It turns out that if $\Sigma_{ii} > 0$ for all i and $-1 < \Sigma_{ij}/\sqrt{\Sigma_{ii}\Sigma_{jj}} < 1$ for all $i \neq j$, then the bivariate marginal distributions of an elliptically distributed random vector $\mathbf{X} =_d \mu + RA\mathbf{U} \sim E_n(\mu, \Sigma, \phi)$ has tail dependence if and only if R is so-called regularly varying (at ∞). For more details, see Hult and Lindskog (2002), and for details about regular variation in general see Resnick (1987) or Embrechts, Mikosch and Klüppelberg (1997).

Equation (5.2) provides an easy algorithm for random variate generation from the t-copula, $C^t_{\nu,R}$.

Algorithm 5.2.
- Find the Cholesky decomposition A of R.
- Simulate n independent random variates $z_1, \ldots, z_n$ from $\mathcal{N}(0, 1)$.
- Simulate a random variate s from χ^2_ν independent of $z_1, \ldots, z_n$.
- Set $\mathbf{y} = A\mathbf{z}$.
- Set $\mathbf{x} = \frac{\sqrt{\nu}}{\sqrt{s}}\mathbf{y}$.
- Set $u_i = t_\nu(x_i)$, $i = 1, \ldots, n$.
- $(u_1, \ldots, u_n)^{\mathrm{T}} \sim C^t_{\nu,R}$.

Figures 2 and 3 show samples from bivariate distributions with Gaussian and t-copulas. In Figure 2, we have contrasted a real example (BMW-Siemens daily return data) with simulated data using marginal t_4 tails, corresponding Kendall's tau (0.5) and varying copulas. Note that the Gaussian copula does not get the extreme joint tail observations clearly present in the real data. The t_2-copula seems to be able to do a much better job in that respect. Indeed the t_2-generated scatter plot shows most of the graphical features in the real data. Note that these examples were only introduced to highlight the simulation procedures and do not constitute a detailed statistical analysis. Figure 3 (a simulated example) further highlights the difference between the Gaussian and t-copulas, this time with standard normal marginals.

The algorithms presented for the Gaussian and t-copulas are fast and easy to implement. We want to emphasize the potential usefulness of t-copulas as an alternative to Gaussian copulas. Both Gaussian and t-copulas are easily parameterized by the linear correlation matrix, but only t-copulas yield dependence structures with tail dependence.

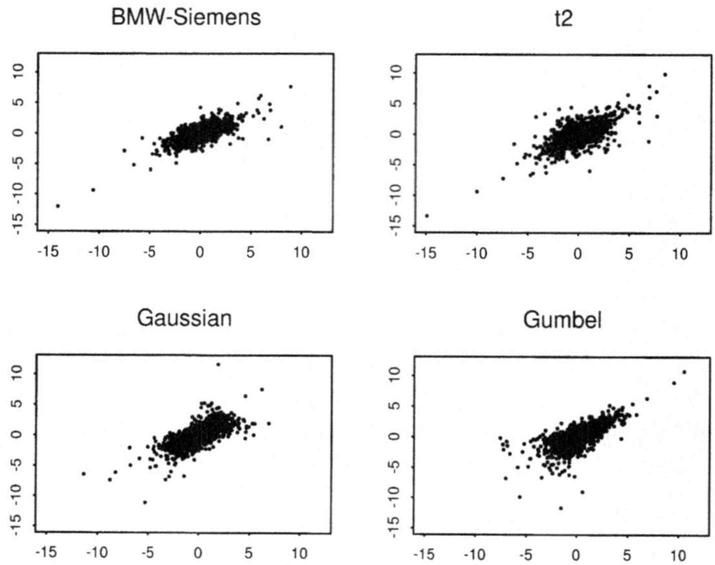

Fig. 2. The upper left plot shows BMW-Siemens daily log returns from 1989 to 1996. The other plots show samples from bivariate distributions with t_4-marginals and Kendall's tau 0.5.

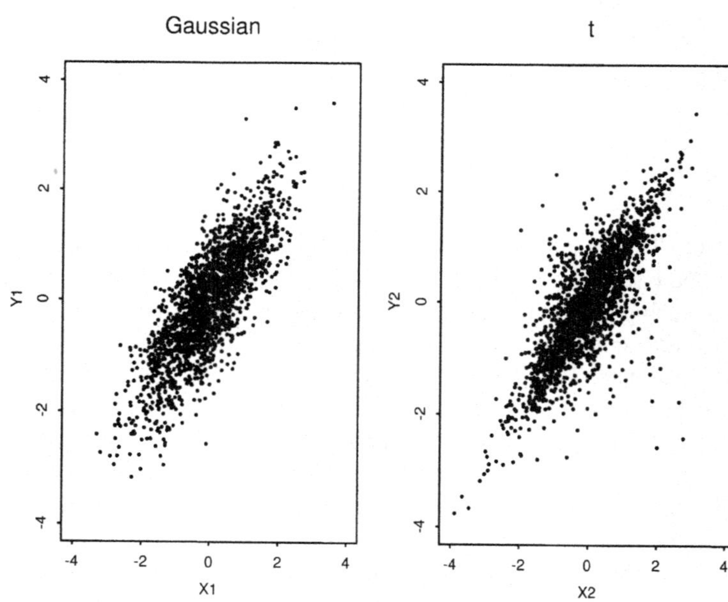

Fig. 3. Samples from two distributions with standard normal marginals, $R_{12} = 0.8$ but different dependence structures. $(X1, Y1)^{\mathrm{T}}$ has a Gaussian copula and $(X2, Y2)^{\mathrm{T}}$ has a t_2-copula.

6. Archimedean copulas

The copula families we have discussed so far have been derived from certain families of multivariate distribution functions using Sklar's Theorem. We have seen that elliptical copulas are simply the distribution functions of componentwise transformed elliptically distributed random vectors. Since simulation from elliptical distributions is easy, so is simulation from elliptical copulas. There are however drawbacks: elliptical copulas do not have closed form expressions and are restricted to have radial symmetry ($C = \widehat{C}$). In many finance and insurance applications it seems reasonable that there is a stronger dependence between big losses (e.g., a stock market crash) than between big gains. Such asymmetries cannot be modelled with elliptical copulas.

In this section we discuss an important class of copulas called Archimedean copulas. This class of copulas is worth studying for a number of reasons. Many interesting parametric families of copulas are Archimedean and the class of Archimedean copulas allow for a great variety of different dependence structures. Furthermore, in contrast to elliptical copulas, all commonly encountered Archimedean copulas have closed form expressions. Unlike the copulas discussed so far these copulas are not derived from multivariate distribution functions using Sklar's Theorem. A consequence of this is that we need somewhat technical conditions to assert that multivariate extensions of Archimedean 2-copulas are proper n-copulas. A further disadvantage is that multivariate extensions of Archimedean copulas in general suffer from lack of free parameter choice in the sense that some of the entries in the resulting rank correlation matrix are forced to be equal. At the end of this section we present one possible multivariate extension of Archimedean copulas. For other multivariate extensions we refer to Joe (1997).

There is much written about Archimedean copulas. For some background on bivariate Archimedean copulas see Genest and MacKay (1986b). For parameter estimation and a discussion on other statistical questions we refer to Genest and Rivest (1993). Good references on Archimedean copulas in general are Genest and MacKay (1986a), Nelsen (1999), Joe (1997). See also the webpage http://www.mat.ulaval.ca/pages/genest/ for further related work.

6.1. Definitions

We begin with a general definition of Archimedean copulas, which can be found in Nelsen (1999, p. 90). As our aim is the construction of multivariate extensions of Archimedean 2-copulas, this general definition will later prove to be a bit more general than needed.

Definition 6.1. Let φ be a continuous, strictly decreasing function from [0, 1] to $[0, \infty]$ such that $\varphi(1) = 0$. The pseudo-inverse of φ is the function $\varphi^{[-1]} : [0, \infty] \to [0, 1]$ given by

$$\varphi^{[-1]}(t) = \begin{cases} \varphi^{-1}(t), & 0 \leqslant t \leqslant \varphi(0), \\ 0, & \varphi(0) \leqslant t \leqslant \infty. \end{cases}$$

Note that $\varphi^{[-1]}$ is continuous and decreasing on $[0, \infty]$, and strictly decreasing on $[0, \varphi(0)]$. Furthermore, $\varphi^{[-1]}(\varphi(u)) = u$ on $[0, 1]$, and

$$\varphi\bigl(\varphi^{[-1]}(t)\bigr) = \begin{cases} t, & 0 \leqslant t \leqslant \varphi(0), \\ \varphi(0), & \varphi(0) \leqslant t \leqslant \infty. \end{cases}$$

Finally, if $\varphi(0) = \infty$, then $\varphi^{[-1]} = \varphi^{-1}$.

Theorem 6.1. *Let φ be a continuous, strictly decreasing function from $[0, 1]$ to $[0, \infty]$ such that $\varphi(1) = 0$, and let $\varphi^{[-1]}$ be the pseudo-inverse of φ. Let C be the function from $[0, 1]^2$ to $[0, 1]$ given by*

$$C(u, v) = \varphi^{[-1]}\bigl(\varphi(u) + \varphi(v)\bigr). \tag{6.1}$$

Then C is a copula if and only if φ is convex.

For a proof, see Nelsen (1999, p. 91).

Copulas of the form (6.1) are called Archimedean copulas. The function φ is called a generator of the copula. If $\varphi(0) = \infty$, we say that φ is a strict generator. In this case, $\varphi^{[-1]} = \varphi^{-1}$ and $C(u, v) = \varphi^{-1}(\varphi(u) + \varphi(v))$ is said to be a strict Archimedean copula.

Example 6.1. Let $\varphi(t) = (-\ln t)^{\theta}$, where $\theta \geqslant 1$. Clearly $\varphi(t)$ is continuous and $\varphi(1) = 0$. $\varphi'(t) = -\theta(-\ln t)^{\theta-1}\frac{1}{t}$, so φ is a strictly decreasing function from $[0, 1]$ to $[0, \infty]$. $\varphi''(t) \geqslant 0$ on $[0, 1]$, so φ is convex. Moreover $\varphi(0) = \infty$, so φ is a strict generator. From (6.1) we get

$$C_\theta(u, v) = \varphi^{-1}\bigl(\varphi(u) + \varphi(v)\bigr) = \exp\bigl(-\bigl[(-\ln u)^{\theta} + (-\ln v)^{\theta}\bigr]^{1/\theta}\bigr).$$

Furthermore $C_1 = \Pi$ and $\lim_{\theta\to\infty} C_\theta = M$ (recall that $\Pi(u, v) = uv$ and $M(u, v) = \min(u, v)$). This copula family is called the Gumbel family. As shown in Example 3.3 this copula family has upper tail dependence.

Example 6.2. Let $\varphi(t) = (t^{-\theta} - 1)/\theta$, where $\theta \in [-1, \infty) \setminus \{0\}$. This gives the Clayton family

$$C_\theta(u, v) = \max\bigl(\bigl[u^{-\theta} + v^{-\theta} - 1\bigr]^{-1/\theta}, 0\bigr).$$

For $\theta > 0$ the copulas are strict and the copula expression simplifies to

$$C_\theta(u, v) = \bigl(u^{-\theta} + v^{-\theta} - 1\bigr)^{-1/\theta}. \tag{6.2}$$

The Clayton family has lower tail dependence for $\theta > 0$, and $C_{-1} = W$, $\lim_{\theta\to 0} C_\theta = \Pi$ and $\lim_{\theta\to\infty} C_\theta = M$. Since most of the following results are results for strict Archimedean copulas we will refer to (6.2) as the Clayton family.

Example 6.3. Let $\varphi(t) = -\ln((\mathrm{e}^{-\theta t} - 1)/(\mathrm{e}^{-\theta} - 1))$, where $\theta \in \mathbb{R} \setminus \{0\}$. This gives the Frank family

$$C_\theta(u, v) = -\frac{1}{\theta} \ln\left(1 + \frac{(\mathrm{e}^{-\theta u} - 1)(\mathrm{e}^{-\theta v} - 1)}{\mathrm{e}^{-\theta} - 1}\right).$$

Frank copulas are strict Archimedean copulas. Furthermore

$$\lim_{\theta \to -\infty} C_\theta = W, \quad \lim_{\theta \to 0} C_\theta = \Pi \quad \text{and} \quad \lim_{\theta \to \infty} C_\theta = M.$$

Members of the Frank family are the only Archimedean copulas which satisfy the equation $C(u, v) = \widehat{C}(u, v)$ for so-called radial symmetry, see Frank (1979) for details.

Example 6.4. Let $\varphi(t) = 1 - t$ for t in $[0, 1]$. Then $\varphi^{[-1]}(t) = 1 - t$ for t in $[0, 1]$, and 0 for $t > 1$; i.e., $\varphi^{[-1]}(t) = \max(1 - t, 0)$. Since $C(u, v) = \max(u + v - 1, 0) =: W(u, v)$, we see that the bivariate Fréchet–Hoeffding lower bound W is Archimedean.

6.2. Properties

The results in the following theorem will enable us to formulate multivariate extensions of Archimedean copulas.

Theorem 6.2. *Let C be an Archimedean copula with generator φ. Then*
(1) *C is symmetric, i.e., $C(u, v) = C(v, u)$ for all u, v in* $[0, 1]$.
(2) *C is associative, i.e., $C(C(u, v), w) = C(u, C(v, w))$ for all u, v, w in* $[0, 1]$.

Proof: The first part follows directly from (6.1). For (2),

$$\begin{aligned} C\big(C(u, v), w\big) &= \varphi^{[-1]}\big(\varphi\big(\varphi^{[-1]}\big(\varphi(u) + \varphi(v)\big)\big) + \varphi(w)\big) \\ &= \varphi^{[-1]}\big(\varphi(u) + \varphi(v) + \varphi(w)\big) \\ &= \varphi^{[-1]}\big(\varphi(u) + \varphi\big(\varphi^{[-1]}(\varphi(v) + \varphi(w))\big)\big) = C\big(u, C(v, w)\big). \quad \square \end{aligned}$$

The associativity property of Archimedean copulas is not shared by copulas in general as shown by the following example.

Example 6.5. Let C_θ be a member of the bivariate Farlie–Gumbel–Morgenstern family of copulas, i.e., $C_\theta(u, v) = uv + \theta uv(1 - u)(1 - v)$, for $\theta \in [-1, 1]$. Then

$$C_\theta\left(\frac{1}{4}, C_\theta\left(\frac{1}{2}, \frac{1}{3}\right)\right) \neq C_\theta\left(C_\theta\left(\frac{1}{4}, \frac{1}{2}\right), \frac{1}{3}\right)$$

for all $\theta \in [-1, 1] \setminus \{0\}$. Hence the only member of the bivariate Farlie–Gumbel–Morgenstern family of copulas that is Archimedean is Π.

Theorem 6.3. *Let C be an Archimedean copula generated by φ and let*

$$K_C(t) = V_C\big(\{(u,v) \in [0,1]^2 \mid C(u,v) \leqslant t\}\big).$$

Then for any t in $[0,1]$,

$$K_C(t) = t - \frac{\varphi(t)}{\varphi'(t^+)}. \tag{6.3}$$

For a proof, see Nelsen (1999, p. 102).

Corollary 6.1. *If $(U,V)^{\mathrm{T}}$ has distribution function C, where C is an Archimedean copula generated by φ, then the function K_C given by* (6.3) *is the distribution function of the random variable $C(U,V)$.*

The next theorem will provide the basis for a general algorithm for random variate generation from Archimedean copulas. Before the theorem can be stated we need an expression for the density of an absolutely continuous Archimedean copula. From (6.1) it follows that

$$\varphi'\big(C(u,v)\big)\frac{\partial}{\partial u}C(u,v) = \varphi'(u),$$

$$\varphi'\big(C(u,v)\big)\frac{\partial}{\partial v}C(u,v) = \varphi'(v),$$

$$\varphi''\big(C(u,v)\big)\frac{\partial}{\partial u}C(u,v)\frac{\partial}{\partial v}C(u,v) + \varphi'\big(C(u,v)\big)\frac{\partial^2}{\partial u \partial v}C(u,v) = 0,$$

and hence

$$\frac{\partial^2}{\partial u \partial v}C(u,v) = -\frac{\varphi''(C(u,v))\frac{\partial}{\partial u}C(u,v)\frac{\partial}{\partial v}C(u,v)}{\varphi'(C(u,v))} = -\frac{\varphi''(C(u,v))\varphi'(u)\varphi'(v)}{[\varphi'(C(u,v))]^3}.$$

Thus, when C is absolutely continuous, its density is given by

$$\frac{\partial^2}{\partial u \partial v}C(u,v) = -\frac{\varphi''(C(u,v))\varphi'(u)\varphi'(v)}{[\varphi'(C(u,v))]^3}. \tag{6.4}$$

Theorem 6.4. *Under the hypotheses of Corollary* 6.1, *the joint distribution function $H(s,t)$ of the random variables $S = \varphi(U)/[\varphi(U) + \varphi(V)]$ and $T = C(U,V)$ is given by $H(s,t) = sK_C(t)$ for all (s,t) in $[0,1]^2$. Hence S and T are independent, and S is uniformly distributed on* $[0,1]$.

Proof: [This proof, for the case when C is absolutely continuous, can be found in Nelsen (1999, p. 104). For the general case, see Genest and Rivest (1993).] The joint density $h(s,t)$ of S and T is given by

$$h(s,t) = \frac{\partial^2}{\partial u \partial v} C(u,v) \left| \frac{\partial(u,v)}{\partial(s,t)} \right|,$$

where $\partial^2 C(u,v)/\partial u \partial v$ is given by (6.4) and $\partial(u,v)/\partial(s,t)$ denotes the Jacobian of the transformation $\varphi(u) = s\varphi(t)$, $\varphi(v) = (1-s)\varphi(t)$. But

$$\frac{\partial(u,v)}{\partial(s,t)} = \frac{\varphi(t)\varphi'(t)}{\varphi'(u)\varphi'(v)},$$

and hence

$$h(s,t) = \left(-\frac{\varphi''(t)\varphi'(u)\varphi'(v)}{[\varphi'(t)]^3}\right)\left(-\frac{\varphi(t)\varphi'(t)}{\varphi'(u)\varphi'(v)}\right) = \frac{\varphi''(t)\varphi(t)}{[\varphi'(t)]^2}.$$

Therefore

$$H(s,t) = \int_0^s \int_0^t \frac{\varphi''(y)\varphi(y)}{[\varphi'(y)]^2}\,\mathrm{d}y\,\mathrm{d}x = s\left[y - \frac{\varphi(y)}{\varphi'(y)}\right]_0^t = sK_C(t),$$

from which the conclusion follows. □

An application of Theorem 6.4 is the following algorithm for generating random variates $(u,v)^{\mathrm{T}}$ whose joint distribution is an Archimedean copula C with generator φ.

Algorithm 6.1.
- Simulate two independent $U(0,1)$ random variates s and q.
- Set $t = K_C^{-1}(q)$, where K_C is the distribution function of $C(U,V)$.
- Set $u = \varphi^{[-1]}(s\varphi(t))$ and $v = \varphi^{[-1]}((1-s)\varphi(t))$.

Note that the variates s and t correspond to the random variables S and T in Theorem 6.4 and from the proof it follows that this algorithm yields the desired result.

Example 6.6. Consider the Archimedean copula family given by

$$C_\theta(u,v) = \left(1 + \left[(u^{-1}-1)^\theta + (v^{-1}-1)^\theta\right]^{1/\theta}\right)^{-1}$$

generated by $\varphi_\theta(t) = (t^{-1}-1)^\theta$ for $\theta \geqslant 1$. To generate a random variate from C we simply apply Algorithm 6.1 with

$$\varphi_\theta(t) = \left(t^{-1} - 1\right)^\theta,$$
$$\varphi_\theta^{-1}(t) = \left(t^{1/\theta} + 1\right)^{-1},$$
$$K_{C_\theta}^{-1}(t) = \frac{\theta_i + 1}{2} - \sqrt{\left(\frac{\theta_i + 1}{2}\right)^2 - \theta_i s}.$$

6.3. Kendall's tau revisited

Recall that Kendall's tau for a copula C can be expressed as a double integral of C. This double integral is in most cases not straightforward to evaluate. However for an Archimedean copula, Kendall's tau can be expressed as an (one-dimensional) integral of the generator and its derivative, as shown in the following theorem from Genest and MacKay (1986a).

Theorem 6.5. *Let X and Y be random variables with an Archimedean copula C generated by φ. Kendall's tau of X and Y is given by*

$$\tau_C = 1 + 4\int_0^1 \frac{\varphi(t)}{\varphi'(t)}\,\mathrm{d}t. \tag{6.5}$$

Proof: Let U and V be $U(0,1)$ random variables with joint distribution function C, and let K_C denote the distribution function of $C(U,V)$. Then from Theorem 3.3 we have

$$\tau_C = 4\mathbb{E}\big(C(U,V)\big) - 1 = 4\int_0^1 t\,\mathrm{d}K_C(t) - 1$$
$$= 4\left(\left[tK_C(t)\right]_0^1 - \int_0^1 K_C(t)\,\mathrm{d}t\right) - 1 = 3 - 4\int_0^1 K_C(t)\,\mathrm{d}t.$$

From Theorem 6.3 and Corollary 6.1 it follows that

$$K_C(t) = t - \frac{\varphi(t)}{\varphi'(t^+)}.$$

Since φ is convex, $\varphi'(t^+)$ and $\varphi'(t^-)$ exist for all t in $(0,1)$ and the set $\{t \in (0,1) \mid \varphi'(t^+) \neq \varphi'(t^-)\}$ is at most countable (i.e., it has Lebesgue measure zero). Hence

$$\tau_C = 3 - 4\int_0^1 \left(t - \frac{\varphi(t)}{\varphi'(t^+)}\right)\mathrm{d}t = 1 + 4\int_0^1 \frac{\varphi(t)}{\varphi'(t)}\,\mathrm{d}t. \quad \square$$

Example 6.7. Consider the Gumbel family with generator $\varphi(t) = (-\ln t)^\theta$, for $\theta \geqslant 1$. Then $\varphi(t)/\varphi'(t) = (t \ln t)/\theta$. Using Theorem 6.5 we can calculate Kendall's tau for the Gumbel family.

$$\tau_\theta = 1 + 4\int_0^1 \frac{t \ln t}{\theta}\,\mathrm{d}t = 1 + \frac{4}{\theta}\left(\left[\frac{t^2}{2}\ln t\right]_0^1 - \int_0^1 \frac{t}{2}\,\mathrm{d}t\right) = 1 - \frac{1}{\theta}.$$

As a consequence, in order to have Kendall's tau equal to 0.5 in Figure 2 (the Gumbel case), we put $\theta = 2$.

Example 6.8. Consider the Clayton family with generator $\varphi(t) = (t^{-\theta} - 1)/\theta$, for $\theta \in [-1, \infty) \setminus \{0\}$. Then $\varphi(t)/\varphi'(t) = (t^{\theta+1} - t)/\theta$. Using Theorem 6.5 we can calculate Kendall's tau for the Clayton family.

$$\tau_\theta = 1 + 4\int_0^1 \frac{t^{\theta+1} - t}{\theta}\,\mathrm{d}t = 1 + \frac{4}{\theta}\left(\frac{1}{\theta+2} - \frac{1}{2}\right) = \frac{\theta}{\theta+2}.$$

Example 6.9. Consider the Frank family presented in Example 6.3. It can be shown that [see, e.g., Genest (1987)] Kendall's tau is $\tau_\theta = 1 - 4(1 - D_1(\theta))/\theta$, where $D_k(x)$ is the Debye function, given by

$$D_k(x) = \frac{k}{x^k}\int_0^x \frac{t^k}{\mathrm{e}^t - 1}\,\mathrm{d}t$$

for any positive integer k.

6.4. *Tail dependence revisited*

For Archimedean copulas, tail dependence can be expressed in terms of the generators.

Theorem 6.6. *Let φ be a strict generator such that φ^{-1} belongs to the class of Laplace transforms of strictly positive random variables. If $\varphi^{-1\prime}(0)$ is finite, then*

$$C(u, v) = \varphi^{-1}\big(\varphi(u) + \varphi(v)\big)$$

does not have upper tail dependence. If C has upper tail dependence, then $\varphi^{-1\prime}(0) = -\infty$ and the coefficient of upper tail dependence is given by

$$\lambda_U = 2 - 2\lim_{s \searrow 0} \frac{\varphi^{-1\prime}(2s)}{\varphi^{-1\prime}(s)}.$$

Proof: [This proof can be found in Joe (1997, p. 103).] Note that

$$\lim_{u\nearrow 1}\frac{\overline{C}(u,u)}{1-u}=\lim_{u\nearrow 1}\frac{1-2u+\varphi^{-1}(2\varphi(u))}{1-u}$$
$$=2-2\lim_{u\nearrow 1}\frac{\varphi^{-1\prime}(2\varphi(u))}{\varphi^{-1\prime}(\varphi(u))}$$
$$=2-2\lim_{s\searrow 0}\frac{\varphi^{-1\prime}(2s)}{\varphi^{-1\prime}(s)}.$$

If $\varphi^{-1'}(0)\in(-\infty,0)$, then the limit is zero and C does not have upper tail dependence. Since $\varphi^{-1'}(0)$ is the negative of the expectation of a strictly positive random variable, $\varphi^{-1'}(0)<0$ from which the conclusion follows. □

The additional condition on the generator φ might seem somewhat strange. It will however prove quite natural when we turn to the construction of multivariate Archimedean copulas. Furthermore, the condition is satisfied by the majority of the commonly encountered Archimedean copulas.

Example 6.10. The Gumbel copulas are strict Archimedean with generator $\varphi(t)=(-\ln t)^{\theta}$. Hence $\varphi^{-1}(s)=\exp(-s^{1/\theta})$ and its derivative $\varphi^{-1'}(s)=-s^{1/\theta-1}\exp(-s^{1/\theta})/\theta$. Using Theorem 6.6 we get

$$\lambda_U=2-2\lim_{s\searrow 0}\frac{\varphi^{-1\prime}(2s)}{\varphi^{-1\prime}(s)}=2-2^{1/\theta}\lim_{s\searrow 0}\frac{\exp(-(2s)^{1/\theta})}{\exp(-s^{1/\theta})}=2-2^{1/\theta},$$

see also Example 3.3.

Theorem 6.7. *Let φ be as in Theorem 6.6. The coefficient of lower tail dependence for the copula $C(u,v)=\varphi^{-1}(\varphi(u)+\varphi(v))$ is equal to*

$$\lambda_L=2\lim_{s\to\infty}\frac{\varphi^{-1\prime}(2s)}{\varphi^{-1\prime}(s)}.$$

The proof is similar to that of Theorem 6.6.

Example 6.11. Consider the Clayton family given by $C_\theta(u,v)=(u^{-\theta}+v^{-\theta}-1)^{-1/\theta}$, for $\theta>0$. This strict copula family has generator $\varphi(t)=(t^{-\theta}-1)/\theta$. It follows that $\varphi^{-1}(s)=(1+\theta s)^{-1/\theta}$. Using Theorems 6.6 and 6.7 shows that $\lambda_U=0$ and that the coefficient of lower tail dependence given by

$$\lambda_L=2\lim_{s\to\infty}\frac{\varphi^{-1\prime}(2s)}{\varphi^{-1\prime}(s)}=2\lim_{s\to\infty}\frac{(1+2\theta s)^{-1/\theta-1}}{(1+\theta s)^{-1/\theta-1}}=2^{-1/\theta}.$$

Example 6.12. Consider the Frank family given by

$$C_\theta(u,v) = -\frac{1}{\theta}\ln\left(1+\frac{(e^{-\theta u}-1)(e^{-\theta v}-1)}{e^{-\theta}-1}\right)$$

for $\theta \in \mathbb{R}\setminus\{0\}$. This strict copula family has generator $\varphi(t) = -\ln((e^{-\theta t}-1)/(e^{-\theta}-1))$. It follows that

$$\varphi^{-1}(s) = -\frac{1}{\theta}\ln\bigl(1-(1-e^{-\theta})e^{-s}\bigr) \quad \text{and} \quad \varphi^{-1\prime}(s) = -\frac{1}{\theta}\frac{(1-e^{-\theta})e^{-s}}{1-(1-e^{-\theta})e^{-s}}.$$

Since

$$\varphi^{-1\prime}(0) = -\frac{e^{\theta}-1}{\theta}$$

is finite, the Frank family does not have upper tail dependence according to Theorem 6.6. Furthermore, members of the Frank family are radially symmetric, i.e. $C = \widehat{C}$, and hence the Frank family does not have lower tail dependence.

6.5. Multivariate Archimedean copulas

In this section we look at the construction of one particular multivariate extension of Archimedean 2-copulas. For other multivariate extensions see Joe (1997). It should be noted that in order to show that other multivariate extensions are proper copulas, we essentially have to go through the same arguments as those given below.

The expression for the n-dimensional product copula Π^n, with $\mathbf{u} = (u_1,\dots,u_n)^{\mathrm{T}}$, can be written as $\Pi^n(\mathbf{u}) = u_1\dots u_n = \exp(-[(-\ln u_1)+\dots+(-\ln u_n)])$. This naturally leads to the following generalization of (6.1):

$$C^n(\mathbf{u}) = \varphi^{[-1]}\bigl(\varphi(u_1)+\dots+\varphi(u_n)\bigr). \tag{6.6}$$

In the 3-dimensional case,

$$C^3(u_1,u_2,u_3) = \varphi^{[-1]}\bigl(\varphi\circ\varphi^{[-1]}\bigl(\varphi(u_1)+\varphi(u_2)\bigr)+\varphi(u_3)\bigr) = C\bigl(C(u_1,u_2),u_3\bigr),$$

and in the 4-dimensional case,

$$\begin{aligned} &C^4(u_1,\dots,u_4)\\ &= \varphi^{[-1]}\bigl(\varphi\circ\varphi^{[-1]}\bigl(\varphi\circ\varphi^{[-1]}\bigl(\varphi(u_1)+\varphi(u_2)\bigr)+\varphi(u_3)\bigr)+\varphi(u_4)\bigr)\\ &= C\bigl(C^3(u_1,u_2,u_3),u_4\bigr) = C\bigl(C\bigl(C(u_1,u_2),u_3\bigr),u_4\bigr). \end{aligned}$$

Whence in general, $n \geqslant 3$, $C^n(u_1, \ldots, u_n) = C(C^{n-1}(u_1, u_2, \ldots, u_{n-1}), u_n)$. This technique of constructing higher-dimensional copulas generally fails. But since Archimedean copulas are symmetric and associative it seems more likely that C^n as defined above, given certain additional properties of φ (and $\varphi^{[-1]}$), is indeed a copula for $n \geqslant 3$.

Definition 6.2. A function $g(t)$ is completely monotone on the interval I if it has derivatives of all orders which alternate in sign, i.e., if it satisfies

$$(-1)^k \frac{\mathrm{d}^k}{\mathrm{d}t^k} g(t) \geqslant 0$$

for all t in the interior of I and $k = 0, 1, 2, \ldots$.

If $g : [0, \infty) \mapsto [0, \infty)$ is completely monotone on $[0, \infty)$ and there is a $t \in [0, \infty)$ such that $g(t) = 0$, then $g(t) = 0$ for all $t \in [0, \infty)$. Hence if the pseudo-inverse $\varphi^{[-1]}$ of an Archimedean generator φ is completely monotone, then $\varphi^{[-1]}(t) > 0$ for all $t \in [0, \infty)$ and hence $\varphi^{[-1]} = \varphi^{-1}$.

The following theorem from Kimberling (1974) gives necessary and sufficient conditions for the function (6.6) to be an n-copula.

Theorem 6.8. *Let φ be a continuous strictly decreasing function from $[0, 1]$ to $[0, \infty]$ such that $\varphi(0) = \infty$ and $\varphi(1) = 0$, and let φ^{-1} denote the inverse of φ. If C^n is the function from $[0, 1]^n$ to $[0, 1]$ given by (6.6), then C^n is an n-copula for all $n \geqslant 2$ if and only if φ^{-1} is completely monotone on $[0, \infty)$.*

This theorem can be partially extended to the case where φ is nonstrict and $\varphi^{[-1]}$ is m-monotone on $[0, \infty)$ for some $m \geqslant 2$, that is, the derivatives of $\varphi^{[-1]}$ alter sign up to and including the mth order on $[0, \infty)$. Then the function C^n given by (6.6) is an n-copula for $2 \leqslant n \leqslant m$. However, for most practical purposes, the class of strict generators φ such that φ^{-1} is completely monotone is a rich enough class.

The following corollary shows that the generators suitable for extensions to arbitrary dimensions of Archimedean 2-copulas correspond to copulas which can model only positive dependence.

Corollary 6.2. *If the inverse φ^{-1} of a strict generator φ of an Archimedean copula C is completely monotone, then $C \succ \Pi$, i.e., $C(u, v) \geqslant uv$ for all u, v in $[0, 1]$.*

For a proof, see Nelsen (1999, p. 122).

While it is simple to generate n-copulas of the form given by (6.6), they suffer from a very limited dependence structure since all k-marginals are identical, they are distribution functions of n exchangeable $U(0, 1)$ random variables. One would like to have a multivariate extension of the Archimedean 2-copula given by (6.1) which allows for nonexchangeability. Such multivariate extensions are discussed in Joe (1997). We will now discuss one such extension in detail. Since any multivariate extension should contain (6.6) as a special case, clearly the necessary conditions for (6.6) to be a copula has to be satisfied. In the light of Theorem 6.8, we restrict ourselves to strict generators.

The expression for the general multivariate extension of (6.1) we will now discuss is notationally complex. For that reason we will discuss sufficient conditions for the 3- and 4-dimensional extensions to be proper 3- and 4-copulas respectively. The pattern and conditions indicated generalize in an obvious way to higher dimensions. The 3-dimensional generalization of (6.1) is

$$\varphi_1^{-1}\big(\varphi_1 \circ \varphi_2^{-1}\big(\varphi_2(u_1) + \varphi_2(u_2)\big) + \varphi_1(u_3)\big), \tag{6.7}$$

where φ_1 and φ_2 are generators of strict Archimedean copulas. The 4-dimensional generalization of (6.1) is

$$\varphi_1^{-1}\big(\varphi_1 \circ \varphi_2^{-1}\big(\varphi_2 \circ \varphi_3^{-1}\big(\varphi_3(u_1) + \varphi_3(u_2)\big) + \varphi_2(u_3)\big) + \varphi_1(u_4)\big), \tag{6.8}$$

where φ_1, φ_2 and φ_3 are generators of strict Archimedean copulas. The expressions (6.7) and (6.8) can be written as

$$C_1\big(C_2(u_1, u_2), u_3\big) \quad \text{and} \quad C_1\big(C_2\big(C_3(u_1, u_2), u_3\big), u_4\big),$$

respectively, where C_i denotes an Archimedean copula generated by φ_i.

If generators φ_i are chosen so that certain conditions are satisfied, then multivariate copulas can be obtained such that each bivariate marginal has the form (6.1) for some i. However, the number of distinct generators φ_i among the $n(n-1)/2$ bivariate marginals is only $n-1$, so that the resulting dependence structure is one of partial exchangeability.

Clearly the generators have to satisfy the necessary conditions for the n-copula given by (6.6) in order to make (6.7) and (6.8) valid copula expressions. What other conditions are needed to make these proper copulas? To answer that question we now introduce function classes $\mathcal{L}_n$ and $\mathcal{L}_n^*$. Let

$$\mathcal{L}_n = \big\{\phi\colon [0, \infty) \to [0, 1] \mid \phi(0) = 1,\ \phi(\infty) = 0,\ (-1)^j \phi^{(j)} \geqslant 0,\\ j = 1, \ldots, n\big\},$$

$n = 1, 2, \ldots, \infty$, with $\mathcal{L}_\infty$ being the class of Laplace transforms of strictly positive random variables.

Also introduce

$$\mathcal{L}_n^* = \big\{\omega\colon [0, \infty) \to [0, \infty) \mid \omega(0) = 0,\ \omega(\infty) = \infty,\ (-1)^{j-1} \omega^{(j)} \geqslant 0,\\ j = 1, \ldots, n\big\},$$

$n = 1, 2, \ldots, \infty$. Note that $\varphi^{-1} \in \mathcal{L}_1$ if φ is the generator of a strict Archimedean copula. The functions in $\mathcal{L}_n^*$ are usually compositions of the form $\psi^{-1} \circ \phi$ with $\psi, \phi \in \mathcal{L}_1$.

Note also that with this notation, the necessary and sufficient conditions for (6.6) to be a proper copula is that $\varphi^{-1} \in \mathcal{L}_n$ and that, if (6.6) is a copula for all n, then φ^{-1} must be completely monotone and hence be a Laplace transform of a strictly positive random variable.

It turns out that if φ_1^{-1} and φ_2^{-1} are completely monotone (Laplace transforms of strictly positive random variables) and $\varphi_1 \circ \varphi_2^{-1} \in \mathcal{L}_\infty^*$, then (6.7) is a proper copula. Note that (6.7) has $(1, 2)$ bivariate marginal of the form (6.1) with generator φ_2 and $(1, 3)$ and $(2, 3)$ bivariate marginals of the form (6.1) with generator φ_1. Also (6.6) is the special case of (6.7) with $\varphi_1 = \varphi_2$. The 3-dimensional copula in (6.7) has a $(1, 2)$ bivariate marginal copula which is larger than the $(1, 3)$ and $(2, 3)$ bivariate marginal copulas (which are identical).

As one would expect, there are similar conditions for the 4-dimensional case. If φ_1^{-1}, φ_2^{-1} and φ_3^{-1} are completely monotone (Laplace transforms of strictly positive random variables) and $\varphi_1 \circ \varphi_2^{-1}$ and $\varphi_2 \circ \varphi_3^{-1}$ are in $\mathcal{L}_\infty^*$, then (6.8) is a proper copula. Note that all 3-dimensional marginals of (6.8) have the form (6.7) and all bivariate marginals have the form (6.1). Clearly the idea underlying (6.7) and (6.8) generalize to higher dimensions.

Example 6.13. Let $\varphi_i(t) = (-\ln t)^{\theta_i}$ with $\theta_i \geqslant 1$ for $i = 1, \ldots, n$, i.e., the generators of Gumbel copulas. What conditions do we have to impose on $\theta_1, \ldots, \theta_n$ in order to obtain an n-dimensional extension of the Gumbel family of the form indicated above (expressions (6.7) and (6.8)). It should first be noted that $\varphi_i^{-1} \in \mathcal{L}_\infty$ for all i, so (6.6) with the above generators gives an n-copula for all $n \geqslant 2$. Secondly, $\varphi_i \circ \varphi_{i+1}^{-1}(t) = t^{\theta_i/\theta_{i+1}}$. If $\theta_i/\theta_{i+1} \notin \mathbb{N}$, then the nth derivative of $\varphi_i \circ \varphi_{i+1}^{-1}(t)$ is given by

$$\frac{\theta_i}{\theta_{i+1}} \cdots \left(\frac{\theta_i}{\theta_{i+1}} - (n-1)\right) t^{\theta_i/\theta_{i+1}-n}.$$

Hence if $\theta_i/\theta_{i+1} \notin \mathbb{N}$, then $\varphi_i \circ \varphi_{i+1}^{-1} \in \mathcal{L}_\infty^*$ if and only if $\theta_i/\theta_{i+1} < 1$. If $\theta_i/\theta_{i+1} \in \mathbb{N}$, then $\varphi_i \circ \varphi_{i+1}^{-1} \in \mathcal{L}_\infty^*$ if and only if $\theta_i/\theta_{i+1} = 1$. Hence an n-dimensional extension of the Gumbel family of the form indicated above, given by

$$\exp\left\{-\left(\left[(-\ln u_1)^{\theta_2} + (-\ln u_2)^{\theta_2}\right]^{\theta_1/\theta_2} + (-\ln u_3)^{\theta_1}\right)^{1/\theta_1}\right\}$$

in the 3-dimensional case, is a proper n-copula if $\theta_1 \leqslant \cdots \leqslant \theta_n$.

Example 6.14. Consider the Archimedean copula family given by

$$C_\theta(u, v) = \left(1 + \left[\left(u^{-1} - 1\right)^\theta + \left(v^{-1} - 1\right)^\theta\right]^{1/\theta}\right)^{-1}$$

generated by $\varphi_\theta(t) = (t^{-1} - 1)^\theta$ for $\theta \geqslant 1$. Set $\varphi_i(t) = \varphi_{\theta_i}(t)$ for $i = 1, \ldots, n$. Can the above copulas be extended to n-copulas of the form indicated by (6.7) and (6.8), and if so under what conditions on $\theta_1, \ldots, \theta_n$? By calculating derivatives of φ_i^{-1} and $\varphi_i \circ \varphi_{i+1}^{-1}$ it follows that $\varphi_i^{-1} \in \mathcal{L}_\infty$ and $\varphi_i \circ \varphi_{i+1}^{-1} \in \mathcal{L}_\infty^*$ if and only if $\theta_i/\theta_{i+1} \leqslant 1$. Hence the n-dimensional extension of the above copulas are n-copulas if $\theta_1 \leqslant \cdots \leqslant \theta_n$.

Copulas of the above form have upper and lower tail dependence, with coefficients of upper and lower tail dependence given by $2 - 2^{1/\theta}$ and $2^{-1/\theta}$ respectively. One limiting factor for the usefulness of this copula family might be that they only allow for a limited range of positive dependence, as seen from the expression for Kendall's tau given by $\tau = 1 - 2/(3\theta)$, for $\theta \geqslant 1$.

Note that the results presented in this section hold for strict Archimedean copulas. With some additional constraints most of the results can be generalized to hold also for nonstrict Archimedean copulas. However for practical purposes it is sufficient to only consider strict Archimedean copulas. This basically means (there are exceptions such as the Frank family) that we consider copula families with only positive dependence. Furthermore, risk models are often designed to model positive dependence, since in some sense it is the "dangerous" dependence: assets (or risks) move in the same direction in periods of extreme events.

7. Modelling extremal events in practice

7.1. *Insurance risk*

Consider a portfolio consisting of n risks $X_1, \ldots, X_n$, representing potential losses in different lines of business for an insurance company. Suppose that the insurance company, in order to reduce the risk in its portfolio, seeks protection against simultaneous big losses in different lines of business. One suitable reinsurance contract might be the one which pays the excess losses $X_i - k_i$ for $i \in B \subseteq \{1, \ldots, n\}$ (where B is some prespecified set of business lines), given that $X_i > k_i$ for all $i \in B$. Hence the payout function f is given by

$$f\big((X_i, k_i);\ i \in B\big) = \Bigg(\prod_{i \in B} 1_{\{X_i > k_i\}}\Bigg)\Bigg(\sum_{i \in B}(X_i - k_i)\Bigg). \tag{7.1}$$

In order to price this contract the seller (reinsurer) would typically need to estimate $\mathbb{E}(f((X_i, k_i);\ i \in B))$. Without loss of generality let $B = \{1, \ldots, l\}$ for $l \leqslant n$. If the joint distribution H of $X_1, \ldots, X_l$ could be accurately estimated, calculating the expected value of (7.1) (possibly by using numerical methods) would not be difficult. Unfortunately, accurate estimation of H is seldom possible due to lack of reliable data. It is more realistic, and we will assume this, that the data available allow for estimation of the marginals $F_1, \ldots, F_n$ of H and pairwise rank correlations. The probability of payout is given by

$$\overline{H}(k_1, \ldots, k_l) = \widehat{C}\big(\overline{F}_1(k_1), \ldots, \overline{F}_l(k_l)\big), \tag{7.2}$$

where $\overline{H}$ and $\widehat{C}$ denotes the joint survival function and survival copula of $X_1, \ldots, X_l$. If the thresholds are chosen to be quantiles of the X_is, i.e., if $k_i = F_i^{-1}(\alpha_i)$ for all i, then the right-hand side of (7.2) simplifies to $\widehat{C}(1-\alpha_1,\ldots,1-\alpha_l)$. In a reinsurance context, these quantile levels are often given as return periods and are known to the underwriter.

For a specific copula family, Kendall's tau estimates can typically be transformed into an estimate of the copula parameters. For Gaussian (elliptical) n-copulas this is due to the relation $R_{ij}=\sin(\pi\tau(X_i,X_j)/2)$, where $R_{ij}=\Sigma_{ij}/\sqrt{\Sigma_{ii}\Sigma_{jj}}$ with Σ being the dispersion matrix of the corresponding normal (elliptical) distribution. For the multivariate extension of the Gumbel family presented in Section 6.5 this is due to the relation $\theta=1/(1-\tau(X_i,X_j))$, where θ denotes the copula parameter for the bivariate Gumbel copula of $(X_i,X_j)^{\mathrm{T}}$. Hence, once a copula family is decided upon, calculating the probability of payout or the expected value of the contract is easy. However there is much uncertainty in choosing a suitable copula family representing the dependence between potential losses for the l lines of business. The data may give indications of properties such as tail dependence but it should be combined with careful consideration of the nature of the underlying loss causing mechanisms. To show the relevance of good dependence modelling, we will consider marginal distributions and pairwise rank correlations to be given and compare the effect of the Gaussian and Gumbel copula on the probability of payout and expected value of the contract. To be able to interpret the results more easily, we make some further simplifications: let $X_i\sim F$ for all i, where F is the distribution function of the standard Lognormal distribution LN(0, 1), let $k_i=k$ for all i and let $\tau(X_i,X_j)=0.5$ for all $i\neq j$. Then,

$$\overline{H}(k,\ldots,k)=1+(-1)\binom{l}{1}C_1\bigl(F(k)\bigr)+\cdots+(-1)^l\binom{l}{l}C_l\bigl(F(k),\ldots,F(k)\bigr),$$

where C_m, for $m=1,\ldots,l-1$, are m-dimensional marginals of $C=C_l$ (the copula of $(X_1,\ldots,X_l)$). In the Gaussian case,

$$C_m\bigl(F(k),\ldots,F(k)\bigr)=\Phi^m_{R_m}\bigl(\Phi^{-1}\bigl(F(k)\bigr),\ldots,\Phi^{-1}\bigl(F(k)\bigr)\bigr),$$

where $\Phi^m_{R_m}$ denotes the distribution function of m multivariate normally distributed random variables with linear correlation matrix R_m with off-diagonal entries $\sin(\pi 0.5/2)=1/\sqrt{2}$. $\Phi^m_{\rho_l}(\Phi^{-1}(F(k)),\ldots,\Phi^{-1}(F(k)))$ can be calculated by numerical integration using the fact that [see Johnson and Kotz (1972, p. 48)]

$$\Phi^m_{\rho_l}(a,\ldots,a)=\int_{-\infty}^{\infty}\phi(x)\left[\Phi\left(\frac{a-\sqrt{\rho_l}x}{\sqrt{1-\rho_l}}\right)\right]^m \mathrm{d}x,$$

where ϕ denotes the univariate standard normal density function. In the Gumbel case,

$$\begin{aligned}C_m\bigl(F(k),\ldots,F(k)\bigr)&=\exp\bigl\{-\bigl[\bigl(-\ln F(k)\bigr)^{\theta}+\cdots+\bigl(-\ln F(k)\bigr)^{\theta}\bigr]^{1/\theta}\bigr\}\\&=F(k)^{m^{1/\theta}},\end{aligned}$$

where $\theta=1/(1-0.5)=2$.

For illustration, let $l = 5$, i.e., we consider 5 different lines of business. Figure 4 shows payout probabilities (probabilities of joint exceedances) for thresholds $k \in [0, 15]$, when the dependence structure among the potential losses are given by a Gaussian copula (lower curve) and a Gumbel copula (upper curve). If we let $k = F^{-1}(0.99) \approx 10.25$, i.e., payout occurs when all 5 losses exceed their respective 99% quantile, then Figure 5 shows that if one would choose a Gaussian copula when the true dependence structure between

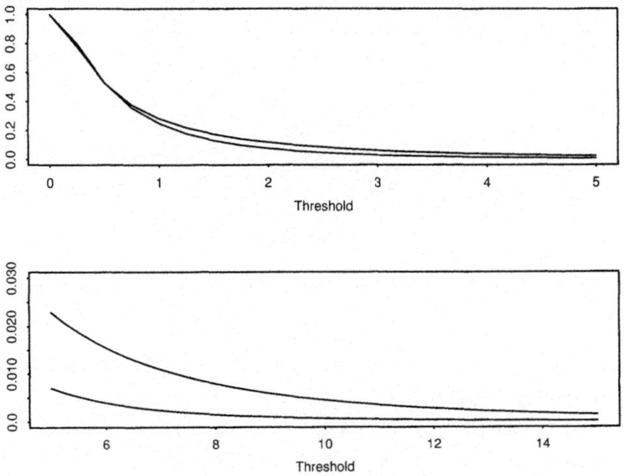

Fig. 4. Probability of payout for $l = 5$ when the dependence structure is given by a Gaussian copula (lower curve) and Gumbel copula (upper curve).

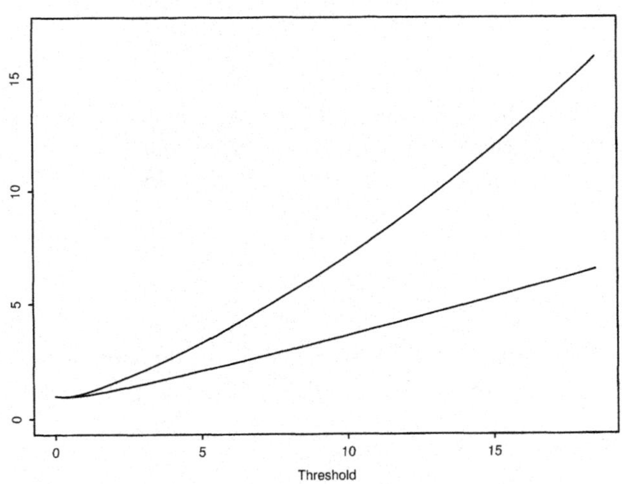

Fig. 5. Ratios of payout probabilities (Gumbel/Gaussian) for $l = 3$ (lower curve) and $l = 5$ (upper curve).

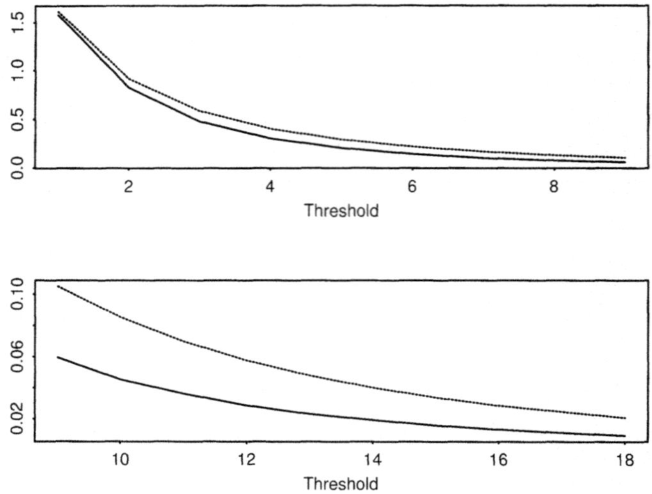

Fig. 6. Estimates of $\mathbb{E}(f(X_1, X_2, k))$ for Gaussian (lower curve) and Gumbel (upper curve) copulas.

the potential losses $X_1, \ldots, X_5$ is given by a Gumbel copula, the probability of payout is underestimated almost by a factor 8.

Figure 6 shows estimates of $\mathbb{E}(f(X_1, X_2, k))$ for $k = 1, \ldots, 18$. The lower curve shows estimates for the expectation when $(X_1, X_2)^{\mathrm{T}}$ has a Gaussian copula and the upper curve when $(X_1, X_2)^{\mathrm{T}}$ has a Gumbel copula. The estimates are sample means from samples of size 150000. Since $F^{-1}(0.99) \approx 10.25$, Figure 6 shows that if one would choose a Gaussian copula when the true dependence between the potential losses X_1 and X_2 is given by a Gumbel copula, the expected loss to the reinsurer is underestimated by a factor 2.

7.2. Market risk

We now consider the problem of measuring the risk of holding an equity portfolio over a short time horizon (one day, say) without the possibility of rebalancing. More precisely, consider a portfolio of n equities with current value given by

$$V_t = \sum_{i=1}^{n} \beta_i S_{i,t},$$

where β_i is the number of units of equity i and $S_{i,t}$ is the current price of equity i. Let $\Delta_{t+1} = -(V_{t+1} - V_t)/V_t$, the (negative) relative loss over time period $(t, t+1]$, be our aggregate risk. Then

$$\Delta_{t+1} = \sum_{i=1}^{n} \gamma_{i,t} \delta_{i,t+1}$$

where $\gamma_{i,t} = \beta_i S_{i,t}/V_t$ is the portion of the current portfolio value allocated to equity i, and $\delta_{i,t+1} = -(S_{i,t+1} - S_{i,t})/S_{i,t}$ is the (negative) relative loss over time period $(t, t+1]$ of equity i.

We will highlight the techniques introduced by studying the effect of different distributional assumptions for $\delta := (\delta_{1,t+1}, \ldots, \delta_{n,t+1})^{\mathrm{T}}$ on the aggregate risk $\Delta := \Delta_{t+1}$. The classical distributional assumption on δ, widely used within market risk management, is that of multivariate normality. However, in general the empirical distribution of δ has (one-dimensional) marginal distributions which are heavier tailed than the normal distribution. Furthermore, there is an even more critical problem with multivariate normal distributions in this context. Extreme falls in equity prices are often joint extremes, in the sense that a big fall in one equity price is accompanied by simultaneous big falls in other equity prices. This is for instance seen in Figure 7, an example already encountered in Figure 2. Loosely speaking, a problem with the multivariate normal distributions (or models based on them) is that they do not assign a high enough probability of occurrence to the event in which many thing go wrong a the same time – the "perfect storm" scenario. More precisely, daily equity return data often indicate that the underlying dependence structure has the property of tail dependence, a property which we know Gaussian copulas lack.

Suppose δ is modelled by a multivariate normal distribution $\mathcal{N}_n(\mu, \Sigma)$, where μ and Σ are estimated from historical prices of the equities in the portfolio. There seems to be much agreement on the fact that the quantiles of $\Delta = \gamma^{\mathrm{T}}\delta \sim \mathcal{N}(\gamma^{\mathrm{T}}\mu, \gamma^{\mathrm{T}}\Sigma\gamma)$ do not

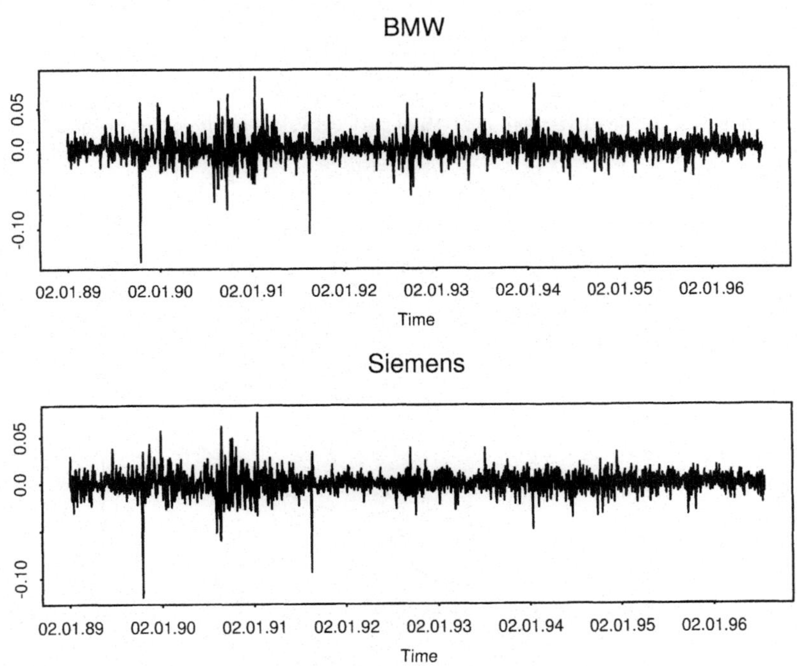

Fig. 7. Daily log returns from 1989 to 1996.

capture the portfolio risk due to extreme market movements; see for instance Embrechts, Mikosch and Klüppelberg (1997), Embrechts (2000) and the references therein. Therefore, different stress test solutions have been proposed. One such "solution" is to choose μ_s and Σ_s in such a way that $\delta_s \sim \mathcal{N}_n(\mu_s, \Sigma_s)$ represents the distribution of the relative losses of the different equities under more adverse market conditions. The aim is that the quantiles of $\Delta_s = \gamma^{\mathrm{T}}\delta_s \sim \mathcal{N}(\gamma^{\mathrm{T}}\mu_s, \gamma^{\mathrm{T}}\Sigma_s\gamma)$ should be more realistic risk estimates. To judge this approach we note that

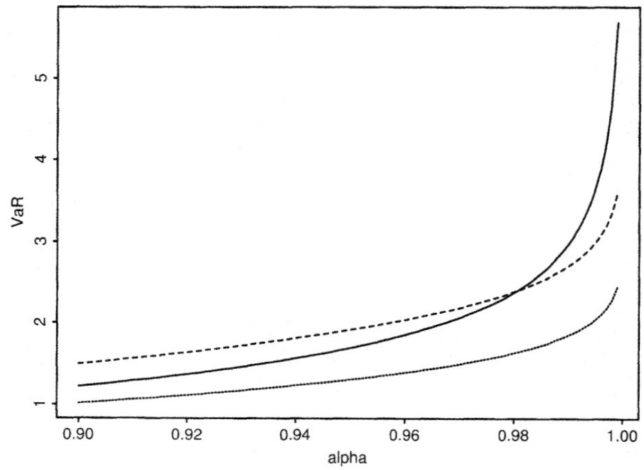

Fig. 8. Quantile curves: $\mathrm{VaR}_{\Delta}(\alpha)$, $\mathrm{VaR}_{\Delta_s}(\alpha)$ and $\mathrm{VaR}_{\Delta^*}(\alpha)$ from lower to upper.

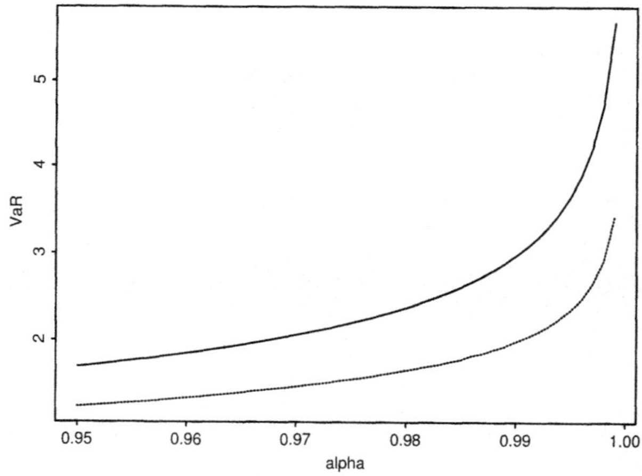

Fig. 9. Quantile curves: $\mathrm{VaR}_{\Delta'}(\alpha)$ and $\mathrm{VaR}_{\Delta^*}(\alpha)$ from lower to upper.

$$\frac{F_s^{-1}(\alpha) - \gamma^{\mathrm{T}}\mu_s}{F^{-1}(\alpha) - \gamma^{\mathrm{T}}\mu} = \sqrt{\frac{\gamma^{\mathrm{T}}\Sigma_s\gamma}{\gamma^{\mathrm{T}}\Sigma\gamma}},$$

where F and F_s denotes the distribution functions of Δ and Δ_s respectively. Hence the effect of this is simply a translation and scaling of the quantile curve $F^{-1}(\alpha)$. As a comparison, let δ^* have a t_4-distribution with mean μ and covariance matrix Σ and let Δ^* be the corresponding portfolio return. Furthermore let $n = 10$, $\mu_i = \mu_{s,i} = \mu_i^* = 0$, $\gamma_i = 1/10$ for all i and let $\tau(\delta_i, \delta_j) = \tau(\delta_i^*, \delta_j^*) = 0.4$, $\tau(\delta_{s,i}, \delta_{s,j}) = 0.6$, $\Sigma_{ij} = \sin(\pi\tau(\delta_i, \delta_j)/2)$, $\Sigma_{s,ij} = 1.5\sin(\pi\tau(\delta_{s,i}, \delta_{s,j})/2)$ for all i, j. Then Figure 8 shows from lower to upper the quantile curves of Δ, Δ_s and Δ^* respectively. If Δ^* were the true portfolio return, Figure 8 shows that the approach described above would eventually underestimate the quantiles of the portfolio return. It should be noted that this is not mainly due to the heavier tailed t_4-marginals. This can be seen in Figure 9 which shows quantile curves of Δ^* and $\Delta' = \gamma^{\mathrm{T}}\delta'$, where δ' is a random vector with t_4-marginals, a Gaussian copula, $\mathbb{E}(\delta') = \mathbb{E}(\delta)$ and $\mathrm{Cov}(\delta') = \mathrm{Cov}(\delta)$.

There are of course numerous alternative applications of copula techniques to integrated risk management. Besides the references already quoted, also see Embrechts, Hoeing and Juri (2001) where the calculation of Value-at-Risk bounds for functions of dependent risks is discussed. The latter paper also contains many more relevant references to this important topic.

References

Barlow, R., Proschan, F., 1975. Statistical Theory of Reliability and Life Testing. Hoult, Rinehart & Winston, New York.

Cambanis, S., Huang, S., Simons, G., 1981. On the theory of elliptically contoured distributions. Journal of Multivariate Analysis 11, 368–385.

Capéraà, P., Genest, C., 1993. Spearman's rho is larger than Kendall's tau for positively dependent random variables. Journal of Nonparametric Statistics 2, 183–194.

Denneberg, D., 1994. Non-Additive Measure and Integral. Kluwer Academic, Boston.

Embrechts, P., 2000. The bell curve is wrong: so what? In: Embrechts, P. (Ed.), Extremes and Integrated Risk Management. Risk Waters Group, pp. xxv–xxviii.

Embrechts, P., Hoeing, A., Juri, A., 2001. Using copulae to bound the value-at-risk for functions of dependent risks. ETH preprint.

Embrechts, P., McNeil, A., Straumann, D., 2002. Correlation and dependence in risk management: Properties and pitfalls. In: Dempster, M.A.H. (Ed.), Risk Management: Value at Risk and Beyond. Cambridge University Press, Cambridge, pp. 176–223.

Embrechts, P., Mikosch, T., Klüppelberg, C., 1997. Modelling Extremal Events for Insurance and Finance. Springer, Berlin.

Fang, K.-T., Kotz, S., Ng, K.-W., 1987. Symmetric Multivariate and Related Distributions. Chapman & Hall, London.

Frank, M.J., 1979. On the simultaneous associativity of $f(x, y)$ and $x + y - f(x, y)$. Aequationes Mathematicae 19, 194–226.

Fréchet, M., 1951. Sur les tableaux de corrélation dont les marges sont données. Annales de l'Université de Lyon, Sciences Mathématiques et Astronomie 14, 53–77.

Fréchet, M., 1957. Les tableaux de corrélation dont les marges et des bornes sont données. Annales de l'Université de Lyon, Sciences Mathématiques et Astronomie 20, 13–31.
Genest, C., 1987. Frank's family of bivariate distributions. Biometrika 74, 549–555.
Genest, C., MacKay, J., 1986a. The joy of copulas: Bivariate distributions with uniform marginals. The American Statistician 40, 280–283.
Genest, C., MacKay, R.J., 1986b. Copules archimédiennes et familles de lois bidimensionell dont les marges sont données. The Canadian Journal of Statistics 14, 145–159.
Genest, C., Rivest, L.-P., 1993. Statistical inference procedures for bivariate Archimedean copulas. Journal of the American Statistical Association 88, 1034–1043.
Hoeffding, W., 1940. Massstabinvariante Korrelationstheorie. Schriften des Mathematischen Seminars und des Instituts für Angewandte Mathematik der Universität Berlin 5, 181–233.
Hult, H., Lindskog, F., 2002. Multivariate extremes, aggregation and dependence in elliptical distributions. Advances in Applied Probability 34, 587–608.
Joe, H., 1997. Multivariate Models and Dependence Concepts. Chapman & Hall, London.
Johnson, N.L., Kotz, S., 1972. Distributions in Statistics: Continuous Multivariate Distributions. Wiley, New York.
Kendall, M., Stuart, A., 1979. Handbook of Statistics. Griffin, London.
Kimberling, C.H., 1974. A probabilistic interpretation of complete monotonicity. Aequationes Mathematicae 10, 152–164.
Kruskal, W.H., 1958. Ordinal measures of association. Journal of the American Statistical Association 53, 814–861.
Lehmann, E.L., 1975 Nonparametrics: Statistical Methods Based on Ranks. Holden-Day, San Francisco.
Lindskog, F., McNeil, A., 2001. Common Poisson shock models: Applications to insurance and credit risk modelling. ETH preprint.
Lindskog, F., McNeil, A., Schmock, U., 2001. A note on Kendall's tau for elliptical distributions. ETH preprint.
Marshall, A.W., 1996. Copulas, marginals and joint distributions. In: Rüschendorff, L., Schweizer, B., Taylor, M.D. (Eds.), Distributions with Fixed Marginals and Related Topics. Institute of Mathematical Statistics, Hayward, CA, pp. 213–222.
Marshall, A.W., Olkin, I., 1967. A multivariate exponential distribution. Journal of the American Statistical Association 62, 30–44.
Mikusinski, P., Sherwood, H., Taylor, M., 1992. The Fréchet bounds revisited. Real Analysis Exchange 17, 759–764.
Muliere, P., Scarsini, M., 1987. Characterization of a Marshall–Olkin type class of distributions. Annals of the Institute of Statistical Mathematics 39, 429–441.
Nelsen, R., 1999. An Introduction to Copulas. Springer, New York.
Resnick, S.I., 1987. Extreme Values, Regular Variation and Point Processes. Springer, New York.
Scarsini, M., 1984. On measures of concordance. Stochastica 8, 201–218.
Schweizer, B., Sklar, A., 1983. Probabilistic Metric Spaces. North-Holland, New York.
Schweizer, B., Wolff, E., 1981. On nonparametric measures of dependence for random variables. The Annals of Statistics 9, 879–885.
Sklar, A., 1959. Fonctions de répartition à *n* dimensions et leurs marges. Publications de l'Institut de Statistique de l'Université de Paris 8, 229–231.
Sklar, A., 1996. Random variables, distribution functions, and copulas – a personal look backward and forward. In: Rüschendorff, L., Schweizer, B., Taylor, M.D. (Eds.), Distributions with Fixed Marginals and Related Topics. Institute of Mathematical Statistics, Hayward, CA, pp. 1–14.

Chapter 9

PREDICTION OF FINANCIAL DOWNSIDE-RISK WITH HEAVY-TAILED CONDITIONAL DISTRIBUTIONS

STEFAN MITTNIK*

Institute of Statistics, University of Munich, Akademiestr. 1, D-80799 Munich, Germany
Ifo Institute for Economic Research, Munich, Germany
Center for Financial Studies, Frankfurt, Germany

MARC S. PAOLELLA

Institute of Statistics and Econometrics, University of Kiel, Olshausenstr. 40, D-24098 Kiel, Germany

Contents

* The research of S. Mittnik was supported by the Deutsche Forschungsgemeinschaft.

Handbook of Heavy Tailed Distributions in Finance, Edited by S.T. Rachev

Abstract

The use of GARCH models with stable Paretian innovations in financial modeling has been recently suggested in the literature. This class of processes is attractive because it allows for conditional skewness and leptokurtosis of financial returns without ruling out normality. This contribution illustrates their usefulness in predicting the downside risk of financial assets in the context of modeling foreign exchange-rates and demonstrates their superiority over use of normal or Student's t GARCH models.

1. Introduction

Risk managers of financial institutions are particularly interested in the left – i.e., downside – tail of the return distribution of financial assets. To assess the short-term exposure to market risks, they are required to evaluate future shortfall probabilities or value-at-risk levels of financial investments. Such estimates can be based on the distribution of the returns themselves. For example, ever since the pioneering works of Mandelbrot (1963) and Fama (1965) there have been numerous studies investigating the appropriateness of the stable Paretian distribution for modeling the unconditional distribution of asset returns [for an overview, see, for example, Mittnik and Rachev (1993), McCulloch (1997)].

However, short-term prediction often benefits substantially when taking conditional volatility into account. The GARCH class of conditional models has been widely and – both from an academic and applied perspective – successfully used to model returns on financial assets [see Palm (1997), Gouriéroux (1997), for surveys]. Although a stationary GARCH model with normally distributed innovations gives rise to an unconditional distribution with higher (possibly nonexistent) kurtosis than the normal, it is often found that residuals from estimated GARCH models of financial return data still tend to exhibit nonnegligible kurtosis. To allow for this, other fatter tailed distributions for GARCH innovations have been considered in the literature, most notably the Student's t. Only very recently has the stable Paretian distribution been considered in the context of modeling the conditional heteroscedastic distribution of asset returns. Special cases of the model considered herein were developed by McCulloch (1985), Nelson (1990), Panorska, Mittnik and Rachev (1995), and Mittnik, Rachev and Paolella (1998), while a more general case was examined in Liu and Brorsen (1995), Paolella (1999) and Mittnik, Paolella and Rachev (2000, 2002).

Like the Student's t, the stable Paretian distribution includes the normal distribution as a special, limiting case and permits heavy-tailed distributions for GARCH innovations. However, the stable Paretian distribution allows for skewness, an attractive property in financial applications not shared by the Student's t. In addition to this practical aspect, the stable Paretian distribution also has the appealing theoretical property that it is the only valid distribution that arises as a limiting distribution of sums of independently, identically distributed (iid) random variables. This is highly desirable, given that error terms in econometric models are usually interpreted as random variables that represent the sum of the external effects not being captured by the model.

This contribution investigates the use of asymmetric stable Paretian power GARCH models for modeling downside risk and demonstrates that this model class is more suitable than the class of Student's t GARCH models, particularly when one uses a goodness-of-fit criterion that focuses on the tails of the conditional distribution.

The remainder is organized as follows. Section 2 discusses GARCH processes with stable Paretian innovations and stationarity conditions. Section 3 reconsiders the empirical analysis of the five exchange-rate series in Liu and Brorsen (1995) using the appropriate measure for persistence of volatility and compares the goodness of fit of the estimated stable Paretian and Student's t GARCH models. The problem of out-of-sample conditional density prediction with particular focus on predicting downside market risk is considered in Section 4. Section 5 concludes.

2. GARCH-stable processes

Sequence y_t is said to be a stable Paretian power GARCH process or, in short, an $\mathrm{S}^{\delta}_{\alpha,\beta}\mathrm{GARCH}(r,s)$ process [see Panorska, Mittnik and Rachev (1995), Paolella (1999), Rachev and Mittnik (2000)], if

$$y_t = \mu + c_t \varepsilon_t, \qquad \varepsilon_t \overset{\text{iid}}{\sim} S_{\alpha,\beta}(0,1) \tag{1}$$

and

$$c_t^{\delta} = \theta_0 + \sum_{i=1}^{r} \theta_i |y_{t-i} - \mu|^{\delta} + \sum_{j=1}^{s} \phi_j c_{t-j}^{\delta}, \tag{2}$$

where $S_{\alpha,\beta}(0,1)$ denotes the standard asymmetric stable Paretian distribution with stable index α, skewness parameter $\beta \in [-1,1]$, zero location parameter, and unit scale parameter. There exist several notational varieties of the stable Paretian distribution; we use the same as in Samorodnitsky and Taqqu (1994) and Rachev and Mittnik (2000), whereby

$$\int_{-\infty}^{\infty} \mathrm{e}^{\mathrm{i}tx}\,\mathrm{d}H(x) = \begin{cases} \exp\left\{-c^{\alpha}|t|^{\alpha}\left[1 - \mathrm{i}\beta\,\mathrm{sign}(t)\tan\dfrac{\pi\alpha}{2}\right] + \mathrm{i}\delta t\right\}, & \text{if } \alpha \neq 1, \\ \exp\left\{-c|t|\left[1 + \mathrm{i}\beta\dfrac{2}{\pi}\mathrm{sign}(t)\ln|t|\right] + \mathrm{i}\delta t\right\}, & \text{if } \alpha = 1, \end{cases} \tag{3}$$

is the characteristic function and H denotes the distribution function corresponding to $S_{\alpha,\beta}(\delta,c)$. The density is symmetric for $\beta = 0$ and skewed to the right (left) for $\beta > 0$ ($\beta < 0$). Stable index α, which, in general, assumes values in interval (0,2], determines the tail-thickness of the distribution. As α approaches 2, tails become thinner; and for $\alpha = 2$ the standard stable Paretian distribution coincides with normal distribution $N(0,2)$. For $\alpha < 2$, ε_t does not possess moments of order α or higher.

Mittnik, Paolella and Rachev (2002) derived sufficient conditions under which the $\mathrm{S}^{\delta}_{\alpha,\beta}\mathrm{GARCH}(r,s)$ process has a unique strictly stationary solution. These are given by $1 < \alpha \leqslant 2$, $0 < \delta < \alpha$, $c_0 > 0$, $c_i \geqslant 0$, $i = 1,\dots,r$, $r \geqslant 1$, $d_j \geqslant 0$, $j = 1,\dots,s$, $s \geqslant 0$, and that the *volatility persistence*, V_S, defined by

$$V_S := \mathrm{E}|Z|^{\delta} \sum_{i=1}^{r} \theta_i + \sum_{j=1}^{s} \phi_j \tag{4}$$

for $Z \sim S_{\alpha,\beta}(0, 1)$, satisfies

$$V_S \leqslant 1. \tag{5}$$

If $1 < \alpha \leqslant 2$ and $0 < \delta < \alpha$, they also showed that

$$\lambda_{\alpha,\beta,\delta} := \mathrm{E}|Z|^{\delta} = \frac{1}{\psi_{\delta}}\Gamma\left(1 - \frac{\delta}{\alpha}\right)\left(1 + \tau_{\alpha,\beta}^{2}\right)^{\delta/(2\alpha)} \cos\left(\frac{\delta}{\alpha} \arctan \tau_{\alpha,\beta}\right), \tag{6}$$

where $\tau_{\alpha,\beta} := \beta \tan(\alpha\pi/2)$ and

$$\psi_{\delta} = \begin{cases} \Gamma(1-\delta)\cos\dfrac{\pi\delta}{2}, & \text{if } \delta \neq 1, \\ \dfrac{\pi}{2}, & \text{if } \delta = 1. \end{cases} \tag{7}$$

Restrictions $1 < \alpha \leqslant 2$ and $0 < \delta < \alpha$ not only appear to be satisfied for the data sets used below, but also for other, even more volatile series, such as stock price indices and East Asian currencies [see Mittnik, Rachev and Paolella (1998), Mittnik, Paolella and Rachev (2000), respectively].

Analogous to the ordinary normal GARCH model (Engle and Bollerslev, 1986), we say that y_t is an *integrated* $\mathrm{S}^{\delta}_{\alpha,\beta}\mathrm{GARCH}(r, s)$ process, denoted $\mathrm{S}^{\delta}_{\alpha,\beta}\mathrm{IGARCH}(r, s)$, if, in (5), $V_S = 1$. In practice, the estimated volatility persistence, $\widehat{V}_S$, tends to be quite close to one for highly volatile series, so that an integrated model might offer a reasonable data description. Because both finite sample and even asymptotic properties of $\widehat{V}_S$ and the associated likelihood ratio test statistics are not known [see, however, Mittnik, Paolella and Rachev (2000)], it is not immediately clear how one can test for an integrated process. Instead of formally testing, we suggest fitting both models and examining the change in various goodness-of-fit statistics, most notably the Anderson–Darling statistic, which is particularly relevant for assessing the models' ability to successfully model the value-at-risk (see Section 3.3 below).

3. Modeling exchange-rate returns

To examine the appropriateness of the stable GARCH hypothesis, we model returns[1] on five daily spot foreign exchange rates against the U.S. dollar, namely the British pound, Canadian dollar, German mark, Japanese yen, and the Swiss franc. The choice of exchange rate allows us to compare our more general GARCH specification to that used by Liu and Brorsen (1995), who set $\alpha = \delta$ in (2). However, our sample is somewhat larger than theirs,

[1] We define the return r_t in period t by $r_t = 100 \times (\ln P_t - \ln P_{t-1})$, where P_t is the exchange rate at time t.

covering the period January 2, 1980 to July 28, 1994, yielding series of lengths 3681, 3682, 3661, 3621, and 3678, respectively. Serial correlation was found to be negligible, and, as is common in practice, a GARCH(r, s) specification with $r = s = 1$ was sufficient to capture serial correlation in the absolute returns. Therefore, we specify a model of the form

$$r_t = \mu + c_t \varepsilon_t, \tag{8}$$

$$c_t^\delta = \theta_0 + \theta_1 |r_{t-1} - \mu|^\delta + \phi_1 c_{t-1}^\delta \tag{9}$$

for each of the five currencies.

3.1. Approximate maximum likelihood estimation

Evaluation of the probability density function (pdf) and, thus, the likelihood function of the $S_{\alpha,\beta}$ distribution is nontrivial, because it lacks an analytic expression. The maximum likelihood (ML) estimate of parameter vector $\boldsymbol{\theta} = (\mu, c_0, \theta_0, \theta_1, \phi_1, \alpha, \beta, \delta)'$ for the $S_{\alpha,\beta}^\delta$GARCH(1, 1) models (8), (9) is obtained by maximizing the logarithm of the likelihood function

$$L(\boldsymbol{\theta}; r_1, \ldots, r_T) = \prod_{t=1}^{T} c_t^{-1} S_{\alpha,\beta}\left(\frac{r_t - \mu}{c_t}\right), \tag{10}$$

where c_0 denotes the unknown initial value of c_t.

The ML estimation we conduct is *approximate* in the sense that the stable Paretian density function $S_{\alpha,\beta}((r_t - \mu)/c_t)$ needs to be approximated. To do so, we follow the algorithm of Mittnik et al. (1999), which approximates the stable Paretian density via fast Fourier transform of the characteristic function. DuMouchel (1973) shows that the ML estimator of the parameters of the stable density is consistent and asymptotically normal with the asymptotic covariance matrix being given by the inverse of the Fisher information matrix. Approximate standard errors of the estimates can be obtained via numerical approximation of the Hessian matrix.

Below, we will demonstrate that – for the five series under consideration – the $S_{\alpha,\beta}^\delta$GARCH(r, s) model outperforms its Student's t counterpart. However, it is of practical interest to know at least three things before adopting a new and more complex method: first, how easy the stable ML estimation routine is to implement; second, whether it is numerically well-behaved; and third, how fast it performs. When implemented in high-level software which provide both FFT and linear interpolation routines (such as Matlab and Splus), the algorithm becomes a straightforward programming exercise. Our experience has shown that the method is extremely well behaved, giving rise to numerical problems only for grossly misspecified and/or overspecified models (for which the Student's t GARCH model also has difficulties) or, in the case of the more general class of ARMA-GARCH models, when there is near zero-pole cancellation in the ARMA structure – a well-known difficulty in ARMA estimation.

The satisfactory behavior of the algorithm is actually not surprising for at least two reasons. First, there is no explicit numerical integration involved [as in the approach of Liu and Brorsen (1995)] and, second, the method can be made arbitrarily accurate by the choice of several tuning constants [recommendations for which are given in Mittnik et al. (1999)]. Nevertheless, it is clear that the method will take longer than the (essentially closed form) evaluation of the Student's t density. For the series considered in this paper, use of a quasi-Newton minimization algorithm (BFGS, as implemented in Matlab 5.2) with convergence tolerance of 10^{-4} resulted in convergence after about 150 to 350 function evaluations (including gradient calculations). Rather contrary to our initial expectations – and fears –, the choice of initial values is of surprisingly little importance. Given any "reasonable" set of values, say $\alpha > 1.4$, $|\beta| < 0.7$, $|\mu| < 0.2$, $\theta_0 > 0$, $\theta_1 > 0$ and $\phi_1 > 0.2$, convergence to the same respective maxima occurred for all five exchange-rate series under consideration, and also for the vast majority of trials from simulation experiments. From a purely numerical standpoint then, the method appears both highly reliable and "stable".

Evaluation of the GARCH recursion requires presample values ε_0 and c_0. Following Nelson and Cao (1992), one could set those to their unconditional expected values, i.e.,

$$\hat{c}_0 = \frac{\hat{\theta}_0}{1 - \lambda_{\hat{\alpha},\hat{\beta},\hat{\delta}} \sum_{i=1}^{r} \hat{\theta}_i - \sum_{j=1}^{s} \hat{\phi}_j} \quad \text{and} \quad \hat{\varepsilon}_0 = \hat{\lambda}\hat{c}_0. \tag{11}$$

In the IGARCH case, (11) will be invalid, so we instead *estimate* c_0 as an additional parameter. In fact, we chose to do this for all models considered here, as (11) will clearly be problematic for nearly integrated GARCH models.

For the integrated model $\mathrm{S}^{\delta}_{\alpha,\beta}$IGARCH(1, 1), the restriction $\phi_1 = 1 - \lambda_{\alpha,\beta,\delta}\theta_1$ needs to be imposed. Notice that this entails evaluation of (4) at each iteration, as ϕ_1 is also dependent on values $\hat{\alpha}$, $\hat{\beta}$ and $\hat{\delta}$.

We compare the $S^{\delta}_{\alpha,\beta}$GARCH model to the most commonly used heavy-tailed variant of the GARCH model, the Student's t-GARCH models in power form, say t^{δ}_{ν}-GARCH(r, s), given by

$$r_t = \mu + c_t\varepsilon_t, \quad \varepsilon_t \overset{\text{iid}}{\sim} t(\nu), \tag{12}$$

$$c_t^{\delta} = \theta_0 + \sum_{i=1}^{r} \theta_i |r_{t-i} - \mu|^{\delta} + \sum_{j=1}^{s} \phi_j c_{t-j}^{\delta}, \tag{13}$$

where $t(\nu)$ refers to the Student's t distribution with ν degrees of freedom, i.e.,

$$f(x;\nu) = K_{\nu}\left(1 + \frac{x^2}{\nu}\right)^{-(\nu+1)/2} \tag{14}$$

and

$$K_\nu = \frac{\Gamma((\nu+1)/2)\nu^{-1/2}}{\sqrt{\pi}\Gamma(\nu/2)}. \tag{15}$$

Assuming $0 < \delta < \nu$ and $\nu > 1$,[2] taking unconditional expectations of c_t^δ in (13) shows that $\mathrm{E}c_t^\delta$ exists if $\mathrm{E}|T|^\delta \sum_{i=1}^r \theta_i + \sum_{j=1}^s \phi_j < 1$, where $T \sim t(\nu)$ and

$$\lambda_{\nu,\delta} := \mathrm{E}|T|^\delta = \sqrt{\frac{\nu^\delta}{\pi}}\Gamma\left(\frac{\delta+1}{2}\right)\Gamma\left(\frac{\nu-\delta}{2}\right)\Gamma^{-1}\left(\frac{\nu}{2}\right). \tag{16}$$

Analogous to (4), the measure of volatility persistence for t_ν^δ-GARCH(r, s) models is defined to be

$$V_t := \lambda_{\nu,\delta}\sum_{i=1}^r \theta_i + \sum_{j=1}^s \phi_j. \tag{17}$$

Similar remarks regarding treatment of presample values and the imposing of the IGARCH constraint apply to the Student's t model as well.

3.2. Estimation results and volatility persistence

The parameter estimates of the models are presented in Table 1. Noteworthy are the estimates of the skewness parameter β: all $\hat{\beta}$ values are (statistically) significantly different from zero, although those for the British pound and German mark series are quite close to zero. In addition, when $|\beta| < 0.3$ and α is over 1.8, the amount of skewness is, for practical purposes, slight. Skewness is most pronounced for the Japanese yen, for which $\hat{\alpha} = 1.81$ and $\hat{\beta} = -0.418$.

The persistence-of-volatility measure given in the last column of Table 1 reflects the speed with which volatility shocks die out. A $\widehat{V}$-value near one is indicative of an integrated GARCH process, in which volatility shocks have persistent effects. Under the $S_{\alpha,\beta}$ assumption, the models for the Canadian dollar ($\widehat{V_S} = \lambda_{\hat{\alpha},\hat{\beta},\hat{\delta}}\hat{\theta}_1 + \hat{\phi}_1 = 1.001$) and Japanese yen ($\widehat{V_S} = 1.002$) series would suggest that they are very close to being integrated. Under the Student's t assumption, $\widehat{V_t} = \lambda_{\hat{\nu},\hat{\delta}}\hat{\theta}_1 + \hat{\phi}_1 = 0.992$ for the Canadian dollar, which is also rather close to being integrated, while $\widehat{V_t}$ is only 0.972 for the Japanese yen. Thus, for these two currencies, the indications regarding persistence of volatility differ under the two distributional assumptions. For the other currencies, the measures are strikingly close, most notably for the German mark ($\widehat{V_S} = \widehat{V_t} = 0.969$) and the Swiss franc

[2] The condition $\nu > 1$ is analogous to requiring $\alpha > 1$ in the stable Paretian case and implies existence of a finite first moment of the innovations.

Table 1
GARCH parameter estimates[a]

	Intercept	GARCH parameters				Distribution parameters		Persistence measure[b]
	μ	θ_0	θ_1	ϕ_1	δ	Shape	Skew	$\widehat{V}$
British								
$S_{\alpha,\beta}$	−9.773e−3	8.085e−3	0.04132	0.9171	1.359	1.850	−0.1368	0.984
	(0.012)	(2.39e−3)	(6.42e−3)	(0.0118)	(0.0892)	(0.0245)	(0.0211)	
t	−2.312e−3	0.01190	0.06373	0.9071	1.457	6.218	–	0.976
	(0.010)	(3.56e−3)	(0.0115)	(0.0200)	(0.167)	(0.615)		
Canadian								
$S_{\alpha,\beta}$	5.167e−3	1.034e−3	0.04710	0.9164	1.404	1.823	0.3577	1.001
	(0.0614)	(3.12e−4)	(6.63e−3)	(0.0118)	(0.0143)	(0.0104)	(0.0209)	
t	−2.240e−3	7.774e−4	0.06112	0.9118	1.793	5.900	–	0.992
	(3.83e−3)	(6.90e−4)	(5.98e−3)	(7.27e−3)	(0.0150)	(0.0801)		
German								
$S_{\alpha,\beta}$	2.580e−3	0.01525	0.05684	0.8971	1.101	1.892	−0.06779	0.969
	(0.016)	(1.61e−3)	(3.44e−3)	(7.42e−3)	(9.78e−3)	(0.0216)	(0.0184)	
t	6.643e−3	0.01812	0.07803	0.8938	1.261	7.297	–	0.969
	(9.21e−4)	(2.25e−3)	(6.45e−3)	(4.43e−3)	(0.147)	(0.186)		
Japanese								
$S_{\alpha,\beta}$	−0.01938	4.518e−3	0.06827	0.8865	1.337	1.814	−0.4175	1.002
	(0.0166)	(1.12e−3)	(7.91e−3)	(0.0124)	(0.0132)	(0.0107)	(8.80e−3)	
t	5.318e−3	9.949e−3	0.07016	0.8756	1.816	5.509	–	0.972
	(8.87e−3)	(3.03e−3)	(0.0119)	(0.0205)	(0.162)	(0.461)		
Swiss								
$S_{\alpha,\beta}$	−2.677e−3	0.01595	0.04873	0.9115	1.041	1.902	−0.2836	0.971
	(0.0124)	(3.30e−3)	(6.84e−3)	(0.0132)	(0.144)	(0.0206)	(0.0722)	
t	8.275e−3	0.02099	0.06825	0.9061	1.159	8.294	–	0.968
	(0.0118)	(3.91e−3)	(6.85e−3)	(7.25e−3)	(0.179)	(0.933)		

[a] Estimated models: $r_t = \mu + c_t \varepsilon_t$, $c_t^{\delta} = \theta_0 + \theta_1 |r_{t-1} - \mu|^{\delta} + \phi_1 c_{t-1}^{\delta}$. "Shape" denotes the degrees of freedom parameter ν for the Student's t distribution and stable index α for the stable Paretian distribution; "Skew" refers to the stable Paretian skewness parameter β. Standard deviations resulting from ML estimation are given in parentheses.
[b] $\widehat{V}$ corresponds to $\widehat{V}_S$ in the stable Paretian and $\widehat{V}_t$ in the Student's t case. $V = 1$ implies an IGARCH model.

($\widehat{V}_S = 0.971$, $\widehat{V}_t = 0.968$). It is interesting to note that, for each of these two currencies, the log-likelihood values $\mathcal{L}_t$ and $\mathcal{L}_S$ are also extremely close. These are discussed further in the next section.

For all five series, we also estimated the models with the IGARCH condition imposed. Table 2 shows the resulting parameter estimates. Not surprisingly, for those models for which the persistence measure was close to unity, the IGARCH-restricted parameter estimates differ very little. For the remaining models, the greatest changes occur with the

Table 2
IGARCH parameter estimates[a]

	Intercept	IGARCH parameters				Distribution parameters	
	μ	θ_0	θ_1	ϕ_1	δ	Shape	Skew
British							
$S_{\alpha,\beta}$	−0.01023	7.050e−3	0.03781	0.9114	1.598	1.846	−0.1340
	(0.0103)	(1.79e−3)	(5.64e−3)	–	(0.0677)	(0.0224)	(0.0147)
t	−3.033e−3	4.237e−3	0.05774	0.9130	1.949	5.543	–
	(0.0101)	(1.68e−3)	(9.83e−3)	–	(0.264)	(0.484)	
Canadian							
$S_{\alpha,\beta}$	5.148e−3	1.115e−3	0.04689	0.9154	1.404	1.823	0.3578
	(3.65e−3)	(2.14e−4)	(5.71e−3)	–	(0.0143)	(0.0105)	(0.0209)
t	−2.098e−3	4.998e−4	0.06468	0.9146	1.796	5.890	–
	(3.48e−3)	(1.37e−4)	(7.54e−3)	–	(0.0226)	(0.0838)	
German							
$S_{\alpha,\beta}$	8.959e−3	9.666e−3	0.04518	0.8896	1.676	1.881	0.03944
	(0.0113)	(1.85e−3)	(6.10e−3)	–	(0.0662)	(0.0217)	(0.0930)
t	8.851e−3	5.505e−3	0.08124	0.9003	1.741	6.560	–
	(0.0106)	(1.60e−3)	(0.0106)	–	(0.231)	(0.676)	
Japanese							
$S_{\alpha,\beta}$	−0.01932	4.814e−3	0.06768	0.8858	1.336	1.814	−0.4175
	(8.44e−3)	(9.75e−4)	(7.68e−3)	–	(0.0751)	(0.0226)	(0.0151)
t	6.136e−3	5.611e−3	0.06036	0.8689	2.314	5.066	–
	(8.57e−3)	(1.31e−3)	(0.0112)	–	(0.224)	(0.410)	
Swiss							
$S_{\alpha,\beta}$	3.823e−3	0.01111	0.03700	0.9009	1.724	1.889	−0.1703
	(0.0127)	(2.65e−3)	(5.40e−3)	–	(0.0419)	(0.0169)	(0.137)
t	9.130e−3	2.047e−3	0.07125	0.9347	1.166	8.194	–
	(0.0119)	(8.34e−4)	(9.13e−3)	–	(9.79e−3)	(0.0996)	

[a] Estimated models: $r_t = \mu + c_t\varepsilon_t$, $c_t^\delta = \theta_0 + \theta_1|r_{t-1} - \mu|^\delta + (1 - \lambda\theta_1)c_{t-1}^\delta$ with IGARCH condition $\hat{\phi}_1 = 1 - \hat{\lambda}\hat{\theta}_1$ imposed. See footnote to Table 1 for further details.

power parameter δ and, to a lesser extent, the shape parameters α and ν. The former increase, while the latter decrease under IGARCH restrictions.

It should also be noted that the restriction $\alpha = \delta$, imposed by Liu and Brorsen (1995) when estimating GARCH-stable models for the same five currencies, is not supported by our results. This is important because, if $\delta \geqslant \alpha$, the unconditional first moments of c_t is infinite for any $\alpha < 2$. The knife-edge specification $\delta = \alpha$ does not only induce conceptual difficulties, but also leads to a highly volatile evolution of the c_t series in practical work. For our estimates, we obtain $\hat{\delta} < \hat{\alpha}$, which suggest that conditional volatility c_t^δ is a well-defined quantity in the sense that $\mathrm{E}(c_t^\delta \mid r_{t-1}, r_{t-2}, \ldots) < \infty$ for $V_S < 1$.

3.3. Goodness of fit

We employ three likelihood-based and one empirical CDF-based criteria for comparing the goodness of fit of the candidate models. The first is the maximum log-likelihood value obtained from ML estimation. This value may be viewed as an overall measure of goodness of fit and allows us to judge which candidate is more likely to have generated the data. The second is the AICC [Hurvich and Tsai (1989), see also Brockwell and Davis (1991), Equation (9.3.7)] given by

$$\mathrm{AICC} = -2\mathcal{L} + \frac{2T\,(k+1)}{T-k-2}, \tag{18}$$

where k denotes the number of estimated parameters and T the number of observations. This is the bias-corrected information criterion of Akaike (1973), which corrects the latter's tendency to overfit. Similarly, the SBC (Schwartz, 1978), defined as

$$\mathrm{SBC} = -2\mathcal{L} + \frac{k\log(T)}{T}, \tag{19}$$

is a similar penalizing strategy which is commonly used.

The fourth criterion is the Anderson–Darling statistic [Anderson and Darling (1952), see also Press et al. (1991), and Tanaka (1996)], given by

$$\mathrm{AD} = \sup_{x\in\mathbf{R}} \frac{|F_s(x) - \widehat{F}(x)|}{\sqrt{\widehat{F}(x)(1-\widehat{F}(x))}}, \tag{20}$$

where $\widehat{F}(x)$ denotes the cdf of the estimated parametric density, and $F_s(x)$ is the empirical sample distribution, i.e.,

$$F_s(x) = \frac{1}{T}\sum_{t=1}^{T} \mathcal{I}_{(-\infty,x]}\left(\frac{r_t - \hat{\mu}}{\hat{c}_t}\right),$$

where $\mathcal{I}(\cdot)$ is the usual indicator function. The AD statistic weights discrepancies appropriately across the whole support of the distribution. This is especially important if one is interested in determining conditional shortfall probabilities, i.e., the probability of large investment losses, or so-called value-at-risk measures, where one focuses on the left tail of the conditional return distribution.

Table 3 displays the aforementioned goodness-of-fit measures for the estimated models. In both the unrestricted and IGARCH restricted cases, the inference suggested from the maximum log-likelihood value $\mathcal{L}$, and the AICC and SBC are identical. This is not too surprising, given the large ratio of observations to parameters, and the fact that there is only one parameter difference between the Student's t and stable Paretian GARCH models.

Table 3
Goodness-of-fit measures of estimated models[a]

	$\mathcal{L}$		AICC		SBC		AD	
	$S_{\alpha,\beta}$	t	$S_{\alpha,\beta}$	t	$S_{\alpha,\beta}$	t	$S_{\alpha,\beta}$	t
Britain:								
GARCH	−3842.0	−3828.6	7700.0	7671.2	7684.0	7657.2	0.0375	0.0244
IGARCH	−3842.3	−3837.1	7698.6	7686.2	7684.6	7674.2	0.0417	0.0420
Canada:								
GARCH	−159.92	−152.25	0335.9	0318.5	0319.9	0304.5	0.0532	0.0571
IGARCH	−159.97	−153.71	0334.0	0319.4	0320.0	0307.4	0.0529	0.0633
Germany:								
GARCH	−3986.5	−3986.2	7989.0	7986.4	7973.0	7972.4	0.0368	0.345
IGARCH	−3989.9	−3999.4	7993.8	8010.8	7979.8	7998.8	0.0506	0.200
Japan:								
GARCH	−3178.7	−3333.7	6373.4	6681.4	6357.4	6667.4	0.0401	0.0986
IGARCH	−3178.8	−3334.6	6371.6	6681.2	6357.6	6669.2	0.0394	0.0793
Switzerland:								
GARCH	−4308.6	−4308.1	8633.2	8630.2	8617.2	8616.2	0.0457	0.287
IGARCH	−4314.2	−4325.0	8642.4	8662.0	8628.4	8650.0	0.0460	0.278

[a] $\mathcal{L}$ refers to the maximum log-likelihood value; AICC is the corrected AIC criteria (18); SBC is the Schwarz Bayesian criteria (19); and AD is the Anderson–Darling statistic (20).

It appears that $\mathcal{L}$ significantly favors the Student's t distribution for the British pound (with values, in obvious notation, $\mathcal{L}_t = -3828.6$ and $\mathcal{L}_S = -3842.0$) and the Canadian dollar ($\mathcal{L}_t = -152.25$, $\mathcal{L}_S = -159.92$). For the German mark ($\mathcal{L}_t = -3896.2$, $\mathcal{L}_S = -3896.5$) and the Swiss franc ($\mathcal{L}_t = -4308.1$, $\mathcal{L}_S = -4308.6$), the log-likelihood values, AICC and SBC are very close, albeit larger for the Student's t. On the other hand, the $S_{\alpha,\beta}$ assumption is favored quite strongly for the Japanese yen with $\mathcal{L}_S = -3178.7$ as compared to $\mathcal{L}_t = -3331.7$.

For the British pound, the AD statistic ($\mathrm{AD}_t = 0.0244$, $\mathrm{AD}_S = 0.0375$) slightly favors the Student's t model, in agreement with $\mathcal{L}$, although the difference is relatively small. The AD statistics for the remaining countries all favor the stable Paretian model, particularly for the German mark ($\mathrm{AD}_t = 0.345$, $\mathrm{AD}_S = 0.0368$), the Japanese yen ($\mathrm{AD}_t = 0.0986$, $\mathrm{AD}_S = 0.0401$) and the Swiss franc ($\mathrm{AD}_t = 0.287$, $\mathrm{AD}_S = 0.0457$). The usual caveat applies, in that, statistically speaking, it is not clear to what extent these differences are significant. However, given virtually identical log-likelihood values, but AD statistics which are several times smaller for the $S_{\alpha,\beta}$ distribution, one might safely conclude that, particularly in the tails of the conditional distribution, the $S_{\alpha,\beta}$ model offers a distinct advantage, irrespective of its desirable theoretical properties which are not shared by the Student's t distribution.

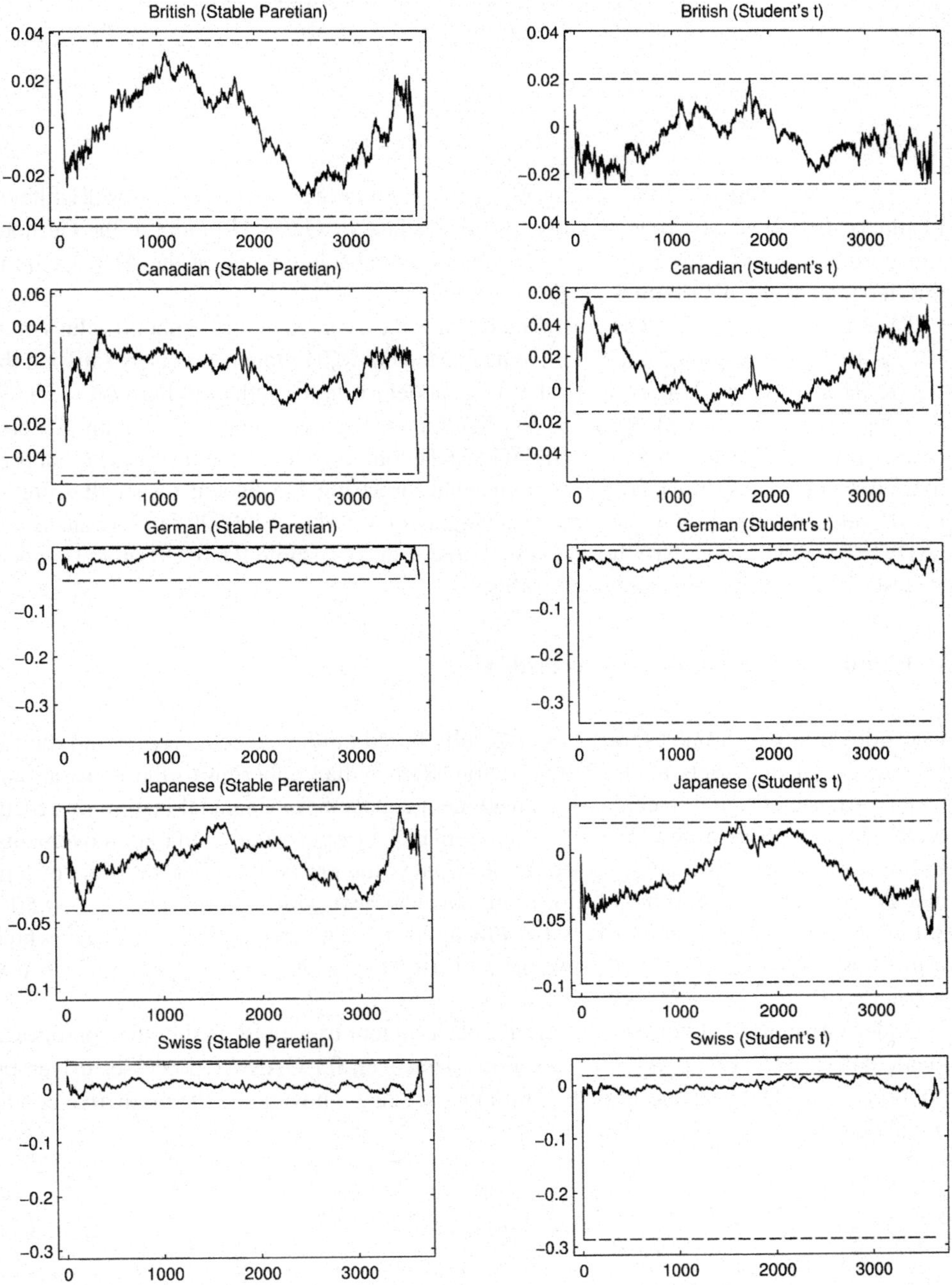

Fig. 1. Comparison of the variance adjusted differences between the sample and fitted distribution functions.

For each currency and both distributional assumptions, Figure 1 plots the values

$$\mathrm{AD}_t = \frac{|F_s(\hat{\varepsilon}_{t:T}) - \widehat{F}(\hat{\varepsilon}_{t:T})|}{\sqrt{\widehat{F}(\hat{\varepsilon}_{t:T})(1 - \widehat{F}(\hat{\varepsilon}_{t:T}))}},$$

$t = 1, \ldots, T$, where T is the sample size and $\hat{\varepsilon}_{t:T}$ denotes the *sorted* GARCH-filtered residuals. In most cases, most notably for the Student's t GARCH model of the German, Japanese and Swiss currency returns, the maximum absolute value of the AD_t occurs in the (left) tail of the distribution.

Turning now to the IGARCH-restricted fits, it is clear that the log-likelihood values must necessarily decrease, since none of the unrestricted GARCH models precisely satisfied the IGARCH restrictions. However, for the $S_{\alpha,\beta}$ model of the Canadian dollar ($\mathcal{L} = 159.97$) and Japanese yen ($\mathcal{L} = 3178.8$), the log-likelihoods are very close to their unrestricted counterparts. This was expected, as the IGARCH condition for the unrestricted models of these two currencies were nearly met. Somewhat surprising, however, is the small *de*crease in AD values for the $S_{\alpha,\beta}$ model of the Canadian dollar ($\mathrm{AD}_S = 0.0529$) and Japanese yen ($\mathrm{AD}_S = 0.0394$). Particularly for the latter two currencies, stable IGARCH models appear to describe the daily returns quite plausibly.

4. Prediction of densities and downside risk

Decisions on financial investments are typically based on the expected return and the expected risk of the assets under consideration. Rather than adhering to the conventional mean-variance criterion, recent risk management concepts for financial institutions focus on the downside risk or the value-at-risk of a financial position due to market movements. In this context, a typical question would be: what is the probability that the value of a financial position will drop by 50% or more over the next period, i.e., $\Pr(r_{t+1} < -0.50)$? Alternatively, one may ask what is the threshold or value-at-risk, $-z(\gamma)$, under which a position will not fall with a probability of $100(1-\gamma)\%$; i.e., find $-z(\gamma)$ such that $\Pr(r_{t+1} < -z(\gamma)) = \gamma$.

Under unconditional normality, it would be sufficient to simply predict the conditional mean and variance to answer such questions. However, for GARCH processes driven by nonnormal, asymmetric and, possibly, infinite-variance innovations, the predictive conditional density

$$\hat{f}_{t+1|t}(r_{t+1}) = f\left(\frac{r_{t+1} - \mu(\hat{\boldsymbol{\theta}}_t)}{c_{t+1}(\hat{\boldsymbol{\theta}}_t)} \middle| r_t, r_{t-1}, \ldots\right), \tag{21}$$

needs to be computed. In (21), $\hat{\boldsymbol{\theta}}_t$ refers to the estimated parameter vector using the sample information up to and including period t; and $c_{t+1}(\cdot)$ is obtained from the conditional-scale recursion (2) using $\hat{\boldsymbol{\theta}}_t$.[3] Multistep density predictions,

$$\hat{f}_{t+n|t}(r_{t+n}) = f\left(\frac{r_{t+n} - \mu(\hat{\boldsymbol{\theta}}_t)}{c_{t+n}(\hat{\boldsymbol{\theta}}_t)}\middle| r_t, r_{t-1}, \ldots\right), \tag{22}$$

are obtained by recursive application of (2) with unobserved quantities being replaced by their conditional expectations.

For each of the five currencies under consideration, we evaluate $\hat{f}_{t+1|t}(r_{t+1})$, $t = 2000, \ldots, T-1$, for the $S^{\delta}_{\alpha,\beta}$GARCH(1, 1) and t^{δ}_{ν}GARCH(1, 1) models, as well as the conventional GARCH(1,1) model with normal innovations.[4] We re-estimate (via ML estimation) the model parameters at each step, as would typically be done in actual applications.

The overall density forecasting performance of competing models can be compared by evaluating their conditional densities at the future observed value r_{t+1}, i.e., $\hat{f}_{t+1|t}(r_{t+1})$. A model will fare well in such a comparison if realization r_{t+1} is near the mode of $\hat{f}_{t+1|t}(\cdot)$ and if the mode of the conditional density is more peaked. The conditional densities are determined not only by the specification of the mean and GARCH equations, but also by the distributional choice for the innovations.

Table 4 presents the means, standard deviations and medians of the density values $\hat{f}_{t+1|t}(r_{t+1})$, $t = 2000, \ldots, T-1$, for each currency. Based on the means, values cor-

Table 4
Comparison of overall predictive performance[a]

	British	Canadian	German	Japanese	Swiss
			Mean		
Normal	0.4198	1.1248	0.4064	0.4796	0.3713
t	0.4429	1.1871	0.4258	0.5207	0.3851
$S_{\alpha,\beta}$	0.4380	1.1798	0.4213	0.5173	0.3820
			Standard deviation		
Normal	0.1934	0.5697	0.1888	0.1988	0.1620
t	0.2325	0.6802	0.2151	0.2782	0.1840
$S_{\alpha,\beta}$	0.2189	0.6482	0.2016	0.2662	0.1771
			Median		
Normal	0.4291	1.0824	0.4178	0.5172	0.3942
t	0.4483	1.1500	0.4452	0.5261	0.4069
$S_{\alpha,\beta}$	0.4493	1.1730	0.4477	0.5242	0.4041

[a] The entries represent average predictive likelihood values, $\sum_{t=2000}^{T-1} \hat{f}_{t+1|t}(r_{t+1})$.

[3] A conditionally varying location parameter, μ_t, would be handled analogously.

[4] Since the sample sizes, T, of the five currencies vary, the number of forecasts ranges from 1,621 to 1,682.

responding to the $S_{\alpha,\beta}$ and Student's t assumptions are extremely close, with the Student's t values nevertheless larger in each case. Based on the medians, however, the stable Paretian model is (slightly) favored by the British, Canadian and German currencies. Notice that this is contrary to the model selection based on the goodness of fit measures; both AICC and AD statistics favored use of stable Paretian innovations for the Japanese yen and Student's t innovations for the British pound.

Next, we examine how well the models predict the downside risk. Consider the value-at-risk implied by a particular model, M, namely

$$\Pr\left(r_{t+1} \leqslant -z_{t+1}^{M}(\gamma)\right) = \gamma, \quad t = 2000, \ldots, T-1. \tag{23}$$

For a correctly specified model we expect $100\gamma\%$ of the observed r_{t+1}-values to be less than or equal to the implied threshold-values $-z_{t+1}(\gamma)$. If the observed frequency

$$\hat{\gamma}^{M} := \frac{1}{T-2000} \sum_{t=2000}^{T-1} \mathcal{I}_{(-\infty, -z_{t+1}^{M}(\gamma)]}(r_{t+1})$$

is less (higher) than γ, then model M tends to overestimate (underestimate) the risk of the currency position; i.e., the implied absolute $z_{t+1}^{M}(\gamma)$-values tend to be too large (small).

The predictive performance for assessing the downside risk achieved by the normal, Student's t and stable Paretian models are compared in Table 5 for the shortfall probabilities $\gamma = 0.01, 0.025, 0.05, 0.10$. A comparison of the stable Paretian and Student's t

Table 5
Comparison of predictive performance for downside risk[a]

100γ	Model	British	German	Canadian	Japanese	Swiss
	Normal	1.9036	1.5051	1.3674	1.9124	1.4899
1.0	t	1.3682	0.9031	0.7134	1.4189	1.3707
	$S_{\alpha,\beta}$	1.3682	0.9031	1.3080	1.3572	1.2515
	Normal	3.0339	2.6490	2.3187	2.8994	3.2777
2.5	t	2.8554	2.9500	2.1403	3.2079	3.3969
	$S_{\alpha,\beta}$	2.9149	2.9500	2.4970	2.5910	3.1585
	Normal	4.7591	4.5756	3.6266	4.9969	4.7676
5.0	t	5.1160	5.2378	3.9834	5.7372	5.0656
	$S_{\alpha,\beta}$	5.1160	5.2378	5.0535	5.2437	5.0656
	Normal	8.3879	9.2113	8.5612	8.0814	8.9392
10.0	t	9.8751	10.6562	9.9287	10.3023	10.8462
	$S_{\alpha,\beta}$	9.6966	10.4154	10.2259	9.8088	10.2503

[a] The entries show the observed frequencies $\hat{\gamma}^{M} = (T-2000)^{-1} \sum_{t=2000}^{T-1} \mathcal{I}_{(-\infty, -z_{t+1}^{M}(\gamma)]}(r_{t+1})$ multiplied by 100. For a correctly specified model, we expect $\hat{\gamma}^{M} \approx \gamma$.

GARCH models over the five currencies and four cutoff values, γ, shows that, in 4 out the 20 cases, the Student's t GARCH model outperforms that of the stable Paretian, while the latter is more accurate in 11 cases, sometimes considerably so (as for the Canadian dollar with $\gamma = 0.025$ and 0.05). The remaining 5 cases are tied.

Table 6 presents summary measures[5] for the predictive performance of the three models across all five currencies in the form of the mean error

$$ME_M(\gamma) = \frac{1}{5}\sum_{i=1}^{5} 100(\hat{\gamma}_i^M - \gamma),$$

mean absolute error

$$MAE_M(\gamma) = \frac{1}{5}\sum_{i=1}^{5} 100|\hat{\gamma}_i^M - \gamma|$$

Table 6
Summary measures for the predictive performance[a]

100γ	Model	$ME(\gamma)$	$MAE(\gamma)$	$MSE(\gamma)$
1.0	Normal	0.6357	0.6357	0.4558
	t	0.1549	0.3083	0.1080
	$S_{\alpha,\beta}$	0.2376	0.2764	0.0861
2.5	Normal	0.3357	0.4083	0.2209
	t	0.4101	0.5540	0.3527
	$S_{\alpha,\beta}$	0.3223	0.3235	0.1633
5.0	Normal	−0.4548	0.4548	0.4357
	t	0.0280	0.4346	0.3302
	$S_{\alpha,\beta}$	0.1433	0.1433	0.0273
10.0	Normal	−1.3638	1.3638	2.0195
	t	0.3217	0.4002	0.2517
	$S_{\alpha,\beta}$	0.0794	0.2772	0.0830
Aggregate	Normal	−0.2118	0.7156	0.7830
	t	0.2287	0.4243	0.2607
	$S_{\alpha,\beta}$	0.1956	0.2551	0.0899

[a] Shown are the mean error (*ME*), mean absolute error (*MAE*) and mean squared error (*MSE*) of the observed extreme-tail frequencies from Table 5 across the five currencies. The bottom panel is the aggregate over all γ-values considered.

[5] The measures are evaluated for 100γ rather than γ because the resulting scales of the reported values enhance readability.

and the mean squared error

$$MSE_M(\gamma) = \frac{1}{5}\sum_{i=1}^{5} 100^2(\hat{\gamma}_i^M - \gamma)^2.$$

The *ME*'s for the normal show that it underestimates the probability of extreme downturns ($ME_{\text{Normal}}(\gamma) > 0$ for $\gamma = 0.01, 0.025$) and overestimates the probability of moderate downturns ($ME_{\text{Normal}}(\gamma) < 0$ for $\gamma = 0.05, 0.10$). With one exception, the *ME*'s of the stable Paretian and Student's t GARCH models are smaller (in absolute terms) than those for the normal. However, they are always positive, indicating, on average, slight underprediction of the downturn probabilities. For $\gamma = 0.01$ and $\gamma = 0.05$, the Student's t model has smaller *ME* than the stable Paretian model. This is due to the Student's t model's offsetting prediction error for the Canadian dollar for these γ-values.

While the *ME*'s indicate possible systematic prediction bias, the *MAE*s and *MSE*s reflect the size of the prediction error. With respect to both measures, the stable Paretian model dominates those of both the normal and the Student's t for all γ-values considered. This is also evident from the bottom panel of Table 6, which aggregates the summary measures over all γ-values considered. In the aggregate, the model using the stable Paretian innovation assumption outperforms those using the normal and Student's t in terms of all three summary measures.

5. Conclusions

Power GARCH processes driven by either stable Paretian or Student's t innovations have been evaluated and compared in the context of predicting downside market risk, an activity which is particularly important for risk managers of financial institutions. For all five exchange-rate series considered, the asymmetric stable Paretian distributional assumption was found to be superior.

While there exist several popular model classes designed to parsimoniously and effectively fit financial return data, the GARCH class of models is arguably the most common. Furthermore, the usual assumption, and that which is implemented in popular software packages, is that the driving innovations are either normally or Student's t distributed. The former is the "standard" assumption in financial and even most econometric or statistical models, but fails demonstrably in empirical applications [see, e.g., Palm (1997), Gouriéroux (1997), and the references therein]. Indeed, normality is a special, limiting case of the stable Paretian distribution, which, otherwise, allows for fatter-than-normal tails and skewness, these being precisely two of the typical "stylized facts" associated with financial returns data. The Student's t assumption does allow for fatter tails, but is restricted to being symmetric. The latter restraint can actually be overcome if more general Student's t-like distributions are used (Paolella, 1999; Mittnik and Paolella, 2000), but these suggestions, while often providing admirable in- and out-of-sample fits, do not possess the theoretical property of summability, common only to the stable Paretian (and, thus, normal) class of distributions.

With respect to the summability property, one might argue that the value of stable Paretian models is, as shown here, their improved forecasting ability as compared to competing models, with such "theoretical niceties" as summability being largely irrelevant. In a larger context, however, the summability property can often be judiciously used when building more complex financial models such as those used in portfolio analysis. In such models, the ad hoc nature of, say, the Student's t distribution can become quite problematic. Further discussion along these lines and a test for the summability property in the context of GARCH models has been proposed in Paolella (2001) and further applied in Mittnik, Paolella and Rachev (2000).

References

Akaike, H., 1973. Information theory and an extension of the maximum likelihood principle. In: Petrov, B., Csaki, F. (Eds.), 2nd International Symposium on Information Theory. Akademiai Kiado, Budapest, pp. 267–281.

Anderson, T., Darling, D., 1952. Asymptotic theory of certain "goodness of fit" criteria based on stochastic processes. The Annals of Mathematical Statistics 23, 193–212.

Brockwell, P., Davis, R., 1991. Time Series: Theory and Methods. Springer-Verlag.

DuMouchel, W., 1973. On the asymptotic normality of the maximum-likelihood estimate when sampling from a stable distribution. The Annals of Statistics 1, 948–957.

Engle, R.F., Bollerslev, T., 1986. Modelling the persistence of conditional variances. Econometric Reviews 5, 1–50; 81–87.

Fama, E., 1965. The behavior of stock market prices. Journal of Business 38, 34–105.

Gouriéroux, C., 1997. ARCH Models and Financial Applications. Springer, New York.

Hurvich, C., Tsai, C., 1989. Regression and time series model selection in small samples. Biometrika 76, 297–307.

Liu, S., Brorsen, B.W., 1995. Maximum likelihood estimation of a GARCH-stable model. Journal of Applied Econometrics 10, 273–285.

Mandelbrot, B., 1963. The variation of certain speculative prices. Journal of Business 36, 394–419.

McCulloch, J.H., 1985. Interest-risk sensitive deposit insurance premia: stable ACH estimates. Journal of Banking and Finance 9, 137–156.

McCulloch, J.H., 1997. Financial applications of stable distributions. In: Maddala, G., Rao, C. (Eds.), Handbook of Statistics, Vol. 14. Elsevier, Amsterdam.

Mittnik, S., Paolella, M.S., 2000. Conditional density and value-at-risk prediction of Asian currency exchange rates. Journal of Forecasting 19, 313–333.

Mittnik, S., Paolella, M.S., Rachev, S.T., 2000. Diagnosing and treating the fat tails in financial returns data. Journal of Empirical Finance 7, 389–416.

Mittnik, S., Paolella, M.S., Rachev, S.T., 2002. Stationarity of stable power-GARCH processes. Journal of Econometrics 106, 97–107.

Mittnik, S., Rachev, S., 1993. Modeling asset returns with alternative stable models. Econometric Reviews 12, 261–330.

Mittnik, S., Rachev, S.T., Doganoglu, T., Chenyao, D., 1999. Maximum likelihood estimation of stable Paretian models. Mathematical and Computer Modelling 29, 275–293.

Mittnik, S., Rachev, S.T., Paolella, M.S., 1998. Stable Paretian modeling in finance. In: Adler, R.J., Feldman, R.E., Taqqu, M.S. (Eds.), A Practical Guide to Heavy Tails: Statistical Techniques for Analyzing Heavy Tailed Distributions. Birkhäuser, Boston.

Nelson, D., 1990. Stationarity and persistence in the GARCH(1,1) model. Econometric Theory 6, 318–344.

Nelson, D.B., Cao, C.Q., 1992. Inequality constraints in the univariate GARCH model. Journal of Business and Economic Statistics 10 (2), 229–235.

Palm, F.C., 1997. GARCH models of volatility. In: Maddala, G., Rao, C. (Eds.), Handbook of Statistics, Vol. 14. Elsevier, Amsterdam.

Panorska, A.K., Mittnik, S., Rachev, S.T., 1995. Stable GARCH models for financial time series. Applied Mathematics Letters 8 (5), 33–37.

Paolella, M., 1999. Tail estimation and conditional modeling of heteroscedastic time-series. Ph.D. Thesis. Institute of Statistics and Econometrics, Christian Albrechts University at Kiel. Pro Business, Berlin.

Paolella, M.S., 2001. Testing the stable Paretian assumption. Mathematical and Computer Modelling 34, 1095–1112.

Press, W.H., Teukolsky, S.A., Vetterling, W.T., Flannery, B.P., 1991. Numerical Recipes in Fortran: The Art of Scientific Computing, 2nd edition. Cambridge University Press, New York.

Rachev, S.T., Mittnik, S., 2000. Stable Paretian Models in Finance. Wiley, Chichester.

Samorodnitsky, G., Taqqu, M., 1994. Stable Non-Gaussian Random Processes, Stochastic Models with Infinite Variance. Chapman & Hall, London.

Schwartz, G., 1978. Estimating the dimension of a model. The Annals of Statistics 6, 461–464.

Tanaka, K., 1996. Time Series Analysis, Nonstationary and Noninvertible Distribution Theory. Wiley, New York.

Chapter 10

STABLE NON-GAUSSIAN MODELS FOR CREDIT RISK MANAGEMENT

BERNHARD MARTIN

Institute of Statistics and Mathematical Economics, University of Karlsruhe, Germany

SVETLOZAR T. RACHEV

Department of Statistics and Applied Probability, University of California, Santa Barbara, USA
Institute of Statistics and Mathematical Economics, University of Karlsruhe, Germany
e-mail: rachev@lsoe-4.wiwi.uni-karlsruhe.de

EDUARDO S. SCHWARTZ

Anderson School of Management, University of California, Los Angeles, USA

Contents

Handbook of Heavy Tailed Distributions in Finance, Edited by S.T. Rachev

Abstract

Unlike the credit risk models based on the normal assumption, the model in this chapter assumes credit returns to follow a stable distribution. As empirical studies show, the daily returns of a bond and its credit spread obey a stable law, exhibiting peaked, heavy tailed, and skewed distributions. This implies the application of stable Credit Value-at-Risk (CVaR) in order to obtain a more precise measure for a bond's risk compared to normal Credit Value-at-Risk.

Describing the returns of a financial instrument subject to credit risk, our model is based on the one-factor model proposed by Rachev, Schwartz and Khindanova (2000). It separates the risky asset into a default free component subject to interest risk and a component that represents the default risk. For this model, we change the definition of the bonds' returns and derive the risk of individual corporate bonds directly from their historical market prices. Thus, we avoid to construct yield curves mapping the individual credit risk of the observed risky bonds.

In an empirical example consisting of a portfolio with two corporate bonds, we compare the stable and normal Credit Value-at-Risk. Furthermore, we analyze the effects of considering the dependence among different instruments compared to the independence assumption.

The second part of the chapter analyzes the presence of long-range dependence (LRD) in credit returns using time series of corporate bond indices. For the detection of LRD, we apply the classical R/S analysis by Mandelbrot and Wallis (1968), the statistic of Lo (1991), and the Mansfield–Rachev–Samorodnitsky (MRS) statistic (1999). Our results show that the Hurst-Exponent is greater than 0.5 for all four time series. Under the Gaussian assumption, the LRD hypothesis is significant for two of the four time series. Allowing a tail-index α of less than 2, $1 < \alpha \leqslant 2$, the applied MRS-statistic also indicates significant LRD for these two series.

1. Stable modeling in credit risk – recent advances

Academics and practitioners[1] have examined the application of stable distributions for modeling asset returns. As it is well documented in the literature on empirical finance,[2] changes in the value of a financial asset are heavy-tailed and peaked, whereas the mass of the commonly used normal distribution is located around its center. Therefore, the normal distribution[3] fails to model crashes and strong upturns in financial markets.

Recent research has also examined the returns of instruments subject to credit risk. Those studies[4] found that credit returns are also peaked and heavy-tailed. Moreover, they turned out to be skewed.

Rachev, Schwartz and Khindanova (2000) suggested the application of stable distributions for credit instruments to meet those properties. As explained above, for stable distributions, the peakedness and the heavy tails are determined by the stability index α, whereas, the parameter β is responsible for skewness or asymmetry.

In the following, we propose a model that describes the returns of individual corporate bonds assuming those to follow a stable law. We are especially interested in determining the Value-at-Risk (VaR) of such financial instruments subject to credit risk for a given time horizon. VaR is a measure for the riskiness of an asset and determines the economic capital required for holding the asset.[5] VaR models seek to measure the maximum loss of value on a given asset or liability over a given time period at a given confidence level (e.g., 95%). The VaR is defined as a threshold regarding the price change of the instrument over the observed time horizon. The return over time horizon τ is expected to fall below that threshold with a probability of $1-c$. It says, with a probability of $1-c$ the returns are expected to be less than $-VaR_c$.[6] The VaR is expressed as

$$P\left[\Delta p(\tau) \leqslant -VaR_c\right] = 1 - c, \tag{1}$$

with

- $\Delta p(\tau)$: price change over time horizon τ;
- c: confidence level of VaR, e.g., 95%;
- The probability that losses exceed VaR_c is $1-c$.

1 See the work of Mandelbrot (1963), Fama (1965a, b), Fama and Roll (1971).

2 For example, see Rachev and Mittnik (2000).

3 The application of the normal distribution for financial returns goes back to the work of Bachelier (1900).

4 Federal Reserve System Task Force on Internal Credit Risk Models (1998), Basel Committee on Banking Supervision (1999).

5 Saunders (1999).

6 The VaR is defined as a positive number.

2. A one-factor model for stable credit returns

In their model for credit returns, Rachev, Schwartz and Khindanova (2000) assumed a linear relationship between the returns of a risky credit instrument and the returns of a comparable risk-free credit instrument.

For such a credit instrument i, the returns are described by

$$R_i = a_i + b_i Y_i + U_i, \tag{2}$$

where

- R_i are the log-returns of an asset i that is subject to credit risk;
- Y_i are the log-returns of a risk-free asset;
- U_i is the disturbance. It represents the spread or the premium for the credit risk;
- a_i and b_i are constants which are obtained by ordinary least squares (OLS) estimation.

In this linear model, the returns of both the risky (R_i) and the risk-free (Y_i) credit instrument are assumed to follow a strictly stable law. Moreover, the disturbance term (U_i) is a strictly stable random variable:

- $U_i \sim S_\alpha(\sigma_\alpha, \beta_\alpha, \mu_\alpha)$, $1 < \alpha < 2$;
- $Y_i \sim S_\gamma(\sigma_\gamma, \beta_\gamma, \mu_\gamma)$, $1 < \gamma < 2$.

For credit instruments, the log-return $R_{i,t}$ at time t is defined as

$$R_{i,t} = \log\left(\frac{P_{i,t,T}}{P_{i,t-1,T-1}}\right), \tag{3}$$

where $P_{i,t,T}$ is the price of an instrument i subject to credit risk with maturity date T evaluated at time t. The log-returns of the riskless asset $Y_{i,t}$ are determined by

$$Y_{i,t} = \log\left(\frac{B_{i,t,T}}{B_{i,t-1,T-1}}\right), \tag{4}$$

with $B_{i,t,T}$ as the price of the risk-free asset with maturity date T evaluated at time t. This means that all prices used for the calculation of the returns are determined on the basis of constant *time-to-maturity*. Therefore, the time series of log-returns (both $Y_{i,t}$ and $R_{i,t}$) is calculated such that the time-to-maturity is the same for all t.

It must be noted that $Y_{i,t}$ and $R_{i,t}$ are not directly observable for individual bonds whose market price movements are recorded on a daily basis. The prices $B_{i,t,T}$, $B_{i,t-1,T-1}$, $B_{i,t-2,T-2}, \ldots$ are calculated from the yield curve of riskless treasury bonds and $P_{i,t,T}$, $P_{i,t-1,T-1}$, $P_{i,t-2,T-2}, \ldots$ are derived from a yield curve generated from risky bonds representing a similar level of credit risk (e.g., having equal credit ratings). Such an approach enables us to deal with constant time-to-maturity. This is crucial, because for the prices of individual bonds, time-to-maturity decreases with increasing time t. However, a decreasing time to maturity does have an effect on the credit returns. Thus, the advantage of the approach in (3) and (4) is that we do not have to pay attention on the influence of a changing time-to-maturity.

The effect of changing time-to-maturity on credit returns can be demonstrated by a small example with two riskless zero-bonds: the first one has a time-to-maturity of one year, the other one has a time to maturity of two years. Furthermore, the term structure is assumed to be flat, and therefore, both securities have equal yields. If the yield of both increases by the same percentage, then the price of the two-year bond reacts more sensitively, compared to the one-year bond.

However, the approach of modelling the returns as in (3) and (4) is very difficult to implement in practice. Historical data of daily yield curves is available for treasury bonds, but it is practically impossible to observe a time series $P_{t,T}, P_{t-1,T-1}, P_{t-2,T-2}, \ldots$ for an individual bond. We would have to define a number of different credit risk categories and assign individual bonds with different maturities to those.[7] We would use the prices of the bonds assigned to the same risk category in order to generate the corresponding yield curve.[8]

In order to avoid such difficulties, we look for a more practical way to define the credit returns. Obviously, a risk manager would prefer to deal with the observed real prices of a bond to fit a model, rather than deriving prices from yield curves that have to be generated before. Moreover, each yield curve only represents an average credit risk level. Our approach proposed in the following paragraph determines the individual credit risk of the analyzed bond.

A new approach to define the returns R_i and Y_i. From the historical yield curve data of treasury bonds, we can construct daily prices for any riskless bond with given coupon, coupon dates, and maturity. Thus, we can generate a corresponding riskless bond i with identical specifications for each risky corporate bond i. We define the return $R_{i,t}$ of a risky corporate bond as its actual (observable) daily price movement:

$$R_{i,t} = \log\left(\frac{P_{i,t,T}}{P_{i,t-1,T}}\right). \tag{5}$$

Here, time-to-maturity is no longer kept fixed. The return $R_{i,t}$ is that of an individual bond i with fixed maturity date T. The riskless returns $Y_{i,t}$ are defined the same way:

$$Y_{i,t} = \log\left(\frac{B_{i,t,T}}{B_{i,t-1,T}}\right). \tag{6}$$

This riskless bond i has the same specifications (maturity, coupon, coupon dates), as the risky bond i.

With this new approach, the original linear risk-return relation $R_i = a_i + b_i Y_i + U_i$ remains, but its components R_i, Y_i, and U_i now have a different meaning. R_i and Y_i are individual bond returns, and the disturbance U_i incorporates both credit spread and the risk of time-to-maturity.

7 For example, one can use the rating grades assigned by Standard & Poor's or Moody's to define a risk category.

8 For example, see McCulloch (1971, 1975).

For all empirical examinations in this chapter, we used the model with the returns defined in (5) and (6). In the following, we present a brief summary of advantages and disadvantages of both approaches:

The model whose returns are defined by Equations (3) and (4), abandons the problem of changing time-to-maturity. This is its main advantage. The disadvantage of such an approach is that yield curves have to be modelled for a number of different risk levels (e.g., corporate credit ratings), and for the risk free (treasury) bonds. After fitting the parameters a and b of Equation (2), we can simulate future scenarios for each yield curve integrating a model for the riskless returns. Such a framework would enable us to simulate future daily returns for each time-to-maturity. With the simulated yield curves, we would then be able to calculate the future returns of individual bonds.

With our model defined by the returns in (5) and (6) we neither need to construct yield curves for a number of risk levels (of risky bonds), nor do we have to simulate future representations of those yield curves by applying a complex term structure model. Thus, we can directly simulate future returns of individual bonds by generating representations of Y_i and U_i according to their fitted distributions.

The advantage of the chosen approach is that we can work with the actual historical price data and spread information of the individual bonds, instead of generating yield curves, each for a certain risk grade. Such yield curves only represent the average of a risk grade. Studies found that in some cases a higher rated bond can even have a larger credit spread than bonds with a lower rating grade. This is due to the fact that the range of credit spreads within a given rating grade can be relatively wide and that the spread ranges of neighboring grades are usually overlapping. A reason for this effect could be that the market values the creditworthiness of an issuer differently than the rating agencies do. Sometimes the market can anticipate a change in the credit quality of an issuer before the rating agencies react.

The construction of a yield curve for a given credit grade usually requires data from a large number of bonds with various issuers. The yield curve of a single issuer is calculable even only for large corporations with many issued bonds.

2.1. *Credit risk evaluation for single assets*

In order to obtain the Credit Value-at-Risk (CVaR) for a bond i over a time horizon of one period, we perform the following steps:

- We create a corresponding risk-free treasury bond with equal maturity, coupon, and coupon dates.
- The estimates for a_i and b_i are calculated with OLSE.

  As in Rachev, Khindanova and Schwartz, the estimates are given by

$$\hat{a}_i = \frac{\sum_{t=1}^{T} Y_{it}^2 \sum_{t=1}^{T} R_{it} - \sum_{t=1}^{T} Y_{it} \sum_{t=1}^{T} R_{it} Y_{it}}{T \sum_{t=1}^{T} Y_{it}^2 - (\sum_{t=1}^{T} Y_{it})^2}, \tag{7}$$

$$\hat{b}_i = \frac{T \sum_{t=1}^{T} R_{it} Y_{it} - \sum_{t=1}^{T} Y_{it} \sum_{t=1}^{T} R_{it}}{T \sum_{t=1}^{T} Y_{it}^2 - (\sum_{t=1}^{T} Y_{it})^2}, \tag{8}$$

where $i = 1, \ldots, N$; $t = 1, \ldots, T$.

With the estimates $\hat{a}_i$ and $\hat{b}_i$, we obtain the residuals $\widehat{U}_i$,

$$\widehat{U}_i = R_i - \hat{a}_i - \hat{b}_i Y_i. \tag{9}$$

- Finally, we perform a stable fit for $\widehat{U}_i$ and Y_i.
- In order to calculate the CVaR of asset i for one period, we simulate 1000 representations of $R_i = a_i + b_i Y_i + U_i$.

2.2. *A stable portfolio model with independent credit returns*

Suppose there are n different credit instruments i (bonds) in a portfolio, and let v_i be the weight of security i within the portfolio.[9] The return of the portfolio is given by

$$R_p = \sum_{i=1}^{n} v_i R_i, \tag{10}$$

with

$$R_p = \sum_{i=1}^{n} v_i (a_i + b_i Y_i + U_i) = \sum_{i=1}^{n} v_i a_i + \sum_{i=1}^{n} v_i b_i Y_i + \sum_{i=1}^{n} v_i U_i, \tag{11}$$

and

$$\sum_{i=1}^{n} v_i = 1. \tag{12}$$

R_p can be expressed by

$$R_p = \sum_{i=1}^{n} v_i a_i + Y_p + U_p, \tag{13}$$

with Y_P and U_p given by

$$Y_p = \sum_{i=1}^{n} v_i b_i Y_i \tag{14}$$

[9] v_i can also be negative in case short-selling is allowed.

and

$$U_p = \sum_{i=1}^{n} v_i U_i. \tag{15}$$

The constant a_p of the total portfolio is

$$a_p = \sum_{i=1}^{n} v_i a_i. \tag{16}$$

As we assume the R_i to be driven by independent α-stable distributions, this also means that both the U_i and the Y_i, $i = 1, \ldots, n$, are independent of each other. We further assume that both the U_i and the Y_i, $i = 1, \ldots, n$, are characterized by a common index of stability (α for the U_i, γ for the Y_i). A common stability index allows an easy analytical solution for the parameters of the distributions for U_p and Y_p. For the properties of stable distributions, see Samorodnitsky and Taqqu (1994).

The common index of stability is calculated as an average from the stability indices of the distributions of the individual U_i and Y_i, weighted according to formula (10):

$$\alpha = \frac{\sum_{i=1}^{n} |v_i| \alpha_i}{\sum_{i=1}^{n} |v_i|} \tag{17}$$

and

$$\gamma = \frac{\sum_{i=1}^{n} |v_i| \gamma_i}{\sum_{i=1}^{n} |v_i|}. \tag{18}$$

With the common stability index, the parameters β, σ, μ have to be re-estimated for the individual U_i and Y_i first.

The assumption of independent returns gives us an analytical solution for the portfolio's U_p and Y_p.

The parameters of U_p and Y_p are then determined by the following expressions:

$$\sigma_{U_p} = \left[\sum_{i=1}^{n} \left(|v_i| \sigma_{U_i} \right)^{\alpha} \right]^{1/\alpha}, \tag{19}$$

$$\beta_{U_p} = \frac{\sum_{i=1}^{n} [\operatorname{sign}(v_i) \beta_{U_i} (|v_i| \sigma_{U_i})^{\alpha}]}{\sum_{i=1}^{n} (|v_i| \sigma_{U_i})^{\alpha}}, \tag{20}$$

$$\mu_{U_p} = \sum_{i=1}^{n} v_i \mu_{U_i}, \tag{21}$$

$$\sigma_{Y_p} = \left[\sum_{i=1}^{n} (|v_i \hat{b}_i| \sigma_{Y_i})^{\gamma} \right]^{1/\gamma}, \tag{22}$$

$$\beta_{Y_p} = \frac{\sum_{i=1}^{n} [\mathrm{sign}(v_i \hat{b}_i) \beta_{Y_i} (|v_i \hat{b}_i| \sigma_{Y_i})^{\gamma}]}{\sum_{i=1}^{n} (|v_i \hat{b}_i| \sigma_{Y_i})^{\gamma}}, \tag{23}$$

$$\mu_{Y_p} = \sum_{i=1}^{n} v_i \mu_{Y_i}. \tag{24}$$

The portfolio's returns R_p are given by (13).

2.3. A stable portfolio model with dependent credit returns

This section introduces a solution for modelling the dependence between credit returns on the one hand, and integrating the skewness-property of their distributions on the other hand.

Each variable U_i and Y_i is split into a dependent symmetric and into an independent skewed component. Both components are independent of each other;

$$U_i = U_i^{(1)} + U_i^{(2)}, \tag{25}$$

$$Y_i = Y_i^{(1)} + Y_i^{(2)}. \tag{26}$$

By the example of U_i, we show the derivation of the parameters for the two independent components. Both components are defined to have identical stability indices:

$$U_i^{(1)} \sim S_\alpha(\sigma_1, \beta_1, 0), \tag{27}$$

$$U_i^{(2)} \sim S_\alpha(\sigma_2, \beta_2, 0). \tag{28}$$

Because of the independence of $U_i^{(1)}$ and $U_i^{(2)}$, the parameters' values of U_i are calculated as follows:

$$\sigma = (\sigma_1^\alpha + \sigma_2^\alpha)^{1/\alpha}, \tag{29}$$

$$\beta = \frac{\beta_1 \sigma_1^\alpha + \beta_2 \sigma_2^\alpha}{\sigma_1^\alpha + \sigma_2^\alpha}. \tag{30}$$

$U_i^{(1)}$ is symmetric, therefore $\beta_1 = 0$. We also set equal values for the scale parameters, $\sigma_1 = \sigma_2 = \sigma^*$.

Thus, the parameters of U_i are:

$$\sigma = 2^{1/\alpha} \sigma^*, \tag{31}$$

$$\beta = \frac{1}{2}\beta_2. \tag{32}$$

Summing up the results for the parameters, we have: $\sigma_1 = \sigma_2 = \sigma^* = 2^{-1/\alpha}\sigma$, $\beta_2 = 2\beta$ (β_2 is for the skewed component $U_i^{(2)}$), and $\beta_1 = 0$ (β_1 is for the symmetrical component $U_i^{(1)}$),

$$U_i^{(1)} \sim S_\alpha\left(2^{-1/\alpha}\sigma, 0, 0\right), \tag{33}$$

$$U_i^{(2)} \sim S_\alpha\left(2^{-1/\alpha}\sigma, 2\beta, 0\right). \tag{34}$$

Analogously, Y_i is split into $Y_i^{(1)} + Y_i^{(2)}$, and their parameters are obtained the same way. The return of the credit instrument i is then given by

$$R_{i,t} = a + b\left(Y_{i,t}^{(1)} + Y_{i,t}^{(2)}\right) + \left(U_{i,t}^{(1)} + U_{i,t}^{(2)}\right). \tag{35}$$

The symmetric components $Y_{i,t}^{(1)}$ and $U_{i,t}^{(1)}$ are used to incorporate the dependence among the n assets. The dependence structure of the $S\alpha S$[10] vectors $(U_1^{(1)}, U_2^{(1)}, \ldots, U_n^{(1)})$ and $(Y_1^{(1)}, Y_2^{(1)}, \ldots, Y_n^{(1)})$ is modelled by representing them as sub-Gaussian vectors. Thus, $(U_1^{(1)}, U_2^{(1)}, \ldots, U_n^{(1)})$ is represented as

$$\left(U_1^{(1)}, U_2^{(1)}, \ldots, U_n^{(1)}\right) \sim \left(A^{1/2}G_1, A^{1/2}G_2, \ldots, A^{1/2}G_n\right), \tag{36}$$

where A is a totally skewed $\alpha/2$-stable random variable with

$$A \sim S_{\alpha/2}\left(\left(\cos\frac{\pi\alpha}{4}\right)^{2/\alpha}, 1, 0\right)$$

and $G = (G_1, G_2, \ldots, G_n)$ is an n-dimensional Gaussian zero mean random vector. Let $R_{ij} = EG_iG_j$, $i, j = 1, \ldots, n$, be the covariances within the vector $G = (G_1, G_2, \ldots, G_n)$. Then $(U_1^{(1)}, U_2^{(1)}, \ldots, U_n^{(1)})$ is generated by simulating a representation of the Gaussian vector G with correlated elements $G_1, G_2, \ldots, G_n$ and an independent representation of the $\alpha/2$-stable random variable A.[11]

The generation of vector $(Y_1^{(1)}, Y_2^{(1)}, \ldots, Y_n^{(1)})$ is performed analogously.

[10] A $S\alpha S$ vector is a symmetrically stable random vector.

[11] There are various ways to model the dependence. For example, see Rachev, Khindanova and Schwartz (2000).

3. Comparison of empirical results

In order to illustrate the effects of the different assumptions on our stable credit modelling, we perform an empirical example. We chose a portfolio consisting of two corporate bonds and calculated its daily Credit Value-at-Risk (CVaR) for the independent and the dependent case.

3.1. The observed portfolio data

As our sample-portfolio, two corporate bonds (country: US market; currency: US-Dollars) were selected. Both bonds pay coupons twice a year. Historical prices were obtained from Bloomberg[12] for the past four years (March 14th 1996 up to March 13th 2000). According to their credit ratings, the bonds exhibit considerable credit risk. For our portfolio, we assume to have one unit of each security. Both have a nominal value of 100 US \$ (see Table 1).

3.2. Generating comparable risk-free bonds from the yield curve

First, we calculate the daily returns of the above listed bonds using market prices. Then, for each bond a corresponding riskless bond was generated in order to derive the values for the Y_i. The corresponding riskless bond has the same specifications (maturity, coupon, coupon date) as the risky corporate bond. The history of daily prices of these artificial treasury bonds were calculated from the daily yield curves. The treasury-yield curve for each day was approximated by prices of 9 risk-free zero bonds with maturities: 0.25, 0.5, 1, 2, 3, 4, 5, 7, 10 years. These 9 points were interpolated by a natural cubic spline algorithm.[13]

Table 1
Bonds selected for sample-portfolio

Corporation	Coupon	Rating (S&P/Moodys)	Maturity
Pennzoil (Bond 1)	10.25	BBB+/Baa2	11/05
United Airlines (Bond 2)	9.0	BB+/Baa2	12/03

Table 2
Estimates for a and b

Corporation	$\hat{a}_i$	$\hat{b}_i$
Pennzoil (Bond 1)	0.0000	0.9262
United Airlines (Bond 2)	0.0000	0.9878

[12] Bloomberg Information System, Corporate Bonds Section.
[13] Burden and Faires (1997).

With the obtained daily yield curves, we can generate historical prices for our artificial treasury bonds and calculate their daily returns according to (6).

Next, we perform the linear regressions to estimate the parameters a and b of the equations $R_i = a_i + b_i Y_i + U_i$. The resulting $\hat{a}_i$ and $\hat{b}_i$ are OLS-estimates (see Equations (7) and (8), and Table 2).

With the values $\hat{b}$ and $\hat{a}$ the estimates for the disturbances U_i can be calculated: $\widehat{U}_i = R_i - \hat{a} - \hat{b}Y_i$. We now have the empirical distributions for R_i, Y_i, and $\widehat{U}_i$. Next, we apply both a stable and a normal fit to those.

During the available sample period from 1996 to 2000, time-to-maturity for the observed bonds reduces by 41% and 52%. The question rises if this has a systematic effect on the fitted parameters of the Y_i over time. However, in our case empirical analysis found no evidence for this.

3.3. *Fitting the empirical time series for R_i, Y_i, and $\widehat{U}_i$*

For the stable fit, we applied the Maximum Likelihood Estimation (MLE) to obtain the four parameters of the distribution. The stable densities were approximated via Fast Fourier Transformation.[14] The procedure was implemented with Matlab 5.3.

Table 3
Parameters for R fitted with stable and normal distribution

Corporation	Stable				Normal	
	alpha	beta	sigma	mu	mean	std-dev
Pennzoil (Bond 1)	1.5451	−0.0690	0.0019	0.0000	−0.0001	0.0041
United Airlines (Bond 2)	1.5199	−0.0744	0.00164	0.0000	−0.0001	0.0035

Table 4
Parameters for Y fitted with stable and normal distribution

Corporation	Stable				Normal	
	alpha	beta	sigma	mu	mean	std-dev
Pennzoil (Bond 1)	1.3639	−0.0297	0.0012	0.0000	−0.0001	0.0027
United Airlines (Bond 2)	1.2811	0.0062	0.0009	0.0000	−0.0001	0.0022

Table 5
Parameters for the disturbance U fitted with stable and normal distribution

Corporation	Stable				Normal	
	alpha	beta	sigma	mu	mean	std-dev
Pennzoil (Bond 1)	1.0348	−0.0247	0.0006	0.0000	0.0001	0.0027
United Airlines (Bond 2)	1.1663	0.0117	0.0008	0.0000	0.0000	0.0032

[14] For example, see Rachev and Mittnik (2000).

The parameters of the stable and Gaussian distributions fitted for the R_i, Y_i, and U_i are shown in Tables 3, 4, and 5.

3.4. CVaR-results for the independence assumption

The assumption of independence between the bonds in our portfolio leads to the application of the equations in Section 2.2. We perform the stable fit for both the Y_i and the U_i based on a common stability index, and select $\alpha = 1.10$ for the U_i, and $\gamma = 1.32$ for the Y_i. Re-estimating the parameters by a stable fit applying common stability indices, we obtain the results presented in Tables 6 and 7.

The parameters of the portfolio's U_p and Y_p, given by

$$U_p = v_1 U_1 + v_2 U_2 \quad \text{and} \quad Y_p = v_1 \hat{b}_1 Y_1 + v_2 \hat{b}_1 Y_2, \tag{37}$$

are determined by the relationships presented in Section 2.2. With $v_1 = v_2 = 0.5$, we have

$$U_p = 0.5 U_1 + 0.5 U_2 \quad \text{and} \quad Y_p = 0.5 \hat{b}_1 Y_1 + 0.5 \hat{b}_1 Y_2. \tag{38}$$

Table 6
Parameters for the disturbance U fitted with stable and normal distribution assuming $\alpha = 1.10$

Corporation	Stable			
	alpha	beta	sigma	mu
Pennzoil (Bond 1)	1.1000	−0.0047	0.0007	0.0000
United Airlines (Bond 2)	1.1000	0.0828	0.0011	0.0000

Table 7
Parameters for Y fitted with stable and normal distribution assuming $\gamma = 1.32$

Corporation	Stable			
	alpha	beta	sigma	mu
Pennzoil (Bond 1)	1.3200	0.0089	0.0013	0.0001
United Airlines (Bond 2)	1.3200	−0.0430	0.0010	0.0000

Table 8
Stable parameters for portfolio U_p and Y_p

	Stable			
	alpha	beta	sigma	mu
Y_p	1.3200	−0.0137	0.0009	0.0000
U_p	1.1000	0.0497	0.0008	0.0000

The results for the parameters of U_p and Y_p are printed in Table 8. Their calculation is performed according to Equations (19)–(24).

The resulting equation describing the portfolio's returns is

$$R_p = 0.5(\hat{a}_1 + \hat{a}_2) + Y_p + U_p = 0.0000 + Y_p + U_p. \tag{39}$$

Based on this, we can simulate 1000 daily returns. This provides us the daily Credit Value-at-Risk of the portfolio. For the stable model with independence assumption, we obtain a one-day CVaR of 0.67% at the 95% level and a one-day CVaR of 2.24% at the 99% level.

So far, we have assumed both bonds to be independent of each other. However, empirical examinations exhibit strong dependence among the Y_i and among the U_i. Therefore, the following section presents the results of the model in Section 2.3 incorporating dependence among the U_i and dependence among the Y_i.

3.5. *CVaR-results for the dependence assumption*

Calculating the Gaussian covariances and correlations between the U_i and Y_i of our example portfolio, the results are presented in Tables 9–12.

The modelling of the dependent case – as demonstrated in the former section – is performed by splitting both the Y_i and the U_i into two components. The first component includes the dependence which is modelled by a sub-Gaussian random vector. The second component exhibits the skewness (see Section 2.3).

Table 13 provides the Credit-Value-at-Risk (CVaR) for the 95% and 99% level with horizon one day, comparing both stable models (independent and dependent case) with the empirical data.

Table 9
$\mathrm{cov}(Y_i, Y_j),\ i, j = 1, 2$

	$\mathrm{cov}(Y_i, Y_j) * 10^{-4}$	
	Y_1	Y_2
Y_1	0.7699	0.5821
Y_2	0.5821	0.4785

Table 10
$\mathrm{cov}(U_i, U_j),\ i, j = 1, 2$

	$\mathrm{cov}(U_i, U_j) * 10^{-4}$	
	U_1	U_2
U_1	0.1038	0.0850
U_2	0.0850	0.0748

Table 11
$\mathrm{cor}(Y_i, Y_j),\ i, j = 1, 2$

	$\mathrm{cov}(Y_i, Y_j)$	
	Y_1	Y_2
Y_1	1.0000	0.9591
Y_2	0.9591	1.0000

Table 12
$\mathrm{cor}(U_i, U_j),\ i, j = 1, 2$

	$\mathrm{cov}(U_i, U_j)$	
	U_1	U_2
U_1	1.0000	0.9653
U_2	0.9653	1.0000

Table 13
Stable portfolio Credit Value-at-Risk (one-day) as log-price and percental price changes

	95%		99%	
	log-price change	perc. change (%)	log-price change	perc. change (%)
Empirical	0.0054	0.54	0.0108	1.08
Dependent	0.0060	0.60	0.0242	2.40
Independent	0.0068	0.67	0.0226	2.24

Table 14
Gaussian portfolio Credit Value-at-Risk (one-day) as log-price and percentile price changes

	95%		99%	
	log-price change	perc. change (%)	log-price change	perc. change (%)
Dependent	0.0061	0.61	0.0087	0.87
Independent	0.0044	0.44	0.0063	0.63

For comparison, Table 14 presents the CVaR assuming the returns to follow a Gaussian law.

The results for CVaR confirm the earlier findings[15] that for credit returns the Gaussian VaR is only acceptable for the 95% level, but does underestimate the 99% level. The stable VaR is also appropriate for the 95% level, but it is a more conservative measure for the 99% level. This is actually good because the empirical VaR tends to underestimate the true VaR due to the low number of observations in the tails.[16]

Calculating the CVaR also for longer horizons, e.g., 10 days, we would have to build the 10-day returns for both the corporate bonds and their corresponding treasury bonds from the empirical data, and fit the above models with those data. It has to be pointed out that longer horizons cannot be calculated by taking the one-day return model and extend it to the desired horizon by simply applying a Lévy process with independent increments. Subsequent observations of the returns are not i.i.d. as volatility clustering can be observed and long-memory effects might occur. Thus, the volatility for a multiple-day horizon cannot be obtained by a simple scaling approach.[17] Longer forecast horizons require new types of models while sample data should be available for longer periods.

So far we have dealt with the phenomenon of heavy tails in credit returns. Two others, volatility clustering and long-range dependence, have already been mentioned. The following part of the chapter explains the phenomenon of long-range dependence. It introduces the theory and possible ways of detection. Finally, we examine such behavior for credit return data.

[15] Rachev, Schwartz and Khindanova (2000).

[16] See Rachev and Mittnik (2000).

[17] See Christoffersen, Diebold and Schuermann (1998).

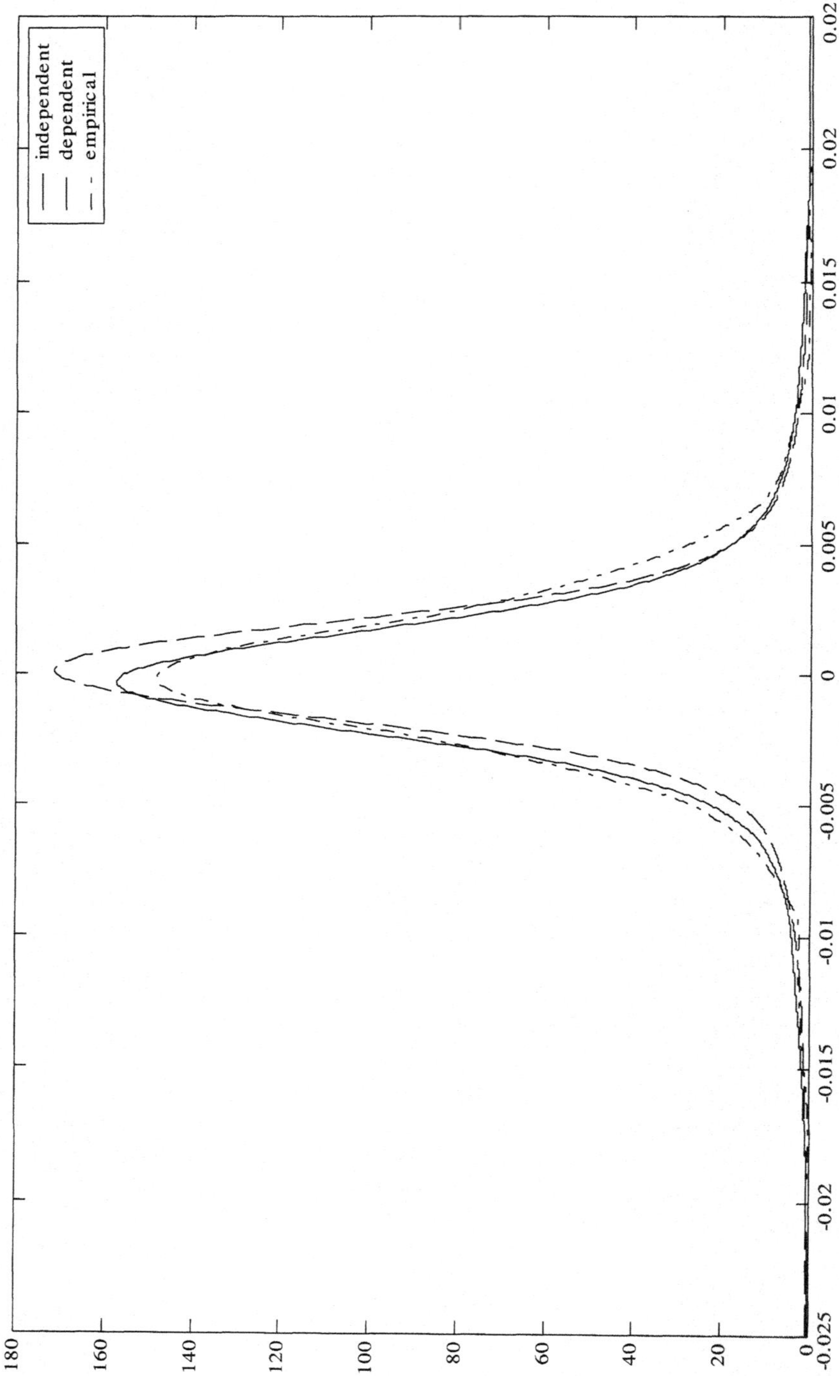

Fig. 1. Stable models for independent and dependent case: density of daily log-returns (portfolio).

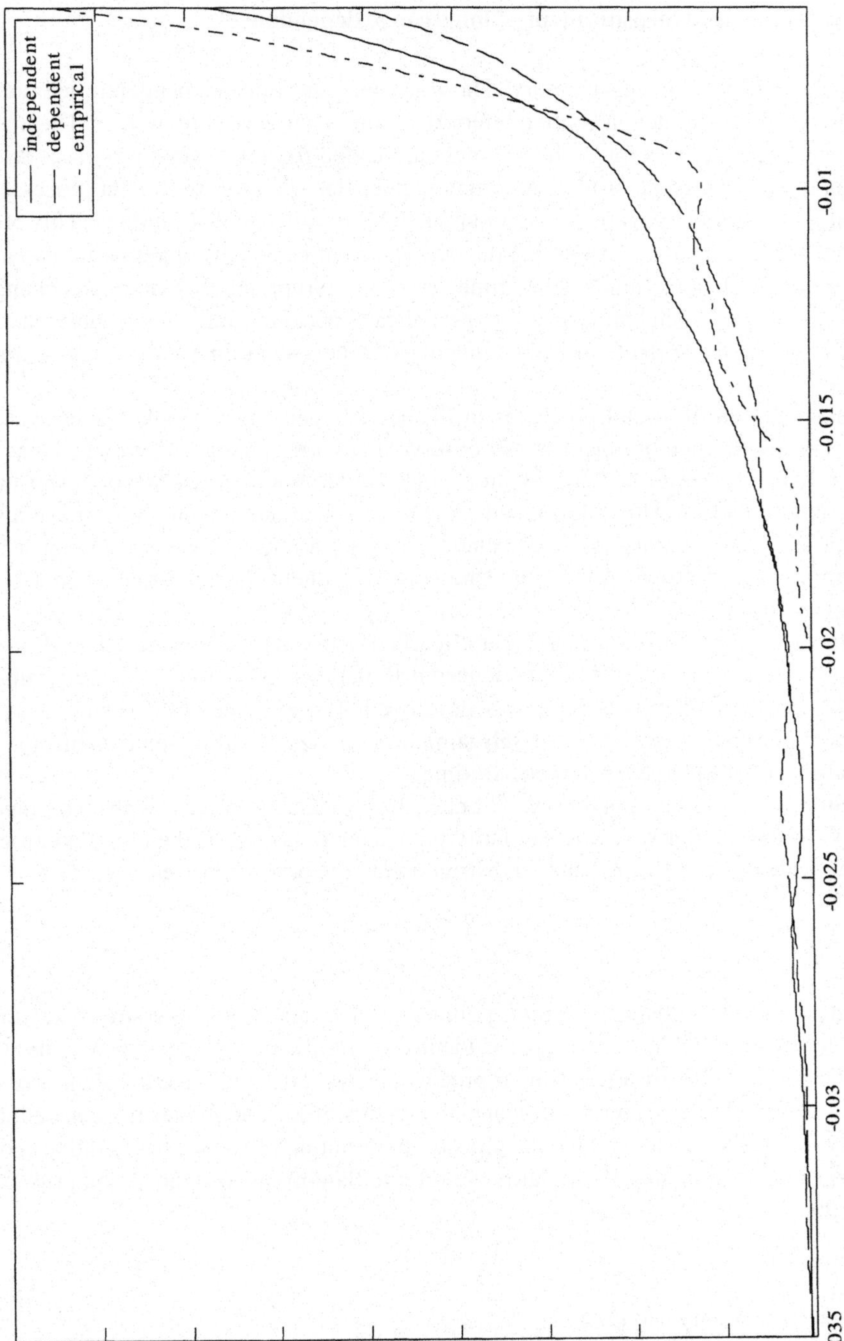

Fig. 2. Stable models for independent and dependent case: density of daily log-returns (portfolio) – tails on left-hand side.

4. The detection and measurement of long-range dependence

Time series can have a long memory. Those systems are not independently identically distributed. This phenomenon is often referred as *burstiness* in the literature.[18] The underlying stochastic processes for such burstiness are called *fractal*. Fractal processes with a long memory are called persistent. A common characteristic of those fractal processes is that their space time is governed parsimoniously by power law distributions. This effect is called the "Noah-Effect", explaining the occurrence of heavy tails and infinite variance. It can be observed as the tendency of time series for abrupt and discontinuous changes. Another property of fractal processes is hyperbolically decaying autocorrelations, which is known as the "Joseph-Effect". It is the tendency of a persistent time series to have trends and cycles.

The examination of fractal processes in finance has become a popular topic over the years.[19] For a long-memory process, we observe that larger-than-average representations are more likely followed by larger-than-average representations instead of lower-than-average representations. Hurst developed a statistic to examine the long memory of a stochastic process. As significant autocorrelations are often not visible, he came up with a new methodology to provide a measure (the Hurst-Exponent) for long-range dependence within a time series.

Due to the failures of traditional capital market theory which is largely based on the theory of martingales, researchers experienced that markets do not follow a purely random walk. The fractal market hypothesis was developed. The existence of self-similar structures is a major component of it. For self-similar processes, small increments of time are statistically similar to larger increments of time.

Self-Similarity is defined as follows:[20] Let X_t be a stochastic process with a continuous time t. X_t is self-similar with self-similarity parameter H (H-ss), if the re-scaled process with time scale ct, $c^{-H}X_{ct}$, is equal in distribution to the original process X_t,

$$X_t \stackrel{\mathrm{d}}{=} c^{-H} X_{ct}. \tag{40}$$

This means, for a sequence of time points $t_1, \ldots, t_k$ and a positive stretch factor c, the distribution of $c^{-H}(X_{ct_1}, \ldots, X_{ct_k})$ is identical with the one of $X_{t_1}, \ldots, X_{t_k}$. In other words, the path covered by a self-similar process always looks the same, regardless of the scale it is observed with. In terms of financial data this means: no matter if we have intraday, daily, weekly, or monthly data, the plots of the resulting processes have similar looks. For further information on self-similarity we refer to Samorodnitsky and Taqqu (1994), or Beran (1994).

[18] Willinger, Taqqu and Erramilli (1996).

[19] For example, we refer to Mandelbrot (1997a, b, 1999), and Peters (1994).

[20] Beran (1994).

4.1. Fractal processes and the Hurst-Exponent

First, we consider a process without a long memory. A perfect example is Standard Brownian Motion, which is characterized as a standard random walk. [21] Commonly known is Einstein's "to the one-half" – rule, describing the distance covered by a particle driven by Standard Brownian Motion. It states that the distance between consecutive values of the observed time series of this particle is proportional to the square root of time:[22]

$$R \sim T^{0.5}. \tag{41}$$

The power of 0.5 refers to the Hurst-Exponent which is already known as the self-similarity parameter. For Standard Brownian Motion, the Hurst-Exponent H is equal to 0.5 which means that we have an unbiased random walk. A process with a Gaussian limiting distribution but a Hurst-Exponent H different from 0.5 is called Fractional Brownian Motion. Fractional Brownian Motion differs from Standard Brownian Motion by the fact that it is a biased random walk. The odds are biased in one direction or the other.

Definition of Fractional Brownian Motion. [23] Let us assume a self-similar Gaussian process with X_t, $t \in R$, having mean zero and the autocovariance function

$$\mathrm{Cov}(X_{t_1}, X_{t_2}) = \frac{1}{2}\left(|t_1|^{2H} + |t_2|^{2H} - |t_1 - t_2|^{2H}\right) \mathrm{Var}\, X(1), \tag{42}$$

where H is the self-similarity parameter and $H \in (0, 1)$.

Such a process is called a Fractional Brownian Motion. For $H = 1/2$ it becomes a Standard Brownian Motion.

The increments of Fractional Brownian Motion, $Y_j = B_H(j+1) - B_H(j)$, $j \in Z$, form a stationary sequence Y_j which is called Fractional Gaussian Noise.

Fractional Gaussian Noise. A sequence of Fractional Gaussian Noise has the following properties:

(i) its mean is zero,
(ii) its variance $EY_j^2 = EB_H^2(1) = \sigma_0^2$, and
(iii) its autocovariance function is

$$r(j) = \frac{\sigma_0^2}{2}\left[(j+1)^{2H} - 2j^{2H} + (j-1)^{2H}\right],$$

where $j \in Z$, $j \geqslant 0$, and $r(j) = r(-j)$ for $j < 0$.

[21] See Campbell, Lo and McKinlay (1997).
[22] Peters (1994).
[23] Samorodnitsky and Taqqu (1994).

For $j \to \infty$, $r(j)$ behaves like a power function.

$$\lim_{j\to\infty} r(j) \to 0. \tag{43}$$

The autocorrelations are given by

$$\rho(j) = \frac{1}{2}\left[(j+1)^{2H} - 2j^{2H} + (j-1)^{2H}\right], \tag{44}$$

where $j \geqslant 0$ and $\rho(j) = \rho(-j)$ for $j < 0$. As j tends to infinity, $\rho(j)$ is equivalent to $H(2H-1)j^{2H-2}$.[24]

In the presence of long memory, $1/2 < H < 1$, the correlations decay to zero so slowly that they are no more summable:

$$\sum_{j=-\infty}^{\infty} \rho(j) = \infty. \tag{45}$$

For $H = 1/2$, i.e., a Gaussian i.i.d. process, all correlations at non-zero lags are zero. For $0 < H < 1/2$, the correlations are summable, and it holds:

$$\sum_{j=-\infty}^{\infty} \rho(j) = 0. \tag{46}$$

$H = 1$ implies $\rho(j) = 1$. For $H > 1$, the condition $-1 \leqslant \rho(j) \leqslant 1$ is violated.

For $0 < H < 1$, a Gaussian process with mean zero and the given autocovariance function is self-similar and has stationary increments (H-sssi). The above autocovariance function is shared by all Gaussian H-sssi processes.

Fractional processes with stable innovations. There are many different extensions of the Fractional Brownian motion to the α-stable case with $\alpha < 2$. Most common is the so called Linear Fractional Stable Motion or, Linear Fractional Lévy Motion.

In an analogy to the Gaussian case with $\alpha = 2$, the increments of Linear Fractional Stable Motion[25] show long-range dependence for $H > 1/\alpha$. LRD for $\alpha < 1$ does not exist, as H must lie in $(0, 1)$. Processes with $H = 1/\alpha$ are called α-stable Lévy Motion whose increments $X(t_{j+1}) - X(t_j)$ are all mutually independent.

For α-stable Lévy processes with infinite variance, we carefully have to interpret the value obtained for H and how it is related to the parameter d measuring the degree of long-range dependence.

[24] Beran (1994).

[25] Samorodnitsky and Taqqu (1994).

H, the Hurst-Exponent, is the scaling parameter and describes asymptotical self-similarity:

For finite variance processes, the relation between H and d is

$$H = d + \frac{1}{2}. \tag{47}$$

For processes with infinite variance ($\alpha < 2$), the relation is

$$H = d + \frac{1}{\alpha}. \tag{48}$$

If $d > 0$, the time series is governed by a long-memory process.

There is a number of methods to distinguish a purely random time series from a fractal one. For example, the classical R/S analysis[26] determines the parameter H of a time series. The resulting graph is called pox-plot of R/S or rescaled-adjusted-range plot.

Before the classical R/S method will be described, we briefly explain two other methods to derive the Hurst-Exponent H, the Aggregated Variance Method and the similar method Absolute Values of Aggregated Series.[27]

4.2. The Aggregated Variance Method

The original time series $X = (X_i,\ i = 1, \ldots, N)$ is divided into blocks. Each block has the size m elements. The index k labels the block. The aggregated series is calculated as the mean of each block:

$$X^{(m)}(k) = \frac{1}{m} \sum_{i=(k-1)m+1}^{km} X_i \quad \text{with } k = 1, 2, \ldots, \left[\frac{N}{m}\right]. \tag{49}$$

After building the aggregated series, we get the sample variance of $X^{(m)}(k)$ as

$$\widehat{\text{Var}} X^{(m)} = \frac{1}{N/m} \sum_{k=1}^{N/m} \left(X^{(m)}(k)\right)^2 - \left(\frac{1}{N/m} \sum_{k=1}^{N/m} X^{(m)}(k)\right)^2. \tag{50}$$

The procedure is repeated for different values of m $\{m_i,\ i \geqslant 1\}$. The chosen values for m should be equidistant on a log-scale, i.e., $m_{i+1}/m_i = C$.

As $X^{(m)}$ scales like $m^{(H-1)}$, the sample variance $\widehat{\text{Var}} X^{(m)}$ behaves like $m^{(2H-2)}$. Thus, plotting a log-log representation of m and $\widehat{\text{Var}} X^{(m)}$, the plots form a straight line with slope $2H - 2$.

[26] Mandelbrot and Wallis (1968).

[27] Teverovsky, Taqqu and Willinger (1995) as well as Teverovsky, Taqqu and Willinger (1998).

4.3. Absolute Values of the Aggregated Series

This method is similar to the Method of Aggregated Variance explained above. Starting again with the aggregated series, we calculate the sum of the absolute values of the aggregated series.

$$\frac{1}{(N/m)} \sum_{k=1}^{N/m} \left|X^{(m)}(k)\right|. \tag{51}$$

If the original series has a long-range dependence parameter H, the log–log-plot of m versus the corresponding values of the statistic provides us with a line of slope $H - 1$.

4.4. Classical R/S analysis

Let us assume we have a time series of N consecutive values. $Y(n) = \sum_{i=1}^{n} X_i$, $n \geqslant 1$, is the partial sum and $S^2(n) = \frac{1}{n}\sum_{i=1}^{n}[X_i - n^{-1}Y(n)]^2$, $n \geqslant 1$, is the corresponding sample variance.

We define $Z(t) = Y(t) - \frac{t}{n}Y(n)$. The *rescaled-adjusted-range* statistic or R/S statistic is defined by

$$\frac{R}{S}(n) = \frac{1}{S(n)}\Big[\max_{0 \leqslant t \leqslant n} Z(t) - \min_{0 \leqslant t \leqslant n} Z(t)\Big]. \tag{52}$$

R/S is called the rescaled adjusted range as its mean is zero, and it is expressed in terms of the local standard deviation. For large n, the expected value of the statistic approaches $c_1 n^H$:

$$E\big[R/S(n)\big] \sim c_1 n^H, \tag{53}$$

where c_1 is a positive, finite constant and does not depend on n. In case of long-range-dependence in a Gaussian process, the values for H range in the interval (0.5, 1.0). For an i.i.d. Gaussian process (i.e., pure random walk) or a short-range dependent process, the value of $R/S(n)$ approaches $c_2 n^{0.5}$. c_2 is independent of n, finite, and positive.

$$E\big(R/S(n)\big) \sim c_2 n^{0.5}. \tag{54}$$

The practical application of the R/S analysis is performed graphically. It is described in Mandelbrot and Wallis (1968).

With this procedure K different estimates of $R/S(n)$ are obtained. It starts with dividing the total sample of N consecutive values into K blocks, each of size N/K. We define

$$k_{(m)} = \frac{(m-1)N}{K} + 1 \tag{55}$$

as the starting points of each block, where K is the total number of blocks and $m = 1, \ldots, K$ is the current block number. Now we compute the $R(n, k_{(m)})/S(n, k_{(m)})$ for each lag n such that $k_{(m)} + n < N$. All data points before $k_{(m)}$ are ignored in order to avoid the influence of particular short-range dependence in the data.

Plotting the $\log(R(n, k_{(m)})/S(n, k_{(m)}))$ for each block versus $\log(n)$, we can estimate the slope of the fitted straight line. The classical R/S analysis is quite robust against variations in the marginal distribution of the data. This is also true for data with infinite variance.

Calculating the Hurst-Exponent H and the stability index α of the process innovations, the long-range dependence parameter d is obtained by

$$d = H - \frac{1}{2}, \tag{56}$$

for finite variance ($\alpha = 2$), and b

$$d = H - \frac{1}{\alpha}, \tag{57}$$

for infinite variance ($\alpha < 2$).

Long-range dependence occurs if d is greater than 0.

The R/S analysis is a nonparametric tool for examining long-memory effects. There is no requirement for the time series' underlying limiting distribution. In case of an underlying Gaussian process ($\alpha = 2$), a Hurst-Exponent of $H = 0.5$ implies that there is no long-range dependence among the elements of the time series.

For $0.5 < H < 1$, a Gaussian time series is called *persistent*.[28] A persistent time series is characterized by long-memory effects. If long memory is present, the effects occur regardless of the scale of the time series. All daily changes are correlated with all future daily changes, and all weekly changes are correlated with all future weekly changes. The fact that there is no characteristic time scale is an important property of fractal time series.

$0 < H < 0.5$ signals an *antipersistent* system for finite variance. Such a system reverses itself more frequently than a purely random one. At the first glance, it looks like a mean-reverting process. But this would actually require a stable mean, which is not the case in such systems.

4.5. The modified approach by Lo

Hurst's R/S statistic turned out to react sensitively towards short-memory processes. Thus, Lo (1991) modified the classical R/S statistic, now showing robustness towards short-range dependence.[29] Lo's statistic only focuses on lag $n = N$, the length of the series.[30] Multiple lags are not analyzed, the statistic does not vary n over several lags $< N$.

[28] Peters (1994).

[29] Lo (1991).

[30] Teverovsky, Taqqu and Willinger (1998).

Compared to the graphical R/S method, which delivers an estimate of the parameter H, Lo's modified statistic just indicates the presence of long-range dependence, but does not deliver an estimate of the Hurst-Exponent. The statistic performs a test of the hypotheses

- H_0: no long-range dependence.

Instead of the ordinary sample standard deviation S for normalization, there is an adjusted standard deviation S_q in the denominator. S_q considers the elimination of short term memory to the statistic. As it is known that the R/S statistic responds very sensitively towards short range dependence, the influence of short range dependence can be offset by normalizing R with a weighted sum of short-lag autocovariances. To the variance S^2 Lo added weighted autocovariances up to order q.[31] His modified statistic $V_q(N)$ is defined by

$$V_q(N) = N^{-1/2} \frac{R(N)}{S_q(N)}, \tag{58}$$

with

$$S_q(N) = \sqrt{S^2 + 2\sum_{j=1}^{q} w_j(q)\hat{\gamma}_j}, \tag{59}$$

where $\hat{\gamma}_j$ is the autocovariance of order j for the observed time series. $w_j(q)$ is defined as

$$w_j(q) = 1 - \frac{j}{q+1} \quad \text{with } q < N. \tag{60}$$

The statistic $V_q(N)$ is applied for a hypothesis test. It checks if the null hypothesis of the test can be rejected or not, given a certain confidence level. The two hypotheses are:

- H_0: no long-range dependence present in the observed data, $0 < H \leqslant 0.5$.
- H_1: long-range dependence is present in the data, $0.5 < H < 1$.

The statistic assumes a Gaussian process ($\alpha = 2$). In cases where the value of $V_q(N)$ lies inside the interval [0.809, 1.862], H_0 is accepted as the statistic is in the 95% acceptance region. For $V_q(N)$ outside the interval [0.809, 1.862], H_0 is rejected.

Lo's results are asymptotic assuming $N \to \infty$ and $q = q(N) \to \infty$.[32] However, in practice the sample size is finite and the value of the statistic depends on the chosen q. Thus, the question arises, what would be the proper value for q in order to perform the hypothesis test? Andrews (1991) has developed a data driven method for choosing q:[33]

$$q_{\text{opt}} = \left[\left(\frac{3N}{2}\right)^{1/3} \left(\frac{2\hat{\rho}}{1-\hat{\rho}^2}\right)^{2/3}\right], \tag{61}$$

[31] Peters (1994).

[32] Teverowsky, Taqqu and Willinger (1998).

[33] See Lo (1991).

here $[\cdot]$ stands for the greatest integer smaller than the value between. $\hat{\rho}$ is the first order autocorrelation coefficient. Therefore, choosing Andrews' q assumes that the true underlying process is AR(1).

Critique of Lo's statistic. Lo's statistic is applied by calculating V_q for a number of lags q, plotting those values against q. The confidence interval for accepting H_0 at the 95% confidence level is plotted as well.

Simulations have shown that the acceptance of H_0 (and therefore the value of $V_q(N)$) varies significantly with q. Taqqu, Willinger and Teverowsky (1998) found that the larger the time series and the larger the value for q, the less likely H_0 is rejected.

Whereas, Lo's statistic just checks for the significance of long-range dependence, the graphical method of the classical R/S provides relatively good estimates of H.

For small q the results of V_q usually vary strongly. Then a range of stability follows after the so called "extra" short-range dependence has been eliminated, and the only effect measurable for the statistic would be long-range dependence.

Applying the statistic to Fractional Brownian Motion with $H > 0.5$, which is a purely long-range dependent process without short memory effects, V_q is expected to stabilize at very low values of q. Unfortunately this could not be confirmed by the testing of Taqqu, Willinger and Teverowsky (1998). Moreover, they demonstrate that – if q is large enough – the following holds for $V_q(N)$ and $q^{1/2-H}$:

$$V_q(N) \simeq q^{1/2-H}. \tag{62}$$

For $H > 0.5$, V_q decreases with increasing q. Even for strongly fractional processes with time series containing 10000 samples, Taqqu, Willinger and Teverowsky found that, with increasing values for q, the probability of V_q lying inside the H_0 95% confidence interval and accepting the null-hypothesis grows. To mention three cases only: for $q = 500$ and $H = 0.9$ the null-hypothesis (no long-range dependence) is accepted with 90% for Fractional Brownian Motion, with 92% for FARIMA$(0.5, d, 0)$, and with 94% for FARIMA$(0.9, d, 0)$.[34]

Lo's test is very conservative in rejecting the null-hypothesis. It works for short-range dependence, but in cases of long-range dependence it mostly accepts the null-hypothesis. The statistic of Lo is certainly an improvement compared to the short-range sensitive classical R/S, but should not be used isolated without comparing its results with other tests for LRD.

In practical applications, the question for a proper choice of q remains. The value of Andrews' data driven q_{opt} depends on the econometric model underlying the observed time series, but, the appropriate model is not known in advance. Andrews' choice bears the assumption that the time series obeys an AR(1) process.

[34] FARIMA$(0.5, d, 0)$ means a fractional ARIMA process with an AR(1) coefficient of 0.5 and an MA(1) coefficient of 0.

It used to be a common way to asses long-range dependence by looking at the rate at which the autocorrelations decay. With a Hurst-Exponent H different from 0.5 the correlations are no longer summable. Such non-summability of autocorrelations used to be seen as a comfortable way of assuming long-range dependence. But there are pitfalls: if the underlying process is considered to follow a stable law with $\alpha < 2$, a second moment does not exist and therefore autocorrelations do not exist either.

It can be concluded that – if testing for long-range dependence – the application of a single technique is insufficient.

4.6. The statistic of Mansfield, Rachev and Samorodnitsky (MRS)

Long-range dependence means that a time series exhibits a certain kind of order over a long comprehensive period. Instead of pure chaos with no rule in the price movements of an asset, we can find periods of time with its sample mean significantly different from the theoretical mean. The stronger the long-memory effects in the time series, the longer an interval of the series whose mean deviates from the expected value.

Mansfield, Rachev and Samorodnitsky (1999) concentrate on this property of LRD-exhibiting time series. This property of LRD is valid regardless of the assumed underlying stochastic model.

The authors define a statistic that delivers the length of the longest interval within the time series, where the sample mean lies beyond a certain threshold. The threshold is set greater than the finite mean EX_i of the whole time series. Furthermore, the time series is assumed to follow a stationary ergodic process.

Expressed in mathematical terms, the statistic is defined as

$$R_n(A) = \sup\left\{ j - i\colon 0 \leqslant i < j \leqslant n,\ \frac{X_{i+1} + \cdots + X_j}{j - i} \in A \right\}, \tag{63}$$

which is defined for every $n = 1, 2, \ldots$. If the supremum is taken over the empty set, the statistic is defined to be equal to zero.

The set A is defined either as

$$A = (\theta, \infty) \quad \text{with } \theta > \mu, \tag{64}$$

or as

$$A = (-\infty, \theta) \quad \text{with } \theta < \mu, \tag{65}$$

where μ is the theoretical mean of the time series.

$R_n(-\infty, \theta)$ and $R_n(\theta, \infty)$ are interpreted as "greatest lengths of time intervals when the system runs under effective load that is different from the nominal load".[35] In the following, the examination is restricted to $R_n(\theta, \infty)$.

[35] Mansfield, Rachev and Samorodnitsky (1999).

A theoretical way to examine a time series for long-range dependence would be the log–log plot of $R_n(\theta, \infty)$ versus n. In the case of long-range dependence, the slope of the plot would be expected to be greater than $1/\alpha$ with α as the tail index. However, α is not known in advance. Therefore, Mansfield, Rachev and Samorodnitsky developed a statistic that does not rely on an a-priori tail index. They defined

$$W_n(\theta) = \frac{R_n(\theta, \infty)}{M_n}, \tag{66}$$

where $M_n = \max(X_1, \ldots, X_n)$ is the largest of the first n observations, $n \geqslant 1$. This statistic has a self-normalizing nature, and because of the denominator it has the ability to compensate for the effects of the tail-index α.

In case of short-range dependence, the ratio $W_n(\theta)$ approaches a weak limit as $n \to \infty$. In case of long-range dependence, R_n grows faster than M_n and the statistic diverges.

For visualization, the statistic $\theta W_n(\theta)$ is plotted against θ. Its limiting distribution is independent of θ. A difficult task is the selection of the proper range of θ. It has to be determined empirically by looking where the values for $\theta W_n(\theta)$ stabilize.

Once the value of the statistic is at least 19 for a certain θ then long-range dependence is present at a significance level of 0.05.

4.7. Empirical results for long-range dependence in credit data

For our empirical examination of long-memory effects in daily credit return data, we use the returns of bond indices provided by Merill Lynch.[36] We have selected four indices with time series of daily index prices from January 1988 to April 2000. Each index represents a number of bonds with similar properties (see explanation in Table 15). As the analysis of long-memory effects requires large data samples, an important criterion for the selection of an index was the available sample size. The sample sizes are listed in Table 16.

We apply three different methods for estimating the self-similarity parameter H and two methods performing a hypothesis test regarding the presence of LRD. As explained before, we have chosen

(i) the "Aggregated Variance Method",
(ii) the method "Absolute Values of Aggregated Series",
(iii) the classical R/S analysis developed by Mandelbrot and Wallis,
(iv) Lo's modified R/S statistic, and
(v) the statistic of Mansfield, Rachev and Samorodnitsky (MRS).

All these methods have been implemented with Matlab 5.3. Methods (i)–(iii) provide an estimate of the Hurst-Exponent H. Method (iv) tests if the null hypothesis "no long-range dependence" has to be accepted or rejected at a given confidence level. Method (v) is also a hypothesis test, however, contrary to Lo's test it works independently of the tail index.

[36] The time series were obtained via Bloomberg's Index Section.

Table 15
Explanation of the selected indices

Index	Explanation
X0H0	High Yield 175
C8B0	Corporates C rated, cash pay
J0A3	AAA-AA rated corporates, time to maturity 15 yrs
C0A0	US Corporate master

Table 16
Data sets used for testing LRD

Index	No. of observations	Starting date	Ending date
X0H0	3083	10–31–86	04–30–00
C8B0	3470	10–31–86	04–30–00
J0A3	2920	08–04–88	04–30–00
C0A0	3472	01–04–88	04–30–00

Table 17
Results for Aggregated Variance and Absolute Values of the Aggregated Series

Index	H for Aggreg. Variance	H for Abs. Values of Aggreg. Ser.
X0H0	0.7632	0.7596
C8B0	0.5527	0.5511
J0A3	0.8070	0.8022
C0A0	0.5856	0.5838

Testing the index-returns for long-range dependence, we computed the daily changes of the index log-prices

$$r_t = \log(p_t) - \log(p_{t-1}). \tag{67}$$

The results of the methods "Aggregated Variance" and "Absolute Values of the Aggregated Series". For methods (i) and (ii), we plotted the values of the statistic over m (number of elements in each block), with m ranging from 10 up to 40. Finally, we determined the slope of the data points in order to obtain H. The values for H are printed in Table 17. Both methods calculate Hurst-Exponents greater than 0.5 for all observed indices. Thus, under the Gaussian assumption, the underlying processes are long-memory processes. X0H0 and J0A3 show strong LRD, whereas C8B0 and C0A0 have a weaker long memory.

The results of classical R/S and Lo's statistic. As we only have about 3000 observations for each time series, we do not divide the data set into several blocks for the classical R/S statistic. Thus, we choose $K = 1$.

The results of classical R/S and the values of Lo's statistic V_q (for q we chose a range of $1, \ldots, 50$) are presented in Table 18. We plotted both the $\log(R/S)$–$\log(n)$, and the

Table 18
Results for the classical R/S statistic and Lo's test

Index	Classical R/S	Lo's statistic	
	Fitted H	Range of V_q ($q = 1, \ldots, 50$)	Optimal q (Andrews)
XOHO	0.7579	[1.74, 3.44]	11
C8B0	0.4874	[1.33, 1.40]	6
J0A3	0.9213	[2.04, 4.29]	10
COA0	0.4493	[1.23, 1.40]	–

V_q–q graphs for our observed indices X0H0, C8B0, J0A3, and C0A0 (see Figures 3–6 for classical R/S, and Figure 7 for Lo's test). The second column of Table 18 presents the Hurst-Exponent estimated with the R/S statistic. In the third column the table presents the intervals in which the values of Lo's V_q are located for $q = 1, \ldots, 50$. The fourth column provides the optimal lag q, determined by Andrews' data driven method.[37] The results of the R/S-statistic are similar to the ones obtained by the Aggregated Variance Method and the Absolute Values of the Aggregated Series. The time series of X0H0 and J0A3 exhibit strong LRD according to their Hurst-Exponent H. This is supported by the result of Lo's test that rejects the null-hypothesis "no LRD" at the 95% level. However, for C8B0 and C0A0, the Hurst-Exponent appears already in the area of antipersistence. Another interesting finding is that for C0A0 – which has the lowest value for H – the sample autocorrelation of order 1 is negative. Therefore, we cannot calculate the optimal q for C0A0.

4.7.0.1. The results of the statistic by Mansfield, Rachev and Samorodnitsky (MRS). Figures 8–11 show the plots of $\theta W_n(\theta)$ over the range of θ. For the time series of the index X0H0, we found that the statistic $\theta W_n(\theta)$ is linearly increasing with θ in the range of $[0.5\,e^{-4}, 3.5\,e^{-4}]$ (the empirical mean of the whole series is $0.497\,e^{-4}$). The value of $\theta W_n(\theta)$ reaches levels of about 19 and then declines until it stabilizes at a level of about 1 (see Figure 8). This result clearly indicates the presence of LRD. The presence of long memory is significant at the 0.05 level once the value of the statistic is at least 19. Thus, the MRS-statistic supports the LRD-hypothesis for X0H0. Lo's statistic and classical R/S also indicate long-range dependence for the index X0H0, but this was based on the assumption that the underlying process of the time series follows a Gaussian law, i.e., that $\alpha = 2$. However, the MRS-statistic is independent of α.

The second bond index that exhibits strong LRD in its returns with the former tests, was the J0A3-index (C rated corporates). Its empirical mean is $-3.2\,e^{-4}$. We observe a sharp increase of $\theta W(\theta)$ for $\theta \in [0, 6.5\,e^{-4}]$ up to a value of about 15, and it finally drops to a level of about 1 as well. Thus, the hypothesis of long-range dependence can also be confirmed for the J0A3-series as the MRS-statistic also exhibits significant values (see Figure 9). However, the significance is not as strong as for the X0H0-series.

[37] See Lo (1991).

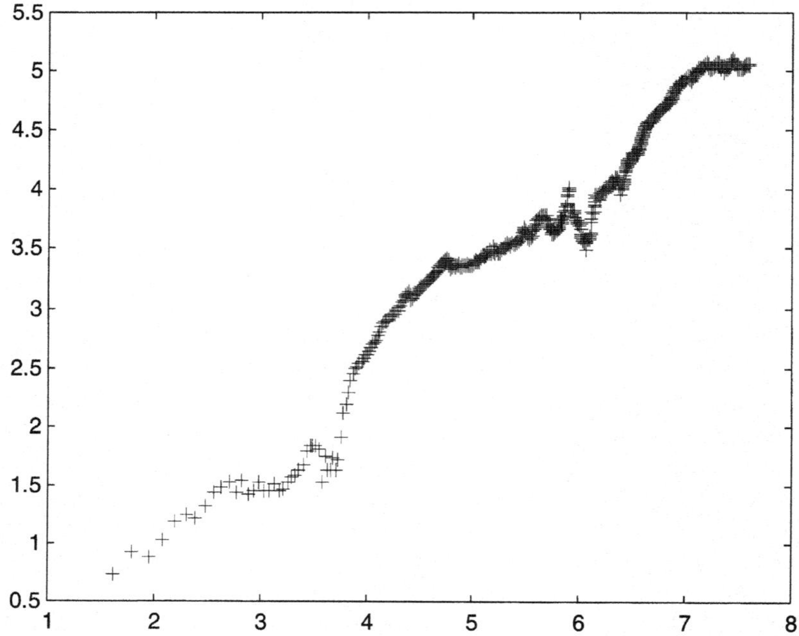

Fig. 3. Plot of log(R/S)–log(n) for X0H0.

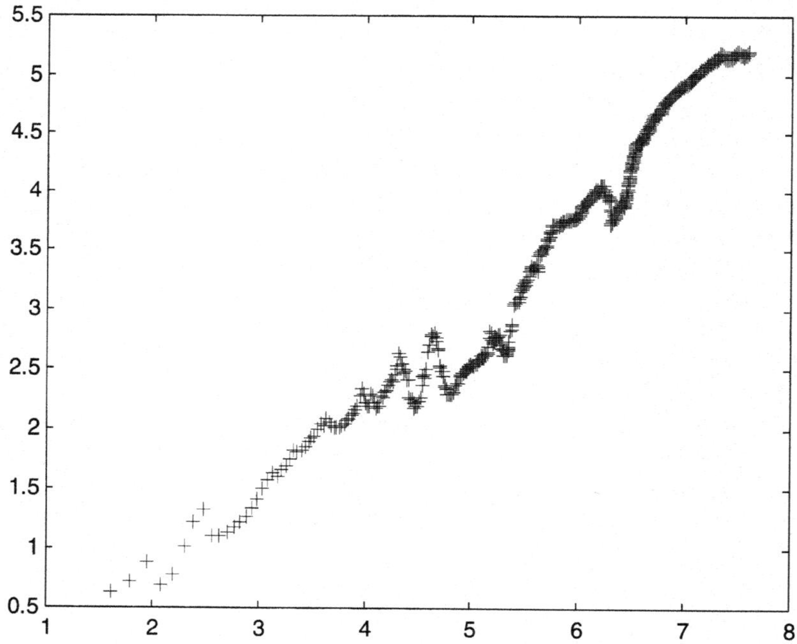

Fig. 4. Plot of log(R/S)–log(n) for J0A3.

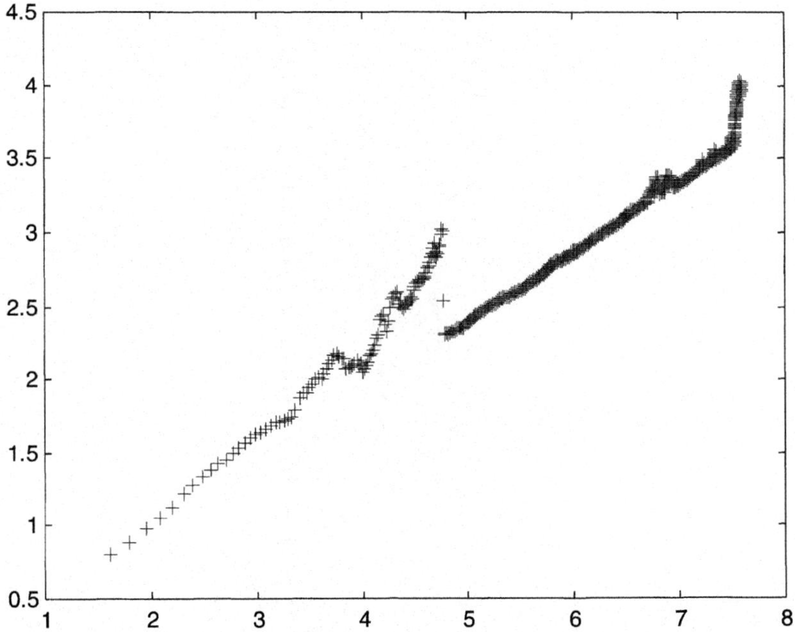

Fig. 5. Plot of log(R/S)–log(n) for C0A0.

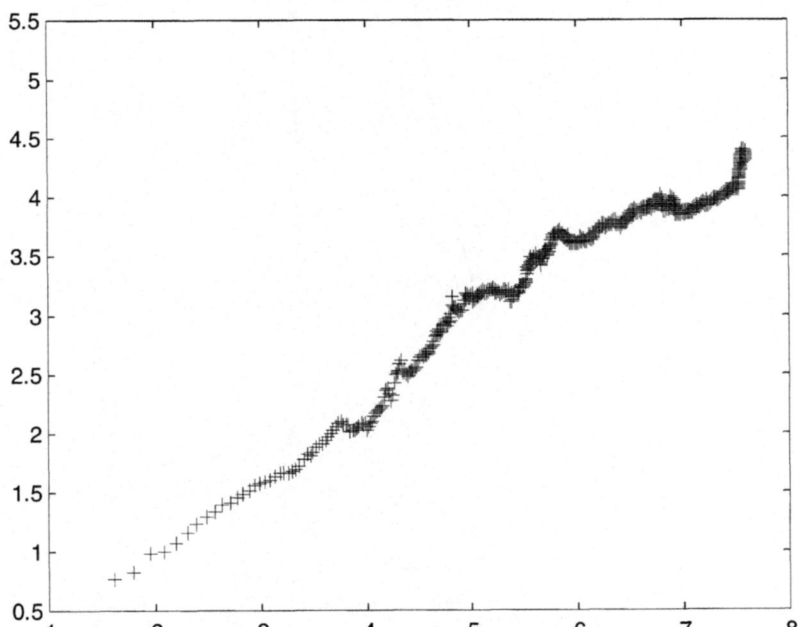

Fig. 6. Plot of log(R/S)–log(n) for C8B0.

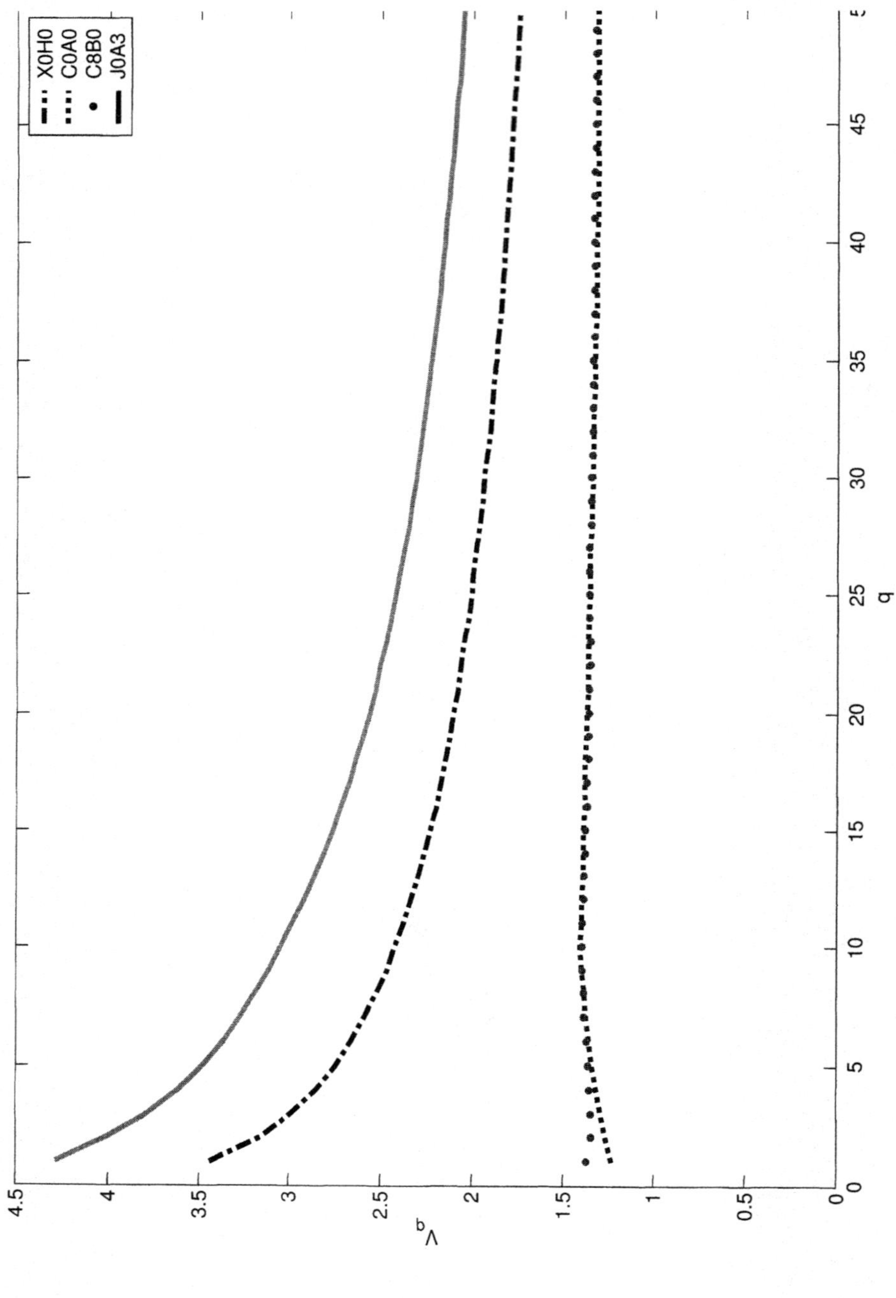

Fig. 7. The plots of Lo's statistic for X0H0, J0A3, C0A0, and C8B0.

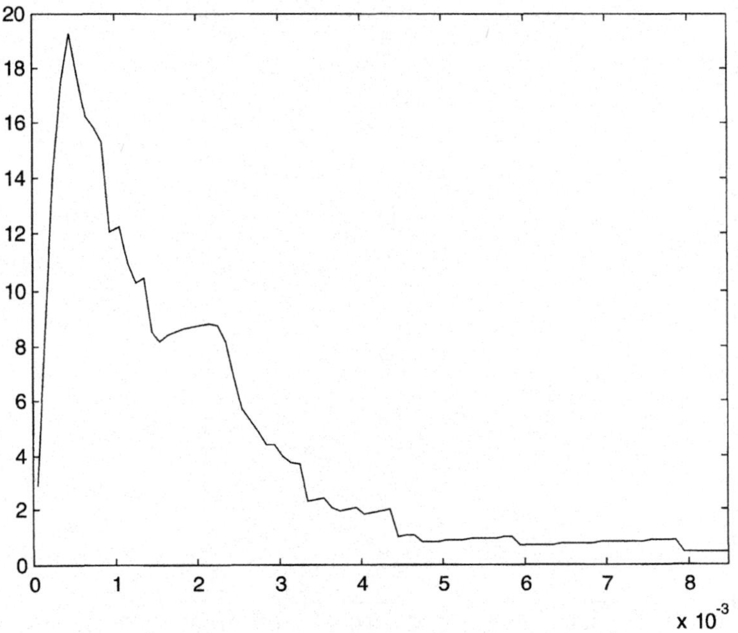

Fig. 8. Plot of $\theta W(\theta)-\theta$ for X0H0.

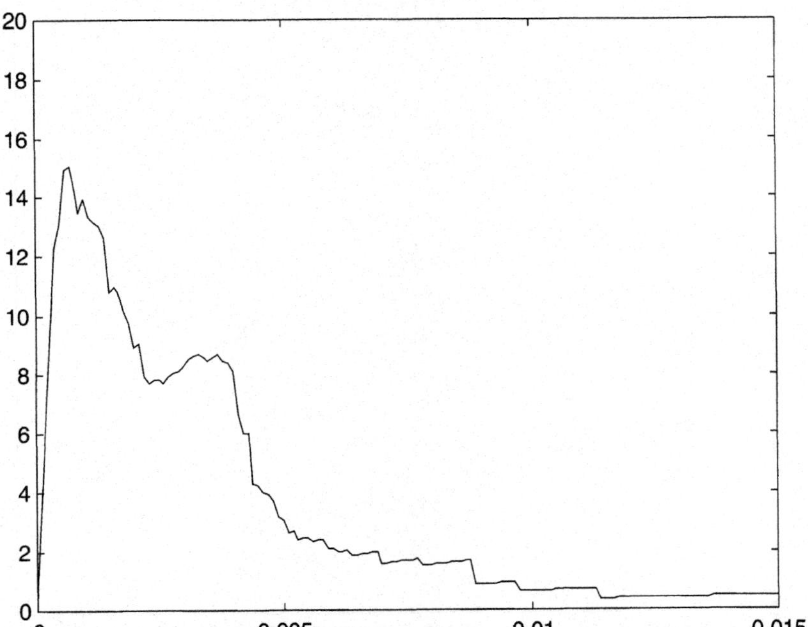

Fig. 9. Plot of $\theta W(\theta)-\theta$ for J0A3.

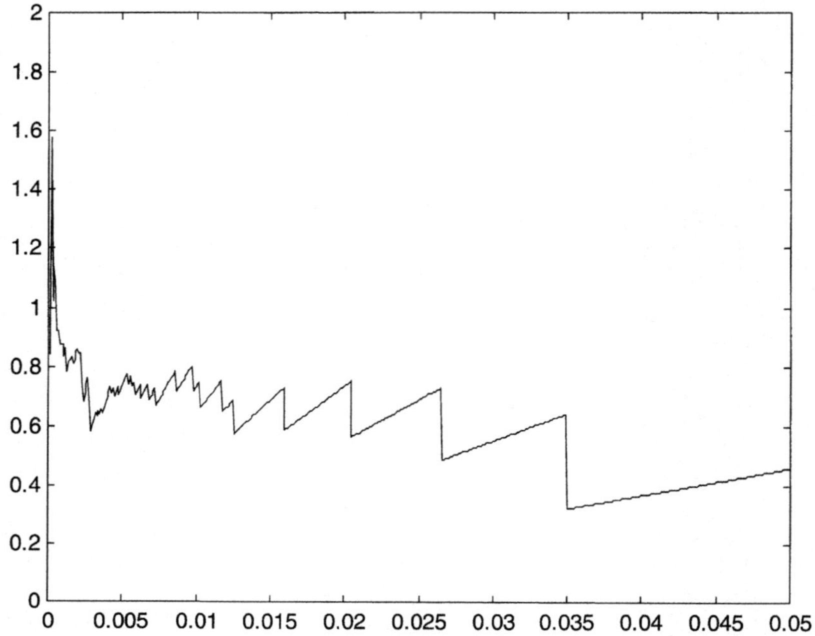

Fig. 10. Plot of $\theta W(\theta)-\theta$ for C0A0.

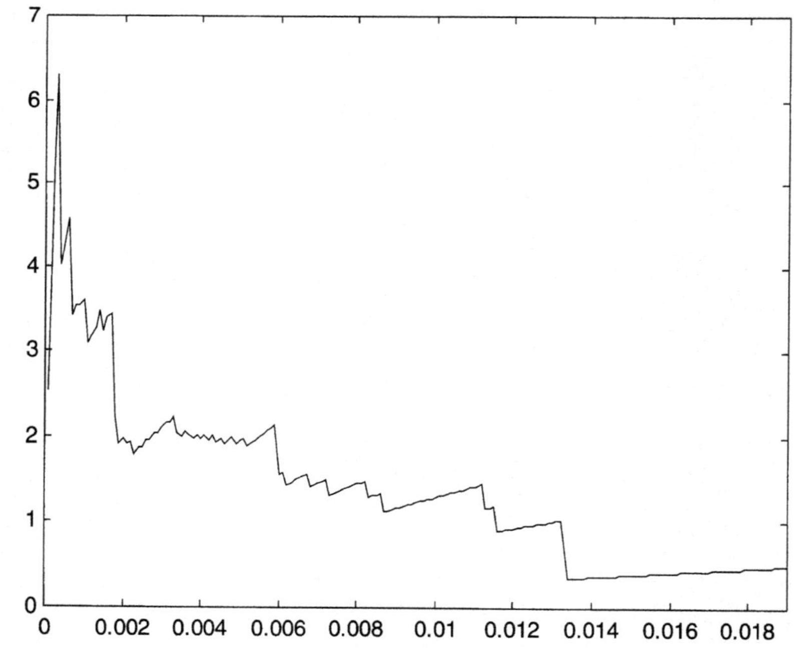

Fig. 11. Plot of $\theta W(\theta)-\theta$ for C8B0.

The returns of the two other indices, C0A0 and C8B0, do not exhibit long-range dependence with the $\theta W(\theta)$-statistic, and this is consistent with the results of the formerly applied tests. The returns of the C8B0-index show a higher probability for the LRD-hypothesis than the returns of C0A0, however, both are not significant. Thus, for both indices C0A0 and C8B0, there is no significant indication for long-range dependence with the MRS-statistic (see Figures 10 and 11).

5. Conclusion

In the first part of this chapter we have shown the predominant performance of Value-at-Risk models based on stable distributions compared to Gaussian models. Furthermore, we have presented a modified model for credit returns which makes practical implementation easier.

In the second part of the chapter we have studied long-range dependence in credit return data.

In Section 3 we have illustrated that the stable distribution much better approximates the tail of the empirical distribution of credit returns. This is especially important for Value-at-Risk (VaR) applications. VaR has become increasingly important for risk management. The stable VaR exhibits excellent performance for the high quantiles (i.e., 99% VaR). While the Gaussian 99% VaR underestimates the empirical VaR, the stable 99% VaR slightly overestimates it. Thus, the heavy-tailedness property of time series of credit returns is captured very uniquely by the application of non-Gaussian stable distributions, as well as the skewness property. The stability indices of the fitted corporate bond returns lie in the range of 1.5–1.6, which clearly indicates heavy-tailedness.

In this context, a slightly modified model for credit returns has also been presented which can be implemented without the building of yield curves for various rating grades. It makes a practical application less burdensome.

The other phenomenon that has been analyzed in this work is the long-memory property of credit returns (Section 4). A sign of long memory is the "burstiness" of plotted time series. Long-range dependence is characterized by hyperbolically decaying autocorrelations and the property that large (small) representations are more likely followed by large (small) representations than small (large) representations.

While three of the chosen tests measure the Hurst-Exponent, the other two are hypothesis-tests checking the significance of the LRD-hypothesis.

Applying the methods "Aggregated Variance" and "Absolute Values of Aggregated Series", all four analyzed time series exhibit a Hurst-Exponent H greater than 0.5, which means long-range dependence under the Gaussian assumption. For two of the four credit return series, the modified R/S statistic of Lo confirms LRD to be significant. This is remarkable because Lo's test tends to confirm the null-hypothesis "no LRD" for large sample sizes and increasing lag q, even when the actual process is strongly long-range dependent.[38] Also allowing infinite variance ($\alpha < 2$), we apply the MRS-statistic. It analyzes a

[38] Teverowsky, Taqqu and Willinger (1998).

process for LRD without relying on the tail-index. For the X0H0 and J0A3 series which have been confirmed for LRD by Lo's test, the MRS-statistic $\theta W(\theta)$ also indicates significant long memory. This is probably the strongest result of our LRD studies which states that long-range dependence in credit returns is also found to be significant in combination with the non-Gaussian stable assumption.

Our examinations have only focused on the returns. However, for other financial series – such as stock prices – LRD has also been discovered in the trading time process, as demonstrated by Marinelli et al. (1999).

The use of bond indices for the empirical examination, instead of individual bonds, is advantageous in two respects: First, each index incorporates numerous bonds of a certain market segment. Thus, the obtained results can then be considered a widespread phenomenon. If only a small number of bonds within the observed indices would exhibit such an effect, it would probably fade away. Second, LRD-analysis requires large samples which are more readily available for indices than for single bonds.

Finally we can conclude that the issue of long memory cannot be neglected for time series of credit returns. The increments of the underlying stochastic process are not i.i.d.

With the proven LRD in the time series of credit returns and by demonstrating that the distribution of credit returns is better captured with stable non-Gaussian models, we obtain a powerful tool to generate accurate forecasts of Value-at-Risk for longer horizons.

References

Andrews, D., 1991. Heteroskedasticity and autocorrelation consistent covariance matrix estimation. Econometrica 59, 817–858.

Bachelier, L., 1900. Theorie de la spéculation. Annales de École Normale Superieure Series 3 (17), 28–86. English translation: Coonter, P.H., (Ed.), The Random Character of Stock Market Prices. MIT Press, Cambridge, MA, 1964.

Basle Committee on Banking Supervision, 1988. International Convergence of Capital Measurement and Capital Standards. Bank of International Settlements.

Basle Committee on Banking Supervision, 1999. Credit Risk Modeling: Current Practices and Applications. Bank of International Settlements.

Beran, J., 1994. Statistics for Long-Memory Processes. In: Monographs on Statistics and Applied Probability, Vol. 61. Chapman and Hall.

Burden, R.C., Faires, J.D., 1997. Numerical Analysis. Sixth edition. Brooks/Cole.

Campbell, J.Y., Lo, A.W., MacKinlay, A.C., 1997. The Econometrics of Financial Markets. Princeton University Press, Princeton, NJ.

Christoffersen, P.F., Diebold, F.X., Schuermann, T., 1998. Horizon problems and extreme events in financial risk management. Economic Policy Review, Federal Reserve Bank of New York, 4.

Fama, E.F., 1965a. The behavior of stock market prices. Journal of Business 38, 34–105.

Fama, E.F., 1965b. Portfolio analysis in a stable Paretian market. Management Science 11, 404–419.

Fama, E.F., 1970. Risk, return, and equilibrium. Journal of Political Economy 78, 30–55.

Fama, E.F., Roll, R., 1971. Parameter estimates for symmetric stable distributions. Journal of the American Statistical Association 66, 331–338.

Federal Reserve System Task Force on Internal Credit Models, 1998. Credit risk models at major US banking institutions: Current state of the art and implications for assessment of capital adequacy.

Gamrowski, B., Rachev, S.T., 1994. Stable models in testable asset pricing. In: Anastassiou, G., Rachev, S.T., (Eds.), Approximation, Probability, and Related Fields. Plenum, New York, pp. 223–236.

Hamilton, J.D., 1994. Time Series Analysis. Princeton University Press.

Hull, J.C., 2000. Options, Futures, and Other Derivatives. Prentice-Hall.

Hurst, S.R., Platen, E., Rachev, S.T., 1997. Subordinated market index models: A comparison. Financial Engineering and the Japanese Markets 4, 97–124.

Lo, A.W., 1991. Long-term memory in stock market prices. Econometrica 59 (5), 1279–1313.

Lux, T., 1999. Multi-fractal processes as models for financial returns: A first assessment. Discussion Paper. University of Bonn.

Mandelbrot, B., 1963. The variation on certain speculative prices. Journal of Business 36, 394–419.

Mandelbrot, B., 1997a. Fractals and Scaling in Finance, Discontinuity, Concentration, Risk. Springer, New York.

Mandelbrot, B., 1997b. Fractals, Hasard, et Finance. Flammarion, Paris.

Mandelbrot, B., 1999. Survey of multifractality in Finance. Cowles Foundation Discussion Paper No. 1238. Yale University, New Haven.

Mandelbrot, B., Wallis, J., 1968. Noah, Joseph, and operational hydrology. Water Resources Research 4, 909–918.

Mansfield, P., Rachev S.T., Samorodnitsky, G., 1999. Long strange segments of a stochastic process and long range dependence. The Annals of Applied Probability, to appear.

Marinelli, C., Rachev, S.T., Roll, R., 1999. Subordinated exchange rate models: Evidence for heavy-tailed distributions and long-range dependence.

Marinelli, C., Rachev, S.T., Roll, R., Göppl, H., 1999. Subordinated stock price models: Heavy tails and long-range dependence in the high-frequency Deutsche Bank price record.

McCulloch, H., 1971. Measuring the term structure of interest rates. Journal of Business 44 (1), 19–31.

McCulloch, H., 1975. The tax-adjusted yield curve. Journal of Finance 30 (3), 811–830.

Mittnik, S., Rachev, S.T., Paolella, M.S., 1997. Stable Paretian models in finance: Some theoretical and empirical aspects. In: Adler, R., et al. (Eds.), A Practical Guide to Heavy Tails: Statistical Techniques for Analyzing Heavy Tailed Distributions, Birkhäuser. Boston.

Peters, E.E., 1994. Fractal Market Analysis: Applying Chaos Theory to Investment and Economics. Wiley.

Rachev, S.T., Mittnik S., 2000. Stable Paretian Models in Finance. In: Wiley Series in Financial Economics. Wiley, New York.

Rachev, S.T., Samorodnitsky, G., 1999. Long strange segments in long range dependent moving average. Stochastic Processes and their Applications, to appear.

Rachev, S.T., Schwartz, E., Khindanova, I., 2000. Stable modeling of Value-at-Risk. In: Stable Models in Finance. Pergamon Press, to appear.

Samorodnitsky, G., Taqqu, M.S., 1994. Stable Non-Gaussian Random Processes. Chapman & Hall, New York.

Saunders, A., 1999. Credit Risk Measurement. New Approaches to Value at Risk and Other Paradigms. In: Wiley Frontiers in Finance. Wiley.

Taqqu, M.S., Teverovsky, V., 1998. On estimating the intensity of long-range dependence in finite and infinite variance time series. In: Adler, R.J., Feldman, R., Taqqu, M.S. (Eds). A Practical Guide to Heavy Tails: Statistical Techniques and Applications. Springer-Verlag.

Teverovsky, V., Taqqu, M.S., Willinger, W., 1995. Estimators for long-range dependence: An empirical study. Fractals 3 (4) 785–788.

Teverovsky, V., Taqqu, M.S., Willinger, W., 1998. A critical look at Lo's modified R/S statistic. Journal of Statistical Planning and Inference, to appear.

Willinger, W., Taqqu, M.S., Erramilli, A., 1996. A bibliographical guide to self-similar traffic and performance modeling for modern high speed networks. In: Kelly, F.P., Zachary, S., Ziedins, I. (Eds.), Stochastic Networks: Theory and Applications. Claredon Press, Oxford, pp. 339–366.

Willinger, W., Taqqu, M.S., Teverovsky, V., 1999. Stock Market prices and Long Range Dependence. Online Publication, Springer-Verlag, Berlin.

Chapter 11

MULTIFACTOR STOCHASTIC VARIANCE MODELS IN RISK MANAGEMENT: MAXIMUM ENTROPY APPROACH AND LÉVY PROCESSES*

ALEXANDER LEVIN

Group Risk Management, TD Bank Financial Group, Toronto
e-mail: Alex.Levin@td.com

ALEXANDER TCHERNITSER

Bank of Montreal, Toronto
e-mail: Alexander.Tchernitser@bmo.com.

Contents

* The views expressed in this chapter are those of the authors and not necessarily of the Bank of Montreal.

Handbook of Heavy Tailed Distributions in Finance, Edited by S.T. Rachev

Abstract

This chapter investigates a class of multifactor non-normal models for Market Risk Management, and, specifically, for Value-at-Risk (VaR) calculations, with stochastic variance (SV) driven by Lévy processes. Relevant statistical and dynamic properties for the risk factors are discussed. A short review of the Market Risk Management requirements and stochastic models for VaR is presented.

In the case of one asset, a broad class of pure jump Generalized Gamma processes for the SV is derived from the Maximum Entropy principle. The corresponding family of Lévy processes for the risk factors (RF) possesses skewed leptokurtic marginal distributions with a wide range of heavy tails, from exponential and sub-exponential (stretched exponential) to polynomial. The introduced extended Generalized Gamma Variance family is a two shape parameter class of conditionally normal symmetric distributions (there is the third shape parameter in the case of non-zero skewness) with the SV represented as an arbitrary power (positive, zero or negative) of a gamma distribution. It includes normal, Variance Gamma (Generalized Laplace), Student t, and Weibull Variance Mixture distributions as special cases. Ornstein–Uhlenbeck type processes for the SV driven by positive Lévy noise and the corresponding term structure of the RF kurtosis and quantiles are considered for the purpose of modelling non-linear dependence in the asset returns.

A general framework for constructing multidimensional conditionally Gaussian stochastic processes with the correlated multivariate stochastic variances that follow Lévy processes is considered. This methodology allows for different shape and tail behavior of the marginal RF and linear sub-portfolio distributions, exact fit into the RF correlation structure, and proper non-linear scaling of VaR for different holding periods. Presented empirical evidence for different markets confirms a good agreement between the model and historical RF distributions. Effective numerical calibration and Monte Carlo simulation procedures are developed.

1. Review of market risk models

1.1. Market risk management and Value-at-Risk

Market Risk Management deals with the risk of potential portfolio losses due to adverse changes in the price of financial instruments caused by stochastic fluctuations of the market variables (JP Morgan, 1996; Basle Committee on Banking Supervision, 1997; Jorion, 2001; Crouhy, Galai and Mark, 2001). The are many types of general market and specific risk factors (RF) with different distributional properties and stochastic behavior in the foreign exchange, interest rate, commodity and equity markets. Market variables include, for example, stock prices, equity indices, spot foreign exchange rates, commodity prices, as well as complex aggregate structures: interest rate curves, commodity futures price curves, credit spread curves, implied volatility surfaces (e.g., European option implied volatility as a function of strike and maturity) or "cubes" (e.g., swaption implied volatility as a function of underlying swap tenor, swaption maturity and strike). Also, there are such "wild" and "exotic" market variables as, for example, electricity prices and interest rate or foreign exchange rate cross-correlations (the changes of latter variables effect the spread and cross-currency option prices).

Proper modelling of the multivariate future RF distributions is important for financial institutions for the purpose of accurate estimation of the market risk, identification of the risk concentration, developing of trading and hedging strategies, portfolio optimization, consistent measurement of the risk adjusted performance for different units (Risk Adjusted Return On Capital (RAROC) and Capital-at-Risk methodologies), setting up the trading limits, calculating of the regulatory capital (Basle Committee on Banking Supervision, 1997), back-testing of the market risk models required by regulators (Basle Committee on Banking Supervision, 1996). Many financial institutions need to consistently estimate market risk for large portfolios and sub-portfolios (aggregation levels) that comprise hundreds of thousands of instruments dependent on thousands of risk factors in all markets. These portfolios usually include sub-portfolios of options, which magnify and non-linearly transform deviations of the underlyings. Modern Market Risk Management is interested in comprehensive modelling of the multidimensional risk factor stochastic processes and marginal distributions for different time horizons rather than static multivariate distributions for some fixed holding period. This interest comes from the requirements to capture liquidity risk for many instrument types with varying liquidation periods [see Crouhy, Galai and Mark (2001)], estimate intraday risk for some frequently rebalanced positions, consistently evaluate VaR for one-day and ten-day time horizons prescribed by BIS documents (Basle Committee on Banking Supervision, 1996, 1997) for back-testing and regulatory capital calculations respectively, and actively dynamically manage risk. This problem points out on the importance of adequate modelling of a non-linear dependence in the underlying returns observed in the market to capture a proper VaR term profile.

Along with the RF volatilities (standard deviations of daily changes) and correlations combined with the portfolio sensitivities [Greeks, Hull (1999)], the most widely accepted methodology for measuring market risk is the Value-at-Risk approach. The VaR can be defined as the worst possible loss in the portfolio value over a given holding period (1 or 10 days) at the 99% confidence level (Jorion, 2001; Crouhy, Galai and Mark, 2001). Essentially, a mathematical model for VaR consists of two main parts: (1) modelling of proper multivariate risk factor distributions (processes) for the required time horizons; (2) evaluation of the portfolio (linear instruments, options and other derivatives) changes for the risk factor scenarios to produce a portfolio distribution. The evaluation part can be based on a full revaluation for the prices of instruments or partial revaluation methodologies [for example, Delta–Gamma–Vega approximation (Hull, 1999)]. Regulators also require complementing the VaR analysis with stress testing (scenarios for crashes, extreme movements in the market, stresses of volatilities and correlations, etc.). Traditional methods of the VaR calculation are analytical (variance–covariance) method (JP Morgan, 1996), historical simulation [combined with some bootstrapping procedures or other non-parametric methods (Crouhy, Galai and Mark, 2001)], and parametric Monte Carlo simulation approach [see Duffie and Pan (1997)]. Primarily developed for the "normal" market conditions (multivariate Gaussian distribution for the risk factors), the variance–covariance method can be applied only for linear portfolios. The variance–covariance method can be extended from multivariate normal to the non-normal elliptical RF distributions (see Section 3.3). VaR for option portfolios is usually calculated based on simulation approaches. In this chapter, we concentrate on the parametric modelling of the RF distributions based on the Monte Carlo simulation procedures given an appropriate portfolio valuation methodology.

There are some market risk measures other than VaR closely related to the tails of the RF probability distributions, for example, Expected Shortfall [see Mausser and Rosen (2000)]. The Expected Shortfall is defined as an average loss calculated from the losses that exceed VaR. The Expected Shortfall, as a conditional mathematical expectation, is an example of so-called coherent risk measures [see Artzner et al. (1999)] that, contrary to VaR, possess a natural subadditivity property (total risk of entire portfolio should be less or equal to a sum of risks of all sub-portfolios). In some cases, Expected Shortfall reflects the market risk better than VaR (it gives an answer to the question, what is the average of the worst case losses that occur at the corresponding confidence level). This market risk measure is more sensitive to the tail behavior than VaR. In general, it is wrong to say that only tails of the underlying RF distributions are important for the VaR or other risk measures. For example, a left tail for the portfolio of some barrier options or even European near at-the-money options may mostly depend on the central part of the underlying distribution. Therefore, it is a necessity to accurately model all parts of the RF distributions, including peaks at the origin and tails.

Due to short time horizons utilized in Market Risk Management (1–10 business days) contrary to Credit Risk Management with usual time horizons of years (Crouhy, Galai and Mark, 2001; Duffie and Pan, 2001), the market risk factors are defined as daily log-returns, relative or absolute changes in the underlying prices, rates or implied volatilities, rather than these underlyings themselves. Such long-term effects as mean-reversion in the interest rate, commodity price, and implied volatility dynamics (with characteristic times 1–20 years) are not taken into account in the VaR modelling. Most of financial variables are positive (although, spreads and interest rate differentials may be negative). Except some rare situations (e.g., Japanese interest rates), daily changes for the underlyings are much less than 100% of the notional values, and, therefore, there is no need to apply any positive transformations to the market variables, like exponential or square transformations. Heuristically, this means that in most cases one can use "linear" RF simulation models for the VaR calculation.

1.2. Statistical properties of the market risk factors

There is extensive empirical evidence that historical daily return distributions for different underlyings in the foreign exchange, interest rate, commodity, and equity markets have high peaks, "fat" tails (excess kurtosis, Figures 1 and 2) and skewness (right graph on Figure 2) contrary to the normal distribution [see, for example, Mandelbrot (1960), Fama (1965), Duffie and Pan (1997), Müller, Dacorogna and Pictet (1998), Barndorff-Nielsen and Shephard (2000b), Rachev and Mittnik (2000), Bouchaud and Potters (2000), Cont (2001)]. Also, it is well known that the volatility of these financial variables varies stochastically with clustering (Bollerslev, Engle and Nelson, 1994) (see Figure 3). These distributional properties have significant impact on Risk Management, specifically on VaR. A standard methodology usually used for the VaR calculation (JP Morgan, 1996) exploits a multivariate normal distribution as a proxy for the RF distributions. The standard model corresponds to stable market conditions when one can neglect large jumps of the underlyings and volatility fluctuations. This results in underestimating of the actual VaR by the standard methodology and breaching the back-testing. A comprehensive RF simulation model should additionally capture the following important features observed in the market:

- different distributional shapes for different risk factors and markets (for example, short interest rates have much heavier tails, higher peaks and kurtosis than long term rates even for the same interest rate curve, Figure 1; some commodity price distributions deviate more from normal than others);
- anomalously small normalization effect for large diversified portfolios contrary to the one predicted by the Central Limit Theorem (for example, S&P 500 Industrial Index or TSE 300 Index (Figure 2), viewed as large portfolios of stocks, have markedly non-normal distributions with kurtosis about ten). This phenomenon points to a non-linear dependence between different risk factors [see also Embrechts, McNeil and Straumann (1999)];
- normalization of the risk factor distributions for longer holding periods [for example, ten-day return distributions are significantly closer to normal than daily return distributions, on the other hand, intraday change distributions are clearly more distant from normal than daily ones (Müller, Dacorogna and Pictet, 1998; Cont, Potters and Bouchaud, 1997; Mantegna and Stanley, 2000)]. A decreasing term structure of kurtosis points out to the same effect (Duffie and Pan, 1997; Bouchaud and Potters, 2000);
- volatility clustering and non-linear time dependence in risk factor returns (for example, statistically significant autocorrelation in squares of virtually uncorrelated daily returns, see top graph on Figure 3 and Figure 10 in Section 2.3).

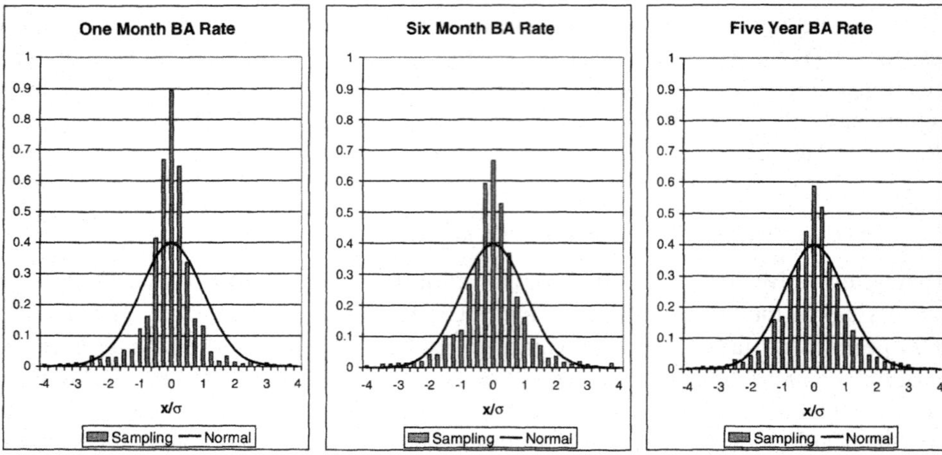

Fig. 1. Variety of distributional shapes for CAD BA interest rate daily returns.

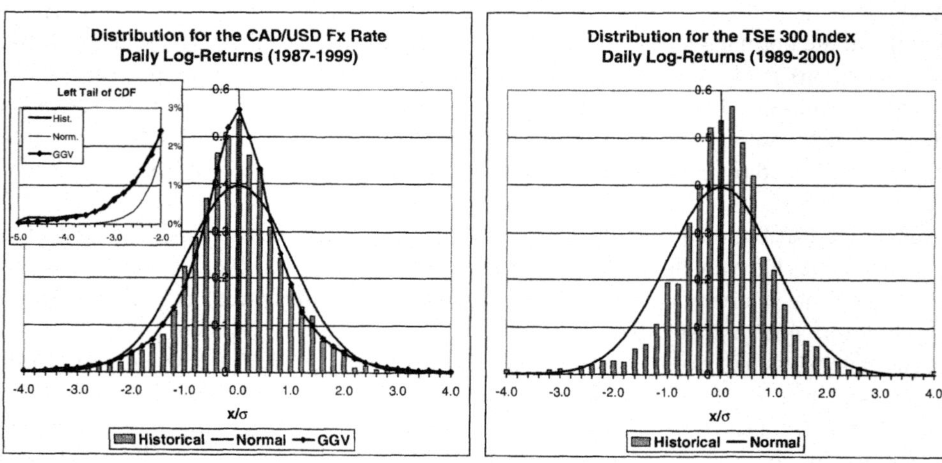

Fig. 2. Distributions for the CAD/USD FX and TSE 300 daily log-returns.

1.3. A short review of stochastic volatility models

In this chapter we restrict consideration of the SV models to the case of continuous time models. Time series approaches (ARCH, GARCH, etc.) (Bollerslev, 1986; Bollerslev, Engle and Nelson, 1994) are beyond the scope of the chapter.

L. Bachelier introduced the normal distribution and Brownian motion in finance in his Ph.D. Thesis (Bachelier, 1900) more than one hundred years ago. Brownian motion [that corresponds to a standard model for VaR (JP Morgan, 1996)] was rediscovered in finance

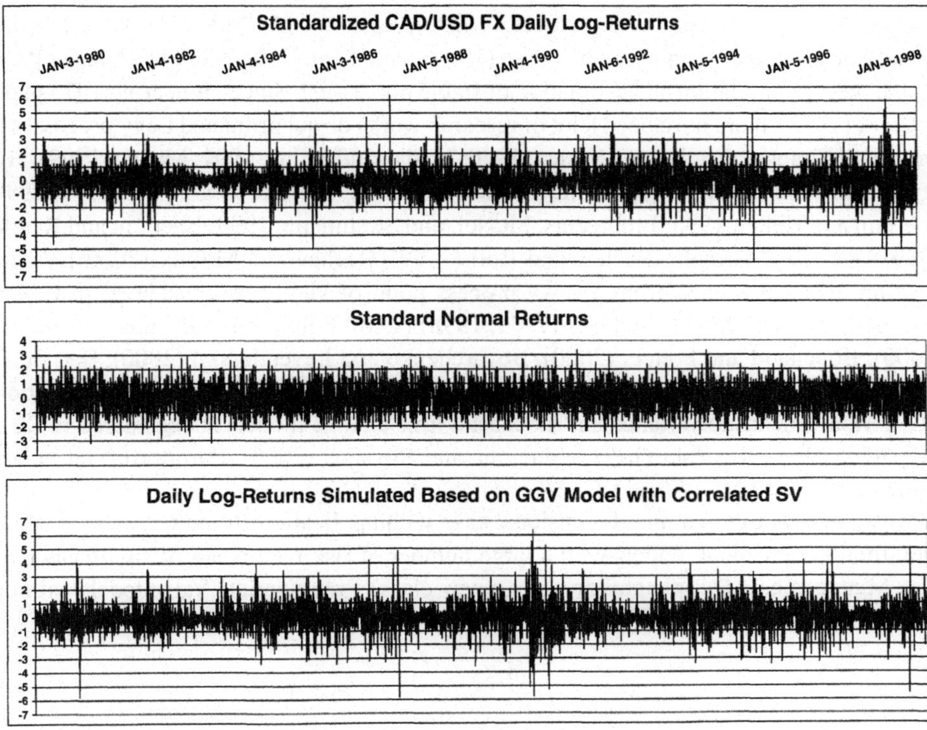

Fig. 3. Volatility clustering and large deviations in CAD/USD FX rate daily returns.

in Osborne (1959), and then replaced by a Geometric Brownian motion for modelling of the stock dynamics (Samuelson, 1965). Without any doubt, the Black–Scholes–Merton (Black and Scholes, 1973) option pricing model has become a main tool in modern finance. Since well-known investigations of Mandelbrot (1960, 1963) and Fama (1965) on stable processes in the market, researchers have developed different approaches for modelling the abnormal behavior of the market variables. Fat-tailed distributions and jumps in the risk factors have been usually modelled by jump-diffusion processes (Merton, 1976, 1990; Bates, 1996; Kou, 2000), processes with diffusion stochastic volatility (Hull and White, 1987; Heston, 1993; Stein and Stein, 1991; Bates, 1991; Melino and Turnbull, 1990), mixtures of normal and other distributions (Duffie and Pan, 1997; Rachev and SenGupta, 1993; Albanese, Levin and Ching-Ming Chao, 1997), and other methods (Hull and White, 1998; Sornette, Simonetti and Andersen, 2000). Also, different types of non-Gaussian Lévy processes were used to describe the dynamics of underlyings [we refer to Bertoin (1996), Feller (1966), Lukacs (1970) and Sato (1999) for the theory of infinitely divisible distributions and Lévy processes]. Stable Paretian models in Finance were considered in Madelbrot (1960, 1963), Fama (1965), McCulloch (1978, 1996), Mittnik and Rachev

(1989), Willinger, Taqqu and Teverovsky (1999), Rachev and Mittnik (2000), and other works [see also Samorodnitsky and Taqqu (1994), Janicki and Weron (1994), Nolan (1998) for the theory, simulation and estimation of stable processes]. Since pioneering 1973 paper of Clark (1973), there have been a lot of research works on subordinated Lévy processes in finance: VG model (Madan and Seneta, 1990; Madan and Milne, 1991; Madan, 1999); Hyperbolic and Generalized Hyperbolic models (Barndorff-Nielsen, 1977, 1978, 1997, 1998; Eberlein and Keller, 1995; Embrechts, McNeil and Straumann, 1999; Eberlein and Raible, 1999) [see also Marinelli, Rachev and Roll (1999), Rachev and Mittnik (2000)]. A fine structure of asset returns from a Lévy process point of view was considered in Carr et al. (2000), Geman, Madan and Yor (1999, 1998) (CGMY model), Mantegna and Stanley (2000), Bouchaud and Potters (2000), Boyarchenko and Levendorskii (2000), Barndorff-Nielsen and Levendorskii (2001) (Truncated Lévy Flight). A general theory of conditionally normal stochastic variance and stochastic time change models is considered in Steutel (1970, 1973), Rosiński (1991), Maejima and Rosiński (2000), Barndorff-Nielsen and Pérez-Abreu (2000).

Most papers discuss a one-dimensional case with applications to option pricing. However, multidimensional models with a large number of risk factors are of significance for Risk Management. This chapter presents a new class of multivariate VaR models with the SV driven by Lévy processes.

2. Single-factor stochastic variance model

2.1. *Maximum entropy approach and Lévy processes*

Let a risk factor X denote a t-day absolute return, relative return, or log-return of the underlying market variable. Value-at-Risk over a given holding period t with a specified confidence level q (usually, $q = 1\%$) is defined as a q-quantile of the distribution for the portfolio changes during the period t. For the standard model, a RF probability density function is normal with given constant mean and variance. We consider a class of conditional normal models where the variance V of the risk factor X is stochastic rather than constant. The stochastic variance of the underlying returns is not directly observable in the market. Generally, the most reliable information about the SV is its average value over some period of time. It can be estimated from the sampling variance of the underlying returns. Under conditions of uncertainty, it is reasonable to adopt a conservative approach, i.e., choose a probability distribution for the SV that provides the most uncertain outcomes given only information about the average value. A well-known measure of uncertainty associated with probability distributions is entropy (Kagan, Linnik and Rao, 1973). Therefore, it is reasonable to determine the SV distribution from the Maximum Entropy principle.

A proposed single-factor SV model is based on the following assumptions (Levin and Tchernitser, 1999a):

Assumption 1. The density function, $p_X(x,T)$, of the risk factor $X = X(T)$ for some holding period T is normal conditional upon the stochastic variance $V = V(T)$ that possesses a probability density function $p_V(v,T)$, $v \geqslant 0$, i.e.,

$$p_X(x,T) = \int_0^\infty \frac{1}{\sqrt{2\pi v}} \exp\left(-\frac{(x-\theta v-\mu T)^2}{2v}\right) p_V(v,T)\,dv. \tag{1}$$

Parameter μT specifies a constant part of the mean for the conditional normal distribution, and parameter θ defines a shift in the mean proportional to the SV. As is shown later, θ determines the correlation between the RF and SV that results in a skewed RF distribution. The case $\theta = 0$ corresponds to a symmetric distribution. Linear dependence of the shift term θv from v in the mean of normal density is important for further construction of a Lévy process for the RF. The stochastic representation for X is as follows:

$$X(T) = \sqrt{V(T)}Z + \theta V(T) + \mu T, \quad Z \sim N(0,1), \tag{2}$$

with Z being a standard normal random variable independent of $V(T)$.

Assumption 2. The average variance $E\{V(T)\}$ for the holding period T is known and equal to $\overline{V}$:

$$E\{V(T)\} = \int_0^\infty v p_V(v,T)\,dv = \overline{V}. \tag{3}$$

Assumption 3. The probability density function $p_V(v,T)$ of the stochastic variance $V(T)$ is defined by the Maximum Entropy principle given the average variance (3):

$$H(p_V) = -\int_0^\infty p_V(v) \ln p_V(v)\,dv \to \max_{p_V(v)\geqslant 0}. \tag{4}$$

The optimization problem (4) for the SV density $p_V(v)$ subject to the constraint on the average variance (3) and standard normalization constraint $\int_0^\infty p_V(v)\,dv = 1$ has the exponential density

$$p_V(v) = \frac{1}{\overline{V}} \exp\left(-\frac{v}{\overline{V}}\right)$$

as a solution calculated by the Lagrange multiplier method (Kagan, Linnik and Rao, 1973). According to the Law of Total Probability, the unconditional density (1) of the risk factor $X(T)$ has the following density:

$$p_X(x,T) = \frac{\lambda}{\overline{V}} \exp\left(-\frac{|x-\mu T|}{\lambda} + \theta(x-\mu T)\right), \quad \lambda = \sqrt{\frac{\overline{V}}{2+\theta^2\overline{V}}}. \tag{5}$$

Distribution (5) is known as the skewed double exponential (Laplace) distribution (Kotz, Kozubowski and Podgórski, 2001). This distribution has a sharp peak, exponential tails and non-zero skewness for $\theta \neq 0$. Kurtosis of a symmetric Laplace distribution is always equal to 6, in contrast to 3 for a normal distribution. Historical distributions of daily returns for many market variables, such as CAD/USD FX rate (Figure 2), JPY/USD FX rate, S&P 500 Index, TSE 300 Index (Figure 2), NYMEX Natural Gas futures prices, some LIBOR rates, etc., have a similar leptokurtic shape (Levin and Tchernitser, 1999a; Kotz, Kozubowski and Podgórski, 2001).

In the case of a linear portfolio and symmetric Laplace distribution for the RF, the impact of non-normality on VaR can be estimated as

$$\frac{VaR_{\text{Laplace}}}{VaR_{\text{Normal}}} = \frac{\ln(2q)}{\sqrt{2}z_q},$$

where z_q is a standardized normal quantile for the confidence level q. For the case $q = 1\%$ ($z_q = 2.3263$), VaR_{Laplace} for a linear portfolio is 19% higher than the standard VaR_{Normal}. The impact on VaR is even more pronounced for non-linear instruments. For example, for a non-linear perfectly delta-hedged option portfolio, $\Pi(x)$, within Delta–Gamma approximation for the portfolio changes, $\delta\Pi(x) = 0.5\Gamma x^2$, the corresponding formulas for VaR are as follows:

$$VaR_{\text{Laplace}} = -\overline{V}\frac{\Gamma}{4}\ln^2(q), \qquad VaR_{\text{Normal}} = -\overline{V}\frac{\Gamma}{2}(z_{q/2})^2.$$

This results in 60% higher VaR_{Laplace} number than VaR_{Normal} (Levin and Tchernitser, 1999a).

The exponential distribution for the SV was derived from the Maximum Entropy principle for some unspecified holding period T. To calculate VaR for different holding periods t, a stochastic process for the risk factor X is required. The standard normal model assumes that the risk factor X follows a Wiener process with independent stationary Gaussian increments. The simplest extension of this assumption is that the RF follows a Lévy process, i.e., a stochastic process with independent stationary (not necessarily Gaussian) increments. It can be shown (Rosiński, 1991) that within the class of conditionally normal models (2) this assumption is equivalent to the following assumption on the SV:

Assumption 4. The total stochastic variance $V(t)$ in (2) follows a positive increasing Lévy process.

The exponential distribution for the $V(T)$ is infinitely divisible. It uniquely determines a positive increasing pure jump Gamma process [see Sato (1999)] for the total stochastic variance $V(t)$, $t > 0$, with a Gamma probability density function

$$p_{V(t)}(v) = \frac{v^{\alpha t-1}}{\Gamma(\alpha t)\beta^{\alpha t}}\exp\left(-\frac{v}{\beta}\right), \tag{6}$$

where $\alpha = 1/T$, $\beta = \overline{V}$. Assumptions 1–4 define the corresponding Lévy process for the risk factor $X(t)$ with the following probability density function:

$$p_X(x,t) = \sqrt{\frac{2}{\pi}}\frac{\lambda^{2\alpha t-1}\exp(\theta\lambda y)}{\Gamma(\alpha t)\beta^{\alpha t}}|y|^{\alpha t-1/2}K_{\alpha t-1/2}\big(|y|\big), \quad y = \frac{x-\mu t}{\lambda}. \tag{7}$$

Here λ is defined in (5), $\Gamma(\nu)$ is a gamma function, and $K_\nu(y)$ is a modified Bessel function of the third kind of the order ν, $K_{-\nu}(y) = K_\nu(y)$ (Abramowitz and Stegun, 1972). Distribution (7) is known as a Bessel K-function distribution (Johnson, Kotz and Balakrishnan, 1994) or as a Generalized Laplace distribution (Kotz, Kozubowski and Podgórski, 2001). Essentially, the SV model derived from the Maximum Entropy principle is equivalent to the Variance Gamma (VG) model [Gamma stochastic time change model, see Madan and Seneta (1990), Madan and Milne (1991), Geman and Ané (1996)]. The tail asymptotic behavior and behavior at the origin for the density (7) follows from known asymptotics for the modified Bessel function $K_\nu(y)$ (Abramowitz and Stegun, 1972)

$$K_\nu(y) \underset{y\to\infty}{\sim} \sqrt{\frac{\pi}{2y}}\,\mathrm{e}^{-y}, \qquad K_\nu(y) \underset{y\to 0}{\sim} \Gamma(\nu)2^{\nu-1}y^{-\nu}, \qquad \nu > 0, \qquad K_0(y) \underset{y\to 0}{\sim} -\ln(y).$$

The RF density (7) has exponential tails for all t and a wide range of shapes at the origin, from almost normal "bell" shape (for large $\alpha \gg 1$) to a highly peaked ($0.5 < \alpha \leqslant 1$) and even unbounded shape ($0 < \alpha \leqslant 0.5$) (see Figure 4). A skewed Laplace density (5) is a special case of (7) for $t = T$. The Bessel K-function family of distributions possesses finite moments of all orders. The characteristic function for the Gamma process has a simple form

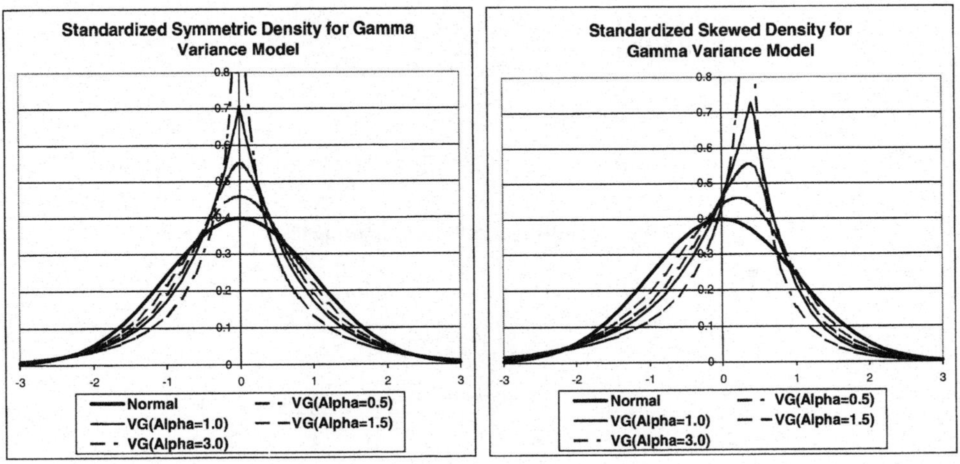

Fig. 4. Probability densities for the Gamma SV model.

$$\phi_{X(t)}(\omega) = E\{e^{i\omega X(t)}\} = \frac{\exp(i\omega\mu t)}{(1 - i\omega\beta\theta + \omega^2\beta/2)^{\alpha t}}. \tag{8}$$

The Lévy density from the Lévy–Khintchine representation of $\phi_{X(t)}(\omega)$ that characterizes the intensity of jumps of different sizes x has the following closed form [see Sato (1999)]:

$$k(x) = \frac{\alpha}{|x|}\exp\left(-\sqrt{\frac{2}{\beta} + \theta^2}|x| + \theta x\right).$$

The RF distribution (7) tends to a normal distribution for $t \to +\infty$. This normalization effect is important for a proper VaR scaling from short holding periods to longer ones. The total variance $D_{X(t)}$ is proportional to time, as it is for any Lévy process with finite variance (Feller, 1966) (a "square root of time" rule for the volatility is valid). However, contrary to the Gaussian case, the ratios of q-quantiles and standard deviation for the RF distributions (7) are not constant for different holding periods t. For example, the standardized 1%-quantile (VaR_{VG}) is higher for shorter holding period than the same 1%-quantile for longer holding period (Figure 5).

The entropy for the SV distribution standardized by time t (the mean of a standardized SV is equal to 1 for all t) has the maximum at $t = T$ (Figure 6) that corresponds to the exponential distribution. This property may be explained by transition of the standardized Gamma density from the delta-function at 0 to the delta-function at 1 as time t passes. Heuristically, this evolution of shape for the SV density corresponds to a transition from the state of maximum certainty at time 0 to the limiting state of maximum certainty at $t = \infty$ (with the limiting normal density for the standardized RF).

The following expressions provide a connection between the first four moments of the RF distribution and those of the SV distribution (Levin and Tchernitser, 2000a):

$$\begin{aligned} m_{X(t)} &= \mu t + \theta m_{V(t)}, \\ D_{X(t)} &= m_{V(t)} + \theta^2 D_{V(t)}, \\ m_{3,X(t)} &= \theta\left(3D_{V(t)} + \theta^2 m_{3,V(t)}\right), \\ m_{4,X(t)} &= 3m_{V(t)}^2 + 3D_{V(t)} + 6\theta^2 m_{V(t)} D_{V(t)} + 6\theta^2 m_{3,V(t)} + \theta^4 m_{4,V(t)}. \end{aligned} \tag{9}$$

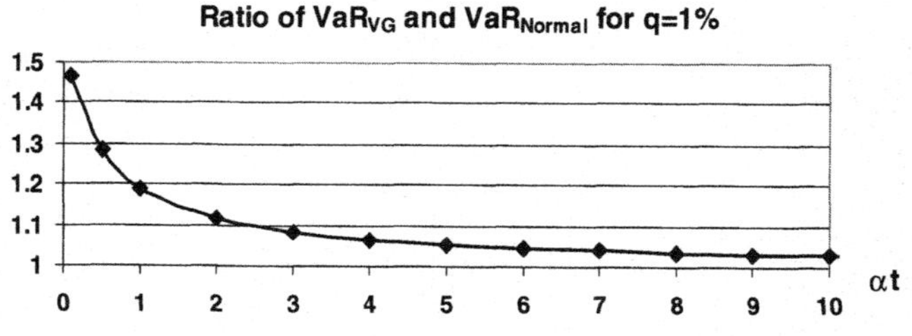

Fig. 5. VG model 1%-VaR term structure with respect to 1% Normal $VaR = 2.33$.

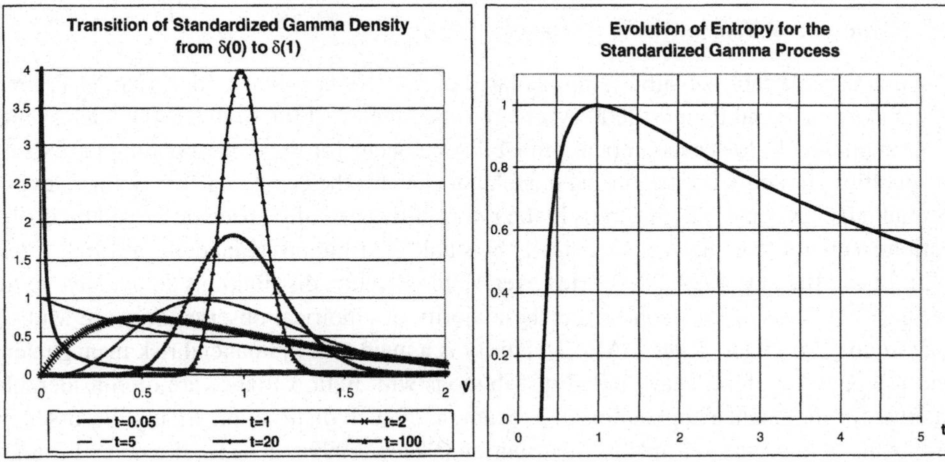

Fig. 6. Evolution of standardized Gamma SV density and entropy.

The expressions (9) for the moments are valid for conditional normal models of the form (1) provided that the distribution $p_{V(t)}(v)$ for the stochastic variance $V(t)$ possesses moments up to the fourth order. Parameter θ controls skewness of the RF distribution and defines the correlation $\rho_{X,V}$ between the risk factor X and its stochastic variance V:

$$\rho_{X,V} = \theta \sqrt{\frac{D_V}{m_V + \theta^2 D_V}}.$$

A parameter estimation procedure (model calibration), with respect to the four parameters, μ, θ, β, and α can be based either on the Maximum Likelihood approach or the method of moments given four sampling central moments for the T_1-day underlying returns and analytical expressions for the moments of the Gamma stochastic variance (Johnson, Kotz and Balakrishnan, 1994)

$$m_{V(T_1)} = \alpha T_1 \beta, \qquad D_{V(T_1)} = \alpha T_1 \beta^2,$$
$$m_{3,V(T_1)} = 2\alpha T_1 \beta^3, \qquad m_{4,V(T_1)} = 3\alpha T_1 \beta^4 (\alpha T_1 + 2).$$

Equations (9) can be used for the model calibration by the method of moments. Note that time T, corresponding to the maximum entropy for the SV density, can be recovered from the calibrated parameter α as $T = 1/\alpha$.

It follows from (6) and (9) that the term structure of the RF variance and kurtosis for the symmetric case of the Gamma-SV model ($\theta = 0$) is:

$$D_{X(t)} = \alpha \beta t, \qquad Kurt_{X(t)} - 3 = \frac{3}{\alpha t}. \tag{10}$$

2.2. Generalized Gamma Variance model

Some market variables exhibit jumps as large as 5 to 10 daily standard deviations (Fama, 1965; Bouchaud and Potters, 2000; Mantegna and Stanley, 2000; Cont, 2001). Such events have significantly lower theoretical probability to occur for the corresponding periods of observations not only for the normal model, but even for the Gamma SV model with exponential tails. Extremely large jumps in the risk factors have often been described by distributions with polynomial tails, specifically by stable distributions (Mandelbrot, 1960, 1963; Mittnik and Rachev, 1989, 2000). However, stable Paretian distributions do not have finite variance (volatility). This contradicts the majority of empirical observations [see Müller, Dacorogna and Pictet (1998)]. Also, volatility is a main tool in financial risk management and pricing. Therefore, heavy tailed distributions with finite variance are of considerable interest for the finance applications. An example of such distribution widely discussed in the financial literature is Student t-distribution (Platen, 1999; Albanese, Levin and Ching-Ming Chao, 1997; Rachev and Mittnik, 2000). A new family of the RF distributions introduced below includes t-distribution as a special case. The symmetric Gamma SV model considered above has only one shape parameter, α, that controls both the tails and central part of the distribution. It seems that one shape parameter is insufficient to distinguish between sources of high kurtosis: whether it comes from heavy tails or high peak. It is possible to show that for a class of conditional normal models the tail asymptotics of the RF distribution depends upon the tail asymptotics of the corresponding SV distribution. Therefore, a more general SV model that allows for separate control for the tails and peak should more successfully describe large deviations of the risk factors.

Note, that the Gamma SV density (6) can be formally derived from the Maximum Entropy principle (4) without Assumption 4. Instead, one can use a constraint on the logarithmic moment $E\{\ln(V)\}$ in addition to the condition on the average variance $E\{V\}$ (Kagan, Linnik and Rao, 1973). Essentially, this logarithmic constraint defines a power behavior of the SV density at the origin, while the constraint on $E\{V\}$ defines the exponential tail behavior. The condition on average variance can be replaced by a more flexible condition to accommodate information on a generalized moment of any power for the SV (Levin and Tchernitser, 2000a, b). For example, one can assume that the average volatility is known instead of average variance. This approximately corresponds to a constraint on the fractional moment $E\{\sqrt{V}\}$ instead of $E\{V\}$. Hence, we can formally define the entropy maximization problem (4) with two essential constraints

$$\int_0^\infty \ln(v) p_V(v)\,\mathrm{d}v = c_0, \qquad \int_0^\infty v^{1/\nu} p_V(v)\,\mathrm{d}v = c_1 \tag{11}$$

and a standard normalization constraint for a probability density function. The use of the Maximum Entropy approach with a constraint on the generalized moment $E\{V^{1/\nu}\}$, $\nu \in \mathbb{R}^1$, allows for a desirable generalization of the Gamma SV model to a broad class of models with a wide range of heavy tails, from exponential and sub-exponential (stretched exponential) to polynomial (Levin and Tchernitser, 2000a, b). A solution of the maximization problem (4), (11) is the Generalized Gamma density for the stochastic variance V:

$$p_V(v) = \frac{v^{\alpha/\nu-1}}{|\nu|\Gamma(\alpha)\beta^\alpha} \exp\left(-\frac{v^{1/\nu}}{\beta}\right). \tag{12}$$

The corresponding stochastic representation for V is a ν-th power of the Gamma distributed random variable γ with the density (6) [see Johnson, Kotz and Balakrishnan (1994)]:

$$V = \gamma^\nu. \tag{13}$$

Stochastic representations (2) and (13) allow for an effective Monte Carlo simulation procedure for the RF given well-known simulation procedures for normal and gamma random variables (Fishman, 1996).

The Generalized Gamma distribution is a very flexible class of distributions with two shape parameters α and ν. This class includes Gamma ($\nu = 1$), Inverse Gamma ($\nu = -1$), and Weibull ($\alpha = 1$, $\nu > 0$) distributions as special cases. It is known that the Generalized Gamma distribution is infinitely divisible for these three representatives [see Grosswald (1976), Ismail (1977), Sato (1999)] and for positive $\nu \geqslant \max(\alpha, 1)$ (Ismail and Kelker, 1979). Therefore, for these cases the Generalized Gamma distribution produces Lévy processes for the SV. We do not know if the Generalized Gamma distribution is infinitely divisible for arbitrary values of $\nu \in \mathbb{R}^1$, nor whether there is a closed form representation for the characteristic function. Hence, we apply the distribution (12) to describe the returns for the shortest holding period available, say one day, and then construct an additive SV process for a longer holding period, say 10 days, by summing up the independent Generalized Gamma distributed random variables. An analytical formula for the moments of the Generalized Gamma distribution is readily available

$$E\{V^k\} = \frac{\beta^{k\nu}\Gamma(\alpha + k\nu)}{\Gamma(\alpha)}$$

(the condition for the k-moments to exist is $(\alpha + k\nu) > 0$).

The corresponding RF density $p_X(x)$ is given by the integral (1) with SV density $p_V(v)$ being of the form (12). We call this density a Generalized Gamma Variance density (GGV). Unfortunately, in the general case there is no closed analytical form for the density $p_X(x)$. However, we consider an effective numerical procedure for calculating the integral (1) to be as good as, for example, a "closed form" formula (7) involving special K-Bessel functions. Effective asymptotic expansion methods (Olver, 1974; Abramowitz and Stegun, 1972) can be applied for the numerical calculations.[1] In the case of a symmetric GGV density, there is an analytical formula for the moments of any fractional order k (finite for $\alpha + k\nu/2 > 0$):

[1] Effective numerical procedure and software for the GGV density calculation was developed by Xiaofang Ma.

$$E\{|X|^k\} = \frac{2^{k/2}\beta^{k\nu/2}\Gamma((k+1)/2)\Gamma(\alpha+k\nu/2)}{\sqrt{\pi}\Gamma(\alpha)}. \tag{14}$$

The moments cease to exist for some combinations of negative values of ν and $\alpha > 0$ because of polynomial tails for the GGV density.

Below, we provide some results for a symmetric density $p_X(x)$. There are some known special analytical cases for $p_X(x)$:

(i) $\nu = -1$ corresponds to the t-distribution with 2α degrees of freedom;
(ii) $\nu = 0$ corresponds to the Gaussian distribution;
(iii) $\nu = +1$ corresponds to the K-Bessel function distribution (7).

Table 1 presents a summary of results for the Generalized Gamma Variance model, including a constraint on the generalized moment in Maximum Entropy principle (column 1), SV stochastic representation (column 2), corresponding RF density (column 3), and asymptotics for the tails of the RF density (column 4).

Some market variables are better described by distributions with polynomial tails, while others are better described by distributions with semi-heavy tails (exponential and sub-exponential) [see Platen (1999), Rachev and Mittnik (2000), Duffie and Pan (1997)]. The GGV model is capable of accommodating both types of behavior. A range of values $\nu < 0$ corresponds to a power low tails. GGV density is finite at zero for all $\nu < 0$. A range of values $\nu > 0$ corresponds to exponential and sub-exponential tails. In this case, tails are far lighter and the moments of all orders exist. The range $\nu > 1$ corresponds to a class of stretched exponential densities $p_X(x)$. The specific class of the stretched exponential distributions based on a modified Weibull density was considered in Sornette, Simonetti and Andersen (2000). Figure 7 shows the RF GGV density $p_X(x)$ for different values of parameters ν and α. Parameter ν brings an extra flexibility to the GGV density: it is seen that GGV model can accommodate a wide variety of shapes and tail behavior.

A statistical investigation of different SV models from a Generalized Hyperbolic family based on historical data for 15 stock market indices was presented in the paper by Platen (1999). The class of Generalized Hyperbolic distributions developed in Barndorff-Nielsen (1978, 1998), Eberlein and Keller (1995), Eberlein, Keller and Prause (1998) is also a two shape parameter family in symmetric case. All members of this family have exponential

Table 1
GGV model summary

Constr. $E\{V^{1/\nu}\}$	SV density & Stoch. rep.	RF density	RF asymptotics $x \to \infty$
$E\{V\}$, $\nu = 1$	Gamma, $V = \gamma$	K-Bessel	$\sim x^{\alpha-1}\,\mathrm{e}^{-cx}$
$E\{\sqrt{V}\}$, $\nu = 2$	Square of Gamma, $V = \gamma^2$	GGV$(2, \alpha)$	$\sim x^{2\alpha/3-1}\,\mathrm{e}^{-cx^{2/3}}$
$E\{1/V\}$, $\nu = -1$	Inverse Gamma, $V = 1/\gamma$	t-Distribution	$\sim x^{-(2\alpha+1)}$
$E\{V^{1/\nu}\}$, $\nu > 0$	Generalized Gamma, $V = \gamma^{\nu}$	GGV(ν, α)	$\sim x^{2\alpha/(1+\nu)-1}\,\mathrm{e}^{-cx^{2/(1+\nu)}}$
$E\{V^{1/\nu}\}$, $\nu < 0$	Generalized (Inv.) Gamma, $V = \gamma^{\nu}$	GGV(ν, α)	$\sim x^{-(2\alpha/\lvert\nu\rvert+1)}$
$\nu = 0$	SV degenerates to $V \equiv 1$	Normal	$\sim \mathrm{e}^{-x^2/2}$

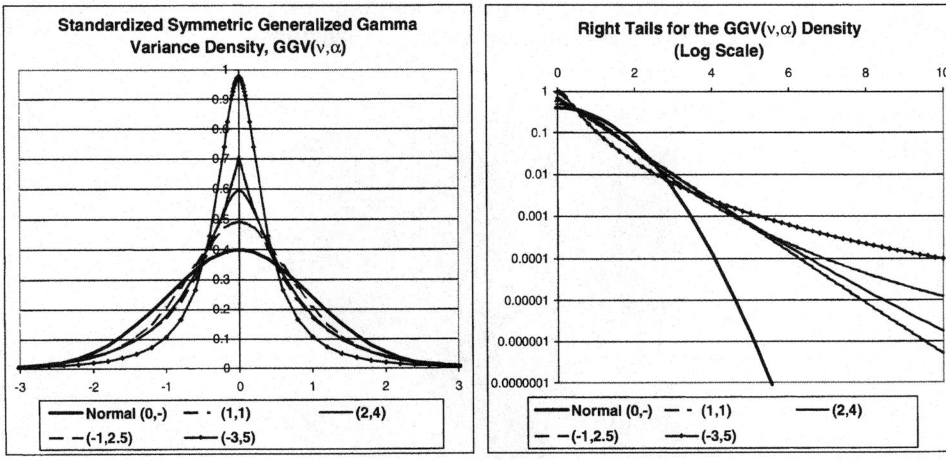

Fig. 7. Generalized Gamma Variance densities.

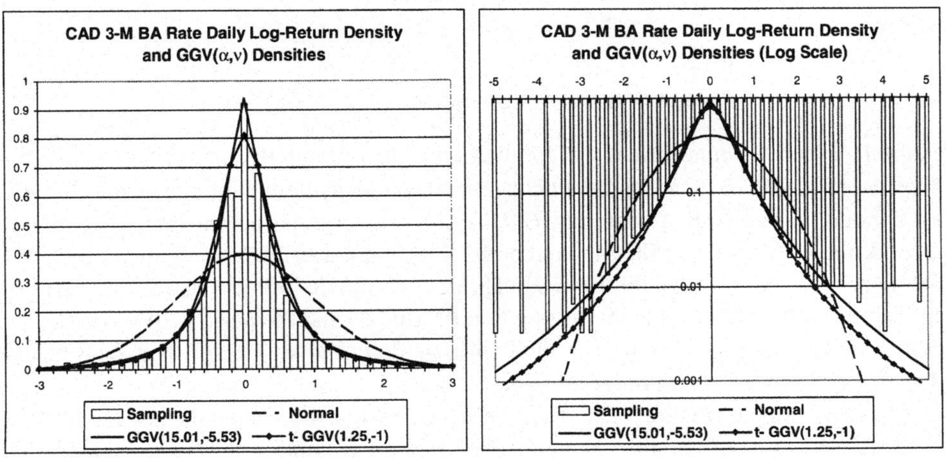

Fig. 8. Historical and calibrated GGV densities for the CAD 3-month BA interest rate daily log-returns.

tails except the Student t-distribution, which has polynomial tails. For this specific case, the class of Generalized Hyperbolic distributions collapses to a one shape parameter (number of degrees of freedom) family. Four representatives from a Generalized Hyperbolic class (Student t-distribution, Normal Inverse Gaussian, Variance Gamma, and Hyperbolic distributions) were compared based on the Maximum Likelihood criteria. The last three of these distributions have exponential tails. Results presented in Platen (1999) show that all distributions having exponential tails fail to satisfy the Pearson χ^2 test. In contrast, the t-distribution has not been rejected on a 99% confidence level for ten of the fifteen indices. Two-parameter Paretian tail GGV distributions perform better than the t-distribution. As an

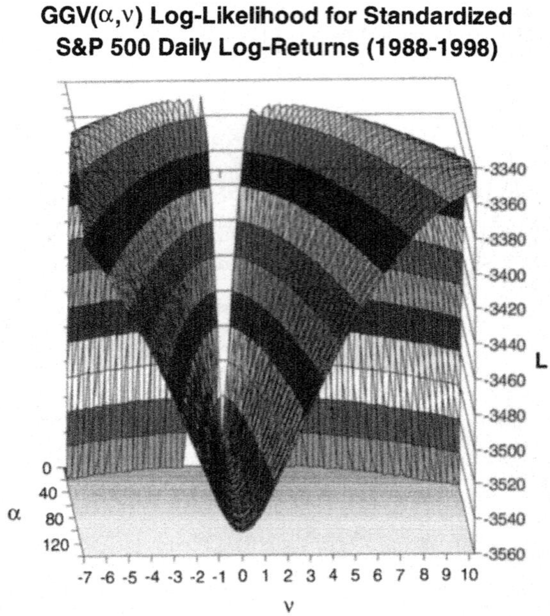

Fig. 9. GGV model log-likelihood surface for S&P 500.

example, Figure 8 demonstrates a fit for the Canadian 3-month BA interest rate daily return density (1992–1998) by normal, Student-t, and GGV densities calibrated using Maximum Likelihood approach. It is seen that $GGV(\alpha, \nu)$ with optimal parameters $\nu = -5.5$ and $\alpha = 15$ outperforms t-distribution, and both GGV and t-distributions significantly outperform normal. The χ^2 value for the $GGV(15, -5.5)$ is about 80% less than χ^2 value for the calibrated t-distribution. It is interesting to note, that during the period 1992–1998, Canadian 3-month BA interest rate exhibited 14 large daily moves greater than four standard deviations (about 1% of all observations).

Another example (Figure 9) shows a GGV model log-likelihood surface for S&P 500 Index as a function of parameters ν and α. A deep minimum for $\nu = 0$ corresponds to the normal distribution, while two wings correspond to the power law ($\nu < 0$) and stretched exponential ($\nu > 1$) tailed distributions. For this example, a stretched exponential sub-class produces almost the same maximum likelihood value as a power law sub-class.

2.3. *Mean-reverting stochastic variance model*

So far, we have considered a class of the SV models driven by Lévy processes with independent, identically distributed, but not necessarily Gaussian increments. The model explains non-normality of the RF distributions. For any conditional normal SV model, expressions (9) provide an exact answer for the term structures of the risk factor variance and kurtosis

$$D_{X(t)} = m_{V(t)}, \qquad Kurt_{X(t)} - 3 = 3\frac{D_{V(t)}}{m_{V(t)}^2}. \tag{15}$$

Here $V(t)$ is a total variance. Since $m_{V(t)}$ and $D_{V(t)}$ are linear functions of time for any Lévy process for $V(t)$, the above expressions predict linear increase of the RF variance and hyperbolic decrease of its excess kurtosis.

However, empirical investigations show that the underlying returns are almost uncorrelated, but not independent [see Bouchaud and Potters (2000), Cont (2001), Müller, Dacorogna and Pictet (1998)]. The easiest way to demonstrate this dependence is to consider the empirical correlations for the absolute values or squares of the returns. It is seen (Figure 10) that autocorrelations of squares are statistically significant. This phenomenon is connected with a known volatility clustering effect (Figure 3). Also, it is known that empirical term structure of kurtosis decreases slower than is predicted by a "Lévy term structure" model (15) [so called "anomalous decay", see Bouchaud and Potters (2000), Cont (2001)]. All this suggests that a better model for the instantaneous stochastic volatility is not a "white noise" kind of process, but rather a process with autocorrelation. One way to account for the autocorrelation structure of the SV is to consider regime-switching SV processes [see Konikov and Madan (2000)]. We will follow another approach to introduce the SV autocorrelation by considering Ornstein–Uhlenbeck (OU) type processes for the instantaneous SV (Levin and Tchernitser, 1999a, 2000a). Such class of non-Gaussian OU type processes driven by positive Lévy noise was investigated in detail in Barndorff-Nielsen and Shephard (2000a, b). In this section we will only demonstrate that the empirically observed term structure of kurtosis can be consistently described by such models.

Consider a stationary non-negative process $\xi(t)$ with autocorrelation function $R_\xi(\tau)$ that describes the instantaneous stochastic variance. For the total variance $V(t)$ being $V(t) = \int_0^t \xi(t')\,\mathrm{d}t'$, it follows that

$$m_{V(t)} = m_{V(1)}t, \qquad D_{V(t)} = 2\int_0^t (t-\tau)R_\xi(\tau)\,\mathrm{d}\tau.$$

The above expressions in conjunction with (15) can be used to calculate a term structure of the RF kurtosis. In particular, assume a mean-reverting process for the instantaneous stochastic variance $\xi(t)$ be a Ornstein–Uhlenbeck type process

$$\mathrm{d}\xi(t) = -\lambda\xi(t)\,\mathrm{d}t + \lambda\,\mathrm{d}G(t), \tag{16}$$

where $G(t)$ is, for example, a Gamma process, $\lambda > 0$ is a mean-reversion speed parameter. Expressions for $R_\xi(\tau)$ and variance $D_{V(t)}$ are as follows

$$R_\xi(\tau) = \frac{\lambda\alpha\beta^2}{2}\mathrm{e}^{-\lambda|\tau|}, \qquad D_{V(t)} = \frac{\alpha\beta^2 t}{2}\left(1 - \frac{1-\mathrm{e}^{-\lambda t}}{\lambda t}\right).$$

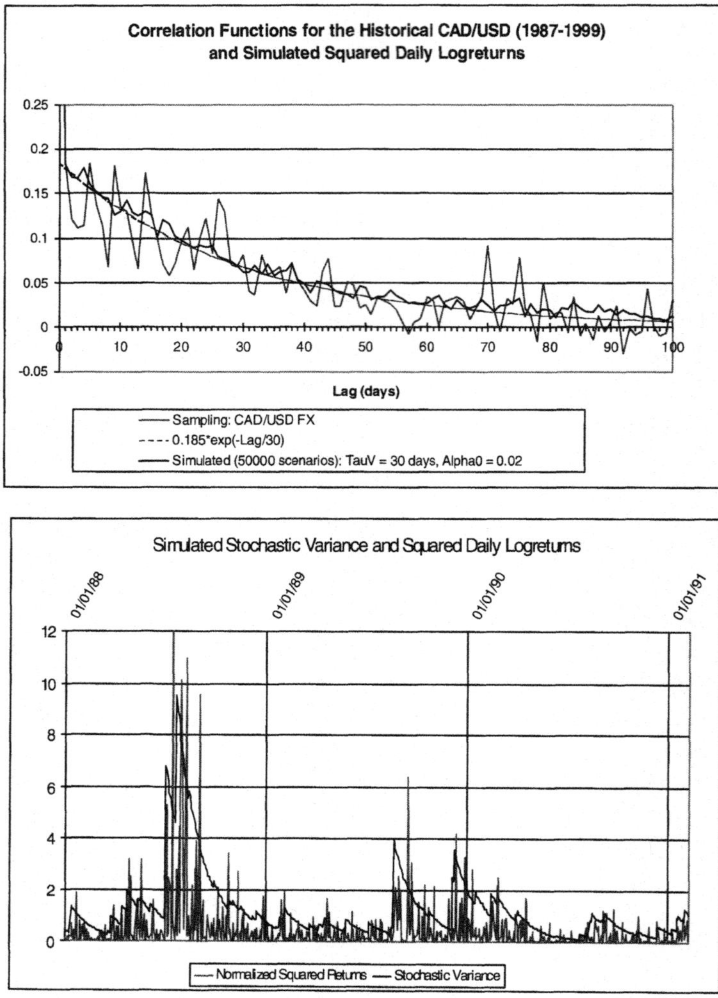

Fig. 10. Autocorrelation in the squared CAD/USD FX daily returns and in the SV.

The autocorrelation function $R_\xi(\tau)$ is an exponential function for any OU model (16). It is seen that $D_{V(t)}$ is not a linear function of time contrary to the Lévy case (10). Previous formulas and formulas (15) result in the following term structure of the RF kurtosis:

$$Kurt_{X(t)}^{\mathrm{OU}} - 3 = \frac{3}{\alpha t}\left(1 - \frac{1 - \mathrm{e}^{-\lambda t}}{\lambda t}\right).$$

Figure 11 shows a term structure of the RF kurtosis for different values of the mean-reversion speed parameter λ. As expected, the OU stochastic variance process provides

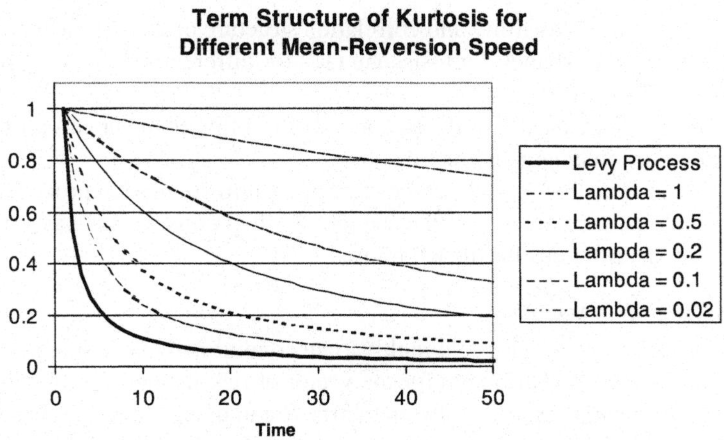

Fig. 11. Term structure of the RF kurtosis for the model with autocorrelated SV.

slower decay of kurtosis vs. Lévy SV process. Reduction in decay can be significant depending upon the mean-reversion speed λ. This is equivalent to slower "normalization" effect. The bottom graph in Figure 10 presents the time series for the simulated SV and empirical CAD/USD FX rate squared daily log-returns. The bottom graph in Figure 3 presents the simulated RF time series. It is evident that the model produces large deviations for the FX rate and volatility clustering effect that is very similar to the one observed in the market (top graph in Figure 3).

3. Multifactor stochastic variance model

3.1. Requirements for multifactor VaR models

A realistic multifactor VaR model should consistently describe not only the correlation and volatility structure for the risk factors, but also different shapes of the marginal risk factor distributions and distributions in other "diagonal" directions. Also, a principal component analysis for daily returns in different markets (interest rate curves, commodity futures prices, implied volatility curves and surfaces), clearly indicates the presence of non-linear dependence between risk factors (principal components). For example, the squared daily changes of the principal components are significantly correlated, while daily changes themselves are uncorrelated. This non-linear dependence breaks conditions of the Central Limit Theorem and has an important impact on VaR calculation: even for well-diversified linear portfolios with a large number of instruments there is no full normalization of the portfolio return distributions (Levin and Tchernitser, 1999a, b). An example of such large diversified portfolio is the S&P 500 Index. Its distribution is quite far from normal despite the portfolio averaging effect. Hence, a comprehensive model for multiple risk factors should additionally capture the following important features observed in the market:

- exact match of a given volatility and correlation structure of the risk factors;
- approximate match of shapes, kurtosis, and tails for different risk factors (marginal distributions);
- approximate match of shapes, kurtosis, and tails for different linear sub-portfolios (marginal distributions in diagonal directions).

The model should also allow for an effective Monte Carlo simulation procedure. To facilitate further multivariate analysis, in the sequel, we shall consider the case of symmetric joint probability distributions for the RF returns.

3.2. "Naïve" multifactor model

A very simple idea for constructing a multivariate conditionally Gaussian stochastic variance model is to define a distribution for the vector of risk factors $X(t) \in \mathbb{R}^N$ as a multivariate normal with some fixed correlation matrix R and independent stochastic variances $V_i(t)$, $i = 1, \ldots, N$. A symmetric multivariate probability density function for the vector of risk factors is represented as:

$$p_{X(t)}(x) = \int_{V_1} \cdots \int_{V_N} \frac{1}{\sqrt{(2\pi)^N \det(C)}} \times \exp\left(-\frac{x'C^{-1}x}{2}\right) p_V(V_1, \ldots, V_N)\, \mathrm{d}V_1 \cdots \mathrm{d}V_N, \tag{17}$$

$$C = \Sigma R \Sigma', \quad \Sigma = \operatorname{diag}\left(\sqrt{V_1(t)}, \ldots, \sqrt{V_N(t)}\right). \tag{18}$$

Here $p_V(V_1, \ldots, V_N) = \prod_{i=1}^N p_{V_i}(V_i)$ is a probability density for independent stochastic variances $V_i(t)$, x' is transpose of x. The corresponding stochastic representation for the risk factors $X(t)$ is

$$X(t) = \operatorname{diag}\left(\sqrt{V_1(t)}, \ldots, \sqrt{V_N(t)}\right) A Z, \quad AA' = C, \; Z \sim N(0, I), \tag{19}$$

where $Z \sim N(0, I)$ is independent of V standard normal vector with identity covariance matrix I. This representation allows for modelling marginal distributions with different leptokurtic shapes.

However, it can be shown that this "naïve" approach reduces the correlations between risk factors because of "randomization" for the covariance matrix (Levin and Tchernitser, 1999a). Due to independence of the stochastic variances V_i, absolute values of the model correlations $\operatorname{Corr}(X_i, X_j)$ are less than absolute values of the correlations R_{ij} used in (17):

$$\begin{aligned}\operatorname{Cov}(X_i, X_j) &= \iint x_i x_j p_X(x)\, \mathrm{d}x_i\, \mathrm{d}x_j = R_{ij} \iint \sqrt{V_i}\sqrt{V_j}\, p_V(V_i, V_j)\, \mathrm{d}V_i\, \mathrm{d}V_j \\ &= R_{ij} \int \sqrt{V_i}\, p_{V_i}(V_i)\, \mathrm{d}V_i \int \sqrt{V_j}\, p_{V_j}(V_j)\, \mathrm{d}V_j \\ &= f_{ij} \sigma_{X,i} \sigma_{X,j} R_{ij}, \quad i \neq j.\end{aligned} \tag{20}$$

Table 2
Correlation reduction factors

$\alpha_i = \alpha_j$	0.5	1	2	5	10
$f(\alpha_i, \alpha_j)$	0.64	0.79	0.88	0.95	0.98

Reduction factors $f_{ij}, i \neq j$, are less than one, because

$$\int \sqrt{V_i}\, 1 p_{V_i}(V_i)\,\mathrm{d}V_i < \sqrt{\int V_i p_{V_i}(V_i)\,\mathrm{d}V_i}\sqrt{\int 1 p_{V_i}(V_i)\,\mathrm{d}V_i} = \sqrt{E\{V_i\}} = \sigma_{X,i}.$$

It means that the sampling correlation matrix cannot be used as the matrix R in (17). For example, the reduction factors $f_{ij} < 1, i \neq j$, calculated explicitly for the case of the Gamma stochastic variances (6) are as follows:

$$\mathrm{Corr}(X_i, X_j) = f_{ij} R_{ij}, \quad f_{ij} = f(\alpha_i, \alpha_j) = \frac{\Gamma(\alpha_i + 1/2)\Gamma(\alpha_j + 1/2)}{\Gamma(\alpha_i)\Gamma(\alpha_j)\sqrt{\alpha_i \alpha_j}}, \quad i \neq j.$$

The underestimation of the correlations can be significant for some values of parameters α_i, α_j as it is shown in Table 2.

The randomization effect exists for any probability density functions $p_{V_i}(V_i)$ for independent stochastic variances. Usually, equations $\mathrm{Corr}(\widetilde{X}_i, \widetilde{X}_j) = \underset{\sim}{f_{ij}} \underset{\sim}{R_{ij}}$ cannot be resolved with respect to correlations R_{ij} given sampling correlations $\mathrm{Corr}(\widetilde{X}_i, \widetilde{X}_j)$ while preserving the necessary conditions $|R_{ij}| \leqslant 1$ or non-negative definite matrices R. Hence, this "naïve" model does not allow to preserve historical correlations between the risk factors.

Remark. Equation (20) and the inequality

$$\iint \sqrt{V_i}\sqrt{V_j}\, p_V(V_i, V_j)\,\mathrm{d}V_i\,\mathrm{d}V_j < \sqrt{\mathrm{Cov}(V_i, V_j) + E\{V_i\}E\{V_j\}}$$

imply that the class of the SV models with the stochastic representation (18) for the covariance matrix preserves the RF correlation structure only if

$$\iint \sqrt{V_i}\sqrt{V_j}\, p_V(V_i, V_j)\,\mathrm{d}V_i\,\mathrm{d}V_j = \sqrt{E\{V_i\}}\sqrt{E\{V_j\}},$$

which requires dependent stochastic variances with positive correlations. We do not investigate this direction in the chapter.

3.3. Elliptical stochastic variance model

The simplest extension of a single-factor SV model to the multifactor case is an elliptical stochastic variance model. Elliptical models are widely used for representing nonnormal multivariate distributions in finance [see Eberlein, Keller and Prause (1998), Kotz, Kozubowski and Podgórski (2001)]. This class of models preserves the observed RF correlation structure. The model is similar by construction to the one-dimensional variance mixture of normals. An elliptical N-dimensional symmetric process $X^E(t)$ for N risk factors has a stochastic representation as a single variance mixture of multivariate normals with a given covariance matrix C:

$$X^E(t) = \sqrt{V(t)}Z^C, \quad Z^C \sim N(0, C). \tag{21}$$

Here $V(t)$ is a univariate SV process, Z^C is a multivariate normal N-dimensional vector independent of $V(t)$. The covariance matrix C is estimated from historical T_1-day returns (e.g., daily returns), while the SV is normalized to satisfy a condition $m_V(T_1) = E\{V(T_1)\} = 1$. The unconditional density for the random vector of risk factors $X^E(t)$ is:

$$p_{X^E(t)}(x) = \int_0^\infty \frac{1}{\sqrt{(2\pi V)^N \det(C)}} \exp\left(-\frac{x'C^{-1}x}{2V}\right) p_{V(t)}(V)\,\mathrm{d}V.$$

As an example, consider the case of Gamma stochastic variance $V(t)$. A closed analytical form for the unconditional elliptical Bessel K-function density for $X^E(T)$ is available in Kotz, Kozubowski and Podgórski (2001). A characteristic function $\phi_{X^E(t)}(\omega)$ for the elliptical Lévy process $X^E(t)$ is represented as:

$$\phi_{X^E(t)}(\omega) = \left(1 + \frac{\beta}{2}\omega' C \omega\right)^{-\alpha t}, \tag{22}$$

where ω is N-dimensional vector, ω' is a vector transposed to ω. Due to known properties of elliptical distributions [see Fang, Kotz and Ng (1990)], all marginal one-dimensional distributions for the risk factors are univariate Bessel K-function distributions with the same shape parameter αt and the same kurtosis. They differ only by the standard deviations. The same property holds for all one-dimensional distributions of linear combinations $X_\Delta = \Delta' X^E(t)$ of the risk factors. These linear combinations correspond to the linear portfolios defined by Δ. The kurtosis of $X_\Delta(T_1)$ for any arbitrary Δ is equal to $k_\Delta = 3(1 + D_V(T_1)/m_V^2(T_1)) = 3(1 + D_V(T_1))$. Therefore, within the class of elliptical models there is no normalization effect at all for the distributions of large diversified portfolios. This is a result of violation of the conditions for the Central Limit Theorem: the risk factors are dependent through the common stochastic variance V. Such property is a drawback for all elliptical models. It is clear that the actual RF fluctuations are not driven by a single stochastic variance ("global market activity"). More realistic SV model should include a multidimensional processes for the SV to model different distributional shapes for the risk factors and linear sub-portfolios. Since sampling marginal RF distributions have different shapes, the calibration of elliptical model is restricted to fitting a distribution of some preselected portfolio. Hence, the calibration of elliptical models is portfolio dependent.

3.4. Independent stochastic variances for the principal components

One of the possible ways to model different shapes for the RF distributions while preserving a given correlation structure was considered in Levin and Tchernitser (1999a, b). An N-dimensional vector of the risk factors is represented as a linear combination of principal components (PC) with independent one-dimensional stochastic variances. The corresponding stochastic representation is as follows:

$$X^L(t) = \tilde{A} Z^I(t), \quad Z_i^I(t) = \sqrt{V_i(t)} Z_i, \quad Z_i \sim N(0,1), \quad i = 1, \ldots, M.$$

Here Z_i are independent standard normal variables, $V_i(t)$ are independent SV processes with a unit mean and some variances D_{Vi} for a specified time horizon T_1. The columns of a constant matrix $\tilde{A}$ are the principal components of a given covariance matrix C. The covariance matrix C is estimated from the historical T_1-day returns. Matrix $\tilde{A}$ is calculated based on eigenvalue decomposition of the covariance matrix C [see Wilkinson and Reinsch (1971)]:

$$C = UDU', \quad U' = U^{-1}, \quad D = \mathrm{diag}(d_1, \ldots, d_N), \tag{23}$$

$$\tilde{A} = U_M D_M^{1/2}, \quad D_M = \mathrm{diag}(d_1, \ldots, d_M), \ M \leqslant N, \quad C = \tilde{A}\tilde{A}'. \tag{24}$$

Matrix U_M consists of the first M columns of the orthogonal matrix U, which correspond to the first M largest eigenvalues $d_1, \ldots, d_M$ of the matrix C. Number M may be chosen less than N if the matrix C is singular and has only M non-zero eigenvalues. Some numerical issues related to singularity of the matrix C were considered in Kreinin and Levin (2000). It follows from the construction of the process $X^L(t)$ that $\mathrm{Cov}(X^L(T_1)) = C$. This ensures an exact match of the sampling covariance matrix C. One can keep even a smaller number M of the principal components in (24) and recover the matrix C with the required accuracy.

A characteristic function for the model is a product of the characteristic functions of one-dimensional processes for the PCs. For example, a characteristic function for the Gamma SV model with independent SV has a form

$$\phi_{X^E(t)}(\omega) = \prod_{i=1}^{M} \left(1 + \frac{(\omega'(\tilde{A})_i)^2}{2}\right)^{-\alpha_i t},$$

where $(\tilde{A})_i$ is i-th column of the matrix $\tilde{A}$.

The matrix $\tilde{A}$ can be defined up to an arbitrary orthogonal transformation H without change of the covariance matrix C since Z_i are independent standard normal variables, Z_i and $V_i(T_1)$ are independent and $E\{V_i(T_1)\} = 1$. Hence, $E\{Z^I(T_1)Z^I(T_1)'\} = I$ and

$$E\{X^L(T_1)X^L(T_1)'\} = \tilde{A}HE\{Z^I(T_1)Z^I(T_1)'\}H'\tilde{A}' = \tilde{A}HH^{-1}\tilde{A}' = C$$

for any orthogonal matrix H. However, the matrix H influences a matrix of the fourth moments of $X^L(t)$, $K_{ij} = E\{(X^L)_i^2(X^L)_j^2\}$. The orthogonal matrix H and shape parameters for the V_i can be determined to approximate a given sampling matrix $\{K_{ij}\}$ of the fourth moments for the RF distribution (all moments $E\{(X^L)_i^3(X^L)_j\}$ are equal to zero for symmetric distributions). An explicit calculation yields:

$$K_{ij} = E\{(X^L)_i^2(X^L)_j^2\} = 3\sum_{k=1}^{M} a_{ik}^2 a_{jk}^2 D_{Vk} + C_{ii}C_{jj} + 2C_{ij}^2,$$
$$i, j = 1, \ldots, N, \tag{25}$$

where a_{ik} are the elements of the matrix $A = \tilde{A}H$. An effective method for calculating the matrix H and shape parameters is discussed in Section 3.6 below.

The model provides an exact match of the RF correlation and volatility structures and approximates different shapes and kurtosis of the marginal RF distributions contrary to the Elliptical model. However, there is a significant drawback for this model. Since the stochastic variances V_i are independent, there is a strong normalization effect in any "diagonal" direction. This means that some linear portfolios $X_\Delta(t) = \Delta' X^L(t)$ have almost normal distributions whenever the portfolio Delta, Δ, is not a marginal direction and the number of principal risk factors M is large enough. Described effect presents a real danger, because the non-normal marginal RF distributions may be well-approximated, while the modelled portfolio distributions (contrary to the actual sampling distributions) may be almost normalized and the VaR underestimated.

3.5. A model with correlated stochastic variances

As it was pointed out above, a more general and realistic market model should incorporate the correlated stochastic variances that can correct the deficiencies of both Elliptical model and the model with independent SV for the principal components. The correlated SV structure should allow modeling of some general economic factors as well as idiosyncratic components that drive the SV processes for different risk factors and markets.

The model is defined via stochastic representation of the following form (Levin and Tchernitser, 2000a, b):

$$X^{CV}(t) = AZ^I(t), \quad Z_i^I(t) = \sqrt{V_i(t)}Z_i, \quad Z_i \sim N(0,1), \quad i = 1, \ldots, M. \tag{26}$$

Here Z_i are independent standard normal variables, $V_i(t)$ are the correlated stochastic variance processes with a unit mean for a specified time horizon T_1. The matrix $A \in \mathbb{R}^{N\times M}$ is defined as in the previous section through the eigenvalue decomposition for the covariance matrix C up to an arbitrary orthogonal transformation $H \in \mathbb{R}^{M\times M}$:

$$C = \mathrm{Cov}(X^{CV}(T_1)) = AA' = \tilde{A}\tilde{A}', \quad A = \tilde{A}H, \; H' = H^{-1}.$$

Stochastic variances $V_i(t)$ are correlated to each other due to the following stochastic representation:

$$V_i(t) = \sum_{k=1}^{L} b_{ik}\xi_k(t), \quad \sum_{k=1}^{L} b_{ik} = 1, \ b_{ik} \geqslant 0, \ B \in \mathbb{R}^{M\times L}, \tag{27}$$

where $\xi_k(t)$ are independent positive increasing Lévy processes with unit mean for the time horizon T_1 and different shape parameters, and B is a constant matrix with non-negative elements. The processes $\xi_k(t)$ are the drivers for the SV processes $V_i(t)$. For example, each driver ξ_k can be a Gamma process or Generalized Gamma process. Linear structure in (27) with $b_{ik} \geqslant 0$ ensures that $V_i(t)$ are positive increasing Lévy processes. The normalization conditions $E\{\xi_k(T_1)\} = 1$' and $\sum b_{ik} = 1$ ensure, as in Section 3.4, exact recovering of the sampling covariance matrix for the risk factors. It follows, that the vector of stochastic variances $V(T_1)$ has covariance matrix C_V equal to

$$\begin{aligned} C_V &= \operatorname{Cov}\bigl(V(T_1)\bigr) = B D_\xi B', \\ D_\xi &= \operatorname{diag}(D_{\xi 1}, \dots, D_{\xi L}), \ D_{\xi k} = \operatorname{Var}\bigl\{\xi_k(T_1)\bigr\}. \end{aligned} \tag{28}$$

The multivariate Generalized Stochastic Variance (GSV) model (26), (27) has two levels of correlations. First level defines usual correlations across the risk factors described by the covariance matrix C. Second level defines the correlations across the stochastic variances described by the covariance matrix C_V. The second level of correlations provides a possibility to obtain an approximate, but consistent match of the higher order moments and shape of the RF multivariate distribution. The elliptical model and the model with independent stochastic variances are the special cases of the above GSV model. Elliptical model corresponds to the matrix B being equal to one column with all unit entries, $B = [1, \dots, 1]'$. The model with independent SV corresponds to the case when the matrix B is equal to the identity matrix, $B = I$.

There is no analytical form for the probability density function of the vector $X^{CV}(t)$ even for the Gamma drivers $\xi_k(t)$. However, a characteristic function $\phi_{X^{CV}(t)}(\omega)$ can be calculated as

$$\begin{aligned} \phi_{X^{CV}(t)}(\omega) &= \int_{\xi\in\mathbb{R}_+^L} \exp\biggl(-\frac{1}{2}\omega' A \operatorname{diag}(B\xi) A'\omega\biggr) p_{\xi(t)}(\xi)\, \mathrm{d}\xi \\ &= \prod_{j=1}^{L} \int_0^{+\infty} \exp\Biggl(-\frac{\xi_j}{2}\sum_{i=1}^{M} b_{ij}\Biggl[\sum_{k=1}^{N} A_{ki}\omega_k\Biggr]^2\Biggr) p_{\xi_j(t)}(\xi_j)\, \mathrm{d}\xi_j. \end{aligned}$$

The expression for the characteristic function above is equivalent to

$$\phi_{X^{CV}(t)}(\omega) = \prod_{j=1}^{L} \int_0^{+\infty} \exp\biggl(-\frac{\xi_j}{2}\omega' \widehat{C}_j\, \omega\biggr) p_{\xi_j(t)}(\xi_j)\, \mathrm{d}\xi_j,$$

where $\widehat{C}_j$, $j = 1, \ldots, L$, are certain positive semi-definite matrices. The latter expression for the characteristic function allows for a different interpretation of the GSV model. It shows that the process for the risk factors $X^{CV}(t)$ can be presented as a sum of L independent elliptical Lévy processes. In turn, each of these elliptical processes has a multivariate conditional normal distribution with a covariance matrix proportional to $\widehat{C}_j$ and the corresponding stochastic variance $\xi_j(t)$.

The kurtosis k_Δ of a linear combination of the risk factors $X_\Delta(T_1) = \Delta' X^{CV}(T_1)$ for any given direction Δ can be calculated analytically:

$$k_\Delta - 3 = \frac{E\{X_\Delta^4(T_1)\}}{E\{X_\Delta^2(T_1)\}^2} - 3 = 3\eta' C_V \eta = 3\eta' B D_\xi B' \eta,$$

$$\eta \in \mathbb{R}^M, \ \eta_k = \frac{(\Delta' A)_k^2}{\|\Delta' A\|^2}, \ k = 1, \ldots, M. \tag{29}$$

The above expression provides a link between the covariance matrix C_V and the kurtosis k_Δ, that characterizes the shape of the RF multivariate distribution for the linear portfolio with Delta equal to Δ. Another useful quantity that clarifies the role of the correlated variances V_i is a standardized matrix of the fourth moments. This matrix, $\{k_{ij}\}$, is a multidimensional analog for kurtosis

$$k_{ij} = \frac{E\{(X^L)_i^2 (X^L)_j^2\}}{E\{(X^L)_i^2\} E\{(X^L)_i^2\}}. \tag{30}$$

The matrix $\{k_{ij}\}$ incorporates kurtosis in all marginal and all pair-wise diagonal directions in the original risk factor space. It is expressed as

$$k_{ij} - (1 + 2\rho_{ij}^2) = \sum_{k=1}^M \sum_{l=1}^M \lambda_{ik}^2 \lambda_{jl}^2 \operatorname{Cov}(V_k V_l) + 2 \sum_{k=1}^M \sum_{l=1}^M \lambda_{ik} \lambda_{jk} \lambda_{il} \lambda_{jl} \operatorname{Cov}(V_k V_l),$$

$$\lambda_{ik} = \frac{a_{ik}}{\|a_i\|}, \ \|a_i\|^2 = \sum_{k=1}^M a_{ik}^2, \tag{31}$$

where ρ_{ij} is a correlation between i-th and j-th risk factors. Formulas (29) and (31) clearly indicate that the correlation structure of the SV is embedded into the correlation structure of the fourth moments of the RF distribution. This connection will be used as the base for the GSV model calibration. A number L of the SV drivers can be chosen significantly smaller than a number of stochastic variances M and risk factors N. These SV drivers may be thought as "stochastic activities" for different countries, industries, sectors, etc.

The GSV model with the correlated stochastic variances is, in fact, a general framework. It can incorporate any reasonable processes to represent the SV drivers $\xi_k(t)$, $k = 1, \ldots, L$. Some examples of suitable one-dimensional SV driver distributions are: Inverse Gaussian distribution (Barndorff-Nielsen, 1997), Gamma distribution (Madan and Seneta, 1990; Levin and Tchernitser, 1999a), Lognormal distribution (Clark, 1973), or considered above class of Generalized Gamma distributions. The GSV model is practical in terms of effective Monte Carlo simulation: it is based on the simulation of one-dimensional SV processes and standard multivariate normal variables.

3.5.1. Example 1. Joint distribution for DEM/USD and JPY/USD FX rates

The first example presents a bivariate GSV model applied to the foreign exchange market data. Four bivariate models were examined for DEM/USD and JPY/USD FX rate daily returns: Standard Gaussian model, Elliptical Gamma Variance model, model with independent stochastic variances for PCs, and the model with correlated SV. Figures 12 and 13 show a 3-D plot and a contour plot of the joint probability density for the historical data and four types of the models considered. All three SV models provide a far better fit than the Gaussian distribution. However, the most convincing fit is provided by the GSV model with the correlated stochastic variances. Marginal distributions for DEM/USD and JPY/USD FX rates have kurtosis 5.2 and 6.9 respectively. Figures 14 and 15 show that the latter model is able to capture kurtosis and shape of marginal distributions in different directions.

3.5.2. Example 2. Twenty risk factors

The second example examines a 20-dimensional GSV model with correlated SV applied to the data from the interest rate, FX rate, and equity markets. The USD and CAD zero interest rate curves each consisting of nine interest rates, CAD/USD FX rate, and S&P 500 Index were chosen as a representative set of the risk factors. There were 5 years (1994–1999) of daily historical data used for the model calibration (about 1,250 data points). Figure 16 presents statistical results for principal component analysis and the correlation matrix for squares of the first three PCs. These results indicate that uncorrelated PCs neither are normal nor independent. The first three "largest" PCs per zero curve were used for the GSV model calibration and simulation. Three Gamma distributed drivers ξ_k, $k = 1, 2, 3$, with different shape parameters were utilized to represent each stochastic variance V_i, $i = 1, 2, 3$, for PCs. Therefore, the following values for parameters were assigned: number of risk factors $N = 9 \times 2 + 1 + 1 = 20$, number of principal risk factors $M = 3 \times 2 + 1 + 1 = 8$, number of SV drivers $L = 3$.

The model was calibrated to match kurtosis (in the least squares sense) for all 20 risk factors and kurtosis for 15 additional linear sub-portfolios. Sampling kurtosis varies within a wide range from 5 to 25. Typically, kurtosis for short-term interest rates is much higher than kurtosis for long-term rates. It is seen (Figure 17) that the GSV model reproduces this typical decreasing kurtosis term structure quite well. It is also seen that the model adequately matches kurtosis of the FX rate and S&P 500 Index, as well as kurtosis of different linear sub-portfolios. To compare, the standard multi-dimensional Gaussian model produces a flat kurtosis term structure identically equal to three.

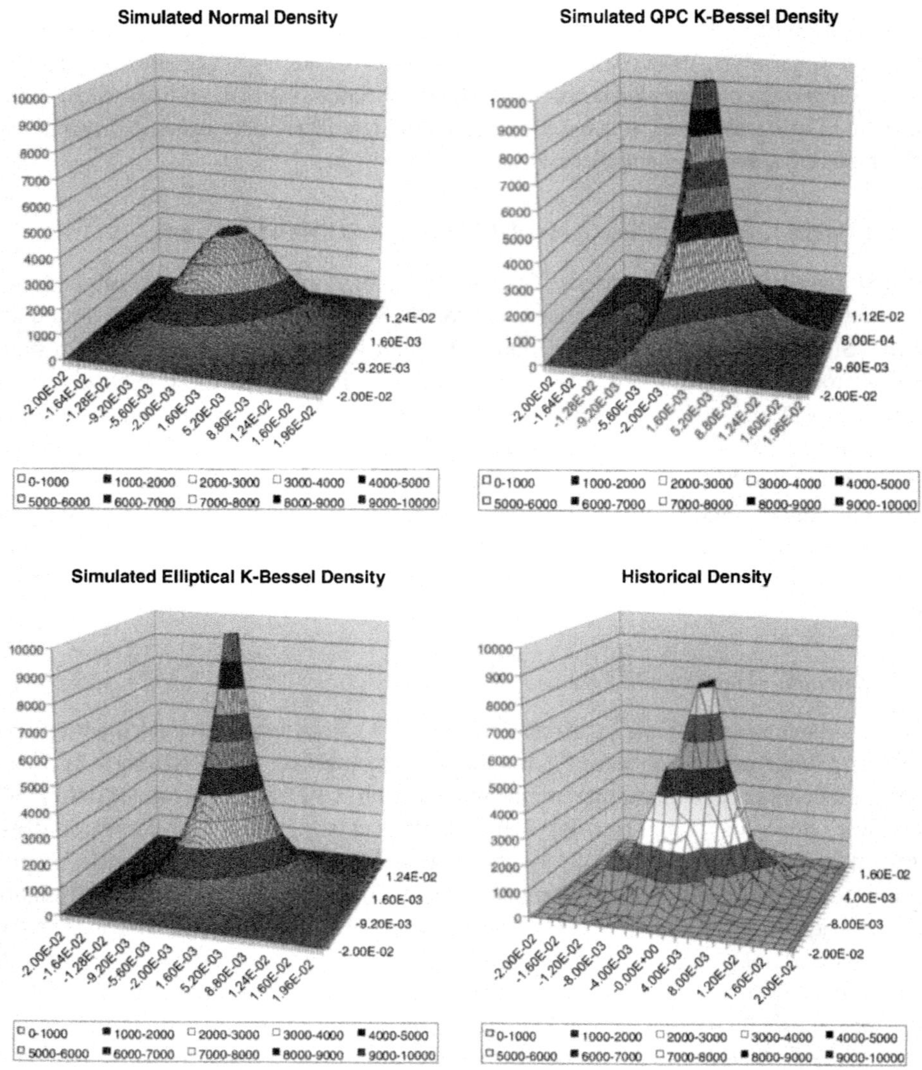

Fig. 12. Joint density for DEM/USD and JPY/USD FX rate.

3.6. *Calibration for the GSV model*

The GSV model is a two-level model that incorporates a traditional variance–covariance structure of the risk factors and novel variance–covariance structure of the RF stochastic variances. The GSV model with correlated SV automatically preserves the RF covariance matrix C. At the second level, it is necessary to calibrate the SV covariance matrix C_V to approximate the fourth moments of the multivariate RF distribution.

The main steps of the model calibration procedure are as follows:

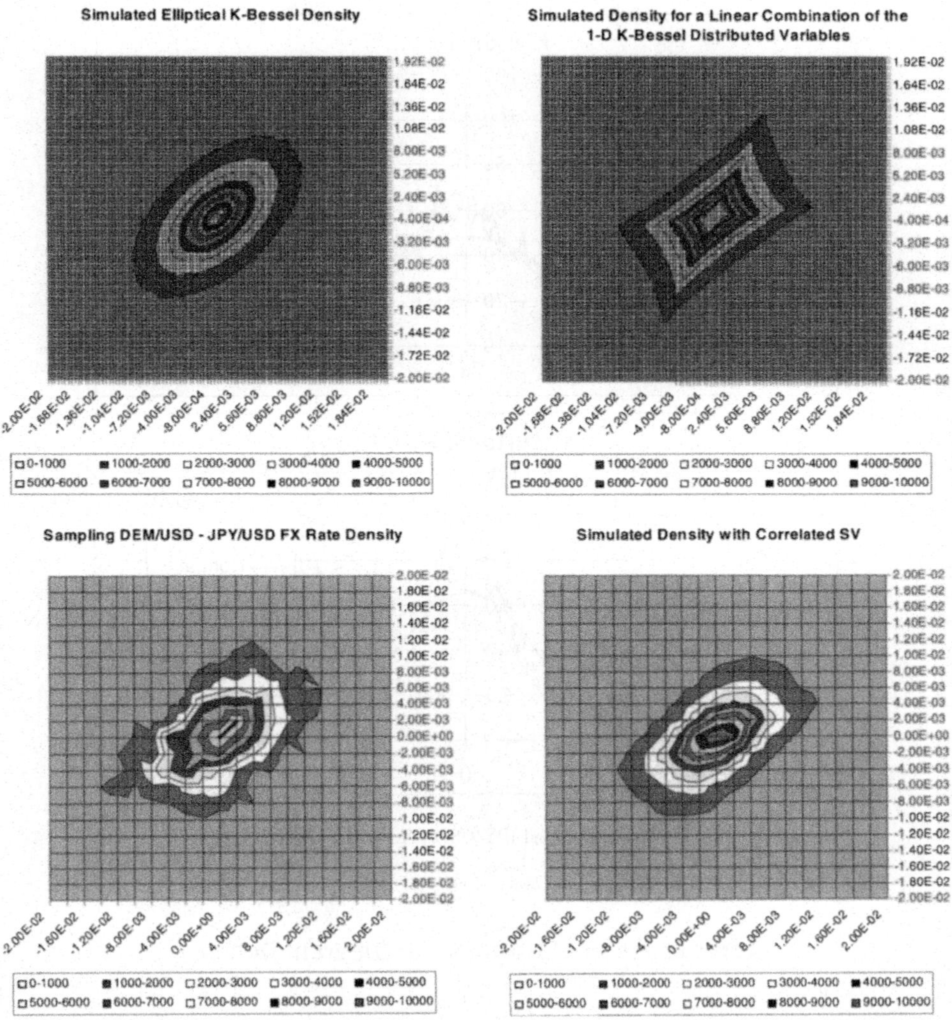

Fig. 13. Contour plot for the DEM/USD–JPY/USD FX rate joint density.

1. Calculate a sampling covariance matrix $C \in \mathbb{R}^{N\times N}$ for a given set of risk factors X. The time window usually used for calibration of the covariance matrix C is about 1–2 years. Exponentially weighted averaging or uniform sliding window are the usual methods for the covariance matrix calculation (JP Morgan, 1996).
2. Decompose the sampling covariance matrix C using a standard eigenvalue decomposition procedure and form a matrix $\tilde{A} \in \mathbb{R}^{N\times M}$ from a set of M eigenvectors corresponding to M largest eigenvalues. Number M has to be chosen to recover the matrix C with a required accuracy.

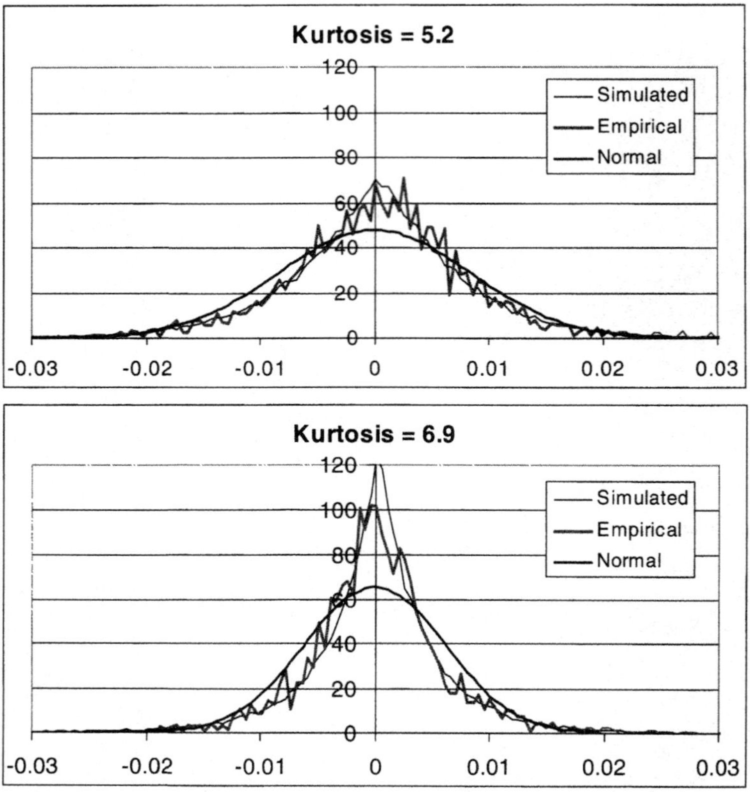

Fig. 14. DEM/USD and JPY/USD FX marginal distributions.

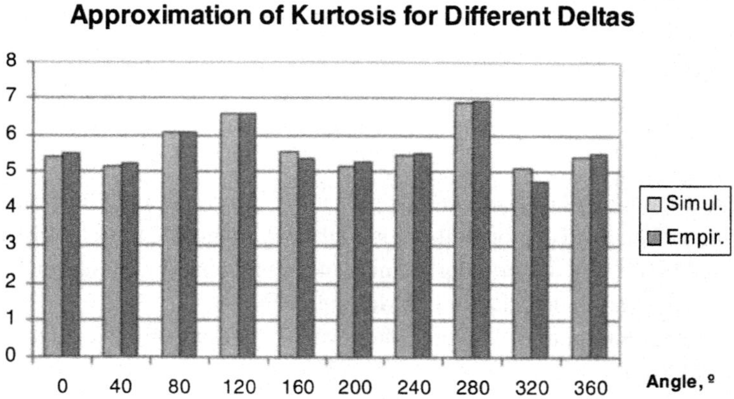

Fig. 15. Fit of the kurtosis for different sub-portfolios.

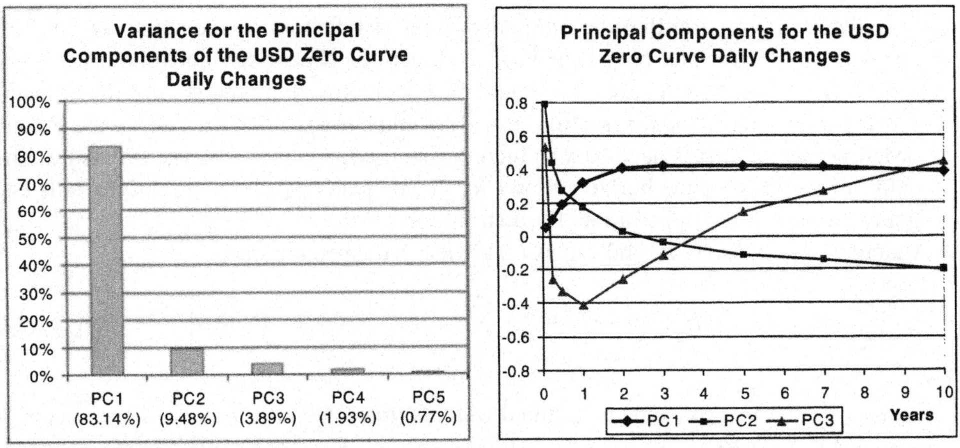

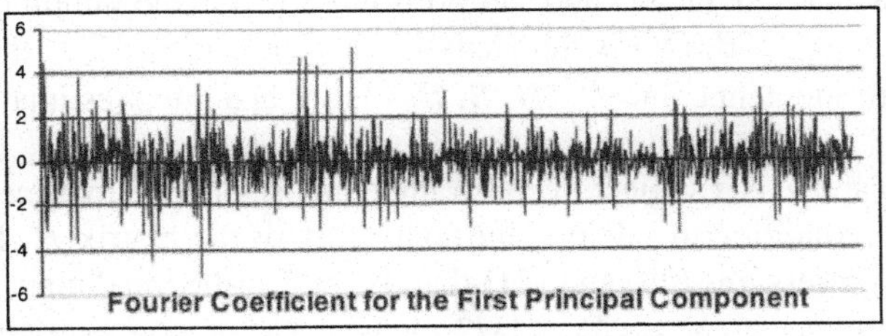

Correlation Matrix for the Fourier Coefficient Squared Daily Changes

	PC1	PC2	PC3
PC1	1.00	0.11	0.26
PC2	0.11	1.00	0.46
PC3	0.26	0.46	1.00

Fig. 16. PCA for USD zero curve.

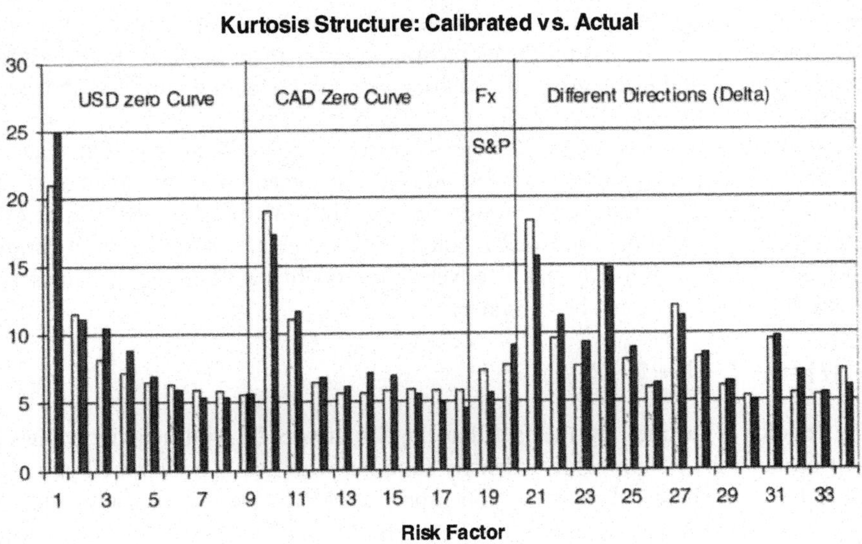

Fig. 17. Fit of the kurtosis.

3. Calculate sampling fourth order moments for the risk factors X (the matrix k_{ij} in (30)) and kurtosis k_Δ for any preselected set of directions (linear sub-portfolios) $\{\Delta\}$. The time window typically required for calculation of the fourth moments is of the order 5–10 years. This period of observations has to be much longer than the one for the second order moments. This is necessary to incorporate relatively rare extreme events into the calibration. Longer time horizon allows for an adequate approximation of the tails and general shape of the multivariate RF distribution.
4. Calculate matrices H, B, and D_ξ using the least squares approach:

$$\sum_i w_i \left(\hat{k}_{\Delta i} - k_{\Delta i}(H, B, D_\xi)\right)^2 + \sum_i \sum_j w_{ij} \left(\hat{k}^e_{ij} - k^e_{ij}(H, B, D_\xi)\right)^2 \to \min_{H,B,D_\xi}, \quad (32)$$

where w_i, and w_{ij} are some predefined weights (these weights may be chosen depending on the importance of particular risk factors and sub-portfolios), $\hat{k}_{\Delta i}$ is the sampling kurtosis for the direction Δ_i, $k_{\Delta i}(H, B, D_\xi)$ is the analytical estimate (29), $\hat{k}^e_{ij}$ is the sampling matrix of the fourth moments, and $k^e_{ij}(H, B, D_\xi)$ is its analytical estimate (31).

The minimization problem above is a subject to constraints imposed on the matrices H, B, D_ξ. The most difficult condition to satisfy is orthogonality of the matrix H. It follows from the analysis of expressions (29) and (31) that $M \times M$ orthogonal matrix H can be constructed as a product of $M \times (M-1)/2$ elementary rotation matrices (Wilkinson and Reinsch, 1971). It can be shown that for the problem (29), a representation for the orthogonal matrix H does not require reflections. The diagonal matrix D_ξ is subject to simple non-negativity constraints. The matrix B is subject to constraints (27). Hence, the non-linear optimization problem (32) can be re-formulated with respect to $M \times (M-1)/2$ angles φ_m for the elementary rotation matrices with simple constraints $-\pi \leqslant \varphi_m \leqslant \pi$ and elements of the matrices B and D_ξ with mentioned above simple constraints.
5. If the Gamma Variance model for the SV drivers ξ_k is adopted, the diagonal matrix D_ξ and conditions $E\{\xi_k(T_1)\} = 1$ determine the shape and scale parameters α_k and β_k in (6). For the GGV model, the powers $\nu_k \in \mathbb{R}^1$ have to be additionally specified. As a practical approach, the following methodology has been adopted: a set of parameters $\{\nu_k\}$ is fixed in such a way that it covers a reasonably wide range of values ν_k. For example, the set of ν_k can be chosen as

$$\{\nu_k\} = \{-2, -1, +1, +2\}.$$

This choice is justified by the fact that the SV drivers ξ_k with negative values of ν_k will produce the RF probability density function with heavy polynomial tails. On the other hand, positive values of ν_k can produce the RF distributions with semi-heavy exponential and sub-exponential tails, but with unbounded peaks at the origin. However, it is quite possible that a more flexible and adjustable structure for the set of parameters $\{\nu_k\}$ is more beneficial for the model calibration.

Acknowledgment

Authors thank C. Albanese, O. Barndorff-Nielsen, D. Duffie, P. Embrechts, H. Geman, D. Madan, J. Nolan, and, especially, M. Taqqu for interesting discussions and useful comments related to presented models.

References

Abramowitz, M., Stegun, I.A. (Eds.), 1972. Handbook of Mathematical Functions with Formulas, Graphs, and Mathematical Tables. National Bureau of Standards, Washington, DC.

Albanese, C., Jaimungal, S., Rubisov, D.H., 2001. A jump model with binomial volatility. Working Paper. University of Toronto–Math-Point, Toronto.

Albanese, C., Levin, A., Ching-Ming Chao, J., 1997. Bayesian Value at Risk, back-testing and calibration. Working paper. Bank of Montreal–University of Toronto, Toronto.

Anderson, T.W., 1984. An Introduction to Multivariate Statistical Analysis, 2nd edition. Wiley, New York.

Ané, T., Geman, H., 1999. Stochastic volatility and transaction time: an activity-based volatility estimator. Journal of Risk 2, 57–69.

Ané, T., Geman, H., 2000. Order flow, transaction clock and normality of asset returns. Journal of Finance 55, 2259–2284.

Artzner, P., Delbaen, F., Eber, J.-M., Heath, D., 1999. Coherent measures of risk. Mathematical Finance 9 (3), 203–228.

Bachelier, L., 1900. Theory of speculation. Ph.D. Thesis. English translation. In: Cootner, P.H. (Ed.), The Random Character of Stock Market Prices. MIT Press, Cambridge, MA, 1964.

Barndorff-Nielsen, O., 1977. Exponentially decreasing distributions for the logarithm of particle size. Proceedings of the Royal Society London. Series A 353, 401–419.

Barndorff-Nielsen, O., 1978. Hyperbolic distributions and distributions on hyperbolae. Scandinavian Journal of Statistics 5, 151–157.

Barndorff-Nielsen, O., 1997. Normal Inverse Gaussian distributions and stochastic volatility modelling. Scandinavian Journal of Statistics 24, 1–14.

Barndorff-Nielsen, O., 1998. Processes of Normal Inverse Gaussian type. Finance and Stochastics 2, 41–68.

Barndorff-Nielsen, O., Levendorskii, S., 2001. Feller processes on Normal Inverse Gaussian type. Quantitative Finance 1, 318–331.

Barndorff-Nielsen, O., Pérez-Abreu, V., 2000. Multivariate type **G** distributions. Working Paper. Centre for Mathematical Physics and Stochastics, University of Aarhus, Aarhus.

Barndorff-Nielsen, O., Shephard, N., 2000a. Non-Gaussian OU based models and some of their use in financial economics. Working Paper. Centre for Mathematical Physics and Stochastics, University of Aarhus, Aarhus.

Barndorff-Nielsen, O., Shephard, N., 2000b. Modelling by Lévy processes for financial econometrics. Working Paper. Centre for Mathematical Physics and Stochastics, University of Aarhus, Aarhus.

Basle Committee on Banking Supervision, 1996. Supervisory framework for the use of backtesting in conjunction with the internal models approach to market risk capital requirements. January, http://www.bis.org.

Basle Committee on Banking Supervision, 1997. International Convergence of Capital Measurements and Capital Standards. July 1988, amended in April 1997, http://www.bis.org.

Bates, D., 1991. The crash of '87: was it expected? The evidence from option markets. Journal of Finance 46, 1009–1044.

Bates, D., 1996. Jumps and stochastic volatility: exchange rate processes implicit in Deutsche Mark options. Review of Financial Studies 9, 69–107.

Bertoin, J., 1996. Lévy Processes. Cambridge University Press, Cambridge.

Black, F., Scholes, M., 1973. The pricing of options and corporate liabilities. Journal of Political Economy 81, 637–654.

Bollerslev, T., 1986. Generalized autoregressive conditional heteroskedasticity. Journal of Econometrics 31, 307–327.

Bollerslev, T., Engle, R., Nelson, D., 1994. ARCH models. In: Engle, R., McFadden, D. (Eds.), Handbook of Econometrics, Vol. 4. Elsevier, Amsterdam.

Bouchaud, J.-P., Potters, M., 2000. Theory of Financial Risks. Cambridge University Press, Cambridge.

Boyarchenko, S., Levendorskii, S., 2000. Option pricing for truncated Lévy processes. International Journal of Theoretical and Applied Finance 3, 549–552.

Buchen, P., Kelly, M., 1996. The maximum entropy distribution of an asset inferred from option prices. Journal of Financial and Quantitative Analysis 31, 143–159.

Carr, P., Geman, H., Madan, D., Yor, M., 2000. The fine structure of asset returns: an empirical investigation. Working paper. University of Maryland, College Park.

Clark, P., 1973. A subordinated stochastic process model with finite variance for speculative prices. Econometrica 41, 135–155.

Cont, R., 2001. Empirical properties of asset returns: stylized facts and statistical issues. Quantitative Finance 1, 223–236.

Cont, R., Potters M., Bouchaud J.-P., 1997. Scaling in stock market data: stable laws and beyond. In: Dubrulle, B., Graner, F., Sornette, D. (Eds.), Scale Invariance and Beyond, Proceedings of the CNRS Workshop on Scale Invariance. Springer.

Crouhy, M., Galai, D., Mark, R., 2001. Risk Management. McGraw-Hill, New York.

Duffie, D., Pan, J., 1997. An overview of Value at Risk. The Journal of Derivatives 4, 7–49.

Duffie, D., Pan, J., 2001. Analytical Value-at-Risk with jumps and credit risk. Finance and Stochastics 5, 155–180.

Eberlein, E., Keller, U., 1995. Hyperbolic distributions in finance. Bernoulli 1, 281–299.

Eberlein, E., Keller, U., Prause, K., 1998. New insights into smile, mispricing and Value at Risk: the hyperbolic model. Journal of Business 71, 371–405.

Eberlein, E., Raible, S., 1999. Term structure models driven by general Lévy processes. Mathematical Finance 9 (1), 31–53.

Embrechts, P., McNeil, A., Straumann, D., 1999. Correlation and dependency in risk management: properties and pitfalls. ETH Preprint, Zürich.

Fama, E., 1965. The behavior of stock market prices. Journal of Business 38, 34–105.

Fang, K., Kotz, S., Ng, K., 1990. Symmetric Multivariate and Related Distributions. Chapman and Hall, London.

Feller, W., 1966. An Introduction to Probability Theory and its Applications, Vol. 2. Wiley, New York.

Feuerverger, A., McDunnough, P., 1981. On efficiency of empirical characteristic function procedures. Journal of the Royal Statistical Society. Series B 43 (1), 20–27.

Fishman, G.S., 1996. Monte Carlo: Concepts, Algorithms and Applications. Springer-Verlag, New York.

Geman, H., Ané, T., 1996. Stochastic subordination. Risk 9 (September), 12–16.

Geman, H., Madan, D., Yor, M., 1998. Asset prices are Brownian motion: only in business time. Working paper. University of Maryland, College Park.

Geman, H., Madan, D., Yor, M., 1999. Time changes for Lévy processes. Working paper. University of Maryland, College Park.

Grosswald, E., 1976. The Student *t*-distribution of any degree of freedom is infinitely divisible. Zeitschrift für Wahrscheinlichkeitstheorie und Verwandte Gebiete 36, 103–109.

Heston, S., 1993. A closed-form solution for options with stochastic volatility, with applications to bond and currency options. Review of Financial Studies 6, 327–344.

Hull, J.C., 1999. Options, Futures, and Other Derivatives, 4th edition. Prentice-Hall, Upper Saddle River, NJ.

Hull, J., White, A., 1987. The pricing of options on assets with stochastic volatilities. Journal of Finance 42, 281–300.

Hull, J., White, A., 1998. Value at Risk when daily changes in market variables are not normally distributed. The Journal of Derivatives 5, 9–19.

Ismail, M., 1977. Bessel functions and the infinite divisibility of the Student *t*-distribution. The Annals of Probability 5, 582–585.

Ismail, M., Kelker, D., 1979. Special functions, Stieltjes transforms and infinite divisibility. SIAM Journal on Mathematical Analysis 10, 884–901.

Janicki, A., Weron, A., 1994. Simulation and Chaotic Behavior of α-Stable Stochastic Processes. Marcel Dekker, New York.

Johnson, N.L., Kotz, S., Balakrishnan, N., 1994. Continuous Univariate Distributions, Vol. 1, 2nd edition. Wiley, New York.

Johnson, N.L., Kotz, S., Balakrishnan, N., 1995. Continuous Univariate Distributions, Vol. 2, 2nd edition. Wiley, New York.

Jorion, P., 2001. Value-at-Risk: The New Benchmark for Managing Financial Risk, 3rd edition. McGraw-Hill, New York.

JP Morgan, 1996. RiskMetrics™ Technical Document, 4th edition. JP Morgan, New York.

Kagan, A.M., Linnik, Yu.V., Rao, C.R., 1973. Characterization Problems in Mathematical Statistics. Wiley, New York.

Kelker, D., 1971. Infinite divisibility and variance mixtures of the normal distributions. The Annals of Mathematical Statistics 42, 802–808.

Khindanova, I., Rachev, S., Schwartz, E., 2000. Stable modelling of Value at Risk. Working Paper. University of California, Santa Barbara.

Konikov, M., Madan, D., 2000. Pricing options of all strikes and maturities using a generalization of the VG model. Working paper, University of Maryland, College Park.

Kotz, S., Balakrishnan, N., Johnson, N.L., 2000. Continuous Multivariate Distributions, Vol. 1, 2nd edition. Wiley, New York.

Kotz, S., Kozubowski, T., Podgórski, K., 2001. The Laplace Distribution and Generalizations: A Revisit with Applications to Communications, Economics, Engineering, and Finance. Birkhäuser, Boston.

Kou, S.G., 2000. A jump diffusion model for option pricing with three properties: leptokurtic feature, volatility smile, and analytical tractability. Working Paper. Columbia University, New York.

Kreinin, A., Levin, A., 2000. Robust Monte Carlo simulation for approximate covariance matrices and VaR analysis. In: Uryasev, S. (Ed.), Probabilistic Constrained Optimization: Methodology and Applications. Kluwer Academic, Boston, pp. 166–178.

Levin, A., Tchernitser, A., 1999a. Multifactor stochastic variance Value-at-Risk model. Presentation at the Workshop Probability in Finance, Toronto, January 26–30, 1999. The Fields Institute, Toronto.

Levin, A., Tchernitser, A., 1999b. Multifactor Gamma stochastic variance Value-at-Risk model. Presentation at the Conference Applications of Heavy Tailed Distributions in Economics, Engineering and Statistics. Washington, DC, June 3–5, 1999. American University, Washington, DC.

Levin, A., Tchernitser, A., 2000a. A class of multifactor stochastic variance VaR models: Maximum Entropy approach and jump processes. Presentation at the First World Congress of the Bachelier Finance Society, Paris, June 28–July 1, 2000.

Levin, A., Tchernitser, A., 2000b. Stochastic volatility and jump Lévy processes in Value-at-Risk modelling. Presentation at the Risk Conference Math Week '2000, London, November 27–December 1, 2000.

Lukacs, E., 1970. Characteristic Functions. Griffin, London.

Madan, D., 1999. Purely discontinuous asset price processes. Working paper. University of Maryland, College Park.

Madan, D., Milne, F., 1991. Option pricing with VG martingale components. Mathematical Finance 1 (4), 39–55.

Madan, D., Seneta, E., 1990. The Variance Gamma (VG) model for share market returns. Journal of Business 63, 511–524.

Maejima, M., Rosiński, J., 2000. Type G distributions on $\mathbb{R}^d$. Working Paper. Keio University.

Mandelbrot, B., 1960. The Pareto–Lévy law and the distribution of income. International Economic Review 1, 79–106.

Mandelbrot, B., 1963. The variation of certain speculative prices. Journal of Business 36, 394–419.

Mandelbrot, B., 1997. Fractal and Scaling in Finance: Discontinuity, Concentration, Risk. Springer, New York.

Mantegna, R.N., Stanley, H.E., 1994. Stochastic process with ultraslow convergence to a Gaussian: the truncated Lévy flight. Physical Reviews Letters 73, 2946–2949.

Mantegna, R.N., Stanley, H.E., 2000. An Introduction to Econophysics: Correlations and Complexity in Finance. Cambridge University Press, Cambridge.

Marinelli, C., Rachev, S., Roll, R., 1999. Subordinated exchange rate models: evidence for heavy tailed distributions and long-range dependence. Working Paper. University of Padova.

Mausser, H., Rosen, D., 2000. Managing risk with expected shortfall. In: Uryasev, S. (Ed.), Probabilistic Constrained Optimization: Methodology and Applications. Kluwer Academic, Boston, pp. 204–225.

McCulloch, J., 1978. Continuous time processes with stable increments. Journal of Business 36, 394–419.

McCulloch, J., 1996. Financial applications of stable distributions. In: Maddala, G., Rao, C. (Eds.), Statistical Methods in Finance, Elsevier, Amsterdam, pp. 393–425.

Melino, A., Turnbull, S., 1990. Pricing foreign currency options with stochastic volatility. Journal of Econometrics 45, 239–265.

Merton, R., 1976. Option pricing when underlying stock returns are discontinuous. Journal of Financial Economics 3, 125–144.

Merton, R., 1990. Continuous-Time Finance. Blackwell, Cambridge.

Mittnik, S., Rachev, S., 1989. Stable distributions for asset returns. Applied Mathematics Letters 213, 301–304.

Müller, U., Dacorogna, M., Pictet, O., 1998. Heavy tails in high-frequency financial data. In: Adler, R., Feldman, R., Taqqu, M. (Eds.), A Practical Guide to Heavy Tails. Birkhäuser, Boston.

Nolan, J., 1998. Multivariate stable distributions: approximation, estimation, simulation and identification. In: Adler, R., Feldman, R., Taqqu, M. (Eds.), A Practical Guide to Heavy Tails. Birkhäuser, Boston, pp. 509–525.

Olver, F., 1974. Introduction to Asymptotics and Special Functions. Academic Press, New York.

Osborne, M., 1959. Brownian motion in the stock market. Operations Research 7, 145–173.

Platen, E., 1999. A Minimal Share Market model with stochastic volatility. Research Paper 21. University of Technology, Sydney.

Prudnikov, A., Brychkov, Yu., Marichev, O., 1986. Integrals and Series, Vol. 1. Gordon & Breach, New York.

Rachev, S., Mittnik, S., 2000. Stable Paretian Models in Finance. Wiley, Chichester, New York.

Rachev, S., SenGupta, A., 1993. Laplace–Weibull mixtures for modelling price changes. Management Science 39, 1029–1038.

Rosiński, J., 1991. On a class of infinitely divisible processes represented as mixtures of Gaussian processes. In: Cambanis, S., Samorodnitsky, G., Taqqu, M. (Eds.), Stable Processes and Related Topics. Birkhäuser, Boston, pp. 27–41.

Rosiński, J., 2001. Series representations of Lévy processes from the perspective of point processes In: Barndorff-Nielsen, O., Mikosch, T., Resnick, S. (Eds.), Lévy Processes – Theory and Applications. Birkhäuser, Boston, pp. 27–41.

Samorodnitsky, G., Taqqu, M., 1994. Stable Non-Gaussian Random Processes: Stochastic Models with Infinite Variance. Chapman & Hall, New York.

Samuelson, P., 1965. Rational theory of warrant pricing. Industrial Management Review 6, 13–32.

Sato, K.-I., 1999. Lévy Processes and Infinitely Divisible Distributions. Cambridge University Press, Cambridge.

Schoenberg, I., 1938. Metric spaces and completely monotonic functions. Annals of Mathematics 39, 811–841.

Sornette, D., Simonetti, P., Andersen, J.V., 2000. ϕ^q-field theory for portfolio optimization: "fat tails" and non-linear correlations. Physics Reports 335 (2), 19–92.

Stein, E., Stein, J., 1991. Stock price distributions with stochastic volatility: an analytic approach. Review of Financial Studies 4, 727–752.

Steutel, F., 1970. Preservation of infinite divisibility under mixing and related topics. In: Math. Centre Tracts, No. 33. Math. Centrum, Amsterdam.

Steutel, F., 1973. Some recent results in infinite divisibility. Stochastic Processes and their Applications 1, 125–143.

Wiggins, J., 1987. Option values under stochastic volatility: theory and empirical estimates. Journal of Financial Economics 19, 351–372.

Wilkinson, J.H., Reinsch, C., 1971. Linear Algebra, Vol. 2. In: Handbook for Automatic Computation. Springer-Verlag, New York.

Willinger, W., Taqqu, M., Teverovsky, V., 1999. Stock market prices and long-range dependence. Finance and Stochastics 3, 1–13.

Chapter 12

MODELLING THE TERM STRUCTURE OF MONETARY RATES*

LUISA IZZI

Credit Risk Department of Banca Nazionale del Lavoro, Department of Mathematics, Statistics and Information Technology, Faculty of Economics, and Centro Interdipartimentale Vito Volterra, University of Rome Tor Vergata, Italy

Contents

* The analysis and conclusions set forth in this chapter are those of the author and do not indicate concurrence by other members of the Banca Nazionale del Lavoro.

Handbook of Heavy Tailed Distributions in Finance, Edited by S.T. Rachev

Abstract

The chapter addresses the general problem of modeling and estimating the term structure of interest rates by adopting the use of jump-diffusion mean-reverting and stable Paretian models. The chapter proposes a new procedure to recursively compute interest rates subject to both Brownian and Poissonian noises. This procedure is consistent with the absence of arbitrage, non-negativity of interest rates, the mean-reverting hypothesis and the recombining condition, and can be calibrated with respect to any term structure which can be observed in the market. The numerical study shows that the proposed model is particularly suited to describe the behavior of European money market rates.

1. Introduction

The term structure of interest rates provides a characterization of interest rates as a function of maturity, which is mainly used in the pricing of fixed-income securities and for the valuation of contingent claims. The chapter addresses the general problem of the modeling and estimation of the term structure of interest rates. The problem is indeed of great importance, in view of its possible application to the analysis of Economic and Monetary Union monetary policy. It is this specific application that makes the problem of modeling interest rates difficult; in fact, the narrowness of the time horizons of interest rates time series has been one of the main obstacles in developing a solid quantitative analysis of the short-term interest rates in the Euro-zone.

The term structure has been modeled extensively using primarily pure-Gaussian models. Notable examples of these kinds of models are those of Merton (1973), Cox (1975), Cox and Ross (1976), Vasicek (1977), Dothan (1978), Brennan and Schwartz (1980), Cox, Ingersoll and Ross (1980, 1985) and Black, Derman and Toy (1990). Recent empirical works show that models which accommodate skewness and kurtosis, such as stochastic volatility [see, e.g., Bates (1996) and Britten-Jones and Neuberger (2000)] and jump-diffusion models (Ahn and Thompson, 1988; Ho, Perraudin and Sorensen, 1996; Scott, 1997), appear to fit the time series of interest rates better than pure-Gaussian type ones.[1] In fact, unlike jump-diffusion processes, stochastic volatility models are not capable of generating high levels of skewness and kurtosis at short maturities under reasonable parameterizations. As a consequence, stochastic volatility models cannot generate as sharp implied volatility "smiles" as those typically observed empirically.[2]

It should also be noted that conditional skewness and kurtosis in stochastic volatility models are always hump-shaped in the length of horizon: indeed, for plausible parameter values, both quantities must be increasing over short to moderate maturities. This implies that the smile does not flatten out appreciably as maturity increases. Moreover, the kurtosis of changes in interest rates increases when the data is sampled daily instead of monthly, and interest rates often display discontinuous behavior, partly on account of the discrete changes the Central Bank makes in the official rates [see Baz and Das (1996), Das and Foresi (1996), Björk, Kabanov and Runggaldier (1997), Attari (1999)]. Modeling the term structure using a jump-diffusion process captures these market features better, because with these processes the variance shrinks with the time interval, yet the size of jumps remains

[1] The class of jump-diffusion models augments the Black–Scholes (1973) return distribution with a Poisson-driven jump process; while the class of stochastic volatility models extends the Black–Scholes model by allowing the volatility of the return process itself to evolve randomly over time.

[2] The "smile" is the plot of implied volatilities from a range of options of the same maturity across different strike prices. It can be observed that at-the-money options seem to trade at the lowest implied volatilities and the in-the-money and out-of-the-money options trade at higher volatilities. Since the options are all written on the same underlying variable, there should be no plausible reason for this, other than the fact that the model is inexact. Since the observed distribution of interest rates has fatter tails than that assumed by a pure-Gaussian model, such an effect is intuitively obvious. When plotted against the strike price, the graph of implied volatilities appears U-shaped like a smile. Hence the terminology.

the same, enhancing kurtosis (which is the relative size of the outliers to the variance of the process).

In this chapter a jump-diffusion mean-reverting model for estimating monetary rates is introduced and related to the class of models driven by infinitely divisible processes [see Gnedenko (1992)] to which Gaussian, Poissonian and Stable ones [see Samorodnitsky and Taqqu (1994)] belong. Relating interest rate models to the class of Stable processes is an attempt to unify, from the probabilistic point of view, the extensive literature on pure-diffusion, pure-Poissonian and jump-diffusion processes. It completes recent works on the stock [see Schumacher (1997), Rieken (1999)] and exchange markets [see Akgiray and Booth (1988), Rachev and Mittnik (2000)] concerning the same topic.

We also propose a new numerical procedure to recursively compute interest rates subject to both Brownian and Poissonian noises. This procedure – which generalizes the trinomial approach proposed by Hull and White (1994, 1996) in the presence of pure-diffusion stochastic components – is consistent with the absence of arbitrage, non-negativity of interest rates, the mean-reverting hypothesis and the recombining condition. It can be calibrated with respect to any term structure which can be observed in the market, for instance, the present one.

Numerical experiments demonstrate how the jump-diffusion mean-reverting model is particularly suited to describe the behavior of European money market rates. Interest rates controlled by the monetary authorities behave like jump processes and the term structure, at short maturity, is contingent upon the levels of these official rates.

The rest of the chapter is organized as follows. Section 2 introduces the interest rate model used to describe the behavior of monetary rates both in continuous and in discrete time. To provide a direct application of the model, a new numerical procedure is proposed in Section 3. The new method determines a unique fully specified hexanomial tree, which is consistent with respect to the risk neutral probability. Econometric estimations are related in Section 4. The estimation of the mean-reverting jump-diffusion process is obtained via the indirect inference method. Finally, Section 5 concludes the chapter.

2. The mathematical framework

2.1. Model setup and notation

Let Ω be a (rich enough) sample space, representing the state-of-the-world, and let $X_t : \Omega \to \Re$ be a stochastic process on Ω such that

$$X_t = \sigma B_t + \sum_{i=1}^{n} a_i N_t^{(i)} + \eta, \tag{1}$$

where $t \in [0, T]$ and T is a finite horizon; B_t is a Brownian motion; $\sigma \geqslant 0$ is constant; $n \geqslant 1$ is fixed; $N_t^{(i)}$ are mutually independent Poisson processes, also independent of B_t, with parameters $\lambda_i > 0$, for $i = 1, \ldots, n$; $a_i \sim \mathrm{Bin}(\mu_i, \gamma_i^2)$ is independent of any other variable, with μ_i and $\gamma_i \in \Re$, and represents the jump size of the corresponding Poisson process, for $i = 1, \ldots, n$; η is a shift parameter. Let $\mathcal{F} = \{\mathcal{F}_t\}_{0 \leqslant t \leqslant T}$ be the filtration generated by $\{X_t\}_{0 \leqslant t \leqslant T}$ (the information structure on Ω), that is $\mathcal{F}_t$ is the σ-field generated by $\{X_s\}_{0 \leqslant s \leqslant t}$, where the initial filtration $\mathcal{F}_0$ is trivial, and P a probability measure on $(\Omega, \mathcal{F}_T)$. Let $r_t : \Omega \to \Re_+$ be a stochastic process on the filtered probability space $(\Omega, \{\mathcal{F}_t\}, P)$ described by the stochastic differential equation:

$$\mathrm{d}r_t = \alpha_t \, \mathrm{d}t + \beta_t \, \mathrm{d}X_t \quad (t \in [0, T]), \tag{2}$$

where $\alpha_t = k(\theta_t - r_t)$ is the drift function (or conditional mean); θ_t is a positive bounded real function of time; k is the calling back strength of the θ_t rate on the instantaneous rate r_t; β_t is the diffusion function of the interest rate change (β_t^2 is the conditional variance).

So, we can redefine the model as follows:

$$\mathrm{d}r_t = k(\theta_t - r_t)\, \mathrm{d}t + \beta_t[\sigma \, \mathrm{d}B_t + a \, \mathrm{d}N_t] \quad (t \in [0, T]), \tag{3}$$

where $a \, \mathrm{d}N_t = \sum_{i=1}^{n} a_i \, \mathrm{d}N_t^{(i)}$; $a = [a_1, a_2, \ldots, a_n]$; $N_t = [N_t^{(1)}, N_t^{(2)}, \ldots, N_t^{(n)}]^{\mathrm{T}}$.

Remark 2.1. Let us consider the particular case in which $n = 2$ and $a_1 \sim \mathrm{Bin}(\mu_1, \gamma_1^2)$ and $a_2 \sim \mathrm{Bin}(\mu_2, \gamma_2^2)$ are independent of any other random variable, with $\mu_1 > 0$, $\mu_2 < 0$, $\gamma_1, \gamma_2 \in \Re$. Then, the stochastic process X_t has the form:

$$X_t = \sigma B_t + a_1 N_t^{(1)} + a_2 N_t^{(2)} + \eta \quad (t \in [0, T])$$

and the two Poisson processes $N_t^{(1)}$ and $N_t^{(2)}$ could represent the actions of the European Central Bank ($N_t^{(1)}$ "up" and $N_t^{(2)}$ "down") on the official rates. Moreover, the stochastic differential equation:

$$\mathrm{d}r_t = k(\theta_t - r_t)\, \mathrm{d}t + \beta_t\big[\sigma \, \mathrm{d}B_t + a_1 \, \mathrm{d}N_t^{(1)} + a_2 \, \mathrm{d}N_t^{(2)}\big] \quad (t \in [0, T])$$

is subject to the constraint: $|\theta_t - r_t| \leqslant u_t - d_t$, where θ_t is the rate of the main refinancing operation of the European Central Bank (ECB); u_t is the rate of the marginal lending facility; d_t is the rate of the deposit facility (see Figure 1). In fact, money market rates should not exceed the corridor fixed by u_t (upper bound) and d_t (lower bound), because under normal conditions the ECB gives banks all the money they need at the u_t rate. The only limit is determined by the availability of collateral to be presented to the ECB as a guarantee of the credit received. In such a way, if the collateral available to banks has no limits, also the credit that banks receive from the ECB at the u_t rate has no limits. Similarly, at the end of every working day, banks can deposit an unlimited amount of money in the ECB accounts at the d_t rate.[3]

[3] EONIA (the European overnight rate which is computed as an average value by the ECB) has gone up to the u_t rate only in very few occasions due to market inefficiency problems.

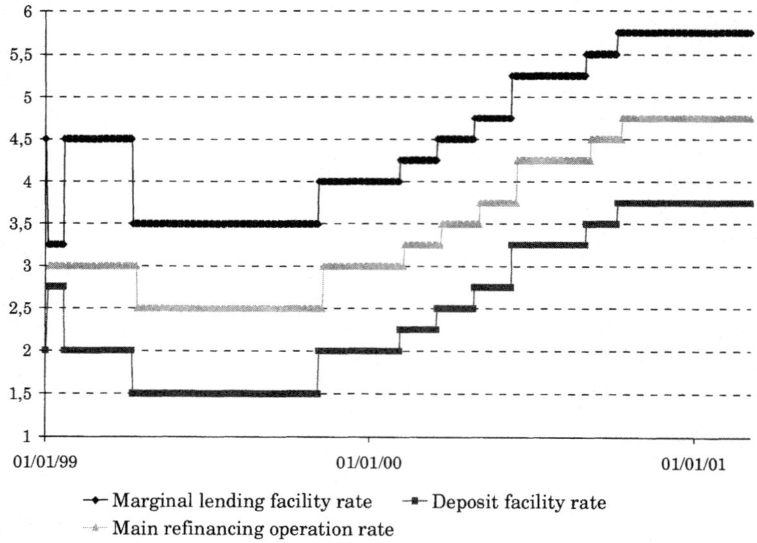

Fig. 1. Official rates during the ten policy periods (percentage values).

2.2. *Regularity conditions on the jump-diffusion process*

The jump-diffusion process defined in (3) can be used to approximate a wide range of Markovian or non-Markovian processes. In particular, the weak convergence of a driving noise process to a Lévy process implies the weak convergence of the driving noise process to a jump-diffusion process [see, for example, Blasikiewicz and Brown (1996)]. In the general case, the Brownian motion B_t and the compound Poisson process $\int_0^t a_\tau \, \mathrm{d}N_\tau(\bar{\lambda})$ are infinitely divisible in time, when appropriately scaled, and have independent increments. It is noted, however, that even though B_t is a martingale process, a compound Poisson process is in general not a martingale. Define the *compensated Poisson process*:

$$\mathrm{d}\widetilde{N}_t(\bar{\lambda}) = \mathrm{d}N_t(\bar{\lambda}) - \bar{\lambda}\,\mathrm{d}t$$

since

$$E\left\{\int_t^{t+\Delta t} a_\tau \, \mathrm{d}\widetilde{N}_\tau(\bar{\lambda})\right\} = E\left\{\int_t^{t+\Delta t} a_\tau \, \mathrm{d}N_\tau(\bar{\lambda}) - \bar{\lambda}\,\mathrm{d}\tau\right\} = 0,$$

i.e., $\int_0^t a_\tau \, \mathrm{d}\widetilde{N}_\tau(\bar{\lambda})$ is a martingale process. Therefore, without loss of generality, we only need to consider the case in which the compound Poisson process is a martingale process.[4] To ensure that the generalized stochastic differential equation defined in (3) is well behaved, it is necessary to make two additional assumptions.

[4] For further details, see Zacks et al. (1999).

Assumption 2.1. The coefficient functions $\alpha(\cdot)$ and $\beta(\cdot)$ are measurable in the product σ-algebra $\mathcal{B} \times \mathcal{F}$, where $\mathcal{B}$ is the σ-field of the Borel sets on $\Re$, and furthermore

$$\int_0^{\mathrm{T}} |\alpha_t| \, \mathrm{d}t < \infty \quad \text{a.s.}, \qquad \int_0^{\mathrm{T}} \beta_t^2 \, \mathrm{d}t < \infty \quad \text{a.s.}$$

and

$$E\left\{\int_0^{\mathrm{T}} \beta_t^2 \, \mathrm{d}t\right\} < \infty \quad \text{a.s.}$$

To ensure the existence and uniqueness of a solution to the stochastic differential equation (3), it is necessary to impose:

Assumption 2.2. The coefficient functions $\alpha(\cdot)$ and $\beta(\cdot)$ satisfy both Lipschitz and linear growth conditions, i.e., there exists a positive constant K for which

$$\begin{aligned} &\left|\alpha(r,t) - \alpha(r',t)\right| + \left|\beta(r,t) - \beta(r',t)\right| \leqslant K|r - r'|, \\ &\left|\alpha(r,t) - \alpha(r,t')\right| + \left|\beta(r,t) - \beta(r,t')\right| \leqslant K|t - t'| \end{aligned}$$

and

$$\alpha^2(r,t) + \beta^2(r,t) \leqslant K\left(1 + r^2\right).$$

Let us consider the model:

$$\mathrm{d}r_t = k(\theta_t - r_t)\,\mathrm{d}t + \sigma\,\mathrm{d}B_t + a\,\mathrm{d}N_t \quad (t \in [0,T]), \tag{4}$$

where the diffusion function β_t is assumed to be equal to 1; $\sigma \geqslant 0$ is the volatility of the interest rate, provided that no jumps occur; $N_t(\bar{\lambda})$ is a Poisson process with an intensity parameter $\bar{\lambda}$; a is the jump size; $\mathrm{d}B_t$ and $\mathrm{d}N_t$ are statistically independent.

2.3. *Interest rates with non-identically distributed jumps*

Let us focus our attention on the stochastic part of the interest rate process, i.e., on X_t, defined by

$$X_t = \sigma B_t + \sum_{i=1}^{n} a_i N_t^{(i)} + \eta \quad (t \in [0,T] \text{ and } n \geqslant 1). \tag{5}$$

Following Rachev and Rüschendorf (1994) and Schumacher (1997), we can study X_t as an *infinitely divisible* random variable.[5]

[5] For further details, see Rachev and Mittnik (2000) and Rieken (1999).

Definition 2.1. A sequence $\{X_{n,k}\}$ of random variables (with $n \geqslant 1$ and $k = 1, \ldots, n$) is *uniformly asymptotically negligible* (u.a.n.) if

$$X_{n,k} \xrightarrow{P} 0 \quad (n \to \infty),$$

uniformly in k, or equivalently if for every $\varepsilon > 0$

$$\max_{1 \leqslant k \leqslant n} P\{|X_{n,k}| \geqslant \varepsilon\} \to 0 \quad (n \to \infty).$$

Let $F_{n,k}$ denote the cumulative distribution function of $X_{n,k}$ and $\phi_{n,k}$ its characteristic function. It can be shown [see Rachev and Mittnik (2000) and the references therein] that the u.a.n. condition is equivalent to

$$\max_{1 \leqslant k \leqslant n} \int \frac{x^2}{1+x^2} \, \mathrm{d}F_{n,k} \to 0 \quad (n \to \infty)$$

or

$$\max_{1 \leqslant k \leqslant n} |\phi_{n,k} - 1| \to 0 \quad (n \to \infty),$$

uniformly on every finite interval.

The use of u.a.n. random variables restricts the class of summands, but this restricted class, known as the class of *infinitely divisible* random variables, is that investigated in the classical central limit problem. It is of primary interest in modeling interest rates and includes the most commonly models, i.e., Normal, Poisson, Stable and others.

Definition 2.2. A random variable X, its distribution F_X, and its characteristic function ϕ_X are said to be *infinitely divisible* (I.D.) if, for every $n \geqslant 1$, there are independent and identically distributed (i.i.d.) random variables $\eta_1, \eta_2, \ldots, \eta_n$, with distribution function F_η and characteristic function ϕ_η each, such that

$$X \stackrel{d}{=} \eta_1 + \cdots + \eta_n,$$

or equivalently if, for every $n \geqslant 1$, F_X is the n-fold convolution of F_η, i.e.,

$$F_X = F_\eta * \cdots * F_\eta = (F_\eta)^{n*},$$

or equivalently if, for every $n \geqslant 1$, its characteristic function ϕ_X is the n-th power of the characteristic function ϕ_η.

From Definition 2.2 it follows immediately that the Normal, the Poisson and the degenerate distributions, i.e., all possible limits of $\sum_{k=1}^{n} X_{n,k}$ in the i.i.d. case, belong to this class.

The I.D. law is uniquely defined by the Lévy–Khintchine representation of its characteristic function [see Rachev and Mittnik (2000)]:

$$\phi_X(u) = E\{e^{iuX}\} = \exp\left\{iu\alpha + \int_{-\infty}^{+\infty}\left(e^{iux} - 1 - \frac{iux}{1+x^2}\right)\frac{1+x^2}{x^2}\,d\psi(x)\right\}, \tag{6}$$

where $\alpha \in \Re$ and ψ is a distribution function up to a multiplicative constant. The characteristic function (6) describes the stochastic process which drives the interest rate in the natural world. Hence, α and ψ can be estimated from the observed data. In the case in which the variance of X is finite, the characteristic function (6) simplifies to:

$$\phi_X(u) = \exp\left\{iu\alpha + \int_{-\infty}^{+\infty}\left(e^{iux} - 1 - iux\right)\frac{1}{x^2}\,d\psi(x)\right\}. \tag{7}$$

The finite variance I.D. model with characteristic function (7) contains the Normal and the Poisson distributions as special cases. In fact, suppose that

$$\psi(x) = \sigma I_{[0,\infty)}(x),$$

where $I_{[a,\infty)}(x)$ denotes the indicator function of the interval $[a,\infty)$, i.e.,

$$I_{[a,\infty)}(x) = \begin{cases} 0, & \text{if } x < a, \\ 1, & \text{if } x \geqslant a, \end{cases}$$

with $a, x \in \Re$. In this case, the integral in (7) becomes $-\sigma^2 t^2/2$ and the characteristic function reduces to:

$$\phi_X(u) = \exp\left\{iu\alpha - \frac{\sigma^2 t^2}{2}\right\}, \tag{8}$$

which is the characteristic function of a Normal random variable with mean α and variance σ^2. The Poisson distribution is obtained from (7) by setting

$$\psi(x) = \lambda K^2 I_{[K,\infty)}(x),$$

with $\lambda > 0$ and $K \neq 0$. In this case, the characteristic function becomes:

$$\phi_X(u) = \exp\{\lambda(e^{iuK} - 1 - iuK)\}. \tag{9}$$

The relation (9) defines the characteristic function of a scaled and shifted Poisson random variable of the form $K(N_\lambda - \lambda)$, where N_λ is a Poisson random variable with parameter λ, i.e.,

$$P\{N_\lambda = k\} = \frac{\lambda\, e^{-\lambda}}{k!},$$

with $\lambda > 0$ and $k = 1, 2, \ldots$. Yet, the characteristic function of N_λ is:

$$\phi_{N_\lambda}(u) = \exp\{\lambda(\mathrm{e}^{\mathrm{i}u} - 1)\}. \tag{10}$$

Hence, the finite-variance I.D. distribution defined by (7) can be interpreted as an infinite mixture of independent Poisson random variables and one independent Gaussian random variable. For practical purposes, the I.D. model can, for example, be specified as a Normal random variable and a finite number n of independent Poisson random variables, as defined in (5), i.e.,

$$X_t = \sigma B_t + \sum_{i=1}^{n} a_i N_t^{(i)} + \eta, \tag{11}$$

where $t \in [0, T]$; $n \geqslant 1$; $\sigma \geqslant 0$; B_t is a standard Normal random variable; $N_t^{(i)}$ are independent Poisson random variables (independent from B_t), respectively, of parameters λ_i, for $i = 1, \ldots, n$; $a_i \sim \mathrm{Bin}(\mu_i, \gamma_i^2)$ are independent of any other variable, with $\mu_i, \gamma_i \in \Re$, for $i = 1, \ldots, n$; $\eta \in \Re$.

2.4. *The smile effect and infinitely divisible distributions*

If the Black–Scholes (1973) model of pricing contingent claims were true, the implied volatility would be constant for all claims. This, however, is not observed in practice. Instead, the phenomenon of a volatility smile is observed on call options on increasing strike prices. That is, deep-in-the-money call options (options with a strike price $K \ll X$, where X is the current price of the underlying asset) and deep-out-of-the-money options have an implied volatility parameter σ (the scale parameter in the Normal distribution) which exceeds the implied volatility parameters of at-the-money options. Said another way, $\sigma(K)$ appears to be a convex function of K. Said still a third way, the Black–Scholes formula underprices deep-in-the-money and deep-out-of-the-money call options, whereas it overprices at the money options. A related issue is the leptokurtic nature of financial asset values. That is, it is also widely observed that asset values tend to be more highly peaked and have fatter tails than is implied by the Normal model. The use of infinitely divisible distributions may help to explain the phenomenon of leptokurtosis as well as the smile effect.[6]

Definition 2.3. The coefficient of kurtosis α_4 of the random variable X is defined as

$$\alpha_4 = \frac{\mu_4}{\mu_2^2},$$

where μ_4 and μ_2 are the fourth and second central moments of X, respectively.

[6] See Schumacher (1997) and Rachev and Mittnik (2000) for further details.

Remark 2.2. The existence of a moment generating function of X is enough to ensure the finiteness of the central moments.

The coefficient of kurtosis is a measure of the peakedness of the distribution. It can be shown that for a Normally distributed random variable X, the coefficient of kurtosis $\alpha_4 = 3$. Given the leptokurtotic nature of asset prices, we might find it desirable that our model of interest rate behavior have a coefficient of kurtosis $\alpha_4 \geqslant 3$.

Theorem 2.1. *Let $\alpha_4(X)$ be the coefficient of kurtosis of an infinitely divisible random variable X. Then $\alpha_4(X) \geqslant 3$.*

Proof: Let $Y_n = \sum_{i=1}^{n}(a_i N_i + b_i)$ where N_i are independent Poisson random variables with rate λ_i. The log characteristic function of Y_n is then

$$\log \phi_{Y_n}(u) = \sum_{i=1}^{n}\left[\lambda_i\left(\mathrm{e}^{\mathrm{i}a_i u} - 1\right) + \mathrm{i}ub_i\right].$$

The k-th derivative (for $k \geqslant 2$) of the log characteristic function is

$$\frac{\mathrm{d}^{(k)} \log \phi_{Y_n}(u)}{\mathrm{d}u^{(k)}} = \sum_{i=1}^{n} \lambda_i (\mathrm{i}a_i)^k \,\mathrm{e}^{\mathrm{i}a_i u}.$$

Hence, we obtain the semi-invariants of order j, for $j \geqslant 2$ for the finite sum of scaled and shifted Poisson random variables:

$$\kappa_j = \sum_{i=1}^{n} a_i^j \lambda_i.$$

The semi-invariants can be expressed in terms of the central moments as

$$\kappa_2 = \mu_2 = \sigma^2, \qquad \kappa_3 = \mu_3, \qquad \kappa_4 = \mu_4 - 3\mu_2^2,$$

and so on. From this it follows that the coefficient of kurtosis for the given sum is given by

$$\alpha_4 = \frac{\mu_4}{\mu_2^2} = \frac{\sum_{i=1}^{n} a_i^4 \lambda_i + 3\left(\sum_{i=1}^{n} a_i^2 \lambda_i\right)^2}{\left(\sum_{i=1}^{n} a_i^2 \lambda_i\right)^2} \geqslant 3.$$

Now by a remark of Gnedenko (1992), all infinitely divisible random variables are limits of sums of scaled and shifted Poisson random variables; so the theorem follows. □

2.5. The discrete-time process

Let us divide the time interval $[0, T]$ into M ($M \geqslant 1$) sub-intervals of equal length Δt, i.e., $[t_i, t_{i+1})$, where $i = 0, 1, \ldots, M-1$, $t_i = i\Delta t$ and $\Delta t = T/M$. Then, for $i = 0, 1, \ldots, M-1$, the stochastic differential equation (4) admits the discretization:

$$r_{t_{i+1}}^{(M)} - r_{t_i}^{(M)} = k\big(\theta_{t_i}^{(M)} - r_{t_i}^{(M)}\big)\Delta t + \sigma\big(B_{t_{i+1}}^{(M)} - B_{t_i}^{(M)}\big) + a\big(N_{t_{i+1}}^{(M)} - N_{t_i}^{(M)}\big), \tag{12}$$

where $\{B_{t_{i+1}}^{(M)} - B_{t_i}^{(M)}\}_{i=0}^{M-1}$ are i.i.d. random variables, distributed as $N(0, \Delta t)$;

$$a\big(N_{t_{i+1}}^{(M)} - N_{t_i}^{(M)}\big) = \sum_{j=1}^{n} a_j\big(N_{t_{i+1}}^{(M),(j)} - N_{t_i}^{(M),(j)}\big);$$

$\{N_{t_{i+1}}^{(M),(j)} - N_{t_i}^{(M),(j)}\}_{i=0}^{M-1}$ are i.i.d. random variables, distributed as Poisson $(\bar{\lambda}^{(j)}\Delta t)$, for any $j = 1, \ldots, n$. As $M \to \infty$, $r_{t_i}^{(M)}$ ($i = 0, 1, \ldots, M$), as defined in (12), converges in weak sense to the process r_t ($t \in [0, T]$), satisfying (4). In the following, the index M will be omitted for simplicity.

Let μ and $\gamma \in \Re$ and set, for notational convenience, $\lambda = \bar{\lambda}\Delta t$. Following Das (1997), let us define the four sequences of random variables:

$$\overline{X}_{t_{i+1}}^{(1)} = \frac{1}{\sqrt{\Delta t}}(B_{t_{i+1}} - B_{t_i}) \quad (i = 0, 1, \ldots, M-1),$$

where $\{\overline{X}_{t_i}^{(1)}\}_{i=0}^{M}$ are i.i.d. standard Normal random variables;

$$X_{t_i}^{(2)} = \begin{cases} \mu + \gamma, & \text{w. prob. } \frac{\lambda}{2}, \\ 0, & \text{w. prob. } 1 - \lambda, \\ \mu - \gamma, & \text{w. prob. } \frac{\lambda}{2}, \end{cases}$$

i.i.d. for $i = 0, 1, \ldots, M$ and with $\{X_{t_i}^{(2)}\}_{i=0}^{M}$ and $\{\overline{X}_{t_i}^{(1)}\}_{i=0}^{M}$ mutually independent;

$$\overline{X}_{t_i}^{(2)} = \begin{cases} \mu + \gamma, & \text{w. prob. } \frac{1}{2}, \\ \mu - \gamma, & \text{w. prob. } \frac{1}{2}, \end{cases}$$

i.i.d. for $i = 0, 1, \ldots, M$;

$$\mathrm{Bin}(\lambda)_{t_i} = \begin{cases} 1, & \text{w. prob. } \lambda, \\ 0, & \text{w. prob. } 1 - \lambda, \end{cases}$$

i.i.d. for $i = 0, 1, \ldots, M$ and with $\{\mathrm{Bin}(\lambda)_{t_i}\}_{i=0}^{M}$ and $\{\overline{X}_{t_i}^{(2)}\}_{i=0}^{M}$ mutually independent. It is easy to show that $\{X_{t_i}^{(2)}\}_{i=0}^{M} \overset{d}{=} \{\overline{X}_{t_i}^{(2)}\,\mathrm{Bin}(\lambda)_{t_i}\}_{i=0}^{M}$.

Let $\mathcal{D}$ be the metric space of right continuous real-valued functions on $[0, T]$, with left limits. Let the interest rate process be defined on the path space $\mathcal{D}[0, T]$. Let $\mathcal{C}[0, T]$ be the subspace of $\mathcal{D}[0, T]$ of all real-valued continuous functions on $[0, T]$. The space $\mathcal{D}$ is restricted to the Skorokhod topology which, when restricted to the space $\mathcal{C}[0, T]$, is the topology of uniform convergence.[7]

By the Functional Limit Theorem for stochastic differential equations, to approximate the spot rate process as defined in (4), in weak sense in $\mathcal{D}[0, T]$, with the process r_{t_i} $(i = 0, 1, \ldots, M)$ as defined in (12), we can replace $\{\sigma \overline{X}_{t_i}^{(1)} \sqrt{\Delta t}\}_{i=0}^{M}$ by the sequence of i.i.d. random variables $\{X_{t_i}^{(1)}\}_{i=0}^{M}$ distributed as follows:

$$X_{t_i}^{(1)} = \begin{cases} +\sigma\sqrt{\Delta t}, & \text{w. prob. } \frac{1}{2}, \\ -\sigma\sqrt{\Delta t}, & \text{w. prob. } \frac{1}{2}, \end{cases}$$

for any $i = 0, 1, \ldots, M$. Then, the discrete spot rate process r_{t_i} $(i = 0, 1, \ldots, M)$ as defined in (12) is equivalent to

$$r_{t_{i+1}} = r_{t_i} + k(\theta_{t_i} - r_{t_i})\Delta t + X_{t_{i+1}}^{(1)} + X_{t_{i+1}}^{(2)} \quad (i = 0, \ldots, M - 1). \tag{13}$$

Remark 2.3. The discrete process above mimics the behavior of a continuous-time jump-diffusion process. Hence for $i = 0, \ldots, M$, the first noise term, $X_{t_i}^{(1)}$, represents the diffusion component, while the second one, $X_{t_i}^{(2)}$, represents the jump, which assumes values $\mu + \gamma$ or $\mu - \gamma$. Therefore, the jump has mean μ and variance γ^2. $\bar{\lambda}$ is the probability parameter of a jump in unit time and hence the probability of a jump in any time interval approximates $\lambda = \bar{\lambda}\Delta t$.

In our process, jumps occur "rarely", which is achieved by choosing a low value for $\lambda \in (0, 1)$. Moreover, as the time interval Δt decreases, the probability of a jump goes to zero. Finally, we may choose the parameters μ and γ to provide the necessary skewness and kurtosis. In particular, the parameter μ governs skewness, while γ drives kurtosis.

Remark 2.4. The moments of the discrete process provide the intuition for why the choice of jump form injects the necessary skewness and kurtosis into the model. The first four moments are as follows:

- mean: $r_t + k(\theta_t - r_t) + \lambda\mu$;
- variance: $\lambda(\mu^2 + \gamma^2) - \lambda^2\mu^2 + \sigma^2\Delta t$;
- skewness: $(1 - \lambda)\lambda\mu(\mu^2 + 3\gamma^2 - 2\lambda\mu^2)$, the sign of which clearly depends on the sign of μ;

[7] See Billingsley (1995).

- kurtosis (expressed for $\sigma = 0$ and $\mu = 0$, to keep the expression simple and to observe the effect of γ): $\lambda\gamma^4$, which demonstrates that the magnitude of kurtosis depends on the size of γ. When $\mu \neq 0$, kurtosis is equal to: $(1-\lambda)\lambda^4\mu^4 + \frac{\lambda}{2}(\mu - \gamma - \lambda\mu)^4 + \frac{\lambda}{2}(\mu + \gamma - \lambda\mu)^4$, which means that the sign and size of μ also affect the degree of kurtosis. However, here too, the degree of kurtosis increases with γ.

3. The tree representation

The diffusion process $X_{t_i}^{(1)}$ $(i = 0, \ldots, M-1)$, as defined above, can be represented on the binomial tree:

$$X_{t_i}^{(1)} \begin{cases} X_{t_{i+1}}^{(1)} = X_{t_i}^{(1)} + \sigma\sqrt{\Delta t}, & \text{w. prob. } \frac{1}{2}, \\ X_{t_{i+1}}^{(1)} = X_{t_i}^{(1)} - \sigma\sqrt{\Delta t}, & \text{w. prob. } \frac{1}{2}. \end{cases}$$

The random term representing the jump in Equation (13) is $X_{t_i}^{(2)}$ and can be represented on the trinomial tree:

$$X_{t_i}^{(2)} \begin{cases} X_{t_{i+1}}^{(2)} = X_{t_i}^{(2)} + \mu + \gamma, & \text{w. prob. } \frac{\lambda}{2}, \\ X_{t_{i+1}}^{(2)} = X_{t_i}^{(2)}, & \text{w. prob. } 1 - \lambda, \\ X_{t_{i+1}}^{(2)} = X_{t_i}^{(2)} + \mu - \gamma, & \text{w. prob. } \frac{\lambda}{2}. \end{cases}$$

Let us consider, at first, only the discrete diffusion process:

$$r_{t_{i+1}} = r_{t_i} + k(\theta_{t_i} - r_{t_i})\Delta t + X_{t_{i+1}}^{(1)} \quad (i = 0, \ldots, M-1), \tag{14}$$

and, following the methodology of Hull and White (1994), let us construct a binomial recombining tree for a variable $r_{t_i}^*$ $(i = 0, \ldots, M)$, which is initially zero, follows the process:

$$r_{t_{i+1}}^* = r_{t_i}^* - kr_{t_i}^*\Delta t + X_{t_{i+1}}^{(1)} \quad (i = 0, \ldots, M-1) \tag{15}$$

and is symmetrical about $r_{t_i}^* = 0$. The variable $[r_{t_{i+1}}^* - r_{t_i}^*]$ is Bernoulli distributed. If terms of higher order than Δt^2 are ignored, it holds:

$$E\{r_{t_{i+1}}^* - r_{t_i}^*\} = -kr_{t_i}^*\Delta t, \qquad \mathrm{Var}\{r_{t_{i+1}}^* - r_{t_i}^*\} = \sigma^2\Delta t + k^2 r_{t_i}^{*2}\Delta t^2.$$

Let j_i be a positive or negative integer, varying in the range $(-\frac{1}{k\Delta t}, +\frac{1}{k\Delta t})$ and depending on i, with $i = 0, \ldots, M$. Recalling that $t_i = i\Delta t$, let denote $r_D^*(i, \cdot)$ the corresponding value of $r_{t_i}^*$ on a lattice scheme. For any (integer) $i \in [0, M]$ and (any integer) $j_i \in (-\frac{1}{k\Delta t}, +\frac{1}{k\Delta t})$

the value of $r_D^*(i, \cdot)$ at "level" j_i, i.e., at node (i, j_i), will be denoted with $r_D^*(i, j_i)$ and similar notation applies to r_{t_i}.

Assumption 3.1. The binomial tree is recombining.

For every $i = 0, \ldots, M$, let n_i be the number of nodes of the tree generated at time $t_i = i\Delta t$. The binomial tree is assumed to be recombining, then, the number of nodes on the tree will be

$$\sum_{k=0}^{i} n_k = \sum_{k=0}^{i} (k+1) = \sum_{k=1}^{i+1} k.$$

Let J_i denote the set of indices j_i generated at time $t_i = i\Delta t$, with $i = 0, \ldots, M$. Then $|J_i| = n_i = i + 1$. Assuming $J_{-2} = J_{-1} = \emptyset$ and recalling that the tree is recombining, we can calculate recursively the set J_i as: $J_i = J_{i-2} \cup \{i, -i\}$, for any $i = 0, \ldots, M$.

For any integer $j_i \in [0, +\frac{1}{k\Delta t})$, let $\Delta(j_i)$ denote the spacing between interest rates on the tree by moving from level $j_i - 1$ to level j_i ("up" movement), or from level j_i to level $j_i - 1$ ("down" movement). In both cases, "up" or "down" movement, the width of spacing between rates depends on the maximum (starting or ending) level reached by moving between the two adjacent nodes, i.e., if the interest rate moves (up or down) from node (i, j_i) to node $(i + 1, j_{i+1})$, with $j_i \geqslant 1$, then

$$\left| r_D^*(i+1, j_{i+1}) - r_D^*(i, j_i) \right| = \Delta\left(\max\{j_i, j_{i+1}\}\right).$$

On the contrary, for any integer $j_i \in (-\frac{1}{k\Delta t}, 0)$ set:

$$\left| r_D^*(i+1, j_{i+1}) - r_D^*(i, j_i) \right| = \Delta\left(\min\{j_i, j_{i+1}\}\right).$$

Then

$$\Delta(j_i) = \Delta(-j_i), \quad \forall j_i \in \left(-\frac{1}{k\Delta t}, +\frac{1}{k\Delta t}\right) \setminus \{0\}.$$

Remark 3.1. Note that, by construction, the quantity $\Delta(0) = 0$ will never be employed in the tree building, i.e., in deriving the interest rate position on the tree.

For any $i = 1, \ldots, M - 1$ and any integer $j_i \in (-\frac{1}{k\Delta t}, +\frac{1}{k\Delta t})$, let $\Delta_u(j_i, j_{i+1})$ and $\Delta_d(j_i, j_{i+1})$ denote, respectively, an "up" and a "down" movement on the binomial tree from level j_i to level j_{i+1}. Then

$$\Delta_u(j_i, j_{i+1}) = -\Delta_d(-j_i, -j_{i+1}).$$

Assumption 3.2. Let us suppose that, for any $i = 1, \ldots, M$ and $j_i \in J_i \cap (-\frac{1}{k\Delta t}, +\frac{1}{k\Delta t})$, the spacing between interest rates on the tree is such that

$$\Delta(j_i) = \frac{|j_i|\sigma\sqrt{\Delta t}}{\sqrt{1 - j_i^2 k^2 \Delta t^2}}.$$

Remark 3.2. The spacing between nodes is not constant, but assumes values increasing with j_i. This makes our model more flexible than the standard Cox, Ross and Rubinstein (1979) binomial tree, which consists of a set of nodes – representing possible future stock prices – with constant logarithmic spacing between them. This spacing is a measure of the future volatility, itself assumed to be constant in the Cox, Ross and Rubinstein framework and in the Black and Scholes model (1973) to which a Cox, Ross and Rubinstein tree, with an "infinite" number of time steps, converges.

The constancy of volatility cannot easily be reconciled with the observed structure of implied volatilities for options traded in most financial markets. Volatility varies with both strike price and expiration. This variation, known as implied volatility "smile", is currently a significant and persistent feature of option markets. Nevertheless, the constant local volatility assumption in the Black and Scholes theory and in the Cox, Ross and Rubinstein tree leads to the absence of a volatility smile, at least as long as market frictions are ignored [see Derman, Kani and Chriss (1996)]. Our tree refers to implied tree theories that extend the Black and Scholes model to make it consistent with the shape of the smile [see Derman and Kani (1997a, b) and Rubinstein (1994)].

Because $r^*_{t_i}$ is initially zero, then $r^*_D(0, 0) = 0$. The dynamics of $r^*_{t_i}$ incorporate a mean reversion to zero, where the strength of the mean reversion is proportional to the value of $r^*_{t_i}$. Therefore, there exist an upper and a lower bound to the interest rate process. Let us formalize this condition as follows:

Assumption 3.3. There exist two integer numbers $j_u \in (0, +\frac{1}{k\Delta t})$ and $j_d \in (-\frac{1}{k\Delta t}, 0)$ such that

- $j_d \leqslant j_i \leqslant j_u$ for any $j_i \in J_i$ $(i = 0, \ldots, M)$;
- if $j_i = j_u$ at step i, then $j_{i+1} = j_u - 1$ at step $i + 1$ $(i = 0, \ldots, M - 1)$;
- if $j_i = j_d$ at step i, then $j_{i+1} = j_d + 1$ at step $i + 1$ $(i = 0, \ldots, M - 1)$.

The parameters j_u and j_d may be estimated from the real values assumed by r_{t_i}, otherwise we can set:

$$j_u = \left[\frac{1}{k\Delta t}\right] - 1 \quad \text{and} \quad j_d = -j_u,$$

where $[\frac{1}{k\Delta t}]$ denotes the integer part of $\frac{1}{k\Delta t}$.

Let $p_u(i, j_i)$ and $p_d(i, j_i)$ denote, respectively, the transition probabilities of "up" and "down" movements starting from node (i, j_i), with $i = 0, \ldots, M - 1$ and $j_i \in J_i \cap (j_d, j_u)$.

The two probabilities must be chosen to match the expected change and variance of the change in $r^*_{t_i}$ over the next time interval Δt. The probabilities must also be non-negative and sum to unity. This leads to the equations below:

$$\begin{cases} p_u(i,j_i)\Delta_u(j_i,j_{i+1}) + p_d(i,j_i)\Delta_d(j_i,j_{i+1}) = -kr^*_D(i,j_i)\Delta t, \\ p_u(i,j_i)\Delta_u^2(j_i,j_{i+1}) + p_d(i,j_i)\Delta_d^2(j_i,j_{i+1}) = \sigma^2\Delta t + k^2 r^*_D(i,j_i)^2\Delta t^2, \\ p_u(i,j_i) + p_d(i,j_i) = 1, \\ 0 \leqslant p_u(i,j_i) \leqslant 1, \\ 0 \leqslant p_d(i,j_i) \leqslant 1. \end{cases}$$

Using Assumptions 3.2 and 3.3, the solution to the system above is given by the following cases:

- for any $i = 0, \ldots, M-1$ and $j_i \in J_i \cap [+1, j_u)$:

$$p_u(i,j_i) = p_u(j_i) = \frac{1}{2} - \sqrt{\frac{1}{4} - \left[\frac{\sigma\sqrt{\Delta t}}{\Delta_u(j_i,j_{i+1}) - \Delta_d(j_i,j_{i+1})}\right]^2},$$

$$p_d(i,j_i) = p_d(j_i) = \frac{1}{2} + \sqrt{\frac{1}{4} - \left[\frac{\sigma\sqrt{\Delta t}}{\Delta_u(j_i,j_{i+1}) - \Delta_d(j_i,j_{i+1})}\right]^2};$$

- for any $i = 0, \ldots, M-1$ and $j_i \in J_i \cap (j_d, -1]$:

$$p_u(i,j_i) = p_u(j_i) = \frac{1}{2} + \sqrt{\frac{1}{4} - \left[\frac{\sigma\sqrt{\Delta t}}{\Delta_u(j_i,j_{i+1}) - \Delta_d(j_i,j_{i+1})}\right]^2},$$

$$p_d(i,j_i) = p_d(j_i) = \frac{1}{2} - \sqrt{\frac{1}{4} - \left[\frac{\sigma\sqrt{\Delta t}}{\Delta_u(j_i,j_{i+1}) - \Delta_d(j_i,j_{i+1})}\right]^2};$$

- for any $i = 0, \ldots, M-1$ and $j_i = 0$:

$$p_u(i,0) = p_u(0) = \frac{1-k\Delta t}{2}, \qquad p_d(i,0) = p_d(0) = \frac{1+k\Delta t}{2};$$

- for any $i = 0, \ldots, M-1$ and $j_i = j_u$:

$$p_u(i,j_u) = p_u(j_u) = 0, \qquad p_d(i,j_u) = p_d(j_u) = 1;$$

- for any $i = 0, \ldots, M-1$ and $j_i = j_d$:

$$p_u(i,j_d) = p_u(j_d) = 1, \qquad p_d(i,j_d) = p_d(j_d) = 0.$$

Remark 3.3. At each node (i, j_i), for any $i = 0, \ldots, M-1$ and $j_i \in J_i \cap [j_d, j_u]$, with $j_d = -j_u$ and $j_i \neq 0$, the transition probabilities depend only on j_i, so the resulting binomial tree, besides being recombining, is also symmetrical around $j_i = 0$, i.e.,

$$p_u(j_i) = p_d(-j_i), \quad p_u(j_u) = p_d(j_d) \quad \text{and} \quad p_u(j_d) = p_d(j_u).$$

Thus, it is not necessary to save all transition probabilities in one large array and the loss in computing time is very small.

Remark 3.4. For every $i = 0, \ldots, M-1$ and $j_i \in J_i \cap [j_d, j_u]$, the transition probabilities $p_u(j_i)$ and $p_d(j_i)$ reflect the mean reversion of the rate $r^*_{t_i}$ and, thus, of $r^*_D(i, j_i)$. In fact,

- if $j_i > 0$, $r^*_{t_i}$ lies above its medium term tendency, $p_u(j_i) < \frac{1}{2}$ and $p_d(j_i) > \frac{1}{2}$; moreover, for any integer $k > 0$, $p_u(j_i) > p_u(j_i + k)$ while $p_d(j_i) < p_d(j_i + k)$, i.e., the larger the j_i, the greater the calling back strength;
- if $j_i < 0$, $r^*_{t_i}$ lies below its medium term tendency, $p_u(j_i) > \frac{1}{2}$ and $p_d(j_i) < \frac{1}{2}$; moreover, for any integer $k < 0$, $p_u(j_i) < p_u(j_i + k)$ while $p_d(j_i) > p_d(j_i + k)$, i.e., the smaller the j_i, the greater the calling back strength.

Following Hull and White (1994), we have to convert the tree for $r^*_{t_i}$ (i.e., for $r^*_D(i, \cdot)$) into a tree for r_{t_i} (i.e., for $r_D(i, \cdot)$). This is accomplished by displacing the nodes on the $r^*_{t_i}$-tree so that the initial term structure is exactly matched. Define:

$$\alpha_{t_i} = r_{t_i} - r^*_{t_i} \quad (i = 0, \ldots, M)$$

and let

$$\alpha(i) = r_D(i, \cdot) - r^*_D(i, \cdot)$$

its lattice version. For a given "time level" $i = 0, \ldots, M$, all nodes are shifted by the same amount $\alpha(i)$. By the definitions of r_{t_i} and $r^*_{t_i}$ (14), (15), the variation of α_{t_i} in the interval Δt is:

$$\Delta\alpha_{t_i} = k[\theta_{t_i} - \alpha_{t_i}]\Delta t.$$

Then, as $\Delta t \to 0$ it holds:

$$\mathrm{d}\alpha_t = k[\theta_t - \alpha_t]\,\mathrm{d}t.$$

The estimation of α_t, i.e., the calibration of the tree, could be done via the "spline–wavelet" method proposed in Cattani and Izzi (2000)[8] or, in turn, by following the procedure pro-

[8] Cattani and Izzi (2000) propose a spline Haar-wavelet interpolation as a flexible tool for interest rate and term structure estimation. The main advantage in employing wavelet functions is that they are a very well-localized representation of the process: the accuracy is reached with only a few sets of data and singularities do not influence the estimation of the entire process.

posed by Hull and White (1994). The solution α_t can be used to create a tree for r_{t_i} from the corresponding tree for $r^*_{t_i}$. The approach is to set the interest rates on the r_D-tree at time $i\Delta t$ to be equal to the corresponding interest rates on the r^*_D-tree plus the value of $\alpha(\cdot)$ at time $i\Delta t$ while keeping the probabilities the same. Then, for any $i = 0, \ldots, M-1$ and $j_i \in J_i \cap [j_d, j_u]$, it holds that:

$$r^*_D(i, j_i) = \mathrm{sgn}(j_i) \cdot \sum_{l=0}^{|j_i|} \frac{|l|\sigma\sqrt{\Delta t}}{\sqrt{1 - l^2k^2\Delta t^2}}$$

and

$$r_D(i, j_i) = r^*_D(i, j_i) + \alpha(i),$$

where

$$\mathrm{sgn}(j_i) = \begin{cases} +1, & \text{if } j_i \geqslant 0, \\ -1, & \text{if } j_i < 0. \end{cases}$$

Remark 3.5. For any $i = 0, \ldots, M-1$ and $j_i \in J_i \cap [j_d, j_u]$, $r^*_D(\cdot,\cdot)$, $p_u(\cdot)$ and $p_d(\cdot)$ do not depend on i, a property referred to as *stationarity*: the behavior of r^*_D depends on its current value via j but not on the date.

In the next instant of time, $t_{i+1} = (i+1)\Delta t$, $r^*_D(\cdot)$ could assume the following values:

$$r^*_D(j_{i+1}) = \begin{cases} r^*_D(j_i) + \Delta_u(j_i, j_{i+1}), & \text{w. prob. } p_u(j_i), \\ r^*_D(j_i) + \Delta_d(j_i, j_{i+1}), & \text{w. prob. } p_d(j_i), \end{cases}$$

while

$$r_D(i+1, j_{i+1}) = \begin{cases} r^*_D(j_i) + \Delta_u(j_i, j_{i+1}) + \alpha(i+1), & \text{w. prob. } p_u(j_i), \\ r^*_D(j_i) + \Delta_d(j_i, j_{i+1}) + \alpha(i+1), & \text{w. prob. } p_d(j_i). \end{cases}$$

Let us consider now the discrete jump process:

$$X^{(2)}_{t_i} = \begin{cases} \mu + \gamma, & \text{w. prob. } \frac{\lambda}{2}, \\ 0, & \text{w. prob. } 1 - \lambda, \\ \mu - \gamma, & \text{w. prob. } \frac{\lambda}{2}. \end{cases}$$

For $i = 0, \ldots, M-1$, let $r_{JD}(i, \cdot)$ denote the value corresponding to

$$r_{t_{i+1}} = r_{t_i} + k(\theta_{t_i} - r_{t_i})\Delta t + X^{(1)}_{t_{i+1}} + X^{(2)}_{t_{i+1}} \quad (i = 0, \ldots, M-1), \tag{16}$$

on a lattice scheme, while $r^*_{JD}(i,\cdot)$ that of

$$r^*_{t_{i+1}} = r^*_{t_i} - k r^*_{t_i}\Delta t + X^{(1)}_{t_{i+1}} + X^{(2)}_{t_{i+1}} \quad (i = 0, \ldots, M-1). \tag{17}$$

The convolution between the jump and the diffusion processes provides the outcome for representing the evolution of the entire term structure. The product space resulting from this convolution could be represented by a hexanomial tree.

With reference to $r^*_{JD}(i,\cdot)$, i.e., to Equation (17), we have:

$$r^*_{JD}(i+1, j_{i+1}) = r^*_{JD}(j_{i+1})$$
$$= \begin{cases} r^*_D(j_i) + \Delta_u(j_i, j_{i+1}) + \mu + \gamma, & \text{w. prob. } \frac{\lambda}{2} p_u(j_i), \\ r^*_D(j_i) + \Delta_u(j_i, j_{i+1}), & \text{w. prob. } (1-\lambda) p_u(j_i), \\ r^*_D(j_i) + \Delta_u(j_i, j_{i+1}) + \mu - \gamma, & \text{w. prob. } \frac{\lambda}{2} p_u(j_i), \\ r^*_D(j_i) + \Delta_d(j_i, j_{i+1}) + \mu + \gamma, & \text{w. prob. } \frac{\lambda}{2} p_d(j_i), \\ r^*_D(j_i) + \Delta_d(j_i, j_{i+1}), & \text{w. prob. } (1-\lambda) p_d(j_i), \\ r^*_D(j_i) + \Delta_d(j_i, j_{i+1}) + \mu - \gamma, & \text{w. prob. } \frac{\lambda}{2} p_d(j_i). \end{cases}$$

Then, the hexanomial tree representing the jump-diffusion process (16) is defined by

$$r_{JD}(i+1, j_{i+1}) = r^*_{JD}(j_{i+1}) + \alpha(i+1),$$

for any $i = 0, \ldots, M-1$ and $j_i \in J_i \cap (j_d, j_u)$.

4. The econometric analysis

4.1. Data

The short-term rate used in estimating our model is the one-month Euribor (EURo Inter-Bank Offered Rate). Data come from the European Central Bank and Datastream and have been sampled daily from 1 January 1999 to 5 March 2001 (567 working days). Weekends and holidays have not been treated specifically (Monday is taken as the next day after Friday). Whereas weekend effects have been documented for stock prices, there does not seem to be a conclusive weekend effect on money market instruments. While similar theoretical and empirical work could be performed on zero-coupon bond yield, the liquidity and default characteristics of such securities are different from those of interbank instruments and the time-varying effects introduced by such features would blur our analysis.

Before defining our stochastic model for the interbank interest rates, a statistical analysis on a wide set of data has been done. The short-term rates used in this preliminary step are the daily interbank euro rates: overnight, 1 week, 1–12 months. The overnight rate is EONIA (Euro OverNight Index Average) daily calculated by the European Central Bank; the

Table 1
Descriptive statistics of the short-term interbank rates

Rate	Mean	Standard deviation	Skewness	Kurtosis
o / n	3.549	0.906	0.256	−1.295
1 w	3.604	0.872	0.257	−1.512
1 m	3.653	0.872	0.216	−1.524
2 m	3.712	0.875	0.175	−1.508
3 m	3.768	0.876	0.133	−1.498
4 m	3.809	0.870	0.126	−1.499
5 m	3.846	0.867	0.114	−1.499
6 m	3.875	0.867	0.099	−1.509
7 m	3.900	0.868	0.083	−1.511
8 m	3.927	0.870	0.062	−1.510
9 m	3.955	0.873	0.036	−1.503
10 m	3.983	0.877	0.015	−1.484
11 m	4.007	0.877	−0.016	−1.469
12 m	4.034	0.881	−0.042	−1.450

interbank rates, from 1 week to 12 months, are the Euribor. Table 1 reports the descriptive statistics of the fourteen rates.[9]

4.2. Estimation results

In the European money market, jumps may arise from intervention by the European Central Bank (ECB) on the official rates. Short rates tend to track these rates – and, above all, the main refinancing operation rate – rather closely. The aim of estimations is to examine the impact of changes in the main refinancing operation rate level θ_t on monetary rates. Equation (4) is specified as follows:

$$\Delta r_t = k[\theta_{t_i} - r_{t_i}]\Delta t + \sigma \Delta B_t + a(\mu, \gamma^2)\Delta N_t(\lambda),$$
$$a_{t+\Delta t} = \alpha_0[\theta_{t+\Delta t} - \theta_t].$$

The process is modelled as if the main refinancing operation rate were actually implemented and perceived by the market. Main refinancing operation rate changes are assumed to be independent from the short-rate process and to be infrequent, with $\lambda \ll 1$ the probability of the main refinancing operation rate change on any day t. In other words, the main refinancing operation rate changes are modelled here as outcomes in Bernoulli trials, where the probability of an event – main refinancing operation rate change on any given day – occurring in a single trial is constant.

[9] It should be noted that monetary policy actions influence the three policy rates, beginning from different dates; however, for the statistical analysis, we consider as simultaneous the ECB influence on the three rates, because of the immediate effect of the announcement on the short-term structure.

The jump-diffusion process is estimated using indirect inference method. The problem of the optimal estimator is solved with respect to both the generalized method of moments (GMM) and the full information maximum likelihood estimation (FIML). In the latter case, parameter estimates are obtained by numerically maximizing the sample log-likelihood function. The standard error is estimated using the Hessian matrix at the optimum point. The optimization algorithm is the Gauss–Newton method.

In a second stage, the generalized method of moments is used to obtain efficient estimates of the parameters of the Poissonian part of the process. These are derived as follows: $\lambda = E\{\theta_{t+\Delta t} - \theta_t\}$, $\mu = E\{a_t\}$, $\gamma^2 = \mathrm{Var}\{a_t\}$. In particular, the estimate for the daily probability λ is given by the empirical frequency of main refinancing operation rate changes. There are 8 main refinancing operation rate changes in our sample of 567 working days.

Equation (4) has been estimated with respect to all the interbank short rates, from overnight to twelve-month maturity. Estimation results indicate that the presence of jump components superimposed on the diffusion process is significant in any of the fourteen interest rates, so that the assumption is consistent with the empirical evidence. Estimations also show that the impact of monetary policy actions is particularly evident in those rates whose maturity lies between one and three months. The estimates performed on these rates produce – from the statistical point of view – the most significant results.

The model shows a good fit with respect to both the GMM and the FIML estimation method. The best performance is obtained with respect to the two-month Euribor (see Figure 2). A T-test on the forecast error has also been applied. The test results show how, in the case of two-month Euribor, the prevision error is in mean statistically equal to zero. In the case of one-month and three-month Euribor, even if the error is not so large, the T-test refuses the hypothesis that the observed and estimated series are statistically equal in the average (see Tables 2 and 3).[10]

[10] The usual T-statistic for testing the equality of averages $\overline{X}_n$ and $\overline{Y}_m$ from two independent samples with n and m observations is:

$$T_{n,m} = \frac{|\overline{X}_n - \overline{Y}_m|}{\sqrt{(n-1)S_n^2 + (m-1)S_m^2}} \sqrt{\frac{nm(n+m-2)}{n+m}} \sim T_{(n+m-2)},$$

where S_n^2 and S_m^2 are the sample variances of the two groups. The T-statistic assumes that $\sigma_n = \sigma_m$, where σ_n and σ_m are the population variances of the two groups.

We use the folded form of the F (Fisher) statistic, F', to test the assumption that the variances are equal, where

$$F' = \frac{\max[S_n^2, S_m^2]}{\min[S_n^2, S_m^2]}.$$

Under the assumption of equal variances, the T-statistic is computed with the formula given above. Under the assumption of unequal variances, the approximate T-statistic is computed as

$$T_{n,m} = \frac{|\overline{X}_n - \overline{Y}_m|}{\sqrt{S_n^2/n + S_m^2/m}}.$$

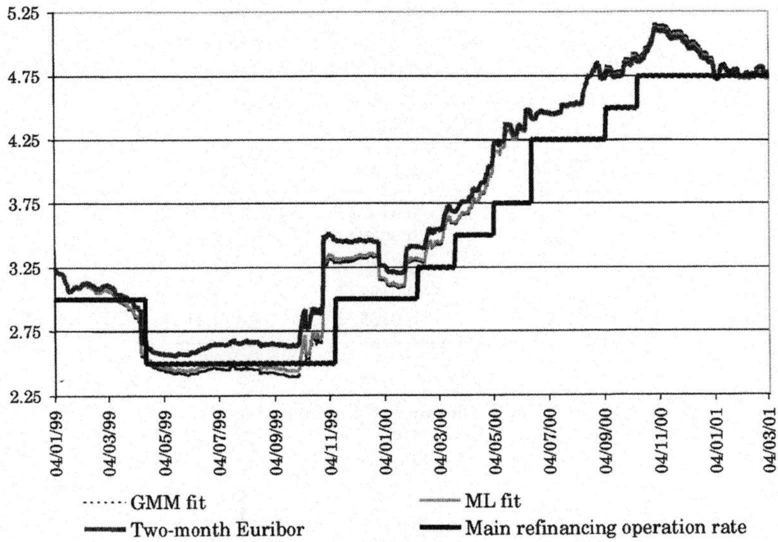

Fig. 2. A comparison among the main refinancing operation rate and actual and fitted two-month Euribor (percentage values).

Table 2

T-test between the actual series of one-, two- and three-month Euribor and their estimates with respect to the full information maximum likelihood method

Series	1m Euribor	2m Euribor	3m Euribor
Mean value of the actual series	3.65312898	3.71241696	3.76756714
Mean value of the fitted series	3.43574164	3.64401606	3.51284076
T-test	4.2619	1.2725	5.0116
Prob > $\|T\|$	0.0001	0.2035	0.0001

Table 3

T-test between the actual series of one-, two- and three-month Euribor and their estimates with respect to the generalized method of moments

Series	1m Euribor	2m Euribor	3m Euribor
Mean value of the actual series	3.65312898	3.71241696	3.76756714
Mean value of the fitted series	3.42677354	3.63484966	3.50756698
T-test	4.4117	1.4274	5.0721
Prob > $\|T\|$	0.0001	0.1538	0.0001

Table 4
Parameter estimates of the jump-diffusion process with respect to the generalized method of moments

One-month Euribor		
Parameter	Estimate	T-statistic
k	−0.010033	−22.08
σ	0.993974	395.68
λ	0.003086	1.70
μ	0.019877	4.41
γ	0.013522	4.41

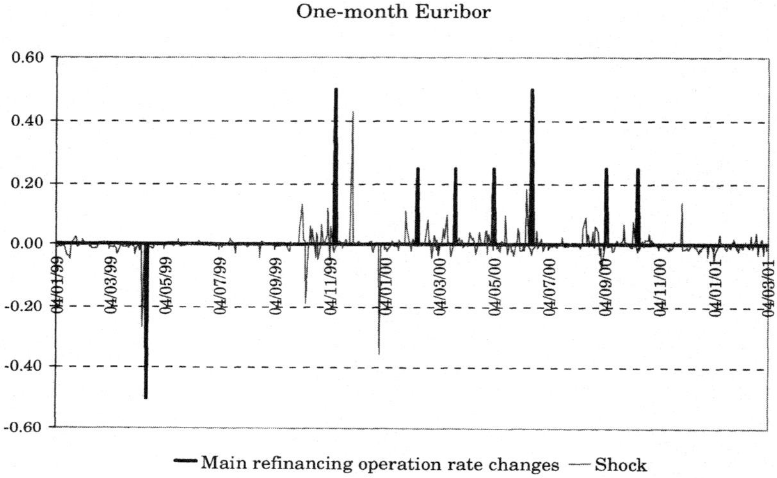

Fig. 3. A comparison between the main refinancing operation rate changes and the shock related to the estimation of the one-month Euribor (percentage values).

The estimates of the parameters of the Poissonian part of the process, which are calculated with respect to the one-month Euribor and become asymptotically more significant as the length of the horizon grows, suffer the narrow size of the sample, which does not allow us to make rolling estimations. This problem could be overcame using Monte Carlo simulations. Parameter estimates are summarized in Table 4.

As a final observation, let us consider the shock of the model, derived from a stochastic perturbation process. As shown in Figures 3–5, the shocks related to the three monetary rates (from one- to three-month Euribor) capture the effects of the approach main refinancing operation rate changes, thus giving a measure of market expectations.

It has to be noted that in all the three cases, the shocks (Figures 3–5, peaks in plain) anticipate the interventions of the ECB (peaks in bold). In particular, in the case of the first intervention – the only downward one in our sample period – the shock magnitude

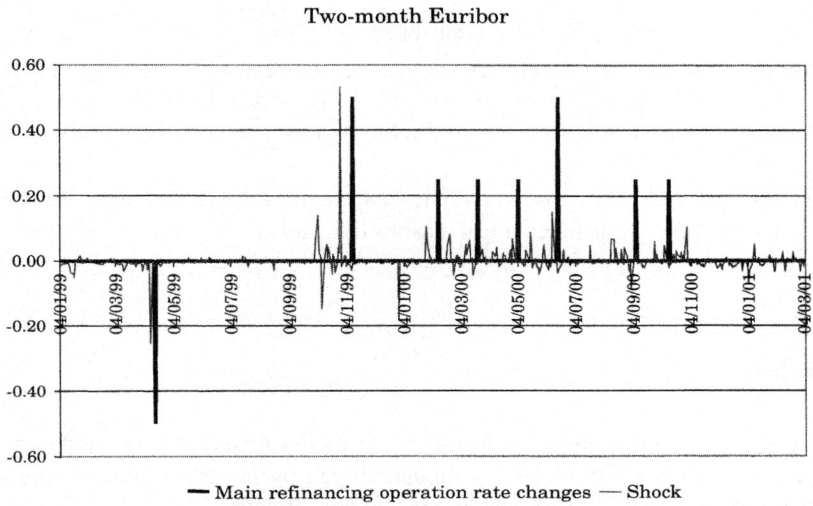

Fig. 4. A comparison between the main refinancing operation rate changes and the shock related to the estimation of the two-month Euribor (percentage values).

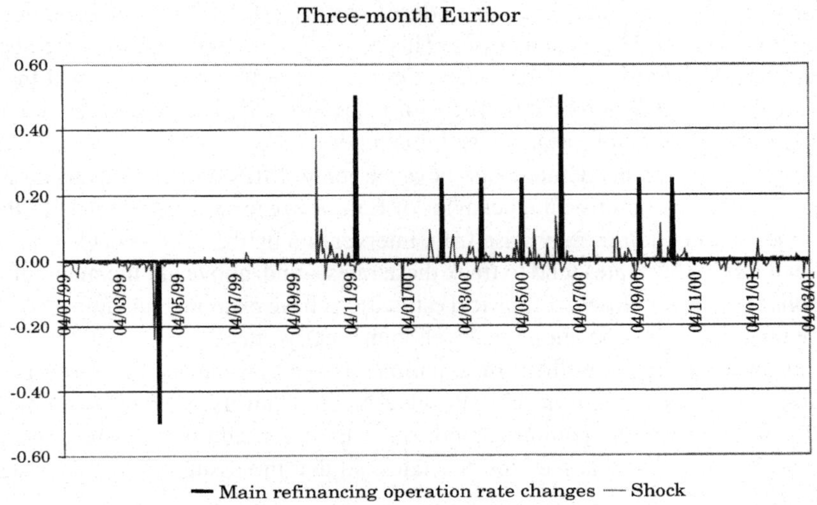

Fig. 5. A comparison between the main refinancing operation rate changes and the shock related to the estimation of the three-month Euribor (percentage values).

is smaller than that observed in the case of upward interventions, taking into account that upward interventions are closer to each other.

The downward movements observed in the three shock series at the end of 1999 (see the plain peaks) show a behavior only apparently contrary to monetary policy indications (bold peaks). In fact, with the intention of contributing to a smooth transition to the year 2000, the European Central Bank injected, via longer-term and main refinancing operations, an amount of liquidity greater than that demanded by the market. Normal liquidity conditions in the money market were restored with the fine-tuning operation conducted by the ECB on 5 January 2000.

Some large positive peaks observed in the shock series far from the main refinancing operation rate changes are due to technical reasons like, first of all, the end of the maintenance period.

5. Conclusions

In this chapter a jump-diffusion mean-reverting model for estimating monetary rates is introduced and related to the class of models driven by infinitely divisible processes to which Gaussian, Poissonian and Stable ones belong. Relating the interest rate models to the class of Stable processes is an attempt to unify, from the probabilistic point of view, the extensive literature of pure-diffusion, pure-Poissonian and jump-diffusion processes and completes recent works on the stock and exchange markets on the same topic.

We also propose a new numerical procedure to recursively compute interest rates subject to both Brownian and Poissonian noises. This procedure is consistent with the absence of arbitrage, non-negativity of interest rates, the mean-reverting hypothesis and the recombining condition, and can be calibrated with respect to any term structure which can be observed in the market, for instance, the initial one.

Numerical experiments demonstrate how the jump-diffusion mean-reverting model is particularly suited to describe the behavior of European money market rates. In the European money market, jumps may arise from intervention by the European Central Bank on the official rates. Short rates tend to track these rates – and, above all, the main refinancing operation rate – rather closely. Empirical estimations have examined the impact of changes in the main refinancing operation rate level on monetary rates.

The mean-reverting jump-diffusion equation has been estimated with respect to all the interbank short rates, from overnight to twelve-month maturity. Statistical results have indicated that the presence of jump components superimposed on the diffusion process is significant in any of the fourteen interest rates, so that the assumption is consistent with empirical evidence.

Estimations have also shown that the influence of monetary policy actions is particularly evident in those rates whose maturity lies between one and three months. With respect to these, the model has shown a good fit with respect to both the generalized method of moments and the full information maximum likelihood estimation method.

Finally, the analysis of the stochastic perturbation of the model has shown that the shocks related to the three monetary rates, from one- to three-month Euribor, capture the effects of the approach of main refinancing operation rate changes, giving a measure of market expectations.

References

Ahn, C.M., Thompson, H.E., 1988. Jump-diffusion processes and the term structure of interest rates. Journal of Finance 43 (1), 155–174.

Akgiray, V., Booth, G.G., 1988. Mixed diffusion-jump process modeling of exchange rate movements. The Review of Economics and Statistics LXX (4), 631–637.

Attari, M., 1999. Discontinuous interest rate processes: An equilibrium model for bond option prices. Journal of Financial and Quantitative Analysis 34 (3), 293–322.

Bates, D.S., 1996. Jump and stochastic volatility: exchange rate processes implicit in Deutsche mark options. The Review of Financial Studies 9 (1), 69–107.

Baz, J., Das, S.R., 1996. Analytical approximations of the term structure for jump-diffusion processes: A numerical analysis. The Journal of Fixed Income 6 (1), 78–86.

Billingsley, P., 1995. Probability and Measure, 3rd edition. Wiley.

Björk, T., Kabanov, Y., Runggaldier, W., 1997. Bond market structure in the presence of market point processes. Mathematical Finance 7 (2), 211–239.

Black, F., Derman, E., Toy, W., 1990. A one-factor model of interest rates and its applications to treasury bond options. Financial Analysts Journal 46, 33–39.

Black, F., Scholes, M., 1973. The pricing of options and other corporate liabilities. Journal of Political Economy 81, 637–654.

Blasikiewicz, M., Brown, T.C., 1996. Poisson approximations for Markov-driven point processes. Stochastic Processes and their Applications 62 (May/June), 179–189.

Brennan, M.J., Schwartz, E.S., 1980. Analyzing convertible bonds. Journal of Financial and Quantitative Analysis 15 (4), 907–929.

Britten-Jones, M., Neuberger, A., 2000. Option prices, implied price processes, and stochastic volatility. Journal of Finance 85, 839–866.

Buttiglione, L., Del Giovane, P., Tristani, O., 1997. Monetary policy actions and the term structure of interest rates: A cross-country analysis. Temi di discussione del Servizio Studi, No. 306. Banca d'Italia.

Cattani, C., Izzi, L., 2000. Spline wavelet interpolation of interest rates. Working Paper, No. 31/2000. Department of Mathematics "G. Castelnuovo", University of Rome "La Sapienza".

Cox, J.C., 1975. Notes on option pricing I: Constant elasticity of variance diffusion. Working Paper. Stanford University.

Cox, J.C., Ingersoll, J.E., Jr., Ross, S.A., 1980. An analysis of variable rate loan contracts. Journal of Finance 35 (2), 389–403.

Cox, J.C., Ingersoll, J.E., Jr., Ross, S.A., 1985. A theory of the term structure of interest rates. Econometrica 53 (2), 385–407.

Cox, J.C., Ross, S.A., 1976. The valuation of options for alternative stochastic processes. Journal of Financial Economics 3 (January/March), 173–202.

Cox, J.C., Ross, S.A., Rubinstein, M., 1979. Option pricing: A simplified approach. Journal of Financial Economics (7), 229–263.

Das, S.R., 1997. An efficient generalized discrete-time approach to Poisson–Gaussian bond option pricing in the Heath–Jarrow–Morton model. Technical Working Paper, No. 212. NBER.

Das, S.R., Foresi, S., 1996. Exact solutions for bond and option prices with systematic jump risk. Review of Derivatives Research 1 (1), 7–24.

Derman, E., Kani, I., 1997a. Riding on a smile. Risk 7 (1), 18–20.

Derman, E., Kani, I., 1997b. Stochastic implied trees: Arbitrage pricing with stochastic term and strike structure of volatility. In: Quantitative Strategies Research Notes. Goldman Sachs, pp. 1–54.

Derman, E., Kani, I., Chriss, N., 1996. Implied trinomial trees of the volatility smile. In: Quantitative Strategies Research Notes. Goldman Sachs, February, pp. 1–24.

Dothan, L.U., 1978. On the term structure of interest rates. Journal of Financial Economics 6 (1), 59–69.

Estrella, A., Mishkin, F.S., 1995. The term structure of interest rates and its role in monetary policy for the European Central Bank. Working Paper, No. 5279. NBER.

European Central Bank, 2001a. Monthly Bulletin, June.
European Central Bank, 2001b. The Monetary Policy of the ECB, August.
Gnedenko, B.V., 1992. Teoria della Probabilitá. Editori Riuniti.
Ho, M.S., Perraudin, W.R.M., Sorensen, B.E., 1996. A continuous-time arbitrage-pricing model with stochastic volatility and jumps. Journal of Business and Economic Statistics 14 (1), 31–43.
Hull, J., White, A., 1994. Numerical procedures for implementing term structure models I: Single-factor models. The Journal of Derivatives, Fall, 7–16.
Hull, J., White, A., 1996. Using the hull-white interest rate trees. The Journal of Derivatives, Spring, 26–36.
Izzi, L., Racheva, B., 2002. The term structure of interest rates in the economic and monetary union. In: Rachev, S.T. (Ed.), Special Issue on Mathematical Models in Market and Credit Risk. Mathematical Methods of Operations Research 55 (2), 187–224.
Merton, R.C., 1973. Theory of rational option pricing. Bell Journal of Economics and Management Science 4, 141–183.
Rachev, S.T., Mittnik, S., 2000. Stable Paretian Models in Finance. Wiley.
Rachev, S.T., Rüschendorf, L., 1994. Models for option pricing. Theory of Probability and its Applications 39 (1) 120–152.
Rieken, S., 1999. Option Pricing Using Subordinated and Infinitely Divisible Return Processes. Pro Business.
Rubinstein, M., 1994. Implied binomial trees. Journal of Finance LXIX (3), 771–818.
Samorodnitsky, G., Taqqu, M.S., 1994. Stable Non-Gaussian Random Processes. Stochastic Models with Infinite Variance. Chapman & Hall.
Santillan, J., Bayle, M., Thygesen, C., 2000. The impact of the Euro on money and bond markets. Occasional Paper, No. 1. European Central Bank.
Schumacher, N., 1997. Option pricing with infinitely divisible returns. Ph.D. Dissertation in Statistics. University of California, Santa Barbara.
Scott, L.O., 1997. Pricing stock options in a jump-diffusion model with stochastic volatility and interest rates: Applications of Fourier inversion methods. Mathematical Finance 7 (4), 413–426.
Vasicek, O., 1977. An equilibrium characterization of the term structure. Journal of Financial Economics 5, 177–188.
Zacks, S., Perry, D., Bshouty, D., Bar-Lev, S., 1999. Distributions of stopping times for compound Poisson processes with positive jumps and linear boundaries. Stochastic Models 15 (1), 89–101.

Chapter 13

ASSET LIABILITY MANAGEMENT: A REVIEW AND SOME NEW RESULTS IN THE PRESENCE OF HEAVY TAILS

YESIM TOKAT

Department of Economics, University of California, Santa Barbara, USA

SVETLOZAR T. RACHEV

Department of Statistics and Applied Probability, University of California, Santa Barbara, USA
Institute of Statistics and Mathematical Economics, University of Karlsruhe, Germany
e-mail: rachev@lsoe-4.wiwi.uni-karlsruhe.de

EDUARDO S. SCHWARTZ

Anderson School of Management, University of California, Los Angeles, USA

Contents

Handbook of Heavy Tailed Distributions in Finance, Edited by S.T. Rachev

Abstract

Asset and liability management is the simultaneous consideration of assets and liabilities in strategic investment planning. In this chapter, asset and liability management models that use stochastic programming framework are reviewed. Most of these models describe the financial uncertainty by a set of representative scenarios. We propose to replace the classical assumption of Gaussian returns in the scenario generation with the stable Paretian distribution, which can capture the leptokurtic nature of financial data. A multistage stochastic asset allocation model with decision rules is analyzed. Optimal asset allocation under the Gaussian and stable Paretian returns are compared. Our computational results suggest that asset allocation may be up to 20% different depending on the utility function and the risk aversion level of the investor. Certainty equivalent return can be increased up to 0.13% and utility can be improved up to 0.72% by switching to the stable Paretian model.

1. Introduction

Managing assets and liabilities is a concern for banks, pension funds and insurance companies. Before the deregulation of interest rates, the market value of liabilities changed very little from year to year. However, after interest rates were deregulated in 1979, they showed much more volatility. This lead the institutional investors mentioned above to consider assets and liabilities simultaneously during their strategic planning. Strategic investment planning is the allocation of portfolio across broad asset classes such as bonds, stocks, cash and real estate considering the legal and policy constraints facing the institution. Empirical evidence by Brinson, Hood and Beebower (1986) suggests that asset allocation is the most important factor in determining investment performance.

Most of the early models in this field are either myopic or represent deterministic formulations of multiperiod problems. Hakansson (1971) show that solving a sequence of single period models optimizes investor's long-run wealth or the expected utility of wealth.[1] They assume absence of transaction costs, market impact costs, and liquidity considerations. However, these assumptions are not justifiable in many situations. Myopic models cannot capture long-term investment goals in the presence of transaction costs. There is considerable evidence of predictability in asset returns[2] and the myopic models do not take this empirical finding into account. These models tend to produce high portfolio turnovers and opportunistic asset trades.

There has been a growing interest in the development of multiperiod stochastic models for asset and liability management (ALM). Kusy and Ziemba (1986) developed a multiperiod stochastic linear programming model for Vancouver City Savings Credit Union for a 5-year planning period. Their work suggests that their stochastic ALM model is superior to 5-year deterministic models. Another successful application of multistage stochastic programming is the Russell–Yasuda Kasai model by Carino et al. (1994). The investment strategy suggested by the model resulted in extra income of $79 million during the first two years of its application (1991 and 1992). An ALM model designed by Mulvey (1994) has been implemented by the Pacific Financial Asset Management Company. Boender (1997) reported the success of a hybrid simulation/optimization scenario model for ALM of pension funds in the Netherlands. The application of the model to a particular pension fund lead to a reduction of the yearly expected contributions of $100 million.

The ALM models that have gained applicability are based on stochastic programming with or without decision rules. In these models, the future economic uncertainty is modeled using discrete scenarios. Most of the models assume that the variables or the innovations of these variables follow normal distribution or the continuous time counterpart, Brownian

[1] Merton (1969) and Samuelson (1969) show that a constant relative risk aversion investor chooses the same asset proportions independent of the investment horizon if the market is frictionless and returns are independent over time.

[2] See for example Hodrick (1992), Bekaert and Hodrick (1992), Kandel and Staumbaugh (1996), and Brandt (1999).

motion. In response to the empirical evidence about the heavy tails, high peak and possible skewness in financial data, Fama (1965) and Mandelbrot (1963, 1967) propose stable Paretian distribution[3] as an alternative model. Among the alternative non-Gaussian distributions in the literature,[4] stable distribution has unique characteristics that make it an ideal candidate. The stable laws are the only possible limit distributions for properly normalized and centered sums of independent identically distributed random variables (Embrechts et al., 1997; Rachev and Mittnik, 2000). If a financial variable can be regarded as the result of many microscopic effects, then it can be described by a stable law. Stable distributions are leptokurtotic. When compared to normal distribution, they typically have fatter tails and higher peak around the center. Asymmetry is allowed. Due to its flexibilities, stable model fits the empirical distribution of the financial data better [see Mittnik et al. (2000)]. Gaussian distribution is a special case of stable distribution. In fact, it is the only distribution in the stable family with a finite second moment. Although autoregressive conditional heteroskedastic models driven by normally distributed innovations imply unconditional distributions that possess heavier tails, there is still considerable kurtosis unexplained by this model. Mittnik, Paolella and Rachev (2000) present empirical evidence favoring stable hypothesis over the normal assumption as a model for unconditional, homoskedastic conditional, and heteroskedastic conditional distributions of several asset return series.

The purpose of this chapter is to review the stochastic programming models in the ALM literature and to analyze an asset allocation problem in the presence of heavy tails. In the first part of the chapter, we review ALM models that utilize stochastic programming methodology. In the second part, a multistage asset allocation model with decision rules is analyzed under the Gaussian and stable returns scenarios. Our computational results suggest that if the investor has very high or low risk aversion, then the normal and stable scenarios result in similar asset allocations. However, when the risk aversion level is between the two cases, the two distributional assumptions may result in considerably different asset allocations depending on the utility function and the risk aversion level of the decision maker. The investor may reduce his equity allocation up to 20%, increase his certainty equivalent wealth up to 0.13% and improve his utility by 0.72% by switching to stable model.

Section 2 reviews the stochastic programming ALM models without imposing any decision rules. Section 3 describes stochastic programming models that assume the investor uses the same decision rule to update the asset allocation every period. In Section 4, we present the scenario generation methods available in the ALM literature. In the second part of the chapter, a stable asset allocation model is described. Section 5 states the reasons for desirability of the stable model, describes the distribution and presents estimation methods. In Section 6, we set up the asset allocation model and report the computational results. Section 7 concludes.

[3] We will call it stable distribution from now on.

[4] A well known alternative to stable model is Student-t distribution. A major drawback of Student-t distribution is its lack of stability with respect to summation, i.e., a portfolio of Student-t distributed asset returns does not have Student-t distribution. It is not supported by a central limit theorem. Student-t distribution is a symmetric distribution and it cannot capture the possible skewness in financial data.

Part I: Review of the stochastic programming ALM literature

2. Stochastic programming ALM models

This method provides a general-purpose modeling framework that conveniently addresses real world concerns such as transaction costs, taxes, legal and policy constraints. The number of decision variables becomes very large resulting in large scale optimization problems. The computational costs make it impractical to test the recommendations out of the sample.

We describe various modeling approaches developed within this framework:

2.1. *Chance-constrained model*

Charnes and Kirby (1966) develop a chance-constrained model that expresses future deposits and loan payments as jointly distributed random variables, and capital adequacy formula by chance-constraints on meeting withdrawal claims. A drawback of the model is that constraint violations are not penalized according to their magnitude.

The methodology has found applications in various areas: Charnes, Gallegos and Yao (1982) applies this methodology to balance sheet management, Li (1995a, b) uses chance-constrained programming in portfolio analysis of insurance companies, and Dert (1998) develops a multistage chance-constrained ALM model for a defined benefit pension fund. As opposed to the original approach of Charnes and Kirby, Dert models the uncertainty using scenarios rather than making distributional assumptions.

Dert's model minimizes the cost of funding while ensuring the stability of contributions and ability to make benefit payments timely with an acceptable level of insolvency risk. The solvency requirement is the asset level being at least equal to the product of required funding level with the value of the remaining liabilities (constraint (7)). The asset value falling below the required level is modeled as a probabilistic constraint. Since uncertainty is modeled through scenarios, binary variables are needed to formulate the chance constraint explicitly (constraints (8)–(10)). It is assumed that remedial contributions are made in case of under-funding (constraint (6)).

The ALM model is formulated as follows:

$$\min A_{01} + \sum_{t=1}^{T-1}\sum_{s=1}^{S_t} P(t,s)\gamma_{ts}Y_{ts} + \lambda\sum_{t=1}^{T}\sum_{s=1}^{S_t} P(t,s)\gamma_{ts}Z_{ts} \quad \text{subject to}$$

$$Y_{ts}^l \leqslant Y_{ts} \leqslant Y_{ts}^u, \tag{1}$$

$$y_{ts}^l \leqslant \frac{Y_{ts}}{W_{ts}} \leqslant y_{ts}^u, \tag{2}$$

$$\frac{Y_{ts}}{W_{ts}} - \frac{Y_{t-1,\hat{s}}}{W_{t-1,\hat{s}}} \leqslant \beta_t, \tag{3}$$

$$A_{ts} + Y_{ts} - l_{ts} = \sum_{i=1}^{N} X_{its}, \tag{4}$$

$$x^{l}_{its}(A_{ts} + Y_{ts} - l_{ts}) \leqslant X_{its} \leqslant x^{u}_{its}(A_{ts} + Y_{ts} - l_{ts}),$$
$$t = 0, \ldots, T-1,\ s = 1, \ldots, S_t, \tag{5}$$

$$A_{ts} = Z_{ts} + \sum_{i=1}^{N} \mathrm{e}^{r_{its}} X_{i,t-1,\hat{s}}, \tag{6}$$

$$A_{ts} \geqslant \alpha L_{ts}, \tag{7}$$

$$Z_{ts} \geqslant f_{ts} M_{ts}, \tag{8}$$

$$\sum_{s=1}^{S_t} P\big[(t,s)|(t-1,\hat{s})\big] f_{ts} \leqslant \Psi_{t-1,\hat{s}}, \tag{9}$$

$$f_{ts} \in \{0,1\}, \quad t = 0, \ldots, T-1,\ s = 1, \ldots, S_t, \tag{10}$$

where,

$t = 0, 1, \ldots, T$ is the time period,
$s = 1, 2, \ldots, S_t$ is the status of the world,
$i = 1, 2, \ldots, N$ is the asset class,
α is the demanded funding level,
β_t is the maximal raise in contribution per period as a fraction of the cost of wages at time t,
γ_{ts} is the discount factor for a cash flow at time t in state s,
l_{ts} is the benefit payments and costs to the fund at time t in state s,
L_{ts} is the actuarial reserve at time t in state s,
λ is the penalty parameter to penalize remedial contributions,
r_{its} is the continuous return on investment of each asset class i during period t in state s,
M_{ts} is the large constant at time t in state s,
W_{ts} is the cost of wages during period t in state s,
A_{ts} is the total asset value before receiving regular contributions and making benefit payments at time t in state s,
f_{ts} is the binary variable for remedial contributions at time t in state s,
Ψ_{ts} is the probability of under-funding at time $t+1$ given the world was in state s at time t,
X_{its} is the amount of money invested in asset class i at time t in state s,
x_{its} is the fraction of asset value invested in asset class i at time t in state s,
Y_{ts} is the regular contribution during period t in state s,
y_{ts} is the regular contribution as a fraction of the cost of wages during period t in state s,
Z_{ts} is the remedial contribution at time t in state s.

The first three constraints, namely (1)–(3), limit the regular contribution amount, regular contribution as a fraction of wages and maximal raise in contribution as a fraction of cost of wages, respectively. After receiving regular contributions and making benefit payments, the assets are reallocated (4) considering the upper and lower bounds on the asset mix (5).

The price inflation, wage inflation, and asset return scenarios are generated using vector autoregressive model. The characteristics of participants are modeled using a Markov chain. More detailed description of a similar model is given in Dert (1998).

2.2. Dynamic programming

The main idea behind dynamic programming is to solve the problem by dealing with one stage at a time. The procedure produces one solution per possible state in each stage. If there are many state variables or the objective function depends in an arbitrary way on the whole history up to the current period, this method is not very appropriate. It can handle small number of financial instruments simultaneously. Therefore it is of limited use in practice.

Eppen and Fama (1971) model a three-asset portfolio problem using this approach. At any point in time, they assume that state of the system is described by two variables: m being the level of cash balance ($m \in N$), and b being the level of bond account ($b \in \{N - N^-\}$). Decisions concerning the state of the system are made at equally spaced discrete points in time. The stochastic changes in the cash balance between the periods are a sequence of independent identically distributed random variables with the discrete probability mass function $p(d)$. The function $p(d)$ is positive only on a finite state space, i.e., there is a finite K such that $p(d) = 0$ if $|d| > K$.

The notation is as follows:

$T(m, b; m', b')$ is the minimum transfer cost involved in changing the state from (m, b) to (m', b'),

c_h is the marginal opportunity cost of starting a period with an additional dollar of cash,

c_p is loss of being a dollar short on cash which is incurred at the beginning of the period,

$L(m')$ is the penalty cost of carrying cash:

$$L(m') = \begin{cases} c_h m', & m' \geqslant 0, \\ -c_p m', & m' < 0, \end{cases}$$

α is the discount factor,

$f_n(m, b)$ is the discounted expected cost for an n period problem whose state at the beginning of period n is (m, b).

The recursive relationship for $f_n(m, b)$ is given by:

$$f_n(m, b) = \min_{m', b'} \left[T(m, b; m', b') + G_n(m', b') \right],$$

where

$$G_n(m,b) = c_h^b \cdot b + L(m) + \alpha \sum_{d=-K}^{K} f_{n-1}(m+d,b) \cdot p(d).$$

$G_n(m,b)$ is the current expected holding penalty cost (the first two terms) plus the discounted expected cost if a decision is made to start period n in state (m,b) and an optimal policy is followed in period $n-1$ and all future periods.

2.3. Sequential decision analysis

This approach uses implicit enumeration to find an optimal solution. It results in extremely large equivalent linear programming problems since it enumerates all possible portfolio strategies for all scenarios in all periods of consideration. The method ensures feasibility of the first period for every possible scenario, this shrinks the feasible set and gives substantial importance to scenarios with low probabilities of occurrence.

Stochastic decision tree model by Bradley and Crane (1972) overcomes the computational difficulties of the approach by using a decomposition algorithm. The objective is the maximization of expected terminal wealth of the firm. Constraint (11) guarantees that the firm cannot purchase assets that cost more than it has funds available. The second set of constraints balance the inventory. The net realized capital losses in a period are controlled by some pre-specified upper bound using (13). Constraint (14) limits the holding of a particular asset.

Their linear programming formulation[5] is

$$\max \sum_{e_N \in E_N} p(e_N) \left\{ \sum_{k=1}^{K} \sum_{m=0}^{N-1} [y_m^k(e_m) + u_{m,N}^k(e_N)] h_{m,N}^k(e_N) \right.$$

$$\left. + [y_N^k(e_N) + u_{N,N}^k(e_N)] b_N^k(e_N) \right\} \quad \text{subject to}$$

$$\sum_{k=1}^{K} b_n^k(e_n) - \sum_{k=1}^{K} \left[\sum_{m=0}^{n-2} y_m^k(e_m) h_{m,n-1}^k(e_{n-1}) + y_{n-1}^k(e_{n-1}) b_{n-1}^k(e_{n-1}) \right]$$

$$- \sum_{k=1}^{K} \sum_{m=0}^{n-1} [1 + g_{m,n}^k(e_n)] s_{m,n}^k(e_n) = f_n(e_n), \tag{11}$$

$$-h_{m,n-1}^k(e_n - 1) + s_{m,n}^k(e_n) + h_{m,n}^k(e_n) = 0, \quad m = 0, 1, \ldots, n-2, \tag{12}$$

[5] The formulation is taken from Kusy and Ziemba (1986).

$$-b^k_{n-1}(e_{n-1}) + s^k_{n-1,n}(e_n) + h^k_{n-1,n}(e_n) = 0,$$

$$h^k_{0,0}(e_0) = h^k_0,$$

$$-\sum_{k=1}^{K}\sum_{m=0}^{n-1} g^k_{m,n}(e_n)s^k_{m,n}(e_n) \leqslant L_n(e_n), \tag{13}$$

$$\sum_{k\in K_i}\left[b^k_n(e_n) + \sum_{m=0}^{n-1} h^k_{m,n}(e_n)\right] \leqslant C^i_n(e_n), \quad i = 1, 2, \ldots, I, \tag{14}$$

$$y^k_{m,n}(e_n) \geqslant 0,$$

$$s^k_{m,n}(e_n) \geqslant 0,$$

$$h^k_{m,n}(e_n) \leqslant 0, \quad m = 1, \ldots, n-1,$$

where

$e_n \in E_n, n = 1, 2, \ldots, N,$
$k = 1, 2, \ldots, K,$
e_n is an economic scenario from period 1 to n having probability $p(e_n)$,
E_n is the set of possible economic scenarios from period 1 to n,
K is the total number of assets,
K_i is the number of assets of type i,
N is the number of time periods,
$y^k_m(e_m)$ is the income yield per dollar of purchase price in period m of asset k conditional on scenario e_n,
$u^k_{m,N}(e_N)$ is the expected terminal value per dollar of purchase price in period m of asset k held at period N, conditional on scenario e_n,
$b^k_n(e_n)$ is the dollar amount of asset k purchased in period n conditional on scenario e_n,
$h^k_{m,n}(e_n)$ is the dollar amount of asset k purchased in period m and held in period n conditional on scenario e_n,
$s^k_{m,n}(e_n)$ is the dollar amount of asset k purchased in period m and sold in period n conditional on scenario e_n
$g^k_{m,n}(e_n)$ is the capital gain or loss per dollar of purchase price of asset k purchased in period m and sold in period n conditional on scenario e_n,
$f_n(e_n)$ is the incremental increase or decrease of funds available for period n,
$L_n(e_n)$ is the dollar amount of maximum allowable net realized capital losses in period n,
$C^i_n(e_n)$ is the upper bound in dollars on the amount of funds invested in asset type i in period n.

They use a decomposition algorithm to breakdown the problem and use an efficient technique to solve the sub-problems of the overall portfolio. However, the solution is still computationally intractable for real life problems.

2.4. Stochastic Linear Programming with Recourse (SLPR)

The basic formulation of the general T-stage SLPR model is

$$\min_{x_1}\left\{c_1'x_1 + E_{w_1}\left[\min_{x_2}\left(c_2(w_1)'x_2 + \cdots + E_{w_{T-1}}\left[\min_{x_T} c_T(w_{T-1})'x_T\right]\right)\right]\right\}$$

subject to

$$\begin{aligned}
&A_1x_1 &&= b_1,\\
&B_2(w_1)x_1 + A_2(w_1)x_2 &&= b_2(w_1),\\
&\qquad B_3(w_2)x_2 + A_3(w_2)x_3 &&= b_3(w_2),\\
& &&\ \vdots\\
&\qquad\qquad B_T(w_{T-1})x_{T-1} + A_T(w_{T-1})x_T &&= b_T(w_{T-1}),\\
&l_t \leqslant x_t \leqslant u_t, \quad t = 1, 2, \ldots, T,
\end{aligned}$$

where

w_t is the random vector that generates the coefficients b_t, c_t, A_t, and B_t of the decision problem at the t-th stage, $t = 2, \ldots, T$,
l_t, u_t are the vector of deterministic bounds on x_t at stage t, $t = 2, \ldots, T$,
b_1, c_1, and A_1 are the deterministic first stage coefficient vectors or matrices, and
x_t is the vector decision variable.

The objective formalizes a sequence of optimization problems corresponding to different stages. At stage 1, the outcome completely depends on future realizations of the uncertainty. After the first period, decisions are allowed to be a function of the observed realization (x_{t-1}, w_t) only. One first decides on x_1, then observes w_1, then decides on x_2, then observes w_2, and so on. The recourse decisions depend on the current state of the system as determined by previous decisions and random events. The uncertainty is modeled by using finite scenarios which have pre-assigned probabilities. In this case, the problem reduces to a large linear program of a special structure:

$$\min\left\{c_1'x_1 + \sum_{k_2=2}^{K_2} p_{k_2}c_{k_2}'x_{k_2} + \sum_{k_3=K_2+1}^{K_3} p_{k_3}c_{k_3}'x_{k_3} + \cdots + \sum_{k_T=K_{T-1}+1}^{K_T} p_{k_T}c_{k_T}'x_{k_T}\right\}$$

subject to

$$\begin{aligned}
&A_1x_1 &&= b_1,\\
&B_{k_2}x_1 + A_{k_2}x_2 &&= b_{k_2}, \quad k_2 = 2, \ldots, K_2,\\
&\qquad B_{k_3}x_{a(k_3)} + A_{k_3}x_{k_3} &&= b_{k_3}, \quad k_3 = K_2 + 1, \ldots, K_3,\\
& &&\ \vdots\\
&\qquad\qquad B_{k_T}x_{a(k_T)} + A_{k_T}x_{k_T} &&= b_{k_T}, \quad k_T = K_{T-1+1}, \ldots, K_T,\\
&l_t \leqslant x_{k_t} \leqslant u_t, \quad k_t = K_{t-1} + 1, \ldots, K_t, \quad t = 1, 2, \ldots, T.
\end{aligned}$$

The scenarios used determine the size, form and optimal solution of the linear program. There are finitely many sequences of possible realizations of the random coefficients

$(c_{k_t}, A_{k_t}, B_{k_t}, b_{k_t})$ with path probabilities p_{k_t} of the subsequences of these realizations, $p_{k_t} > 0$, $\forall k_t$, $\sum_{k_t=K_{t-1}+1}^{K_t} p_{k_t} = 1$, $t = 2, \ldots, T$, that identify the discrete joint probability distribution of $w = \{w_1, \ldots, w_{T-1}\}$. In the program, $a(k_t)$ denotes the immediate predecessor of k_t, for example $a(k_2) = k_1$.

An important application of stochastic linear programming with simple recourse model is given by Kusy and Ziemba (1986). The model was developed for the Vancouver City Savings Credit Union for a 5-year planning period. The formulation has the following features:

(1) Changing yield spreads across time, transaction costs associated with selling assets prior to maturity, and synchronization of cash flows across time are incorporated in a multiperiod context.
(2) Assets and liabilities are considered simultaneously to satisfy basic accounting principles and match liquidities.
(3) Transaction costs are included.
(4) Uncertainty of withdrawal claims and deposits is reflected in uncertain cash flows.
(5) Uncertainty of interest rates is explicitly recognized.
(6) Legal and policy constraints are taken into account.

Their two-stage model did not contain end effects. Three possible scenarios that are independent over time were considered to keep the computations tractable. Their results indicate that their model generates policies that are superior than stochastic decision analysis.

Another milestone after the Kusy and Ziemba model is the Russell–Yasuda Kasai model by Carino et al. (1994). The model builds on the previous research to design a large scale SLPR model with possibly dependent scenarios, end effects, and all the relevant institutional and policy constraints. We present their model next.

Decision variables are

V_t: total fund market value at time t,
X_{nt}: market value in asset n at time t,
w_{t+1}: income shortfall at time $t+1$, and
v_{t+1}: income surplus at time $t+1$.

Random coefficients are

RP_{nt+1}: price return of asset n from end of t to end of $t+1$,
RI_{nt+1}: income returns of asset n from end of t to end of $t+1$,
F_{t+1}: deposit inflow from end of t to end of $t+1$,
P_{t+1}: principal payout from end of t to end of $t+1$,
I_{t+1}: interest payout from end of t to end of $t+1$,
g_{t+1}: rate at which interest is credited to policies from end of t to end of $t+1$,
L_t: liability valuation at t.

The objective is to maximize the expected market value of the firm at the horizon net of penalties for the shortfalls. Expected amount by which goals are not achieved is a more tangible risk measure than variance. The penalty costs of shortfalls may be based on expected financial impact or psychological costs. The piecewise linear convex cost function for the shortfall is denoted by $c_t(w_t)$. (15) is the budget constraint. The return on assets and inflow of deposits net of principal and interest payout gives the total fund market value (16).

Liability balances and cash flows are computed to model liability accumulation (18). If Yasuda does not achieve adequate income, recourse action must be taken at a cost. The income generation is modeled as a soft constraint (17), which permits surpluses or deficits.

$$\max E\left[V_t - \sum_{t=1}^{T} c_t(w_t)\right] \quad \text{subject to}$$

$$\sum_n X_{nt} - V_t = 0, \tag{15}$$

$$V_{t+1} - \sum_n (1 + RP_{nt+1} + RI_{nt+1})X_{nt} = F_{t+1} - P_{t+1} - I_{t+1}, \tag{16}$$

$$\sum_n RI_{nt+1}X_{nt} + w_{t+1} - v_{t+1} = g_{t+1}L_t, \tag{17}$$

$$L_{t+1} = (1 + g_{t+1})L_t + F_{t+1} - P_{t+1} - I_{t+1}, \tag{18}$$

$$X_{nt} \geqslant 0, \quad w_{t+1} \geqslant 0, \quad v_{t+1} \geqslant 0.$$

The abbreviated formulation does not include some elements of the model. There are additional types of shortfalls, indirect investment types, regulatory restrictions, multiple accounts, loan assets, tax effects and end effects that are included in the original model. See Carino and Ziemba (1998) for the details of the formulation. Carino, Myers and Ziemba (1998) discusses the concepts, technical issues and uses of the model.

Korhonen (1987) applies SLPSR to multicriteria decision making. Oguzsoy and Guven (1997) use the SLPSR methodology for a bank ALM model in Turkey. Geyer et al. (2002) describes a pension fund planning model that utilizes this approach.

Some authors argue against linearizing the objective function. Bai, Carpenter and Mulvey (1997) demonstrates that nonlinear programs are not much more difficult than their linear counterparts. Zenios et al. (1998) applies multistage stochastic nonlinear programming with recourse to fixed income portfolio management.

2.5. *Dynamic generalized networks*

Multistage stochastic nonlinear programs with recourse can be represented by generalized network formulations. This framework can be used to account for the dynamic aspects of ALM problems while considering uncertainty in all relevant parameters and accommodating random parameters by means of a moderate number of scenarios.

The network structure is exploited in the solution procedure. The problem is decomposed into its constituent scenario subproblems. Preserving the network structure of each subproblem is challenged by the existence of non-anticipativity constraints. These constraints dictate that scenarios that share common information history up to a specific period must yield the same decision up to that period, i.e., dependence on hindsight is avoided.

The desired decomposition is achieved by dualizing the non-anticipativity constraints. The algorithm by Rockafeller and Wets (1991) operates on the split-variable form of the original problem. The problem is solved by progressively enforcing the non-anticipativity constraints.

Mulvey (1994) utilizes this methodology in designing an asset allocation model for the Pacific Financial Asset Management Company. The single period portfolio model is formulated as a network model. The arcs can be constrained to impose legal or policy constraints. The objective function is the expected utility of surplus at the end of the planning horizon. The model was implemented in a PC environment with acceptable accuracy and efficiency.

Mulvey and Vladimirou (1989, 1991, 1992) present several aspects of stochastic generalized network models. See also the review of Mulvey and Ziemba (1995) which discusses the model in a general context.

2.6. *Scenario optimization*

According to the scenario optimization approach, one computes a solution to the deterministic problem under all scenarios then solves a coordinating model to find a single feasible policy. This approach can be compared to the scenario aggregation method suggested by Rockafeller and Wets (1991). It handles multistage stochastic programming problems, and allows for decisions to depend on future outcomes. On the other hand, scenario optimization is designed for two-period models only. It is assumed that scenario probabilities are functions of time, and estimates of the random parameters in the future stages are poor. Hence one only selects a policy for the immediate future.

Suppose the scenario subproblem is

$$v_s = \min c_s^{\mathrm{T}} x \quad \text{subject to}$$

$$A_s x = b_s, \tag{19}$$

$$A_d x = b_d, \tag{20}$$

$$x \geqslant 0,$$

where the objective function is a particular realization of the uncertainty under scenario s, (19) is also a particular realization of the uncertain constraints under scenario s, and (20) is the deterministic constraints.

A possible coordinating model could be

$$\min \sum_s p_s \left\| c_s^{\mathrm{T}} x - v_s \right\|^2 + \sum_s p_s \| A_s x - b_s \|^2 \quad \text{subject to}$$

$$A_d x = b_d,$$

$$x \geqslant 0.$$

The coordinating model tracks the scenario solution as close as possible while still maintaining feasibility. Alternative coordination models are discussed in Dembo (1991, 1993). Illustrative applications in portfolio immunization and dedication are also presented therein.

2.7. Robust optimization

Robust optimization approach integrates goal programming formulations with a scenario based description of the uncertainty in the data. The aim is to produce solutions that are relatively less sensitive to the realizations of different scenarios. The objective function, in its most general form, is composed of two terms: the first term trades off between mean value and the variability in the mean; the second term is a feasibility penalty function. Consider the following formulation.

$$\min \sigma(x, y_1, \ldots, y_s) + w\rho(z_1, \ldots, z_s) \quad \text{subject to}$$

$$Ax = b,$$

$$B_s x + C_s y_s + z_s = e_s, \quad \forall s \in \Omega,$$

$$x \geqslant 0, \quad y_s \geqslant 0, \quad \forall s \in \Omega,$$

where

x is the vector of decision variables whose value cannot be adjusted once a specific realization of the data is observed,
y is the vector of decision variables that are subject to adjustment once uncertain parameters are observed,
z is the vector of decision variables that measure infeasibility allowed,
$s \in \Omega = \{1, \ldots, S\}$ is the set of possible scenarios,
A, b, B_s, C_s, e_s are the coefficients related to the variables,
w is the goal programming weight that is used to derive a spectrum of answers that trade-off the two objectives.

The inclusion of higher order moments in the objective function reduces the variability of the solution. Hence, few adjustments become necessary as scenarios unfold. The model recognizes that it may not always be possible to find a feasible solution to the problem under all scenarios. The penalty function is used to detect the least amount of infeasibilities to be dealt with outside the model. See Mulvey, Carpenter and Mulvey (1995) for possible objective function choices and their applications. Bai, Vanderbrei and Zenios (1997) argues that linear objective functions fail to identify robust solutions and concave utility functions produce much better results for risk averse decision makers even when penalty term is not used. Both papers compare robust optimization with stochastic linear programming approach (SLP). Since SLP optimizes only the first moment of the distribution of the objective value, more adjustment is needed as scenarios are realized. However, there is no mechanism for choosing w, and the cost of the robust solution may be higher than that of SLP.

3. Multistage stochastic ALM programming with decision rules

In this method, time is discretized into n-stages across the planning horizon, and investments are made using a decision rule, e.g., fixed mix, at the beginning of each time period. The decision rule can easily be tested with out-of-sample scenarios and confidence limits on the recommendations can be constructed. The use of this approach hinges on discovering policies that are intuitive and that will produce superior results. Decision rules may lead to non-convexities and highly nonlinear functions. Some decision rules used in the literature are fixed mix, no rebalancing, life cycle mix (Berger and Mulvey, 1998), constant proportional portfolio insurance (Perold and Sharpe, 1988), target wealth path tracking (Mulvey and Ziemba, 1998).

Boender (1997) and Boender, van Aalst and Heemskerk (1998) describe an ALM model designed for Dutch pension funds. Their goal is to find efficient frontiers of initial asset allocations, which minimize the value of downside risk for given certain values of average contribution rates. The scenarios are generated across time horizon of interest. The management selects a funding policy, an indexation policy of the earned pension rights, and an investment decision rule. These strategies are simulated against generated scenarios. Then, the objective function of the optimization problem is a completely specified simulation model except for the initial asset mix. The hybrid simulation/optimization model has the following three steps:

(1) Randomly generate initial asset mixes, simulate them, and evaluate their contribution rates and downside risks.
(2) Select the best performing initial asset mixes that are located at a minimal critical distance from each other.
(3) Use a local search algorithm to identify the optimal initial asset mix.

Maranas et al. (1997) adopt another approach to stochastic programming with decision rules. They determine the optimal parameters of the decision rule by means of a global optimization algorithm. They propose a dynamically balanced investment policy which is specified by the following parameters:

w_0: initial dollar wealth,
r_{it}^s: percentage return of asset $i \in \{1, 2, \ldots, I\}$ in time period $t \in \{1, 2, \ldots, T\}$ under scenario $s \in \{1, 2, \ldots, S\}$,
p^s: probability of occurrence of scenario s.

The decision variables are:

w_t^s: dollar wealth at time t in scenario s,
λ_i: fraction of wealth invested in asset category i (note that it is constant over time).

The model is a multiperiod extension of mean–variance method. The multi-period efficient frontier is obtained by varying β $(0 \leqslant \beta \leqslant 1)$. The formulation is as follows:

$$\max_{\lambda_i, w_t^s} \ \beta \operatorname{mean}(w_T) - (1-\beta) \operatorname{var}(w_T) \quad \text{subject to}$$

$$w_T^s = w_0 \prod_{t=1}^{T} \left[\sum_{i=1}^{I} (1 + r_{it}^s) \lambda_i \right], \quad s = 1, \ldots, S, \tag{21}$$

$$\sum_{i=1}^{I} \lambda_i = 1, \tag{22}$$

$$0 \leqslant \lambda_i \leqslant 1, \quad i = 1, \ldots, I.$$

The wealth accumulation is governed by (21). When (21) is substituted into the objective function, we get a nonconvex multivariable polynomial function in λ_i involving multiple local minima. A global optimization tool has been developed which obtains the above efficient frontier.

There are other models constructed using similar methodologies. Hiller and Eckstein (1993) design a stochastic dedication model to construct the optimal passive fixed income portfolio. Berger and Mulvey (1998) describe Home Account Advisor™ which assists individual investors in ALM using decision rules. Sweeney et al. (1998) applies a simulation/optimization scenario approach to optimal insurance asset allocation in a multi-currency environment.

4. Scenario generation

A scenario gives a single set of outcomes for the random variables in the model over the planning horizon. A representative set of scenarios describes the possible future financial environmental. Traditional quantitative forecasting methods extrapolate new ideas about future developments based on the knowledge of the past and present. However, the economic environment may change invalidating the past assumptions. Hence, subjective beliefs of the management has become an essential part of scenario building. See Bunn and Salo (1993) for a review of qualitative scenario generation techniques.

The earlier ALM models use a few number of independent scenarios to describe uncertainty. The recent models have become more sophisticated in scenario generation methods. The following are the widely accepted techniques in the literature.

4.1. *Discrete time series model*

4.1.1. *Multivariate approach*

Using Vector Autoregression (VAR) in the analysis of economic systems became popular after an influential paper by Sims (1980). Following Sims, Dert (1998) uses this methodology to generate scenarios for a pension plan. He creates future price inflation, wage inflation, stock returns, bond returns, cash return and real estate returns that are consistent with historical patterns in means, standard deviations, autocorrelations and cross correlations between state variables. A Markov model is used in determining future development of each individual participating in the pension plan. Carino et al. (1994) also employ VAR in generating scenarios for the Yasuda Kasai model. Asset return predictability papers in finance also generally use VAR methodology to predict the return on a stock or bond index.

There are a number of variables used to predict stock returns in various studies. Brennan, Schwartz and Lagnado (1997) use Treasury bill rate, Treasury bond rate and dividend yield as state variables in their model. Brandt (1999) uses lagged excess return on NYSE index over Treasury bill rate as a state variable in addition to dividend yield, default spread and term spread.

VAR may sometimes diverge from long-term equilibrium. Boender, van Aalst and Heemskerk (1998) extend VAR model to a Vector Error Correction Model (VECM) which additionally takes economic regime changes and long term equilibria into account. First, a sub-model generates future economic scenarios. Then, a liability sub-model determines the earned pension rights and payments corresponding to each economic scenario.

The economic scenario sub-module uses time series analysis. The vector of the log-normal transformations of inflation, wage growth, bond return, cash return, equity return, real estate return and nominal GNP growth is y_t. Diagnostic tests revealed that the order of the VAR process as 1.

$$y_t \sim N\big(\mu + \Psi * \{y_{t-1} - \mu\}, \Sigma\big),$$

where $N(\mu, \Sigma)$ denotes a Gaussian distribution with mean μ and covariance matrix Σ.

The extended VECM is given as

$$y_t \sim N\big(\Psi_1 y_{t-1} + \Psi_2 C^T (x_{t-1} - \mu_1 I_{\{T_1\}} - \mu_2 I_{\{T_2\}}), \Sigma\big),$$

where the Ψ_1 corresponds to the short term dynamics and the Ψ_2 corresponds to the long term correction. The index set T_1 specifies the period of an economic regime with growth vector μ_1, and T_2 gives the period of another economic regime with growth vector μ_2. The second term, $C^T(x_{t-1} - \mu_1 I_{\{T_1\}} - \mu_2 I_{\{T_2\}})$ generates the error correction to restore violations of the equilibria, while Ω_2 determines the speed of the response. They estimated the model by row wise ordinary least squares and seemingly unrelated regression methods. Then, scenarios are generated iteratively using the parameter estimates. They report that the VECM improves the explanatory power of the model. The VECM has a more clear economic interpretation which incorporates regime changes and long run equilibrium.

The liability sub-module uses a push Markov model to determine the future status of each individual plan member depending on age, gender, and employee category. Given this information, the pull part of the model is used to determine additional promotions and new employees. Then, the pension rules are applied to compute the guaranteed pension payments and earned pension rights.

4.1.2. Cascade approach

Wilkie (1986) suggests using a cascade structure rather than a multivariate model, in which each variable could affect each of the others. He considers inflation, ordinary shares and fixed interest securities as the main economic determinants of a stochastic investment model. The model includes the following variables: inflation, an index of share of dividends, the dividend yield (the dividend index divided by the corresponding price index)

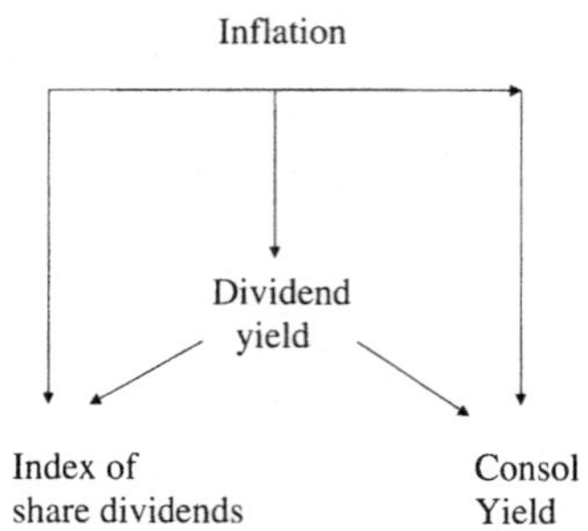

Fig. 1. Wilkie's scenario generation model.

on these share indices, and the yield on consols (as a measure of the general level of fixed interest yields in the market).

Wilkie's investigations and actuarial experience lead him to the conclusion that inflation is the driving force for the other investment variables. Figure 1,[6] where the arrows indicate the direction of influences, depicts the cascade structure of the model.

The inflation is described first using a first order autoregressive model. The dividend yield depends on both the current level of inflation and the previous values of itself. The index of share dividends depends on inflation and the residual of the yield model. The consol yield also depends on inflation and the residual of the yield model along with the previous values of itself. Then, the estimated parameters are used to generate future economic scenarios. Wilkie (1986) improves this basic model.

4.2. Continuous time model

Mulvey (1996) designs an economic projection model for Towers Perrin using stochastic differential equations. The model has a cascade structure as depicted in Figure 2.[7] First the Treasury yield curve, and then government bond returns, price and wage inflation, and large cap returns are generated. Lastly, returns on primary asset categories such as small cap stock and corporate bonds are projected.

It is assumed that short- and long-term interest rates (denoted by r_t and l_t, respectively) are linked through a correlated white noise term. The spread between the two is kept under control by using a stabilizing term. This variant of the two-factor Brennan and Schwartz (1982) model is as follows:

$$\mathrm{d}r_t = a(r_0 - r_t)\,\mathrm{d}t + b\sqrt{r_t}\,\mathrm{d}z_1,$$

$$\mathrm{d}l_t = c(l_0 - l_t)\,\mathrm{d}t + e\sqrt{l_t}\,\mathrm{d}z_2,$$

[6] Source: Wilkie (1986).

[7] Resource: Mulvey (1996).

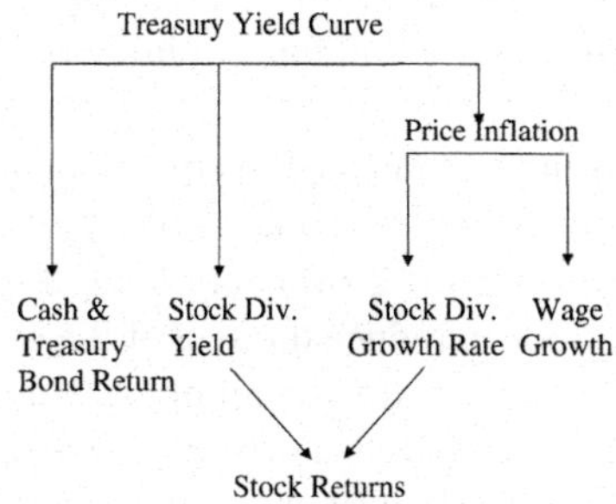

Fig. 2. Mulvey's scenario generation model.

where a and c are functions that depend on the spread between the long and short rates, b and e are constants, and $\mathrm{d}z_1$ and $\mathrm{d}z_2$ are correlated Weiner terms.

The price inflation rate is modeled as a diffusion process that depends on short term interest rate:

$$\mathrm{d}p_t = n\,\mathrm{d}r_t + g(p_0 - p_t)\,\mathrm{d}t + h(v_{pt})\,\mathrm{d}z_3,$$

where p_t is the price inflation at time t, and v_{pt} is the stochastic volatility at time t. Since the volatility of inflation persists, it is represented using Autoregressive Conditional Heteroskedasticity (ARCH) model. The equation for the stochastic volatility is given by:

$$\mathrm{d}v_{pt} = k(v_{p0} - v_{pt})\,\mathrm{d}t + m\sqrt{v_{pt}}\,\mathrm{d}z_4,$$

where g and k are functions that handle the independent movement of the underlying prices at time t for the price inflation and stochastic volatility, respectively, and h and m are constants.

Real yields are related to interest rates, current inflation, and expectations for future inflation. The diffusion equation for long-term yield is

$$\mathrm{d}y_t = n(y_u, y_t, l_u, l_t, p_u, p_l)\,\mathrm{d}l + q(y_u, y_t, l_u, l_t, p_u, p_l)\,\mathrm{d}t + u(y_t)\,\mathrm{d}z_5,$$

where y_u is the normative level of real yields, n and q are vector functions that depend upon various economic factors.

The wage inflation is connected to price inflation in a lagged and smoothed fashion. The stock returns are broken down into two components: dividends and capital appreciation, and they are estimated independently. Mulvey reports that the decomposed structure provides more accurate linkages to the key economic factors such as inflation and interest rates.

The parameters of the model are calibrated by considering the overall market trends in the light of historical evidence and subjective beliefs of the management. This model has been in use at Towers Perrin since 1992. Mulvey and Thorlacius (1998) extend the model to a global environment that links the economies of individual countries within a common framework.

Modeling term structure of interest rates is a very essential part of scenario generation. The use of binomial lattice models in the valuation of interest rate contingencies is prevalent. However, the number of scenarios grows very large if the valuation is to be precise. There are some sampling methods to reduce the size of the event tree such as Monte Carlo simulation, antithetic sampling and stratified sampling. However, Klaasen (1997) points out that even if the underlying description is arbitrage-free, a subset of it may include arbitrage opportunities that may lead to spurious profits. Instead of sampling paths, Klaasen (1998a) suggests an aggregation method that can be used to reduce the size of the event tree preserving the arbitrage free description of uncertainty. In Klaasen (1998b), he presents a solution algorithm which iteratively disaggregates the condensed representation towards a more precise description of uncertainty.

Part II: Stable asset allocation

In this part of the chapter, we analyze asset allocation problem for an investor that maximizes isoelastic utility function or an analog of mean-variance objective function at the end of the investment horizon. Stochastic programming with decision rules is used as the solution methodology. The financial uncertainty is represented by using a branching event tree. Each node of the tree represents the joint outcome of all the random variables at that decision stage and each path through the event tree represents a 'scenario'. Financial scenarios are generated by using asset return predictors documented in the literature. The investor perceives the world as modeled by the scenario tree and chooses the fix mix proportions that maximize his objective function. The asset allocation problem is solved under Gaussian and stable return scenarios. Optimal allocations and utility function values under these alternative sets of scenarios are compared.

We state the reasons for desirability of the stable model, describe the distribution and present estimation methods in Section 5. Section 6 sets up the asset allocation model and reports computational results. Section 7 concludes.

5. Stable distribution

There are several important reasons for modeling financial variables using stable distributions. The stable law is supported by a generalized central limit theorem (Embrechts et al., 1997; Rachev and Mittnik, 2000). Stable distributions are leptokurtotic. Since they can accommodate the fat tails and asymmetry, they fit empirical distribution of the financial data better.

Any distribution in the domain of attraction of a specified stable distribution will have properties, which are close to the ones of stable distribution. Even if the observed data does not exactly follow the ideal distribution specified by the modeler, in principle, the resulting decision is not affected.

Each stable distribution has an index of stability, which remains the same regardless of the sampling interval adopted. The index of stability can be regarded as an overall parameter that can be employed in inference and decision making. However, we should note that for some financial data empirical analysis shows that the index of stability increases as the sampling interval increases.

It is possible to check whether a distribution is in the domain of attraction of a stable distribution or not by examining the tails of the distribution. The tails dictate the properties of the distribution.

This section describes the properties of stable distribution and addresses the estimation issues.

5.1. Description of stable distributions

If the sums of linear functions of independent identically distributed (iid) random variables belong to the same family of distributions, the family is called stable. Formally, a random variable r has stable distribution if for any $a > 0$ and $b > 0$ there exists constants $c > 0$ and $d \in R$ such that

$$ar_1 + br_2 \stackrel{d}{=} cr + d,$$

where r_1 and r_2 are independent copies of r, and $\stackrel{d}{=}$ denotes equality in distribution. The distribution is described by the following parameters: $\alpha \in (0, 2]$ (index of stability), $\beta \in [-1, 1]$ (skewness parameter), $\mu \in R$ (location parameter), and $\sigma \in [0, \infty)$ (scale parameter). The variable is then represented as $r \sim S_{\alpha,\beta}(\mu, \sigma)$. Gaussian distribution is actually a special case of stable distribution when $\alpha = 2$, $\beta = 0$. The smaller the stability index is, the stronger the leptokurtic nature of the distribution becomes, i.e., with higher peak and fatter tails. If the skewness parameter is equal to zero, as in the case of Gaussian distribution, the distribution is symmetric. When $\beta > 0$ ($\beta < 0$), the distribution is skewed to the right (left). If $\beta = 0$ and $\mu = 0$, then the stable random variable is called symmetric α-stable ($S\alpha S$). The scale parameter generalizes the definition of standard deviation. The stable analog of variance is variation, v_α, which is given by σ^α.

Stable distributions generally do not have closed form expressions for density and distribution functions. They are more conveniently described by characteristic functions. The characteristic function of random variable r, $\Phi_r(\theta) = E[\exp(\mathrm{i}r\theta)]$, is given by

$$\Phi_r(\theta) = \begin{cases} \exp\left\{-\sigma^\alpha |\theta|^\alpha \left(1 - \mathrm{i}\beta \operatorname{sign}(\theta) \tan\frac{\pi\alpha}{2}\right) + \mathrm{i}\mu\theta\right\}, & \text{if } \alpha \neq 1, \\ \exp\left\{-\sigma |\theta| \left(1 - \mathrm{i}\beta \frac{2}{\pi} \operatorname{sign}(\theta) \ln\theta\right) + \mathrm{i}\mu\theta\right\}, & \text{if } \alpha = 1. \end{cases}$$

The p-th absolute moment of r, $E|X|^p = \int_0^\infty P(|X|^p > y)\,\mathrm{d}y$, is finite if $0 < p < \alpha$, and infinite otherwise. Hence, when $\alpha < 1$ the first moment is infinite, and when $\alpha < 2$ the second moment is infinite. The only stable distribution that has finite first and second moments is the Gaussian distribution.

In models that use financial data, it is generally assumed that $\alpha \in (1, 2]$. There are several reasons for this:

(1) When $\alpha > 1$, the first moment of the distribution is finite. It is convenient to be able to speak of expected returns.
(2) Empirical studies support this parametrization.
(3) Although the empirical distributions of the financial data sometimes depart from normality, the deviation is not "too much".

In scenario generation, one may need to use multivariate stable distributions. The extension to the multivariate case is nontrivial. Although most of the literature concentrates on the univariate case, recently some new results have become available. See for example Samorodnitsky and Taqqu (1994), Rachev and Mittnik (2000).

If R is a stable d-dimensional stable vector, then any linear combination of the components of R is also a stable random variable. However, the converse is true under certain conditions (Samorodnitsky and Taqqu, 1994). The characteristic function of R is given by:

$$\Phi_Y(\theta) = \begin{cases} \exp\left\{-\int_{S_d} |\theta^{\mathrm{T}} s| \left(1 - \mathrm{i}\,\mathrm{sign}\,(\theta^{\mathrm{T}} s) \tan \frac{\pi\alpha}{2}\right) \Gamma(\mathrm{d}s) + \mathrm{i}\theta^{\mathrm{T}}\mu\right\}, & \text{if } \alpha \neq 1, \\ \exp\left\{-\int_{S_d} |\theta^{\mathrm{T}} s| \left(1 + \mathrm{i}\frac{2}{\pi}\,\mathrm{sign}\,(\theta^{\mathrm{T}} s) \ln|\theta^{\mathrm{T}} s|\right) \Gamma(\mathrm{d}s) + \mathrm{i}\theta^{\mathrm{T}}\mu\right\}, & \text{if } \alpha = 1, \end{cases}$$

where Γ is the spectral measure which replaces the scale and skewness parameters that enter the description of the univariate stable distribution. It is a bounded nonnegative measure on the unit sphere S_d, and $s \in S_d$ is the integrand unit vector. The index of stability is again α, and μ is the vector of locations.

Stable distributions have infinite variances. The stable equivalent of covariance for $S\alpha S$ variables is covariation:

$$[R_1, R_2]_\alpha = \int_{S_d} s_1 s_2^{\langle\alpha-1\rangle} \Gamma(\mathrm{d}s),$$

where (R_1, R_2) is a $S\alpha S$ vector ($\alpha \in (1, 2)$), and $x^{\langle\alpha-1\rangle} = |x|^k\,\mathrm{sign}(x)$. The matrix of covariations determines the dependence structure among the individual variables.

5.2. *Financial modeling and estimation*

Financial modeling frequently involves information on past market movements. Examples include technical analysis to derive investment decisions, or researchers assessing the efficiency of financial markets. In such cases, it is not the unconditional return distribution which is of interest, but the conditional distribution, which is conditioned on information contained in past return data, or a more general information set. The class of autoregressive moving average (ARMA) models is a natural candidate for conditioning on the past of a return series. These models have the property that the conditional distribution is homoskedastic. In view of the fact that financial markets frequently exhibit volatility clusters, the homoskedasticity assumption may be too restrictive. As a consequence, conditional heteroskedastic models, such as Engle's (1982) autoregressive conditional heteroskedastic (ARCH) models and the generalization (GARCH) of Bollerslev (1986), possibly in combination with an ARMA model, referred to as an ARMA-GARCH model, are now common in empirical finance. It turns out that ARCH-type models driven by normally distributed innovations imply unconditional distributions which themselves possess heavier tails. Thus, in this respect, ARCH models and stable distributions can be viewed as competing hypotheses.

Mittnik, Rachev and Paolella (1997) present empirical evidence favoring stable hypothesis over the normal assumption as a model for unconditional, homoskedastic conditional, and heteroskedastic conditional distributions of several asset return series.

5.2.1. Maximum likelihood estimation

We will describe an approximate conditional maximum-likelihood (ML) estimation procedure suggested by Mittnik et al. (1999). The unconditional ML estimate of $\theta = (\alpha, \beta, \mu, \sigma)$ is obtained by maximizing the logarithm of the likelihood function

$$L(\theta) = \prod_{t=1}^{T} S_{\alpha,\beta}\left(\frac{r_t - \mu}{\sigma}\right)\sigma^{-1}.$$

One needs to use conditional ML to estimate ARMA and ARMA-GARCH models. The ML estimation is conditional, in the sense that, when estimating, for example, an ARMA(p, q) model, one conditions on the first p realizations of the sample, r_p, $r_{p-1}, \ldots, r_1$, and, when $\alpha > 1$ holds, sets innovations $\varepsilon_p, \varepsilon_{p-1}, \ldots, \varepsilon_{p-q+1}$ to their unconditional mean $\mathbf{E}(\varepsilon_t) = 0$. The estimation of all stable models is approximate in the sense that the stable density function, $S_{\alpha,\beta}(\mu, \sigma)$, is approximated via fast Fourier transformation (FFT) of the stable characteristic function,

$$\int_{-\infty}^{\infty} e^{itx}\, dH(x) = \begin{cases} \exp\left\{-\sigma^{\alpha}|t|^{\alpha}\left[1 - i\beta \operatorname{sign}(t)\tan\dfrac{\pi\alpha}{2}\right] + i\mu t\right\}, & \text{if } \alpha \neq 1, \\ \exp\left\{-\sigma|t|\left[1 + i\beta\dfrac{2}{\pi}\operatorname{sign}(t)\ln|t|\right] + i\mu t\right\}, & \text{if } \alpha = 1, \end{cases}$$

where H is the distribution function corresponding to $S_{\alpha,\beta}(\mu, \sigma)$.

This ML estimation method essentially follows that of DuMouchel (1973), but differs in that the stable density is approximated numerically by an FFT of the characteristic function rather than some series expansion. As DuMouchel shows, the resulting estimates are consistent and asymptotically normal with the asymptotic covariance matrix of $T^{1/2}(\hat{\theta} - \theta_0)$

being given by the inverse of the Fisher information matrix. The standard errors of the estimates are obtained by evaluating the Fisher information matrix at the ML point estimates. For details on stable ML estimation see Mittnik et al. (1999), Mittnik and Rachev (1993), and Paulauskas and Rachev (1999).

5.2.2. *Comparison of estimation methods*

When the residuals of the ARMA model have Gaussian distribution, Least Squares (LS) estimation is equivalent to conditional ML estimation. Furthermore, Whittle estimator is asymptotically equivalent to LS and ML estimation methods. However, when the innovations have stable distribution, the properties of conventional estimation methods may change due to the infinite variance property. In the stable case, ML estimates are still consistent and asymptotically normal (DuMouchel, 1973); LS and Whittle estimates are consistent but they are not asymptotically normal. The LS and Whittle estimates have infinite variance limits with a convergence rate that is faster than that of the Gaussian case (Mikosch et al., 1995). Calder and Davis (1998) compare LS, Least Absolute Deviation (LAD), and ML methods for the estimation of ARMA model with stable innovations. Their simulations reveal that the difference between the estimates of the three methods is insignificant when the index of stability of the residuals is 1.75. However, when $\alpha = 1$ or $\alpha = 0.75$, they report that the LAD and ML estimation procedures are superior to LS estimation. ML estimation has desirable properties in both the Gaussian and stable setting, but it is computationally very demanding. Since the variables of interest in this paper have indices of stability greater than 1.5, nonlinear LS estimation method has been utilized in this study. Our parameter estimates are consistent, but they are not asymptotically normal. However, due to the high index of stability, the parameter estimates are comparable to those that would be achieved if ML estimation were to be used.

6. Multistage stable asset allocation model with decision rules

The asset allocation problem for an investor that maximizes isoelastic utility function or an analog of mean-variance objective function at the end of the investment horizon is formulated as follows:

$$\max E\big[u\big(\acute{R}^i_{s,T}\big)\big] \quad \text{subject to}$$

$$\acute{R}^i_{s,T} = \prod_{t=1}^{T} \left(1 + R^i_{s,t}\right) - 1,$$

$$R^i_{s,t} = \sum_{j=1}^{J} w^i_j r_{jst},$$

$$w^i_j \geqslant 0,$$

where w_j^i is the proportion of funds[8] of portfolio i invested in asset j,

$\acute{R}_{s,T}^i$ is compound return of allocation i in time period of 1 through T under scenario $s \in \{1, 2, \ldots, S\}$,

$R_{s,t}^i$ is the return of the portfolio i under scenario $s \in \{1, 2, \ldots, S\}$ in time period $t \in \{1, 2, \ldots, T\}$, and

r_{jst} is the percentage return of asset $j \in \{1, 2, \ldots, J\}$ under scenario s in time period t.

The restrictions on the model are that there are no short sales and the asset allocation is updated every month according to fixed mix decision rule.[9] In general, fixed mix strategy requires the purchase of stocks as they fall in value, and the sale of stocks as they rise in value. Fixed mix strategy does not have much downside protection, and tends to do very well in flat but oscillating markets. However, it tends to do relatively poorly in bullish markets (Perold and Sharpe, 1988).

We use two alternative objective functions: the first one is power utility function and the second one is an analog of mean–variance analysis. The power utility function, which has constant relative risk aversion, is calculated as follows:[10]

$$U(W^i) = \frac{1}{S} \sum_{s=1}^{S} \frac{1}{(1-\gamma)} (W_s^i)^{(1-\gamma)}, \quad \gamma > -1,$$

where γ is the coefficient of relative risk aversion, and W_s^i is the final wealth. Assuming that the initial wealth is 1, we compute the final wealth as follows: $W_s^i = 1 \cdot (1 + \acute{R}_{s,T}^i)$.

A constant relative risk aversion investor chooses the same investment proportions independent of the investment horizon if the market is frictionless and returns are independent over time. Fix mix is the optimal portfolio choice in this setting. However, if the returns are predictable, which is the conjecture of this paper, then the portfolio choice depends on the investment horizon. Although the fix mix strategy is no longer optimal in this economic environment, the investor is assumed to follow this decision strategy for computational simplicity.

The second objective function trades off between mean final return and a measure of risk:

$$U(\acute{R}_T^i) = E[\acute{R}_T^i] - c \cdot MD(\acute{R}_T^i),$$

where c is the coefficient of risk aversion.

[8] Fix mix rule requires that w_j^i does not depend on time.

[9] Perold and Sharpe (1988) suggest constant proportion portfolio insurance as an alternative strategy. In this strategy, one sells stocks as they fall in value and buy stocks as they rise in value.

[10] Note that $U(W^i)$ is finite if $(1-\gamma) < 2$.

The mean compound portfolio return of fixed mix rule $i \in \{1, 2, \dots, I\}$ at the final date is:

$$E[\acute{R}^i_T] = \frac{1}{S} \sum_{s=1}^{S} \acute{R}^i_{s,T}.$$

We consider the following risk measure which gives less importance to outliers than variance does:

$$MD(\acute{R}^i_T) = \frac{1}{S} \sum_{s=1}^{S} \left|\acute{R}^i_{s,T} - E[\acute{R}^i_T]\right|^r, \quad \text{where } 1 < r < 2.$$

Notice that when $r = 2$, the above risk measure becomes the variance. Since variance is not defined for non-Gaussian stable variables, we use those values of $r < 2$ for which $MD(\acute{R}^i_T)$ is finite, such as $r = 1.5$.

The scenario generation module generates asset return scenarios, r_{jst}, for each time period. At each stage, n new offspring scenarios are generated from the parent scenarios. If the horizon of interest is T periods, then we produce n^T alternative asset return scenarios for the final date. Optimal asset allocation is calculated for this scenario tree. The scenario tree is repeated 100 times and the sample average of optimal allocations is reported as the optimal asset allocation.

6.1. Scenario generation

The portfolio we analyze is composed of Treasury bill and S&P 500. The monthly return on Treasury bill is assumed to be constant at 6% annualized rate of return. The main challenge is predicting the return scenarios for S&P 500. The financial variables that are used to generate the return scenarios for S&P 500 are modeled in a cascade structure similar to Mulvey[11] (1996) (see Figure 3). However, the analysis is done in discrete time as in Wilkie[12] (1995). Monthly data from 2/1965 through 12/1999 is used for the estimation of the time series models.3-month Treasury bill rate and 10-year Treasury bond rate are modeled first as measures of short term and long term interest rates. The price inflation depends on the Treasury bond rate and the previous values of inflation. Following Wilkie's and Mulvey's approaches, stock returns are analyzed in two components: dividend growth and dividend yield growth.[13]

The relationship of economic variables does not denote a one way casual relationship, but rather indicates the sequencing of the modules. The economic variables are modeled

[11] See Section 4.2 for a brief description.

[12] See Section 4.1.2 for a brief review.

[13] Tokat, Rachev and Schwartz (2002) gives the details of the time series analysis.

using Box–Jenkins methodology. The standard Gaussian Box–Jenkins techniques carry over to the stable setting with some possible changes.

We do not model the time varying volatility of the economic variables. Fitting ARMA-GARCH models may reduce the kurtosis in the residuals. However, Balke and Fomby (1994) show that even after estimating GARCH models, significant excess kurtosis and/or skewness still remains. Mittnik, Rachev and Paolella (1997) present empirical evidence favoring stable hypothesis over the normal assumption as a model for ARMA-GARCH residuals. We postpone modeling the time varying volatility to another paper.

Future economic scenarios are simulated at monthly intervals. One set of scenarios is generated by assuming that the residuals of each variable is identical normally distributed. This is the classical assumption made in the literature. Another set of scenarios is generated by assuming that the residuals are identical stable distributed. The estimated normal and stable parameters[14] for the innovations of the time series models are given in Table 1. See Figures 4–8 for graphical comparison of stable and normal fit to the residuals.

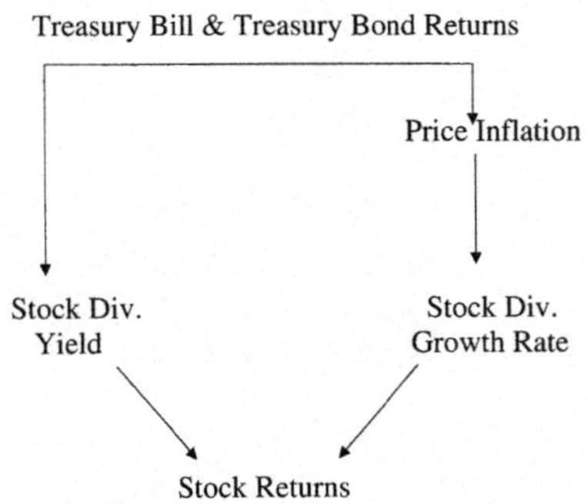

Fig. 3. The scenario generation model.

Table 1
The estimated normal and stable parameters for the innovations

Innovations of	Normal distribution		Stable distribution			
	μ	σ	α	β	μ	σ
Price inflation (*Inf*)	6.15e−06	0.0021	1.7072	0.1073	6.15e−06	0.0012
Dividend gr. (*Divg*)	9.89e−4	0.0195	1.7505	−0.0229	9.89e−4	0.0114
Dividend yield (d(*Divy*))	−0.002551	0.0407	1.8076	0.2252	−0.002551	0.0239
Treasury bill (d(*Tbill*))	0.000336	0.0579	1.5600	0	0	0.0308
Treasury bond (d(*Tbond*))	0.000818	0.0339	1.9100	0	0	0.0230

[14] Stable parameters are estimated using maximum likelihood estimation method.

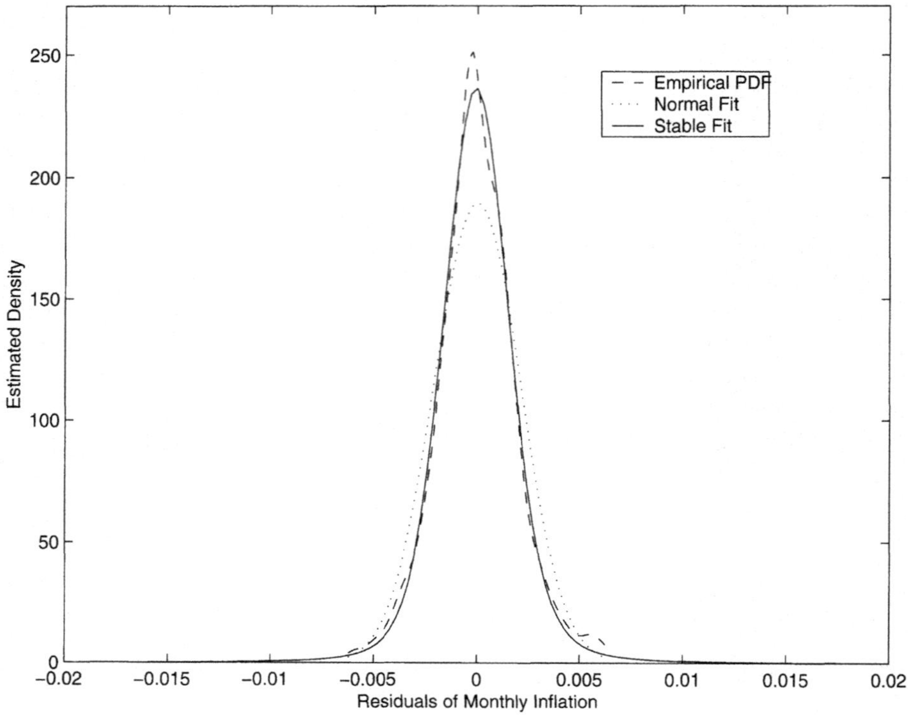

Fig. 4. The empirical pdf of the residuals of inflation, the Gaussian fit and the stable fit.

The scenarios have a tree structure. At each stage (month) we generate n possible scenarios. For each scenario, we first generate a normal or stable residual for Treasury bill, and calculate the corresponding Treasury bill rate for the proceeding month. Then, given this short rate, we generate Treasury bond rate, price inflation, dividend growth rate and dividend yield for that month according to the cascade structure and the time series models we have built. For instance, the inflation rate for next month is generated by using the Treasury bond rate, inflation rate and the surprise to expected inflation this month, and the normal or stable innovation of inflation rate next month. Note that we allow for innovation of each economic variable in each simulated month.

At the next stage, n new offspring scenarios are generated from the parent scenarios. This continues until the final time of interest. In this study, we generate 2 scenarios for each month, so 512 possible economic scenarios are considered over the next three quarters.

6.2. *Valuation of assets*

The monthly return of S&P 500 is derived using the dividend yield and the dividend index. Dividend index is calculated by multiplying price index with the dividend yield:

$$DI_t = P_t \times DY_t,$$

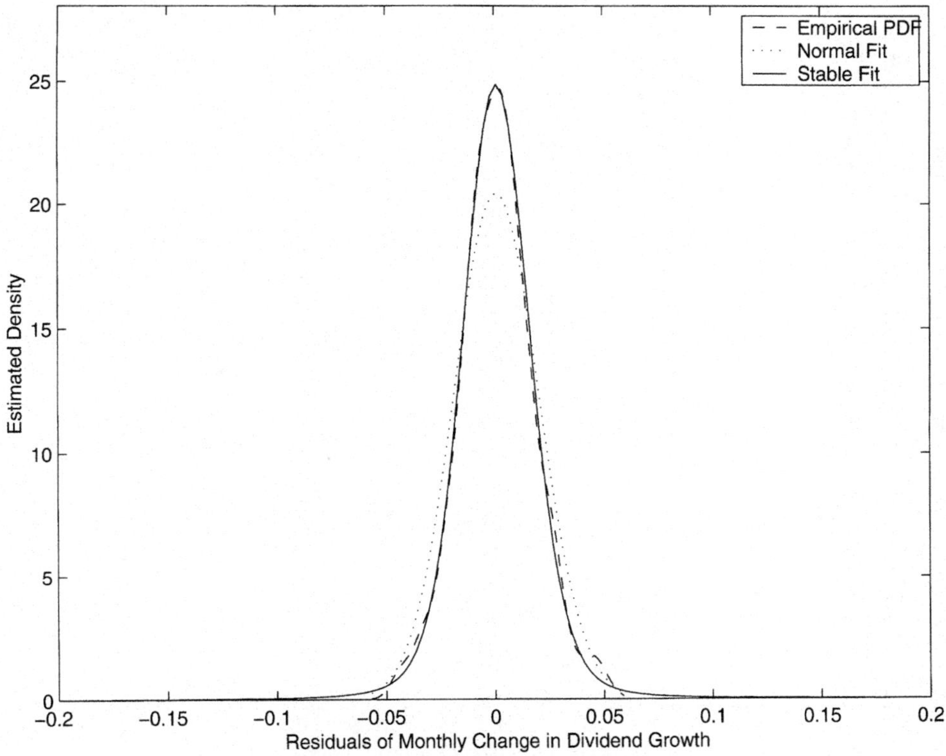

Fig. 5. The empirical pdf of the residuals of dividend growth, the Gaussian fit and the stable fit.

where DI_t is the dividend index for period t, P_t is the price index for period t, and DY_t is the dividend yield for period t. The dividend growth is just log differences of dividend indices.

The dividend yield and dividend growth rate are simulated as explained in the previous section. Hence, we can get back simulated future price index in period t under scenario s from the simulated dividend growth and dividend yield indices by

$$P_{st} = DI_{st}/DY_{st}.$$

Then, we can calculate the return for holding S&P 500 for a month under scenario s as

$$r_{st} = (P_{st} - P_{s(t-1)} + DI_{st})/P_{s(t-1)}.$$

6.3. Computational results

We first present the mean annualized return of S&P 500 in 100 repetitions of the scenario tree generated by using the Gaussian and stable distribution models (see Table 2).

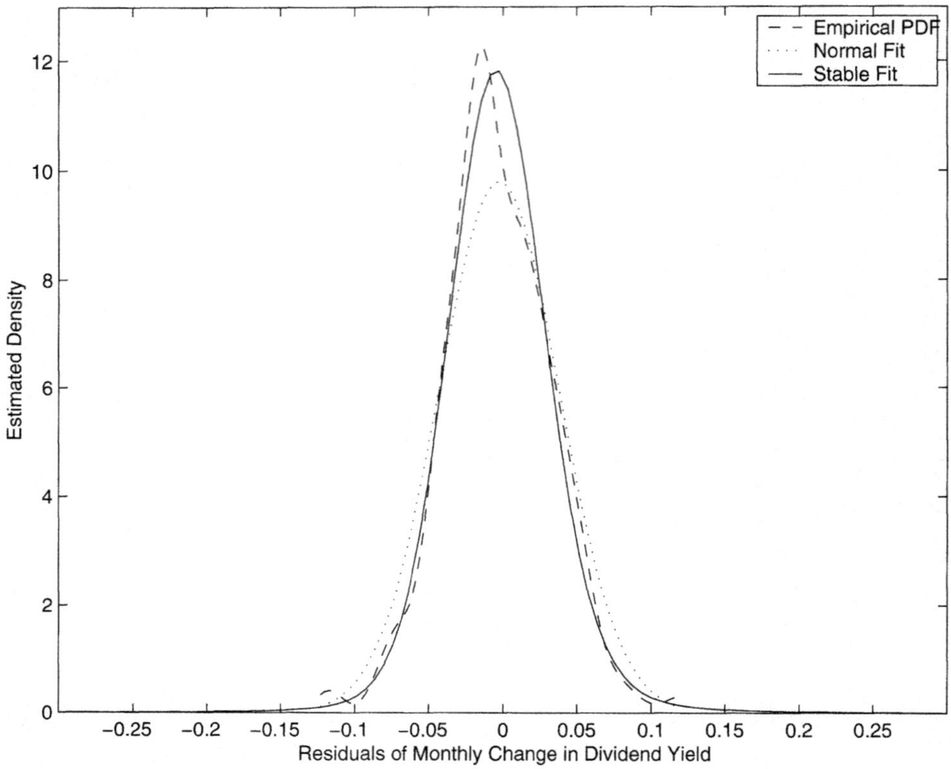

Fig. 6. The empirical pdf of the residuals of dividend yield, the Gaussian fit and the stable fit.

Table 2
Annualized return scenarios on S&P 500

	Mean	1%	2.5%	25%	75%	97.5%	99%
Normal scenarios	9.07	−122.34	−103.03	−31.97	48.54	129.66	152.90
Stable scenarios	10.20	−149.17	−107.29	−27.45	44.96	128.68	171.16

The table also depicts the percentiles of these return scenarios. It should be noted that the S&P 500 returns generated by stable scenarios have fatter tails than those of Gaussian scenarios. Hence, stable scenarios consider more extreme scenarios than Gaussian scenarios do. Khindanova, Rachev and Schwartz (2001) report similar observations in their paper where they compute value at risk employing Gaussian and stable distributed daily returns. They state that 5% percentile of normal and stable distribution are very close, but the 1% percentile of stable distribution is greater than that of the Gaussian.

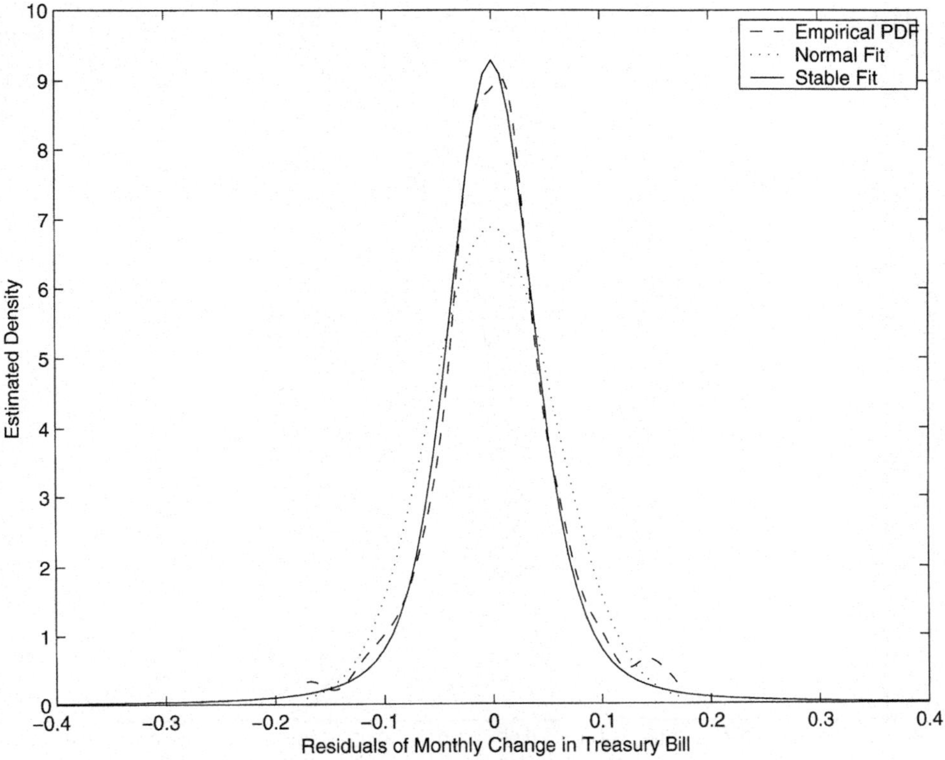

Fig. 7. The empirical pdf of the residuals of Treasury bill, the Gaussian fit and the stable fit.

The asset allocation problem has been solved for an investor that maximizes the power utility of final wealth. The optimal asset allocation depends on the risk aversion level of the agent. If his relative risk aversion coefficient is very low, such as 0.80, or very high, such as 10.00, then the Gaussian and stable scenarios result in similar asset allocations (see Table 3). The intuitive explanation for this is that, the investor who has very low risk aversion, does not mind the risk very much. Therefore, his decision does not change when the extreme events are modeled more realistically. Similarly, the investor who has very high risk aversion, is already scared away from the risky asset. The fatter tails do not affect his decision much either. On the other hand, an investor who would put 60% in S&P 500 if he were to use normal scenarios, will put only 48% in S&P 500 if he uses stable scenarios. The fact that stable scenarios model the extreme events more realistically, results in stable investor putting less in the risky asset than Gaussian investor does.

The time series models which generate the Gaussian and stable scenarios are the same except for the residuals being Gaussian or stable, respectively. In our computations, the

[15] Note that when $\gamma = 1$ the power utility function reduces to logarithmic utility function.

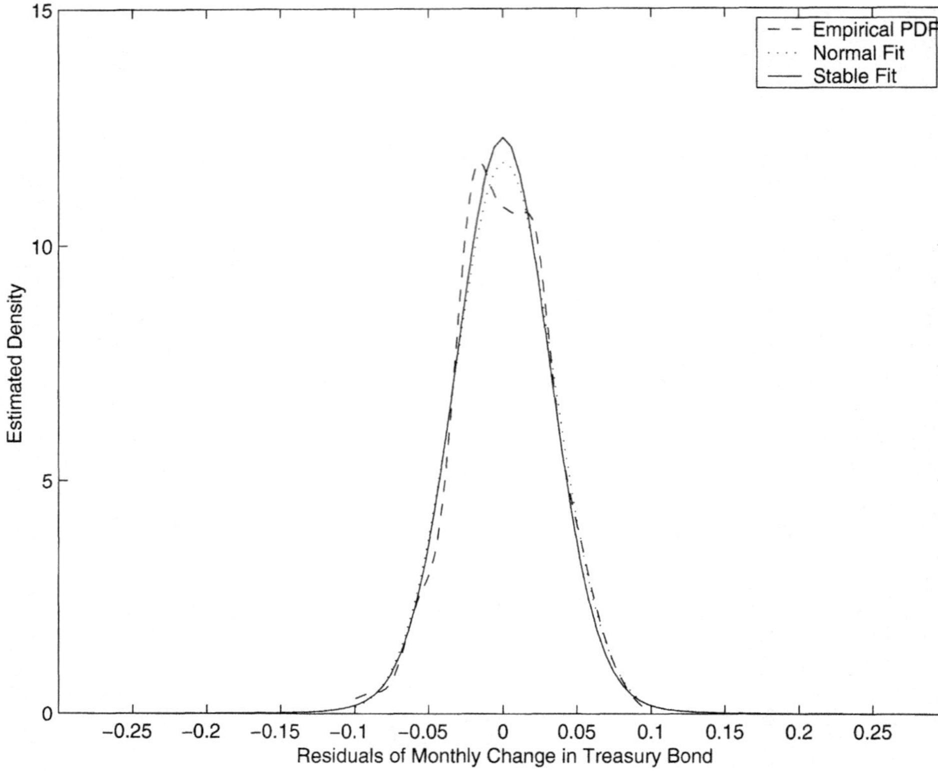

Fig. 8. The empirical pdf of the residuals of Treasury bond yield, the Gaussian fit and the stable fit.

Table 3
Optimal allocations under normal and stable scenarios ($T = 3$ quarters)

γ	Normal scenarios Optimal percentage invested		Stable scenarios Optimal percentage invested	
	S&P 500 (%)	Treasury Bill (%)	S&P 500 (%)	Treasury Bill (%)
0.80	100	0	100	0
1.00[15]	100	0	88	12
1.50	86	14	66	44
2.30	60	40	48	52
2.70	52	48	42	58
10.00	14	86	12	88

mean return of Gaussian S&P 500 scenarios came out to be less than stable S&P 500 scenarios. The equity premium is 3.07% in the normal scenarios and 4.20% in the stable scenarios. Since the premium on equity is higher in stable scenarios, the equity is more attractive. However, the fact that the stable scenarios also have heavier tails outweighs this, and consequently the investor puts considerably less money in the stock index. If the equity premium were the same in both sets of scenarios, we contemplate that the allocation difference would be even more pronounced.

Table 4 depicts the change in the utility[16] if the investor uses stable scenarios rather than Gaussian scenarios. The improvement can be as large as 0.72% depending on the risk aversion level of the investor. Table 5 reports the improvement in the certainty equivalent final wealth (CEFW) if an investor uses stable scenarios rather than Gaussian scenarios.[17] The computations show a 6 basis point improvement in the certainty equivalent wealth of the investor who would put 60% in S&P 500. The difference could get larger or smaller depending on the risk aversion level of the decision maker.

The other 'utility' function we consider is an analog of mean-variance criterion. The computational results achieved are very similar to the constant relative risk aversion utility. The investor who has very low or very high risk aversion, does not gain much from using

Table 4

Comparison of utility achieved from normal and stable scenarios ($T = 3$ quarters)

γ	Normal scenarios		Stable scenarios		% Change in utility
	% in S&P 500	Utility	% in S&P 500	Utility	
0.80	100	5.0633	100	5.0633	0.00
1.00	100	0.0600	88	0.0604	0.72
1.50	86	−1.9458	66	−1.9445	0.06
2.30	60	−0.7188	48	−0.7181	0.09
2.70	52	−0.5391	42	−0.5386	0.09
10.00	14	−0.0728	12	−0.0728	0.03

Table 5

Comparison of certainty equivalent wealth achieved from normal and stable scenarios ($T = 3$ quarters)

γ	Normal scenarios		Stable scenarios		% Change in CEFW
	% in S&P 500	CEFW	% in S&P 500	CEFW	
0.80	100	1.0650	100	1.0650	0.00
1.00	100	1.0618	88	1.0623	0.04
1.50	86	1.0565	66	1.0579	0.13
2.30	60	1.0536	48	1.0543	0.07
2.70	52	1.0526	42	1.0532	0.05
10.00	14	1.0480	12	1.0481	0.00

[16] Note that the utility value becomes negative when $\gamma > 1$. Although negative utility does not make much sense, it can be made positive by monotonic transformations.

[17] Since Gaussian distribution is a special case of stable distribution, the stable model encompasses the Gaussian model. Therefore, the certainty equivalency comparison is made under the assumption that stable is the correct model.

Table 6
Optimal allocations under normal and stable scenarios ($T = 3$ quarters)

c	Normal scenarios Optimal percentage invested		Stable scenarios Optimal percentage invested	
	S&P 500 (%)	Treasury Bill (%)	S&P 500 (%)	Treasury Bill (%)
0.35	100	0	100	0
0.40	90	10	80	20
0.52	60	40	54	46
0.59	50	50	44	66
1.00	20	80	18	82

Table 7
Percentage change in utility achieved from normal and stable scenarios ($T = 3$ quarters)

c	Normal scenarios		Stable scenarios		% Change in utility
	% in S&P 500	Utility	% in S&P 500	Utility	
0.35	100	0.0583	100	0.0583	0.00
0.40	90	0.0561	80	0.0562	0.28
0.52	60	0.0526	54	0.0527	0.10
0.59	50	0.0513	44	0.0514	0.12
1.00	20	0.0479	18	0.0480	0.08

the stable model. However, the stable model makes a difference for the investors in the middle. Table 6 depicts that an investor who would put 60% in S&P 500 if he were to use normal scenarios, will put only 56% in S&P 500 if he uses stable scenarios. Table 7 reports the percentage improvement in the 'utility' function[18] if one uses stable model as opposed to Gaussian model. If there is any percentage improvement in the utility function, an investor can reduce the risk for a given level of mean return or increase the mean return for a given level of risk. This can be achieved by switching from Gaussian scenario generation to stable scenario generation.

7. Conclusion

The ALM models that are based on stochastic programming with or without decision rules are starting to gain applicability in the industry. In these models, the future uncertainty is modeled using discrete scenarios. A representative set of scenarios describes the possible future economic situations facing the institution.

[18] Since the risk corresponding to certainty equivalent return is zero, the certainty equivalent return is equal to the utility of return. Hence, the percentage improvement in the utility of return is equivalent to the percentage improvement in the certainty equivalent return.

Generating scenarios that realistically represent the future uncertainty is important for the validity of the results of stochastic programming based ALM models. The assumption underlying the scenario generation models used in the literature is the normal distribution. The validity of normal distribution has been questioned in the finance and macroeconomics literature. The leptokurtic (heavy tailed and peaked), and asymmetric nature of the economic variables can be better captured by using stable distribution as opposed to normal distribution.

We analyze the effects of the distributional assumptions on optimal asset allocation. A multistage dynamic asset allocation model with decision rules has been set up. The optimal asset allocations found under normal and stable scenarios are compared. The analysis suggests that the normal scenarios greatly underestimate risks. Stable scenario modeling leads to asset allocations that are up to 20% different than those of normal scenario modeling.

Although the financial data exhibit time varying volatility and long range dependence as well as heavy tails, this study has only considered explicit modeling of heavy tails in the financial data. The conditional heteroskedastic models (ARMA-GARCH) utilizing stable distributions can be used to describe the time varying volatility along with the asymmetric and leptokurtic behavior. In addition to these, the long-range dependence can also be modeled if fractional-stable GARCH models are employed. These aspects of financial data will be considered in a later paper.

References

Bai, D., Carpenter, T., Mulvey, J.M., 1997. Making a case for robust optimization models. Management Science 43 (7), 895–907.

Balke, N.S., Fomby, T.B., 1994. Large shocks, small shocks, and economic fluctuations: outliers in macroeconomic time series. Journal of Applied Econometrics 9, 181–200.

Bekaert, G., Hodrick, R.J., 1992. Characterizing predictable components in excess returns on equity and foreign exchange markets. Journal of Finance 47 (2), 467–509.

Berger, A.J., Mulvey, J.M., 1998. The home account advisor: asset and liability management for individual investors. In: Ziemba, W.T., Mulvey, J.M. (Eds.), Worldwide Asset and Liability Modeling. Cambridge University Press, Cambridge, UK, pp. 634–665.

Boender, G.C.E., 1997. A hybrid simulation/optimization scenario model for asset/liability management. European Journal of Operational Research 99, 126–135.

Boender, G.C.E., van Aalst, P., Heemskerk, F., 1998. Modeling and management of assets and liabilities of pension plans in the Netherlands. In: Ziemba, W.T., Mulvey, J.M. (Eds.), Worldwide Asset and Liability Modeling. Cambridge University Press, Cambridge, UK, pp. 561–580.

Bollerslev, T., 1986. Generalized autoregressive conditional homoskedasticity. Journal of Econometrics 52, 5–59.

Bradley, S.P., Crane, D.B., 1972. A dynamic model for bond portfolio management. Management Science 19, 139–151.

Brandt, M.W., 1999. Estimating portfolio and consumption choice: a conditional Euler equations approach. Journal of Finance 54, 1609–1646.

Brennan, M.J., Schwartz, E.S., 1982. An equilibrium model of bond pricing and a test of market efficiency. Journal of Financial and Quantitative Analysis 17 (3), 301–329.

Brennan, M.J., Schwartz, E.S., 1998. The use of Treasury bill futures in strategic asset allocation programs. In: Ziemba, W.T., Mulvey, J.M. (Eds.), Worldwide Asset and Liability Modeling. Cambridge University Press, Cambridge, UK, pp. 205-228.

Brennan, M.J., Schwartz, E.S., Lagnado, R., 1997. Strategic asset allocation. Journal of Economic Dynamics and Control 21, 1377–1403.

Brinson, G.P., Hood, L.R., Beebower, G.L., 1986. Determinants of portfolio performance. Financial Analysts Journal 42 (4), 39–48.

Bunn, D.W., Salo, A.A., 1993. Forecasting with scenarios. European Journal of Operational Research 68 (3), 291–303.

Calder, M., Davis, R.A., 1998. Inference for linear processes with stable noise. In: Adler, R.J., Feldman, R.E., Taqqu, M.S. (Eds.), A Practical Guide to Heavy Tails. Birkhäuser, pp. 159–176.

Carino, D.R., Kent, T., Myers, D.H., Stacy, C., Sylvanus, M., Turner, A.L., Watanabe, K., Ziemba, W.T., 1994. The Russell–Yasuda Kasai model: An asset/liability model for a Japanese insurance company using multistage stochastic programming. Interfaces 24, 29–49.

Carino, D.R., Myers, D.H., Ziemba, W.T., 1998. Concepts, technical issues, and uses of the Russel–Yasuda Kasai financial planning model. Operations Research 46 (4), 450–462.

Carino, D.R., Ziemba, W.T., 1998. Formulation of the Russell–Yasuda Kasai financial planning model. Operations Research 46 (4), 433–449.

Carpenter, T., Lustig, I., Mulvey, J.M., 1991. Formulating stochastic programs for interior point methods. Operations Research 39, 757–770.

Charnes, A., Gallegos, Yao, S., 1982. A chance constrained approach to bank dynamic balance sheet management. Center for Cybernetic Studies CCS 428, University of Texas, Austin.

Charnes, A., Kirby, M.J.L., 1966. Optimal decision rules for the E-model of chance-constrained programming. Cahiers du Centre d'Études de Recherche Operationelle 8, 5–44.

Chopra, V.K., Ziemba, W.T., 1993. The effect of errors in means, variances, and covariances on optimal portfolio choice. Journal of Portfolio Management, Winter, 6–11.

Culp, C., Tanner, K., Mensink, R., 1997. Risk, returns and retirement. RISK 10 (10), 63–69.

Dembo, R.S., 1991. Scenario optimization. Annals of Operations Research 30, 63–80.

Dembo, R.S., 1993. Scenario immunization. In: Zenios, S.T. (Ed.), Financial Optimization. Cambridge University Press, Cambridge, UK, pp. 290-308.

Dert, C.L., 1995. Asset and liability management for pension funds, a multistage chance constrained programming approach. Ph.D. Thesis. Erasmus University, Rotterdam, The Netherlands.

Dert, C.L., 1998. A dynamic model for asset liability management for defined benefit pension funds. In: Ziemba, W.T., Mulvey, J.M. (Eds.), Worldwide Asset and Liability Modeling. Cambridge University Press, Cambridge, UK, pp. 501–536.

DuMouchel, W.H., 1973. On the asymptotic normality of the maximum likelihood estimate when sampling from a stable distribution. The Annals of Statistics 1, 948–957.

Embrechts, P., Klüppelberg, C., Mikosch, T., 1997. Modelling Extremal Events for Insurance and Finance. Springer-Verlag, Berlin.

Engle, R., 1982. Autoregressive conditional heteroskedasticity with estimates of the variance of United Kingdom inflation. Econometrica 50, 987–1007.

Eppen, G., Fama, E., 1971. Three asset cash balance and dynamic portfolio problems. Management Science 17 (5), 311–319.

Fama, E., 1965. The behavior of stock market prices. Journal of Business 38, 34–105.

Geyer, A., Herold, W., Kontriner, K., Ziemba, W.T., 2002. The innovest Austrian pension fund financial planning model InnoALM. Working paper. University of Economics, Vienna, Austria.

Hakansson, N.H., 1971. On the optimal myopic portfolio policies, with and without serial correlation of yields. Journal of Business 44, 324–334. Reprinted in: Ziemba, W.T., Vickson, R.G. (Eds.), 1975. Stochastic Optimization Models in Finance. Academic Press, pp. 401–411.

Hiller, R.S., Eckstein, J., 1993. Stochastic dedication: designing fixed income portfolios using massively parallel Benders decomposition. Management Science 39 (11), 1422–1438.

Hodrick, R.J., 1992. Dividend yields and expected stock returns: alternative procedures for inference and measurement. Review of Financial Studies 5 (3), 357–386.

Kandel, S., Staumbaugh, R.F., 1996. On predictability of stock returns: an asset allocation perspective. Journal of Finance 51, 385–424.

Khindanova, I., Rachev, S.T., Schwartz, E., 2001. Stable modeling of value at risk. Mathematical and Computer Modelling 34 (9–11), 1223–1259.

Kim, J.-R., 1999. Another look on the asymmetric behavior of unemployment rates over the business cycle. Working paper. Technische Universität, Dresden.

Klaasen, P., 1997. Discretized reality and spurious profits in stochastic programming models for asset/liability management. European Journal of Operational Research 101, 374–392.

Klaasen, P., 1998a. Financial asset-pricing theory and stochastic programming models for asset/liability management: a synthesis. Management Science 44 (1), 31–48.

Klaasen, P., 1998b. Solving stochastic programming models for asset/liability management using iterative disaggregation. In: Ziemba, W.T., Mulvey, J.M. (Eds.), Worldwide Asset and Liability Modeling. Cambridge University Press, Cambridge, UK, pp. 427–463.

Korhonen, A., 1987. A dynamic bank portfolio planning model with multiple scenarios, multiple goals and changing priorities. European Journal of Operational Research 30, 13–23.

Kusy, M.I., Ziemba, W.T., 1986. A bank asset and liability management model. Operations Research 35, 356–376.

Li, S.X., 1995a. An insurance and investment portfolio model using chance constrained programming. Omega 23 (5), 577–585.

Li, S.X., 1995b. A satisficing chance constrained model in the portfolio selection of insurance lines and investments. Journal of the Operational Research Society 46 (9), 1111–1120.

Mandelbrot, B.B., 1963. The variation of certain speculative prices. Journal of Business 26, 394–419.

Mandelbrot, B.B., 1967. The variation of some other speculative prices. Journal of Business 40, 393–413.

Maranas, C.D., Androulakis, I.P., Floudas, C.A., Berger, A.J., Mulvey, J.M., 1997. Solving long-term financial planning problems via global optimization. Journal of Economic Dynamics and Control 21, 1405–1425.

Merton, R.C., 1969. Lifetime portfolio selection under uncertainty: the continuous-time case. Review of Economics and Statistics 51 (3), 247–257.

Mikosch, T., Gadrich, T., Klüppelberg, C., Adler, R.J., 1995. Parameter estimation for ARMA models with infinite variance innovations. The Annals of Statistics 23 (1), 305–326.

Mittnik, S., Paolella, M.S., Rachev, S.T., 2000. Diagnosing and treating the fat tails in financial returns data. Journal of Empirical Finance 7, 389–416.

Mittnik, S., Rachev, S.T., 1993. Modeling asset returns with alternative stable distribution. Econometric Review 12 (3), 261–330.

Mittnik, S., Rachev, S.T., Doganoglu, T., Chenyao, D., 1999. Maximum likelihood estimation of stable Paretian models. Mathematical and Computer Modelling 29 (10–12), 275–293.

Mittnik, S., Rachev, S.T., Paolella, M.S., 1997. Stable Paretian modeling in finance: some empirical and theoretical aspects. Unpublished manuscript. Institute of Statistics and Econometrics, Christian Albrechts University at Kiel.

Mulvey, J.M., 1994. An asset-liability investment system. Interfaces 24, 22–33.

Mulvey, J.M., 1996. Generating scenarios for the Towers Perrin investment system. Interfaces 26, 1–15.

Mulvey, J.M., 1998. Generating scenarios for global financial planning systems. International Journal of Forecasting 14, 291–298.

Mulvey, J.M., Thorlacius, A.E., 1998. The Tower Perrin global capital market scenario generation system. In: Ziemba, W.T., Mulvey, J.M. (Eds.), Worldwide Asset and Liability Modeling. Cambridge University Press, Cambridge, UK, pp. 286–312.

Mulvey, J.M., Vanderbrei, R.J., Zenios, S.A., 1995. Robust optimization of large-scale systems. Operations Research 43 (2), 264–281.

Mulvey, J.M., Vladimirou, H., 1989. Stochastic network optimization models for investment planning. Annals of Operations Research 20, 187–217.

Mulvey, J.M., Vladimirou, H., 1991. Applying the progressive hedging algorithm to stochastic generalized networks. Annals of Operations Research 31, 399–424.

Mulvey, J.M., Vladimirou, H., 1992. Stochastic network programming for financial planning problems. Management Science 38 (11), 1642–1664.

Mulvey, J.M., Zenios, S.A., 1994a. Capturing correlations of fixed income instruments. Management Science 40 (10), 1329–1342.

Mulvey, J.M., Zenios, S.A., 1994b. Dynamic diversification of fixed income portfolios. Financial Analysts Journal, January–February, 30–38.

Mulvey, J.M., Ziemba, W.T., 1995. Asset and liability allocation in a global environment. In: Jarrow, R., et al. (Eds.), Handbooks in OR and MS, Vol. 9, pp. 435–463.

Mulvey, J.M., Ziemba, W.T., 1998. Asset and liability management systems for long-term investors: discussion of the issues. In: Ziemba, W.T., Mulvey, J.M. (Eds.), Worldwide Asset and Liability Modeling. Cambridge University Press, Cambridge, UK, pp. 3–38.

Oguzsoy, C.B., Guven, S., 1997. Bank asset and liability management under uncertainty. European Journal of Operational Research 102, 575–600.

Ortobelli, L., Rachev, S.T., Schwartz, E., 2002. On the problem of optimal portfolio with stable distributed returns. Unpublished manuscript. Anderson Graduate School of Business, University of California, Los Angeles.

Paulauskas, V., Rachev, S.T., 1999. Maximum likelihood estimators in regression models with infinite variance innovation. Working paper. Vilnius University, Lithuania.

Perold, J.M., Sharpe, W.F., 1988. Dynamic strategies for asset allocation. Financial Analysts Journal, January, 16–27.

Rachev, S.T., Han, S., 1999. Optimization problems in mathematical finance. Working paper. University of Karlsruhe, Germany.

Rachev, S.T., Kim, J.-R., Mittnik, S., 1999. Stable Paretian models in econometrics: Part I. Working paper. University of Karlsruhe, Germany.

Rachev, S.T., Mittnik, S., 2000. Stable Paretian Models in Finance. Wiley, New York.

Rockafeller, R.T., Wets, R.J.-B., 1991. Scenarios and policy aggregation in optimization under uncertainty. Mathematics of Operations Research 16, 119–147.

Samorodnitsky, G., Taqqu, M.S., 1994. Stable Non-Gaussian Random Variables. Chapman and Hall, New York.

Samuelson, P.A., 1969. Lifetime portfolio selection by dynamic stochastic programming. Review of Economics and Statistics 51 (3), 239–246.

Sims, C., 1980. Macroeconomics and reality. Econometrica 48 (1), 1–48.

Sweeney, J.C., Sonlin, S.M., Correnti, S., Williams, A.P., 1998. Optimal insurance asset allocation in a multi-currency environment. In: Ziemba, W.T., Mulvey, J.M. (Eds.), Worldwide Asset and Liability Modeling. Cambridge University Press, Cambridge, UK, pp. 341–368.

Tokat, Y., Rachev, S.T., Schwartz, E., 2002. The stable non-Gaussian asset allocation: a comparison with the classical Gaussian approach. Journal of Economic Dynamics and Control. Special Issue on High Performance Computing in Finance, forthcoming.

Von Neumann, J., Morgenstern, O., 1947. Theory of Games and Economic Behavior. Princeton University, Princeton, NJ.

Wilkie, A.D., 1986. A stochastic investment model for actuarial use. Transactions of the Faculty of Actuaries 39, 391–403.

Wilkie, A.D., 1995. More on a stochastic asset model for actuarial use. Presented to the Institute Actuaries and Faculty of Actuaries, London.

Zenios, S.A., 1993. Financial Optimization. Cambridge University Press, Cambridge, UK.

Zenios, S.A., Holmer, M.R., McKendall, R., Vassiadou-Zeniou, C., 1998. Dynamic models for fixed income portfolio management under uncertainty. Journal of Economic Dynamics and Control 22, 1517–1541.

Ziemba, W.T., 1974. Choosing investment portfolios when the returns have stable distributions. In: Hammer, P.L., Zoutendijk, G. (Eds.), North-Holland, Amsterdam, pp. 443–482.

Ziemba, W.T., Mulvey, J.M., 1998. Worldwide Asset and Liability Modeling. Cambridge University Press, Cambridge.

Chapter 14

PORTFOLIO CHOICE THEORY WITH NON-GAUSSIAN DISTRIBUTED RETURNS

SERGIO ORTOBELLI

University of Bergamo, Italy

ISABELLA HUBER

University of Karlsruhe, Germany

SVETLOZAR T. RACHEV

Department of Statistics and Applied Probability, University of California, Santa Barbara, USA
Institute of Statistics and Mathematical Economics, University of Karlsruhe, Germany
e-mail: rachev@lsoe-4.wiwi.uni-karlsruhe.de

EDUARDO S. SCHWARTZ

Anderson School of Management, University of California, Los Angeles, USA

Contents

Handbook of Heavy Tailed Distributions in Finance, Edited by S.T. Rachev

Abstract

This chapter discusses the parametric distributions of asset returns and proposes portfolio choice models consistent with the maximization of the expected utility. We analyze multi-parameter models to select nonstochastically dominated portfolios when short sales are allowed and when short sales are not allowed. We also concentrate our attention on the stable distributional approach in order to derive optimal portfolios with heavy-tailed distributed financial returns. Finally, we examine and compare optimal allocations obtained with the multivariate normal model and the sub-Gaussian stable one.

1. Introduction

The purpose of this chapter is to describe the admissible classes of parametric distribution functions of return portfolios and to analyze their consistency with the maximization of the expected utility, in order to choose the optimal portfolios. In particular, we present a general theory and a unifying framework with the following aims: (1) Understanding the distributional approach applied to the portfolio choice theory; (2) Studying the implications of the classical market restrictions on the portfolio distributions; (3) Considering the asymptotic behavior of return data. We conclude our analysis comparing the normal multivariate approach with the sub-Gaussian approach here presented.

The theory of portfolio choice is based on the assumption that investors allocate their wealth across the available assets in order to maximize their expected utility. Markowitz (1952, 1959, 1987) and Tobin (1958, 1965) were among the first to give rigorous results approximating to the portfolio selection problem in terms of the mean and the variance. Their analysis was extended to an equilibrium theory by Sharpe (1964), Lintner (1965) and Mossin's (1966) capital asset pricing model (CAPM). In a mean–variance world the investor is concerned with only two parameters of the probability distribution of total returns on investment: the mean of the return and the variance of the return. The simplicity of the equilibrium theory and the intuitive appeal of the mean–variance analysis attracted and directed the attention toward determining their generality and extensibility. The foundation of the whole theory lays in the arbitrage pricing theory (APT) and in the stochastic dominance analysis. As a matter of fact, both theories are strictly grounded on the equilibrium theory [see, among others, Ross (1975), Dybvig and Ross (1987), Jarrow (1986)]. The arbitrage pricing theory and the fund separation theorems [see Ross (1976, 1978a)] justify and extend CAPM to multi-parameter linear models. Whereas the stochastic dominance analysis justifies the partial consistency of the mean–variance framework with the expected utility maximization when the portfolios are elliptically distributed [see, among others, Bawa (1975), Chamberlein (1983), Owen and Rabinovitch (1983)]. In the same years and subsequently, the theory was further generalized to intertemporal finance and to consumption-based model. Since the space of feasible consumption bundles is quite generally a linear space [as Ross (1978b), Cox and Leland (1982), Rubinstein (1976) and many others have emphasized], the original dynamic problem can be replaced with an equivalent one-period problem, which has appropriate terminal state prices, if all consumption takes place at the end. More generally, if preferences are time separable and if we treat consumption at each date separately, the analysis is unchanged. For this reason, here we propose a static approach. A first generalization to an intertemporal approach can be found in Ortobelli, Rachev and Schwartz (2002).

The mean–variance theory has survived theoretical criticism and empirical rejection, such as that of Samuelson (1967, 1969) and Samuelson and Merton (1975) who have underlined the limits of the approximation given only by the mean and the variance of a portfolio. Later, Roll (1977, 1978, 1979a, b) was the first to understand clearly the weaknesses of the theory and the empirical deficiencies. Then, Dybvig and Ingersoll (1982) verified that the mean–variance pricing and the complete market hypothesis still can lead to arbitrage opportunities. They also proved that the standard mean–variance separation theorem holds in a complete market only if all investors have quadratic utility. Further, Dybvig and Ross (1982) have demonstrated that efficient sets generally are not convex and the market portfolio could be inefficient. Bawa (1976) considered the case of a market with no short sale opportunities and jointly normally distributed returns. Under these assumptions, if there are some investors with increasing nonconcave utility functions [for example Friedman and Savage type utility functions (Friedman and Savage, 1948)], the market portfolio could be inefficient. Moreover, Dybvig and Ross (1985a) observed that assuming symmetric information and an inefficient index, the security market line analysis can be grossly misleading, since in general efficient and inefficient portfolios can plot above and below the security market line. In another paper [see Dybvig and Ross (1985b)] they also argued that differential information disrupts the validity of the security market line analysis, since it takes us outside the domain of the mean–variance analysis. On the other side, the fundamental work of Mandelbrot (1963a, b, 1967), Mandelbrot and Taylor (1967) and Fama (1963, 1965a, b) has sparked considerable interest in studying the empirical distribution of financial assets. The excess kurtosis, found in Mandelbrot's and Fama's investigations, led them to reject the normal assumption (generally used to justify the mean–variance approach) and to propose the stable Paretian distribution as a statistical model for asset returns. Other relevant empirical studies on postwar US data have shown that the slope of the mean–standard deviation frontier or of the expected return-beta lines is much higher than the reasonable risk aversion and consumption volatility estimates suggest. This is the equity premium puzzle (Merha and Prescott, 1985; Hansen and Jagannathan, 1991) that could be generated by one or more of the following conditions: (a) the investors are much more risk averse than the academics might have thought; (b) the stock returns of the last 50 years are due to good luck rather than an equilibrium compensation for risk; (c) something is deeply wrong with the model [see Cochrane (1999)].

The many lacks and contrasting results on the empirical and theoretical mean–variance analysis represented the main justifications and reasons of the alternative mean–dispersion models proposed in the last decades [see Markowitz' (1959) mean–semi-variance approach, Yitzhaki (1982) and Shalit and Yitzhaki's (1984) mean–Gini portfolio theory, Dybvig's (1988a, b) distributional approach for complete market, Speranza (1993) and Konno and Jamazaki's mean–absolute deviation approach (MAD) (Konno and Yamazaki, 1991), the Ogryczak and Ruszczynski's (1999, 2001) mean–semi-deviation models]. However, from a conceptual point of view, the stochastic dominance theory has shown that the variance, as any other dispersion measure, cannot always be considered as a risk measure. As a matter of fact, given a lottery X with a given dispersion measure, we can always find another lottery Y which has a greater, lower or equal dispersion and it is preferred to the first one by some non-satiable or risk averse investors [see among others Levy (1992)]. Then, the implicit problems to solve are the following: "When can we use a dispersion measure as a risk measure? Which relations there exist between the dispersion measure and the other parameters that characterize the parametric family of portfolio distribution functions?". To answer these questions, we recall a distributional stochastic dominance analysis of the parametric families consistent with the maximization of the expected utility, see Ortobelli (2001). The distributional analysis proposed is formally different from Dybvig's (1988a) distributional one. Dybvig's model for complete markets has been applied successfully in theoretical work to value the magnitude of inefficiency of some dynamic portfolio strategies [see Dybvig (1988b)]. However, Dybvig's model is not easily applicable from an empirical point of view. In this chapter we classify clearly the multi-parameter optimization problems to solve so as to obtain optimal portfolios. The proposed multi-parameter approach is alternative to Ross' one and it is a unifying and generalizing extension of the classic moment analysis in portfolio choice theory [see among others Jean (1971), Fishburn (1980), Ingersoll (1987)]. As a matter of fact, when the family of return portfolios belongs to a parametric family uniquely determined by a finite number of moments, we show what kind of optimization problems we have to solve in order to find frontiers of admissible choices. As a consequence, we obtain a further delimitation of the admissible portfolio choices. Therefore, we could classify the distributions for which Markowitz and Tobin's mean–variance rule is an optimal selection rule in a market with and without institutional restrictions (no short sale and limited liability). Moreover, this analysis justifies the stochastic dominance properties of the mean–dispersion models in the works recalled above and of some classic mean–dispersion–skewness approaches [see, for example, Kraus and Litzenberger (1976), Simaan (1993), Ingersoll (1987)].

In second analysis we study the asymptotic distributional behavior of data. The behavior, generally stationary over time of returns, and the Central Limit Theorem and Central Pre-limit Theorem for normalized sums of i.i.d. random variables [see Zolotarev (1986), Klebanov, Rachev and Szekely (2001), Klebanov, Rachev and Safarian (2000)] theoretically justify the stable Paretian approach proposed by Mandelbrot and Fama. As a matter of fact, their conjecture was supported by numerous empirical investigations in the subsequent years [see Mittnik, Rachev and Paolella (1997), Rachev and Mittnik (2000)]. The practical and theoretical appeal of the stable non-Gaussian approach is given by its attractive properties that are almost the same as the normal one. A relevant desirable property of a stable distributional assumption is that stable distributions have domain of attraction. Therefore, any distribution in the domain of attraction of a specified stable distribution will have properties close to those of the stable distribution. Another attractive aspect of the stable Paretian assumption is the stability property, i.e., stable distributions are stable with respect to summation of i.i.d. random stable variables. Hence, the stability governs the main properties of the underlying distribution [detailed accounts for theoretical aspects of stable distributed random variables can be found in Samorodnitsky and Taqqu (1994), Janicki and Weron (1994)]. Here, we adapt the above mentioned multi-parameter approach to portfolio choice problems using stable laws. We find an equivalent parameterization of the stable laws (in terms of some moments) that characterizes the stable laws generally used. Then, we recall three admissible fund separation models where the asset returns are in the domain for attraction of stable laws [see Ortobelli, Rachev and Schwartz (2002)].

We first consider the portfolio allocation among n α-stable sub-Gaussian distributed risky assets (with $1 < \alpha < 2$) and the riskless one. The joint stable sub-Gaussian family is an elliptical family. Hence, as argued by Owen and Rabinovitch (1983), in this case, we can use a mean–dispersion analysis. The resulting efficient frontier is formally the same as Markowitz–Tobin's mean–variance analysis, but, instead of considering the variance as a risk parameter, we have to consider the scale parameter of the stable distributions. All the stable parameters can be estimated. In order to consider the possible asymmetry of asset returns, we describe a three-fund separation model for returns in the domain of attraction of a stable law. In case of asymmetry, the model results from a new stable version of the Simaan's model, see Simaan (1993). In case of symmetry of returns, we obtain a version of a model recently studied by Götzenberger, Rachev and Schwartz (1999), that can also be viewed as a particular version of the two-fund separation of Fama's (1965b) model. In this case too, it is possible to estimate all parameters. Finally, the last model proposed deals with the case of optimal allocation among stable distributed portfolios with different indexes of stability. To overcome the difficulties of the most general case of the stable law, we introduce a $k+1$ fund separation model. Then, we show how to express the model's multi-parameter admissible frontier.

Finally, we analyze an investment allocation problem. It consists of the maximization of the mean minus a measure of portfolio risk. We propose a mean–risk analysis that facilitates the interpretation of the results. In the allocation problem, we consider as the risk measure the expected value of a power absolute deviation. When the power is equal to two, we obtain the classic quadratic utility functional. We examine the optimal allocation among a riskless return and 23 risky returns, then we compare the allocation obtained with the Gaussian and the stable sub-Gaussian distributional assumption for the risky returns. We choose the 6% annual rate as riskless return. The model parameters are estimated using the methodology based on the moment method. We show that there are significant differences in the allocation when the data fit the stable sub-Gaussian or the normal distributions. By comparing the joint normal distribution with the joint stable sub-Gaussian law one, it has occurred that the results performed under the examined optimal allocation problems are substantially different. In particular, the stable market portfolio is generally less risky than the Gaussian market portfolio. This intuitive result is confirmed by the comparison of the optimal allocations when different distributional hypotheses are assumed. Therefore, the investors who fit the data with the stable distributions are generally more risk preserving than the investors who fit the data with the normal laws because they consider the component of risk due to the heavy tails.

Section 2 presents a first classification of the parametric distributions consistent with the maximization of the expected utility. Section 3 analyzes the asymptotic distributional assumption. In Section 4 we compare the stable sub-Gaussian multivariate approach with the normal multivariate one. In the last section, we briefly summarize the results.

2. Choices determined by a finite number of parameters

In this section we propose a distributional analysis of the optimal portfolio choice problem among $n+1$ assets: n of those assets are risky with gross returns[1] $Z=[Z_1,\ldots,Z_n]'$, and the $(n+1)$-th asset has risk-free gross return Z_0. When unlimited short selling is allowed, every portfolio of gross returns is a linear combination of the constant riskless gross return Z_0, and the risky returns Z_i, i.e.:

$$x_0 Z_0 + \sum_{i=1}^{n} x_i Z_i, \tag{1}$$

where $(x_0, x) \in \mathbb{R}^{n+1}$, $x \in \mathbb{R}^n$. Therefore, the distribution functions of all admissible investments belong to a translation and scale invariant family[2] determined by a finite number of parameters.

Assume price taker agents have preferences depending only on the probability distribution of terminal wealth. This assumption allows von Neumann–Morgenstern's preferences (1953) over wealth or more generally Machina's preferences (1982) over wealth but it precludes state dependent preferences.

Assume that the market faced by a decision maker comes from a standard model of perfect market (no transaction costs, taxes, asymmetric information, or arbitrage opportunities and all securities are perfectly divisible) which may not be complete.

Thus, in order to classify the parametric portfolio distribution functions consistent with the expected utility maximization, we distinguish and analyze the differences in portfolio allocation when:

(1) institutional restrictions (no short sales, limited liability) are allowed; or,
(2) unlimited short selling is allowed without penalty.

2.1. Portfolio choice with institutional restrictions

When limited liability and no short sales are allowed, portfolios of gross returns (i.e., $x'Z \geqslant 0$ where $Z_i > 0$ and $x_i \geqslant 0$, $\forall i$) are positive random variables. Thus, we assume that the portfolios of gross returns are positive random variables belonging to a scale invariant family, denoted with $\sigma\tau_k^+(\bar{a})$, that admits positive translations and it has the following characteristics:

[1] Generally, we assume the standard definition of i-th gross return between time t and time $t+1$, $Z_i = (P_{t+1,i} + d_{[t,t+1],i})/P_{t,i}$, where $P_{t,i}$ is the price of the i-th asset at time t and $d_{[t,t+1],i}$ the total amount of cash generated by the instrument between t and $t+1$. We distinguish the definition of gross return (with the capital letter) from the definition of return denoted $z_i = Z_i - 1$ (or the alternative definition of continuously compounded return $r_i = \log Z_i$).

[2] Recall that a parametric family $\Im$ of distribution functions is *translation invariant* if whenever the distribution $F_X(x) = P(X \leqslant x)$ belongs to $\Im$, then for every $t \in \mathbb{R}$, $F_{X+t} \in \Im$ as well. Similarly, we say that a family $\Im$ is *scale invariant* if whenever the distribution F_X belongs to $\Im$, then for every $\alpha > 0$, $F_{\alpha X}$ belongs to $\Im$ as well.

(1) Every distribution F_X belonging to $\sigma\tau_k^+(\bar{a})$ is associated to a positive random variable X and is identified from k parameters

$$(m_X, \sigma_X, a_{1,X}, \dots, a_{k-2,X}) \in A \subseteq \mathbb{R}^k,$$

where m_X is the mean of X, σ_X is the positive scale parameter of X.[3] We assume that the class $\sigma\tau_k^+(\bar{a})$ is *weakly determined* from its parametrization. That is, the equality

$$(m_X, \sigma_X, a_{1,X}, \dots, a_{k-2,X}) = (m_Y, \sigma_Y, a_{1,Y}, \dots, a_{k-2,Y})$$

implies that $F_X \overset{d}{=} F_Y$, but the converse is not necessarily true.

(2) For every admissible real $t \geqslant 0$, the distribution function $F_X \in \sigma\tau_k^+(\bar{a})$ has the same parameters as $F_{X+t} \in \sigma\tau_k^+(\bar{a})$, except the mean and the dispersion measure. In particular, the application $f(t) = \sigma_{X+t}$ is a nonincreasing continuous function.

(3) For every admissible positive α, the distribution function $F_X \in \sigma\tau_k^+(\bar{a})$ has, the same parameters of the distribution $F_{\alpha X}$ except for the mean that is αm_X and the scale parameter that is $\alpha\sigma_X$ (where m_X and σ_X are respectively the mean and the scale parameter of the random variable X).

When portfolios belong to a $\sigma\tau_k^+(\bar{a})$ class, we can identify stochastic dominance relations[4] among portfolios and the following theorem holds.

Theorem 1. *Assume all random admissible portfolios of gross returns belong to a $\sigma\tau_k^+(\bar{a})$ class. Let $w'Z$ and $y'Z$ be a couple of portfolios respectively determined by the parameters*

$$(m_{w'Z}, \sigma_{w'Z}, a_{1,p}, \dots, a_{k-2,p}) \quad \textit{and} \quad (m_{y'Z}, \sigma_{y'Z}, a_{1,p}, \dots, a_{k-2,p}).$$

Then, the following implications hold:

(1) *Suppose* $\frac{m_{w'Z}}{\sigma_{w'Z}} = \frac{m_{y'Z}}{\sigma_{y'Z}}$, *then $w'Z$ FSD $y'Z$ if and only if* $\sigma_{w'Z} > \sigma_{y'Z}$.

(2) $\frac{m_{w'Z}}{\sigma_{w'Z}} \geqslant \frac{m_{y'Z}}{\sigma_{y'Z}}$ *and* $\sigma_{w'Z} \geqslant \sigma_{y'Z}$ *with at least one inequality strict, implies $w'Z$ FSD $y'Z$.*

[3] In our context we use the mean as location parameter but the analysis can be extended to translation invariant families which do not admit finite the first moment. Moreover, we recall Pitman's seminal work (1939) on the estimation of location and scale parameters.

[4] Recall that the portfolio $x'Z$ first order stochastically deaminates (FSD) $y'Z$ if and only if for every increasing utility functions u, $E(u(x'Z)) \geqslant E(u(y'Z))$ and the inequality is strict for some u. Equivalently $x'Z$ FSD $y'Z$ if and only if $P(x'Z \leqslant t) \leqslant P(y'Z \leqslant t)$ for every real t and strictly for some t. Analogously, we say that $x'Z$ second order stochastically dominates (SSD) $y'Z$, if and only if for every increasing, concave utility function u, $E(u(x'Z)) \geqslant E(u(y'Z))$ and the inequality is strict for some u. Equivalently, $x'Z$ SSD $y'Z$, if and only if $\int_{-\infty}^{t} F_{x'Z}(v)\,dv \leqslant \int_{-\infty}^{t} F_{y'Z}(v)\,dv$ for every real t and strictly for some t [see, among others, Quirk and Saposnik (1962), Fishburn (1964), Hanoch and Levy (1969), Hadar and Russel (1969)]. We also say that $x'Z$ Rothschild Stiglitz stochastically dominates (R–S) $y'Z$ if and only if for every concave utility functions u, $E(u(x'Z)) \geqslant E(u(y'Z))$ and the inequality is strict for some u. Equivalently $x'Z$ R–S $y'Z$ if and only if $E(x'Z) = E(y'Z)$ and $x'Z$ SSD $y'Z$ [see Rothschild and Stiglitz (1970)]. However, there exist many other stochastic orders used in Economics and Finance, see, among others, Levy (1992), Shaked and Shanthikumar (1994).

(3) $w'Z$ *R–S* $y'Z$, *if and only if* $m_{w'Z} = m_{y'Z}$ *and* $\sigma_{w'Z} < \sigma_{y'Z}$.

(4) $\frac{m_{w'Z}}{\sigma_{w'Z}} \geqslant \frac{m_{y'Z}}{\sigma_{y'Z}}$ *and* $m_{w'Z} \geqslant m_{y'Z}$ *with at least one inequality strict, implies* $w'Z$ *SSD* $y'Z$.

(5) $\sigma_{w'Z} \geqslant \sigma_{y'Z}$ *and* $w'Z$ *SSD* $y'Z$, *implies* $w'Z$ *FSD* $y'Z$.

(6) $m_{w'Z} \geqslant m_{y'Z}$ *and* $\sigma_{w'Z} \leqslant \sigma_{y'Z}$ *with at least one inequality strict, implies* $w'Z$ *SSD* $y'Z$.

The proofs of Theorem 1 and of the next results are given in the appendix.

Observe that there exist counterexamples to the converse of implications (2), (4), (5) and (6) in Theorem 1. Thus, in order to obtain the converse of these implications, we need additional hypotheses [see Ortobelli (2001)]. Theorem 1 stresses the limits of mean–variance rule. In fact, suppose the portfolios of gross returns (without considering the riskless gross return) belong to a $\sigma\tau_2^+(\bar{a})$ class uniquely determined by the mean and the variance. Then, all non-satiable investors will choose portfolio solutions of the following constrained system

$$\max_x x'Qx \quad \text{subject to}$$

$$\frac{E(x'Z)}{\sqrt{x'Qx}} = h, \quad \text{where } x'e = 1, \tag{2}$$

$$x_i \geqslant 0, \quad i = 1, \ldots, n,$$

for some h, where $e = [1, \ldots, 1]'$, Q is the variance–covariance matrix of the vector of gross returns $Z = [Z_1, \ldots, Z_n]'$. Let σ^* be the maximum standard deviation of all admissible portfolios. Let us denote with μ^* the portfolio mean of gross returns with maximum variance. As a consequence of Theorem 1 and Bawa's results (1976), when the variance–covariance matrix Q is not singular and h varies in the following interval:

$$\frac{\mu^*}{\sigma^*} \leqslant h \leqslant \sqrt{\mu_Z' Q^{-1} \mu_Z} = \max_x \frac{x'\mu_Z}{\sqrt{x'Qx}}, \tag{3}$$

where μ_Z is the mean of the vector of gross returns Z, then the solutions of optimization problem (2) describe a set that contains the efficient frontier for agents with utility functions monotonically increasing in wealth. Moreover, under our assumptions, there exists a nonempty neighborhood U of the global minimum variance portfolio $Z'Q^{-1}e/(e'Q^{-1}e)$ such that every admissible portfolio belonging to $U(Z'Q^{-1}e/(e'Q^{-1}e))$ is not a solution of optimization problem (2).

With reference to the portfolio selection problem, recall that Markowitz (1952, 1987) and Tobin (1958, 1965) proposed the following selection rule for non-satiable risk averse investors: "From among a given set of investment alternatives (which includes the set of securities available in the market as well as all possible linear combinations of those basic securities), the admissible set of alternatives is obtained by discarding those investments with a lower mean and higher variance than a member of the given set". On the basis of

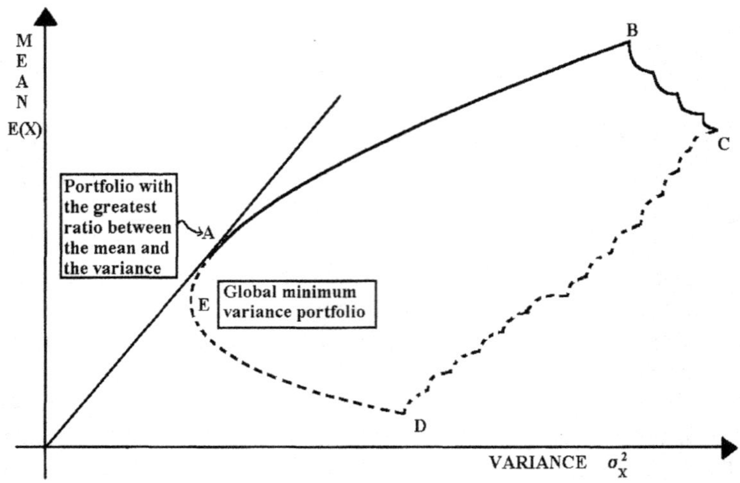

Fig. 1. The continuous curve represents efficient portfolios for non-satiable investors considering restrictions of nonnegative wealth; - - - dominated portfolios. The class of non-satiable investor's optimal choices (which are contained in the arc ABC) is different from the class of risk averse investor's optimal choices (which are contained in the arc DEAB), even if they generally have many common choices. Therefore, when we consider the risk averse non-satiable investor's optimal choices, we obtain the feasible optimal portfolios (which are contained in the arc AB) that are only a part of portfolios given by the classic Markowitz and Tobin's rule (arc EAB).

Theorem 1, we find that Markowitz–Tobin's selection rule is not optimal for non-satiable risk averse investors. In this context it is necessary to underline that no short sales or limited liability restrictions are imposed in a market where no riskless return is allowed. As a consequence, all portfolios are random variables uniformly bounded from below.[5] As a matter of fact, Theorem 1 cannot be extended to nonpositive random variables. Markowitz, Tobin, Bawa and many other authors left behind this observation in their considerations using normal distributions for returns. They have considered as efficient the portfolios on the upper neighborhood of global minimum variance (EA in Figure 1) but the same portfolios whose domain is under this restriction are not all efficient. Therefore, we proved that Markowitz and Tobin selection rule cannot be optimal even when portfolios belong to a family uniquely determined from the mean and the variance. It is well known that a lower variance does not imply a better choice for a non-satiable risk averse investor [see, example, in Hanoch and Levy (1969)]. Moreover, in an opportune neighborhood of global minimum variance portfolio, optimal portfolios for non-satiable investors do not exist. However, when riskless borrowing or lending is allowed, the mean–variance rule provides a sharper decision which permits to derive the efficient set for decision making with increasing and concave utility functions. In fact, if riskless asset is allowed, the global

[5] Recall that a random variable X is *bounded from below* (*above*) if there exists a real t such that $P(X \leqslant t) = 0$ ($P(X \geqslant t) = 0$). Analogously, a parametric family $\Im$ is *uniformly bounded from below* (*above*) if there exists a real t such that, for every random variable $X \in \Im$, $P(X \leqslant t) = 0$ ($P(X \geqslant t) = 0$).

minimum dispersion portfolio is the riskless asset itself. Thus, as shown by Levy and Kroll (1976) and Kroll and Levy (1979), the classification of the efficient frontiers given by the stochastic dominance analysis assumes a simpler form.

Under institutional restriction on the market (no short sales, limited liability), we can assume that the family of all admissible portfolios of gross returns $x'Z$ belongs to a scale invariant family $\Im$ which admits positive translations. If every distribution function $F_X \in \Im$ is associated to a positive random variable X uniquely determined[6] by $(m_X, \sigma_X, p_{1,X}, \ldots, p_{k-2,X})$, where m_X is the mean, σ_X is the standard deviation and

$$p_{i,X} = \frac{E((X - E(X))^{i+2})}{\sigma_X^{i+2}} \quad \text{for } i = 1, \ldots, k-2$$

are the first $k-2$ nontrivial *fundamental ratios*, then, the family $\Im$ is a particular $\sigma\tau_k^+(\bar{a})$. Note for $i = 1$ and $i = 2$ the i-th fundamental ratios are respectively Pearson's asymmetry and kurtosis coefficients of the random variable X. Thus, all risk averse investors will choose non-R–S stochastically dominated portfolios among the solutions of the following constrained optimization problem

$$\begin{aligned}
&\min_x x'Qx \quad \text{subject to} \\
&E(x'Z) = m, \quad x'e = 1, \\
&\frac{E((x'Z - E(x'Z))^i)}{(x'Qx)^{i/2}} = q_i, \quad i = 3, \ldots, k, \\
&x_j \geqslant 0, \quad j = 1, \ldots, n,
\end{aligned} \tag{4}$$

for some m and q_i, $i = 3, \ldots, k$, where $e = [1, \ldots, 1]'$, Q is the variance–covariance matrix of the vector of gross returns $Z = [Z_1, \ldots, Z_n]'$. Moreover, all non-satiable investors will choose portfolio weights, solutions of the following optimization problem

$$\begin{aligned}
&\max_x x'Qx \quad \text{subject to} \\
&\frac{E(x'Z)}{\sqrt{x'Qx}} \geqslant h, \quad x'e = 1, \\
&\frac{E((x'Z - E(x'Z))^i)}{(x'Qx)^{i/2}} = q_i, \quad i = 3, \ldots, k, \\
&x_j \geqslant 0, \quad j = 1, \ldots, n,
\end{aligned} \tag{5}$$

[6] Recall that a class of distributions is *uniquely determined* from k parameters when the equality $(m_X, \sigma_X, a_{1,X}, \ldots, a_{k-2,X}) = (m_Y, \sigma_Y, a_{1,Y}, \ldots, a_{k-2,Y})$ implies the equality of the respective distributions, i.e., $F_X \stackrel{d}{=} F_Y$, and vice versa.

for some q_i, $i = 3, \ldots, k$, and $h \geqslant \mu^*/\sigma^*$, where σ^* is the maximum standard deviation of all admissible portfolios and μ^* is the mean of that portfolio of gross returns. Similarly, all non-satiable risk averse investors will choose portfolio weights among the solutions of the following optimization problem

$$\begin{aligned} &\max_x E(x'Z) \quad \text{subject to} \\ &\frac{E(x'Z)}{\sqrt{x'Qx}} \geqslant h, \quad x'e = 1, \\ &\frac{E((x'Z - E(x'Z))^i)}{(x'Qx)^{i/2}} = q_i, \quad i = 3, \ldots, k, \\ &x_j \geqslant 0, \quad j = 1, \ldots, n, \end{aligned} \tag{6}$$

for some q_i, $i = 3, \ldots, k$, and $h \geqslant \mu_{j^*}/\sigma_{j^*}$, where μ_{j^*} is the maximum mean of all primary gross returns and σ_{j^*} is the standard deviation of that return. We obtain optimization problems analogous to (4), (5) and (6) when we consider the riskless asset. In this case, the mean is given by $E(x'Z) + (1 - x'e)Z_0$ and we require that $0 \leqslant x'e \leqslant 1$ instead of requiring $x'e = 1$. Theorem 1 is used for positive random variables. However, the above results can be generalized to families of random variables, which are uniformly bounded from below. In fact, without loss of generality, we can consider a translation that makes all random variables positive.

2.2. *Portfolio choice when unlimited short sales are allowed*

In the last fifty years the researchers of portfolio choice theory often used unbounded random variables for portfolio of returns, typically: the Gaussian laws. They also used to study continuously compounded portfolio of returns, say $x'r = \sum_{i=1}^n x_i \log Z_i$, where[7] $r_i = \log Z_i$.

In particular, we assume that the distribution functions of portfolios belong to a translation and scale invariant family denoted with $\sigma\tau_k(\bar{a})$ with the following characteristics:

[7] The continuously compounded portfolio of returns $x'r$ represents an approximation to the portfolio of returns $x'z$ (i.e., $\sum_{i=1}^n x_i \log Z_i \sim x'z$, where $z_i = Z_i - 1$). Thus, continuously compounded portfolio of returns $x'r$ are equivalently identified and called portfolio of returns. However, observe that

$$X \text{ FSD } Y \quad \text{if and only if} \quad \log X \text{ FSD } \log Y,$$

while

$$X \text{ SSD } Y \quad \text{implies} \quad \log X \text{ SSD } \log Y$$

but the converse is not necessarily true (you can find a simple counterexample with the log-normal class). Hence, when we study the optimal choices by considering the approximation $\sum_{i=1}^n x_i \log Z_i \sim x'z$, we find a set of choices that would be closer to the efficient set as well as the approximation would be right. (The approximation is good enough when we consider daily – or weekly – data in the empirical analysis).

(1) Every distribution F_X belonging to $\sigma\tau_k(\bar{a})$ is identified from k parameters $(m_X, \sigma_X, a_{1,X}, \ldots, a_{k-2,X}) \in A \subseteq \mathbb{R}^k$ where m_X is the mean of X, σ_X is the positive scale parameter of X. We assume that the class $\sigma\tau_k(\bar{a})$ is *weakly determined* from its parameterization. That is the equality $(m_X, \sigma_X, a_{1,X}, \ldots, a_{k-2,X}) = (m_Y, \sigma_Y, a_{1,Y}, \ldots, a_{k-2,Y})$ implies that $F_X \stackrel{d}{=} F_Y$ but the converse is not necessarily true.
(2) For every admissible real t, the distribution function $F_X \in \sigma\tau_k(\bar{a})$ has the same parameters, except the mean, as $F_{X+t} \in \sigma\tau_k(\bar{a})$ (the translated of F_X).
(3) For every admissible positive α, the distribution function $F_X \in \sigma\tau_k(\bar{a})$ has the same parameters of the distribution $F_{\alpha X} \in \sigma\tau_k(\bar{a})$ except for the mean that is αm_X and the scale parameter that is $\alpha\sigma_X$ (where m_X and σ_X are respectively the mean and the scale parameter of the random variable X).

The random variables associated to the distribution functions of a $\sigma\tau_k(\bar{a})$ class are not uniformly bounded from below because every $\sigma\tau_k(\bar{a})$ class is translation invariant. When portfolios belong to a $\sigma\tau_k(\bar{a})$ class, we can identify a stochastic dominance relation among portfolios unbounded from below and the following theorem holds.

Theorem 2. *Suppose the distribution functions of all random portfolios belong to the same class $\sigma\tau_k(\bar{a})$. Let $w'r$ and $y'r$ be a couple of random portfolios unbounded from below respectively determined by the parameters*

$$(m_{w'r}, \sigma_{w'r}, a_{1,p}, \ldots, a_{k-2,p}) \quad \text{and} \quad (m_{y'r}, \sigma_{y'r}, a_{1,p}, \ldots, a_{k-2,p}).$$

Then, the following properties are equivalent
(1) $E(w'r) \geqslant E(y'r)$, $\sigma_{w'r} \leqslant \sigma_{y'r}$ *with at least one inequality strict.*
(2) $w'r$ *SSD* $y'r$ *and* $y'r \stackrel{d}{=} w'r - (E(w'r) - E(y'r)) + \varepsilon$ *and* $E(\varepsilon / w'r) = 0$.

As for Theorem 2, when all portfolios are random variables unbounded from below and their distribution functions belong to a $\sigma\tau_2(\bar{a})$ class, two portfolios X and Y such that $\sigma_X > \sigma_Y$ and X SSD Y cannot exist. On the contrary, when the random portfolios considered in Theorem 2 are random variables bounded from below, we need further assumptions to get the above equivalence [see Ortobelli (2001)].

According to Theorem 2, it follows that when all portfolios are unbounded random variables belonging to a $\sigma\tau_k(\bar{a})$ class, it is easier to characterize their stochastic dominance properties. In this sense, the continuously compounded portfolios of returns $x'r$, are natural candidates for a simpler stochastic dominance analysis.

Samuelson (1969), Samuelson and Merton (1975) were among the first to investigate the conditions for the mean–variance criterion to provide an approximate optimum. Chamberlein (1983) has shown that when the riskless return is allowed, the families of elliptical distributions with finite variance are necessary and sufficient for the expected utility of final wealth to be a function only of the mean and the variance. Hence, when the portfolios are unbounded random variables[8] with distribution functions belonging to the same elliptical distribution family having finite variance, we can use Markowitz and Tobin's rule to individuate the optimal portfolios. Similarly, assuming that:

(a) there is no riskless asset;
(b) the portfolios of returns are unbounded random variables;
(c) the last $n-1$ components of the return random vector are elliptically distributed (with finite variance) conditional on the first component which has an arbitrary distribution with finite variance [see Chamberlein (1983)];

then, Markowitz and Tobin's rule can be used to individuate the optimal portfolios. Thus, Theorems 1 and 2 underline a further limitation (the above point (b)) of the previous studies on this issue.

We can now find optimal portfolios when all returns are unbounded random variables uniquely determined by a finite number of moments. Thus, if short sale is allowed, all risk averse investors will choose non-R–S stochastically dominated portfolios that are solutions of the constrained optimization problem (4) without the constrain $x_j \geqslant 0$, $j = 1, \ldots, n$. Similarly, we obtain optimal solutions for non-satiable investors maximizing the mean for some fixed central moments.

2.3. *Relations with Ross' multi-parameter models*

Consider the problem of optimal allocation among $n+1$ assets: n of those assets are risky with non-redundant returns $r = [r_1, \ldots, r_n]'$, and the $(n+1)$-th asset return is z_0 risk-free. Then,we are interested in the cases of portfolio distributions belonging to a $\sigma\tau_k(\bar{a})$ family with $k < n$. As argued by Ross (1978a), in order to reduce the variables of the portfolio choice problem, we have to assume some restrictions on the vector $\varepsilon = (\varepsilon_1, \varepsilon_2, \ldots, \varepsilon_n)'$ in the following representation of the returns:

$$r_i = \sum_{p=1}^{q} b_{i,p}\big(Y_p - E(Y_p)\big) + \varepsilon_i, \quad i = 1, \ldots, n, \tag{7}$$

where Y_i and ε_i are random variables and $b_{i,j}$ are scalars. Differently from Ross, we propose to study the case where all random variables $x'\varepsilon$ belong to a $\sigma\tau_s(\bar{a})$ family. Then, the scale parameter $\sigma_{x'\varepsilon}$ of random variable $x'\varepsilon$, has to verify the properties relatively to the $\sigma\tau_s(\bar{a})$ class. Thus, consider the parameterization given by:

(1) the parameters of the $\sigma\tau_s(\bar{a})$ family, and
(2) the parameters $c_j^* = x'b_{\cdot,j}/\sigma_{x'\varepsilon}$, for $j = 1, \ldots, q$.

This parameterization verifies the properties of a $\sigma\tau_k(\bar{a})$ family with $k = s + q$. In fact, for every positive real α the parameters of $\alpha x'(r - \mu)$ do not change except for the scale

[8] The elliptical families with finite variance are symmetric around the mean and are not necessarily associated to unbounded random variables, see Ingersoll (1987), Owen and Rabinovitch (1983). Then, following Theorem 2 we need to specify when elliptically distributed random variables have to be unbounded.

parameter that becomes $\alpha\sigma_{x'\varepsilon}$. Then, all portfolios belong to a $\sigma\tau_k(\bar{a})$ family with $k = s+q$ when the returns admit the form (7) and all admissible random variables $x'\varepsilon$ belong to a $\sigma\tau_s(\bar{a})$ family. Typical examples of this approach are the stable Paretian models presented in the next section.

Note that the above proposed moment analysis generalizes many of the three moment models presented in the last decades [see, for example, Kraus and Litzenberger (1976), Ingersoll (1987), Simaan (1993)]. Moreover, we do not need to require that portfolio returns verify the fund separation conditions as happened in the three moment models. Therefore, the above theorems represent a first classification of portfolio distribution functions which is alternative to those proposed from Ross (1976, 1978a). In fact, we underline the following differences from Ross' models:

(1) We express necessary and sufficient conditions to identify optimal portfolios. We can derive the efficient frontiers solving a constrained optimization problem.
(2) We do not require the closure of the random law under addition.
(3) The above theorems are an unifying and generalizing extension of moment analysis in the portfolio selection theory. In particular, the previous analysis describes further restrictions in using Markowitz and Tobin's selection rule as optimal portfolio selection rule.
(4) We express a portfolio choice theory dependent on a finite number of parameters consistent with expected utility maximization. We do not specify which parameters identify the distribution functions of asset returns. We only require very general properties which determine the existence of a scale parameter and a shift parameter.
(5) The above results can be applied to every economic choice in uncertainty conditions when the distribution functions are weakly determined by a finite number of parameters and verify properties of $\sigma\tau_k(\bar{a})$ or of $\sigma\tau_k^+(\bar{a})$ classes. Besides, this classification of choices under uncertainty conditions implies a first classification of the admissible dispersion measures [see Ortobelli (2001), Giacometti and Ortobelli (2001)].

As it follows from the previous considerations, the models introduced here can be theoretically improved and empirically tested. However, a more general theoretical and empirical analysis with further discussion, studies and comparison of the above models does not enter in the objective of this chapter and it will be the subject of future research.

3. The asymptotic distributional classification of portfolio choices

In this section we study the portfolio choice problem analyzing the asymptotic behavior of data. In particular, we consider unbounded random portfolios of stable distributed returns, $x'r$, that, with abuse of notation, we continue to call as portfolios of stable distributed returns.[9]

[9] If $\log Z_i$ is stable distributed, then $Z_i = 1 + z_i$ is log-stable distributed.

The recent crashes observed in the stock market showed that the stock returns are more volatile than those predicted by the models with finite variance of the asset returns. In the empirical financial literature, it is well documented that the asset returns have a distribution whose tail is heavier than that of the distributions with finite variance, i.e.,

$$P(|r_i| > x) \sim x^{-\alpha_i} L_i(x) \quad \text{as } x \to \infty, \tag{8}$$

where $0 < \alpha < 2$ and $L_i(x)$ is a slowly varying function at infinity, i.e.,

$$\lim_{x\to\infty} \frac{L_i(cx)}{L_i(x)} \to 1 \quad \text{for all } c > 0,$$

see Rachev and Mittnik (2000) and the references therein. In particular, in the data observed until now $1 < \alpha < 2$. The constrain $1 < \alpha < 2$ and the relation (8) imply that returns r_i admit finite mean and non-finite variance. The tail condition in (8) also implies that the vector of returns $r = [r_1, \dots, r_n]'$ is in the domain of attraction of $(\alpha_1, \dots, \alpha_n)$-stable law. That is, given T i.i.d (independent and identically distributed) observations on r, namely

$$r^{(t)} = \left[r_1^{(t)}, \dots, r_n^{(t)}\right]', \quad t = 1, 2, \dots, T,$$

then, there exist normalizing constants

$$a^{(T)} = \left(a_1^{(T)}, \dots, a_n^{(T)}\right) \in R_+^n \quad \text{and} \quad b^{(T)} = \left(b_1^{(T)}, \dots, b_n^{(T)}\right) \in \mathbb{R}^n,$$

such that

$$\left(\sum_{i=1}^{T} \frac{r_1^{(i)}}{a_1^{(T)}} + b_1^{(T)}, \dots, \sum_{i=1}^{T} \frac{r_1^{(i)}}{a_n^{(T)}} + b_n^{(T)}\right) \xrightarrow{d} S(\alpha_1, \dots, \alpha_n) \quad \text{as } T \to \infty, \tag{9}$$

where $S(\alpha_1, \dots, \alpha_n)$ is $(\alpha_1, \dots, \alpha_n)$-stable random variable. This convergence result is a consequence of the stationary behavior of returns and of the Central Limit Theorem for normalized sums of i.i.d. random variables which determines the domain of attraction of each stable law [see Zolotarev (1986)]. Therefore, any distribution in the domain of attraction of a specified stable distribution will have properties close to those of the stable distribution. The constants $a_j^{(T)}$ in (9) have the form

$$a_j^{(T)} = T^{1/\alpha_j} L_j(T),$$

where $L_j(T)$ are slowly varying functions as $T \to \infty$.

Each component of $S(\alpha_1,\dots,\alpha_n)=(s_1,\dots,s_n)$ has a Pareto–Lévy stable distribution, i.e., its characteristic function is given by

$$\Phi_j(t)=\begin{cases}\exp\left[-\sigma_j^{\alpha_j}|t|^{\alpha_j}\left(1-\mathrm{i}\beta_j\operatorname{sgn}(t)\tan\left(\frac{\pi\alpha_j}{2}\right)\right)+\mathrm{i}\mu_j t\right] & \text{if } \alpha_j\neq 1,\\ \exp\left[-\sigma_j|t|\left(1+\mathrm{i}\beta_j\frac{2}{\pi}\operatorname{sgn}(t)\log|t|\right)+\mathrm{i}\mu_j t\right] & \text{if } \alpha_j=1,\end{cases}\tag{10}$$

where $\alpha_j\in(0,2)$ is the so-called stable (tail) index of s_j, $\sigma>0$ is the scale (or dispersion) parameter, $\beta\in[-1,1]$ is a skewness parameter and μ is a location parameter. Moreover, for every fixed α, the Pareto–Lévy α-stable law is a $\sigma\tau_3(\bar{a})$ class. When $\alpha>1$ the location parameter μ is the mean. However, there is a considerable debate in literature concerning the applicability of α-stable distributions as they appear in Lévy's central limit theorems. A serious drawback of Lévy's approach is that in practice one can never know whether the underlying distribution is heavy tailed, or just has a long but truncated tail. Limit theorems for stable laws are not robust with respect to truncation of the tail or with respect to any change from light to heavy tail, or conversely. Based on finite samples, one can never justify the specification of a particular tail behavior. Hence, one cannot justify the applicability of classical limit theorems in probability theory. Therefore, instead of relying on limit theorems, we can use the so-called pre-limit theorem which provides an approximation for distribution functions in case the number of observation T is "large" but not too "large" [see Klebanov, Rachev and Szekely (2001), Klebanov, Rachev and Safarian (2000)]. In particular the "pre-limiting" approach helps to overcome the drawback of Lévy-type central limit theorems. As a matter of fact, we can assume that returns are bounded "far away", say daily returns cannot be outside the interval $[-0.5,0.5]$. Thus, considering the empirical observation on asset returns, we can assume that the asset returns r_i are truncated α_i-stable distributed with support, $[-0.5,0.5]$. Even if the returns will be attracted by the CLT to the Gaussian law, pre-limit theorems show that for any reasonable T the truncated stable laws will be attracted to the stable laws. Therefore, it is plausible assuming that the vector of returns $r=[r_1,\dots,r_n]'$ is in the domain of attraction of a n-dimensional $(\alpha_1,\dots,\alpha_n)$-stable law.

In order to express a multi-parameter choice in portfolio selection theory coherent with the empirical evidence and consistent with the expected utility maximization, we need the *asymptotic distributional assumption* consisting in:

(1) (*Heavy tailedness assumption*) Portfolios $x'r$ are unbounded random variables belonging to L^p with $1<p\leqslant 2$ and the return vector $r=[r_1,\dots,r_n]'$ is in the domain of attraction of $(\alpha_1,\dots,\alpha_n)$-stable law ($1<\alpha_i\leqslant 2$, $i=1,\dots,n$). The assumption $1<\alpha_i\leqslant 2$ is supported by increasing empirical results as shown by Mandelbrot (1963a, b, 1967), Fama (1963, 1965, b), Mittnik, Rachev and Paolella (1997), Rachev and Mittnik (2000).

(2) (*Consistency with the expected utility maximization*) The distributions of the portfolio returns $x'r$ belong to the same $\sigma\tau_k(\bar{a})$ class of distribution functions.

Under these assumptions, as for Theorem 2, we obtain an admissible frontier for non-satiable and non-satiable risk averse investors.

A simpler way to express the asymptotic behavior of data consists in considering every portfolio in the domain of attraction of a Pareto–Lévy α stable distribution with $\alpha > 1$. Given that, we implicitly assume that all optimal choices are identified by four parameters of the underlined stable law. Therefore, every portfolio $x'r$ can be well approximated by a stable distribution, i.e., we can assume:

$$x'r + (1 - x'e)z_0 \stackrel{d}{=} S_{\alpha(x)}\big(\sigma(x), \beta(x), \mu(x)\big), \tag{11}$$

where z_0 is the riskless return, $\alpha(x) \in (\min_{1 \leqslant i \leqslant n} \alpha_i, 2)$ is the index of stability, $\alpha_j > 1$ is the index of stability of the j-th asset return, $\sigma(x)$ is the scale parameter, $\mu(x) = x'E(r) + (1 - x'e)z_0$ is the mean and $\beta(x)$ is the skewness parameter. Properties of $\sigma\tau_4(\bar{a})$ class are verified with this parameterization, so according to Theorem 2 every risk averse investor will choose a portfolio weight, solution of the following constrained problem

$$\begin{aligned} &\min_x \sigma(x) \quad \text{subject to} \\ &x'E(r) + (1 - x'e)z_0 = m, \\ &\beta(x) = \beta^*, \\ &\alpha(x) = \alpha^* \end{aligned} \tag{12}$$

for some m, β^*, α^*. In this case, we are not able to find a closed form of the efficient frontier because we do not know a priori the joint distribution of the asset returns. In order to overcome this problem, we could consider another admissible parameterization of stable distribution for problem (11). For example, we can prove that the mean $\mu(x) = x'E(r) + (1 - x'e)z_0$, the scale parameter $s(x) = E(|x'r - x'E(r)|)$ and the fundamental ratios

$$\rho_1(x) = \frac{E(|x'r - x'E(r)|^{q_1})}{(s(x))^{q_1}} \quad \text{and} \quad \rho_2(x) = \frac{E((x'r - x'E(r))^{\langle q_2 \rangle})}{(s(x))^{q_2}},$$

where $q_1, q_2 \in (1, \min_{1 \leqslant i \leqslant n} \alpha_i)$; represent a parameterization which verifies the properties of $\sigma\tau_4(\bar{a})$ class.[10] In fact, first observe that $\rho_1(x)$ and $\rho_2(x)$ do not depend on $\mu(x)$ and $\sigma(x)$ because

$$\big|x'r - x'E(r)\big|^{q_1} \stackrel{d}{=} \sigma(x)^{q_1} \big|S_{\alpha(x)}(1, \beta(x), 0)\big|^{q_1},$$

[10] The symbology $x^{\langle t \rangle}$ stands for $\operatorname{sgn}(x)|x|^t$.

and also

$$\left(x'r - x'E(r)\right)^{\langle q_2\rangle} \stackrel{d}{=} \sigma(x)^{q_2}\left(S_{\alpha(x)}(1, \beta(x), 0)\right)^{\langle q_2\rangle}.$$

Thus, as a consequence of Property 1.2.17 in Samorodnisky and Taqqu (1994)

$$\begin{aligned}\rho_1(x) &= \frac{E(|x'r - x'E(r)|^{q_1})}{(s(x))^{q_1}} \\ &= K\frac{\Gamma(1 - q_1/\alpha(x))\cos(\arctan(\beta(x)\tan(\pi\alpha(x)/2))q_1/\alpha(x))}{(\Gamma(1 - 1/\alpha(x))\cos(\arctan(\beta(x)\tan(\pi\alpha(x)/2))1/\alpha(x)))^{q_1}},\end{aligned}$$

where K is a constant that depends only on q_1. Hence, for every $q_1 \in (1, \min_{1\leqslant i\leqslant n}\alpha_i)$ and for every fixed $\beta(x)$, $\rho_1(x)$ is a decreasing function of $\alpha(x)$ on the existence interval. Moreover, $\rho_1(x)$ is an even function of $\beta(x)$ and it decreases in $|\beta(x)|$ for fixed $\alpha(x) \in (\min_{1\leqslant i\leqslant n}\alpha_i, 2)$. Instead, $\rho_2(x)$ is an increasing odd function of β for every $q_2 \in (1, \min_{1\leqslant i\leqslant n}\alpha_i)$ and for every fixed $\alpha(x) \in (\min_{1\leqslant i\leqslant n}\alpha_i, 2)$. These relations imply that $\rho_1(x)$ and $\rho_2(x)$ uniquely determinate $\alpha(x)$ and $\beta(x)$. Then, under the assumption (11), every risk averse investor will choose a portfolio weight, solution of the following constrained problem

$$\begin{aligned}&\min_x E\left(\left|x'r - x'E(r)\right|\right) \quad \text{subject to} \\ &x'E(r) + (1 - x'e)z_0 = m, \\ &\frac{E(|x'r - x'E(r)|^{q_1})}{(s(x))^{q_1}} = \rho_1, \\ &\frac{E((x'r - x'E(r))^{\langle q_2\rangle})}{(s(x))^{q_2}} = \rho_2\end{aligned} \tag{13}$$

for some m, ρ_1, ρ_2. Differently from problem (12), problem (13) does not require the knowledge of the joint distribution of asset returns but it is still computationally too complex. Generally, in order to identify the efficient frontier and reduce the number of parameters, we assume that $\alpha_i = \alpha$ for all $i = 1, \ldots, n$. Observe that stable distributions are stable with respect to summation of i.i.d. random stable variables and the vector of returns $r = [r_1, \ldots, r_n]'$ is α-stable distributed with $\alpha > 1$ if and only if all linear combinations are stable [see Samordinsky and Taqqu (1994, Theorems 2.1.2 and 2.1.5)]. In this case the joint characteristic function of returns is given by

$$\Phi_r(t) = \exp\left(-\int_{S_n} |t's|^{\alpha}\left(1 - \mathrm{i}\,\mathrm{sgn}(t's)\tan\left(\frac{\pi\alpha}{2}\right)\right)\gamma(\mathrm{d}s) + \mathrm{i}t'\mu\right),$$

where α is the index of stability, $\gamma(\mathrm{d}s)$ is the spectral measure concentrated on $S_n = \{s \in \mathbb{R}^n \mid \|s\| = 1\}$.

Thus, when the vector of returns is α stable distributed (with $\alpha > 1$), every portfolio $x'r + (1 - x'e)z_0$ (except the riskless return, i.e., $x = 0$) is distributed as

$$x'r + (1 - x'e)z_0 \overset{d}{=} S_\alpha\big(\sigma(x), \beta(x), \mu(x)\big),$$

where

$$\mu(x) = x'E(r) + (1 - x'e)z_0,$$

$$\sigma(x) = \left(\int_{S_n} |x's|^\alpha \gamma(\mathrm{d}s)\right)^{1/\alpha} \quad \text{and} \quad \beta(x) = \frac{\int_{S_n} |x's|^\alpha \operatorname{sgn}(x's)\gamma(\mathrm{d}s)}{(\sigma(x))^\alpha}$$

are respectively the mean, the scale parameter and the skewness parameter of the portfolio $x'r + (1 - x'e)z_0$. Under this distributional assumption, every risk averse investor will choose a portfolio weight, solution of the following constrained problem

$$\begin{aligned} &\min_x \sigma(x) \quad \text{subject to} \\ &x'E(r) + (1 - x'e)z_0 = m, \\ &\beta(x) = \beta^* \end{aligned} \tag{14}$$

for some m and β^*. In order to determine estimates of the scale parameter and of the skewness parameter, we can consider the tail estimator for the index of stability α and the estimator for the spectral measure $\gamma(\mathrm{d}s)$ proposed by Rachev and Xin (1993) and Cheng and Rachev (1995). However, even if the estimates of the scale parameter and the skewness parameter are computationally feasible, they require numerical calculations. Thus, model (14) does not present an easy applicability from an empirical point of view. Similarly to problem (13), we can fix $q < \alpha$ and propose a different representation based on the moments type constrains. Therefore, instead of model (14), we obtain the following constrained problem

$$\begin{aligned} &\min_x E\big(|x'r - x'E(r)|\big) \quad \text{subject to} \\ &x'E(r) + (1 - x'e)z_0 = m, \\ &\frac{E((x'r - x'E(r))^{\langle q_2 \rangle})}{(E(|x'r - x'E(r)|))^{q_2}} = \rho_2 \end{aligned} \tag{15}$$

for some m and ρ_2. Optimization problems (15) and (13) can be used in a more general setting than optimization problems (12), (14). In fact, a priori other classes of distribution functions (not only stable distributions) for returns uniquely determined by the parameters $m(x)$, $s(x)$, $\rho_1(x)$ and $\rho_2(x)$ could exist. Next, in order to overcome the intrinsic difficulties of the problems (12)–(14) and (15), we analyze different fund separation models that consider the asymptotic distributional assumption.

3.1. The sub-Gaussian stable model

Assume the vector of returns $r = [r_1, \ldots, r_n]'$ is sub-Gaussian α-stable distributed with $1 < \alpha < 2$. Then, the characteristic function of r has the following form

$$\Phi_r(t) = E\big(\exp(\mathrm{i}t'r)\big) = \exp\big(-(t'Qt)^{\alpha/2} + \mathrm{i}t'\mu\big), \tag{16}$$

where $Q = [R_{i,j}/2]$ is a positive definite $(n \times n)$-matrix, $\mu = E(r)$ is the mean vector, and $\gamma(\mathrm{d}s)$ is the spectral measure with support concentrated on $S_n = \{s \in \mathbb{R}^n \mid \|s\| = 1\}$. The term $R_{i,j}$ is defined by

$$\frac{R_{i,j}}{2} = [\tilde{r}_i, \tilde{r}_j]_\alpha \|\tilde{r}_j\|_\alpha^{2-\alpha}, \tag{17}$$

where $\tilde{r}_j = r_j - \mu_j$ are the centralized return, the covariation $[\tilde{r}_i, \tilde{r}_j]_\alpha$ between two jointly symmetric stable random variables $\tilde{r}_i$ and $\tilde{r}_j$ is given by

$$[\tilde{r}_i, \tilde{r}_j]_\alpha = \int_{S_2} s_i |s_j|^{\alpha-1} \operatorname{sgn}(s_j)\gamma(\mathrm{d}s),$$

in particular, $\|\tilde{r}_j\|_\alpha = (\int_{S_2} |s_j|^\alpha \gamma(\mathrm{d}s))^{1/\alpha} = ([\tilde{r}_j, \tilde{r}_j]_\alpha)^{1/\alpha}$. Here the spectral measure $\gamma(\mathrm{d}s)$ has support on the unit circle S_2.

This model can be considered as a special case of Owen–Rabinovitch's elliptical model [see Owen and Rabinovitch (1983)]. However, no estimation procedure of the model parameters is given in the elliptical models with non-finite variance. In our approach we use (16) and (17) to provide a statistical estimator of the stable efficient frontier. To estimate the efficient frontier for returns given by (16), we need to consider an estimator for the mean vector μ and an estimator for the dispersion matrix Q. The estimator of μ is given by the vector $\hat{\mu}$ of sample averages. Using Lemma 2.7.16 in Samorodnitsky and Taqqu (1994) we can write for every p such that $1 < p < \alpha$

$$\frac{[\tilde{r}_i, \tilde{r}_j]_\alpha}{\|\tilde{r}_j\|_\alpha^\alpha} = \frac{E(\tilde{r}_i \tilde{r}_j^{\langle p-1\rangle})}{E(|\tilde{r}_j|^p)}, \tag{18}$$

where the scale parameter σ_j can be written $\|\tilde{r}_j\|_\alpha = \sigma_j$. It can be approximated by the moment method suggested by Samorodnitsky and Taqqu (1994) (Property 1.2.17) in the case $\beta = 0$

$$\sigma_j^p = \|\tilde{r}_j\|_\alpha^p = \frac{E(|\tilde{r}_j|^p)p\int_0^{+\infty} u^{-p-1}\sin^2 u \,\mathrm{d}u}{2^{p-1}\Gamma(1 - p/\alpha)}. \tag{19}$$

It follows

$$\frac{R_{i,j}}{2} = \sigma_j^2 \frac{E(\tilde{z}_i \tilde{z}_j^{\langle p-1\rangle})}{E(|\tilde{z}_j|^p)} = \sigma_j^{2-p} \frac{p\int_0^{+\infty} u^{-p-1}\sin^2 u \,\mathrm{d}u}{2^{p-1}\Gamma(1 - p/\alpha)} E\big(\tilde{z}_i \tilde{z}_j^{\langle p-1\rangle}\big).$$

The above suggests the following estimator $\widehat{Q} = [\widehat{R}_{i,j}/2]$ for the entries of the unknown covariation matrix Q

$$\frac{\widehat{R}_{i,j}}{2} = \hat{\sigma}_j^{2-p} \frac{p \int_0^{+\infty} u^{-p-1} \sin^2 u \, du}{2^{p-1}\Gamma(1 - p/\alpha)} \frac{1}{N} \sum_{k=1}^{N} \tilde{z}_i^{(k)} \left(\tilde{z}_j^{(k)}\right)^{\langle p-1 \rangle}, \tag{20}$$

where the σ_j^2 is estimated as follows

$$\hat{\sigma}_j^2 = \frac{\widehat{R}_{j,j}}{2} = \left(\frac{\frac{1}{N} \sum_{k=1}^{N} |\tilde{r}_j^{(k)}|^p p \int_0^{+\infty} u^{-p-1} \sin^2 u \, du}{2^{p-1}\Gamma(1 - p/\alpha)} \right)^{2/p}. \tag{21}$$

The moment estimator makes most sense for each fixed $p \in (1, \alpha)$. The rate of convergence of the empirical matrix $\widehat{Q} = [\widehat{R}_{i,j}/2]$ to the unknown matrix Q (to be estimated), will be faster, if p is as large as possible, see Rachev (1991).

Now, let us recall that our portfolio satisfies the relation

$$x'r \stackrel{d}{=} S_\alpha(\sigma_{x'r}, \beta_{x'r}, m_{x'r})$$

and furthermore, $W = z_0$ when $x = 0$, otherwise

$$W = x'r + (1 - x'e)z_0 \stackrel{d}{=} S_\alpha(\sigma_{x'r}, \beta_{x'r}, m_W),$$

where α is the index of stability, $\sigma_{x'r} = \sqrt{x'Qx}$ is the scale (dispersion) parameter, $\beta_{x'r} = 0$ is the skewness parameter and $m_W = x'E(r) + (1 - x'e)z_0$ is the mean of W. In particular, every sub-Gaussian α-stable family is a particular $\sigma\tau_2(m, \sigma)$ class.

In view of what stated before, when the returns $r = [r_1, \ldots, r_n]'$ are jointly sub-Gaussian α-stable distributed, every risk averse investor will choose an optimal portfolio among all portfolio solutions of the following optimization problem:

$$\min_x x'Qx \quad \text{subject to} \quad x'\mu + (1 - x'e)z_0 = m_W \tag{22}$$

for some given mean m_W, where $W = x'r + (1 - x'e)z_0$. Thus, every optimal portfolio that maximizes a given concave utility function u,

$$\max_x E\big(u\big(x'r + (1 - x'e)z_0\big)\big)$$

belongs to the mean–dispersion frontier

$$\sigma = \begin{cases} \dfrac{m - z_0}{\sqrt{(\mu - ez_0)'Q^{-1}(\mu - ez_0)}} & \text{if } m \geqslant z_0, \\[2ex] \dfrac{z_0 - m}{\sqrt{(\mu - ez_0)'Q^{-1}(\mu - ez_0)}} & \text{if } m < z_0, \end{cases} \tag{23}$$

where $\mu = E(r)$; $m = x'\mu + (1 - x'e)z_0$; $e = [1, \ldots, 1]'$; and $\sigma^2 = x'Qx$. Besides, the optimal portfolio weights x satisfy the following relation:

$$x = Q^{-1}(\mu - z_0 e)\frac{m - z_0}{(\mu - ez_0)'Q^{-1}(\mu - ez_0)}. \tag{24}$$

Note that (23) and (24) have the same forms as the mean–variance frontier. In particular, they assume a more general form for nonnecessarily symmetric dispersion matrix Q. As a matter of fact, even if Q is a symmetric matrix (it is definite positive) the estimator proposed in the sub-Gaussian cases (21) and (22) generally is not necessarily symmetric. Therefore, in some extreme cases we could obtain the inconsistent situation of stable distribution associated to a portfolio x with square scale parameter equal or lower than zero.[11] Moreover, (24) exhibits the two fund separation property for both the stable and the normal case but the matrix Q and the parameter have different meaning. In the normal case, Q is the variance–covariance matrix and σ is the standard deviation, while in the stable case Q is a dispersion matrix and σ is the scale (dispersion) parameter, $\sigma = \sqrt{x'Qx}$. According to the two-fund separation property of the sub-Gaussian α-stable approach, we can assume that the market portfolio is equal to the risky tangent portfolio under the equilibrium conditions (as in the classical mean–variance Capital Asset Pricing Model (CAPM)). Therefore, every optimal portfolio can be seen as the linear combination between the market portfolio

$$\bar{x}'r = \frac{r'Q^{-1}(\mu - z_0 e)}{e'Q^{-1}\mu - e'Q^{-1}ez_0}, \tag{25}$$

and the riskless asset return z_0. Following the same arguments as in Sharpe, Lintner, Mossin's mean–variance equilibrium model, the return of asset i is given by:

$$E(r_i) = z_0 + \beta_{i,m}\big(E(\bar{x}'r) - z_0\big), \tag{26}$$

[11] Observe that for every $x \in \mathbb{R}^n$, we get $x'\widehat{Q}x > 0$ if and only if $(\widehat{Q} + (\widehat{Q})')/2$ is a definite positive matrix. Thus, we can verify that $(\widehat{Q} + (\widehat{Q})')/2$ is definite positive in order to avoid stable portfolios $x'z$ with negative scale parameter estimators. Moreover, we observe that the symmetric matrix $(\widehat{Q} + (\widehat{Q})')/2$ is an alternative estimator of the dispersion matrix Q whose statistical properties have to be proved. In particular, if we want to simulate the vector $\tilde{z} = [\tilde{z}_1, \ldots, \tilde{z}_n]'$ of the centred α stable sub-Gaussian return distributions, we generally use the dispersion matrix $\widehat{\Sigma} = (\widehat{Q} + (\widehat{Q})')/2$. As a matter of fact, we first generate the vector $G = [G_1, \ldots, G_n]'$ of the joint Gaussian distribution $G = N(0, \widehat{\Sigma})$ using the Cholesky decomposition matrix. Then, the vector of returns [see Samorodnitsky and Taqqu (1994)] is given by:

$$\tilde{z} = \sqrt{A}\,G,$$

where $A \stackrel{d}{=} S_{\alpha/2}(2(\cos(\pi\alpha/4))^{2/\alpha}, 1, 0)$ is an $\alpha/2$ stable random variable independent of the Gaussian vector G.

where $\beta_{i,m} = \bar{x}'Qe^i/\bar{x}'Q\bar{x}$, with e^i the vector with 1 in the i-th component and zero in all the other components. As a consequence of Ross' necessary and sufficient conditions of two-fund separation [see Ross (1978a)], the above model admits the form

$$r_i = \mu_i + b_i Y + \varepsilon_i, \quad i = 1, \dots, n,$$

where $\mu_i = E(r_i)$, $E(\varepsilon/Y) = 0$, $\varepsilon = (\varepsilon_1, \varepsilon_2, \dots, \varepsilon_n)'$, $b = [b_1, \dots, b_n]'$ and the vector $bY + \varepsilon$ is sub-Gaussian α-stable distributed with zero mean.

Hence, our sub-Gaussian α-stable version of CAPM is not much different from Gamrowski–Rachev's (1999) version of the two-fund separation α-stable model. As a matter of fact, Gamrowski and Rachev (1999) propose a generalization of Fama's α-stable model (1965b) assuming $r_i = \mu_i + b_i Y + \varepsilon_i$, for every $i = 1, \dots, n$, where ε_i and Y are α-stable distributed and $E(\varepsilon/Y) = 0$. In view of their assumptions,

$$E(r_i) = z_0 + \tilde{\beta}_{i,m}\big(E(\bar{x}'r) - z_0\big),$$

where

$$\tilde{\beta}_{i,m} = \frac{1}{\alpha\|\bar{x}'\tilde{r}\|_\alpha^\alpha}\frac{\partial\|\bar{x}'\tilde{r}\|_\alpha^\alpha}{\partial\bar{x}_i} = \frac{[\tilde{r}_i, \bar{x}'\tilde{r}]_\alpha}{\|\bar{x}'\tilde{r}\|_\alpha^\alpha}.$$

Furthermore, the coefficient $[\tilde{r}_i, \bar{x}'\tilde{r}]_\alpha/\|\bar{x}'\tilde{r}\|_\alpha^\alpha$ can be estimated using the above formula (18).

Now, we see that in the above sub-Gaussian symmetric α-stable model $\bar{x}'Q\bar{x} = \|\bar{x}'\tilde{r}\|_\alpha^2$ and $\bar{x}'Qe^i = \frac{1}{2}\partial\|\bar{x}'\tilde{r}\|_\alpha^2/\partial\bar{x}_i$. Thus, we get the equivalence between the coefficient $\beta_{i,m}$ of model (26) and $\tilde{\beta}_{i,m}$ of Gamrowski–Rachev's model, i.e.:

$$\beta_{i,m} = \frac{\bar{x}'Qe^i}{\bar{x}'Q\bar{x}} = \frac{1}{\sigma_{\bar{x}'r}}\frac{\partial\sigma_{\bar{x}'r}}{\partial\bar{x}_i} = \frac{[\tilde{r}_i, \bar{x}'\tilde{r}]_\alpha}{\|\bar{x}'\tilde{r}\|_\alpha^\alpha} = \tilde{\beta}_{i,m},$$

where $\sigma_{\bar{x}'r}$ is the scale parameter of market portfolio.

3.2. A three fund separation model in the domain of attraction of a stable law

Let us assume that the vector $r = [r_1, \dots, r_n]'$ describes the following three-fund separating stable model of security returns:

$$r_i = \mu_i + b_i Y + \varepsilon_i, \quad i = 1, \dots, n, \tag{27}$$

where the random vector $\varepsilon = (\varepsilon_1, \varepsilon_2, \dots, \varepsilon_n)'$ is independent from Y and follows a joint sub-Gaussian α_1-stable distribution ($1 < \alpha_1 < 2$), with zero mean and characteristic function

$$\Phi_\varepsilon(t) = \exp\big(-|t'Qt|^{\alpha_1/2}\big),$$

where Q is the definite positive dispersion matrix. On the other hand, $Y \stackrel{d}{=} S_{\alpha_2}(\sigma_Y, \beta_Y, 0)$ is α_2-stable distributed random variable independent from ε, with $1 < \alpha_2 < 2$ and zero mean. Under these assumptions, the portfolios are in the domain of attraction of a α stable law with $\alpha = \min(\alpha_1, \alpha_2)$ and belong to a $\sigma\tau_3(\bar{a})$ family. A testable case in which Y is α_2-stable symmetric distributed (i.e., $\beta_Y = 0$), was recently studied by Götzenberger, Rachev and Schwartz (1999). When $\beta_Y = 0$ and $\alpha_1 = \alpha_2$, our model can lead to the two-fund separation Fama's model. The characteristic function of the vector of returns $r = [r_1, r_2, \ldots, r_n]'$ is given by:

$$\begin{aligned}\Phi_r(t) &= \Phi_\varepsilon(t)\Phi_Y(t'b)\,\mathrm{e}^{\mathrm{i}t'\mu} \\ &= \exp\left(-|t'Qt|^{\alpha_1/2} - |t'b\sigma_Y|^{\alpha_2}\left(1 - \mathrm{i}\beta_Y \operatorname{sgn}(t'b)\tan\left(\frac{\pi\alpha_2}{2}\right)\right) + \mathrm{i}t'\mu\right), \quad (28)\end{aligned}$$

where $b = [b_1, \ldots, b_n]'$ is the coefficient vector and $\mu = [\mu_1, \ldots, \mu_n]'$ is the mean vector.

Next we shall estimate the parameter in model (27), (28). First, the estimator of μ is given by the vector $\hat{\mu}$ of sample average. Then, we consider as factor Y a centralized index return (for example the market portfolio (25) given by the above sub-Gaussian model). Therefore, given the sequence of observations $Y^{(k)}$, we can estimate its stable parameters. Observe that the random vector ε admits a representation as a product of random variable V and Gaussian vector G:

$$\varepsilon = VG.$$

$V = \sqrt{A}$, where A is an $\alpha_1/2$-stable subordinator, that is

$$A \stackrel{d}{=} S_{\alpha_1/2}\left(\left(\cos\left(\frac{\pi\alpha_1}{4}\right)\right)^{2/\alpha_1}, 1, 0\right);$$

G is a $(n \times 1)$-Gaussian vector with null mean and variance–covariance matrix Q and it is independent from A. We can generate values A_k, $k = 1, \ldots, N$, of A independent from G. We address to Paulauskas and Rachev's work (1999) the problem of generating such values A_k. Using the centralizing returns $\tilde{r}_j = r_j - \mu_j$ on Y we write the following OLS estimators[12] for $b = [b_1, \ldots, b_n]'$ and Q:

$$\hat{b}_i = \frac{\sum_{k=1}^N Y^{(k)}\tilde{r}_i^{(k)}/A_k}{\sum_{k=1}^N (Y^{(k)})^2/A_k}, \qquad i = 1, \ldots, n,$$

and

$$\widehat{Q} = \frac{1}{N}\sum_{k=1}^N \frac{(\tilde{r}^{(k)} - \hat{b}Y^{(k)})(\tilde{r}^{(k)} - \hat{b}Y^{(k)})'}{A_k}.$$

[12] For a discussion see Tokat, Rachev and Schwartz (2002).

The selection of α_1 is a separate problem. A possible way to estimate α_1 is to consider the OLS estimator $\tilde{b}_i = \sum_{k=1}^{N} Y^{(k)}\tilde{r}_i^{(k)}/\sum_{k=1}^{N}(Y^{(k)})^2$ and then to evaluate the sample residuals $\tilde{\varepsilon}^{(k)} = \tilde{r}^{(k)} - \tilde{b}Y^{(k)}$. If these residuals are heavy tailed, one can take the tail exponent as an estimator for α_1. The asymptotic properties of the above estimator can be derived arguing similarly with Paulauskas and Rachev (1999) and Götzenberger, Rachev and Schwartz (1999).

In order to determine portfolios R–S non-dominated when unlimited short selling is allowed, we have to minimize the scale parameter $\sigma_W = \sqrt{x'Qx}$ for some fixed mean $m_W = x'\mu + (1 - x'e)z_0$ and $\tilde{b} = x'b/\sqrt{x'Qx}$. Alternatively, as shown by Ortobelli, Rachev and Schwartz (2002), we can obtain these portfolios from the solution of the following quadratic programming problem:

$$\begin{aligned} &\min_x x'Qx \quad \text{subject to} \\ &x'\mu + (1 - x'e)z_0 = m_W, \\ &x'b = b^* \end{aligned} \tag{29}$$

for some m_W and b^*. Thus, under our assumptions, every portfolio that maximizes the expected value of a given concave utility function u,

$$\max_x E\big(u(x'r)\big)$$

belongs to the following frontier

$$(1 - \lambda_2 - \lambda_3)z_0 + \lambda_2 \frac{r'Q^{-1}(\mu - z_0 e)}{e'Q^{-1}(\mu - z_0 e)} + \lambda_3 \frac{r'Q^{-1}b}{e'Q^{-1}b} \tag{30}$$

spanned by the riskless return z_0, and the two risky portfolios

$$u^{(1)} = \frac{r'Q^{-1}(\mu - z_0 e)}{e'Q^{-1}(\mu - z_0 e)} \quad \text{and} \quad u^{(2)} = \frac{r'Q^{-1}b}{e'Q^{-1}b}.$$

Observe in (28) that when $\alpha = \alpha_1 = \alpha_2 > 1$, every portfolio $x'r$ is an α-stable distribution and satisfies the relation

$$W = (1 - x'e)z_0 + x'r \stackrel{d}{=} S_\alpha\big(\sigma_{x'r}, \beta_{x'r}, (1 - x'e)z_0 + m_{x'r}\big)$$

and $W = z_0$ when $x = 0$, where

$$\sigma_{x'r}^{\alpha} = (x'Qx)^{\alpha/2} + |x'b\sigma_Y|^{\alpha}, \quad \beta_{x'r} = \frac{|x'b\sigma_Y|^{\alpha}\,\mathrm{sgn}(x'b)\beta_Y}{\sigma_{x'r}^{\alpha}}, \quad m_{x'r} = x'E(r).$$

Hence, this jointly α-stable model is a fund separation model whose solutions are given by the optimization problem (14) and these solutions satisfy the quadratic programming problem (29).

3.3. A $k+1$ fund separation model in the domain of attraction of a stable law

As empirical studies show in the stable case one of the most severe restrictions of performance measurement and asset pricing is the assumption of a common index of stability for all assets – individual securities and portfolio alike. It is well understood that asset returns are not normally distributed. We also know that the return distributions do not have the same index of stability. However, under the assumption that returns have different indexes of stability, it is not generally possible to find a closed form to the efficient frontier. Generalizing the above model instead, we get the following $k+1$ fund separation model [for details on k fund separation models see Ross (1978a)]:

$$r_i = \mu_i + b_{i,1}Y_1 + \cdots + b_{i,k-1}Y_{k-1} + \varepsilon_i, \quad i = 1, \ldots, n. \tag{31}$$

Here, $n \geqslant k \geqslant 2$, the vector $\varepsilon = (\varepsilon_1, \varepsilon_2, \ldots, \varepsilon_n)'$ is independent from $Y_1, \ldots, Y_{k-1}$ and follows a joint sub-Gaussian symmetric α_k-stable distribution with $1 < \alpha_k < 2$, zero mean and characteristic function $\Phi_\varepsilon(t) = \exp(-|t'Qt|^{\alpha_k/2})$, and the random variables $Y_j \stackrel{d}{=} S_{\alpha_j}(\sigma_{Y_j}, \beta_{Y_j}, 0)$, $j = 1, \ldots, k-1$, are mutually independent[13] α_j-stable distributed with $1 < \alpha_j < 2$ and zero mean. Under these assumptions, the portfolios belong to a $\sigma\tau_{k+1}(\bar{a})$ class. If we need to insure the separation obtained in situations where the above model degenerates into a p-fund separation model with $p < k+1$, we require the rank condition [see Ross (1978a)]. In order to determine portfolios R–S non-dominated when unlimited short selling is allowed, we have to minimize the scale parameter $\sigma_W = \sqrt{x'Qx}$ for some fixed mean $m_W = x'\mu + (1 - x'e)z_0$ and $\tilde{b}_j = x'b_{\cdot,j}/\sqrt{x'Qx}$, $j = 1, \ldots, k-1$. Alternatively, as shown by Ortobelli, Rachev and Schwartz (2002), we can obtain these portfolios from the solution of the following quadratic programming problem:

$$\begin{aligned} &\min_x x'Qx \quad \text{subject to} \\ &x'\mu + (1 - x'e)z_0 = m_W, \\ &x'b_{\cdot,j} = c_j, \quad j = 1, \ldots, k-1. \end{aligned} \tag{32}$$

[13] In order to estimate the parameters, we need to know the joint law of the vector $(Y_1, \ldots, Y_{k-1})$. Therefore, we assume independent random variables Y_j, $j = 1, \ldots, k-1$. Then the characteristic function of the vector of returns $r = [r_1, \ldots, r_n]'$ is given by

$$\Phi_r(t) = \Phi_\varepsilon(t) \prod_{j=1}^{k-1} \Phi_{Y_j}(t'b_{\cdot,j}) \mathrm{e}^{\mathrm{i}t'\mu}.$$

Under this additional assumption, we can approximate all parameters of any optimal portfolio using a similar procedure of the previous three fund separation model. However, if we assume a given joint $(\alpha_1, \ldots, \alpha_{k-1})$ stable law for the vector $(Y_1, \ldots, Y_{k-1})$, we can generally determine estimators of the parameters studying the characteristics of the multivariate stable law.

By solving the optimization problem (32), we obtain that the riskless portfolio and other k risky portfolios span the efficient frontier for the risk averse investors given by

$$\left(1-\sum_{j=1}^{k}\lambda_j\right)z_0+\lambda_1\frac{r'Q^{-1}(\mu-z_0e)}{e'Q^{-1}(\mu-z_0e)}+\sum_{j=1}^{k-1}\lambda_{j+1}\frac{r'Q^{-1}b_{\cdot,j}}{e'Q^{-1}b_{\cdot,j}}.$$

The above multivariate models are motivated by arbitrage considerations as in the Arbitrage Pricing Theory (APT) [see Ross (1976)]. Without going into details, it should be noted that there are two versions of the APT for α-stable distributed returns, a so-called equilibrium [see Chen and Ingersoll (1983), Dybvig (1983), Grinblatt and Titman (1983)] and an asymptotic version [see Huberman (1982)]. Connor (1984) and Milne (1988) introduced a general theory which encompassed the equilibrium APT as well as the mutual fund separation theory for returns belonging to any normed vector space (hence also symmetric α-stable distributed returns). While Gamrowski and Rachev (1999) provide the proof for the asymptotic version of α-stable distributed returns. Hence, it follows from Connor and Milne's theory that the above model in the domain of attraction of a stable law of the return is coherent with the classic arbitrage pricing theory and the mean returns can be approximated by the linear pricing relation

$$\mu_i \sim z_0 + b_{i,1}\delta_1 + \cdots + b_{i,k-1}\delta_{k-1},$$

where δ_p, for $p = 1, \ldots, k-1$, are the risk premiums relative to the different factors. The above $k+1$ fund separation model concludes the examples of models in the domain of attraction of stable laws. In the next section we compare the Gaussian multivariate approach with the sub-Gaussian stable one.

4. A first comparison between the normal multivariate distributional assumption and the stable sub-Gaussian one

In this section we examine and compare the stable sub-Gaussian assumption with the normal distributional one. Thus, we implicitly assume that returns belong to a $\sigma\tau_2(m,\sigma)$ class where m is the mean and σ is either the scale parameter of stable distributions or the standard deviation of normal distributions.

In a recent work Ortobelli, Rachev and Schwartz (2002) compare the stable non-Gaussian assumption and the normal one by analyzing optimal allocations between a riskless return and a benchmark index. Three different indexes have been taken into consideration: CAC40, DAX30 and S&P 500. Their analysis has indicated that either the heavy tails of data or a greater centralization of data around the mean can have a significant impact on the approximation of the investors' choices. However, the stable non-Gaussian allocation is generally more risk preserving than the normal one. Precisely, the stable approach considers a further component of risk which is due to the fat tails of the return distributions. This fact does not surprise us excessively. As a matter of fact, also Mehra and Prescott's empirical analysis (1985) underlines that asset pricing puzzles can be justified thinking of people much more risk averse. Clearly, we do not believe that the equity premium puzzle can be explained only considering the sub-Gaussian stable distribution instead of the Gaussian one. However, we believe that the distributional differences between the data and the classic model used in finance can help to understand asset pricing puzzles. This conjecture is partly confirmed by assuming the stable distributions in place of the Gaussian one [see, for example, Kocherlakota's (1997) test on CCAPM with heavy-tailed pricing errors].

Next, we extend Ortobelli, Rachev and Schwartz's comparison to the multivariate case. This comparison is formally and theoretically different from the previous one because here the benchmark index is given by the market portfolio which generally will change, if the distributional assumptions change too. Thus, as a consequence of Roll (1977, 1978, 1979a, b), Dybvig and Ross' (1985a, b) analysis, we observe that:

(a) an investor, who fits the return distributions with a joint α_1-stable sub-Gaussian distribution, will consider as inefficient the choice of another investor who fits the return distributions with a joint α_2-stable sub-Gaussian distribution with $\alpha_1 \neq \alpha_2$; and
(b) the stable CAPM is still subject of some of the criticism already addressed to the classical one.

Nevertheless, it seems that the stable case explains better the empirical data. This is the main reason why here we interpret and analyze the different behavior between the investor who fits the data with joint stable sub-Gaussian distribution and the investor who fits the data with the joint normal distribution.

4.1. An optimal allocation problem

First, we consider the optimal allocation among 24 assets: 23 of those assets are risky assets with returns $r = [r_1, r_2, \dots, r_{23}]'$ and the 24th is riskfree with annual rate 6%. We analyze the portfolio choice problems when short sales are allowed and when short sales are not allowed. In view of this comparison, we discuss and study the differences in portfolio choice problems without examining them so as to choose one of the two assumptions (Gaussian or sub-Gaussian).[14]

In our comparison we use daily data taken from 23 international risky indexes valued in USD and quoted from January 1995 to January 1998. In the analysis proposed we first consider the maximum likelihood estimation of the stable parameters and of the Gaussian ones for every risky asset. Thus, Tables 1 and 2 assembles the approximating parameters obtained from using the program STABLE.[15]

In order to compare the different stable sub-Gaussian joint distributions and the joint normal distributions for the asset returns, we assume that the vector r is sub-Gaussian

[14] On this topic, recent studies [see Ortobelli et al. (2001), Ortobelli, Huber and Schwartz (2002)] have shown that sub-Gaussian multivariate models present a superior performance with respect to the mean–variance model.
[15] See Nolan (1997) and the web site www.ca.american.edu/~jpnolan.

α-stable distributed, with $\alpha = \alpha_k$, $k = 1, 2$ where $\alpha_1 = 1.7488$ represents the average of the indexes of stability and $\alpha_2 = 1.8856$ represents the maximum of the indexes of stability (see Table 2).[16] Moreover, when in the following tables we consider the index of stability $\alpha = 2$, we implicitly assume that the returns are jointly normal distributed. Thus, every portfolio of risky assets is stable distributed in the following way:

$$x'r \stackrel{d}{=} S_{\alpha_k}(\sigma_{x'r}, \beta_{x'r}, m_{x'r}),$$

where α_k is one of the considered index of stability $k = 1, 2$, $\sigma_{x'r} = (x'Q_k x)^{1/2}$ is the respective scale parameter, $Q_k = [R_{ij}/2]_k$ is the dispersion matrix, with $k = 1, 2$, $\beta_{x'r} = 0$ is the skewness parameter, and $m_{x'r}$ represents the mean of $x'r$. Observe that the matrix Q_k is estimated with the method defined in the previous section and thus it depends on the index of stability α_k for $k = 1, 2$. As observed previously, the rate of convergence of the empirical matrix $\widehat{Q}_k$ to the unknown matrix Q_k will be faster, if p is as large as possible. In our estimations we use $p_1 = 1.7$ (relative to $\alpha_1 = 1.7488$) and $p_2 = 1.8$ (relative to $\alpha_2 = 1.8856$).

We assume the investors wish to maximize the following utility functional:

$$U(W) = E(W) - cE(|W - E(W)|^q), \tag{33}$$

where c and q are positive real numbers, $W = \lambda z_0 + (1 - \lambda)\bar{x}'r$ is the return on the portfolio, z_0 is the risk-free asset return, and

$$\bar{x}'r = \frac{r'Q_k^{-1}(\mu - z_0 e)}{e'Q_k^{-1}\mu - e'Q_k^{-1}ez_0}$$

is the tangent portfolio of returns (25). With reference to the allocation problem (33), we observe:

(1) Problem (33) is equivalent to the following maximization of the utility functional

$$aE(W) - bE(|W - E(W)|^q), \tag{34}$$

assuming $c = b/a$ in (33) for every $a, b > 0$. Thus, $E(|W - E(W)|^q)$ represents a particular risk measure of portfolio loss, which satisfies (under the opportune standardization) the main characteristics of the typical dispersion measures. Solving the optimal allocation problem (33), the investor implicitly maximizes the expected mean of the increment wealth aW as well as minimizes the individual risk $bE(|W - E(W)|^q)$.

(2) Furthermore, when $q = 2$, the maximization of utility functional (33) motivates the mean–variance approach in terms of preference relations.

[16] We consider different indexes of stability, in order to value the effects of heavy-tailedness on the portfolio selection problems.

Suppose X dominates Y in the sense of R–S. Since $E(X) = E(Y)$ and $f(x) = c|x - E(X)|^q$ is a concave utility function, for every $q \in [1, \alpha)$, it follows that:

$$U(X) = E(X) - cE(|X - E(X)|^q) \geqslant U(Y), \quad \forall q \in [1, \alpha).$$

The above inequality implies that every risk averse investor with utility functional (34) should choose a portfolio $W = \lambda z_0 + (1 - \lambda)\bar{x}'r$ that maximizes the utility functional (33) for some real λ and some $q \in [1, \alpha)$.

We know that for $\lambda \neq 1$, all the portfolio returns $W = \lambda z_0 + (1 - \lambda)\bar{x}'r$ admits stable distribution

$$S_{\alpha_k}(|1 - \lambda|\sigma_{\bar{x}'r}, 0, \lambda z_0 + (1 - \lambda)m_{\bar{x}'r}), \quad k = 1, 2,$$

and $W = z_0$ when $\lambda = 1$. Now, in order to solve the asset allocation problem

$$\max_{\lambda} E(W) - cE(|W - E(W)|^q),$$

notice first that, for all $q \in [1, \alpha)$ and $1 < \alpha < 2$, we get

$$\begin{aligned} U(W) &= E(W) - cE(|W - E(W)|^q) \\ &= \lambda z_0 + (1 - \lambda)m_{\bar{x}'r} - c(H(\alpha, 0, q))^q |1 - \lambda|^q \sigma_{\bar{x}'r}^q, \end{aligned}$$

where

$$(H(\alpha, 0, q))^q = \frac{2^{q-1}\Gamma(1 - q/\alpha)}{q\int_0^\infty u^{-q-1}\sin^2 u\,du}$$

[see Samorodnitsky and Taqqu (1994), Hardin (1984)]. The above relation analyzes the stable non-Gaussian case. When the vector r admits a joint normal distribution (i.e., $\alpha = 2$), then for all $q > 0$,

$$\begin{aligned} U(W) &= E(W) - cE(|W - E(W)|^q) \\ &= \lambda z_0 + (1 - \lambda)m_{\bar{x}'r} - c\frac{2^{q/2}\Gamma((q + 1)/2)}{\sqrt{\pi}}|1 - \lambda|^q \sigma_{\bar{x}'r}^q. \end{aligned}$$

Hence, the real optimal solution of the problem in the important case $q \in (1, \alpha)$, is given by

$$\bar{\lambda} = 1 - \operatorname{sgn}(1 - \bar{\lambda})\left(\frac{\operatorname{sgn}(1 - \bar{\lambda})(m_{\bar{x}'r} - z_0)}{qc\sigma_{\bar{x}'r}^q V(\alpha, 0, q)}\right)^{1/(q-1)} \tag{35}$$

and

$$x = (1 - \bar{\lambda})\bar{x}, \tag{36}$$

where $\bar{x}$ is given by (25) and

$$V(\alpha, 0, q) = \begin{cases} \left(H(\alpha, 0, q)\right)^q & \text{in the stable case } (1 < \alpha < 2), \\ \dfrac{2^{q/2}\Gamma((q+1)/2)}{\sqrt{\pi}} & \text{in the normal case } (\alpha = 2). \end{cases}$$

Again, one would expect that the optimal allocation was different because the constant $V(\alpha, 0, q)$ and the matrix Q are different in the stable sub-Gaussian and in the normal case.

4.2. *Stable versus normal optimal allocation: a first comparison*

We analyze the differences in optimal allocations with reference to problem (33) when the investor chooses:
(1) joint normal distribution, or,
(2) joint α_k stable sub-Gaussian distribution ($k = 1, 2$) where $\alpha_1 = 1.7488$, $\alpha_2 = 1.8856$
as a model for the asset returns in his/her portfolio. Under these distinctive assumptions, the investors with utility functional (33) have different information about the distributional behavior of data. In particular, we examine the different market portfolio composition and the different investor's wealth allocation in the riskless asset.

First, when short sales are allowed and when short sales are not allowed, we examine optimal allocation among the riskless return and 23 index-daily returns: DAX 30, DAX 100 Performance, CAC 40, FTSE all share, FTSE 100, FTSE actuaries 350, Reuters Commodities, Nikkei 225 simple average, Nikkei 300 weighted stock average, Nikkei 300 simple stock average, Nikkei 500, Nikkei 225 stock average, Nikkei 300, Brent Crude Physical, Brent current month, Corn No 2 Yellow cents, Coffee Brazilian, Dow Jones Futures 1, Dow Jones Commodities, Dow Jones Industrials, Fuel Oil No 2, Goldman Sachs Commodity, S&P 500. We use the riskless return 6% p.a.

Using the estimated daily index parameters, we can compute the dispersion matrixes and the approximating "market" portfolios. The dispersion matrix Q is given by either the variance–covariance matrix (in the normal case) or the matrix Q_k (in the stable cases) which depends on the index of stability α_k for $k = 1, 2$ ($\alpha_1 = 1.7488$ and $\alpha_2 = 1.8856$). Therefore, as shown by Tables 3, 4, the market portfolio weights

$$\bar{x} = \frac{Q^{-1}(\mu - z_0 e)}{e' Q^{-1}\mu - e' Q^{-1} e z_0}$$

change under the different distributional assumptions. We observe that the market portfolio composition does not change excessively when we use either the asymmetric estimator (20) and (21) of matrix Q_k or the symmetric one $(\widehat{Q}_k + (\widehat{Q}_k)')/2$. However, using daily data the elements of the dispersion matrixes are of orders 10^{-6}. Thus the approximation in using data could be determinant to express elements of the matrixes. In particular, Table 3 presents the market portfolio weights when we consider all 23 asset returns and short sales are allowed. Table 4 gives the market portfolio weights when no short sales are allowed. Under this constraint, we value the market portfolio weights in terms of the risky portfolio compositions which maximize the extended Sharpe ratio, i.e., the market portfolio weights are the solution of the following optimization problem

$$\max_x \frac{E(x'r) - z_0}{\sigma_{x'r}},$$

$$x'e = 1,$$

$$x_i \geqslant 0, \quad i = 1, \ldots, n.$$

In this case the optimal allocation is reduced only among the four risky assets: DAX 100 Performance, FTSE all share, Nikkei 300 weighted stock average, Dow Jones Industrials and the riskless one. As argued by Roll (1977, 1978), Dybvig and Ross (1985a), different market portfolios imply a completely different security market line analysis. Thus, the approach which takes into account short sales presents more opportunities of earning than the approach with no short sales constraint. Therefore, it dominates the other approaches. Besides, if the returns are jointly α_k stable sub-Gaussian distributed (for some determined $k = 1, 2$), then the Gaussian approach is inefficient. Since, in general, efficient and inefficient portfolios can plot above and below the "real" security market line.

The analysis of Tables 3 and 4 points out that the composition of the market portfolio is strictly linked to the index of stability. In fact, we see that the allocation of the market portfolio in each asset component is generally monotone with respect of the stability index. Then the intuition suggests that the stable sub-Gaussian approaches take more into consideration the component of risk because of the fat tails. Recall that the tail behavior of every stable non-Gaussian distribution $X \stackrel{d}{=} S_\alpha(\sigma, \beta, \mu)$, with $1 < \alpha < 2$, is given by

$$\lim_{\lambda \to +\infty} \lambda^\alpha P(\pm X > \lambda) = C_\alpha \frac{1 \pm \beta}{2} \sigma^\alpha, \tag{37}$$

where $C_\alpha = (1 - \alpha)/(\Gamma(2 - \alpha) \cos \frac{\pi\alpha}{2})$. Therefore, the fat tails of smaller stability indexes underline the risk of the loss component of every portfolio. In particular, under the diverse distributional assumption, we distinguish the different perception of risk in the market portfolio components. This issue can be easily analyzed in the market portfolio weights with reference to the 23 returns when no short sales are allowed. In fact, Table 2 shows that the index of stability of FTSE all share is greater than the other indexes of stability (of the assets DAX 100 Performance, Nikkei 300 weighted stock average, Dow Jones Industrials). Observe that in Table 4 the component of the FTSE all share in the market portfolio increases with the index of stability α_k of the sub-Gaussian approach and the component of the other assets (DAX 100 Performance, Nikkei 300 weighted stock average,

Dow Jones Industrials) decreases with the index of stability. Thus, the market portfolios obtained under Gaussian and sub-Gaussian distributional hypotheses consider the risks due to heavy tails differently. On the other hand the mean of market portfolios decreases with the index of stability. However, if we accept the idea that the market portfolios represent in some sense the market behavior, then according to the classic mean–risk interpretation, an optimal portfolio that has a greater mean, it has also a greater risk. This fact appears clear enough when we consider and compare the dispersion measures $\sqrt{\bar{x}_k' Q_j \bar{x}_k}$ in every mean–risk plane for every market portfolio weights

$$\bar{x}_k = \frac{Q_k^{-1}(\mu - z_0 e)}{e' Q_k^{-1}\mu - e' Q_k^{-1} e z_0},$$

for every k and j. Observe that $\tilde{\sigma}_{j,k} = \sqrt{\bar{x}_k' Q_j \bar{x}_k}$ is the dispersion measure of market portfolio $\bar{x}_k' r$ considering the α_j stable Paretian approach. Therefore, for every fixed mean–risk plane (i.e., for every fixed α_j stable distributional approach) we can compare the market portfolio risk positions considering their risk position $\tilde{\sigma}_{j,k}$ (varying k). According to a mean–risk interpretation, we could observe that market portfolio with greater mean admits also a greater dispersion measure $\tilde{\sigma}_{j,k}$ in any mean–risk plane (see Tables 5 and 6).

As a consequence of relation (37) it follows that every stable non-Gaussian distribution $X \stackrel{d}{=} S_\alpha(\sigma, \beta, \mu)$, with $1 < \alpha < 2$, admits

$$E\left(|X - E(X)|^q\right) < \infty \quad \text{for } q < \alpha \quad \text{and} \tag{38}$$

$$E\left(|X - E(X)|^q\right) = \infty \quad \text{for } q \geqslant \alpha. \tag{39}$$

Hence, the weight of the risk measure $E(|X - E(X)|^q)$ in optimization problem (33) is generally greater for the investors who use the stable laws for asset returns when q is quite close to the index of stability α.

In Tables 7, 8 we listed the optimal allocation $\bar{\lambda}$ for the normal and the stable fit. Recall that $\bar{\lambda}$ is the optimal proportion of funds invested in the risk free asset which maximizes $E(W) - cE(|W - E(W)|^q)$, where $W = \lambda z_0 + (1 - \lambda)\bar{x}' r$. We have chosen $q = 1.45$ in Table 7 and $q = 1.55$ in Table 8, so that q is strictly less than all indexes of stability in the data set. On the other hand, we want to value and compare the different effects of q distant or closer to the stability parameters α_k. For any given allocation problem, we remark in bold character and in italics respectively the greatest and the smallest allocation in the riskless asset. Both tables show the greatest diversity among the optimal allocations considering small risk aversion coefficients c. Instead, the very risk averse investors assume a less risky position with every distributional hypothesis and the allocations in the riskless asset do not change very much.

As we see from these tables, when $q = 1.45$ and $q = 1.55$ the investors who fit the data with the Gaussian approach generally assume a less risky position than the investors who fit the data with the sub-Gaussian approach. Thus, if the stable sub-Gaussian approximation presents greater performances than Gaussian one (as observed by many empirical analysis) the "stable investors" have more opportunities of earning than the "Gaussian investors". In particular, the investors with $\alpha_1 = 1.7488$ stable sub-Gaussian approach invest less in the riskless asset than the investors who fit the data with the other approaches. However, if we consider q very closer to α_1 in optimization problem (33), then, as a consequence of (38) and (39), the investors who fit the data with α_1 stable sub-Gaussian approach assume a less risky position than the investors who fit the data with the Gaussian approach. In this case, the "stable investor" has a very risk preserving behavior because he prefers not allocating too much wealth in the risky asset. In this sense, intuition suggests that the stable approaches with lower indexes of stability generally are more risk preserving than those with greater indexes of stability because they consider the component of risk due to the fat tails of asset returns. Therefore, the stability index plays a strategic role in the stable optimal portfolio selection. Conversely, q in the above optimization problem can be an opportune measure of the magnitude to be given to the component of risk due to the heavy-tailedness of the asset returns. The importance given to q is intuitively linked to the conditions of the market in which the investor operates.

5. Conclusions

Firstly, we study, analyze and discuss portfolio choice models depending on a finite number of parameters. The distributional analysis presented permits to classify the admissible parametric families of returns. Moreover, by the interrelation between the parameters of each parametric family, we can order the portfolio choices using the basic principles of the stochastic dominance analysis. Thus, we can identify a dispersion measure which has some basic characteristics and represents the implicit measure of the return portfolio risk. In view of the classification of parametric portfolio choices, that is alternative to Ross' multiparameter one, we can distinguish the different efficient frontiers for investors who are non-satiable, risk averse or both (non-satiable and risk averse). In particular, we distinguish further restrictions to the classic Markowitz–Tobin's efficient frontier when no short sales are allowed. Besides, we can identify the optimization problems we have to solve in order to determine more accurate estimations of the investor's optimal allocations. In this sense, the analysis presented represents a general theory and a unifying framework to understand the parametric distributional approach to the portfolio choice theory.

Secondly, we show a simple classification of the portfolio choices considering the asymptotic behavior of returns with heavy tailed distributions. As a matter of fact, when returns have a stationary behavior they are in the domain of attraction of a stable law. Therefore, we present some examples of models in the domain of attraction of stable laws. The first distributional model considered is the case of the sub-Gaussian stable distributed returns. It permits a mean risk analysis pretty similar to Markowitz–Tobin's mean–variance one. In fact, this model admits the same analytical form for the efficient frontier but the parameters have a different meaning in the two models. Thus, the most important difference is given by the way of estimating the parameters. In order to present heavy tailed models that consider the asymmetry of returns, we study a three fund separation model where the portfolios are in the domain of attraction of an (α_1, α_2) stable law. Next, we analyze the case of $k + 1$ fund separation model with portfolios in the domain of attraction of an $(\alpha_1, \dots, \alpha_k)$ stable law. In all models we explicate the efficient frontier for the risk averse investors. In this context, we have shown that if the stable optimal portfolio analysis is stable, our approach is theoretically and empirically possible. Indeed, this work should be viewed only as a starting point for new empirical and theoretical studies on the topic of optimal allocation.

Finally, the comparison made between the stable sub-Gaussian and the normal approach in terms of the allocation problems has indicated that the stable sub-Gaussian allocation is more risk preserving than the normal one and can give more opportunities of earning. Precisely, the stable approach, differently from the normal one, considers the component of risk due to the fat tails. Therefore, we find that the tail behavior of sub-Gaussian and Gaussian approaches could imply substantial differences in the asset allocation. Taken into account that the stable approach is more adherent to the reality of the market, then, as argued by Götzenberger, Rachev and Schwartz (1999), we can obtain models that improve the performance measurements with the stable distributional assumption.

Acknowledgment

We are grateful to Stoyan Stoyanov and Boryana Racheva-Jotova (Sofia University and Bravo Risk Management Group, Santa Barbara) for the computational analysis and helpful comments.

Appendix A: Proofs

In order to prove the following results, we use some Hanoch and Levy's results [in particular see Theorems 3 and 4 in Hanoch and Levy (1969)].

Proof of Theorem 1: *Implication 1*. According to definition of $\sigma\tau_k^+(\bar{a})$ family, it follows

$$\frac{w'Z}{\sigma_{w'Z}} \stackrel{d}{=} \frac{y'Z}{\sigma_{y'Z}}$$

because the two random variables have the same parameters. If $\sigma_{w'Z} > \sigma_{y'Z}$, then for every $t \geqslant 0$

$$P(w'Z \leqslant t) = P\left(\frac{w'Z}{\sigma_{w'Z}} \leqslant \frac{t}{\sigma_{w'Z}}\right) \leqslant P\left(\frac{w'Z}{\sigma_{w'Z}} \leqslant \frac{t}{\sigma_{y'Z}}\right) = P(y'Z \leqslant t)$$

and the above inequality is strict for some t. Conversely, if $w'Z$ FSD $y'Z$, then $E(w'Z) > E(y'Z)$. $\square$

Implication 2. As a consequence of the assumptions, it follows

$$q := E(w'Z) - E(y'Z) \geqslant 0 \quad \text{and} \quad \sigma_{w'Z} \geqslant \sigma_{y'Z} \geqslant \sigma_{y'Z+t},$$

for every $t \geqslant 0$. Moreover, for every $t \geqslant 0$ the function $g(t) = (E(y'Z) + t)/\sigma_{y'Z+t}$ is an increasing continuous positive function that tends to infinity as $t \to \infty$. As a consequence of definition of $\sigma\tau_k^+(\bar{a})$ family there exists $t \leqslant q$ such that the random variable $(w'Z)/\sigma_{w'Z}$ has the same parameters of $(y'Z + t)/\sigma_{y'Z+t}$ and hence $(w'Z)/\sigma_{w'Z} \stackrel{d}{=} (y'Z + t)/\sigma_{y'Z+t}$. Then, for every $\lambda \geqslant 0$:

$$P(w'Z \leqslant \lambda) \leqslant P\left(\frac{y'Z + t}{\sigma_{y'Z+t}} \leqslant \frac{\lambda}{\sigma_{y'Z+t}}\right) \leqslant P(y'Z \leqslant \lambda). \tag{40}$$

Observe that at least one of the two inequalities $\sigma_{w'Z} \geqslant \sigma_{y'Z}$ and $q \geqslant 0$ is strict by hypothesis. Then, at least one of the previous inequalities (40) is strict for some real $\lambda \geqslant 0$. Therefore, $w'Z$ FSD $y'Z$. □

Implication 3. First assume $m_{w'Z} = m_{y'Z}$ and $\sigma_{w'Z} < \sigma_{y'Z}$. Then, $m_{w'Z}/\sigma_{w'Z} > m_{y'Z}/\sigma_{y'Z}$ and there exists $t > 0$ such that $m_{w'Z}/\sigma_{w'Z} = (m_{y'Z} + t)/\sigma_{y'Z+t}$. Therefore,

$$m_{w'Z} < m_{y'Z} + t, \quad 1 < \frac{\sigma_{y'Z+t}}{\sigma_{w'Z}} =: \alpha \quad \text{and} \quad w'Z \stackrel{d}{=} \frac{y'Z + t}{\alpha}.$$

Hence, for every $\lambda \leqslant M := t/(\alpha - 1)$;

$$P(w'Z \leqslant \lambda) = P(y'Z \leqslant \lambda\alpha - t) \leqslant P(y'Z \leqslant \lambda)$$

and for every $\lambda \geqslant M$,

$$P(w'Z \leqslant \lambda) \geqslant P(y'Z \leqslant \lambda).$$

By hypothesis $m_{w'Z} = m_{y'Z}$, thus we cannot have $w'Z$ FSD $y'Z$ or $y'Z$ FSD $w'Z$. Therefore, from Hanoch and Levy Theorem 3 (1969) $w'Z$ SSD $y'Z$ and $m_{w'Z} = m_{y'Z}$; that is $w'Z$ R–S stochastically dominates $y'Z$. Conversely, if $w'Z$ R–S stochastically dominates $y'Z$, then $m_{w'Z} = m_{y'Z}$ and $\sigma_{w'Z} \neq \sigma_{y'Z}$. Thus, by previous demonstration it follows that $\sigma_{w'Z} < \sigma_{y'Z}$ because the converse is absurd. □

Implication 4. From Implication 2, $\sigma_{w'Z} \geqslant \sigma_{y'Z}$, implies $w'Z$ FSD $y'Z$ which implies $w'Z$ SSD $y'Z$. Next, assume $\sigma_{w'Z} < \sigma_{y'Z}$. Therefore, there exists $t \geqslant 0$ such that $m_{w'Z}/\sigma_{w'Z} = (m_{y'Z} + t)/\sigma_{y'Z+t}$ and $w'Z \stackrel{d}{=} (y'Z + t)/\alpha$ where $\alpha = \sigma_{y'Z+t}/\sigma_{w'Z}$. Thus, we can distinguish two cases:

(1) $m_{w'Z} \geqslant m_{y'Z} + t$ and $\sigma_{w'Z} \geqslant \sigma_{y'Z+t}$. As a consequence of Implication 1, $w'Z$ FSD $y'Z$.

(2) $m_{y'Z} \leqslant m_{w'Z} < m_{y'Z} + t$ and $\sigma_{w'Z} < \sigma_{y'Z+t}$. Then, as proved in Implication 3, $w'Z$ SSD $y'Z$. □

Implication 5. First, assume $\sigma_{w'Z} = \sigma_{y'Z}$. As a consequence of stochastic dominance $w'Z$ SSD $y'Z$, the inequality $E(w'Z) > E(y'Z)$ holds. Thus, as a consequence of Implication 2, $w'Z$ FSD $y'Z$. Secondly, assume $\sigma_{w'Z} > \sigma_{y'Z}$. Therefore, we can distinguish two cases:

(1) $m_{w'Z}/\sigma_{w'Z} \geqslant m_{y'Z}/\sigma_{y'Z}$. Thus, as a consequence of Implication 2, $w'Z$ FSD $y'Z$.

(2) $m_{w'Z}/\sigma_{w'Z} < m_{y'Z}/\sigma_{y'Z}$. Then, there exists $t > 0$ such that $(m_{w'Z} + t)/\sigma_{w'Z+t} = m_{y'Z}/\sigma_{y'Z}$ and $(w'Z + t)/\sigma_{w'Z+t} \overset{d}{=} y'Z/\sigma_{y'Z}$. Observe that $m_{w'Z} + t > m_{y'Z}$. Therefore, $\sigma_{w'Z+t} > \sigma_{y'Z}$ and for every $\lambda \leqslant M := t/(\beta - 1)$ where $\beta = (\sigma_{w'Z+t})/\sigma_{y'Z}$, it follows $F_{w'Z}(\lambda) \geqslant F_{y'Z}(\lambda)$. Similarly, for every $\lambda > M$, $F_{w'Z}(\lambda) \leqslant F_{y'Z}(\lambda)$. However, a $\lambda \leqslant M$ such that $F_{w'Z}(\lambda) > F_{y'Z}(\lambda)$ cannot exist, because distribution functions are right continuous and $\int_{-\infty}^{M} F_{w'Z}(u)\,du > \int_{-\infty}^{M} F_{y'Z}(u)\,du$, against the assumption $w'Z$ SSD $y'Z$.

Then, $w'Z$ FSD $y'Z$. □

Implication 6. The assumptions $\sigma_{w'Z} \leqslant \sigma_{y'Z}$ and $m_{w'Z} \geqslant m_{y'Z}$ with at least one inequality strict imply $m_{w'Z}/\sigma_{w'Z} > m_{y'Z}/\sigma_{y'Z}$. As a consequence of Implication 4, it follows $w'Z$ SSD $y'Z$. □

Proof of Theorem 2: From the assumptions of the theorem it follows

$$\frac{w'r - E(w'r)}{\sigma_{w'r}} \overset{d}{=} \frac{y'r - E(y'r)}{\sigma_{y'r}}.$$

First suppose $w'r$ SSD $y'r$. Therefore, $E(w'r) \geqslant E(y'r)$. If $E(w'r) = E(y'r)$ and $\sigma_{w'r} = \sigma_{y'r}$ the equality in distribution $w'r \overset{d}{=} y'r$ holds, against the hypothesis. Suppose for absurd that $\sigma_{w'r} > \sigma_{y'r}$. Then, for every $t < M := (m_{w'r}\sigma_{y'r} - m_{y'r}\sigma_{w'r})/(\sigma_{y'r} - \sigma_{w'r})$ it follows $(t - E(w'r))/\sigma_{w'r} > (t - E(y'r))/\sigma_{y'r}$ and $F_{w'r}(t) \geqslant F_{y'r}(t)$. The inequality is strict for some t because $w'r$ is a random variable unbounded from below. This is a contradiction because by hypothesis $w'r$ SSD $y'r$. Therefore, $\sigma_{w'r} \leqslant \sigma_{y'r}$ and $E(w'r) \geqslant E(y'r)$ with at least one inequality strict.

Conversely, suppose $E(w'r) > E(y'r)$ and $\sigma_{w'r} = \sigma_{y'r}$ we obtain $w'r$ FSD $y'r$ from the properties of $\sigma\tau_k(\bar{a})$ class [see Ortobelli (2001)]. Then, assume $E(w'r) \geqslant E(y'r)$ and $\sigma_{w'r} < \sigma_{y'r}$. The distributions of the random variables $w'r$ and $X \overset{d}{=} w'r - E(w'r) + E(y'r)$ belong to the same $\sigma\tau_k(\bar{a})$ family and $E(X) = E(y'r)$. Moreover, from the properties of $\sigma\tau_k(\bar{a})$ family, the random variables X and $w'r$ have the same scale parameter $\sigma_{w'r}$ and the same other $k-2$ parameters $(a_{1,p}, \ldots, a_{k-2,p})$. Therefore, $Y := (X - E(y'r))/\sigma_{w'r} \overset{d}{=} (y'r - E(y'r))/\sigma_{y'r}$ and for every real u:

$$\int_{-\infty}^{u} \big(P(X \leqslant \lambda) - P(y'r \leqslant \lambda)\big)\,d\lambda$$

$$= \int_{-\infty}^{u} \left(P\left(Y \leqslant \frac{\lambda - E(y'r)}{\sigma_{w'r}}\right) - P\left(Y \leqslant \frac{\lambda - E(y'r)}{\sigma_{y'r}}\right)\right)d\lambda \leqslant 0.$$

The above inequality is strict for some real u because random variable Y is unbounded from below and $\sigma_{w'r} < \sigma_{y'r}$. Thus, X SSD $y'r$. If $E(w'r) = E(y'r)$ we have $X \stackrel{d}{=} w'r$. Instead, if $E(w'r) > E(y'r)$, from the properties of $\sigma\tau_k(\bar{a})$ class the following stochastic relation holds

$$w'r \text{ FSD } X \text{ SSD } y'r.$$

Therefore $w'r$ SSD $y'r$. However, considering the stochastic dominance equivalences for "SSD" [see Levy (1992)] the following equality in distribution holds $y'r \stackrel{d}{=} w'r - E(w'r) + E(y'r) + \varepsilon$ with $E(\varepsilon/w'r) = 0$. □

Appendix B: Tables

Table 1
Maximum likelihood estimations of the normal asset return parameters considering daily data from 1/3/95 to 1/30/98

Assets	Gaussian parameters	
	Mean μ	Standard deviation σ
DAX 30	0.0007	0.0113
DAX 100 Performance	0.0007	0.0106
CAC 40	0.0005	0.011
FTSE all share	0.0007	0.007
FTSE 100	0.0008	0.0078
FTSE actuaries 350	0.0007	0.0072
Reuters Commodities	−0.0002	0.0072
Nikkei 225 simple average	0.0005	0.0157
Nikkei 300 weighted stock average	0.0006	0.0137
Nikkei 300 simple stock average	0.0004	0.0129
Nikkei 500	0.0003	0.0128
Nikkei 225 stock average	−0.0005	0.0158
Nikkei 300	−0.0005	0.0138
Brent Crude Physical	0.0000	0.0185
Brent current month	0.0000	0.0186
Corn No 2 Yellow cents	0.0002	0.0152
Coffee Brazilian	0.0002	0.0270
Dow Jones Futures 1	−0.0001	0.0055
Dow Jones Commodities	−0.0001	0.0079
Dow Jones Industrials	0.0013	0.0086
Fuel Oil No 2	−0.0001	0.0201
Goldman Sachs Commodity	0.0000	0.0092
S&P 500	0.0009	0.0083

Table 2
Maximum likelihood estimators of the stable asset return parameters considering daily data from 1/3/95 to 1/30/98

Assets	Stable parameters			
	Index of stability α	Stable skewness β	Stable mean μ	Stable scale σ
DAX 30	1.8148	−0.6682	0.0005	0.0069
DAX 100 performance	1.7996	−0.6389	0.0004	0.0064
CAC 40	1.8381	−0.1852	0.0004	0.0071
FTSE all share	1.8418	−0.5726	0.0006	0.0045
FTSE 100	1.8856	−0.5192	0.0007	0.0052
FTSE actuaries 350	1.8521	−0.5666	0.0006	0.0047
Reuters Commodities	1.7959	−0.2075	−0.0003	0.0045
Nikkei 225 simple average	1.663	−0.0483	0.0004	0.009
Nikkei 300 weighted stock average	1.6962	0.0869	0.0006	0.0079
Nikkei 300 simple stock average	1.7064	0.085	0.0004	0.0075
Nikkei 500	1.7253	0.0334	0.0003	0.0076
Nikkei 225 stock average	1.6798	−0.0721	−0.0006	0.0091
Nikkei 300	1.6994	0.0303	−0.0005	0.008
Brent Crude Physical	1.7423	−0.229	−0.0003	0.0112
Brent current month	1.7405	−0.2039	−0.0001	0.0112
Corn No 2 Yellow cents	1.6869	−0.1565	0.0002	0.0083
Coffee Brazilian	1.5876	−0.0153	0.0007	0.0144
Dow Jones Futures 1	1.8063	−0.4641	−0.0002	0.0035
Dow Jones Commodities	1.6806	−0.1389	−0.0001	0.0037
Dow Jones Industrials	1.7368	−0.2886	0.0012	0.0049
Fuel Oil No 2	1.7338	−0.1961	−0.0002	0.0117
Goldman Sachs Commodity	1.8036	−0.2663	−0.0002	0.0058
S&P 500	1.7052	−0.0881	0.0010	0.0047

Table 3
Stable sub-Gaussian and Gaussian market portfolio weights when short sales are allowed

Assets	Weights for $\alpha = 1.7488$	Weights for $\alpha = 1.8856$	Gaussian weights ($\alpha = 2$)
DAX 30	0.3929	0.7182	1.2784
DAX 100 Performance	−0.0704	−0.4594	−1.1205
CAC 40	−0.0694	−0.1217	−0.2022
FTSE all share	9.1802	10.6690	14.0729
FTSE 100	−3.0513	−2.5989	−1.6515
FTSE actuaries 350	−4.5968	−6.3873	−10.4307
Reuters Commodities	−1.2833	−1.3451	−1.4679
Nikkei 225 Simple average	0.9302	0.0114	−1.8360
Nikkei 300 weighted stock average	19.7594	19.4478	19.7910
Nikkei 300 simple stock average	−12.7675	−12.0352	−11.2504
Nikkei 500	15.8399	15.5310	15.5214
Nikkei 225 stock average	−2.1322	−1.2088	0.6395
Nikkei 300	−21.2774	−21.3744	−22.4705
Brent Crude Physical	0.2125	0.1850	0.1653
Brent current month	0.0044	0.0260	0.0460
Corn No 2 Yellow cents	0.0295	0.0032	−0.0306
Coffee Brazilian	0.0175	0.0170	0.0201
Dow Jones Futures 1	−0.2376	−0.2771	−0.4155
Dow Jones Commodities	−0.7123	−0.6422	−0.5327
Dow Jones Industrials	3.7278	3.7832	4.0654
Fuel Oil No 2	−0.2806	−0.2740	−0.2714
Goldman Sachs Commodity	0.2473	0.2356	0.2143
S&P 500	−2.8630	−2.9032	−3.1345

Table 4
Stable sub-Gaussian and Gaussian market portfolio weights when no short sales are allowed

Assets	Weights for $\alpha = 1.7488$	Weights for $\alpha = 1.8856$	Gaussian weights ($\alpha = 2$)
DAX 100 performance	0.0746	0.0742	0.0732
FTSE all share	0.2329	0.2379	0.2496
Nikkei 300 weighted stock average	0.0509	0.0502	0.0485
Dow Jones Industrials	0.6416	0.6377	0.6287
Other assets	0	0	0

Table 5
Stable sub-Gaussian and Gaussian market portfolio parameters for every mean–dispersion plane when short sales are allowed

Parameters	Market portfolio $\alpha = 1.7488$	Market portfolio $\alpha = 1.8856$	Gaussian market portfolio
Mean	0.0243	0.0236	0.0231
Dispersion if $\alpha = 1.7488$	0.0104	0.0101	0.0101
Dispersion if $\alpha = 1.8856$	0.0191	0.0185	0.0183
Standard deviation	0.0451	0.0433	0.0418

Table 6
Stable sub-Gaussian and Gaussian market portfolio parameters for every mean–dispersion plane when no short sales are allowed

Parameters	Market portfolio $\alpha = 1.7488$	Market portfolio $\alpha = 1.8856$	Gaussian market portfolio
Mean	0.001079	0.001077	0.001072
Dispersion if $\alpha = 1.7488$	0.001553	0.001549	0.001541
Dispersion if $\alpha = 1.8856$	0.002807	0.002800	0.002783
Standard deviation	0.006393	0.006376	0.006339

Table 7
Optimal allocation for the optimization problem $\max_\lambda E(W) - cE(|W - E(W)|^{1.45})$ when different distributional assumptions are considered

Allocation $\bar{\lambda}$ in the riskless asset considering the market portfolio on 23 assets when unlimited short sales are allowed			
Coefficient c of the optimization problem	Optimal allocation $\bar{\lambda}$ when $\alpha = 1.7488$	Optimal allocation $\bar{\lambda}$ when $\alpha = 1.8856$	Optimal allocation $\bar{\lambda}$ when $\alpha = 2$
1.8	*−10.2855*	−2.3803	**−0.0715**
2.2	*−6.2252*	−1.1642	**0.3140**
3	*−2.6268*	−0.0863	**0.6557**
4.2	*−0.7171*	0.4857	**0.8370**
5	*−0.1655*	0.6509	**0.8893**
6	*0.2227*	0.7672	**0.9262**
7	*0.4482*	0.8347	**0.9476**
10	*0.7502*	0.9252	**0.9763**
15	*0.8985*	0.9696	**0.9904**
21	*0.9520*	0.9856	**0.9954**
Allocation $\bar{\lambda}$ in the riskless asset considering the market portfolio on 23 assets when short sales are not allowed			
Coefficient c of the optimization problem	Optimal allocation $\bar{\lambda}$ when $\alpha = 1.7488$	Optimal allocation $\bar{\lambda}$ when $\alpha = 1.8856$	Optimal allocation $\bar{\lambda}$ when $\alpha = 2$
1.4	*0*	0	**0.3788**
1.5	*0*	0	**0.4671**
1.8	*0*	0	**0.6446**
2.2	*0*	0.3024	**0.7725**
3	*0*	0.6498	**0.8858**
4.2	*0.4559*	0.8342	**0.9459**
5	*0.6306*	0.8875	**0.9633**
6	*0.7537*	0.9250	**0.9755**
7	*0.8251*	0.9467	**0.9826**
10	*0.9208*	0.9759	**0.9921**

This table computes the optimal allocation $\bar{\lambda}$ in the riskless return 6% annual rate (daily $z_0 = 0.00016$) for different risk aversion coefficient c of the optimization problem $\max_\lambda E(W) - cE(|W - E(W)|^{1.45})$ where $W = \lambda z_0 + (1-\lambda)\bar{x}'r$ and $\bar{x}'r$ is either the Gaussian market portfolio (for $\alpha = 2$) or the sub-Gaussian market portfolio (for $\alpha = 1.7488$ or $\alpha = 1.8856$).

Table 8

Optimal allocation for the optimization problem $\max_{\lambda} E(W) - cE(|W - E(W)|^{1.55})$ when different distributional assumptions are considered

Allocation $\bar{\lambda}$ in the riskless asset considering the market portfolio on 23 assets when unlimited short sales are allowed			
Coefficient c of the optimization problem	Optimal allocation $\bar{\lambda}$ when $\alpha = 1.7488$	Optimal allocation $\bar{\lambda}$ when $\alpha = 1.8856$	Optimal allocation $\bar{\lambda}$ when $\alpha = 2$
1.5	*−11.4475*	−4.5346	**−1.2292**
1.8	*−7.9354*	−2.9730	**−0.6003**
2.2	*−5.2038*	−1.7584	**−0.1111**
3	*−2.5298*	−0.5695	**0.3678**
4.2	*−0.9146*	0.1487	**0.6571**
5	*−0.3944*	0.3800	**0.7503**
6	*−0.0010*	0.5549	**0.8207**
7	*0.2437*	0.6637	**0.8646**
10	*0.6046*	0.8242	**0.9292**
15	*0.8108*	0.9159	**0.9661**

Allocation $\bar{\lambda}$ in the riskless asset considering the market portfolio on 23 assets when short sales are not allowed			
Coefficient c of the optimization problem	Optimal allocation $\bar{\lambda}$ when $\alpha = 1.7488$	Optimal allocation $\bar{\lambda}$ when $\alpha = 1.8856$	Optimal allocation $\bar{\lambda}$ when $\alpha = 2$
1.3	*0*	0	**0**
1.5	*0*	0	**0**
1.8	*0*	0	**0.0859**
2.2	*0*	0	**0.3654**
3	*0*	0.1239	**0.6389**
4.2	*0*	0.5248	**0.8042**
5	*0.2305*	0.6539	**0.8574**
6	*0.4476*	0.7515	**0.8976**
7	*0.5826*	0.8123	**0.9226**
10	*0.7818*	0.9019	**0.9595**

This table computes the optimal allocation $\bar{\lambda}$ in the riskless return 6% annual rate (daily $z_0 = 0.00016$) for different risk aversion coefficient c of the optimization problem $\max_{\lambda} E(W) - cE(|W - E(W)|^{1.55})$where $W = \lambda z_0 + (1-\lambda)\bar{x}'r$ and $\bar{x}'r$ is either the Gaussian market portfolio (for $\alpha = 2$) or the sub-Gaussian market portfolio (for $\alpha = 1.7488$ or $\alpha = 1.8856$).

References

Bawa, V.S., 1975. Optimal rules for ordering uncertain prospects. Journal of Financial Economics 2, 95–121.

Bawa, V.S., 1976. Admissible portfolio for all individuals. Journal of Finance 31, 1169–1183.

Chamberlein, G., 1983. A characterization of the distributions that imply mean–variance utility functions. Journal of Economic Theory 29, 975–988.

Chen, N.F., Ingersoll, J., 1983. Exact pricing in linear factor models with finitely many assets: a note. Journal of Finance 38, 985–988.
Cheng, B., Rachev, S., 1995. Multivariate stable futures prices. Mathematical Finance 5, 133–153.
Cochrane, J.H., 1999. Asset pricing. Manuscript. University of Chicago.
Connor, G., 1984. A unified beta pricing theory. Journal of Economic Theory 34, 13–31.
Cox, J., Leland, H., 1982. On dynamic investment strategies. In: Proceeding of Seminar on the Analysis of Security Prices 26. Center for Research in Security Prices, University of Chicago.
Dybvig, P., 1983. An explicit bound on individual assets' deviations from APT pricing in a finite economy. Journal of Financial Economics 12, 483–496.
Dybvig, P., 1988a. Distributional analysis of portfolio choice. Journal of Business 61, 369–393.
Dybvig, P., 1988b. Inefficient dynamic portfolio strategies or how to throw away a million dollars in the stock market. Review of Financial Studies 1, 67–88.
Dybvig, P., Ingersoll, J., 1982. Mean–variance theory in complete markets. Journal of Business 55, 233–251.
Dybvig, P., Ross, S., 1982. Portfolio efficient sets. Econometrica 50, 1525–1546.
Dybvig, P., Ross, S., 1985a. The analytics of performance measurement using a security market line. Journal of Finance 40, 401–416.
Dybvig, P., Ross, S., 1985b. Differential information and performance measurement using a security market line. Journal of Finance 40, 383–399.
Dybvig, P., Ross, S., 1987. Arbitrage. In: The New Palgrave: A Dictionary of Economics. Stockton Press, New York.
Fama, E., 1963. Mandelbrot and the stable Paretian hypothesis. Journal of Business 36, 420–429.
Fama, E., 1965a. The behavior of stock market prices. Journal of Business 38, 34–105.
Fama, E., 1965b. Portfolio analysis in a stable Paretian market. Management Science 11, 404–419.
Fishburn, P., 1964. Decision and Value Theory. Wiley, New York.
Fishburn, P., 1980. Stochastic dominance and moments distributions. Mathematical Operation Research 5, 94–100.
Friedman, M., Savage, L.J., 1948. The utility analysis of choices involving risk. Journal of Political Economy 56, 279–304.
Gamrowski, B., Rachev, S., 1994. Stable models in testable asset pricing. In: Approximation, Probability and Related Fields. Plenum Press, New York.
Gamrowski, B., Rachev, S., 1999. A testable version of the Pareto-stable CAPM. Mathematical and Computer Modeling 29, 61–81.
Giacometti, R., Ortobelli, S., 2001. A comparison between dispersion measures for the asset allocation problem. Technical Report. Department of Mathematics, Statistics, Computer Science and Applications, University of Bergamo, Italy.
Götzenberger, G., Rachev, S., Schwartz, E., 1999. Performance measurements: the stable Paretian approach. In: Applied Mathematics Reviews, Vol. 1. World Scientific, to appear.
Grinblatt, M., Titman, S., 1983. Factor pricing in a finite economy. Journal of Financial Economics 12, 497–508.
Hadar, J., Russel, W., 1969. Rules of ordering uncertain prospects. American Economic Review 59, 25–34.
Hanoch, G., Levy, H., 1969. The efficiency analysis of choices involving risk. Review of Economic Studies 36, 35–46.
Hansen, L.P., Jagannathan, R., 1991. Implications of security market data for models of dynamic economies. Journal of Political Economy 99, 225–262.
Hardin, Jr., 1984. Skewed stable variable and processes. Technical Report 79. Center for Stochastic Processes at the University of North Carolina, Chapel Hill.
Huberman, G., 1982. A simple approach to arbitrage pricing theory. Journal of Economic Theory 28, 183–191.
Ingersoll, J., Jr., 1987. Theory of Financial Decision Making. Rowman & Littlefield, Totowa.
Janicki, A., Weron, A., 1994. Simulation and Chaotic Behavior of Stable Stochastic Processes. Marcel Dekker, New York.
Jarrow, R., 1986. The relationship between arbitrage and first order stochastic dominance. Journal of Finance 41, 915–921.

Jean, W.H., 1971. The extension of portfolio analysis to three or more parameters. Journal of Financial and Quantitative Analysis, 505–515.

Klebanov, L.B., Rachev, S., Safarian, G., 2000. Local pre-limit theorems and their applications to finance. Applied Mathematics Letters 13, 70–73.

Klebanov, L.B., Rachev, S., Szekely, G., 2001. The central pre-limit theorem and its applications. In: Stable Models in Finance. Pergamon Press, to appear.

Kocherlakota, N.R., 1997. Testing the consumption CAPM with heavy-tailed pricing errors. Macroeconomic Dynamics 1, 551–567.

Konno, H., Yamazaki, H., 1991. Mean–absolute deviation portfolio optimization model and its application to Tokyo stock market. Management Science 37, 519–531.

Kraus, A., Litzenberger, R., 1976. Skewness preference and the valuation of risk assets. Journal of Finance 31, 1085–1100.

Kroll, Y., Levy, H., 1979. Stochastic dominance with a riskless asset: an imperfect market. Journal of Financial and Quantitative Analysis 14, 179–204.

Levy, H., 1990. Stochastic dominance. In: Utility and Probability. The Macmillan Press, United Kingdom.

Levy, H., 1992. Stochastic dominance and expected utility: survey and analysis. Management Science 38, 555–593.

Levy, H., Kroll, Y., 1976. Stochastic dominance with riskless assets. Journal of Financial and Quantitative Analysis 11, 743–773.

Lintner, J., 1965. The valuation of risk assets and the selection of risky investments in stock portfolios and capital budgets. Review of Economics and Statistics 47, 13–37.

Machina, M., 1982. Expected utility analysis without the independence axiom. Econometrica 50, 277–323.

Mandelbrot, B., 1963a. New methods in statistical economics. Journal of Political Economy 71, 421–440.

Mandelbrot, B., 1963b. The variation of certain speculative prices. Journal of Business 26, 394–419.

Mandelbrot, B., 1967. The variation of some other speculative prices. Journal of Business 40, 393–413.

Mandelbrot, B.B., Taylor, H.M., 1967. On the distribution of stock price differences. Operations Research 15, 1057–1062.

Markowitz, H., 1952. Portfolio selection. Journal of Finance 7, 77–91.

Markowitz, H., 1959. Portfolio Selection; Efficient Diversification of Investment. Wiley, New York.

Markowitz H., 1977. An algorithm for finding undominated portfolios. In: Levy, H., Sarnat, M. (Eds.), Financial Decision Making under Uncertainty. Academic Press, New York.

Markowitz, H., 1987. Mean–Variance Analysis in Portfolio Choice and Capital Markets. Blackwell, Oxford.

Mehra, R., Prescott, E.C., 1985. The equity premium: a puzzle. Journal of Monetary Economy 15, 145–161.

Milne F., 1988. Arbitrage and diversification a general equilibrium asset economy. Econometrica 56, 815–840.

Mittnik, S., Rachev, S., Paolella, M., 1997. Stable Paretian modelling in finance: some empirical and theoretical aspects. In: Adler, R., et al. (Eds.), A Practical Guide to Heavy Tails: Statistical Techniques for Analyzing Heavy Tailed Distributions. Birkhäuser, Boston.

Mossin, J., 1966. Equilibrium in a capital asset market. Econometrica 34, 768–783.

Nolan, J., 1997. Numerical computation of stable densities and distribution functions. Communications in Statistics. Stochastic Models 13, 759–774.

Ogryczak, W., Ruszczynski, A., 1999. From stochastic dominance to mean–risk models: semideviations as risk measures. European Journal of Operational Research 116, 33–50.

Ogryczak, W., Ruszczynski, A., 2001. On consistency of stochastic dominance and mean–semideviation models. Mathematical Programming 89, 217–232.

Ortobelli, S., 2001. The classification of parametric choices under uncertainty: analysis of the portfolio choice problem. Theory and Decision 51, 297–327.

Ortobelli, S., Huber, I., Höchstötter, M., Rachev, S., 2001. A comparison among Gaussian and non-Gaussian portfolio choice models. In: Proceeding IFAC SME 2001.

Ortobelli, S., Huber, I., Schwartz, E., 2002. Portfolio selection with stable distributed returns. Mathematical Methods of Operation Research 55, 265–300.

Ortobelli, S., Rachev, S., Schwartz, E., 2002. The problem of optimal asset allocation with stable distributed returns. Technical Report. University of Karlsruhe.

Owen, J., Rabinovitch, R., 1983. On the class of elliptical distributions and their applications to the theory of portfolio choice. Journal of Finance 38, 745–752.

Paulauskas, V., Rachev, S., 1999. Maximum likelihood estimators in regression models with infinite variance innovations. Technical Report. Department of Statistics, UCSB, Santa Barbara, CA 93106.

Pitman, 1939. The estimation of the location and scale parameters of a continuous population of any given form. Biometrica 30, 391–421.

Quirk, J., Saposnik, R., 1962. Admissibility and measurable utility function. Review of Economic Studies 29, 140–146.

Rachev, S., 1991. Probability Metrics and the Stability of Stochastic Models. Wiley, New York.

Rachev, S., Mittnik, S., 2000. Stable Model in Finance. Wiley, New York.

Rachev, S., Xin, H., 1993. Test on association of random variables in the domain of attraction of multivariate stable law. Probability and Mathematical Statistics 14, 125–141.

Roll, R., 1977. A critique of the asset pricing theory's tests. Journal of Financial Economics 4, 129–176.

Roll R., 1978. Ambiguity when performance is measured by the security market line. Journal of Finance 33, 1031–1069.

Roll R., 1979a. Testing a portfolio for ex ante mean/variance efficiency. In: Elton, E., Gruber, M. (Eds.), TIMS Studies in the Management Sciences. North-Holland, Amsterdam.

Roll R., 1979b. A reply to Meyer and Rice (1979). Journal of Financial Economics 7, 391–400.

Ross, S., 1975. Return, risk and arbitrage. In: Friend, I., Bicksler, J. (Eds.), Studies in Risk and Return. Ballinger, Cambridge, MA.

Ross, S., 1976. The arbitrage theory of capital asset pricing. Journal of Economic Theory 13, 341–360.

Ross, S., 1978a. Mutual fund separation in financial theory – the separating distributions. Journal of Economic Theory 17, 254–286.

Ross, S., 1978b. A simple approach to the valuation of risky streams. Journal of Business 51, 453–475.

Rothschild, M., Stiglitz, J., 1970. Increasing risk: I. definition. Journal of Economic Theory 2, 225–243.

Rubinstein, M., 1976. The valuation of uncertain income streams and the pricing of options. Bell Journal of Economics 7, 407–425.

Samorodnitsky, G., Taqqu, M.S., 1994. Stable Non-Gaussian Random Processes: Stochastic Models with Infinite Variance. Chapman and Hall, New York.

Samuelson, P.A., 1967. General proof that diversification pays. Journal of Financial Quantitative Analysis 2, 1–13.

Samuelson, P.A., 1969. The fundamental approximation theorem of portfolio analysis in terms of means, variances and higher moments. Review of Economic Studies 37, 537–542.

Samuelson, P.A., Merton, R., 1975. Generalized mean–variance tradeoffs for best perturbation corrections to approximate portfolio decisions. Journal of Finance 30, 27–40.

Shaked, M., Shanthikumar, G., 1994. Stochastic Orders and their Applications. Academic Press – Harcourt, Brace & Company, New York.

Shalit, H., Yitzhaki, S., 1984. Mean-Gini, portfolio theory, and the pricing of risky assets. Journal of Finance 39, 1449–1468.

Sharpe, W., 1964. The capital asset prices: a theory of market equilibrium under condition of risk. Journal of Finance 19, 425–442.

Simaan, Y., 1993. Portfolio selection and asset pricing – three parameter framework. Management Science 5, 568–577.

Speranza, M.G., 1993. Linear programming models for portfolio optimization. Finance 14, 107–123.

Tobin, J., 1958. Liquidity preference as behavior toward risk. Review of Economic Studies 25, 65–86.

Tobin, J., 1965. The theory of portfolio selection. In: Hahn, F.H., Brechling, F.P.R. (Eds.), The Theory of Interest Rates. Macmillan, London.

Tokat, Y., Rachev, S., Schwartz, E., 2002. The stable non-Gaussian asset allocation: a comparison with the classical Gaussian approach. Journal of Economic Dynamics and Control, to appear.

von Neumann, J., Morgenstern, O., 1953. Theory of Games and Economic Behavior, 3rd edition. Princeton University Press, Princeton, NJ.

Yitzhaki, S., 1982. Stochastic dominance, mean, variance and Gini's mean difference. American Economic Review 72, 178–185.

Zolotarev, V.M., 1986. One-Dimensional Stable Distributions. In: Translation of Mathematical Monographs, Vol. 65. American Mathematical Society, Providence, RI. Translation of the original 1983 in Russian.

Chapter 15

PORTFOLIO MODELING WITH HEAVY TAILED RANDOM VECTORS

MARK M. MEERSCHAERT

Department of Mathematics, University of Nevada, Reno, USA
e-mail: mcubed@unr.edu

HANS-PETER SCHEFFLER

Fachbereich Mathematik, University of Dortmund, 44221 Dortmund, Germany
e-mail: hps@math.uni-dortmund.de

Contents

Handbook of Heavy Tailed Distributions in Finance, Edited by S.T. Rachev

Abstract

Since the work of Mandelbrot in the 1960s there has accumulated a great deal of empirical evidence for heavy tailed models in finance. In these models, the probability of a large fluctuation falls off like a power law. The generalized central limit theorem shows that these heavy-tailed fluctuations accumulate to a stable probability distribution. If the tails are not too heavy then the variance is finite and we find the familiar normal limit, a special case of stable distributions. Otherwise the limit is a nonnormal stable distribution, whose bell-shaped density may be skewed, and whose probability tails fall off like a power law. The most important model parameter for such distributions is the tail thickness α, which governs the rate at which the probability of large fluctuations diminishes. A smaller value of α means that the probability tails are fatter, implying more volatility. In fact, when $\alpha < 2$ the theoretical variance is infinite. A portfolio can be modeled using random vectors, where each entry of the vector represents a different asset. The tail parameter α usually depends on the coordinate. The wrong coordinate system can mask variations in α, since the heaviest tail tends to dominate. A judicious choice of coordinate system is given by the eigenvectors of the sample covariance matrix. This isolates the heaviest tails, associated with the largest eigenvalues, and allows a more faithful representation of the dependence between assets.

Keywords

multivariable regular variation, moment estimates, moving averages, generalized domains of semistable attraction, R–O varying measures

1. Introduction

In order to construct a useful probability model for an investment portfolio, we must consider the dependence between assets. If we accept the premise that price changes are heavy tailed, then we are lead to consider random vectors with heavy tails. In this chapter, we survey those portions of the theory of heavy tailed random vectors that seem relevant to portfolio analysis. The most flexible models recognize the possibility that the thickness of probability tails varies in different directions, implying the need for matrix scaling. A judicious change of coordinates often simplifies the model, and may uncover features masked by the original coordinates. The original coordinates are the price changes (or returns) for each asset. The new coordinates can be interpreted as market indices, chosen to capture certain features of the market. In some popular heavy-tailed finance models, the tails are so heavy that the theoretical variance of price changes is undefined. For these models, the theoretical covariance matrix is also undefined. Of course the sample variance and the sample covariance matrix can always be computed for any data set, but these statistics are not estimating the usual model parameters. One of the most interesting discoveries in heavy tailed modeling is that, in the infinite variance case, the sample covariance matrix actually contains quite a bit of important information about the underlying distribution. In fact, the eigenvectors of this matrix provide a very useful coordinate system. We illustrate the application of this principle, and we also include a previously unpublished proof, extending the method to more general heavy tailed vector models with time dependence.

2. Heavy tails

A probability distribution has heavy tails if some of its moments fail to exist. Suppose that X is a random variable with density $f(x)$ so that

$$P(a \leqslant X \leqslant b) = \int_a^b f(x)\,\mathrm{d}x.$$

The k-th moment of the random variable X is defined by an improper integral

$$\mu_k = E\big(X^k\big) = \int_{-\infty}^{\infty} x^k f(x)\,\mathrm{d}x.$$

The mean $\mu = \mu_1$, variance $\sigma^2 = \mu_2 - \mu_1^2$, skewness and kurtosis depend on these moments. Because μ_k is an improper integral, it may not exist. If $f(x)$ is a normal density, a lognormal density, or any other density whose tails fall off exponentially then all of the moments μ_k exist. But if $f(x)$ has heavy tails that fall off like a power law, then some of the moments μ_k will not exist. The simplest example of a heavy tailed distribution is a Pareto, invented to model the distribution of incomes. A Pareto random variable satisfies

$P(X > x) = Cx^{-\alpha}$ so that the probability of large outcomes falls off like a power law. The Pareto density is defined by

$$f(x) = \begin{cases} C\alpha x^{-\alpha-1} & \text{for } x > C^{1/\alpha}, \\ 0 & \text{otherwise} \end{cases}$$

so that

$$\mu_k = \int_{C^{1/\alpha}}^{\infty} C\alpha x^{k-\alpha-1}\,\mathrm{d}x = \alpha C^{k/\alpha} \int_1^{\infty} y^{k-\alpha-1}\,\mathrm{d}y = \alpha C^{k/\alpha}\left[\frac{y^{k-\alpha}}{k-\alpha}\right]_{y=1}^{\infty}$$

using the substitution $x = C^{1/\alpha}y$. If $k < \alpha$ then the limit at infinity is zero and $\mu_k = \alpha C^{k/\alpha}/(\alpha - k)$, but if $k \geqslant \alpha$ then this improper integral diverges, so that the k-th moment does not exist.

Pareto distributions are closely related to some other familiar distributions. If U has a uniform distribution on $(0, 1)$, then $X = U^{-1/\alpha}$ has a Pareto distribution with tail parameter α. To check this, write

$$P(X > x) = P\big(U^{-1/\alpha} > x\big) = P\big(U < x^{-\alpha}\big) = x^{-\alpha}.$$

If X is Pareto with $P(X > x) = x^{-\alpha}$, then $Y = \ln X$ has an exponential distribution with rate α. To see this, note that

$$P(Y > y) = P(\ln X > y) = P\big(X > \mathrm{e}^y\big) = \big(\mathrm{e}^y\big)^{-\alpha} = \mathrm{e}^{-\alpha y}.$$

Some other familiar distributions have Pareto-like power law tails, causing some moments to diverge. If Y has a Student-t distribution with ν degrees of freedom, then $P(|Y| > y) \sim Cy^{-\alpha}$ where $\alpha = \nu$.[1] Then $E(Y^k)$ exists only for $k < \nu$. If Y has a Gamma distribution with density proportional to $y^{p-1}\mathrm{e}^{-qy}$ then the log-Gamma random variable X defined by $Y = \ln X$ satisfies $P(X > x) \sim Cx^{-\alpha}$ for x large, where $\alpha = q$. Some other distributions with Pareto-like tails are the stable and operator stable distributions, which will be discussed later in this chapter.

Heavy tailed random variables with $P(|X| > x) \sim Cx^{-\alpha}$ are observed in many real world applications. Estimation of the tail parameter α is important, because it determines which moments exist. Anderson and Meerschaert (1998) find heavy tails in a river flow with $\alpha \approx 3$, so that the variance is finite but the fourth moment is infinite. Tessier et al. (1996) find heavy tails with $2 < \alpha < 4$ for a variety of river flows and rainfall accumulations. Hosking and Wallis (1987) find evidence of heavy tails with $\alpha \approx 5$ for annual flood levels of a river in England. Benson, Wheatcraft and Meerschaert (2000), Benson et al. (2001) model concentration profiles for tracer plumes in groundwater using

[1] Here $f(x) \sim g(x)$ means that $f(x)/g(x) \to 1$ as $x \to \infty$.

stochastic models whose heavy tails have $1 < \alpha < 2$, so that the mean is finite but the variance is infinite. Heavy tail distributions with $1 < \alpha < 2$ are used in physics to model anomalous diffusion, where a cloud of particles spreads faster than classical Brownian motion predicts (Blumen, Zumofen and Klafter, 1989; Klafter, Blumen and Shlesinger, 1987; Shlesinger, Zaslavsky and Frisch, 1994). More applications to physics with $0 < \alpha < 2$ are cataloged in Uchaikin and Zolotarev (1999). Resnick and Stărică (1995) examine the quiet periods between transmissions for a networked computer terminal, and find heavy tails with $0 < \alpha < 1$, so that the mean and variance are both infinite. Several additional applications to computer science, finance, and signal processing appear in Adler, Feldman and Taqqu (1998). More applications to signal processing can be found in Nikias and Shao (1995).

Mandelbrot (1963) and Fama (1965a) pioneered the use of heavy tail distributions in finance. Mandelbrot (1963) presents graphical evidence that historical daily price changes in cotton have heavy tails with $\alpha \approx 1.7$, so that the mean exists but the variance is infinite. Jansen and de Vries (1991) argue that daily returns for many stocks and stock indices have heavy tails with $3 < \alpha < 5$, and discuss the possibility that the October 1987 stock market plunge might be just a heavy tailed random fluctuation. Loretan and Phillips (1994) use similar methods to estimate heavy tails with $2 < \alpha < 4$ for returns from numerous stock market indices and exchange rates. This indicates that the variance is finite but the fourth moment is infinite. Both daily and monthly returns show heavy tails with similar values of α in this study. Rachev and Mittnik (2000) use different methods to find heavy tails with $1 < \alpha < 2$ for a variety of stocks, stock indices, and exchange rates. McCulloch (1996) uses similar methods to re-analyze the data in Jansen and de Vries (1991), Loretan and Phillips (1994), and obtains estimates of $1.5 < \alpha < 2$. This is important because the variance of price returns is finite if $\alpha > 2$ and infinite if $\alpha < 2$. While there is disagreement about the true value of α, depending on which model is employed, all of these studies agree that financial data is typically heavy tailed, and that the tail parameter α varies between different assets.

Portfolio analysis involves the joint probability distribution of several prices or returns $X_1, \ldots, X_d$, where d is the number of assets in the portfolio. It is natural to model this set of numbers as a d-dimensional random vector $\boldsymbol{X} = (X_1, \ldots, X_d)'$. We say that $\boldsymbol{X}$ has heavy tails if $E(\|\boldsymbol{X}\|^k)$ is undefined for some $k = 1, 2, 3, \ldots$. Let us consider the practical problem of portfolio modeling. We choose d assets and research historical performance to obtain data of the form $X_i(t)$ where $i = 1, \ldots, d$ is the asset and $t = 0, \ldots, n$ is the time variable. Typically the distribution of values $X_i(0), \ldots, X_i(n)$ has a heavy tail whose parameter α_i can be estimated from this data. The research of Jansen and de Vries (1991), Loretan and Phillips (1994), and Rachev and Mittnik (2000) indicates, not surprisingly, that α_i will vary depending on the asset. Then the random vectors $\boldsymbol{X}_t = (X_1(t), \ldots, X_d(t))'$ will have heavier tails in some directions than in others. Despite this well known fact, most existing research on heavy tailed portfolio modeling has assumed that the probability tails are the same in every direction. Nolan, Panorska and McCulloch (2001) consider such a model, based on the multivariable stable distribution, for a vector of two exchange rates.

They argue that α is the same for both.[2] Rachev and Mittnik (2000) use a multivariable stable model for portfolio analysis, so that α is the same for every asset. The same approach was also applied to portfolio analysis by Bawa, Elton and Gruber (1979), Belkacem, Véhel and Walter (2000), Chamberlain, Cheung and Kwan (1990), Fama (1965b), Gamba (1999), Press (1982), Rachev and Han (2000), and Ziemba (1974). If this modeling approach can be enhanced to allow α_i to vary with the asset, a more realistic and flexible representation of financial portfolios can be achieved. The goal of this chapter is to show how this can be accomplished, using modern central limit theory.

3. Central limit theorems

Normal and log-normal models are popular in finance because of their simplicity and familiarity. Their use can also be justified by the central limit theorem. If $X, X_1, X_2, X_3, \ldots$ are independent and identically distributed (IID) random variables with mean $m = E(X)$ and finite variance $\sigma^2 = E[(X-m)^2]$ then the central limit theorem says that

$$\frac{(X_1 + \cdots + X_n) - nm}{n^{1/2}} \Rightarrow Y, \tag{3.1}$$

where Y is a normal random variable with mean zero and variance σ^2, and $\Rightarrow$ means convergence of probability distributions. Essentially, (3.1) means that $X_1 + \cdots + X_n$ is approximately normal (with mean nm and variance $n\sigma^2$) for n large. If the summands X_i represent independent price shocks, then their sum is the price change over a period of time. If price changes are accumulations of many IID shocks, then they should be normally distributed. If price changes accumulate multiplicatively, taking logs changes the product into a sum, leading to a log-normal model.

For portfolio analysis, we need to consider a vector of prices. Suppose that $\boldsymbol{X}, \boldsymbol{X}_1, \boldsymbol{X}_2, \boldsymbol{X}_3, \ldots$ are IID random vectors on a d-dimensional Euclidean space $\mathbb{R}^d$. If $\boldsymbol{X} = (X_1, \ldots, X_d)'$ then the mean $\boldsymbol{m} = E(\boldsymbol{X})$ is a vector with i-th entry $m_i = E(X_i)$, the covariance matrix C is a $d \times d$ matrix with ij entry

$$c_{ij} = \mathrm{Cov}(X_i, X_j) = E\big[(X_i - m_i)(X_j - m_j)\big],$$

and the central limit theorem says that

$$\frac{(\boldsymbol{X}_1 + \cdots + \boldsymbol{X}_n) - n\boldsymbol{m}}{\sqrt{n}} \Rightarrow \boldsymbol{Y}, \tag{3.2}$$

[2] Example 8.1 gives an alternative operator stable model for the same data set.

where $\boldsymbol{Y}$ is a normal random vector with mean zero and covariance matrix $C = E[\boldsymbol{Y}\boldsymbol{Y}']$. In this case, it simplifies the analysis to change coordinates. If the matrix P defines the change of coordinates then it follows from (3.2) that

$$\frac{(P\boldsymbol{X}_1 + \cdots + P\boldsymbol{X}_n) - nP\boldsymbol{m}}{\sqrt{n}} \Rightarrow P\boldsymbol{Y}, \tag{3.3}$$

where $P\boldsymbol{Y}$ is multivariate normal with mean zero and covariance matrix $PCP' = E[(P\boldsymbol{Y})(P\boldsymbol{Y})']$. If we take the new coordinate system defined by the eigenvectors of the covariance matrix C, then the limit $P\boldsymbol{Y}$ has independent normal marginals. The eigenvalues of C determine the variance of each marginal, so their square roots measure volatility. The corresponding marginals of $P\boldsymbol{X}$ are all linear combinations of the original assets, chosen to be asymptotically independent. This coordinate system is one of the cornerstones of Markowitz's theory of optimal portfolios, see for example Elton and Gruber (1995).

For heavy tailed random variables, the central limit theorem may not hold, because the second moment might not exist. An extended central limit theorem applies in this case. If $X, X_1, X_2, X_3, \ldots$ are IID random variables we say that X belongs to the domain of attraction of some random variable Y, and we write $X \in \mathrm{DOA}(Y)$, if

$$\frac{(X_1 + \cdots + X_n) - b_n}{a_n} \Rightarrow Y. \tag{3.4}$$

For mathematical reasons we exclude the degenerate case where $Y = c$ with probability one. The limits in (3.4) are called stable. If $E(X^2)$ exists then the classical central limit theorem shows that Y is normal, a special case of stable. In this case, we can take $a_n = n^{1/2}$ and $b_n = nE(X)$. If X has heavy tails with $P(|X| > r) \sim Cr^{-\alpha}$ then the situation depends on the tail thickness α. If $\alpha > 2$ then $E(X^2)$ exists and sums are asymptotically normal. But if $0 < \alpha \leqslant 2$ then $E(X^2) = \infty$ and (3.4) holds with $a_n = n^{1/\alpha}$ as long as a tail balancing condition holds:

$$\frac{P(X > r)}{P(|X| > r)} \to p \quad \text{and} \quad \frac{P(X < -r)}{P(|X| > r)} \to q \quad \text{as } r \to \infty \tag{3.5}$$

for some $0 \leqslant p, q \leqslant 1$ with $p + q = 1$.

A proof of the extended central limit theorem can be found in Gnedenko and Kolmogorov (1968), see also Feller (1971) and Meerschaert and Scheffler (2001a). The condition for $X \in \mathrm{DOA}(Y)$ is stated in terms of *regular variation*. A function $f(r)$ varies regularly if

$$\lim_{r\to\infty} \frac{f(\lambda r)}{f(r)} = \lambda^{\rho} \quad \text{for all } \lambda > 0. \tag{3.6}$$

For Y stable with index $0 < \alpha < 2$, so that Y is not normal, a necessary and sufficient condition for $X \in \mathrm{DOA}(Y)$ is that $P(|X| > r)$ varies regularly with index $-\alpha$ and (3.5) holds for some $0 \leqslant p, q \leqslant 1$ with $p + q = 1$. If we have $P(|X| > r) \sim Cr^{-\alpha}$ then it is easy to see that $P(|X| > r)$ varies regularly with index $-\alpha$, but the definition also allows a slightly more general tail behavior. For example, if $P(|X| > r) \sim Cr^{-\alpha} \log r$ then $P(|X| > r)$ still varies regularly with index $-\alpha$. The norming constants a_n in (3.4) can always be chosen according to the formula $nP(|X| > a_n) \to C$. If we have $P(|X| > r) \sim Cr^{-\alpha}$ this leads to $a_n = n^{1/\alpha}$. In practical applications, it is common to assume that $P(|X| > r) \sim Cr^{-\alpha}$ because a practical procedure exists for estimating the parameters C, α for a given heavy tailed data set.[3]

Stable distributions are typically specified in terms of their characteristic functions (Fourier transforms). If Y is stable with density $f(y)$ its characteristic function

$$E\big[e^{ikY}\big] = \int_{-\infty}^{\infty} e^{iky} f(y)\, dy$$

is of the form $e^{\psi(k)}$ where

$$\psi(k) = \begin{cases} ibk - \sigma^\alpha |k|^\alpha \left(1 - i\beta\, \mathrm{sign}(k) \tan\left(\dfrac{\pi\alpha}{2}\right)\right) & \text{for } \alpha \neq 1, \\ ibk - \sigma^\alpha |k|^\alpha \left(1 + i\beta\left(\dfrac{2}{\pi}\right) \mathrm{sign}(k) \ln|k|\right) & \text{for } \alpha = 1. \end{cases} \tag{3.7}$$

The entire class of nondegenerate stable laws on $\mathbb{R}^1$ is given by these formulas with *index* $\alpha \in (0, 2]$, *scale* $\sigma \in (0, \infty)$, *skewness* $\beta \in [-1, +1]$, and *center* $b \in (-\infty, \infty)$. The stable distribution with these parameters will be written as $S_\alpha(\sigma, \beta, b)$ using the notation of Samorodnitsky and Taqqu (1994). The skewness $\beta = p - q$ governs the deviations of the distribution from symmetry, so that $f(y)$ is symmetric if $\beta = 0$. The scale σ and the center b have the usual meaning that if Y has a $S_\alpha(1, \beta, 0)$ distribution then $\sigma Y + b$ has a $S_\alpha(\sigma, \beta, b)$ distribution, except that for $\alpha = 1$ and $\beta \neq 0$ multiplication by σ introduces a nonlinear change in the shift. The stable index α governs the tails of Y, and in fact $P(|Y| > r) \sim Cr^{-\alpha}$ where

$$\sigma^\alpha = \begin{cases} C \dfrac{\Gamma(2-\alpha)}{1-\alpha} \cos\left(\dfrac{\pi\alpha}{2}\right) & \text{for } \alpha \neq 1, \\ C \dfrac{\pi}{2} & \text{for } \alpha = 1 \end{cases} \tag{3.8}$$

in the nonnormal case $0 < \alpha < 2$. The tails are balanced so that

$$\frac{P(Y > r)}{P(|Y| > r)} \to p \quad \text{and} \quad \frac{P(Y < -r)}{P(|Y| > r)} \to q \quad \text{as } r \to \infty. \tag{3.9}$$

[3] See Section 8.

Stable laws belong to their own domain of attraction, but more is true. In fact, if Y_n are IID with Y then

$$\frac{(Y_1 + \cdots + Y_n) - b_n}{n^{1/\alpha}} \stackrel{d}{=} Y \tag{3.10}$$

for some b_n, where $\stackrel{d}{=}$ indicates that both sides have the same probability distribution. Sums of IID stable laws are again stable with the same α, β. Although there is no closed analytical formula for stable densities, the efficient computational method of Nolan (1997, 2002; Resnick and Stărică, 1995) can be used to plot density curves. Nolan (2001) uses these methods to compute maximum likelihood estimators for the stable parameters, see also Mittnik et al. (1999), Mittnik, Doganoglu and Chenyao (1999).

If X_n is the price change on day n then the accumulation of these changes will be approximately stable, assuming that X_n are IID with X and $P(|X| > x) \sim Cx^{-\alpha}$. If $\alpha < 2$, as in the cotton prices considered in Mandelbrot (1963), then the price obtained by adding these changes will be approximately stable with a power law tail. The balancing parameters p and q describe the probability that a large change in price will be positive or negative, respectively. The scale σ (or equivalently, the *dispersion* C) depends on the price units (e.g., US dollars). If $2 < \alpha < 4$ then the sum of these price changes will be asymptotically normal. However, the rule of thumb that sums look normal for $n \geqslant 30$ is no longer reliable. The heavy tails slow the rate of convergence in the central limit theorem. To illustrate the point, we simulated Pareto random variables with $\alpha = 3$, using the fact that if U is uniform on (0, 1) then $U^{-1/\alpha}$ is Pareto with tail parameter α. We summed $n = 50$ of these random variables, and repeated the simulation 100 times to get an idea of the distribution of these sums. The boxplot in Figure 1 indicates that the distribution of the resulting sums is skewed to the right, with some outliers. The normal probability plot in Figure 2 indicates a significant deviation from normality. The moral of this story is that for heavy tailed random variables with $\alpha > 2$, sums eventually converge to a normal limit, but slower than usual.

For heavy tailed random vectors, a generalized central limit theorem applies. If $\boldsymbol{X}, \boldsymbol{X}_1, \boldsymbol{X}_2, \boldsymbol{X}_3, \ldots$ are IID random vectors on $\mathbb{R}^d$ we say that $\boldsymbol{X}$ belongs to the generalized domain of attraction of some full dimensional random vector $\boldsymbol{Y}$ on $\mathbb{R}^d$, and we write $\boldsymbol{X} \in \mathrm{GDOA}(\boldsymbol{Y})$, if

$$A_n(\boldsymbol{X}_1 + \cdots + \boldsymbol{X}_n - \boldsymbol{b}_n) \Rightarrow \boldsymbol{Y} \tag{3.11}$$

for some $d \times d$ matrices A_n and vectors $\boldsymbol{b}_n \in \mathbb{R}^d$. The limits in (3.11) are called operator stable (Jurek and Mason, 1993; Sharpe, 1969). If $E(\|\boldsymbol{X}\|^2)$ exists then the classical central limit theorem shows that $\boldsymbol{Y}$ is multivariable normal, a special case of operator stable. In this case, we can take $A_n = n^{-1/2}I$ and $\boldsymbol{b}_n = nE(\boldsymbol{X})$. If $\boldsymbol{X}$ has heavy tails with $P(\|\boldsymbol{X}\| > r) \sim Cr^{-\alpha}$ then the situation depends on the tail thickness α. If $\alpha > 2$ then

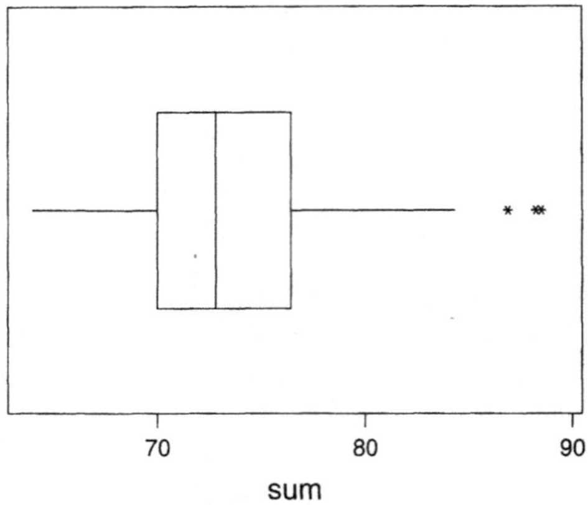

Fig. 1. Sums of 50 Pareto variables with $\alpha = 3$. Their distribution is skewed to the right with several outliers.

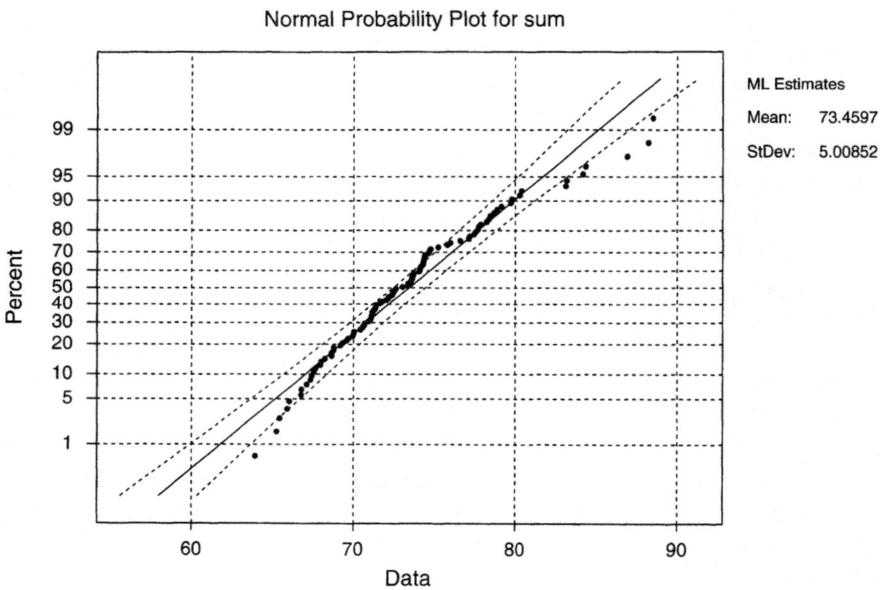

Fig. 2. Sums of 50 Pareto variables with $\alpha = 3$. Upper tail shows systematic deviation from normal distribution.

$E(\|X\|^2)$ exists and sums are asymptotically normal. But if $0 < \alpha < 2$ then $E(\|X\|^2) = \infty$ and (3.11) holds with $A_n = n^{-1/\alpha} I$ as long as a tail balancing condition holds:

$$\frac{P(\|X\| > r,\ X/\|X\| \in B)}{P(\|X\| > r)} \to M(B) \quad \text{as } r \to \infty \tag{3.12}$$

for all Borel subsets[4] B of the unit sphere $S = \{\theta \in \mathbb{R}^d\colon \|\theta\| = 1\}$ whose boundary has M-measure zero, where M is a probability measure on the unit sphere which is not supported on any $d-1$ dimensional subspace of $\mathbb{R}^d$. A proof of the generalized central limit theorem can be found in Rvačeva (1962) or Meerschaert and Scheffler (2001a). In this case, where the tails of $\boldsymbol{X}$ fall off at the same rate in every direction, the limit $\boldsymbol{Y}$ is multivariable stable (Samorodnitsky and Taqqu, 1994), a special case of operator stable.

If $\boldsymbol{Y}$ is multivariable stable with density $f(\boldsymbol{y})$ its characteristic function

$$E\left[e^{i\boldsymbol{k}\cdot\boldsymbol{Y}}\right] = \int e^{i\boldsymbol{k}\cdot\boldsymbol{y}} f(\boldsymbol{y})\,d\boldsymbol{y}$$

is of the form $e^{\psi(\boldsymbol{k})}$ where

$$\psi(\boldsymbol{k}) = i\boldsymbol{b}\cdot\boldsymbol{k} - \sigma^\alpha \int_{\|\theta\|=1} |\theta\cdot\boldsymbol{k}|^\alpha \left(1 - i\,\mathrm{sign}(\theta\cdot\boldsymbol{k}) \tan\left(\frac{\pi\alpha}{2}\right)\right) M(d\theta)$$

for $\alpha \neq 1$ and

$$\psi(\boldsymbol{k}) = i\boldsymbol{b}\cdot\boldsymbol{k} - \sigma^\alpha \int_{\|\theta\|=1} |\theta\cdot\boldsymbol{k}| \left(1 + i\left(\frac{2}{\pi}\right) \mathrm{sign}(\theta\cdot\boldsymbol{k}) \ln|\theta\cdot\boldsymbol{k}|\right) M(d\theta)$$

for $\alpha = 1$. The entire class of multivariable stable laws on $\mathbb{R}^d$ is given by these formulas with index $\alpha \in (0,2]$, scale $\sigma > 0$, mixing measure M and center $\boldsymbol{b} \in \mathbb{R}^d$. We say that $\boldsymbol{Y}$ has distribution $S_\alpha(\sigma, M, \boldsymbol{b})$ in this case. The mixing measure M is a probability distribution on the unit sphere in $\mathbb{R}^d$ that governs the tails of $\boldsymbol{Y}$, so that $f(\boldsymbol{y})$ is symmetric if M is symmetric. The center $\boldsymbol{b}$ and scale σ have the usual meaning that if $\boldsymbol{Y}$ has a $S_\alpha(1, M, 0)$ distribution then $\sigma\boldsymbol{Y} + \boldsymbol{b}$ has a $S_\alpha(\sigma, M, \boldsymbol{b})$ distribution, except when $\alpha = 1$. The stable index α governs the tails of Y in the nonnormal case $(0 < \alpha < 2)$. In fact, $P(\|\boldsymbol{Y}\| > r) \sim Cr^{-\alpha}$ where C is given by (3.8). The mixing measure M is a multivariable analogue of the skewness β. If $d = 1$ then $M\{+1\} = p$ and $M\{-1\} = q$, since the unit sphere on $\mathbb{R}^1$ is the two point set $\{-1, +1\}$. In this case, $\boldsymbol{Y}$ is stable with skewness $\beta = p - q$. The tails of a multivariable stable random vector are balanced so that

$$\frac{P(\|\boldsymbol{Y}\| > r,\ \boldsymbol{Y}/\|\boldsymbol{Y}\| \in B)}{P(\|\boldsymbol{Y}\| > r)} \to M(B) \quad \text{as } r \to \infty. \tag{3.13}$$

If $d = 1$ this reduces to the tail balancing condition (3.9) for stable random variables. Multivariable stable laws belong to their own domain of attraction, and if $\boldsymbol{Y}_n$ are IID with $\boldsymbol{Y}$ then

$$\frac{(\boldsymbol{Y}_1 + \cdots + \boldsymbol{Y}_n) - \boldsymbol{b}_n}{n^{1/\alpha}} \stackrel{d}{=} \boldsymbol{Y} \tag{3.14}$$

[4] The class of Borel subsets is the smallest class that include open sets and is closed under complements and countable unions.

for some $\boldsymbol{b}_n$, so that sums of IID multivariable stable laws are again multivariable stable with the same α. When $\boldsymbol{Y}$ is nonnormal multivariable stable with distribution $S_\alpha(\sigma, M, \boldsymbol{b})$ for some $0 < \alpha < 2$, the necessary and sufficient condition for $\boldsymbol{X} \in \mathrm{DOA}(\boldsymbol{Y})$ is that $P(\|\boldsymbol{X}\| > r)$ varies regularly with index $-\alpha$ and the balanced tails condition (3.12) holds.

Example 3.1. The mixing measure governs the radial direction of large price jumps. Take R_i IID Pareto random variables with $P(R > r) = Cr^{-\alpha}$. Take Θ_i to be IID random unit vectors with distribution M, independent of (R_i). Then $\boldsymbol{X}_i = R_i\Theta_i$ are IID random vectors with $P(\|\boldsymbol{X}_i\| > r) = Cr^{-\alpha}$ and

$$\frac{P(\|\boldsymbol{X}_i\| > r,\ \boldsymbol{X}_i/\|\boldsymbol{X}_i\| \in B)}{P(\|\boldsymbol{X}_i\| > r)} = P(\Theta_i \in B) = M(B)$$

for any Borel subset B of the unit sphere, and so $\boldsymbol{X}_i \in \mathrm{DOA}(\boldsymbol{Y})$ where $\boldsymbol{Y}$ is multivariable stable with distribution $S_\alpha(\sigma, M, \boldsymbol{b})$ for any $\boldsymbol{b} \in \mathbb{R}^d$. We can take $A_n = n^{-1/\alpha}I$ in (3.11), and $\boldsymbol{b}$ depends on the choice of centering $\boldsymbol{b}_n$. We call these heavy tailed random vectors *multivariable Pareto*. If we use a multivariable Pareto model for large jumps in the vector of prices for a portfolio, the parameter α governs the radius and the mixing measure M governs the angle of large jumps. Sums of these IID jumps are asymptotically multivariable stable with the same index α and mixing measure M. The radius $R = \|\boldsymbol{Y}\|$ satisfies $P(R > r) \sim Cr^{-\alpha}$ and the distribution of the radial component $\Theta = \boldsymbol{Y}/\|\boldsymbol{Y}\|$ conditional on $P(\|\boldsymbol{Y}\| > r)$ tends to M as $r \to \infty$ in view of the tail balancing condition (3.13). In other words, multivariable stable random vectors are asymptotically multivariable Pareto on their tails. In a multivariable stable model for price jumps, the mixing measure determines the direction of large jumps. If M is discrete with $M(\theta_i) = p_i$, then it follows from the characteristic function formulas that $\boldsymbol{Y}$ can be represented as the sum of independent stable components laid out along the θ_i directions, and the methods of Nolan (1997, 2002) can be used to plot multivariable stable densities, see Byczkowski, Nolan and Rajput (1993). The same idea is used by Modarres and Nolan (1994) to simulate stable random vectors with discrete mixing measures. For an arbitrary mixing measure, multivariable stable laws can be simulated using sums of independent, identically distributed multivariable Pareto laws. If $0 < \alpha < 1$ then the random vector $n^{-1/\alpha}(\boldsymbol{X}_1 + \cdots + \boldsymbol{X}_n)$ is approximately $S_\alpha(\sigma, M, 0)$ where C is given by (3.8). If $1 < \alpha < 2$ then $n^{-1/\alpha}(\boldsymbol{X}_1 + \cdots + \boldsymbol{X}_n - nE\boldsymbol{X}_1)$ is approximately $S_\alpha(\sigma, M, 0)$ where C is given by (3.8) and

$$E(\boldsymbol{X}_1) = E(R_1)E(\Theta_1) = \frac{\alpha C^{1/\alpha}}{\alpha - 1}\int_{\|\theta\|=1} \theta M(\mathrm{d}\theta).$$

Remark 3.2. Previously a different type of multivariable Pareto distribution was considered by Arnold (1990), see also Kotz, Balakrishnan and Johnson (2000).

4. Matrix scaling

The multivariable stable model is the basis for the work of Nolan, Panorska and McCulloch (2001) on exchange rates, and the portfolio models in Rachev and Mittnik (2000). Under the assumptions of this model, the probability tail of the random vector $\boldsymbol{X}_t$ is assumed to fall off at the same power law rate in every radial direction. Suppose that $\boldsymbol{X}_t = (X_1(t), \ldots, X_d(t))'$ where $X_i(t)$ is the price change of the i-th asset on day t. If $\boldsymbol{X}_t$ belongs to the domain of attraction of some multivariable stable random vector $\boldsymbol{Y} = (Y_1, \ldots, Y_d)'$ with index α, and that (3.11) holds with $A_n = n^{-1/\alpha} I$. Projecting onto the i-th coordinate axis shows that

$$\frac{X_i(1) + \cdots + X_i(n) - b_i(n)}{n^{1/\alpha}} \Rightarrow Y_i, \tag{4.1}$$

where $\boldsymbol{b}_n = (b_1(n), \ldots, b_d(n))'$, so that Y_i is stable with index α and $X_i(t)$ belongs to the domain of attraction of Y_i. According to Jansen and de Vries (1991), Loretan and Phillips (1994), and Rachev and Mittnik (2000), the stable index α_i should vary depending on the asset. Then (4.1) is replaced by

$$\frac{X_i(1) + \cdots + X_i(n) - b_i(n)}{n^{1/\alpha_i}} \Rightarrow Y_i \quad \text{for each } i = 1, \ldots, d \tag{4.2}$$

so that Y_i is stable with index α_i. Mittnik and Rachev (1993) seem to have been the first to apply such models to a problem in finance, see also Section 8.6 in Rachev and Mittnik (2000). Assuming the joint convergence

$$A_n \left[\begin{pmatrix} X_1(1) \\ X_2(1) \\ \vdots \\ X_d(1) \end{pmatrix} + \cdots + \begin{pmatrix} X_1(n) \\ X_2(n) \\ \vdots \\ X_d(n) \end{pmatrix} - \begin{pmatrix} b_1(n) \\ b_2(n) \\ \vdots \\ b_d(n) \end{pmatrix} \right] \Rightarrow \begin{pmatrix} Y_1 \\ Y_2 \\ \vdots \\ Y_d \end{pmatrix} \tag{4.3}$$

and changing to vector-matrix notation we get (3.11) with diagonal norming matrices

$$A_n = \begin{pmatrix} n^{-1/\alpha_1} & 0 & \cdots & 0 \\ 0 & n^{-1/\alpha_2} & & 0 \\ \vdots & & \ddots & \vdots \\ 0 & 0 & \cdots & n^{-1/\alpha_d} \end{pmatrix} \tag{4.4}$$

which we will also write as $A_n = \operatorname{diag}(n^{-1/\alpha_1}, \ldots, n^{-1/\alpha_d})$. The matrix scaling is natural since we are dealing with random vectors, and it allows a more realistic portfolio model. The i-th marginal Y_i of the operator stable limit vector $\boldsymbol{Y}$ is stable with index α_i, so the tail behavior of $\boldsymbol{Y}$ varies with angle. The convergence (3.11) with A_n diagonal was first considered in Resnick and Greenwood (1979), see also Meerschaert (1991).

Matrix notation also leads to a natural analogue of the stable index α. Let $\exp(A) = I + A + A^2/2! + A^3/3! + \cdots$ be the usual exponential operator for $d \times d$ matrices. This operator occurs, for example, in the theory of linear differential equations. If $A = \operatorname{diag}(a_1, \dots, a_d)$ then an easy matrix computation using the Taylor series formula $e^x = 1 + x + x^2/2! + x^3/3! + \cdots$ shows that $\exp(A) = \operatorname{diag}(e^{a_1}, \dots, e^{a_d})$. See Hirsch and Smale (1974) or Section 2.2 of Meerschaert and Scheffler (2001a) for details and additional information. Now define $E = \operatorname{diag}(1/\alpha_1, \dots, 1/\alpha_d)$. Then the norming matrices A_n in (4.4) can also be written in the more compact form $A_n = n^{-E} = \exp(-E \ln n)$, since $-E \ln n = \operatorname{diag}(-(1/\alpha_1)\ln n, \dots, -(1/\alpha_d)\ln n)$ and $e^{-(1/\alpha_i)\ln n} = n^{-1/\alpha_i}$. The matrix E, called an *exponent* of the operator stable random vector $\boldsymbol{Y}$, plays the role of the stable index α. This matrix E need not be diagonal. Diagonalizable exponents involve a change of coordinates, degenerate eigenvalues thicken probability tails by a logarithmic factor, and complex eigenvalues introduce rotational scaling, see Meerschaert (1990). The case of a diagonalizable exponent plays an important role in Example 8.1.

The generalized central limit theorem for matrix scaling can be found in Meerschaert and Scheffler (2001a). Matrix scaling allows for a limit with both normal and nonnormal components. Since $\boldsymbol{Y}$ is infinitely divisible, the Lévy representation [Theorem 3.1.11 in Meerschaert and Scheffler (2001a)] shows that the characteristic function $E[e^{i\boldsymbol{k}\cdot\boldsymbol{Y}}]$ is of the form $e^{\psi(\boldsymbol{k})}$ where

$$\psi(\boldsymbol{k}) = i\boldsymbol{b}\cdot\boldsymbol{k} - \frac{1}{2}\boldsymbol{k}\cdot C\boldsymbol{k} + \int_{\boldsymbol{x}\neq 0}\left(e^{i\boldsymbol{k}\cdot\boldsymbol{x}} - 1 - \frac{i\boldsymbol{k}\cdot\boldsymbol{x}}{1+\|\boldsymbol{x}\|^2}\right)\phi(d\boldsymbol{x})$$

for some $\boldsymbol{b} \in \mathbb{R}^d$, some nonnegative definite symmetric $d \times d$ matrix C and some Lévy measure ϕ. The Lévy measure satisfies $\phi\{\boldsymbol{x}\colon \|\boldsymbol{x}\| > 1\} < \infty$ and

$$\int_{0<\|\boldsymbol{x}\|<1} \|\boldsymbol{x}\|^2 \phi(d\boldsymbol{x}) < \infty.$$

For a multivariable stable law,

$$\phi\left\{\boldsymbol{x}\colon \|\boldsymbol{x}\| > r,\ \frac{\boldsymbol{x}}{\|\boldsymbol{x}\|} \in B\right\} = Cr^{-\alpha}M(B)$$

and the characteristic function formulas for multivariable stable laws follow by a lengthy computation, see Section 7.3 in Meerschaert and Scheffler (2001a) for complete details. If $\phi = 0$ then $\boldsymbol{Y}$ is normal with mean $\boldsymbol{b}$ and covariance matrix C. If $C = 0$ then a necessary and sufficient condition for (3.11) to hold is that

$$nP(A_n\boldsymbol{X} \in B) \to \phi(B) \quad \text{as } n \to \infty \tag{4.5}$$

for Borel subsets B of $\mathbb{R}^d \setminus \{0\}$ whose boundary have ϕ-measure zero, where ϕ is the Lévy measure of the limit $\boldsymbol{Y}$. Proposition 6.1.10 in Meerschaert and Scheffler (2001a) shows that the convergence (4.5) is equivalent to regular variation of the probability distribution $\mu(B) = P(\boldsymbol{X} \in B)$. If (4.5) holds then Proposition 6.1.2 in Meerschaert and Scheffler (2001a) shows that the Lévy measure satisfies

$$t\phi(\mathrm{d}x) = \phi\big(t^{-E}\,\mathrm{d}x\big) \quad \text{for all } t > 0 \tag{4.6}$$

for some $d \times d$ matrix E. Then it follows from the characteristic function formula that $\boldsymbol{Y}$ is operator stable with exponent E, and that for $\boldsymbol{Y}_n$ IID with $\boldsymbol{Y}$ we have

$$n^{-E}(\boldsymbol{Y}_1 + \cdots + \boldsymbol{Y}_n - \boldsymbol{b}_n) \stackrel{d}{=} \boldsymbol{Y} \tag{4.7}$$

for some $\boldsymbol{b}_n$, see Theorem 7.2.1 in Meerschaert and Scheffler (2001a). Hence operator stable laws belong to their own GDOA, so that the probability distribution of $\boldsymbol{Y}$ also varies regularly, and sums of IID operator stable random vectors are again operator stable with the same exponent E. If $E = aI$ then $\boldsymbol{Y}$ is multivariable stable with index $\alpha = 1/a$, and (4.5) is equivalent to the balanced tails condition (3.12).

Example 4.1. Multivariable Pareto random vectors with matrix scaling extend the model in Example 3.1. Suppose $\boldsymbol{Y}$ is operator stable with exponent E and Lévy measure ϕ. Define

$$F_{r,B} = \big\{s^E\theta\colon\ s > r,\ \theta \in B\big\}$$

and let $\lambda(B) = \phi(F_{1,B})$ for any Borel subset B of the unit sphere S whose boundary has λ-measure zero.[5] Let $C = \lambda(S)$ and define the probability measure $M(B) = \lambda(B)/C$. Take R_i IID standard Pareto random variables with $P(R > r) = Cr^{-1}$, Θ_i IID random unit vectors with distribution M and independent of (R_i), and finally let $\boldsymbol{X}_i = R_i^E\Theta_i$. Since $t^E F_{1,B} = F_{t,B}$ we have $\phi(F_{t,B}) = \phi(t^E F_{1,B}) = t^{-1}\phi(F_{1,B}) = Ct^{-1}M(B)$ in view of (4.6). Then

$$\begin{aligned} nP\big(n^{-E}\boldsymbol{X}_i \in F_{t,B}\big) &= nP\big(R_i^E\Theta_i \in F_{nt,B}\big) \\ &= nP(R_i > nt, \Theta_i \in B) \\ &= nC(nt)^{-1}M(B) = \phi(F_{t,B}) \end{aligned}$$

for $n > 1/t$, so that (4.5) holds for the sets $F_{t,B}$ with $A_n = n^{-E}$. Then $\boldsymbol{X}_i \in \mathrm{GDOA}(\boldsymbol{Y})$. Operator stable laws can be simulated using sums of these IID random vectors. If every eigenvalue of E has real part greater than one, then $n^{-E}(\boldsymbol{X}_1 + \cdots + \boldsymbol{X}_n)$ is approximately operator stable with exponent E and Lévy measure ϕ. If every eigenvalue of E has real

5 The measure λ is called the spectral measure of $\boldsymbol{Y}$.

part less than one, then $n^{-E}(\boldsymbol{X}_1+\cdots+\boldsymbol{X}_n-n\boldsymbol{m})$ is approximately operator stable with exponent E and Lévy measure ϕ where

$$\boldsymbol{m}=C\int_{\|\theta\|=1}\int_C^\infty r^E\theta\frac{\mathrm{d}r}{r^2}M(\mathrm{d}\theta)$$

is the mean of $\boldsymbol{X}_1$.

5. The spectral decomposition

The tail behavior of an operator stable random vector $\boldsymbol{Y}$ is determined by the eigenvalues of its exponent E. If $E=(1/\alpha)I$ then $\boldsymbol{Y}$ is multivariable stable and $P(|\boldsymbol{Y}\cdot\theta|>r)\sim C_\theta r^{-\alpha}$ for any $\theta\neq 0$. If $E=\mathrm{diag}(a_1,\dots,a_d)$ then $\boldsymbol{Y}=(Y_1,\dots,Y_d)'$ where Y_i is a stable random variable with index $\alpha_i=1/a_i$. This requires $0<\alpha_i\leqslant 2$ so that $a_i\geqslant 1/2$. For any $d\times d$ matrix E there is a unique *spectral decomposition* based on the real parts of the eigenvalues, see for example Theorem 2.1.14 in Meerschaert and Scheffler (2001a). This decomposition allows us to write $E=PBP^{-1}$ where P is a change of coordinates matrix and B is block-diagonal with

$$B=\begin{pmatrix} B_1 & 0 & \cdots & 0\\ 0 & B_2 & & 0\\ \vdots & & \ddots & \vdots\\ 0 & 0 & \cdots & B_p\end{pmatrix} \tag{5.1}$$

where B_i is a $d_i\times d_i$ matrix, every eigenvalue of B_i has real part equal to a_i, $a_1<\cdots<a_p$, and $d_1+\cdots+d_p=d$. Let $\boldsymbol{e}_1=(1,0,\dots,0)'$, $\boldsymbol{e}_2=(0,1,0,\dots,0)'$, ..., $\boldsymbol{e}_d=(0,\dots,0,1)'$ be the standard coordinates for $\mathbb{R}^d$ and define $\boldsymbol{p}_{ik}=P\boldsymbol{e}_j$ when $j=d_1+\cdots+d_{i-1}+k$ for some $k=1,\dots,d_i$. Then

$$V_i=\mathrm{span}\{\boldsymbol{p}_{i1},\dots,\boldsymbol{p}_{id_i}\}=\left\{\sum_{k=1}^{d_i}t_k\boldsymbol{p}_{ik}\colon t_1,\dots,t_{d_i}\text{ real}\right\}$$

is a d_i-dimensional subspace of $\mathbb{R}^d$. Any vector $\boldsymbol{y}\in\mathbb{R}^d$ can be written uniquely in the form $\boldsymbol{y}=\boldsymbol{y}_1+\cdots+\boldsymbol{y}_p$ with $\boldsymbol{y}_i\in V_i$ for each $i=1,\dots,p$. This is called the spectral decomposition of $\mathbb{R}^d$ with respect to E. Since B is block-diagonal and $E=PBP^{-1}$, every $E\boldsymbol{p}_{ik}$ is a linear combination of $\boldsymbol{p}_{i1},\dots,\boldsymbol{p}_{id_i}$ and therefore $E\boldsymbol{y}_i\in V_i$ for every $\boldsymbol{y}_i\in V_i$. This means that V_i is an E-invariant subspace of $\mathbb{R}^d$. Given a nonzero vector $\theta\in\mathbb{R}^d$, write $\theta=\theta_1+\cdots+\theta_p$ with $\theta_i\in V_i$ for each $i=1,\dots,p$ and define

$$\alpha(\theta)=\min\left\{\frac{1}{a_i}\colon \theta_i\neq 0\right\}. \tag{5.2}$$

Since the probability distribution of $\boldsymbol{Y}$ varies regularly with exponent E, Theorem 6.4.15 in Meerschaert and Scheffler (2001a) shows that for any small $\delta > 0$ we have

$$r^{-\alpha(\theta)-\delta} < P\big(|\boldsymbol{Y} \cdot \theta| > r\big) < r^{-\alpha(\theta)+\delta}$$

for all $r > 0$ sufficiently large. In other words, the tail behavior of $\boldsymbol{Y}$ is dominated by the component with the heaviest tail. This also means that $E(|\boldsymbol{Y} \cdot \theta|^\beta)$ exists for $0 < \beta < \alpha(\theta)$ and diverges for $\beta > \alpha(\theta)$. If we write $\boldsymbol{Y} = \boldsymbol{Y}_1 + \cdots + \boldsymbol{Y}_p$ with $\boldsymbol{Y}_i \in V_i$ for each $i = 1, \ldots, p$, then projecting (4.7) onto V_i shows that $\boldsymbol{Y}_i$ is an operator stable random vector on V_i with some exponent E_i. We call this the spectral decomposition of $\boldsymbol{Y}$ with respect to E. Since every eigenvalue of E_i has the same real part a_i we say that $\boldsymbol{Y}_i$ is spectrally simple, with index $\alpha_i = 1/a_i$. Although $\boldsymbol{Y}_i$ might not be multivariable stable, it has similar tail behavior. For any small $\delta > 0$ we have

$$r^{-\alpha_i-\delta} < P\big(\|\boldsymbol{Y}_i\| > r\big) < r^{-\alpha_i+\delta}$$

for all $r > 0$ sufficiently large, so $E(\|\boldsymbol{Y}_i\|^\beta)$ exists for $0 < \beta < \alpha_i$ and diverges for $\beta > \alpha_i$.

If $\boldsymbol{X} \in \mathrm{GDOA}(\boldsymbol{Y})$ then Theorem 8.3.24 in Meerschaert and Scheffler (2001a) shows that the limit $\boldsymbol{Y}$ and norming matrices A_n in (3.11) can be chosen so that every V_i in the spectral decomposition of $\mathbb{R}^d$ with respect to the exponent E of $\boldsymbol{Y}$ is A_n-invariant for every n, and $V_1, \ldots, V_p$ are mutually perpendicular. Then the probability distribution of $\boldsymbol{X}$ is regularly varying with exponent E and $\boldsymbol{X}$ has the same tail behavior as $\boldsymbol{Y}$. In particular, for any small $\delta > 0$ we have

$$r^{-\alpha(\theta)-\delta} < P\big(|\boldsymbol{X} \cdot \theta| > r\big) < r^{-\alpha(\theta)+\delta}$$

for all $r > 0$ sufficiently large. In this case, we say that $\boldsymbol{Y}$ is spectrally compatible with $\boldsymbol{X}$, and we write $\boldsymbol{X} \in \mathrm{GDOA}_c(\boldsymbol{Y})$.

Example 5.1. If $\boldsymbol{Y}$ is operator stable with exponent $E = aI$ then (4.7) shows that $\boldsymbol{Y}$ is multivariable stable with index $\alpha = 1/a$. Then $p = 1$, $P = I$, and $B = E$. There is only one spectral component, since the tail behavior is the same in every radial direction. If asset price change vectors are IID with $\boldsymbol{X} = (X_1, \ldots, X_d)' \in \mathrm{GDOA}(\boldsymbol{Y})$, then every asset has the same tail behavior. If θ_j measures the amount of the j-th asset in a portfolio, price changes for this portfolio are IID with the random variable $\boldsymbol{X} \cdot \theta = X_1\theta_1 + \cdots + X_d\theta_d$. Since the probability tails of $\boldsymbol{X}$ are uniform in every direction, the probability of a large jump in price falls off like $r^{-\alpha}$ for any portfolio.

Example 5.2. If $\boldsymbol{Y}$ is operator stable with exponent $E = \mathrm{diag}(a_1, \ldots, a_d)$ where $a_1 < \cdots < a_d$ then $p = d$, $P = I$, $B = E$, $B_i = a_i$ and V_i is the i-th coordinate axis. The spectral decomposition of $\boldsymbol{Y} = (Y_1, \ldots, Y_d)'$ with respect to E is $\boldsymbol{Y} = \boldsymbol{Y}_1 + \cdots + \boldsymbol{Y}_d$ with $\boldsymbol{Y}_i = Y_i\boldsymbol{e}_i$, the i-th marginal laid out along the i-th coordinate axis. Projecting (4.7) onto the i-th coordinate axis shows that Y_i is stable with index $\alpha_i = 1/a_i$, so that $P(|Y_i| > r) \sim$

$C_i r^{-\alpha_i}$. If $\theta \neq 0$ then $P(|Y \cdot \theta| > r)$ falls off like $r^{-\alpha(\theta)}$ where $\alpha(\theta) = \min\{\alpha_i \colon \theta_i \neq 0\}$. In other words, the heaviest tail dominates. If asset price change vectors are IID with $X \in \mathrm{GDOA}_c(Y)$, then the assets are arranged in order of increasing tail thickness. If θ_i measures the amount of the i-th asset in a portfolio, the probability of a large jump in price falls off like $r^{-\alpha(\theta)}$.

Example 5.3. If Y is operator stable with exponent $E = \mathrm{diag}(\beta_1, \ldots, \beta_d)$ then $B_i = a_i I$ for some $a_i \geqslant 1/2$ and d_i counts the number of diagonal entries β_j for which $\beta_j = 1/\alpha_i$. The matrix P sorts $\beta_1, \ldots, \beta_d$ in increasing order, and the vectors $\boldsymbol{p}_{ik}$ are the coordinates $\boldsymbol{e}_j$ for which $\beta_j = a_i$. The vectors Y_i are multivariable stable with index $\alpha_i = 1/a_i$, so that $P(\|Y_i\| > r) \sim C_i r^{-\alpha_i}$. For nonzero vectors $\theta \in V_i$ we have

$$P\bigl(|Y \cdot \theta| > r\bigr) = P\bigl(|Y_i \cdot \theta| > r\bigr) \sim C_\theta r^{-\alpha_i}$$

by the balanced tails condition for multivariable stable laws. For any other nonzero vector θ, $P(|Y \cdot \theta| > r) \sim C_\theta r^{-\alpha(\theta)}$ where $\alpha(\theta) = \min\{1/\beta_j \colon \theta_j \neq 0\}$. Again, the heaviest tail dominates. If asset price change vectors are IID with $X \in \mathrm{GDOA}_c(Y)$, then X has essentially the same tail behavior as Y, and P sorts the assets in order of increasing tail thickness.

Example 5.4. Take $B = \mathrm{diag}(a_1, \ldots, a_d)$ where $a_1 < \cdots < a_d$ and P orthogonal, so that $P^{-1} = P'$. If $Y = (Y_1, \ldots, Y_d)'$ is operator stable with exponent $E = PBP^{-1}$ then $p = d$, $B_i = a_i$ and $V_1, \ldots, V_d$ are the coordinate axes in the new coordinate system defined by the vectors $\boldsymbol{p}_i = P\boldsymbol{e}_i$ for $i = 1, \ldots, d$. The spectral component Y_i is the stable random variable $Y \cdot \boldsymbol{p}_i$ with index $\alpha_i = 1/a_i$, laid out along the V_i axis. Since $Y_j = Y \cdot \boldsymbol{e}_j$ is a linear combination of stable laws of different indices, it is not stable. The change of coordinates P rotates the coordinate axes to make the marginals stable. Since $n^{-PBP^{-1}} = Pn^{-B}P^{-1}$ it follows from (4.7) that

$$Pn^{-B}P^{-1}(Y_1 + \cdots + Y_n - \boldsymbol{b}_n) \stackrel{d}{=} Y,$$
$$n^{-B}\bigl(P^{-1}Y_1 + \cdots + P^{-1}Y_n - P^{-1}\boldsymbol{b}_n\bigr) \stackrel{d}{=} P^{-1}Y$$

so that $Y_0 = P^{-1}Y$ is operator stable with exponent B. Then the tail behavior of $Y = PY_0$ follows from Example 5.2 and the change of coordinates. If we write $\theta = \theta_1 \boldsymbol{p}_1 + \cdots + \theta_d \boldsymbol{p}_d$ in these coordinates then $P(|Y \cdot \theta| > r) \sim C_\theta r^{-\alpha(\theta)}$ where $\alpha(\theta) = \min\{\alpha_i \colon \theta_i \neq 0\}$. If asset price change vectors are IID with $X \in \mathrm{GDOA}_c(Y)$, then the tail behavior of X is essentially the same as Y. In particular, taking $\theta = \boldsymbol{p}_1$ gives a portfolio with the lightest probability tails.

Example 5.5. Suppose that Y is operator stable with exponent $E = PBP^{-1}$ where P is orthogonal and B is given by (5.1), with $d_i \times d_i$ blocks $B_i = a_i I$ for some $1/2 \leqslant a_1 < \cdots < a_p$. Let $D_0 = 0$ and $D_i = d_1 + \cdots + d_i$ for $1 \leqslant i \leqslant p$. Then $\boldsymbol{p}_{ik} = P\boldsymbol{e}_j$ when $j = D_{i-1} + k$

for some $k = 1, \ldots, d_i$ and $V_i = \operatorname{span}\{\boldsymbol{p}_{ik}\colon k = 1, \ldots, d_i\}$. To avoid double subscripts we will also write $\boldsymbol{q}_j = P\boldsymbol{e}_j$, so that $\boldsymbol{q}_j = \boldsymbol{p}_{ik}$ when $j = D_{i-1} + k$ for some $k = 1, \ldots, d_i$. The j-th column of the matrix P is the vector $\boldsymbol{q}_j$, and

$$E\boldsymbol{q}_j = PBP^{-1}\boldsymbol{q}_j = PB\boldsymbol{e}_j = Pa_i\boldsymbol{e}_j = a_i P\boldsymbol{e}_j = a_i\boldsymbol{q}_j$$

when $\boldsymbol{q}_j \in V_i$, so that $\boldsymbol{q}_j$ is a unit eigenvector of the matrix E with corresponding eigenvalue a_i. The spectral component

$$\boldsymbol{Y}_i = \sum_{k=1}^{d_i} (\boldsymbol{Y} \cdot \boldsymbol{p}_{ik})\boldsymbol{p}_{ik}$$

is the orthogonal projection of $\boldsymbol{Y}$ onto the d_i-dimensional subspace V_i. The random vector $\boldsymbol{Y}_i$ is multivariable stable with index $\alpha_i = 1/a_i$, so that $P(\|\boldsymbol{Y}_i\| > r) \sim C_i r^{-\alpha_i}$, and every marginal $Y_{ik} = \boldsymbol{Y} \cdot \boldsymbol{p}_{ik}$ is stable with the same index α_i. The change of coordinates P rotates the coordinate axes to find a set of orthogonal unit eigenvectors for E, so that the marginals of $\boldsymbol{Y}$ in the new coordinate system are all stable random variables. The matrix P also sorts the corresponding eigenvalues in increasing order. For any nonzero vector $\theta \in \mathbb{R}^d$,

$$P\big(|\boldsymbol{Y} \cdot \theta| > r\big) \sim C_\theta r^{-\alpha(\theta)},$$

where $\alpha(\theta) = \alpha_i$ for the largest i such that the orthogonal projection of θ onto the subspace V_i is not equal to zero. If asset price change vectors are IID with $\boldsymbol{X} \in \mathrm{GDOA_c}(\boldsymbol{Y})$, then the tail behavior of $\boldsymbol{X}$ is essentially the same as $\boldsymbol{Y}$. If $\theta = \theta_1\boldsymbol{e}_1 + \cdots \theta_d\boldsymbol{e}_d$ so that θ_i measures the amount of the i-th asset in a portfolio, price changes for this portfolio are IID with $\boldsymbol{X} \cdot \theta = X_1\theta_1 + \cdots + X_d\theta_d$. In particular, any $\theta \in V_1$ gives a portfolio with the lightest probability tails.

6. Sample covariance matrix

Given a data set of price changes (or log returns) $X_1, X_2, \ldots, X_n$ for a given asset, the k-th sample moment

$$\hat{\mu}_k = \frac{1}{n}\sum_{t=1}^{n} X_t^k$$

estimates the k-th moment $\mu_k = E(X^k)$. These sample moments are used to estimate the mean, variance, skewness and kurtosis of the data. If X_t are IID with $P(|X_t| > r) \sim Cr^{-\alpha}$, then X_t^k are also IID and heavy tailed with

$$P\big(|X_t^k| > r\big) = P\big(|X_t| > r^{1/k}\big) \sim Cr^{-\alpha/k}$$

so the extended central limit theorem applies. Recall from Section 2 that μ_k exists for $k < \alpha$ and diverges for $k \geqslant \alpha$. If $\alpha > 4$ then $\mathrm{Var}(X_t^2) = \mu_4 - \mu_2^2$ exists and the central limit theorem (3.1) implies that

$$n^{1/2}(\hat{\mu}_2 - \mu_2) = n^{-1/2} \sum_{t=1}^{n} (X_t^2 - \mu_2) \Rightarrow Y, \tag{6.1}$$

where Y is normal. When $2 < \alpha < 4$, the mean $\mu_2 = E(X_t^2)$ of these summands exists but $\mathrm{Var}(X_t^2)$ is infinite, and the extended central limit theorem (3.4) implies that

$$n^{1-2/\alpha}(\hat{\mu}_2 - \mu_2) = n^{-2/\alpha} \sum_{t=1}^{n} (X_t^2 - \mu_2) \Rightarrow Y,$$

where Y is stable with index $\alpha/2$. When $0 < \alpha < 2$ the mean $\mu_2 = E(X_t^2)$ of the squared price change diverges, and the extended central limit theorem implies that

$$n^{1-2/\alpha}\hat{\mu}_2 = n^{-2/\alpha} \sum_{t=1}^{n} X_t^2 \Rightarrow Y,$$

where again Y is stable with index $\alpha/2$. In this case, the sample second moment $\hat{\mu}_2$ exists but the second moment μ_2 does not. When $0 < \alpha < 2$, or when $2 < \alpha < 4$ and $\mu_1 = 0$, the sample variance

$$\hat{\sigma}^2 = \frac{1}{n} \sum_{t=1}^{n} (X_t - \hat{\mu}_1)^2 = \hat{\mu}_2 - \hat{\mu}_1^2 \tag{6.2}$$

is asymptotically equivalent to the sample second moment, see for example Anderson and Meerschaert (1997). Since we can always center to zero expectation when $2 < \alpha < 4$, both have the same asymptotics. If $\alpha > 4$ the sample variance is asymptotically normal, and when $0 < \alpha < 4$ the sample variance is asymptotically stable. Since the variance is a measure of price volatility, the sample variance estimates volatility. Confidence intervals for the variance are based on normal asymptotics when $\alpha > 4$ and stable asymptotics when $2 < \alpha < 4$. When $\alpha < 2$ the variance is undefined, but the sample variance still captures some important features of the data, see Section 8.

Suppose that $\boldsymbol{X}_t = (X_1(t), \ldots, X_d(t))'$ where $X_i(t)$ is the price change of the i-th asset on day t. The covariance matrix characterizes dependence between price changes of different assets over the same day, and the sample covariance matrix estimates the covariance matrix. As before, it is simpler to begin with the uncentered estimate

$$M_n = \frac{1}{n} \sum_{t=1}^{n} \boldsymbol{X}_t \boldsymbol{X}_t', \tag{6.3}$$

where X' denotes the transpose of the vector $X = (X_1, \ldots, X_d)'$ and hence

$$XX' = \begin{pmatrix} X_1 \\ \vdots \\ X_d \end{pmatrix} (X_1, \ldots, X_d) = \begin{pmatrix} X_1X_1 & \cdots & X_1X_d \\ X_2X_1 & \cdots & X_2X_d \\ \vdots & \ddots & \vdots \\ X_dX_1 & \cdots & X_dX_d \end{pmatrix}$$

is an element of the vector space $\mathcal{M}_s^d$ of symmetric $d \times d$ matrices. The ij entry of M_n is

$$M_n(i, j) = \frac{1}{n} \sum_{t=1}^{n} X_i(t) X_j(t)$$

which estimates $E(X_i X_j)$. If X_t are IID with X, then $X_t X_t'$ are IID random matrices and we can apply the central limit theorems from Section 3 [see Section 10.2 in Meerschaert and Scheffler (2001a) for complete proofs]. If the probability distribution of X is regularly varying with exponent E and (4.5) holds with $t\phi\{\mathrm{d}x\} = \phi\{t^{-E}\,\mathrm{d}x\}$ for all $t > 0$, then the distribution of XX' is also regularly varying with

$$nP(A_n XX' A_n' \in B) \to \Phi(B) \quad \text{as } n \to \infty \tag{6.4}$$

for Borel subsets B of $\mathcal{M}_s^d$ that are bounded away from zero and whose boundary has Φ-measure zero. The exponent ξ of the limit measure $\Phi\{d(xx')\} = \phi\{\mathrm{d}x\}$ is defined by $\xi M = EM + ME'$ for $M \in \mathcal{M}_s^d$. Using the matrix norm

$$\|M\| = \left(\sum_{i=1}^{d} \sum_{j=1}^{d} M(i, j)^2 \right)^{1/2}$$

we get

$$\|XX'\|^2 = \sum_{i=1}^{d} \sum_{j=1}^{d} (X_i X_j)^2 = \left(\sum_{i=1}^{d} X_i^2 \right) \left(\sum_{j=1}^{d} X_j^2 \right) = \|X\|^4$$

so that $\|XX'\| = \|X\|^2$. If every eigenvalue of E has real part $a_i < 1/4$, then $E(\|XX'\|^2) = E(\|X\|^4) < \infty$ and the multivariable central limit theorem (3.2) shows that

$$n^{1/2}(M_n - C) = n^{-1/2} \sum_{t=1}^{n} (X_t X_t' - C) \Rightarrow W, \tag{6.5}$$

where W is a Gaussian random matrix and C is the (uncentered) covariance matrix $C = E(XX')$. The estimates of Jansen and de Vries (1991) and Loretan and Phillips (1994) indicate tail estimates in the range $2 < \alpha < 4$. In this case, every eigenvalue of E has real part $1/4 < a_i < 1/2$. Then $E(\|\boldsymbol{X}\boldsymbol{X}'\|^2) = E(\|\boldsymbol{X}\|^4) = \infty$, but $E(\|\boldsymbol{X}\boldsymbol{X}'\|) = E(\|\boldsymbol{X}\|^2) < \infty$ so the covariance matrix $C = E(\boldsymbol{X}\boldsymbol{X}')$ exists. Now the generalized central limit theorem (3.11) gives

$$nA_n(M_n - C)A_n' = A_n\left(\sum_{t=1}^{n}(\boldsymbol{X}_t\boldsymbol{X}_t' - C)\right)A_n' \Rightarrow W, \tag{6.6}$$

where the limit W is a nonnormal operator stable random matrix. The estimates in Rachev and Mittnik (2000) give tail estimates in the range $1 < \alpha < 2$, so that every eigenvalue of E has real part $a_i > 1/2$. Then $E(\|\boldsymbol{X}\boldsymbol{X}'\|) = E(\|\boldsymbol{X}\|^2) = \infty$ and the covariance matrix $C = E(\boldsymbol{X}\boldsymbol{X}')$ diverges. In this case,

$$nA_nM_nA_n' \Rightarrow W \tag{6.7}$$

holds with W operator stable. Since the covariance matrix is undefined, there is no reason to believe that the sample covariance matrix contains useful information. However, we will see in Section 8 that even in this case the sample covariance matrix characterizes the most important distributional features of the random vector $\boldsymbol{X}$.

The centered sample covariance matrix is defined by

$$\Gamma_n = \frac{1}{n}\sum_{i=1}^{n}(\boldsymbol{X}_i - \overline{\boldsymbol{X}}_n)(\boldsymbol{X}_i - \overline{\boldsymbol{X}}_n)',$$

where $\overline{\boldsymbol{X}}_n = n^{-1}(\boldsymbol{X}_1 + \cdots + \boldsymbol{X}_n)$ is the sample mean. In the heavy tailed case $a_i > 1/4$, Theorem 10.6.15 in Meerschaert and Scheffler (2001a) shows that Γ_n and M_n have the same asymptotics, similar to the one dimensional case. In practice, it is common to mean-center the data, so it does not matter which form we choose.

7. Dependent random vectors

Suppose that $\boldsymbol{X}_t = (X_1(t), \ldots, X_d(t))'$ where $X_i(t)$ represents the price change (or log return) of the i-th asset on day t. A model where $\boldsymbol{X}_t$ are IID with $\boldsymbol{X} \in \mathrm{GDOA}(\boldsymbol{Y})$ allows dependence between the price changes $X_i(t)$ and $X_j(t)$ on the same day t, which is commonly observed in practice. If we also want to model dependence between days, we need to relax the IID assumption. A wide variety of time series models can be mathematically reduced to a linear moving average. This reduction may involve integer or fractional differencing, detrending and deseasoning, and nonlinear mappings. Asymptotics for the underlying moving average are established in Section 10.6 of Meerschaert and

Scheffler (2001a). Assume that $\boldsymbol{Z}, \boldsymbol{Z}_1, \boldsymbol{Z}_2, \boldsymbol{Z}_3, \ldots$ are IID random vectors on $\mathbb{R}^d$ whose probability distribution is regularly varying with exponent E, so that

$$nP(A_n\boldsymbol{Z} \in B) \to \phi(B) \quad \text{as } n \to \infty \tag{7.1}$$

for Borel subsets B of $\mathbb{R}^d \setminus \{0\}$ whose boundary have ϕ-measure zero, and $t\phi(\mathrm{d}\boldsymbol{x}) = \phi(t^{-E}\,\mathrm{d}\boldsymbol{x})$ for all $t > 0$. If every eigenvalue of E has real part $a_i > 1/2$ then $\boldsymbol{Z} \in \mathrm{GDOA}(\boldsymbol{Y})$ and

$$A_n(\boldsymbol{Z}_1 + \cdots + \boldsymbol{Z}_n - n\boldsymbol{b}_n) \Rightarrow \boldsymbol{Y}, \tag{7.2}$$

where $\boldsymbol{Y}$ is operator stable with exponent E and Lévy measure ϕ. Define the moving average process

$$\boldsymbol{X}_t = \sum_{j=0}^{\infty} C_j \boldsymbol{Z}_{t-j}, \tag{7.3}$$

where C_j are $d \times d$ real matrices. If every eigenvalue of E has real part $a_i < a_p$ then the moving average (7.3) is well defined as long as

$$\sum_{j=0}^{\infty} \|C_j\|^{\delta} < \infty \tag{7.4}$$

for some $\delta < 1/a_p$ with $\delta \leqslant 1$. If every eigenvalue of E has real part $a_i < 1/2$, then $E(\|\boldsymbol{X}_t\|^2)$ exists and the asymptotics are normal, see Brockwell and Davis (1991). If every eigenvalue of E has real part $a_i > 1/2$, and if for each j either $C_j = 0$, or else C_j^{-1} exists and $A_n C_j = C_j A_n$ for all n, then Theorem 10.6.2 in Meerschaert and Scheffler (2001a) shows that

$$A_n\left(\boldsymbol{X}_1 + \cdots + \boldsymbol{X}_n - n\sum_{j=0}^{\infty} C_j \boldsymbol{b}_n\right) \Rightarrow \sum_{j=0}^{\infty} C_j \boldsymbol{Y}. \tag{7.5}$$

The limit in (7.5) is operator stable with no normal component and Lévy measure $\sum_j C_j\phi$, where $C_j\phi = 0$ if $C_j = 0$ and otherwise $C_j\phi(\mathrm{d}\boldsymbol{x}) = \phi(C_j^{-1}\,\mathrm{d}\boldsymbol{x})$.

If every eigenvalue of E has real part $a_i < 1/2$, then both the mean $\boldsymbol{m} = E(\boldsymbol{X}_t)$ and the lag h covariance matrix

$$\Gamma(h) = E\left[(\boldsymbol{X}_t - \boldsymbol{m})(\boldsymbol{X}_{t+h} - \boldsymbol{m})'\right]$$

exist. The matrix $\Gamma(h)$ tells us when price changes on day t are correlated with price changes (of the same asset or some other asset) h days later. These correlations are useful to identify leading indicators, and they are the basic tools of time series modeling. The sample covariance matrix at lag $h \geqslant 0$ for the moving average $\boldsymbol{X}_t$ is defined by

$$\widehat{\Gamma}_n(h) = \frac{1}{n-h} \sum_{t=1}^{n-h} (\boldsymbol{X}_t - \overline{\boldsymbol{X}})(\boldsymbol{X}_{t+h} - \overline{\boldsymbol{X}})', \tag{7.6}$$

where $\overline{\boldsymbol{X}} = (\boldsymbol{X}_1 + \cdots + \boldsymbol{X}_n)/n$. If every eigenvalue of E has real part $a_i < 1/4$, then $E(\|\boldsymbol{X}_t\|^4) < \infty$ and $\widehat{\Gamma}_n(h)$ is asymptotically normal, see Brockwell and Davis (1991). If every eigenvalue of E has real part $1/4 < a_i < 1/2$, the estimates of Jansen and de Vries (1991) and Loretan and Phillips (1994), then

$$A_n \left(\sum_{t=1}^{n} (\boldsymbol{Z}_t \boldsymbol{Z}_t' - D) \right) A_n' \Rightarrow U \tag{7.7}$$

as in Section 6, where U is a nonnormal operator stable random matrix and $D = E(\boldsymbol{Z}\boldsymbol{Z}')$. Then Theorem 10.6.15 in Meerschaert and Scheffler (2001a) shows that

$$n A_n \big(\widehat{\Gamma}_n(h) - \Gamma(h) \big) A_n' \Rightarrow \sum_{j=0}^{\infty} C_j U C_{j+h}' \tag{7.8}$$

for any $h \geqslant 0$. The asymptotics (7.8) determine which elements of the sample covariance matrix $\widehat{\Gamma}_n(h)$ are statistically significantly different from zero.

If every eigenvalue of E has real part $a_i > 1/2$, as in the estimates of Rachev and Mittnik (2000), then

$$A_n \left(\sum_{t=1}^{n} \boldsymbol{Z}_t \boldsymbol{Z}_t' \right) A_n' \Rightarrow U \tag{7.9}$$

and Theorem 10.6.15 in Meerschaert and Scheffler (2001a) shows that

$$n A_n \widehat{\Gamma}_n(h) A_n' \Rightarrow \sum_{j=0}^{\infty} C_j U C_{j+h}' \tag{7.10}$$

for any $h \geqslant 0$. In this case the covariance matrix $\Gamma(h)$ does not exist, but the sample covariance matrix $\widehat{\Gamma}_n(h)$ still contains useful information about the time series $\boldsymbol{X}_t$ of price changes. In the next section, we will explain this apparent paradox.

8. Tail estimation

Given a set of price changes (or log-returns) $X_1, \ldots, X_n$ for some asset, it is important to estimate the tail behavior. If the price changes X_t are identically distributed[6] with X and $P(X > r) \sim Cr^{-\alpha}$, then the dispersion C and the tail index α determine the central limit behavior, as well as the extreme value behavior, of the price change distribution. Mandelbrot (1963) pioneered a graphical estimation method for C and α. If $y = P(X > r) \approx Cr^{-\alpha}$ then $\log y \approx \log C - \alpha \log r$. Ordering the data so that $X_{(1)} \geqslant X_{(2)} \geqslant \cdots \geqslant X_{(n)}$ we should have approximately that $r = X_{(i)}$ when $y = i/n$. Then a plot of $\log X_{(i)}$ versus $\log(i/n)$ should be approximately linear with slope $-\alpha$ and $\log C$ can be estimated from the vertical axis intercept. If $P(X > r) \approx Cr^{-\alpha}$ for r large, then the upper tail should be approximately linear. We call this a *Mandelbrot plot*. Several Mandelbrot plots for stock market and exchange rate returns appear in Loretan and Phillips (1994) as evidence of heavy tails with $2.5 < \alpha < 3$. Replacing X by $-X$ gives information about the left tail. Least squares estimators for α based on the Mandelbrot plot were proposed by Schultze and Steinebach (1996), see also Csörgő and Viharos (1997).

The most popular numerical estimator for C and α is due to Hill (1975), see also Hall (1982). Sort the data in decreasing order to obtain the *order statistics* $X_{(1)} \geqslant X_{(2)} \geqslant \cdots \geqslant X_{(n)}$. Assuming that $P(X > r) = Cr^{-\alpha}$ for large values of $r > 0$, the maximum likelihood estimates for α and C based on the $m + 1$ largest observations are

$$\hat{\alpha} = \left[\frac{1}{m} \sum_{i=1}^{m} \left(\ln X_{(i)} - \ln X_{(m+1)} \right) \right]^{-1}, \qquad \widehat{C} = \frac{m}{n} X_{(m+1)}^{\hat{\alpha}}, \tag{8.1}$$

where m is to be taken as large as possible, but small enough so that the tail condition $P(X > r) = Cr^{-\alpha}$ remains valid. Replacing X by $-X$ gives estimates for the left tail. Replacing X by $|X|$ gives estimates for the combined tail. This is often advantageous, because it allows us to combine the data from both tails, and increase the number m of order statistics used. Finding the best value of m is a challenge, and creates a certain amount of controversy. Jansen and de Vries (1991) use Hill's estimator with a fixed value of $m = 100$ for several different assets. Loretan and P hillips (1994) tabulate several different values of m for each asset. Hill's estimator $\hat{\alpha}$ is consistent and asymptotically normal with variance α^2/m, so confidence intervals are easy to construct. These intervals clearly demonstrate that the tail parameters in Jansen and de Vries (1991) and Loretan and Phillips (1994) vary depending on the asset.

Aban and Meerschaert (2001) develop a more general Hill's estimator to account for a possible shift in the data. If $P(X > r) = C(r - s)^{-\alpha}$ for r large, the maximum likelihood

[6] Note that we are not assuming IID here.

estimates for α and C based on the $m+1$ largest observations are

$$\hat{\alpha} = \left[\frac{1}{m}\sum_{i=1}^{m}\left(\ln(X_{(i)} - \hat{s}) - \ln(X_{(m+1)} - \hat{s})\right)\right]^{-1},$$
$$\widehat{C} = \frac{m}{n}(X_{(m+1)} - \hat{s})^{\hat{\alpha}}, \tag{8.2}$$

where $\hat{s}$ is obtained by numerically solving the equation

$$\hat{\alpha}(X_{(m+1)} - \hat{s})^{-1} = (\hat{\alpha} + 1)\frac{1}{m}\sum_{i=1}^{m}(X_{(i)} - \hat{s})^{-1} \tag{8.3}$$

over $\hat{s} < X_{(m+1)}$. Once the optimal shift is computed, $\hat{\alpha}$ and $\widehat{C}$ come from Hill's estimator applied to the shifted data. One practical implication is that, since the Pareto model is not shift-invariant, it is a good idea to try shifting the data to get a linear Mandelbrot plot.

If X_t is the sum of many IID price shocks, then it can be argued that the distribution of X_t must be (at least approximately) stable with distribution $S_\alpha(\sigma, \beta, b)$. Maximum likelihood estimation for the stable parameters is now practical, using the efficient method of Nolan (1997) for computing stable densities, see also Mittnik et al. (1999), Mittnik, Doganoglu and Chenyao (1999). Since the stable index $0 < \alpha \leqslant 2$, the stable MLE for α cannot possibly agree with the estimates found in Jansen and de Vries (1991) and Loretan and Phillips (1994). Rachev and Mittnik (2000) use a stable model for price changes, and their estimates yield $1 < \alpha < 2$ for a variety of assets. McCulloch (1997) argues that the $\alpha > 2$ estimates found in Jansen and de Vries (1991) and Loretan and Phillips (1994) are inflated due to a distributional misspecification. The Pareto tail of a stable random variable X disappears as $\alpha \to 2$, so that it may be impossible to take m large enough for a reliable estimate, see Fofack and Nolan (1999) for a more detailed discussion. The estimator in Aban and Meerschaert (2001) corrects for the fact that Hill's $\hat{\alpha}$ is not shift-invariant, and may go some distance towards correcting the problem identified by McCulloch (1997).

Maximum likelihood estimation is quite sensitive to deviations from the proscribed distribution, and it is no surprise that the MLE computations of Jansen and de Vries (1991) and Loretan and Phillips (1994), based on the Pareto model, differ significantly from the estimates of Rachev and Mittnik (2000), based on a stable model. Part of the controversy stems from the fact that the range of α is limited to $(0, 2]$ for the stable model. Akgiray and Booth (1988) interpret the results of Hill's estimator for stock returns as evidence against the stable model. Actual finance data does not exactly fit either the stable or Pareto-tail models, and in our opinion, parameter estimates are only valid with respect to the model used to obtain them, so that Pareto-based estimates of $\alpha > 2$ in no way invalidate the stable model.

Meerschaert and Scheffler (1998) propose a robust estimator

$$\hat{\alpha} = \frac{2\ln n}{\ln n + \ln \hat{\sigma}^2} \tag{8.4}$$

based on the sample variance (6.2). This estimator can be applied whenever $X \in \mathrm{DOA}(Y)$ and Y is stable with index $0 < \alpha < 2$. Then X can be stable or Pareto, or any distribution with balanced power-law tails. The estimator is also applicable to dependent data, since it also applies when $X_t = \sum_j c_j Z_{t-j}$, Z_t is IID with $Z \in \mathrm{DOA}(Y)$, and Y is stable with index $0 < \alpha < 2$. The estimator is based on the simple idea that

$$n^{1-2/\alpha}\hat{\sigma}^2 \Rightarrow Y,$$

$$\ln(n\hat{\sigma}^2) - \frac{2}{\alpha}\ln n \Rightarrow \ln Y,$$

$$2\ln n\left(\frac{\ln(n\hat{\sigma}^2)}{2\ln n} - \frac{1}{\alpha}\right) \Rightarrow \ln Y$$

so that $\ln(n\hat{\sigma}^2)/(2\ln n)$ estimates $1/\alpha$. If X has heavy tails with $\alpha \geqslant 2$ then $\hat{\alpha} \to 2$. In this case, we can apply the estimator to X^k, which also has heavy tails with tail parameter α/k. It is interesting, and even somewhat ironic, that the sample variance can be used to estimate tail behavior, and hence tells us something about the spread of typical values, even in this case $0 < \alpha < 2$ where the variance is undefined.

Portfolio modeling requires a vector model to incorporate dependence between price changes for different assets. In these vector models, the sample variance is replaced by the sample covariance matrix. For heavy tailed price changes with infinite variance, the covariance matrix does not exist. Even so, we will see that the sample covariance matrix is a very useful tool for portfolio modeling. Suppose that $\boldsymbol{X}_t = (X_1(t), \ldots, X_d(t))'$ where $X_i(t)$ is the price change of the i-th asset on day t. If $\boldsymbol{X}_t$ are identically distributed with $\boldsymbol{X}$ and if $\boldsymbol{X}$ has heavy tails with $P(\|\boldsymbol{X}\| > r) \sim Cr^{-\alpha}$ then the vector norms $\|\boldsymbol{X}_1\|, \ldots, \|\boldsymbol{X}_n\|$ can be used to estimate the tail parameter α. Alternatively, we can apply one variable tail estimators to the i-th marginal to get an estimate $\hat{\alpha}_i$ of the tail parameter. If the probability tails of $\boldsymbol{X}$ fall off at the same rate $r^{-\alpha}$ in every radial direction, then these estimates should all be reasonably close. In that case, we might assume that $\boldsymbol{X}$ is multivariable stable with distribution $S_\alpha(\sigma, M, \boldsymbol{b})$. The mean $\boldsymbol{b}$ can be estimated using the sample mean in the usual case $1 < \alpha < 2$. Several estimators now exist for the scale σ and the mixing measure M, or equivalently, for the spectral measure $\lambda(\mathrm{d}\theta) = \sigma^\alpha M(\mathrm{d}\theta)$. Those estimators are surveyed in another chapter in this Handbook (Kozubowski, Panorska and Rachev, 2003), so we will not dwell on them here. If $\alpha > 2$, one might consider the multivariable Pareto laws introduced in Example 3.1. If $P(\|\boldsymbol{X}\| > r) \sim Cr^{-\alpha}$ and the balanced tails condition (3.13) holds for some mixing measure M, then the tail behavior of $\boldsymbol{X}$ is multivariable Pareto. Multivariable stable random vectors have this property with $0 < \alpha < 2$. If $\alpha > 2$ then multivariable Pareto could offer a reasonable alternative, which to our knowledge has not been pursued in the finance literature.

While experts disagree on the range of α for typical assets, there seems to be general agreement that the tail index depends on the asset. Then it is appropriate to assume that the probability distribution of $\boldsymbol{X}$ varies regularly with some exponent E. For IID random vectors, a method for estimating the exponent E can be found in Section 10.4 of Meerschaert and Scheffler (2001a). In Section 9 we show that the same methods also apply to dependent random vectors which are identically distributed. The method is applicable when the eigenvalues of E all have real part $a_i > 1/2$, the infinite variance case. To be concrete, we adopt the model of Example 5.5, which is the simplest model flexible enough for realism. This model assumes that E has a set of d mutually orthogonal unit eigenvectors. Note that if the eigenvalues of E are all distinct then these unit eigenvectors are unique up to a factor of ± 1. On the other hand, if $E = aI$ for some $a > 1/2$ then any set of d mutually orthogonal unit vectors can be used.

Recall the spectral decomposition $E = PBP^{-1}$ from Example 5.5, where P is orthogonal and B is given by (5.1), with $d_i \times d_i$ blocks $B_i = a_i I$ for some $1/2 \leqslant a_1 < \cdots < a_p$. Let $D_0 = 0$ and $D_i = d_1 + \cdots + d_i$ for $1 \leqslant i \leqslant p$. Then $\boldsymbol{q}_j = P\boldsymbol{e}_j$ is a unit eigenvector of the matrix E and the d_i dimensional subspace $V_i = \operatorname{span}\{\boldsymbol{q}_j\colon D_{i-1} < j \leqslant D_i\}$ contains every eigenvector of E with associated eigenvalue a_i. Our estimator for E is based on the sample covariance matrix M_n defined in (6.3). Since M_n is symmetric and nonnegative definite, there exists an orthonormal basis of eigenvectors for M_n with nonnegative eigenvalues. Eigenvalues and eigenvectors of M_n are easily computed using standard numerical routines, see for example Press et al. (1987). Sort the eigenvalues

$$\lambda_1 \leqslant \cdots \leqslant \lambda_d$$

and the associated unit eigenvectors

$$\theta_1, \ldots, \theta_d$$

so that $M_n\theta_j = \lambda_j\theta_j$ for each $j = 1, \ldots, d$. Now Theorem 10.4.5 in Meerschaert and Scheffler (2001a) shows that

$$\frac{\log n + \log \lambda_j}{2\log n} \to a_i \quad \text{as } n \to \infty$$

in probability for any $D_{i-1} < j \leqslant D_i$. This is a multivariable analogue for the one variable tail estimator (8.4). Furthermore, Theorem 10.4.8 in Meerschaert and Scheffler (2001a) shows that the eigenvectors θ_j converge in probability to V_1 when $j \leqslant D_1$, and to V_p when $j > D_{p-1}$. This shows that the eigenvectors estimate the coordinate vectors in the spectral decomposition, at least for the lightest and heaviest tails.

Now we illustrate the practical application of the multivariable tail estimator. Recall that $\boldsymbol{X}_t = (X_1(t), \ldots, X_d(t))'$ where $X_i(t)$ is the price change of the i-th asset on day t. Compute the (uncentered) sample covariance matrix M_n using the formula (6.3) and then compute the eigenvalues $\lambda_1 \leqslant \cdots \leqslant \lambda_d$ and the associated eigenvectors

$$\begin{aligned} \theta_1 &= \big(\theta_1(1), \ldots, \theta_d(1)\big)' \\ &\vdots \\ \theta_d &= \big(\theta_1(d), \ldots, \theta_d(d)\big)' \end{aligned} \tag{8.5}$$

of the matrix M_n. A change of coordinates is essential to the method. Write

$$Z_j(t) = X_t \cdot \theta_j = X_1(t)\theta_1(j) + \cdots + X_d(t)\theta_d(j)$$

for each $j = 1, \ldots, d$. Our portfolio model is based on these new coordinates. Let

$$\hat{\alpha}_j = \frac{2 \log n}{\log n + \log \lambda_j}$$

for each $j = 1, \ldots, n$. Since the eigenvalues are sorted in increasing order we will have $\hat{\alpha}_1 \geqslant \cdots \geqslant \hat{\alpha}_d$. Our model assumes that $Z_j(t)$ are identically distributed with Z_j, and the tail parameter $\hat{\alpha}_j$ governs the j-th coordinate Z_j. If $\hat{\alpha}_j < 2$ then $P(|Z_j| > r)$ falls off like $r^{-\hat{\alpha}_j}$ and if $\hat{\alpha}_j \geqslant 2$ then a finite variance model for Z_j is adequate. We can also use any other one variable tail estimator to get α_j for each of the new coordinates $Z_j(t)$. The new coordinates unmask variations in α that would go undetected in the original coordinates.

Example 8.1. We look at a data set of $n = 2853$ daily exchange rate log-returns $X_1(t)$ for the German Deutsch Mark and $X_2(t)$ for the Japanese Yen, both taken against the US Dollar. We divide each entry by 0.004 which is the approximate median for both $|X_1(t)|$ and $|X_2(t)|$. This has no effect on the eigenvectors but helps to obtain good estimates of the tail thickness. Then we compute

$$M_n = \frac{1}{n} \sum_{t=1}^{n} \begin{pmatrix} X_1(t)^2 & X_1(t)X_2(t) \\ X_1(t)X_2(t) & X_2(t)^2 \end{pmatrix} = \begin{pmatrix} 3.204 & 2.100 \\ 2.100 & 3.011 \end{pmatrix}$$

which has eigenvalues $\lambda_1 = 1.006$, $\lambda_2 = 5.209$ and associated unit eigenvectors $\theta_1 = (0.69, -0.72)'$, $\theta_2 = (0.72, 0.69)'$. Next we compute

$$\begin{aligned} \hat{\alpha}_1 &= \frac{2 \ln 2853}{\ln 2853 + \ln 1.006} = 1.998, \\ \hat{\alpha}_2 &= \frac{2 \ln 2853}{\ln 2853 + \ln 5.209} = 1.656 \end{aligned} \tag{8.6}$$

indicating that $Z_1(t) = 0.69X_1(t) - 0.72X_2(t)$ fits a finite variance model but $Z_2(t) = 0.72X_1(t) + 0.69X_2(t)$ fits a heavy tailed model with $\alpha = 1.656$. Then we can model

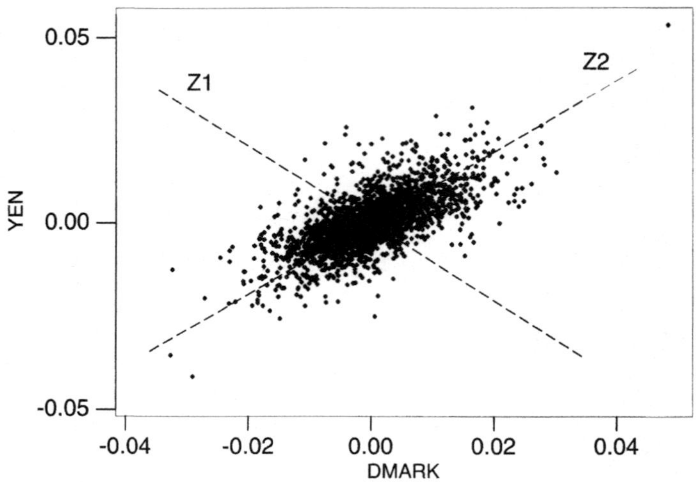

Fig. 3. Exchange rates against the US dollar. The new coordinates uncover variations in the tail parameter α.

$\boldsymbol{Z}_t = (Z_1(t), Z_2(t))'$ as being identically distributed with the random vector $\boldsymbol{Z} = (Z_1, Z_2)'$ where $P(|Z_2| > r) \approx C_1 r^{-1.656}$ and $\mathrm{Var}(Z_1) < \infty$. The simplest model with these properties is to take $Z_1(t)$ normal and $Z_2(t)$ stable with index $\alpha = 1.656$ and independent of $Z_1(t)$.

Next we explain the operator stable model based on these estimates. The random vectors $\boldsymbol{Z}_t$ are operator stable with exponent

$$B = \begin{pmatrix} 0.50 & 0 \\ 0 & 0.60 \end{pmatrix}$$

since $0.50 = 1/1.998$ and $0.60 = 1/1.656$. The change of coordinates matrix

$$P = \begin{pmatrix} 0.69 & -0.72 \\ 0.72 & 0.69 \end{pmatrix}$$

so that

$$\boldsymbol{Z}_t = \begin{pmatrix} Z_1(t) \\ Z_2(t) \end{pmatrix} = \begin{pmatrix} 0.69 & -0.72 \\ 0.72 & 0.69 \end{pmatrix} \begin{pmatrix} X_1(t) \\ X_2(t) \end{pmatrix} = P\boldsymbol{X}_t.$$

Since

$$P^{-1} = \begin{pmatrix} 0.69 & 0.72 \\ -0.72 & 0.69 \end{pmatrix}$$

(rounded off to two decimal places) we also have

$$\boldsymbol{X}_t = P^{-1}\boldsymbol{Z}_t = \begin{pmatrix} X_1(t) \\ X_2(t) \end{pmatrix} = \begin{pmatrix} 0.69 & 0.72 \\ -0.72 & 0.69 \end{pmatrix} \begin{pmatrix} Z_1(t) \\ Z_2(t) \end{pmatrix}$$

so that

$$\begin{aligned} X_1(t) &= 0.69 Z_1(t) + 0.72 Z_2(t), \\ X_2(t) &= -0.72 Z_1(t) + 0.69 Z_2(t). \end{aligned} \tag{8.7}$$

Both exchange rates have a common heavy-tailed stable factor $Z_2(t)$ and so both exchange rates have heavy tails with the same tail index $\alpha = 1.656$. It is tempting to interpret $Z_2(t)$ as the common influence of fluctuations in the US dollar, and the remaining light-tailed factor $Z_1(t)$ as the accumulation of other price shocks independent of the US dollar.

We also take the opportunity to fill in the details of Example 5.4 in this simple case. The original data $\boldsymbol{X}_t = P^{-1}\boldsymbol{Z}_t$ is modeled as operator stable with exponent

$$E = PBP^{-1} = \begin{pmatrix} 0.55 & 0.05 \\ 0.05 & 0.55 \end{pmatrix}.$$

In this case, $Z_1(t)$ and $Z_2(t)$ are independent so the density of $\boldsymbol{Z}_t$ is the product of the two marginal densities, and then the density of $\boldsymbol{X}_t$ can be obtained by a simple change of variables. The columns of the change of variables matrix P are the eigenvectors θ_j of the sample covariance matrix, which estimate the theoretical coordinate system vectors $\boldsymbol{p}_j$ in the spectral decomposition.

Remark 8.2. This exchange rate data in Example 8.1 was also analyzed by Nolan, Panorska and McCulloch (2001) using a multivariable stable model. Since both marginals $X_1(t)$ and $X_2(t)$ have heavy tails with the same α, there is no obvious reason to employ a more complicated model. However, the change of coordinates in Example 8.1 uncovers variations in the tail parameter α, an important modeling insight.

Remark 8.3. Kotz, Kozubowski and Podgórski (2001) employ a very different model for the data in Example 8.1, based on the Laplace distribution. This distribution, and its multivariable analogues, assume exponential probability tails for the data. These models have heavier tails than the Gaussian, but they have moments of all orders.

Remark 8.4. The simplistic model in Example 8.1 assumes that the two factors Z_1 and Z_2 are independent. If we assume that $\boldsymbol{Z}$ is operator stable with Z_1 normal and Z_2 stable then these components must be independent, in view of the general characteristic function formula for operator stable laws. Another alternative is to assume that Z_1 is stable with index $\alpha = 1.998$, very close to a normal distribution. In this case, the two components can be dependent. The dependence is captured by the mixing measure or spectral measure, see Example 4.1. Scheffler (1999) provides a method for estimating the spectral measure from data for an operator stable random vector with a known exponent. This provides a more flexible model including dependence between the two factors.

9. Tail estimator proof for dependent random vectors

In this section, we provide a proof that the multivariable tail estimator of Section 8 is still valid for certain sequences of dependent heavy tailed random vectors. We say that a sequence (B_n) of invertible linear operators is regularly varying with index $-E$ if for any $\lambda > 0$ we have

$$B_{[\lambda n]} B_n^{-1} \to \lambda^{-E} \quad \text{as } n \to \infty.$$

For further information about regular variation of linear operators see Meerschaert and Scheffler (2001a, Chapter 4).

In view of Theorem 2.1.14 of Meerschaert and Scheffler (2001a) we can write $\mathbb{R}^d = V_1 \oplus \cdots \oplus V_p$ and $E = E_1 \oplus \cdots \oplus E_p$ for some $1 \leqslant p \leqslant d$ where each V_i is E invariant, $E_i : V_i \to V_i$ and $\mathrm{Re}(\lambda) = a_i$ for all real parts of the eigenvalues of E_i and some $a_1 < \cdots < a_p$. By Definition 2.1.15 of Meerschaert and Scheffler (2001a) this is called the spectral decomposition of $\mathbb{R}^d$ with respect to E. By Definition 4.3.13 of Meerschaert and Scheffler (2001a) we say that (B_n) is spectrally compatible with $-E$ if every V_i is B_n-invariant for all n. Note that in this case we can write $B_n = B_{1n} \oplus \cdots \oplus B_{pn}$ and each $B_{in} : V_i \to V_i$ is regularly varying with index $-E_i$. [See Proposition 4.3.14 of Meerschaert and Scheffler (2001a).] For the proofs in this section we will always assume that the subspaces V_i in the spectral decomposition of $\mathbb{R}^d$ with respect to E are mutually orthogonal. We will also assume that (B_n) is spectrally compatible with $-E$. Let π_i denote the orthogonal projection operator onto V_i. If we let $P_i = \pi_i + \cdots + \pi_p$ and $L_i = V_i \oplus \cdots \oplus V_p$ then $P_i : \mathbb{R}^d \to L_i$ is a orthogonal projection. Furthermore, $\overline{P_i} = \pi_1 + \cdots + \pi_i$ is the orthogonal projection onto $\bar{L}_i = V_1 \oplus \cdots \oplus V_i$.

Now assume $0 < a_1 < \cdots < a_p$. Since (B_n) is spectrally compatible with $-E$, Proposition 4.3.14 of Meerschaert and Scheffler (2001a) shows that the conclusions of Theorem 4.3.1 of Meerschaert and Scheffler (2001a) hold with $L_i = V_i \oplus \cdots \oplus V_p$ for each $i = 1, \ldots, p$. Then for any $\varepsilon > 0$ and any $x \in L_i \setminus L_{i+1}$ we have

$$n^{-a_i - \varepsilon} \leqslant \|B_n x\| \leqslant n^{-a_i + \varepsilon} \tag{9.1}$$

for all large n. Then

$$\frac{\log \|B_n x\|}{\log n} \to -a_i \quad \text{as } n \to \infty \tag{9.2}$$

and since this convergence is uniform on compact subsets of $L_i \setminus L_{i+1}$ we also have

$$\frac{\log \|\pi_i B_n\|}{\log n} \to -a_i \quad \text{as } n \to \infty. \tag{9.3}$$

It follows that

$$\frac{\log \|B_n\|}{\log n} \to -a_1 \quad \text{as } n \to \infty. \tag{9.4}$$

Since $(B_n')^{-1}$ is regularly varying with index E', a similar argument shows that for any $x \in \bar{L}_i \setminus \bar{L}_{i-1}$ we have

$$n^{a_i - \varepsilon} \leqslant \left\|(B_n')^{-1} x\right\| \leqslant n^{a_i + \varepsilon} \tag{9.5}$$

for all large n. Then

$$\frac{\log \|(B_n')^{-1} x\|}{\log n} \to a_i \quad \text{as } n \to \infty \tag{9.6}$$

and since this convergence is uniform on compact subsets of $\bar{L}_i \setminus \bar{L}_{i-1}$ we also have

$$\frac{\log \|\pi_i (B_n')^{-1}\|}{\log n} \to a_i \quad \text{as } n \to \infty. \tag{9.7}$$

Hence

$$\frac{\log \|(B_n')^{-1}\|}{\log n} \to a_p \quad \text{as } n \to \infty. \tag{9.8}$$

Suppose that $\boldsymbol{X}_t$, $t = 1, 2, \ldots,$ are $\mathbb{R}^d$-valued random vectors and let M_n be the sample covariance matrix of $(\boldsymbol{X}_t)$ defined by (6.3). Note that M_n is symmetric and positive semi-definite. Let $0 \leqslant \lambda_{1n} \leqslant \cdots \leqslant \lambda_{dn}$ denote the eigenvalues of M_n and let $\theta_{1n}, \ldots, \theta_{dn}$ be the corresponding orthonormal basis of eigenvectors.

Basic Assumptions. Assume that for some exponent E with real spectrum $1/2 < a_1 < \cdots < a_p$ the subspaces V_i in the spectral decomposition of $\mathbb{R}^d$ with respect to E are mutually orthogonal, and there exists a sequence (B_n) regularly varying with index $-E$ and spectrally compatible with $-E$ such that:

(A1) The set $\{n(B_n M_n B_n')\colon n \geqslant 1\}$ is weakly relatively compact.

(A2) For any limit point M of this set we have:
- (a) M is almost surely positive definite.
- (b) For all unit vectors θ the random variable $\theta' M \theta$ has no atom at zero.

Now let $\mathbb{R}^d = V_1 \oplus \cdots \oplus V_p$ be the spectral decomposition of $\mathbb{R}^d$ with respect to E. Put $d_i = \dim V_i$ and for $i = 1, \ldots, p$ let $b_i = d_i + \cdots + d_p$ and $\bar{b}_i = d_1 + \cdots + d_i$. Our goal is now to estimate the real spectrum $a_1 < \cdots < a_p$ of E as well as the spectral decomposition $V_1, \ldots, V_p$. In various situation, these quantities completely describe the moment behavior of the X_t.

Theorem 9.1. *Under our basic assumptions, for $i = 1, \ldots, p$ and $\bar{b}_{i-1} < j \leqslant \bar{b}_i$ we have*

$$\frac{\log(n\lambda_{jn})}{2\log n} \to a_i \quad \textit{in probability as } n \to \infty.$$

The proof of Theorem 9.1 is in parts quite similar to the Theorem 2 in Meerschaert and Scheffler (1999b). See also Section 10.4 in Meerschaert and Scheffler (2001a), and Scheffler (1998). We include it here for sake of completeness.

Proposition 9.2. *Under our basic assumptions we have*

$$\frac{\log(n\lambda_{dn})}{2\log n} \to a_p \quad \textit{in probability.}$$

Proof: For $\delta > 0$ arbitrary we have

$$P\left\{\left|\frac{\log(n\lambda_{dn})}{2\log n} - a_p\right| > \delta\right\} \leqslant P\left\{\lambda_{dn} > n^{2(a_p+\delta)-1}\right\} + P\left\{\lambda_{dn} < n^{2(a_p-\delta)-1}\right\}.$$

Now choose $0 < \varepsilon < \delta$ and note that by (9.8) we have $\|(B_n')^{-1}\| \leqslant n^{a_p+\varepsilon}$ for all large n. Using assumption (A1) we obtain for all large n

$$\begin{aligned} P\left\{\lambda_{dn} > n^{2(a_p+\delta)-1}\right\} &= P\left\{\|M_n\| > n^{2(a_p+\delta)-1}\right\} \\ &\leqslant P\left\{\left\|(B_n')^{-1}\right\|^2 \|nB_nM_nB_n'\| > n^{2(a_p+\delta)}\right\} \\ &\leqslant P\left\{\|nB_nM_nB_n'\| > n^{2(\delta-\varepsilon)}\right\} \end{aligned}$$

and the last probability tends to zero as $n \to \infty$.

Now fix any $\theta_0 \in \bar{L}_p \setminus \bar{L}_{p-1}$ and write $(B_n')^{-1}\theta_0 = r_n\theta_n$ for some unit vector θ_n and $r_n > 0$. Theorem 4.3.14 of Meerschaert and Scheffler (2001a) shows that every limit point of (θ_n) lies in the unit sphere in V_p. Then since (9.5) holds uniformly on compact sets we have for any $0 < \varepsilon < \delta$ that $n^{a_p-\varepsilon} \leqslant r_n \leqslant n^{a_p+\varepsilon}$ for all large n. Then for all large n we get

$$\begin{aligned} P\left\{\lambda_{dn} < n^{2(a_p-\delta)-1}\right\} &= P\left\{\max_{\|\theta\|=1} M_n\theta \cdot \theta < n^{2(a_p-\delta)-1}\right\} \\ &\leqslant P\left\{M_n\theta_0 \cdot \theta_0 < n^{2(a_p-\delta)-1}\right\} \end{aligned}$$

$$= P\{nB_nM_nB_n'\theta_n \cdot \theta_n < r_n^{-2}n^{2(a_p-\delta)-1}\}$$
$$\leqslant P\{nB_nM_nB_n'\theta_n \cdot \theta_n < n^{2(\varepsilon-\delta)}\}.$$

Given any subsequence (n') there exists a further subsequence $(n'') \subset (n')$ along which $\theta_n \to \theta$. Furthermore, by assumption (A1) there exists another subsequence $(n''') \subset (n'')$ such that $nB_nM_nB_n' \Rightarrow M$ along (n'''). Hence by continuous mapping [see Theorem 1.2.8 in Meerschaert and Scheffler (2001a)] we have

$$nB_nM_nB_n'\theta_n \cdot \theta_n \Rightarrow M\theta \cdot \theta \quad \text{along } (n''').$$

Now, given any $\varepsilon_1 > 0$ by assumption (A2)(b) there exists a $\rho > 0$ such that $P\{M\theta \cdot \theta < \rho\} < \varepsilon_1/2$. Hence for all large $n = n'''$ we have

$$\begin{aligned} P\{nB_nM_nB_n'\theta_n \cdot \theta_n < n^{2(\varepsilon-\delta)}\} &\leqslant P\{nB_nM_nB_n'\theta_n \cdot \theta_n < \rho\} \\ &\leqslant P\{M\theta \cdot \theta < \rho\} + \frac{\varepsilon_1}{2} \\ &< \varepsilon_1. \end{aligned}$$

Since for any subsequence there exists a further subsequence along which

$$P\{nB_nM_nB_n'\theta_n\theta_n < n^{2(\varepsilon-\delta)}\} \to 0,$$

this convergence holds along the entire sequence which concludes the proof. □

Proposition 9.3. *Under the basic assumptions we have*

$$\frac{\log(n\lambda_{1n})}{2\log n} \to a_1 \quad \textit{in probability.}$$

Proof: Since the set $\mathrm{GL}(\mathbb{R}^d)$ of invertible matrices is an open subset of the vector space of $d \times d$ real matrices, it follows from (A1) and (A2)(a) together with the Portmanteau Theorem [cf., Theorem 1.2.2 in Meerschaert and Scheffler (2001a)] that $\lim_{n\to\infty} P\{M_n \in \mathrm{GL}(\mathbb{R}^d)\} = 1$ holds. Hence we can assume without loss of generality that M_n is invertible for all large n.

Given any $\delta > 0$ write

$$P\left\{\left|\frac{\log(n\lambda_{1n})}{2\log n} - a_1\right| > \delta\right\} \leqslant P\{\lambda_{1n} > n^{2(a_1+\delta)-1}\} + P\{\lambda_{1n} < n^{2(a_1-\delta)-1}\}.$$

To estimate the first probability on the right-hand side of the inequality above choose a unit vector $\theta_0 \in \bar{L}_1$ and write $(B_n')^{-1}\theta_0 = r_n\theta_n$ as above. Then, since (9.5) holds uniformly on the unit sphere in $\bar{L}_1 = V_1$, for $0 < \varepsilon < \delta$ we have $n^{a_1-\varepsilon} \leqslant r_n \leqslant n^{a_1+\varepsilon}$ for all large n. Therefore for all large n

$$\begin{aligned} P\{\lambda_{1n} > n^{2(a_1+\delta)-1}\} &\leqslant P\Big\{\min_{\|\theta\|=1} M_n\theta\cdot\theta > n^{2(a_1+\delta)-1}\Big\} \\ &\leqslant P\{M_n\theta_0\cdot\theta_0 > n^{2(a_1+\delta)-1}\} \\ &\leqslant P\{nB_nM_nB_n'\theta_n\cdot\theta_n > n^{2(\delta-\varepsilon)}\}. \end{aligned}$$

It follows from assumption (A1) together with the compactness of the unit sphere in $\mathbb{R}^d$ and continuous mapping that the sequence $(nB_nM_nB_n'\theta_n\cdot\theta_n)$ is weakly relatively compact and hence by Prohorov's Theorem this sequence is uniformly tight. Since $\delta > \varepsilon$ it follows that $P\{\lambda_{1n} > n^{2(a_1+\delta)-1}\} \to 0$ as $n \to \infty$.

Since the smallest eigenvalue of M_n is the reciprocal of the largest eigenvalue of M_n^{-1} we have

$$\begin{aligned} P\{\lambda_{1n} < n^{2(a_1-\delta)-1}\} &= P\Big\{\frac{1}{\lambda_{1n}} > n^{2(\delta-a_1)+1}\Big\} \\ &= P\Big\{\max_{\|\theta\|=1} M_n^{-1}\theta\cdot\theta > n^{2(\delta-a_1)+1}\Big\} \\ &= P\{\|M_n^{-1}\| > n^{2(\delta-a_1)+1}\} \\ &\leqslant P\Big\{\Big\|\frac{1}{n}(B_n')^{-1}M_n^{-1}B_n^{-1}\Big\| > \|B_n\|^{-2}n^{2(\delta-a_1)}\Big\}. \end{aligned}$$

It follows from (9.4) that for any $0 < \varepsilon < \delta$ there exists a constant $C > 0$ such that $\|B_n\| \leqslant Cn^{-a_1+\varepsilon}$ for all n and hence for some constant $K > 0$ we get $\|B_n\|^{-2} \geqslant Kn^{2(a_1-\varepsilon)}$ for all n. Note that by assumptions (A1) and (A2)(a) together with continuous mapping the sequence

$$\Big(\frac{1}{n}(B_n')^{-1}M_n^{-1}B_n^{-1}\Big)$$

is weakly relatively compact and hence by Prohorov's theorem this sequence is uniformly tight. Hence

$$\begin{aligned} &P\Big\{\Big\|\frac{1}{n}(B_n')^{-1}M_n^{-1}B_n^{-1}\Big\| > \|B_n\|^{-2}n^{2(\delta-a_1)}\Big\} \\ &\quad\leqslant P\Big\{\Big\|\frac{1}{n}(B_n')^{-1}M_n^{-1}B_n^{-1}\Big\| > Kn^{2(\delta-\varepsilon)}\Big\} \to 0 \end{aligned}$$

as $n \to \infty$. This concludes the proof. □

Proof of Theorem 9.1: Let $\mathcal{C}_j$ denote the collection of all orthogonal projections onto subspaces of $\mathbb{R}^d$ with dimension j. The Courant–Fischer Max–Min Theorem [see, Rao (1965, p. 51)] implies that

$$\begin{aligned}\lambda_{jn} &= \min_{P\in\mathcal{C}_j}\max_{\|\theta\|=1} PM_nP\theta\cdot\theta \\ &= \max_{P\in\mathcal{C}_{d-j+1}}\min_{\|\theta\|=1} PM_nP\theta\cdot\theta. \qquad (9.9)\end{aligned}$$

Note that $P_i^2 = P_i$ and that B_n and P_i commute for all n, i. Furthermore $(P_i B_n)$ is regularly varying with index $E_i \oplus \cdots \oplus E_p$. Since

$$n(P_iB_n)P_iM_nP_i(B_nP_i)' = nP_i(B_nM_nB_n')P_i$$

it follows by projection from our basic assumptions that the sample covariance matrix formed from the L_i valued random variables $P_i X_t$ satisfies again those basic assumptions with $E = E_i \oplus \cdots \oplus E_p$ on L_i. Hence if λ_n denotes the smallest eigenvalue of the matrix $P_iM_nP_i$ it follows from Proposition 9.3 that

$$\frac{\log(n\lambda_n)}{2\log n} \to a_i \quad \text{in probability.}$$

Similarly, the sample covariance matrix formed in terms of the $\bar{L}_i$-valued random vectors $\bar{L}_i X_t$ again satisfies the basic assumptions with $E = E_1 \oplus \cdots \oplus E_i$ as above. Then, if $\bar{\lambda}_n$ denotes the largest eigenvalue of the matrix $\overline{P_i}M_n\overline{P_i}$ it follows from Proposition 9.2 above that

$$\frac{\log(n\bar{\lambda}_n)}{2\log n} \to a_i \quad \text{in probability.}$$

Now apply (9.9) to see that

$$\lambda_n \leqslant \lambda_{jn} \leqslant \bar{\lambda}_n$$

whenever $\bar{b}_{i-1} < j \leqslant \bar{b}_i$. The result now follows easily. □

After dealing with the asymptotics of the eigenvalues of the sample covariance in Theorem 9.1 above we now investigate the convergence of the unit eigenvectors of M_n. Recall that $\pi_i : \mathbb{R}^d \to V_i$ denotes the orthogonal projection onto V_i for $i = 1, \ldots, p$. Define the random projection

$$\pi_{in}(\boldsymbol{x}) = \sum_{j=\bar{b}_{i-1}+1}^{\bar{b}_i} (\boldsymbol{x}\cdot\theta_{jn})\theta_{jn}.$$

Theorem 9.4. *Under the basic assumptions we have $\pi_{1n} \to \pi_1$ and $\pi_{pn} \to \pi_p$ in probability as $n \to \infty$.*

Again the proof is quite similar to the proof of Theorem 3 in Meerschaert and Scheffler (1999b) and Theorem 10.4.8 in Meerschaert and Scheffler (2001a). See also Scheffler (1998). We include here a sketch of the arguments.

Proposition 9.5. *Under our basic assumptions we have: If $j > \bar{b}_{p-1}$ and $r < p$ then*

$$\pi_r \theta_{jn} \to 0 \quad \textit{in probability.}$$

Proof: Since $\pi_r \theta_{jn} = (\pi_r M_n / \lambda_{jn})\theta_{jn}$ we get

$$\|\pi_r \theta_{jn}\| \leqslant \left\| \frac{\pi_r M_n}{\lambda_{jn}} \right\| \leqslant \frac{\|\pi_r B_n^{-1}\| \|n B_n M_n B_n'\| \|(B_n')^{-1}\|}{n\lambda_{jn}}.$$

By assumption (A1) together with continuous mapping it follows from Prohorov's theorem that $(n\|B_n M_n B_n'\|)$ is uniformly tight. Also, by (9.7), (9.8) and Theorem 9.1 we get

$$\begin{aligned} &\frac{\log(\|\pi_r B_n^{-1}\| \|n B_n M_n B_n'\| \|(B_n')^{-1}\|)/(n\lambda_{jn})}{\log n} \\ &= \frac{\log \|\pi_r B_n^{-1}\|}{\log n} + \frac{\log \|(B_n')^{-1}\|}{\log n} - \frac{\log(n\lambda_{jn})}{\log n} \\ &\to a_r + a_p - 2a_p < 0 \quad \text{in probability.} \end{aligned}$$

Hence the assertion follows. □

Proposition 9.6. *Under our basic assumptions we have: If $j \leqslant \bar{b}_1$ and $r > 1$ then*

$$\pi_r \theta_{jn} \to 0 \quad \textit{in probability.}$$

Proof: Since $\pi_r \theta_{jn} = (\pi_r M_n^{-1} \lambda_{jn})\theta_{jn}$ we get

$$\|\pi_r \theta_{jn}\| \leqslant \left\|\pi_r M_n^{-1} \lambda_{jn}\right\| \leqslant \|\pi_r B_n'\| \left\| \frac{1}{n} (B_n')^{-1} M_n^{-1} B_n^{-1} \right\| \|B_n\| (n\lambda_{jn}).$$

As in the proof of Proposition 9.3 the sequence $(\frac{1}{n}(B_n')^{-1} M_n^{-1} B_n^{-1}\|)$ is uniformly tight and now the assertion follows as in the proof of Proposition 9.5. □

Proof of Theorem 9.4: The proof is almost identical to the proof of Theorem 3 in Meerschaert and Scheffler (1999b) or Theorem 10.4.8 in Meerschaert and Scheffler (2001a) and therefore omitted. □

Corollary 9.7. *Under our basic assumptions, if* $p \leqslant 3$ *then* $\pi_{in} \to \pi_i$ *in probability for* $i = 1, \ldots, p$.

Proof: Obvious. □

Example 9.8. Suppose that $\boldsymbol{Z}, \boldsymbol{Z}_1, \boldsymbol{Z}_2, \ldots$ is a sequence of independent and identically distributed (IID) random vectors with common distribution μ. We assume that μ is regularly varying with exponent E. That means that there exists a regularly varying sequence (A_n) of linear operators with index $-E$ such that

$$n(A_n\mu) \to \phi \quad \text{as } n \to \infty. \tag{9.10}$$

For more information on regularly varying measures see Meerschaert and Scheffler (2001a, Chapter 6).

Regularly varying measures are closely related to the generalized central limit theorem discussed in Section 3. Recall that if

$$A_n(\boldsymbol{Z}_1 + \cdots + \boldsymbol{Z}_n - n\boldsymbol{b}_n) \Rightarrow \boldsymbol{Y} \quad \text{as } n \to \infty \tag{9.11}$$

for some nonrandom $\boldsymbol{b}_n \in \mathbb{R}^d$, we say that $\boldsymbol{Z}$ belongs to the generalized domain of attraction of $\boldsymbol{Y}$ and we write $\boldsymbol{Z} \in \mathrm{GDOA}(\boldsymbol{Y})$. Corollary 8.2.12 in Meerschaert and Scheffler (2001a) shows that $\boldsymbol{Z} \in \mathrm{GDOA}(\boldsymbol{Y})$ and (9.11) holds if and only if μ varies regularly with exponent E and (9.10) holds, where the real parts of the eigenvalues of E are greater than $1/2$. In this case, $\boldsymbol{Y}$ has an operator stable distribution and the measure ϕ in (9.10) is the Lévy measure of the distribution of $\boldsymbol{Y}$. Operator stable distributions and Lévy measures were discussed in Section 4, where (9.10) is written in the equivalent form $nP(A_n\boldsymbol{Z} \in \mathrm{d}x) \to \phi(\mathrm{d}x)$. The spectral decomposition was discussed in Section 5. Theorem 8.3.24 in Meerschaert and Scheffler (2001a) shows that we can always choose norming operators A_n and limit $\boldsymbol{Y}$ in (9.11) so that $\boldsymbol{Y}$ is spectrally compatible with $\boldsymbol{Z}$, meaning that A_n varies regularly with some exponent $-E$, the subspaces V_i in the spectral decomposition of $\mathbb{R}^d$ with respect to E are mutually orthogonal, and these subspaces are also A_n-invariant for every n. In this case, we write $\boldsymbol{Z} \in \mathrm{GDOA}_c(\boldsymbol{Y})$.

Recall from Section 6 that, since the real parts of the eigenvalues of E are greater than $1/2$,

$$nA_nM_nA_n' \Rightarrow W \quad \text{as } n \to \infty, \tag{9.12}$$

where M_n is the uncentered sample covariance matrix

$$M_n = \frac{1}{n}\sum_{i=1}^{n} \boldsymbol{Z}_i\boldsymbol{Z}_i'$$

and W is a random $d \times d$ matrix whose distribution is operator stable. Theorem 10.2.9 in Meerschaert and Scheffler (2001a) shows that W is invertible with probability one, and

Theorem 10.4.2 in Meerschaert and Scheffler (2001a) shows that for all unit vectors $\theta \in \mathbb{R}^d$ the random variable $\theta \cdot W\theta$ has a Lebesgue density. Then the basic assumptions of this section hold, and hence the results of this section apply.

The tail estimator proven in this section approximates the spectral index function $\alpha(\boldsymbol{x})$ defined in (5.2). This index function provides sharp bounds on the tails and radial projection moments of $\boldsymbol{Z}$. Given a d-dimensional data set $\boldsymbol{Z}_1, \ldots, \boldsymbol{Z}_n$ with uncentered covariance matrix M_n, let $0 \leqslant \lambda_{1n} \leqslant \cdots \leqslant \lambda_{dn}$ denote the eigenvalues of M_n and $\theta_{1n}, \ldots, \theta_{dn}$ the corresponding orthonormal basis of eigenvectors. Writing $x_j = \boldsymbol{x} \cdot \theta_j$ we can estimate the spectral index $\alpha(\boldsymbol{x})$ by

$$\hat{\alpha}(\boldsymbol{x}) = \min\{\hat{\alpha}_j \colon x_j \neq 0\}, \quad \text{where } \hat{\alpha}_j = \frac{2\log n}{\log(n\lambda_{jn})}$$

using the results of this section. Hence the eigenvalues are used to approximate the tail behavior, and the eigenvectors determine the coordinate system to which these estimates pertain. A practical application of this tail estimator appears in Example 8.1.

Example 9.9. The same tail estimation methods used in the previous example also apply to the moving averages considered in Section 7. This result is apparently new. Given a sequence of IID random vectors $\boldsymbol{Z}, \boldsymbol{Z}_j$ whose common distribution μ varies regularly with exponent E, so that (9.10) holds, we define the moving average process

$$\boldsymbol{X}_t = \sum_{j=-\infty}^{\infty} C_j \boldsymbol{Z}_{t-j}, \tag{9.13}$$

where we assume that the $d \times d$ matrices C_j fulfill for each j either $C_j = 0$ or C_j is invertible and $A_n C_j = C_j A_n$ for all n. Moreover if a_p denotes the largest real part of the eigenvalues of E we assume further

$$\sum_{j=-\infty}^{\infty} \|C_j\|^{\delta} < \infty \tag{9.14}$$

for some $\delta < 1/a_p$ with $\delta \leqslant 1$. Recall from Section 7 that under those conditions $\boldsymbol{X}_t$ is almost surely well defined, and that if the real parts of the eigenvalues of E are greater than $1/2$ we have that

$$n A_n \widehat{\Gamma}_n(0) A_n' \Rightarrow M = \sum_{j=-\infty}^{\infty} C_j W C_j' \quad \text{as } n \to \infty, \tag{9.15}$$

where the sample covariance matrix $\widehat{\Gamma}_n(h)$ is defined by (7.6) and W is a random $d \times d$ matrix whose probability distribution is operator stable. Suppose that the norming operators A_n are chosen so that (9.11) holds and $\boldsymbol{Z} \in \mathrm{GDOA}_c(\boldsymbol{Y})$. Then in view of our basic assumptions (A1) and (A2) it remains to show:

Lemma 9.10. *Under the assumptions of the paragraph above the limiting matrix M in* (9.15) *is a.s. positive definite and for any unit vector θ the random variable $M\theta \cdot \theta$ has no atom at zero.*

Proof: Since W in (9.12) is a.s. positive definite we have for any $\theta \neq 0$ that $C_j W C_j' \theta \cdot \theta = W C_j' \theta \cdot C_j' \theta \geqslant 0$ for all j and strictly greater that zero for those j with $C_j \neq 0$. Hence

$$M\theta \cdot \theta = \sum_{j=-\infty}^{\infty} C_j W C_j' \theta \cdot \theta > 0$$

for any $\theta \neq 0$ so M is positive definite.

Moreover if for a given unit vector θ we set $z_j = C_j' \theta$ then $z_{j_0} \neq 0$ for at least one j_0. Since W is almost surely positive definite we have

$$P\{M\theta \cdot \theta < t\} = P\left\{ \sum_{j=-\infty}^{\infty} W z_j \cdot z_j < t \right\} \leqslant P\{W z_{j_0} \cdot z_{j_0} < t\} \to 0$$

as $t \to 0$ using the fact that $W z_{j_0} \cdot z_{j_0}$ has a Lebesgue density as above. Hence $M\theta \cdot \theta$ has no atom at zero. □

It follows from (9.15) together with Lemma 9.10 that the $\boldsymbol{X}_t$ defined above fulfill the basic assumptions of this section. Hence it follows from Theorems 9.1 and 9.4 that the tail estimator used in Example 9.8 also applies to time-dependent data that can be modeled as a multivariate moving average. We can also utilize the uncentered sample covariance matrix (6.3), which has the same asymptotics as long as $E\boldsymbol{Z} = 0$ [cf. Theorem 10.6.7 and Corollary 10.2.6 in Meerschaert and Scheffler (2001a)]. In either case, the eigenvalues can be used to approximate the tail behavior, and the eigenvectors determine the coordinate system in which these estimates apply.

Example 9.11. Suppose now that $\boldsymbol{Z}_1, \boldsymbol{Z}_2, \ldots$ are IID $\mathbb{R}^d$-valued random vectors with common distribution μ. We assume that μ is $\mathrm{ROV}_\infty(E, c)$, meaning that there exist (A_n) regularly varying with index $-E$, a sequence (k_n) of natural numbers tending to infinity with $k_{n+1}/k_n \to c > 1$ such that

$$k_n(A_{k_n}\mu) \to \phi \quad \text{as } n \to \infty. \tag{9.16}$$

See Meerschaert and Scheffler (2001a, Section 6.2) for more information on R–O varying measures.

R–O varying measures are closely related to a generalized central limit theorem. In fact, if μ is $\mathrm{ROV}_\infty(E, c)$ and the real parts of the eigenvalues of E are greater than $1/2$ then (9.16) is equivalent to

$$A_{k_n}(\boldsymbol{Z}_1 + \cdots + \boldsymbol{Z}_{k_n} - k_n \boldsymbol{b}_n) \Rightarrow \boldsymbol{Y} \quad \text{as } n \to \infty,$$

where $\boldsymbol{Y}$ has a so called (c^E, c) operator semistable distribution. See Meerschaert and Scheffler (2001a), Sections 7.1 and 8.2 for details. Once again, a judicious choice of norming operators and limits guarantees that $\boldsymbol{Y}$ is spectrally compatible with $\boldsymbol{Z}$, so that A_n varies regularly with some exponent $-E$, the subspaces V_i in the spectral decomposition of $\mathbb{R}^d$ with respect to E are mutually orthogonal, and these subspaces are also A_n-invariant for every n. It follows from Theorem 8.2.5 of Meerschaert and Scheffler (2001a) that $\boldsymbol{Z}$ has the same moment and tail behavior as for the generalized domain of attraction case considered in Section 5. In particular, there is a spectral index function $\alpha(\boldsymbol{x})$ taking values in the set $\{a_1^{-1}, \ldots, a_p^{-1}\}$ where $a_1 < \cdots < a_p$ are the real parts of the eigenvalues of E. Given $\boldsymbol{x} \neq 0$, for any small $\delta > 0$ we have

$$r^{-\alpha(\boldsymbol{x})-\delta} < P\big(|\boldsymbol{Z} \cdot \boldsymbol{x}| > r\big) < r^{-\alpha(\boldsymbol{x})+\delta}$$

for all $r > 0$ sufficiently large. Then $E(|\boldsymbol{Z} \cdot \boldsymbol{x}|^\beta)$ exists for $0 < \beta < \alpha(\boldsymbol{x})$ and diverges for $\beta > \alpha(\boldsymbol{x})$.

Now let

$$M_n = \frac{1}{n} \sum_{i=1}^{n} \boldsymbol{Z}_i \boldsymbol{Z}_i'$$

denote the sample covariance matrix of $(\boldsymbol{Z}_i)$. Then it follows from Theorem 10.2.3, Corollaries 10.2.4 and 10.2.6, Theorem 10.2.9, and Lemma 10.4.2 in Meerschaert and Scheffler (2001a) that M_n fulfills the basic assumptions (A1) and (A2) of this section. Hence, by Theorems 9.1 and 9.4 we rediscover Theorems 10.4.5 and 10.4.8 of Meerschaert and Scheffler (2001a). See also Scheffler (1998). In other words, the approximation $\hat{\alpha}(\boldsymbol{x})$ from Example 9.8 still functions in this more general case, which represents the most general setting in which sums of IID random vectors can approximated in distribution via a central limit theorem.

10. Conclusions

If one believes that asset price changes (or log-returns) have heavy tails, then there is ample reason to seek a model where the tail thickness parameter α varies with the asset. Operator stable random vectors provide such a model, and are justified by a central limit theorem. Matrix-scaled sums of independent, identically distributed random vectors can only converge (in a distributional sense) to an operator stable limit. Such random vectors have regularly varying probability distributions whose tails are governed by a matrix exponent. Time dependent models can be constructed by taking moving averages of these random vectors. If X_i is the price change in the i-th asset then the vector of price changes $X = (X_1, \ldots, X_d)'$ can be described by such models. If θ_i measures the amount of the i-th asset in a portfolio, price changes for this portfolio are of the form $X \cdot \theta = X_1\theta_1 + \cdots + X_d\theta_d$. The probability of large jumps in price depends on the mix according to a tail index function $\alpha(\theta)$. If $2 < \alpha(\theta) < 4$ we have a finite variance model with infinite fourth moments. Then the sample covariance matrix plays the usual role as a descriptor of dependence between assets, but its asymptotics are operator stable. If $\alpha(\theta) < 2$ indicating heavy tails with infinite variance, the sample covariance matrix still provides some useful information. In particular, the coordinate system that diagonalizes this matrix also identifies the portfolios with the best or worst tail behavior.

References

Aban, I., Meerschaert, M., 2001. Shifted Hill's estimator for heavy tails. Communications in Statistics. Simulation and Computation 30, 949–962.

Adler, R., Feldman, R., Taqqu, M., 1998. A Practical Guide to Heavy Tails. Birkhäuser, Boston.

Akgiray, V., Booth, G., 1988. The stable-law model of stock returns. Journal of Business and Economic Statistics 6, 51–57.

Anderson, P., Meerschaert, M., 1997. Periodic moving averages of random variables with regularly varying tails. The Annals of Statistics 25, 771–785.

Anderson, P., Meerschaert, M., 1998. Modeling river flows with heavy tails. Water Resources Research 34, 2271–2280.

Arnold, B.C., 1990. A flexible family of multivariate Pareto distributions. Journal of Statistics Planning and Inference 24, 249–258.

Bawa, V., Elton, E., Gruber, M., 1979. Simple rules for optimal portfolio selection in stable Paretian markets. Journal of Finance 34 (4), 1041–1047.

Belkacem, L., Véhel, J., Walter, C., 2000. CAPM, risk and portfolio selection in α-stable markets. Fractals 8 (1), 99–115.

Benson, D., Schumer, R., Meerschaert, M., Wheatcraft, S., 2001. Fractional dispersion, Lévy motions, and the MADE tracer tests. Transport Porous Med. 42, 211–240.

Benson, D., Wheatcraft, S., Meerschaert, M., 2000. Application of a fractional advection–dispersion equation. Water Resources Research 36, 1403–1412.

Blumen, A., Zumofen, G., Klafter, J., 1989. Transport aspects in anomalous diffusion: Lévy walks. Physical Review A 40, 3964–3973.

Brockwell, P., Davis, R., 1991. Time Series: Theory and Methods, 2nd edition. Springer-Verlag, New York.

Byczkowski, T., Nolan, J.P., Rajput, B., 1993. Approximation of multidimensional stable densities, Journal of Multivariate Analysis 46, 13–31.

Chamberlain, T.W., Cheung, C.S., Kwan, C.C., 1990. Optimal portfolio selection using the general multi-index model: A stable-Paretian framework. Decision Sciences 21, 563–571.

Csörgő, S., Viharos, L., 1997. Asymptotic normality of least-squares estimators of tail indices. Bernoulli 3, 351–370.

Elton, E., Gruber, M., 1995. Modern Portfolio Theory and Investment Analysis. Wiley, New York.

Embrechts, P., Klüppelberg, C., Mikosch, T., 1997. Modelling Extremal Events. For Insurance and Finance. In: Applications of Mathematics, Vol. 33. Springer-Verlag, Berlin.

Fama, E., 1965a. The behavior of stock market prices. Journal of Business 38, 34–105.

Fama, E.F., 1965b. Portfolio analysis in a stable Paretian market. Management Science 11, 404–419.

Feller, W., 1971. An Introduction to Probability Theory and its Applications, Vol. II, 2nd edition. Wiley, New York.

Fofack, H., Nolan, J., 1999. Tail behavior, modes and other characteristics of stable distribution. Extremes 2, 39–58.

Gamba, A., 1999. Portfolio analysis with symmetric stable Paretian returns. In: Current Topics in Quantitative Finance, Venice, 1997. In: Contribution to Management Science. Physica, Heidelberg, pp. 48–69.

Gnedenko, B., Kolmogorov, A., 1968. Limit Distributions for Sums of Independent Random Variables. Translated from the Russian, annotated, and revised by K.L. Chung. With appendices by J.L. Doob and P.L. Hsu. Revised edition. Addison-Wesley, Reading, MA.

Hall, P., 1981. A comedy of errors: the canonical form for a stable characteristic function. Bulletin of the London Mathematical Society 13, 23–27.

Hall, P., 1982. On some simple estimates of an exponent of regular variation. Journal of the Royal Statistical Society. Series B 44, 37–42.

Hill, B., 1975. A simple general approach to inference about the tail of a distribution. The Annals of Statistics 3, 1163–1173.

Hirsch, M., Smale, S., 1974. Differential Equations, Dynamical Systems and Linear Algebra. Academic Press, New York.

Hosking, J., Wallis, J., 1987. Parameter and quantile estimation for the generalized Pareto distribution. Technometrics 29, 339–349.

Janicki, A., Weron, A., 1994. Simulation and Chaotic Behavior of α-Stable Stochastic Processes. Marcel Dekker, New York.

Jansen, D., de Vries, C., 1991. On the frequency of large stock market returns: Putting booms and busts into perspective. Review of Economical Statistics 23, 18–24.

Jurek, Z., Mason, J.D., 1993. Operator-Limit Distributions in Probability Theory. Wiley, New York.

Klafter, J., Blumen, A., Shlesinger, M., 1987. Stochastic pathways to anomalous diffusion. Physical Review A 35, 3081–3085.

Kotz, S., Balakrishnan, N., Johnson, N.L., 2000. Continuous Multivariate Distributions, Vol. 1. Wiley, New York.

Kotz, S., Kozubowski, T., Podgórski, K., 2001. The Laplace Distribution and Generalizations: A Revisit with Applications to Communications, Economics, Engineering and Finance. Birkhäuser, Boston.

Kozubowski, T.J., Panorska, A.K., Rachev, S.T., 2003. Statistical issues in modeling multivariate stable portfolios. In: Rachev, S. (Ed.), Handbook of Heavy Tailed Distributions in Finance, Vol. 1. Elsevier, Amsterdam, pp. 131–167.

Leadbetter, M., Lindgren, G., Rootzén, H., 1980. Extremes and Related Properties of Random Sequences and Processes. Springer-Verlag, New York.

Loretan, M., Phillips, P., 1994. Testing the covariance stationarity of heavy tailed time series. Journal of Empirical Finance 1, 211–248.

Mandelbrot, B., 1963. The variation of certain speculative prices. Journal of Business 36, 394–419.

Mandelbrot, B., 1982. The Fractal Geometry of Nature. Freeman, San Francisco.

McCulloch, J., 1996. Financial applications of stable distributions. In: Madfala, G., Rao, C.R. (Eds.), Statistical Methods in Finance. In: Handbook of Statistics, Vol. 14. Elsevier, Amsterdam, pp. 393–425.

McCulloch, J., 1997. Measuring tail thickness to estimate the stable index α: A critique. Journal of Business and Economics Statistics 15, 74–81.

Meerschaert, M., 1990. Moments of random vectors which belong to some domain of normal attraction. The Annals of Probability 18, 870–876.

Meerschaert, M., 1991. Regular variation in $\mathbb{R}^k$ and vector-normed domains of attraction. Statistics & Probability Letters 11, 287–289.

Meerschaert, M., Scheffler, H.P., 1998. A simple robust estimator for the thickness of heavy tails. Journal of Statistical Planning and Inference 71, 19–34.

Meerschaert, M., Scheffler, H.P., 1999a. Sample covariance matrix for random vectors with heavy tails. Journal of Theoretical Probability 12, 821–838.

Meerschaert, M., Scheffler, H.P., 1999b. Moment estimator for random vectors with heavy tails. Journal of Multivariate Analysis 71, 145–159.

Meerschaert, M., Scheffler, H.P., 2000. Moving averages of random vectors with regularly varying tails. Journal of Time Series Analysis 21, 297–328.

Meerschaert, M., Scheffler, H.P., 2001a. Limit Distributions for Sums of Independent Random Vectors: Heavy Tails in Theory and Practice. Wiley, New York.

Meerschaert, M., Scheffler, H.P., 2001b. Limit theorems for continuous time random walks. Preprint.

Mittnik, S., Rachev, S., 1993. Reply to comments on Modeling asset returns with alternative stable distributions, and some extensions. Econometric Reviews 12, 347–389.

Mittnik, S., Doganoglu, T., Chenyao, D., 1999. Computing the probability density function of the stable Paretian distribution. Mathematical and Computer Modelling 29, 235–240.

Mittnik, S., Rachev, S., Doganoglu, T., Chenyao, D., 1999. Maximum likelihood estimation of stable Paretian models. Mathematical and Computer Modelling 29, 275–293.

Modarres, R., Nolan, J.P., 1994. A method for simulating stable random vectors. Computational Statistics 9, 11–19.

Nikias, C., Shao, M., 1995. Signal Processing with Alpha Stable Distributions and Applications. Wiley, New York.

Nolan, J.P., 1997. Numerical calculation of stable densities and distribution functions. Heavy tails and highly volatile phenomena. Communications in Statistics. Stochastic Models 13, 759–774.

Nolan, J.P., 2001. Maximum likelihood estimation of stable parameters. In: Barndorff-Nielsen, O., Mikosch, T., Resnick, S. (Eds.), Levy Processes. Birkhäuser, Boston.

Nolan, J.P., 2002. Stable Distributions: Models for Heavy Tailed Data. Birkhäuser, Boston.

Nolan, J.P., Panorska, A., 1997. Data analysis for heavy tailed multivariate samples. Heavy tails and highly volatile phenomena. Communications in Statistics. Stochastic Models 13, 687–702.

Nolan, J.P., Panorska, A., McCulloch, J.H., 2001. Estimation of stable spectral measures. Mathematical and Computer Modelling 34, 1113–1122.

Press, S.J., 1982. Applied Multivariate Analysis, 2nd edition. Krieger, Malabar.

Press, W., Flannery, B., Teukolsky, S., Vetterling, W., 1987. Numerical Recipes. Cambridge University Press, New York.

Rachev, S., Han, S., 2000. Portfolio management with stable distributions. Mathematical Methods of Operations Research 51 (2), 341–352.

Rachev, S., Mittnik, S., 2000. Stable Paretian Models in Finance. Wiley, Chichester.

Rao, C.R., 1965. Linear Statistical Inference and its Applications. Wiley, New York.

Resnick, S., Greenwood, P., 1979. A bivariate characterization and domains of attraction. Journal of Multivariate Analysis 9, 206–221.

Resnick, S., Stărică, C., 1995. Consistency of Hill's estimator for dependent data. Journal of Applied Probability 32, 139–167.

Rvačeva, E., 1962. On domains of attraction of multidimensional distributions. In: Select. Transl. Math. Stat. Prob. 2. American Mathematical Society, Providence, RI, pp. 183–205.

Samorodnitsky, G., Taqqu, M., 1994. Stable Non-Gaussian Random Processes. Chapman & Hall, New York.

Scheffler, H.P., 1998. Multivariable R–O varying measures and generalized domains of semistable attraction. Habilitation thesis. University of Dortmund.

Scheffler, H.P., 1999. On estimation of the spectral measure of certain nonnormal operator stable laws. Statistics & Probability Letters 43, 385–392.

Schultze, J., Steinebach, J., 1996. On least squares estimates of an exponential tail coefficient. Statistics & Decisions 14, 353–372.

Sharpe, M., 1969. Operator-stable probability distributions on vector groups. Translations of the American Mathematical Society 136, 51–65.

Shlesinger, M., Zaslavsky, G., Frisch, U., 1994. Lévy Flights and Related Topics in Physics. In: Lecture Notes in Physics, Vol. 450. Springer-Verlag, Berlin.

Tessier, Y., Lovejoy, S., Hubert, P., Schertzer, D., 1996. Multifractal analysis and modeling of rainfall and river flows and scaling, causal transfer functions. Journal of Geophysical Research 101, 427–440.
Uchaikin, V., Zolotarev, V., 1999. Chance and Stability: Stable Distributions and their Applications. VSP, Utrecht.
Weron, R., 1996. On the Chambers–Mallows–Stuck method for simulating skewed stable random variables. Statistics & Probability Letters 28 (2), 165–171.
Ziemba, W., 1974. Choosing investment portfolios when the returns have stable distributions. In: Mathematical Programming in Theory and Practice, Proc. NATO Adv. Study Inst., Figueira da Foz, 1972. North-Holland, Amsterdam, pp. 443–482.

Chapter 16

LONG RANGE DEPENDENCE IN HEAVY TAILED STOCHASTIC PROCESSES

BORJANA RACHEVA-IOTOVA

Faculty of Economics and Business Administration, University of Sofia, Bulgaria
Bravo Risk Management Group, 1101-C Eugenia Place, Carpinteria, Santa Barbara, CA 93013, USA
e-mail: borjana.racheva@bravo-group.com

GENNADY SAMORODNITSKY*

School of Operations Research and Industrial Engineering, and Department of Statistical Science, Cornell University, Ithaca, NY 14853, USA
e-mail: gennady@orie.cornell.edu

Contents

* G. Samorodnitsky's research was partially supported by NSF grants DMS-0071073 and DMI-9713549 at Cornell University.

Handbook of Heavy Tailed Distributions in Finance, Edited by S.T. Rachev

Abstract

The notion of long range dependence has traditionally been defined through a slow decay of correlations. This approach may be completely inappropriate in the case of a stochastic process with heavy tails. Yet long memory has been reported to be found in various fields where heavy tails are a standard feature of the commonly used stochastic models. Financial and communications networks data are among those often believed to exhibit long memory. We discuss alternative points of view on long range dependence that are applicable in the heavy tailed case. Such alternative approaches may be tailored for a particular applications at hand.

Keywords

heavy tails, long range dependence, rare events, large deviations

1. Introduction

A glance at the plot on Figure 1 describing the annual minima of the water level in the Nile river suggests that the process plotted there has at least 4 distinct time periods when the "level" and "drift" of the process change. This needs not, however, be necessarily taken as an indication that a nonstationary model should be used for the annual minima of water level process. Even though using a nonstationary model is possible in such situations, sometimes a more *parsimonious* model is a *long memory stationary model* (or a *stationary model with long range dependence*). In fact, commonly used long memory models exhibit what Benoit Mandelbrot termed "persistence", or "Joseph effect" (referring to the long stretches of plenty and famine in Egypt of the Bible). Here is what one sees when looking at the increments of a long memory Fractional Brownian motion (also called Fractional Gaussian noise): "Nearly every sample looks like a "random noise" superimposed upon a background that performs several cycles. However, there cycles are *not* periodic, that is, *cannot* be extrapolated as the sample lengthens. In addition, one often sees an underlying trend that need not continue in the extrapolate." (Mandelbrot, 1983, p. 251.)

The Nile river data set is a famous one; arguably, it is the data set that forced us to think about long range dependence in the first place. It was, of course, the same Mandelbrot who with co-workers (Mandelbrot, 1965; Mandelbrot and Van Ness, 1968; Mandelbrot and Wallis 1968, 1969) first realized that a long memory *stationary* Gaussian process may explain the behaviour a particular statistic (the so-called R/S statistic) suggested and applied to the Nile river data by Hurst (1951, 1955).

Today long memory models are still being used in hydrology and related areas. However, new applications have arisen, significantly in finance and communication networks. Often observations from these latter areas feature *heavy tails*, and such data sets sometimes provide extreme illustrations to the Mandelbrot remark on "spurious cycles". For example, Figure 2 describing the load offered by a network server, suggests that the process plotted there has at least 10 distinct time periods when the "nature" of the process changes. Once

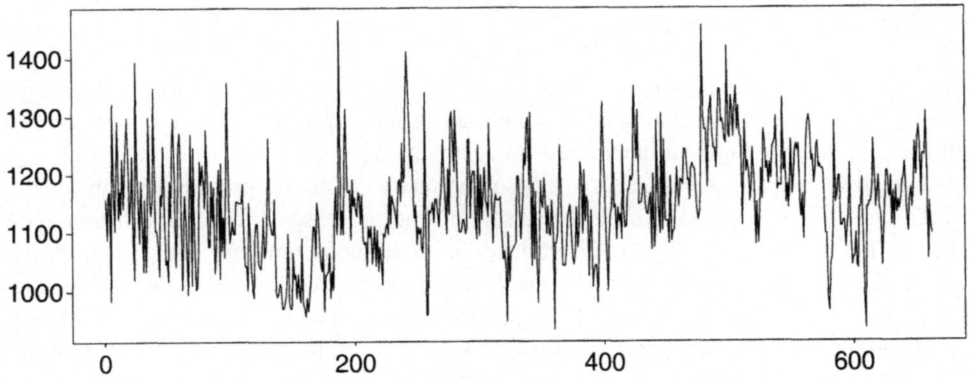

Fig. 1. Annual minima of the water level in the Nile river for the years 622 to 1281, measured at the Roda gauge near Cairo.

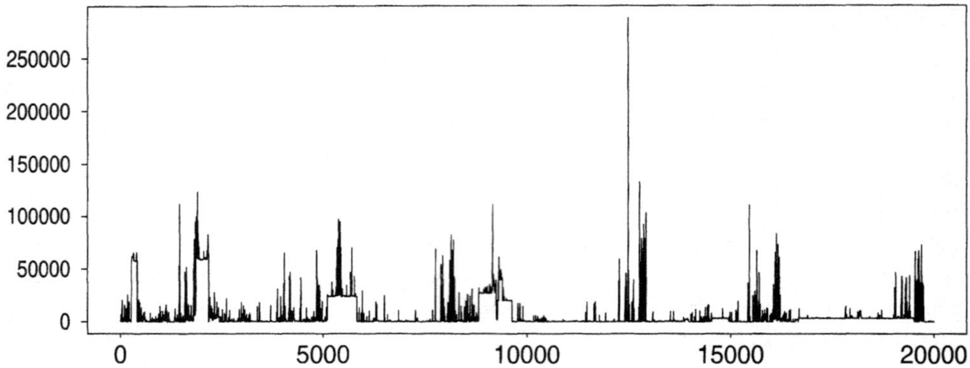

Fig. 2. Amount of information (in bytes) sent by a server from a major telecommunication company in the middle of a workday. Time is measured in seconds.

again, one should not automatically decide to use a nonstationary model because there are perfectly reasonable stationary models that have a similar behavior.

Stationary models are attractive not only because of parsimony but also because it is important to have a reasonably small class of well studied and well understood models that have wide applicability. Hence it is important to study stationary processes that can account for features we saw above; stationary processes with long range dependence.

This chapter is an attempt to survey stationary models with long range dependence and heavy tails. These two features are believed to be present in various data sets of financial and communication networks origin and, hence, attracted recently much attention. Describing long range dependence in the heavy tailed case is especially challenging and most of the work is still ahead of us. Nevertheless, it is an exciting task, and we argue that the insights we hope one will obtain are likely to be useful in other areas of stochastic modeling.

2. What is long range dependence?

The obvious way to measure the length of memory in a stochastic process is by looking at the rate at which its correlations decay with lag. Annoyingly, this requires correlations to make sense, hence finite variance needs to be assumed.

Let, therefore, X_n, $n = 0, 1, 2, \ldots$, be a stationary stochastic process with mean $\mu = EX_0$ and $0 < EX_0^2 < \infty$ (we discuss discrete time processes, but parallel formulations for stationary processes with finite variance in continuous time are entirely clear). Let $\rho_n = \mathrm{Corr}(X_0, X_n)$, $n = 0, 1, \ldots$, be the correlation function. For most "usual" stochastic models: ARMA processes, GARCH processes, many Markov and Markov modulated processes the correlations decay exponentially fast with n; this has a number of important consequences, one of which is $\sum_{n=0}^{\infty} |\rho_n| < \infty$. This, in turn, guarantees that the variance of the partial sums $S_n = X_1 + \cdots + X_n$, $n \geqslant 0$, cannot grow more than linearly fast, which says, heuristically, that we do not expect to see S_n to be more than about $\sqrt{n}$ away from its mean $n\mu$. What Mandelbrot realized two and half decades ago was that the strange behavior of R/S statistic on the Nile river data might be explained if the variance of the partial sums could grow faster than linearly fast. As we already know this implies that

$$\sum_{n=0}^{\infty} |\rho_n| = \infty. \tag{2.1}$$

Hence, (2.1) is often taken as the definition of the long memory; as a definition it seems to originate with Cox (1984).

Looking over the literature on long range dependence, one realizes that the definition (2.1) has not proved to be the most popular one. By far the most widely used definition is the more concrete

$$\rho_n \sim cn^{-d} \quad \text{as } n \to \infty \text{ for some } 0 < d < 1 \text{ and } c > 0. \tag{2.2}$$

A stationary process satisfying (2.2) would have been called long range dependent of index d by Cox (1984). A weaker version of (2.2) is also sometimes mentioned; it replaces the constant c by a slowly varying function (Beran, 1994). Quite often one writes the exponent $d = 2 - 2H$ for some $0.5 < H < 1$ [e.g., Beran (1992)], and the reasons are historical: this is the relation between the exponent H of self-similarity of a Fractional Brownian motion and the rate of decay of correlations of its increments. A misnomer, $H = 1 - d/2$ is at times referred to as the self-similarity parameter even if nothing in the model is self similar.

Less common (but still used) point of view on long range dependence is to allow d in (2.2) to take any positive value or, indeed, a similar assumption of regular variation of correlations (which also allows for the slowly varying case $d = 0$). In fact, one could even draw the line between long and short memory by distinguishing between correlations decaying slower than exponentially fast and those decaying at least exponentially fast.

It is difficult to justify such importance assigned to the rate of decay of correlations (or, almost equivalently, to the rate at which the spectral density of the stationary process grows at the origin), *unless* one deals with a Gaussian model (like the Fractional Gaussian noise, the increment process of a Fractional Brownian motion), or a process that is very close to being Gaussian. Using correlations as a measure of the length of memory becomes untenable in the case of heavy tails. Specifically, let X_n, $n = 0, 1, 2, \ldots$, be a stationary stochastic process. Let F be the distribution function of X_0, and $\overline{F} = 1 - F$ the (right) distribution tail. In a tradition going back, once again, to Mandelbrot in the early 1960s [an exhaustive list of references is in Mandelbrot (1983)] heavy tails are synonymous with infinite variance of X_0. Once again, more concrete views are prevailing in literature; it is common to identify heavy tails with a particular tail behavior of $\overline{F}$. Sometimes one assumes

$$\overline{F}(x) \sim cx^{-\alpha} \quad \text{as } x \to \infty \text{ for some } 0 < \alpha < 2 \text{ and } c > 0. \tag{2.3}$$

See, for example, a recent collection Park and Willinger (2000). Note the parallels between the various definitions of heavy tails and viewing long range dependence via the rate of decay of correlations. Again, one allows sometimes any positive value of α in (2.3) [see, e.g., Müller, Dacorogna and Pictet (1998), Gomez, Selman and Crato (1997)]. Here regular variation of the tails as opposed to power-like (*Pareto-type* is the common expression) is widely accepted. Finally, faster (but still slower than exponentially fast) rate of decay of the distribution tail is sometimes also regarded as being consistent with heavy tails. Here one does not usually go beyond the class of *subexponential distributions*; see Embrechts, Klüppelberg and Mikosch (1997).

Obviously, one cannot use correlations to draw the line between short and long memory if the variance is infinite. Several attempts have been made to use "correlation-like" notions in that case. In the important class of stable processes notions of *covariation* and *codifference* have been introduced and their rate of decay for various classes of stationary stable processes computed; see, e.g., Astrauskas, Levy and Taqqu (1991). Obviously, such "surrogate correlations" can be expected to carry even less information than the "real" correlations do in the case when the latter are defined [although, surprisingly, codifference turns out to characterize mixing of stationary symmetric stable processes, a fact due, essentially, to Maruyama (1970), see also Gross (1994)].

Before finishing this section we remark that when talking about tails we are thinking about the *right tails*. For as long as the left tails do not interfere with the right tails, we will leave it that way. When right and left tails begin to interfere with one another, we will need to say more about the left tails and how heavy they are as well.

3. Tails and rare events

Here is an alternative point of view on long range dependence in heavy tailed processes. Most practitioners using heavy tailed models will agree that the most important feature of such processes is precisely their tails as expressed in probabilities of various *rare events*. Risk analysis, ruin probabilities, congestion and overflow analysis are just some of the key words that name such rare events in various modern applications. To be a bit more concrete here are several specific examples of rare events one usually deals with. Let, once again, X_n, $n = 0, 1, 2, \ldots$, be a stationary stochastic process.

Example 3.1. For large $\lambda > 0$ the event $\{X_0 > \lambda\}$ is a rare event, whose probability is clearly related to the tails. This event is so elementary that it does not tell us anything about the memory in the process.

Example 3.2. For $k \geqslant 1$ and large $\lambda_0, \lambda_1, \ldots, \lambda_k$ the event $\{X_0 > \lambda_0, X_1 > \lambda_1, \ldots, X_k > \lambda_k\}$ is a rare event whose probability can carry very important information about the dependence in finite pieces of the process. Generally, the dependence we can measure using such rare events is a "tail dependence". However, for specific classes of heavy tailed processes (e.g., stable processes, linear processes, etc.) these events can provide even more information.

Example 3.3. For large $n \geqslant 1$ and a positive sequence $(\lambda_j)_{j\geqslant 0}$ that does not converge to zero the event $\{X_j > \lambda_j,\ j = 0, 1, \ldots, n\}$ is a rare event and its probability is a very interesting measure of the length of memory in the process. The case $\lambda_j = \lambda > 0$ for all $j \geqslant 0$ seems to be especially appealing.

A slight generalization of this example uses a triangular array $(\lambda_j^{(n)})_{n\geqslant 1,\ 0\leqslant j\leqslant n}$. Here the case $\lambda_j^{(n)} = \lambda^{(n)}$ for $j \leqslant n$ with various asymptotic rules for $(\lambda^{(n)})$ is very interesting.

Example 3.4. For $k \geqslant 1$ and large λ the event $\{X_1 + \cdots + X_k > \lambda\}$ is a tail event. Similarly to Example 3.2 the probability of this event can be used to clarify the "finite dimensional dependence" in the process.

Example 3.5. Suppose that the mean $\mu = EX_0$ is finite, and that the stationary process X_n, $n = 0, 1, 2, \ldots$, is ergodic. For large $n \geqslant 1$ and $\delta > 0$ the event $\{X_1 + \cdots + X_n > n(\mu + \delta)\}$ is a rare event whose probability measures the length of memory in the sense of a tendency of being over the mean for long stretches of time. It is, obviously, related to the tails. The effect of heavy tails is quite special, as will be discussed below.

Example 3.6. This example has a flavor similar to that of Example 3.5. Let, once again, the process $X_n, n = 0, 1, 2, \ldots$, be ergodic with a finite mean $\mu = EX_0$. Let $\delta > 0$. For large λ the event $\{X_1 + \cdots + X_n > n(\mu + \delta) + \lambda$ for some $n \geqslant 1\}$ is a rare event, whose probability is sometimes referred to as *ruin probability* in the context of risk analysis. In the queuing context various stationary quantities often have expressions of this kind for their probability tails. Adopting the risk analysis term, the ruin probability can be used to measure the length of memory; the effect of heavy tailed case is, once again, very special here.

The list of examples can be continued indefinitely, and we have omitted some very interesting ones. Instead, let us look at some details of the interplay between the tails, memory and rare events in the heavy tailed case, especially in the light of Examples 3.5 and 3.6. The starting point is to adopt the lenses of large deviations: *an unlikely event happens in the most likely way*. We will argue that such lenses provide a powerful way of thinking about the length of memory in a process. It is unfortunate that this idea is not made more explicit in many beautiful texts on large deviations (that also reserve the term "large deviation principle" for something else); see, e.g., Deuschel and Stroock (1989) and Dembo and Zeitouni (1993). The following statement is not a rigorous mathematical statement. Nevertheless, it is often very useful as a guide and, in many ways, it captures the essence of heavy tails:

> *the most likely way tail related rare events happen in a heavy tailed stochastic process is because of the smallest possible number of causes.*

This "smallest possible number of causes" is often equal to one.

Thus, in Example 3.4 it turns out that, if $X_1, \ldots, X_k$ are i.i.d. and heavy tailed, then

$$P(X_1 + \cdots + X_k > \lambda) \sim kP(X_1 > \lambda) \sim P\big(\max(X_1, \ldots, X_k) > \lambda\big) \quad \text{as } \lambda \to \infty. \tag{3.1}$$

That is, the sum $X_1 + \cdots + X_k$ is most likely to be very large due to one of the terms being very large. In this case the possible "causes" are simply the individual terms in the sum. The greatest generality under which (3.1) is valid is that of subexponential distributions, introduced by Chistyakov (1964). See also Chover, Ney and Wainger (1973), and a survey in Goldie and Klüppelberg (1998). Similarly, in Example 3.5, for every $\delta > 0$

$$P\big(X_1 + \cdots + X_n > n(\mu + \delta)\big) \sim nP(X_1 > n\delta) \quad \text{as } n \to \infty \tag{3.2}$$

for exactly the same reason as in (3.1). Indeed, one of the terms ($\equiv$ causes) in the sum $X_1 + \cdots + X_n$ has to be exceptionally large; exactly how large can be determined by realizing that the "nonexceptional" terms in that sum add up to about $n\mu$. While the domain of heavy tails over which (3.2) is valid does not extend to all subexponential distributions, it does extend to all distributions with regularly varying tails of index $\alpha > 1$; see, e.g., Heyde (1968) and Nagaev (1979).

On the other hand, for distributions with "light" tails not only (3.1) and (3.2) fail, even their spirit is false. In fact, in the case of exponentially fast decaying tails the most likely way for the event $\{X_1 + \cdots + X_n > n(\mu + \delta)\}$ to happen is not because of a single cause, or a small number of causes but, rather, because most of the terms in the sum "conspire" to be a bit bigger than they would normally be. This is, in fact, the point of the classical large deviation principle.

When X_n, $n = 0, 1, 2, \ldots$, is a stationary heavy tailed stochastic process *with memory*, it is not, generally, the case that individual observations should be viewed as "causes" of rare events. The nature of such causes depends on the nature of the process and it is, sometimes, a nontrivial problem to figure out what the "right causes" are. We will see several examples below. Moreover, and this is precisely the point why we are interested in rare events, the causes, when found, typically have their effect distributed over time and it is in this way that they make the rare events happen. We argue that *this temporal distribution of the effect of the "causes" on rare events is a useful way of thinking about long range dependence.*

There are two important classes of heavy tailed processes for which progress has been made in understanding the "right causes" of certain rare events and the way the effect of these causes is distributed over time: linear processes and infinitely divisible processes. We discuss these below. Before doing so we would like to introduce another notion related to certain rare events with a potential of being useful, in a similar way, in studying long range dependence.

Certain rare events should be rather viewed as sequences of events that become more and more rare. Examples 3.3 and 3.5 are of this nature. More generally and formally, let $A_j \in \mathbb{R}^j$ be a Borel set, $j = 1, 2, \ldots$, such that

$$p_j := P\big((X_1, \ldots, X_j) \in A_j\big) \to 0 \quad \text{as } j \to \infty. \tag{3.3}$$

For $n \geqslant 1$ define

$$R_n = \max\big\{j - i + 1:\ 1 \leqslant i \leqslant j \leqslant n,\ (X_i, X_{i+1} \ldots, X_j) \in A_{j-i+1}\big\}. \tag{3.4}$$

That is, R_n is the highest dimension of an A_j observed over the first n observations $X_1, \ldots, X_n$. We call R_n the functional associated with the sequence of rare events (A_j).

It is obvious that if X_n, $n = 0, 1, 2, \ldots$, is a mixing stationary process and $p_j > 0$ for an infinite sequence of j's then $R_n \to \infty$ with probability 1 as $n \to \infty$. It appears to be almost obvious that the *rate* at which R_n grows is related to the rate at which p_j decays to zero. Certain rigorous connections are, indeed, possible; other connections seem to require additional information on the process. In any case, the rate of growth of R_n is, in its own right, related to the way rare events happen and, hence, to the memory in the process.

There is a very important reason to concentrate on the probabilities of certain rare events and on functionals associated with sequences of certain rare events, instead of concentrating on correlations, when trying to understand the boundary between short memory and long memory. Such rare events and functionals are often of a direct importance on their own right, as one can see by looking at the examples above and thinking, for instance, of applications in risk analysis and congestion control. On the other hand, nobody is interested in correlations on their own right. We only study correlations hoping that they are significant for whatever application we might have at hand. Unfortunately, the information that the correlations carry is often only indirect and very limited, as anyone familiar, for example, with ARCH and GARCH models realizes.

4. Some classes of heavy tailed processes

4.1. Linear processes

One of the classes of heavy tailed processes we will consider is that of *heavy tailed linear processes*.

Let ε_n, $n \in \mathbb{Z}$, be iid random variables. A (two-sided) linear process with the noise sequence ε_n, $n \in \mathbb{Z}$, is defined by

$$X_n = \sum_{j=-\infty}^{\infty} \varphi_{n-j}\varepsilon_j, \quad n = 0, 1, 2, \ldots, \tag{4.1}$$

where φ_j, $j \in \mathbb{Z}$, is a sequence of (nonrandom) coefficients. We will assume that the noise variables are heavy tailed, but how heavy the tails are will be left open at the moment. It is obvious that the linear process X_n, $n = 0, 1, 2, \ldots$, is a stationary stochastic process as long as it is well defined, meaning that the sum defining it converges. The latter is an assumption on the coefficients φ_j. In particular, if $E\varepsilon_0^2 < \infty$ and $E\varepsilon_0 = 0$, then a necessary and sufficient condition for convergence of the series in (4.1) is

$$\sum_{j=-\infty}^{\infty} \varphi_j^2 < \infty; \tag{4.2}$$

a nonzero mean will require, in addition, the series $\sum_{j=-\infty}^{\infty} \varphi_j$ to converge. Frequently we will assume that the noise variables have regularly varying tails. Unless one is working with constant sign coefficients (an assumption that we will not make in this chapter), it is necessary to control both right and left probability tails of the noise since, say, a negative coefficient will "translate" the left tail of the noise into the right tail of the sum in (4.1). Therefore, a typical assumption is

$$\begin{cases} P\big(|\varepsilon_0| > \lambda\big) = L(\lambda)\lambda^{-\alpha}, \\ \lim\limits_{\lambda\to\infty} \dfrac{P(\varepsilon_0 > \lambda)}{P(|\varepsilon_0| > \lambda)} = p, \quad \lim\limits_{\lambda\to\infty} \dfrac{P(\varepsilon_0 < -\lambda)}{P(|\varepsilon_0| > \lambda)} = q, \end{cases} \tag{4.3}$$

as $\lambda \to \infty$, for some $\alpha \geqslant 0$ and $0 < p = 1 - q \leqslant 1$. Here L is a slowly varying (at infinity) function. If $\alpha > 2$ we are in the case of finite variance, but for $\alpha \leqslant 2$ the precise condition for convergence in (4.1) depends on the slowly varying function, and can be stated through the three series theorem. In particular,

$$\sum_{j=-\infty}^{\infty} |\varphi_j|^{\alpha-\varepsilon} < \infty \tag{4.4}$$

for some $\varepsilon > 0$ is a sufficient condition for convergence if $0 < \alpha \leqslant 1$ or if $1 < \alpha \leqslant 2$ and $E\varepsilon_0 = 0$; a nonzero mean in the latter case will also require, as before, the series $\sum_{j=-\infty}^{\infty} \varphi_j$ to converge.

A rich source of information on linear processes in Brockwell and Davis (1991). This book covers, mostly, the L^2 case. For more information on the infinite variance case see, for example, Cline (1983, 1985) and Mikosch and Samorodnitsky (2000b).

Heavy tailed linear processes are attractive to us because, in this case, the potential "causes" of rare events appear to be evident: those are the individual noise variables ε_n, $n \in \mathbb{Z}$. This intuition has been born out in a number of situations, as will be seen below.

4.2. *Infinitely divisible processes*

A stochastic process X_n, $n = 0, 1, 2, \ldots$, is *infinitely divisible* if for any $k = 1, 2, \ldots$ there is a stochastic process $Y_n^{(k)}$, $n = 0, 1, 2, \ldots$, such that the finite dimensional distributions of X_n, $n = 0, 1, 2, \ldots$, and of $\sum_{i=1}^{k} Y_n^{(k,i)}$, $n = 0, 1, 2, \ldots$, coincide. Here for $i = 1, \ldots, k$, the processes $Y_n^{(k,i)}$, $n = 0, 1, 2, \ldots$, are iid copies of $Y_n^{(k)}$, $n = 0, 1, 2, \ldots$. Many important classes of stochastic processes are, in fact, infinitely divisible. All Gaussian processes, and all stable processes in particular, are infinitely divisible. In general, an infinitely divisible process will have two independent components, a Gaussian one and a non-Gaussian one. Since we are interested in heavy tails, for a vast majority of applications the Gaussian component will have only a negligible effect on the probabilities of rare events we consider. Therefore, we will only consider infinitely divisible processes without a Gaussian component. Such processes have a characteristic function of the form

$$E\exp\left\{\mathrm{i}\sum_{n=0}^{\infty}\theta_n X_n\right\} \tag{4.5}$$

$$= \exp\left\{\int_{\mathbb{R}^\infty}\left(\exp\left\{\mathrm{i}\sum_{n=0}^{\infty}\theta_n x_n\right\} - 1 - \mathrm{i}\sum_{n=0}^{\infty}\theta_n x_n \mathbf{1}\big(|x_n| \leqslant 1\big)\right)\nu(\mathrm{d}\mathbf{x}) + \mathrm{i}\sum_{n=0}^{\infty}\theta_n b_n\right\}$$

for all θ_n, $n = 0, 1, 2, \ldots$, only finitely many of which are different from zero. Here ν is a σ-finite measure on $\mathbb{R}^\infty$ equipped with the product σ-field (the *Lévy measure of the process*) and b_n, $n = 0, 1, 2, \ldots$, is a constant vector in $\mathbb{R}^\infty$.

The Lévy measure of an infinitely divisible process is its most important feature. Often an infinitely divisible process is given in the form of a stochastic integral with respect to an infinitely divisible random measure. In that case there is a natural way to relate the Lévy measure of the process to the basic characteristics of such an integral.

Unlike the linear processes in the previous subsection, it is less obvious what are the potential "causes" of rare events when one deals with infinitely divisible processes as above. There is, however, a point of view on infinitely divisible processes that turns out to be useful here. To be able to see the essence better and not to get bogged in the technical details, let us consider, first, a particular case, when

$$\int_{\mathbb{R}^\infty} x_n \mathbf{1}\big(|x_n| \leqslant 1\big)\nu(\mathrm{d}\mathbf{x}) < \infty \quad \text{for all } n = 0, 1, 2 \ldots. \tag{4.6}$$

In that case one can rewrite (4.5) in the form

$$E\exp\left\{\mathrm{i}\sum_{n=0}^{\infty}\theta_n X_n\right\} = \exp\left\{\int_{\mathbb{R}^\infty}\left(\exp\left\{\mathrm{i}\sum_{n=0}^{\infty}\theta_n x_n\right\} - 1\right)\nu(\mathrm{d}\mathbf{x}) + \mathrm{i}\sum_{n=0}^{\infty}\theta_n b_n'\right\} \tag{4.7}$$

with $b_n' = b_n - \int_{\mathbb{R}^\infty} x_n \mathbf{1}(|x_n| \leqslant 1)\nu(\mathrm{d}\mathbf{x})$ for $n \geqslant 0$.

Let M be a Poisson random measure on $\mathbb{R}^\infty$ with mean measure ν. It is easy to check that the process $\int_{\mathbb{R}^\infty} x_n M(\mathrm{d}\mathbf{x}) - b_n'$ for $n \geqslant 0$ is well defined and has characteristic function given by (4.7). That is, one can represent the process X_n, $n = 0, 1, 2, \ldots$, in the sense of equality of finite dimensional distributions in the form

$$X_n = \int_{\mathbb{R}^\infty} x_n M(\mathrm{d}\mathbf{x}) - b_n', \quad n = 0, 1, 2, \ldots. \tag{4.8}$$

If $(\mathbf{z}^{(j)} = (z_n^{(j)},\ n \geqslant 0),\ j = 1, 2, \ldots)$ is a (measurable) enumeration of the points of the random measure M, then (4.8) means that the process X_n, $n = 0, 1, 2, \ldots$, is the sum of $(\mathbf{z}^{(j)})$, $j = 1, 2, \ldots$, (shifted by the sequence (b_n')). This "discrete" structure of infinitely divisible processes makes the potential "causes" of certain rare events visible, and it is precisely the Poisson points $((\mathbf{z}^{(j)}),\ j=1,2,\ldots)$ that turn out to be such "causes".

Even if the assumption (4.6) does not hold, then a representation similar to (4.8) can still be written, but this time an appropriate centering is required to make the Poisson integral to converge. The important point is that the discrete structure is still here, and the potential causes of rare events are still visible.

There are various ways of summing up Poisson points to get an infinitely divisible process. A very general description is in Rosiński (1989, 1990). Sometimes it is convenient to order the Poisson points according to the value of a particular test functional. If the process is originally given in the form of a stochastic integral with respect to an infinitely divisible random measure, then one can have a more concrete structure of the Poisson points, hence better understanding of the possible causes of rare events.

The literature on infinitely divisible processes is rich. The framework preferred by many authors is that of infinitely divisible probability laws on Banach (or other nice) spaces. See for example Araujo and Giné (1980) and Linde (1986). A very general treatment of stochastic integrals with respect to infinitely divisible random measures as well as representations of infinitely divisible processes as such stochastic integrals is in Rajput and Rosiński (1989).

An important and reasonably well understood class of infinitely divisible processes is that of *α-stable processes*. The latter are characterized by the following scaling property of their Lévy measure:

$$\nu(rA) = r^{-\alpha}\nu(A) \quad \text{for all measurable } A \in \mathbb{R}^{\infty} \text{ and } r > 0. \tag{4.9}$$

Here α is a parameter with the range $0 < \alpha < 2$. See Samorodnitsky and Taqqu (1994) for information on stable processes; the structure of stationary stable processes has been elucidated by J. Rosinski; see, e.g., Rosiński (1998).

5. Rare events, associated functionals and long range dependence

Suppose that we are considering a parametric family of laws of a stationary stochastic process $X_n, n = 0, 1, 2, \ldots$. Let Ξ be the (generally, infinite dimensional) parameter space. We are interested in significant changes ("phase transitions") in the rate of decay of probabilities of certain rare events and/or in the rate of growth of the functionals associated with sequences of rare events that may occur when the parameter ξ crosses the boundary between a subset Ξ_1 of Ξ and its complement. We argue that *certain phase transitions of this kind can be viewed as transitions between short and long range dependence.*

It is clear that it is not useful to view *every* significant change in, say, probabilities of rare events as an indication of interesting and important things happening to the memory of the process. Other factors may be in play as well, most significantly related to the heaviness of the tails. If, for example, one of the components of parameter $\xi \in \Xi$ governs how heavy the tails of X_0 are, one can very easily induce a very significant change in the probabilities of certain rare events by simply changing that particular component of the parameter without doing anything to the memory of the process. In the examples in the sequel we will be careful to look for phase transitions that do not involve changing how heavy the tails are.

We will see several examples of such phase transitions indicating a shift from short to long memory below. We present some known results; these are quite scarce. When appropriate, we supplement those with conjectures. In other cases we have performed numerical studies to try to guess whether a phase transition occurs and, if so, of what kind.

5.1. Unusual sample mean and long strange segments for heavy tailed linear processes

Here we consider the sequence of rare events of the Example 3.5 $A_n = \{X_1 + \cdots + X_n > n(\mu + \delta)\}$ (for a fixed $\delta > 0$) and the corresponding associated functional

$$R_n = \max\left\{ j - i + 1\colon 1 \leqslant i \leqslant j \leqslant n,\ \frac{X_i + X_{i+1} + \cdots + X_j}{j - i + 1} > \mu + \delta \right\}. \tag{5.1}$$

We will keep the distribution of the noise variables ε_n, $n \in \mathbb{Z}$, in the heavy tailed linear processes of Section 4.1 fixed; it is assumed to have the regular variation property (4.3) with $\alpha > 1$. In particular, the parameter α which is responsible for the heaviness of the tails is kept fixed. We will also assume that the $E\varepsilon_0 = 0$. In this case the parameter space is

$$\begin{aligned} \Xi = \big\{ &\varphi = (\ldots, \varphi_{-1}, \varphi_0, \varphi_1, \varphi_2, \ldots) \in \mathbb{R}^{\mathbb{Z}}, \\ &\text{satisfying (4.2) if } \alpha > 2 \text{ or (4.4) if } 1 < \alpha \leqslant 2 \big\}. \end{aligned} \tag{5.2}$$

Let $\Xi_1 \subset \Xi$ be the set of all sequences $\varphi \in \mathbb{R}^{\mathbb{Z}}$ satisfying

$$\sum_{j=-\infty}^{\infty} |\varphi_j| < \infty. \tag{5.3}$$

Note that the set Ξ_1 contains the parameter sequence $\varphi_j = \mathbf{1}(j = 0)$, $j \in \mathbb{Z}$, in which case the linear process is an iid sequence.

It turns out that for any value of the parameters in Ξ_1 the functionals R_n defined by (5.1) grow at the same rate, i.e., at the same rate as for an iid sequence with the same marginal tails. This has been established in Mansfield, Rachev and Samorodnitsky (2001). Specifically, let F be the distribution function of the noise random variable ε_0 and define the usual quantile sequence

$$a_n = \left(\frac{1}{1 - F} \right)^{\leftarrow} (n). \tag{5.4}$$

Here for a function U on $[0,\infty)$, $U^{\leftarrow}$ denotes its generalized inverse

$$U^{\leftarrow}(y)=\inf\{s\colon U(s)\geqslant y\}.$$

Note that, by (4.3), the sequence (a_n) is regularly varying at infinity with exponent $1/\alpha$. See Resnick (1987) for more information on regular varying tails and their quantile functions.

For $\beta>0$ let Z_β be a Fréchet random variable with

$$P(Z_\beta\leqslant z)=\exp\{-z^{-\beta}\},\quad z>0. \tag{5.5}$$

Assume (5.3). Then the numbers

$$\begin{cases} M_+(\boldsymbol{\varphi})=\max\left\{\sup_{-\infty<k<\infty}\left(\sum_{j=-\infty}^{k}\varphi_j\right)_+,\ \sup_{-\infty<k<\infty}\left(\sum_{j=k}^{\infty}\varphi_j\right)_+\right\},\\ M_-(\boldsymbol{\varphi})=\max\left\{\sup_{-\infty<k<\infty}\left(\sum_{j=-\infty}^{k}\varphi_j\right)_-,\ \sup_{-\infty<k<\infty}\left(\sum_{j=k}^{\infty}\varphi_j\right)_-\right\}\end{cases} \tag{5.6}$$

are, obviously, finite. Then

$$a_n^{-1}R_n\Rightarrow\delta^{-1}\left(pM_+(\boldsymbol{\varphi})^\alpha+qM_-(\boldsymbol{\varphi})^\alpha\right)^{1/\alpha}Z_\alpha\quad\text{(weakly) as } n\to\infty, \tag{5.7}$$

once again as long as (5.3) holds. Here p and q are the tail weights in (4.3). See Theorem 2.1 in Mansfield, Rachev and Samorodnitsky (2001).

What happens if $\boldsymbol{\varphi}\in\Xi_1^c$ (i.e., if (5.3) fails)? It is not known whether, in this case, R_n *always* grows at the rate faster than a_n, that is whether the sequence (of the laws of) $(a_n^{-1}R_n,\ n=1,2,\ldots)$ is not tight. However, the following is known. Assume that the coefficients (φ_j) are themselves regularly varying and balanced. That is, there is a function $\varphi\colon[0,\infty)\to[0,\infty)$ such that

$$\varphi(t)=L_2(t)t^{-h} \tag{5.8}$$

as $t\to\infty$ and such that

$$\lim_{j\to\infty}\frac{\varphi_j}{\varphi(j)}=c_+,\qquad\lim_{j\to\infty}\frac{\varphi_{-j}}{\varphi(j)}=c_-, \tag{5.9}$$

for some $c_+,c_-\geqslant 0$, at least one of which is positive. Here

$$1>h>\max\left\{\frac{1}{\alpha},\frac{1}{2}\right\} \tag{5.10}$$

and L_2 is a slowly varying function. Clearly any such parameter vector $\boldsymbol{\varphi}$ is in Ξ_1^c. Define

$$b_n = \left(\frac{1}{\varphi}\right)^{\leftarrow}(a_n), \tag{5.11}$$

$n \geqslant 1$ and note that sequence (b_n) is regularly varying with exponent $1/(\alpha h)$. Then

$$b_n^{-1} R_n \Rightarrow p^{1/\alpha h}\big((1-h)\delta\big)^{-1/h}\big(c_+^{1/h} + c_-^{1/h}\big) Z_{\alpha h} \quad \text{(weakly) as } n \to \infty. \tag{5.12}$$

See Theorem 2.1 in Rachev and Samorodnitsky (2001).

Since b_n grows faster than a_n does, under the assumptions (5.9) the sequence R_n does grow faster than a_n and, hence, faster than in the iid case and, more generally, faster than it is the case for any $\boldsymbol{\varphi} \in \Xi_1$.

Both results (5.7) and (5.12) are, in the final analysis, a consequence of change in the temporal distribution of the effect of the individual "causes": exceptionally large or exceptionally small values of the noise variables (ε_m). In fact, the contribution of each individual noise variable ε_m to the sum $X_i + X_{i+1} + \cdots + X_j$ in (5.1) is $\varepsilon_m \sum_{d=i-m}^{j-m} \varphi_d$. The intuition of heavy tailed large deviations says that it is a single ε_m that is most likely to be responsible for a large value of R_n. Therefore, one would expect that for large x_n

$$\begin{aligned} &P(R_n > x_n) \\ &\quad \sim P\Bigg(\text{for some } m = \ldots, -1, 0, 1, \ldots, \Bigg(\sum_{d=i-m}^{j-m} \varphi_d\Bigg)\varepsilon_m > (j-i+1)(\mu+\delta) \\ &\qquad\qquad \text{for some } 1 \leqslant i \leqslant j \leqslant n,\ j-i+1 \geqslant x_n\Bigg). \end{aligned} \tag{5.13}$$

This turns out to be valid. Moreover, this intuition allows one, in both cases (i.e., under (5.3) and under (5.9)) to select the right rate of growth for x_n in (5.13), which is equivalent to selecting the appropriate normalization to R_n.

It is a bit surprising that less is known about the apparently easier problem of identifying the rate of decay of probabilities $p_n = P(X_1 + \cdots + X_n > (\mu+\delta)n)$ for $\delta > 0$ as $n \to \infty$. It has been checked that under the assumption

$$\sum_{j=-\infty}^{\infty} j|\varphi_j| < \infty \tag{5.14}$$

which defined a proper subset of Ξ_1,

$$p_n \sim n^{-(\alpha-1)} L(n)\delta^{-\alpha}\left(p\Bigg(\sum_{j=-\infty}^{\infty} \varphi_j\Bigg)_+^{\alpha} + q\Bigg(\sum_{j=-\infty}^{\infty} \varphi_j\Bigg)_-^{\alpha}\right) \quad \text{as } n \to \infty, \tag{5.15}$$

where p, q and L are defined in (4.3), and one assumes that $q > 0$ if $\sum_{j=-\infty}^{\infty} \varphi_j < 0$. See Lemma A.5 in Mikosch and Samorodnitsky (2000b). It looks very plausible that (5.15) holds for every parameter $\boldsymbol{\varphi} \in \Xi_1$. The logic of large deviations indicates that, under the assumptions (5.9), p_n is regularly varying with exponent $-(\alpha h - 1)$ at infinity, but nobody has presented a rigorous proof so far.

5.2. *Ruin probability for heavy tailed linear processes*

In this subsection we consider the rare event in Example 3.6, $A = \{X_1 + \cdots + X_n > n(\mu + \delta) + \lambda$ for some $n \geqslant 1\}$, when $\delta > 0$ is fixed and λ is large. Unfortunately, the result for the entire set Ξ_1 is not available here. However, there is a result for the subset of Ξ_1 defined by (5.14). In the latter case, the probability of the event A (commonly referred to as the ruin probability) satisfies

$$P(A) \sim \frac{pM_+^{(1)}(\boldsymbol{\varphi})^\alpha + qM_-^{(1)}(\boldsymbol{\varphi})^\alpha}{\delta(\alpha - 1)} \lambda^{-(\alpha-1)} L(\lambda) \quad \text{as } \lambda \to \infty, \tag{5.16}$$

where

$$M_+^{(1)}(\boldsymbol{\varphi}) = \sup_{-\infty<k<\infty} \left(\sum_{j=-\infty}^{k} \varphi_j \right)_+, \qquad M_-^{(1)}(\boldsymbol{\varphi}) = \sup_{-\infty<k<\infty} \left(\sum_{j=-\infty}^{k} \varphi_j \right)_-, \tag{5.17}$$

compare with (5.6). See Theorem 2.1 in Mikosch and Samorodnitsky (2000b). We conjecture that (5.16) holds whenever $\boldsymbol{\varphi} \in \Xi_1$. Once again, a good way to think of the asymptotic behavior of the ruin probability is to think about the most likely way the "ruin" can happen. Realizing that the ruin is, most likely, due to a single "extraordinary" value of a noise variable ε_m, one would expect that

$$P(A) \sim \sum_{m=-\infty}^{\infty} P\left(\left(\sum_{d=1-m}^{n-m} \varphi_d \right) \varepsilon_m > n\delta + \lambda \text{ for some } n \geqslant 1 \right). \tag{5.18}$$

Once again, this turns out to be valid (at least, under the assumption (5.14)).

The problem of the behaviour of the ruin probability for $\xi \in \Xi_1^c$ has not, to the best of our knowledge, been treated. One can pursue the logic of large deviations, leading to (5.18). This leads us to conjecture that, under the assumptions (5.9), $P(A)$ is, as a function of λ, regularly varying with exponent $-(\alpha h - 1)$ at infinity.

Based on the above discussion (admittedly, some part of it is "hard" results, and another part is conjectures) one can argue that a significant change occurs for heavy tailed linear processes as parameter θ crosses the boundary between Ξ_1 and its complement. Not only the order of magnitude of the probabilities of certain rare events, and of certain functionals associated with sequences of certain rare events, appears to change at that boundary but another interesting phenomenon seems to happen. Various orders of magnitude do not change as the parameter varies inside of Ξ_1; not only these orders of magnitude do change at the boundary but, also, they may keep changing as the parameter varies outside of Ξ_1.

It is important to make a remark at this moment. It does appear that one should, in fact, look at the behavior of a family of related rare events, or a family of sequences of related rare events, if one wants to see what precisely happens at a boundary. For example, the assumptions (5.9) do not cover the entire Ξ_1^c. We conjecture, however, that important changes happen when one moves from Ξ_1 into Ξ_1^c and not, necessarily, into the subset of Ξ_1^c defined by (5.9). It is likely that, in order to see these changes, one should look not only, say, at the event $A_n = \{X_1 + \cdots + X_n > n(\mu + \delta)\}$ but also at some related rare events, for example at the event $B_n = \{|X_1| + \cdots + |X_n| > n(\mu_1 + \delta)\}$, with $\mu_1 = E|X_1|$.

It is also interesting to mention that, in the case $\alpha > 2$, the condition (5.3) also implies the absolute summability of correlations (i.e., (2.1) fails).

5.3. *Rare events for stationary stable processes*

The situation regarding "phase transitions" for general stationary heavy tailed infinitely divisible processes of Section 4.2 has been investigated even less than it is the case with the heavy tailed linear processes. There are several reasons for this, including relatively complicated structure of stationary infinitely divisible processes and its very involved parameter space, which is a space of measures. Most of the known results are for stable processes, whose structure is better understood. We present here the results for a subclass of stationary stable processes, where we will be able to see a "phase transition".

Specifically, let X_n, $n = 0, 1, 2, \ldots$, be the linear fractional symmetric α-stable noise, $1 < \alpha < 2$. For a fixed α the law of the process has an important parameter $H \in (0, 1)$. That is,

$$X_n = \int_{\mathbb{R}} f_n(x) M(\mathrm{d}x), \quad n = 0, 1, 2, \ldots, \tag{5.19}$$

where M is a symmetric α-stable random measure on the real line with the Lebesgue control measure, and $f_n(x) = f(x + n) - f(x + n + 1)$, $n = 0, 1, 2, \ldots$, $x \in \mathbb{R}$, with

$$f(x) = a\big((-x)_+^{H-1/\alpha} - (-x-1)_+^{H-1/\alpha}\big) + b\big((-x)_-^{H-1/\alpha} - (-x-1)_-^{H-1/\alpha}\big) \tag{5.20}$$

if $H \in (0, 1)$, $H \neq 1/\alpha$. Here a and b are real numbers not simultaneously equal to zero. For $H = 1/\alpha$ one has two choices,

$$f(x) = a\mathbf{1}\big([-1, 0]\big)(x) \tag{5.21}$$

and

$$f(x) = a\big(\ln|x| - \ln|x+1|\big). \tag{5.22}$$

In the latter two cases a is a real number different from zero. The resulting symmetric α-stable process in (5.19) is an ergodic stationary process. It is the increment process of the linear fractional symmetric α-stable motion if $H \neq 1/\alpha$, an iid sequence ($\equiv$ the increment process of the symmetric α-stable Lévy motion) under (5.21), and the increment process of the log-fractional symmetric α-stable motion under (5.22). All of these processes are H-self-similar with stationary increments. We refer the reader to Samorodnitsky and Taqqu (1994) for information on stable processes, their integral representations and on self-similar processes. The parameter space Ξ is, then, the collection of all triples (H, a, b) with $H \in (0, 1)$, $H \neq 1/\alpha$, and a, b real, $a^2 + b^2 > 0$, together with the triples (H, a, i) with $H = 1/\alpha$, a real, different from zero, and $i = 1, 2$, depending on the choice between (5.21) and (5.22). Let Ξ_1 be the subset of Ξ corresponding to $0 < H < 1/\alpha$.

We consider, once again, the rare event in the Example 3.6, $A = \{X_1 + \cdots + X_n > n(\mu + \delta) + \lambda \text{ for some } n \geqslant 1\}$, when $\delta > 0$ is fixed and λ is large. Of course $\mu = 0$ here. Then

$$P(A) \sim \begin{cases} \dfrac{K}{\delta}\lambda^{-(\alpha-1)} & \text{if } 0 < H < \dfrac{1}{\alpha} \text{ or under (5.21)}, \\ \dfrac{K}{\delta}\lambda^{-(\alpha-1)}(\log\lambda)^{\alpha} & \text{under (5.22)}, \\ \dfrac{K}{\delta^{\alpha H}}\lambda^{-\alpha(1-H)} & \text{if } \dfrac{1}{\alpha} < H < 1 \end{cases} \tag{5.23}$$

as $\lambda \to \infty$. Here K is a finite positive constant that depends on α, H, a and b, but not on δ. See Proposition 4.4 in Mikosch and Samorodnitsky (2000a).

Observe that the order of magnitude of the ruin probability remains the same as H varies in $(0, 1/\alpha)$. Furthermore, this order of magnitude is the same as under independence. On the other hand, as H varies in the interval $(1/\alpha, H)$, the order of magnitude of the ruin probability is greater than that in the case of independence and, furthermore, *this order of magnitude changes with* H. As we argued earlier, this gives us a reason to say that the range $H \in (0, 1/\alpha)$ corresponds to short memory, and the range $H \in (1/\alpha, 1)$ corresponds to long memory. It is interesting that, in this case, the boundary $H = 1/\alpha$ contains two points, corresponding to (5.21) and to (5.22), and it makes sense to view the latter as corresponding to long memory, while the former is the independent case.

Here is how the intuition of large deviations works here. As mentioned in Section 4.2, the process X_n, $n = 0, 1, 2, \ldots$, can be represented as a sum of Poisson points. In the symmetric stable case this can be done as follows. One can write (in terms of equality of finite dimensional distributions) the process given by (5.19) in the form

$$X_n = C_\alpha^{1/\alpha} \sum_{j=1}^{\infty} \varepsilon_j \Gamma_j^{-1/\alpha} g(V_j)^{-1/\alpha} f_n(V_j), \quad n = 0, 1, 2, \ldots, \tag{5.24}$$

where C_α is a finite positive constant that depends only on α, g a strictly positive measurable function such that $\int_{\mathbb{R}} g(x)\,\mathrm{d}x = 1$, $(\varepsilon_n)_{n \geqslant 1}$ is an iid sequence of Rademacher variables

$(P(\varepsilon_n = -1) = P(\varepsilon_n = 1) = 1/2)$, $(\Gamma_n)_{n\geqslant 1}$ are the points of a unit rate Poisson process on $(0, \infty)$, and $(V_n)_{n\geqslant 1}$ is an iid sequence of real valued random variables with common density g. Moreover, the three sequences are mutually independent. See Samorodnitsky and Taqqu (1994, Section 3.10).

Rewriting

$$P(A) = P\left(C_\alpha^{1/\alpha} \sup_{n\geqslant 1} \left(\sum_{j=1}^{\infty} \varepsilon_j \Gamma_j^{-1/\alpha} g(V_j)^{-1/\alpha} \sum_{k=1}^{n} f_k(V_j) - n\delta \right) > u \right),$$

the intuition of rare events says that it is a single one of the Poisson points (in the function space) $(\varepsilon_j \Gamma_j^{-1/\alpha} g(V_j)^{-1/\alpha} \sum_{k=1}^{n} f_k(V_j),\ n = 1, 2, \ldots)$ that is most likely to cause the ruin. This intuition translates into

$$P(A) \sim \sum_{j=1}^{\infty} P\left(C_\alpha^{1/\alpha} \Gamma_j^{-1/\alpha} g(V_j)^{-1/\alpha} \sup_{n\geqslant 1} \left(\varepsilon_j \sum_{k=1}^{n} f_k(V_j) - n\delta \right) > u \right) \tag{5.25}$$

as $\lambda \to \infty$. It is the equivalence (5.25) that allows one to understand the change in the way the effect of these Poisson points is distributed over time as the parameter H crosses the boundary $1/\alpha$.

Interestingly, the probabilities of the rare events of Example 3.5 $A_n = \{X_1 + \cdots + X_n > n(\mu + \delta)\}$ do not indicate anything interesting happening at the point $H = 1/\alpha$. In fact, since the processes under considerations are the increments of H-self-similar processes,

$$p_n = P(X_1 + \cdots + X_n > \delta n) = P\left(n^H X_1 > \delta n\right) \sim \text{const} \cdot \delta^{-\alpha} n^{-\alpha(1-H)}$$

as $n \to \infty$. Hence the order of magnitude of p_n changes "ordinarily" as H crosses the boundary $1/\alpha$. As mentioned at the end of Section 5.2, one should, probably, look at certain related rare events as well. The behavior of the associated functionals in (5.1) does not seem to have been studied so far.

5.4. High dimensional joint tails for a linear process with stable innovations

We conclude this chapter with a simulation study of a situation in which no analytical results are yet available. Consider a heavy tailed linear process (4.1). For a fixed $\lambda > 0$ we consider the probability of the event $A_n = \{X_j > \lambda,\ j = 0, \ldots, n\}$, when n is large. We are within the framework of Example 3.3. The discussion above makes it possible to conjecture that there is a phase transition at the boundary between the set Ξ_1 in (5.3) and its complement in the set Ξ in (5.2). To check this conjecture we ran a simulation of 10^7 realizations of a linear process with symmetric α-stable innovations with different α. We estimated both the probability $P(A_n)$ as a function of n and the rate of growth of the associated functional

$$R_n = \max\left\{j - i + 1\colon 1 \leqslant i \leqslant j \leqslant n,\ \min(X_i, \ldots, X_j) > \lambda\right\}. \tag{5.26}$$

We simulated first an AR(1) process with $\varphi_j = 0$ for $j \neq 0$ or 1, $\varphi_0 = 1$ and varying φ_1. This choice of coefficients is, clearly, in Ξ_1. Then we simulated a linear process with $\varphi_j = 0$ for $j < 0$ and $\varphi_j = (1+j)^{-0.8}$ for $j \geqslant 0$ (and $\alpha > 1/0.8$). This choice of parameters is in the set Ξ_1^c.

While a simulation study of this type cannot provide a definite answer, it seems to indicate that for the AR(1) process the probabilities $P(A_n)$ decay exponentially fast with n. We plotted in Figure 3 the ratio $-(\log P(A_n))/n$ over the range of n for λ in the set $\{0.1, 0.2, 0.3, 0.4\}$ for the AR(1) process with $\alpha = 1.5$ and $\varphi_1 = 0.5$. Notice how the curves become horizontal.

In comparison, our simulations seem to indicate that for the linear process with $\varphi_j = (1+j)^{-0.8}$, $j \geqslant 0$, the probabilities $P(A_n)$ decay hyperbolically fast with n. We plotted

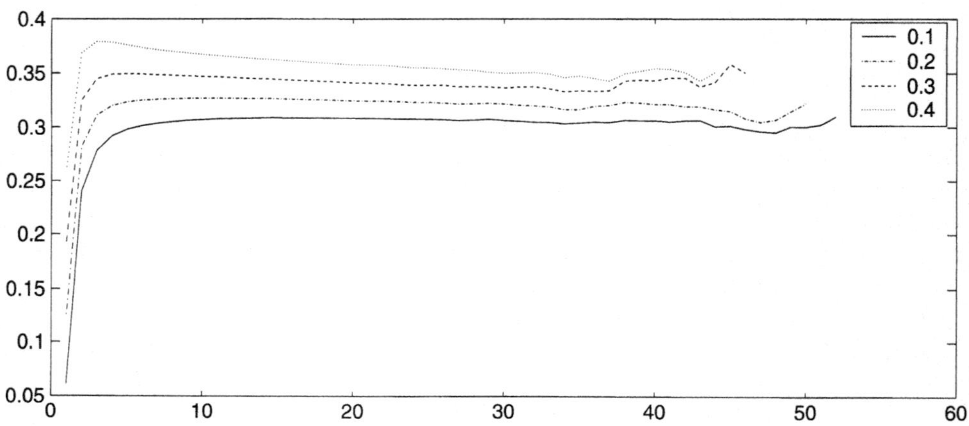

Fig. 3. The ratio $-(\log P(A_n))/n$ for the AR(1) process with $\alpha = 1.5$ and $\varphi_1 = 0.5$.

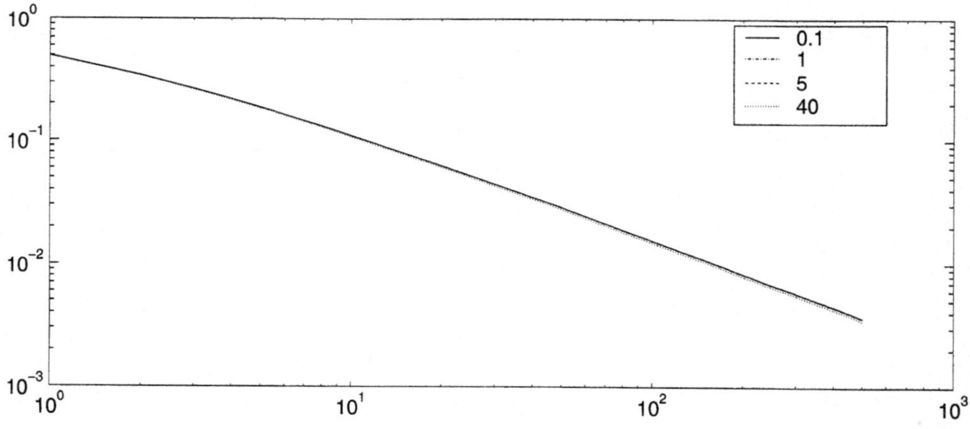

Fig. 4. A plot of $P(A_n)$ against n for a linear process with $\alpha = 1.5$ and $\varphi_j = (1+j)^{-0.8}$, $j \geqslant 0$. Log–log scale.

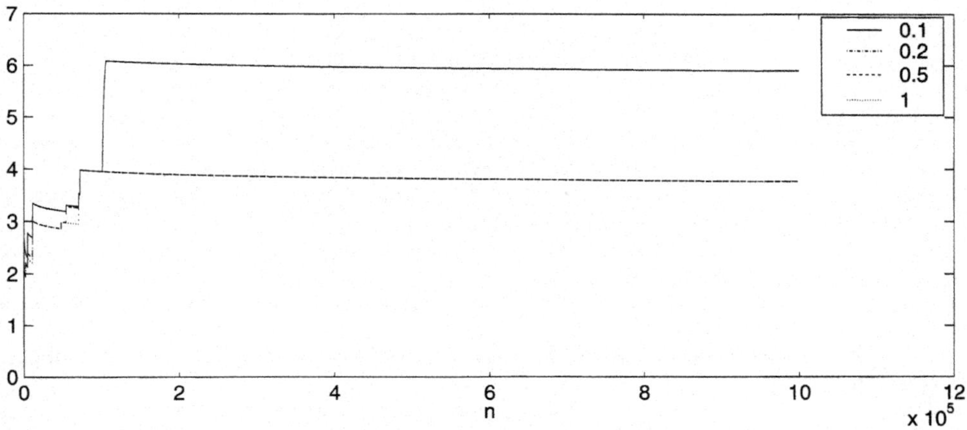

Fig. 5. A plot of $(\log R_n)/\log n$ for a linear process with $\alpha = 1.5$ and $\varphi_j = (1+j)^{-0.8}$, $j \geqslant 0$.

in Figure 4 $P(A_n)$ against n in the log scale, for the case $\alpha = 1.5$. Here we use λ in the set $\{0.1, 1, 5, 40\}$. Notice how linear the plots are. Finally, we present a plot of $(\log R_n)/\log n$ for the long memory process with $\alpha = 1.5$ and $\lambda \in \{0.1, 0.2, 0.5, 1\}$ (Figure 5). Our intuition tells us that in that case R_n should grow polynomially fast with n, and the simulation appears to bear this out.

Once again, even though a simulation study is not a conclusive evidence of a phase transition at the boundary between the set Ξ_1 and its complement, its results are consistent with such a phase transition.

References

Araujo, A., Giné, E., 1980. The Central Limit Theorem for Real and Banach Valued Random Variables. Wiley, New York.

Astrauskas, A., Levy, J., Taqqu, M.S., 1991. The asymptotic dependence structure of the linear fractional Lévy motion. Lietuvos Matematikos Rinkinys (Lithuanian Mathematical Journal) 31, 1–28.

Beran, J., 1992. Statistical methods for data with long-range dependence. Statistical Science 7, 404–416. With discussions and rejoinder, pp. 404–427.

Beran, J., 1994. Statistics for Long-Memory Processes. Chapman & Hall, New York.

Brockwell, P., Davis, R., 1991. Time Series: Theory and Methods, 2nd edition. Springer-Verlag, New York.

Chistyakov, V., 1964. A theorem on sums of independent random variables and its applications to branching random processes. Theory of Probability and its Applications 9, 640–648.

Chover, J., Ney, P., Wainger, S., 1973. Functions of probability measures. Journal d'Analyse Mathématique 26, 255–302.

Cline, D., 1983. Estimation and linear prediction for regression, autoregression and ARMA with infinite variance data. Ph.D. Thesis. Colorado State University.

Cline, D., 1985. Linear prediction of ARMA processes with infinite variance. Stochastic Processes and their Applications 19, 281–296.

Cox, D., 1984. Long-range dependence: a review. In: David, H., David, H. (Eds.), Statistics: An Appraisal. Iowa State University Press, pp. 55–74.

Dembo, A., Zeitouni, O., 1993. Large Deviations Techniques and Applications. Jones and Bartlett, Boston.

Deuschel, J.-D., Stroock, D., 1989. Large Deviations. Academic Press, Boston.

Embrechts, P., Klüppelberg, C., Mikosch, T., 1997. Modelling Extremal Events for Insurance and Finance. Springer-Verlag, Berlin.

Goldie, C., Klüppelberg, C., 1998. Subexponential distributions. In: A Practical Guide to Heavy Tails: Statistical Techniques for Analysing Heavy Tailed Distributions. Birkhäuser, Boston, pp. 435–460.

Gomez, C., Selman, B., Crato, N., 1997. Heavy-tailed probability distributions in combinatorial search. In: Smolka, G. (Ed.), Principles and Practice of Constraint Programming. Springer, Berlin, pp. 121–135.

Gross, A., 1994. Some mixing conditions for stationary symmetric stable stochastic processes. Stochastics Processes and their Applications 51, 277–295.

Heyde, C., 1968. On large deviation probabilities in the case of attraction to a non-normal stable law. Sankyā. Series A 30, 253–258.

Hurst, H., 1951. Long-term storage capacity of reservoirs. Transactions of the American Society of Civil Engineers 116, 770–808.

Hurst, H., 1955. Methods of using long-term storage in reservoirs. Proceedings of the Institution of Civil Engineers, Part I, 519–577.

Linde, W., 1986. Probability in Banach Spaces – Stable and Infinitely Divisible Distributions. Wiley, Chichester.

Mandelbrot, B., 1965. Une classe de processus stochastiques homothetiques a soi; application a loi climatologique de H.E. Hurst. Comptes Rendus l'Académie Sciences Paris 240, 3274–3277.

Mandelbrot, B., 1983. The Fractal Geometry of Nature. Freeman, San Francisco.

Mandelbrot, B., Van Ness, J., 1968. Fractional Brownian motions, fractional noises and applications. SIAM Review 10, 422–437.

Mandelbrot, B., Wallis, J., 1968. Noah, Joseph and operational hydrology. Water Resources Research 4, 909–918.

Mandelbrot, B., Wallis, J., 1969. Robustness of the rescaled range R/S in the measurement of noncyclic long-run statistical dependence. Water Resources Research 5, 967–988.

Mansfield, P., Rachev, S., Samorodnitsky, G., 2001. Long strange segments of a stochastic process and long range dependence. The Annals of Applied Probability 11, 878–921.

Maruyama, G., 1970. Infinitely divisible processes. Theory of Probability and its Applications 15, 1–22.

Mikosch, T., Samorodnitsky, G., 2000a. Ruin probability with claims modeled by a stationary ergodic stable process. The Annals of Probability 28, 1814–1851.

Mikosch, T., Samorodnitsky, G., 2000b. The supremum of a negative drift random walk with dependent heavy–tailed steps. The Annals of Applied Probability 10, 1025–1064.

Müller, U., Dacorogna, M., Pictet, O., 1998. Hill, bootstrap and jacknife estimators for heavy tails. In: A Practical Guide to Heavy Tails: Statistical Techniques for Analysing Heavy Tailed Distributions. Birkhäuser, Boston, pp. 283–310.

Nagaev, S., 1979. Large deviations of sums of independent random variables. The Annals of Probability 7, 745–789.

Park, K., Willinger, W. (Eds.), 2000. Self-Similar Network Traffic and Performance Evaluation. Wiley, New York.

Rachev, S., Samorodnitsky, G., 2001. Long strange segments in a long range dependent moving average. Stochastic Processes and their Applications 93, 119–148.

Rajput, B., Rosiński, J., 1989. Spectral representations of infinitely divisible processes. Probability Theory and Related Fields 82, 451–488.

Resnick, S., 1987. Extreme Values, Regular Variation and Point Processes. Springer-Verlag, New York.

Rosiński, J., 1989. On path properties of certain infinitely divisible processes. Stochastics Processes and their Applications 33, 73–87.

Rosiński, J., 1990. On series representation of infinitely divisible random vectors. The Annals of Probability 18, 405–430.

Rosiński, J., 1998. Structure of stationary stable processes. In: Adler, R., Feldman, R., Taqqu, M. (Eds.), A Practical Guide to Heavy Tails: Statistical Techniques for Analysing Heavy Tailed Distributions. Birkhäuser, Boston, pp. 461–472.

Samorodnitsky, G., Taqqu, M., 1994. Stable Non-Gaussian Random Processes. Chapman & Hall, New York.

AUTHOR INDEX

n indicates citation in footnote.

SUBJECT INDEX

FORTHCOMING TITLES

HANDBOOK OF ASSET AND LIABILITY MANAGEMENT
Edited by S.A. Zenios and W.T. Ziemba

HANDBOOK OF INVESTMENTS:
EFFICIENCY OF SPORTS AND LOTTERY MARKETS
Edited by D. Hausch and W.T. Ziemba

HANDBOOK OF MATHEMATICAL FINANCE:
SEMI-MARTINGALE THEORY OF OPTIONS AND DERIVATIVES
Edited by D. Madan

HANDBOOK OF CORPORATE FINANCE:
EMPIRICAL CORPORATE FINANCE
Edited by E. Eckbo

HANDBOOK OF CORPORATE FINANCE:
CORPORATE GOVERNANCE
Edited by E. Eckbo

HANDBOOK OF FINANCIAL ECONOMETRICS
Edited by Y. Ait-Sahalia and L. Hansen

HANDBOOK OF INVESTMENTS: ASSET PRICING
Edited by G. Connor and R. Korajczyk

HANDBOOK OF INVESTMENTS: EQUITY RISK PREMIUM
Edited by R. Mehra

CPSIA information can be obtained
at www.ICGtesting.com
Printed in the USA
LVOW04*2255010616
490842LV00007B/70/P

9 780444 508966